Fuel Cell Systems

Fuel Cell Systems

Edited by

Leo J. M. J. Blomen
Blomenco B. V.
Rotterdam, The Netherlands
Formerly of
Kinetics Technology International Group B. V.
Zoetermeer, The Netherlands

and

Michael N. Mugerwa
Kinetics Technology International S.p.A.
Rome, Italy

Springer Science+Business Media, LLC

Library of Congress Cataloging-in-Publication Data

Fuel cell systems / edited by Leo J.M.J. Blomen and Michael N. Mugerwa.
p. cm.
Includes bibliographical references and index.

1. Fuel cells. I. Blomen, Leo J. M. J. II. Mugerwa, Michael N.
TK2931.F836 1993
621.31'2429--dc20 93-31687
CIP

10 9 8 7 6 5 4 3 2

ISBN 978-1-4899-2426-1 ISBN 978-1-4899-2424-7 (eBook)
DOI 10.1007/978-1-4899-2424-7

Originally published by Plenum Press, New York in 1993
Softcover reprint of the hardcover 1st edition 1993

For Willy, Jeanne, and Jan
who gave me the energy

— L. J. M. J. B.

To Sabine Namirembe

— M. N. M.

Contributors

Rioji Anahara • Executive Chief Engineer, Technology Development Management Center, Fuji Electric Company, 1-12-1 Yurakucho, Chiyoda-ku, Tokyo 100, Japan

A. J. Appleby • Director, Center for Electrochemical Systems and Hydrogen Research, Texas Engineering Experiment Station, The Texas A&M University System, College Station, Texas 77843-3402

Embrecht Barendrecht • Emeritus Professor, Eindhoven University of Technology, De Naaldenmaker 2, 5506 CD Veldhoven, The Netherlands

Maurizio Bezzeccheri • ICROT S.p.A., Via S. Giovanni d'Acri 6, 16152, Genova, Italy

Leo J. M. J. Blomen • Former Managing Director Energy Activities, Mannesmann Plant Construction, 2700 AB Zoetermeer, The Netherlands. *Present address*: Blomenco B. V., Achtermonde 31, 4156 AD Rumpt, The Netherlands

B. B. Davé • Center for Electrochemical Systems and Hydrogen Research, Texas Engineering Experiment Station, The Texas A&M University System, College Station, Texas 77843-3402. *Present address*: Nalco Chemical Company, Naperville, Illinois 60566

Diane Traub Hooie • President, NORA Management, Inc., P.O. Box 525, Matawan, New Jersey 07747

J. A. A. Ketelaar • Emeritus Professor of Electrochemistry, University of Amsterdam, Park de Eschhorst 6, 7461 BN Rijssen, The Netherlands

Michael N. Mugerwa • Business Development Coordinator, Kinetics Technology International S.p.A., Via Monte Carmelo 5, 00166 Rome, Italy

K. A. Murugesamoorthi • Research Associate, Center for Electrochemical

Systems and Hydrogen Research, Texas Engineering Experiment Station, The Texas A&M University System, College Station, Texas 77843-3402. *Present address*: AT&T Bell Laboratories, Energy Systems, Mesquite, Texas 75149

A. Parthasarathy • Center for Electrochemical Systems and Hydrogen Research, Texas Engineering Experiment Station, The Texas A&M University System, College Station, Texas 77843-3402. *Present address*: Department of Chemistry, Colorado State University, Fort Collins, Colorado 80522

P. Pietrogrande • Bain, Cunio & Associates, Via Lutezia 8, 00198, Rome, Italy

J. R. Selman • Illinois Institute of Technology, Department of Chemical Engineering, Illinois Institute of Technology, Chicago, Illinois 60616-3793

S. Srinivasan • Deputy Director, Center for Electrochemical Systems and Hydrogen Research, Texas Engineering Experiment Station, The Texas A&M University System, College Station, Texas 77843-3402

H. Van den Broeck • Managing Director, Elenco N. V., Gravenstraat 73 bis, B-2480 Dessel, Belgium

David S. Watkins • Director, Product Development, Ballard Power Systems Inc., 980 West 1st Street, North Vancouver, British Columbia, Canada V7P 3N4

Preface

In March 1987, Plenum Publishing Corporation took the initiative to explore the possibility of a new book on fuel cells. At the time, the editors were finalizing a 15,000-hour systematic design study of fuel cell power plants, using a relatively novel approach that essentially considers the plant as a hydrogen production system. At the time, fuel cell efforts worldwide had just resulted in a number of crowning technical achievements, such as the successful demonstration of Gas Research Institute United Technologies Corporation program with approximately 50 fuel cell modular plants of 40 kW. However, fuel cell technology was still unable to achieve commercial success, as was exemplified by the lagging sales of International Fuel Cell's 11-MW (PC-23) plants. As a result, the mood in the market was fairly pessimistic, and many people argued that phosphoric acid fuel cells may have missed their window of opportunity. However, the vast accumulation of experience and developmental effort in the various fuel cell technology areas provided a continuous push forward, not least in Japan where special emphasis on phosphoric acid fuel cell (PAFC) and molten carbonate fuel cell (MCFC) technology development provided a spur to increased fuel cell development activities worldwide.

Early in 1988, the idea of the present multiauthored volume was worked out, and a decision was made to try to combine state-of-the-art overviews of research and development activities on each type of fuel cell with a more general system approach toward fuel cell plant technology, including plant design and economics. We were most fortunate to receive positive acceptance of the concept among the original list of distinguished authors, and the tedious work on the manuscript began—a continuing battle for priority for all involved.

Four years later, the manuscript was completed and book production had begun. The world has changed. Increasing environmental awareness and imminent fossil-fuel-supply shortages are forcing society to develop and adopt a combination of clean and highly efficient power-generation systems. The world has recently witnessed ecological disasters on an unprecedented scale. Greenhouse-effect discussions, despite not having achieved full consensus, have led to the legislation of emission penalties in several countries. The Gulf War has renewed the call for a greater diversification of primary energy sources, and the world has not yet recovered from the Chernobyl disaster. Due to these and other factors, international interest in fuel cells is perhaps stronger than ever before. In Japan, the Ministry of International Trade and Industry (MITI) has set specific targets for fuel cell market penetration which are to be met by the year 2000. Further, major multinationals are securing a portfolio of fuel cell technologies in anticipation of gaining a reasonable share of the market. So, more than 150 years after their invention, fuel cells are finally reaching the market place, and events over the coming decade will be crucial to the successful commercialization of the technology.

This is, therefore, a most exciting moment to publish a book about fuel cells. All the authors, and the editors in particular, hope that this book will contribute to the general knowledge in the field and that the reader will get a balanced, multidisciplinary overview of fuel cell technology and of the systems that can be built with these fascinating electrochemical devices.

Being multidisciplinary in nature, this book is not primarily intended for use as a university textbook for a single course. Nevertheless, much of the material can be used for (under)graduate courses in energy, chemical, mechanical, electrical, and power engineering, materials science, environmental subjects, and (applied) electrochemistry. It will also be of use to professionals in all these fields. In addition, the book will find an audience among the early commercial users and evidently the developers themselves. All chapters are, to a large extent, self-explanatory. A drawback of this approach, noticeable to the reader of the entire text, is a certain amount of repetition and a limited degree of inconsistency in the use of symbols and terminology. The editors have tried to avoid as much inconsistency as possible, with a minimum of distortion to each author's ideas and text.

This book would never have reached its final form if it were not for the great patience and working effort of many. The editors would like to acknowledge the efforts of all the authors. In particular, we thank the authors who provided the previously unforeseen extra chapters on the fast-developing fuel cell types known as solid oxide fuel cell (SOFC) and solid polymer (electrolyte) fuel cell [SP(E)FC], within a relatively short time. We are most grateful for the patience shown by the publisher, since a delay of two years has been difficult for all concerned, and in particular we thank Mr. Ken Derham of Plenum Publishing Corporation in London for his understanding throughout.

All chapters were converted into a single format by Mrs. Nelma Rademaker. She spent many hours transferring manuscripts, correcting errors, arranging correspondence, and coordinating drawing activities without losing her smiles and positive attitude. Her efforts have been invaluable to the editors. In addition, a special thanks is due to Mrs. Jutta Golunski for her unceasing support. A special word of thanks also concerns Mr. J. E. Hille, the reliable support and right hand of the first editor, who accepted many special assignments and therewith enabled said editor, among others, to do his editorial work. Mrs. Karin Onderdelinden combined accuracy and artful design to produce many of the illustrations. The editors would like to acknowledge the support of KTI/Mannesmann management, which permitted us to spend so much of our time and energy on this book. In particular, we would like to acknowledge the encouragement and support of Dr. W. B. Balthasar and Dr. E. Bitterlich. Moreover, special thanks are due to NOVEM, The Netherlands' Agency for Energy and the Environment, for continuing support of our energy activities throughout the years, and who funded most of the design work for Chapters 6 and 12.

Last but not least, Dr. Blomen would like to express his gratitude to his wife, Willy, who had to sacrifice a substantial part of the first 10 years of marriage to let her husband enjoy the pleasure of working on fuel cell technology and the burden of editing this book.

Contents

2. Overview of Fuel Cell Technology

S. Srinivasan, B. B. Davé, K. A. Murugesamoorthi, A. Parthasarathy, and A. J. Appleby

3. Electrochemistry of Fuel Cells

Embrecht Barendrecht

4. Fuel Processing

P. Pietrogrande and Maurizio Bezzeccheri

5. Characteristics of Fuel Cell Systems

A. J. Appleby

6. System Design and Optimization

Michael N. Mugerwa and Leo J. M. J. Blomen

9. Research, Development, and Demonstration of Molten Carbonate Fuel Cell Systems

J. R. Selman

10. Research, Development, and Demonstration of Solid Oxide Fuel Cell Systems

K. A. Murugesamoorthi, S. Srinivasan, and A. J. Appleby

11. Research, Development, and Demonstration of Solid Polymer Fuel Cell Systems

David S. Watkins

Fuel Cell Systems

Introduction

Leo J. M. J. Blomen and Michael N. Mugerwa

I.1. PERSPECTIVE

The outbreak of the Gulf War in January 1991 reemphasized the global importance of the world's most traded commodity, crude oil. Yet again, this valuable feedstock is the focus of attention.

Since the first oil crisis of 1973, the world energy perspective has changed. Most rich nations have attempted to reduce their dependency on oil by diversification of primary energy sources. More significantly, however, has been the trend of increased environmental awareness. Inversion problems in the atmosphere above some of the world's largest cities first led to strict emission legislation in Japan and California, two of the world's largest economies. Over the last few years, discussions on the greenhouse effect have led to general acceptance of the theory that carbon dioxide (CO_2) emissions cause atmospheric heating, and global studies have conclusively shown that the earth's fossil fuel resources should be better maintained in order to secure a sustainable future. In particular, the automobile industry is being strongly affected by the increased interest in cleaner fuels, including the ultimate fuel hydrogen. Legislation is changing rapidly, and as a result the electric power industry worldwide is also undergoing structural change, since people now realize that the limited fossil fuel reserves should be used as efficiently as possible, with minimum losses, during production as well as transmission and end use. Particularly in the last few years, this has resulted in a trend toward decentralization of power production, with a strong emphasis on energy savings at the end-user's level (mainly household appliances and lighting systems).

I.2. FEEDSTOCKS FOR ENERGY AND THE CHEMICAL INDUSTRY

Traditionally, heavy hydrocarbons (wood, coal, and oil) have been used to meet the world's energy needs. A trend toward lighter hydrocarbons for energy

Leo J. M. J. Blomen • Mannesmann Plant Construction, 2700 AB Zoetermeer, The Netherlands. *Present address*: Blomenco B. V., Achtermonde 31, 4156 AD Rumpt, The Netherlands. Michael N. Mugerwa • Kinetics Technology International S.p.A., Via Monte Carmelo 5, 00166 Rome, Italy.

Fuel Cell Systems, edited by Leo J. M. J. Blomen and Michael N. Mugerwa. Plenum Press, New York, 1993.

use only became apparent in the twentieth century, and gradually the carbon-to-hydrogen ratio of the fuels has decreased. On the other hand, hydrocarbon resources are also needed for the production of polymers and other petrochemicals which have relatively high carbon-to-hydrogen ratios. Traditional petrochemical manufacture is almost exclusively based on the lighter fractions of oil, mainly naphtha, which are converted to lighter components, such as ethylene, propylene, butadiene, and aromatics, to be used in turn as building blocks for polymers and other petrochemicals. Figure I.1 illustrates this process.

These traditional production pathways for energy, petrochemicals, and polymers are not very logical from a global perspective. The more logical way would be to use heavy hydrocarbons (high carbon-to-hydrogen ratios) for the production of polymers and petrochemicals. On the other hand, light components (such as natural gas, biogas, and preferably hydrogen) should be used for energy production and, perhaps, for the production of some basic petrochemicals such as ethylene, hence the concept of C_1 chemistry. However, hydrogen is no fossil fuel, and will still be converted from light hydrocarbons or, in the future, produced directly from water by electrolysis or direct electrochemical synthesis. It is expected that global technology will follow these general trends, since they make more sense from an overall perspective.

With respect to natural gas and biogas this will imply an increased use over the next few decades, after which other sources of energy (especially those based on hydrogen) may play a more prominent role. Although the present reserves of natural gas are relatively limited, world annual production at current and future levels is predicted to be less than the expected growth of proven reserves (see Fig. I.2). Such growth in reserves is not expected for oil and coal, as it is generally believed that most of the proven reserves are already known. In fact, as shown in Fig. I.3, this trend is not new. Statistical studies on simple predictive models

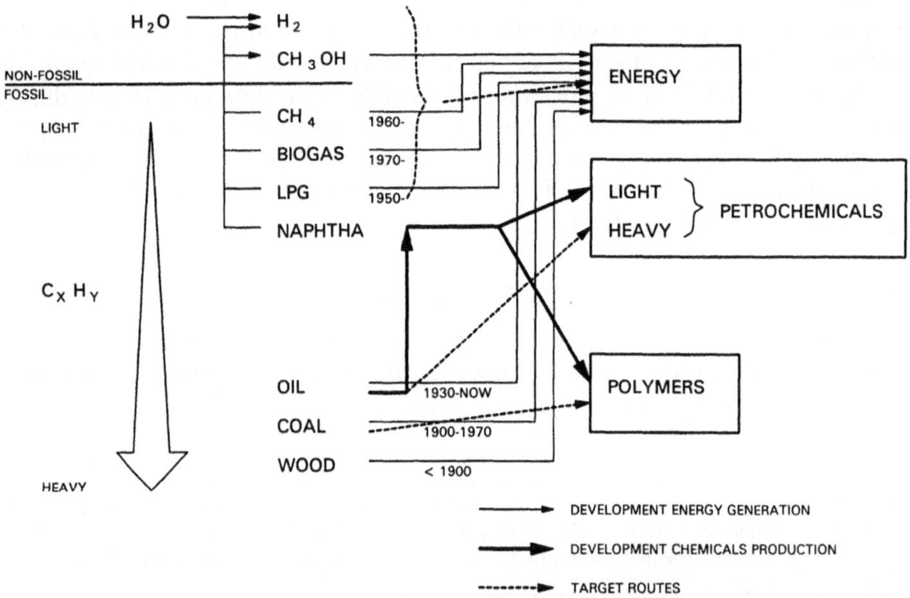

Figure I.1. Trends in hydrocarbon usage for the supply of energy and production of chemicals.

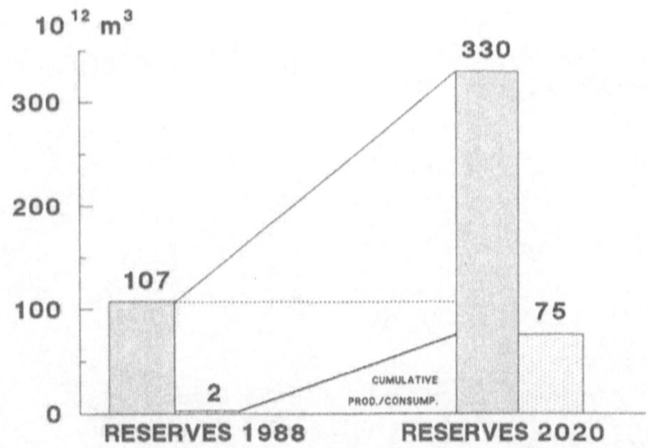

Figure I.2. Natural gas consumption/reserves. (After: World Gas Conference, Washington, 1988).

developed by the Austrian-based International Institute for Applied System Analysis (IIASA) have shown that the relative role played by the heavy primary fuels wood and coal is declining, whereas oil may be considered to have achieved its peak production within the last 10 years. On the other hand, the relative usage of natural gas is still growing. Evidently extrapolation of these models toward the next century is only of limited validity, but the descriptive value of the model over the last 150 years certainly yields some qualitative predictive trends that one would be reckless to ignore. Hydrogen is only expected to play a modest role over the next few decades, but fuel cells, based on natural gas and biogas feed, may "artificially" increase the relative usage of hydrogen in this world energy substitution model (see Fig. I.3.).

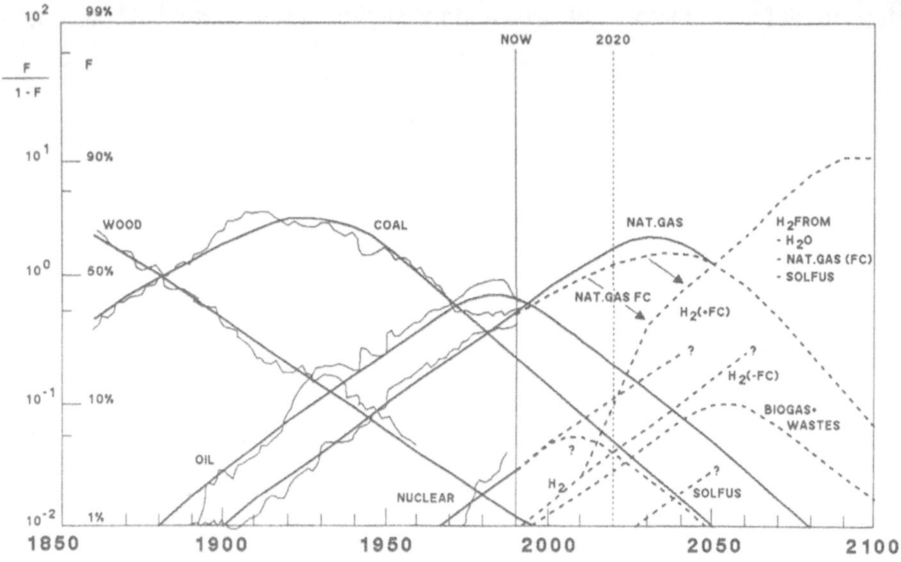

Figure I.3. World primary energy substitution. [After C. Marchetti & N. Nakicenovic, "The dynamics of energy systems and the logistic substitution model" International Institute for Applied Systems Analysis, Austria (1979).]

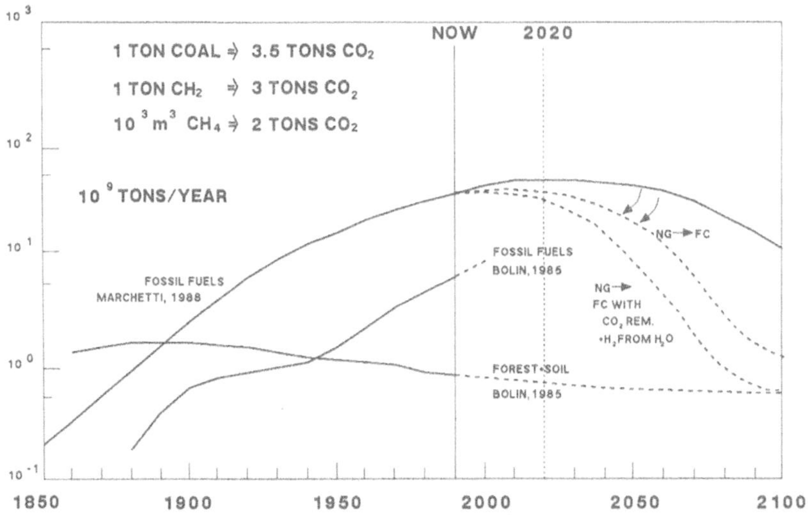

Figure I.4. Total forest, soil, and fossil fuels CO_2 emissions (10^9 t/yr) for assumed 2.3% growth per year in energy consumption.

I.3. EMISSIONS

On the basis of IIASA's energy substitution model, calculations of corresponding carbon dioxide (CO_2) emissions are straightforward, and the results are shown in Fig. I.4. The effect of fuel cells on the normalization of CO_2 emissions can be appreciable, especially if fuel cells in combination with CO_2 removal systems are quickly introduced into the market.

Evidently the greenhouse effect cannot be tackled by installing fuel cell power plants alone. As Fig. I.5 shows, many researchers now attribute the greenhouse effect to different factors, of which only slightly more than 50% is due to

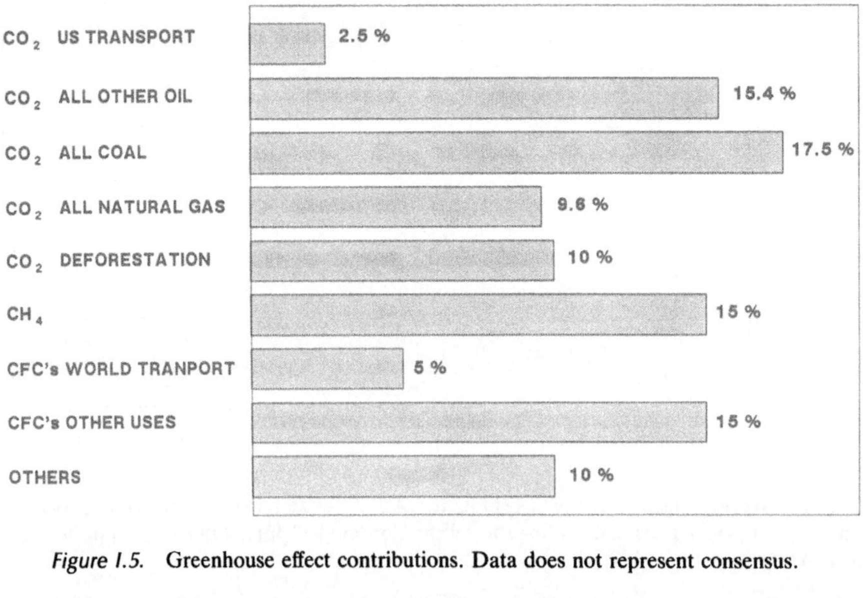

Figure I.5. Greenhouse effect contributions. Data does not represent consensus.

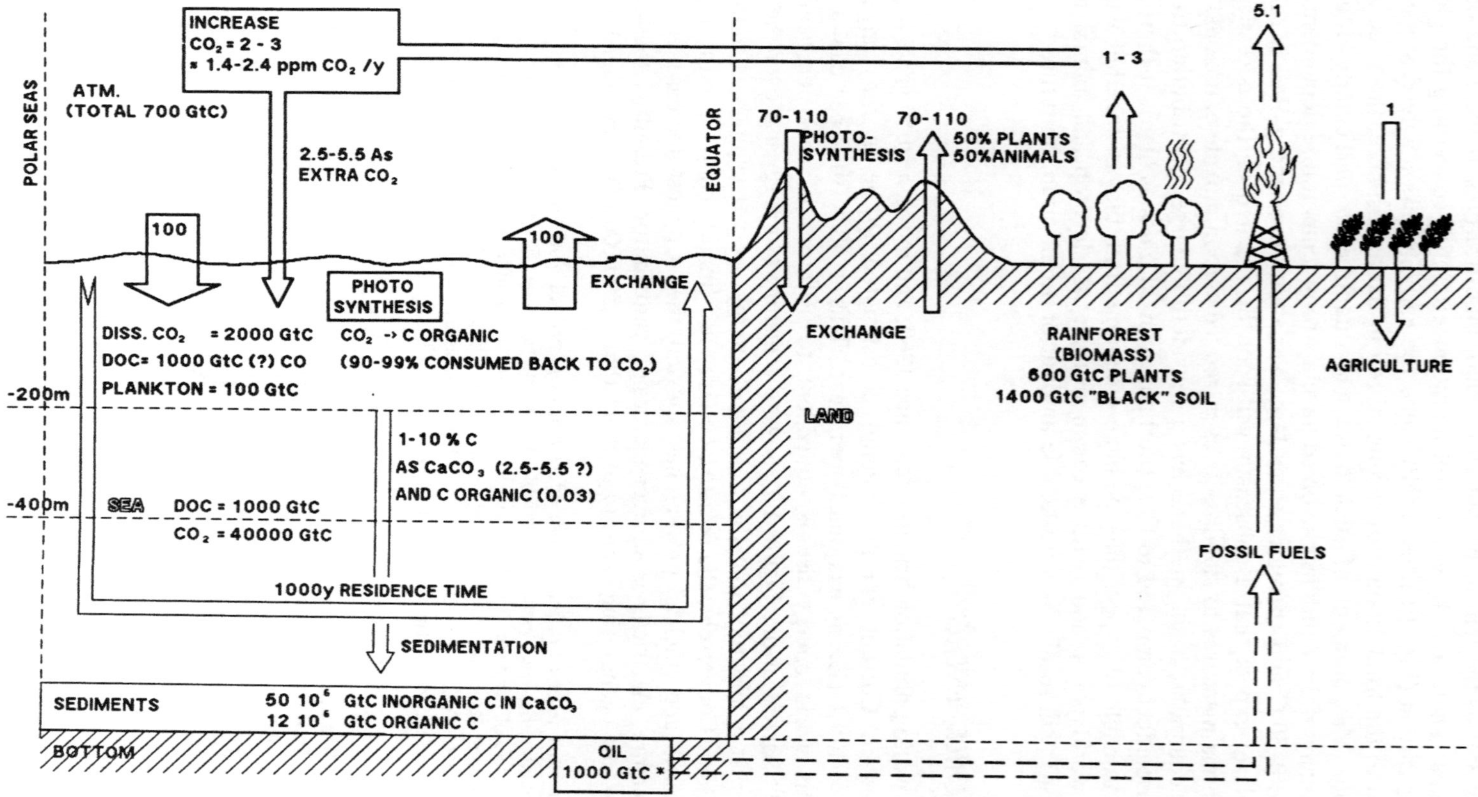

Figure I.6. Global carbon cycling (in Gt C/yr or 10^9 t C/yr). The asterisk denotes total fossil reserves greater than 5000 Gt C.

combustion-sourced CO_2 emissions, including deforestation! Another 15% has its origin in methane emissions from natural processes and pipeline losses. The remaining 35% is thought to be caused by chlorofluorocarbons (CFCs) and other components. There are, however, still many uncertainties concerning the global cycling of carbon (Fig. I.6), especially since the carbon flows involved are small compared to the total amount of carbon stored in the atmosphere, the seas, soil, and plants. Measurement of such flows and quantities is particularly difficult because many of the equilibria involved have a strong temperature dependence in the range 0–30°C and therefore show large variability over the globe. It seems reasonable to expect that a consensus will soon be reached on the necessity of implementing measures to stabilize or even reduce carbon dioxide emissions.

On an equimolar basis, methane has a considerably larger contribution to the greenhouse effect compared to CO_2 (perhaps even 20 times as high). It therefore seems reasonable to expect that methane emissions from biogas, together with a growing world population (with a commensurate rise in crop and animal herd production), will receive a considerable amount of attention in the future.*

I.4. WORLD ENERGY

The world population has doubled since 1960, and in the next 50 years it may double again. Current energy demand is estimated to be in the range of 340–370 × 10^{18} J (EJ or exajoules) per year. This is 68 × 10^9 J per capita per year. This world energy use is equivalent to 10^{10} tons of coal per year or 5.5 × 10^{10} barrels of oil. Figure I.7 shows the history of energy use over the last 100 years. It is apparent that the *traditional* energy use per capita was more or less constant. Conversely, the *industrial* energy use (mainly coal, oil, natural gas, hydroelectric and nuclear power) has showed the strongest increase. In this energy picture, electricity is an ever-increasing contributor. Presently, the world has an installed generating capacity of more than 2,500,000 MW_e (see Fig. I.8), of

*Simplified estimate of extra greenhouse effect due to growth of world population: In 50 years the world population has been assumed to double. For world methane (CH_4) emissions, one may assume that the sources of methane directly related to population will double, with other sources expected to remain constant. Based on current average estimate from literature (B. Giovanni, D. Pain, Proc. 2nd Annual CMDC Conference, Zürich, Dec. 1990), this leads to the following estimated methane emissions [note the values (in million tons CH_4/yr) are still very uncertain]:

Source	Currently	Ater 50 years
Swamps, marshes	100	100
Paddy fields	115	230
Livestock	90	180
Biomass burning	60	120
Other biogenic	400	400+
Natural gas (leaks, etc.)	35	70
Total	800	1100+

A comparison may now be made: 1989 CO_2 emissions from fossil fuel combustion, 7.6 Gt carbon/yr, corresponds to 630 × 10^9 kmol CO_2/yr (see Section I.5). After 50 years, an *extra* amount of 300 mio tons CH_4/yr will be liberated from causes related to population growth. This is equivalent to 20 × 10^9 kmol CH_4/yr. Each methane molecule absorbs 25 times more IR radiation than each CO_2 molecule. Hence, the extra methane will cause a greenhouse effect equivalent to 500 × 10^9 kmol CO_2/yr. *This corresponds to 80% of all CO_2 currently liberated from fossil fuel combustion.* If these data are correct, this will impel humanity to use its biogas.

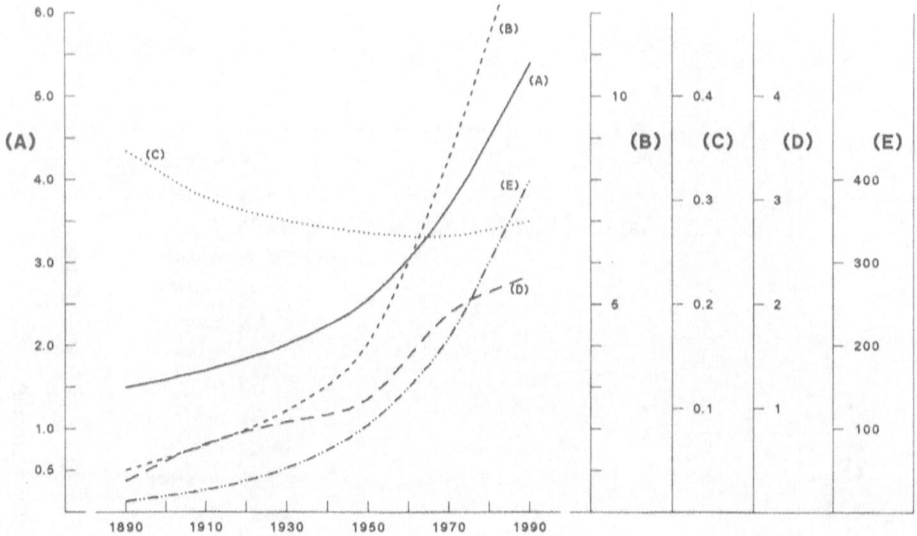

Figure I.7. History of energy use since 1890. (a) World population × 10^9 (—); (b) total world energy use (TW) (- - -); (c) traditional energy use per capita (kW) (wood, crop wastes, dung) (....); (d) industrial energy use per capita (kW) (coal, oil, natural gas, hydro, nuclear) (– – –); (e) cumulative industrial energy use since 1850 (TWy) (–··–··–··). 1 TW ≈ 1 × 10^9 t coal/yr ≈ 5 × 10^9 oil barrels/yr ≈ 31.5 EJ/yr ≈ 31.5 × 10^{18} J/yr.

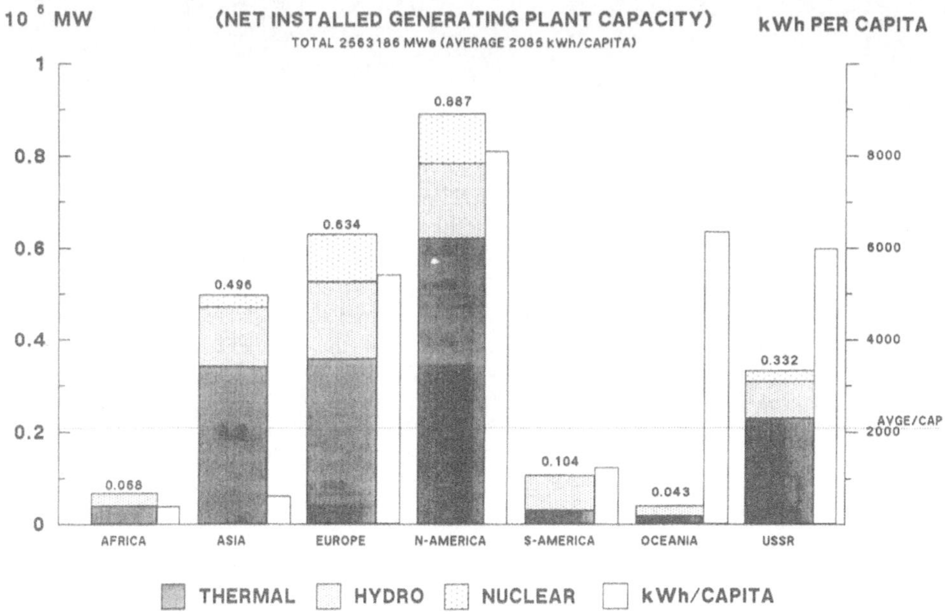

Figure I.8. World electricity supply (1987 UN energy statistics yearbook).

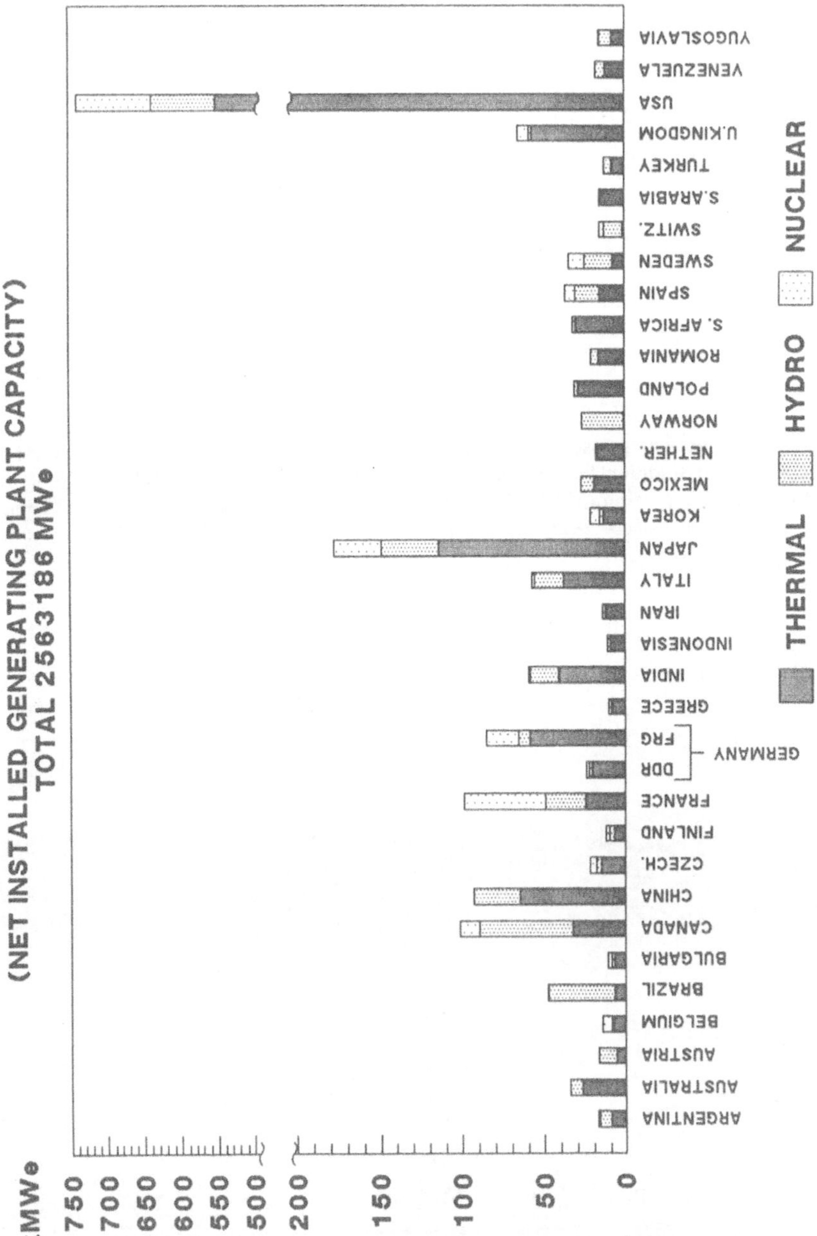

Figure I.9. World electricity supply, selected countries (1987 UN energy statistics yearbook).

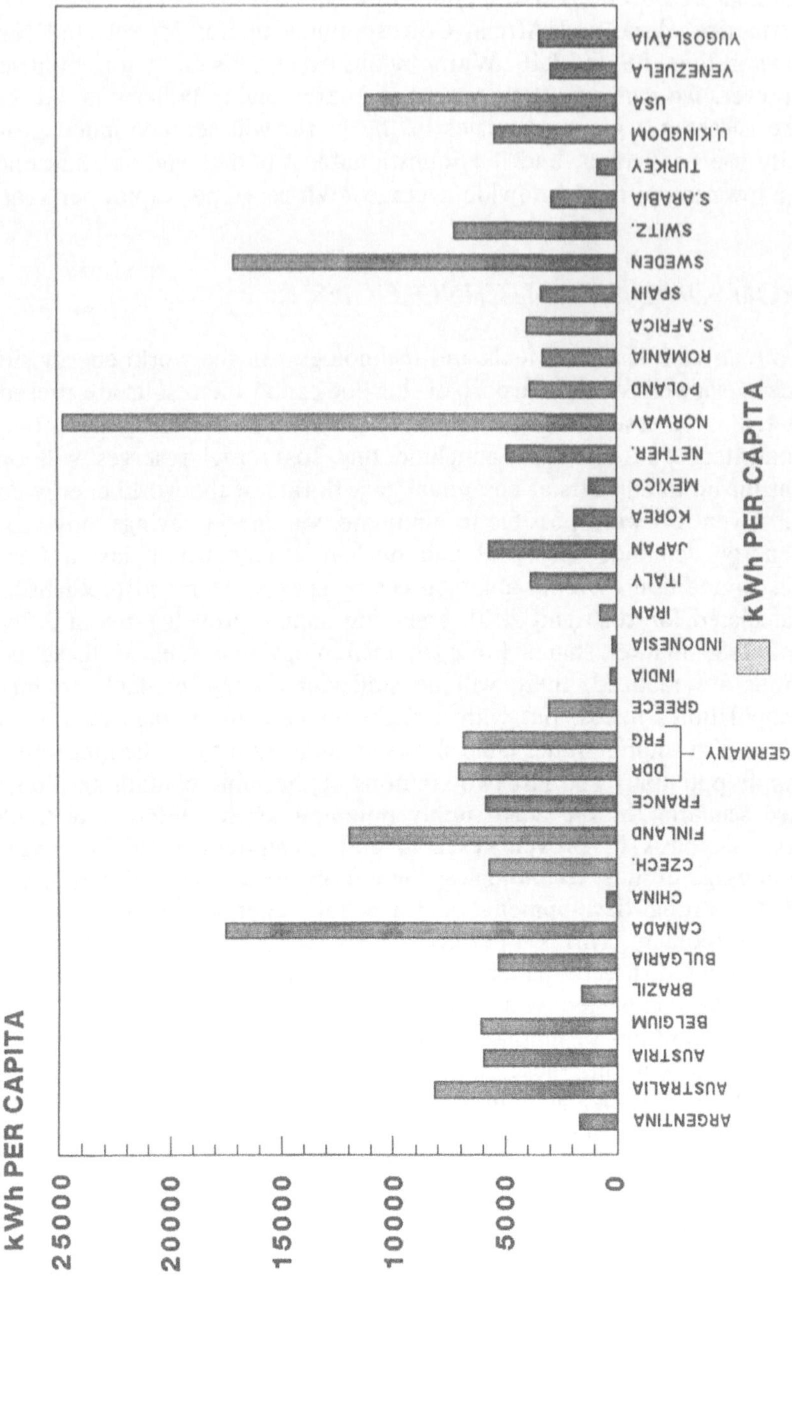

Figure I.10. World electricity supply, selected countries (1987 UN energy statistics yearbook).

which one third is in North America, one fourth in Europe, one fifth in Asia, and the remainder in the rest of the world. The average usage per capita is slightly above 2000 kWh per year, and is four times the average in North America, three times as high in Europe, Russia, and Oceania, and two to four times as low in South America, Asia, and Africa. Corresponding figures for selected countries are shown in Figs. I.9 and I.10. With obvious exceptions made for countries with cheap power, we can expect that even if energy-saving policies at the end-use level are successful, many countries in the world will see continued growth in electricity use per capita, and it is questionable whether end-use efficiency will enable a lowering of the worldwide average kWh usage per capita per year.

I.5. ROLE OF ENERGY TECHNOLOGIES

The relative roles of all fuels and technologies in the world energy situation have been reviewed for the purpose of this book, and the results are presented in Table I.1. Proven reserves (and for renewable sources the yearly potential) have been indicated, and one may conclude that fossil fuel reserves will only be sufficient for up to 80 years at an annual growth rate of the world energy demand of 2.2%. Even if it were possible to eliminate, via energy savings, any growth in world energy demand, only coal and nuclear energy would last a few more centuries in addition to renewable sources of energy. Demand predictions have been calculated for 2020 and 2040, assuming annual growth rates of 2.2%, and assuming 1989 market shares for each technology and fuel. If these demand projections are realized, there will be sufficient energy available to meet the world population's needs, but with a slight increase in average energy use per capita. The fifth column in the table shows an indication from the literature of the 2040 supply potential. The last two columns of the table contain an alternative, gas-based scenario for the 2020 supply potential of the different technologies, made on the basis of proven reserves and technology maturity. Even with maximum usage of most technologies, the energy demands of the future can only be met by strong development of renewable energy sources, with special emphasis on biomass. This is a perfectly plausible assertion, since the amount of biomass generated directly and/or indirectly by the increased world population will need to be combusted as a fuel, or else via natural digestion will generate considerable quantities of methane (and to a lesser extent CO_2). In view of its greenhouse contribution, these huge quantities of CH_4 may become a major problem, which may *force* us to utilize it rather than "waste" it. It may eventually become not just a helpful fuel, but a fuel that must be used in order to sustain our atmosphere (cf. the footnote in Section I.3.).

At the basis of the estimated average supply potential in this study are the expected development of reserves and the expected technological development of renewable potential. For natural gas a rate of development corresponding with the expected increase in reserves is foreseen (Fig. I.2). This is in line with the predicted course of primary energy substitution as described in Fig. I.3. Figure I.3 is also the basis for the current stabilization of oil utilization and a decrease in the use of coal. The rationale behind this is the conservation of coal for use as a strategic chemical feedstock to manufacture petrochemicals and polymers. The enormous potential of oil tar sand and shale oil may be also exploited by 2040.

Table 1.1. World Energy Situation (10^{18} J/yr or EJ/yr)

	Demand 1989	Proven[b] reserves (EJ)/potential 1989	Sufficient for # yrs: At no growth	Sufficient for # yrs: At 2.2% Δ/yr	Literature supply potential[m] (coal–oil scenario) 2040	Demand[n] 2040 at Δ = 2.2%/yr 2040	Demand[n] 2020 at Δ = 2.2%/yr 2020	Average supply potential (gas scenario) 2020	Max. supply potential (gas scenario) 2020
Gas	70–74	4,000–6000 (?)[c,d]	57–81	37–49	120	210	140	220[o]	370[o]
Oil	140	6,000	44	31	400	420	280	150	200
Oil tar/shale	—	30,000[e]	Many	Many	30	—	—	100 (?)	200 (?)
Coal	130	30,000	230	82	450	390	260	75	250
Nuclear[a]	5	875[f,g]	440	73	?	15	10	<5	30
Renew									
solar therm./electr.	—	} 6,000 EJ/yr[h]	Inf.	Inf.	30	—	—	20	25
photovoltaics	—		Inf.	Inf.	10	—	—	30 (?)	50 (?)
wind	—	94 EJ/yr[i]	Inf.	Inf.	25	—	—	25	30
water	21	158 EJ/yr[j]	Inf.	Inf.	100	63	42	40	70
oceans & tidal	—	950 EJ/yr	Inf.	Inf.	20	—	—	10	15
geothermal	—	4,000[k]	Many	Many	20	—	—	10	15
biomass	—	4,200 EJ/yr[l]	Inf.	Inf.	250	—	—	100	200 (?)
Total	340–370	—	—	—	1435 + ?	1098	732	775	1440
World population	5×10^9	—	—	—	10×10^9 (?)	10×10^9 (?)	8.5×10^9 (?)	8.5×10^9 (?)	8.5×10^9 (?)
Energy use/capita (10 J/cap/yr)	68	—	—	—	144 (?)	110 (?)	86 (?)	91 (?)	169 (?)

[a]Nuclear: 400 EJ in 2040 would imply 20,000 plants (= 500 plants/yr) ≡ 3000 under construction at any time → impossible; 30 EJ seems realistic maximum by 2040; note nuclear fusion has not been included within time frame involved.

[b]Data on reserves/potential may vary considerably with source; information on renewable potential was provided by E. Lysen, Leusden, Netherlands.

[c]Excludes several speculative sources, such as Siberian and ocean sediments, gas hydrates/clathrates and high depth reserves. Especially clathrates include huge energy amounts.

[d]Ref: B. de Vries, in *Climate and Energy* (Kluwer, Dordrecht, 1989).

[e]Much of this needed to generate steam to free the oil; tar sand can be exploited more efficiently, however. Of the amount approximately 40% is tar sands, the remainder oil shale.

[f]Based on 12.5 mio ton U_{308}; if all 0.1% ore, 175 TJ_{th}/ton U_{308}; based on Ref. 14, this corresponds to proven reserves at mining costs <80 $/kW uranium metal.

[g]OECD Uranium Resources, Production and Demand, Paris, 1988.

[h]Based on 1% of land surface area @ 130 W/m^2; excl. passive solar energy use (noted as energy savings).

[i]Ref. *Energy in a Finite World, a Global Systems Analysis,* (Ballinger, Cambridge, 1981).

[j]Ref. *Energy for a Sustainable World* (Wiley Eastern, New Delhi, 1988).

[k]Ref. *Renewable Energy* (Academic Press, London, 1979) note reserves in EJ, since reheating takes a very long time.

[l]Only 13 EJ/yr used as food; no growth yet calculated.

[m]R. W. J. Kouffeld (1990), *Delft Integraal* **90** (6), 14–18.

[n]50 yrs time: $(1022)^{50} = 3$; 32 yrs. time: $(1.022)^{32} = 2$; demand extrapolated from 1989 to 2040; and from 1988 to 2020, both at same relative penetration of each technology.

[o]World Gas Conf. 1988 predicted proven reserves in 1988 = 4000 EJ ≡ 53 yr; in 2020 12,300 EJ = 56 @ 220 EJ/yr or 83 yr @ 150 EJ/yr.

Table 1.2. Fuel Exhaustion Calculations and CO_2 Emissions

	Current demand Total CO_2 emitted in 1989 10^9 t CO_2	Reserves remaining in 2040 if linear growth to coal–oil scenario supply potential			Reserves remaining in 2020 if linear growth to gas scenario supply potential		Reserves remaining in 2040 if same linear growth extrapolated	
		Δ/yr	Left	Total CO_2 emitted 1989–2040 10^9 t CO_2	Δ/yr	2020 Left	2040 Left	Total CO_2 emitted 1989–2040 10^9 tCO_2
Gas[a]	94	+1.43%	940	290	+7.14%/+0.62%[b]	−678/+3,600[b]	−26,000/+1,820[b]	1,830
Oil	10	+3.71%	−13,550	1,400	+0.24%/−0.75%[c]	+1,500/+2,100[c]	−1,590/+50[c]	540
Oil tar/shale	—	—	29,250	54	—	28,500	27,500	180
Coal	14	+4.92%	3,500	2,950	−1.41%	26,800	25,250	528
Nuclear	—	?	?	0	<0%	>725	>630	0
Renew								
solar therm./electr.	—	—	} 6,000 yr	0	—	} 6,000/yr	}6,000/yr	0
photovoltaics	—							
wind	—	—	94/yr	0	—	94/yr	94/yr	0
water	—	—	158/yr	0	—	158/yr	158/yr	0
oceans & tidal	—	—	950/yr	0	—	950/yr	950/yr	0
geothermal	—	—	3,990	0	—	3,995	3,995	0
biomass	—	—	4,200/yr	0[d]	—	4,200/yr	4,200	0[d]
Total	28 GtCO_2			4,694 GtCO_2[e]				3,078 GtCO_2[e]

[a]Assumed reserves 6000 EJ (higher limit from Table I.1).
[b]Based on demand growth of 7.14% per year plus reserve growth according to World Gas Conference 1988 (which is 6.48%/yr).
[c]Based on demand growth of 0.24% per year plus reserve growth assumed to be 1.0%/yr.
[d]Net effect biomass production and combustion.
[e]Both CO_2 emissions sufficient for world energy production growing 2.2% per annum from current demand. Converted to 10^9 ton carbon (GtC), the numbers are 1280 GtC and 839 GtC respectivelyr. Averaged over 51 years these numbers correspond to 25 and 16 GtC/yr. Note current CO_2 emissions contribute to 7.6 GtC/yr (in 1989).

However, current technology requires large amounts of energy to raise steam in order to free the oil, especially from shale. Many experts in the field have questioned whether the recovery of energy from these sources will ever be economical.

For nuclear energy, Table I.1 shows a scenario in which the estimated average supply potential in 2020 does not allow for any new plant constructions and only assumes replacement of (part of the) existing capacity. The maximum supply potential scenario assumes that nuclear energy is accepted and stimulated, culminating in a maximum of up to 30 EJ per year in 30 years' time. This implies that at any moment during the next 30 years, approximately 300 new nuclear plants should be under construction (far more than presently considered likely). Renewable energy sources will have to play a major role, particularly in decentral applications. For logistic reasons this is also the most probable, since the largest population growth rates and associated energy requirements will increasingly take place in countries where solar energy, biogas energy, and to a lesser extent, wind and water energy are available.

Table I.2 contains an estimation of the amounts of fuel remaining in 2020 and 2040 for both the coal–oil and gas scenarios. These estimates are based on linear growth for the values given by the average supply potential in Table I.1 (gas scenario) and the coal–oil supply potential (coal–oil scenario).

For the natural gas scenario, assuming no additional supplies are found, natural gas will be exhausted in 2020 and oil in 2040. However, if corrected for the expected reserve growth of gas and oil supplies, these reserves are not expected to become exhausted during the reference period. Coal reserves will remain largely the same and would therefore be freely available for chemicals production. The cumulative amounts of CO_2 liberated from all these fuels from 1989 to 2040 is also indicated.

Table I.2 also shows that, according to the coal–oil scenario, by the year 2040 oil reserves will be completely exhausted, and 90% of the world's coal will be exhausted as well. The amounts of CO_2 liberated in the 50 years period will also be much higher than in the natural-gas-favoring scenario. Both scenarios attribute large importance to biomass.

If CO_2 emissions are viewed as gigatons (Gt) of carbon, the coal–oil scenario (averaged over 51 years) reveals emissions of 25 Gt carbon per year. The natural gas scenario of this study leads to 16 Gt carbon per year, both much higher than current emission levels, which are calculated to be 7.6 Gt carbon per year (cf. Fig. I.6, where 5.1 Gt carbon was measured, corresponding to "slightly" less fossil fuel combustion approximately 5–10 years ago). So both scenarios may even turn out to be unacceptable with respect to CO_2 emissions, necessitating a natural gas scenario with added CO_2 removal from flue gas, or a combination of extra energy end-use savings, higher generating efficiencies, and CO_2 removal.

I.6. CAPACITY ADDITIONS FOR POWER PLANTS

Table I.3 shows the calculated capacity addition and replacement of power plants for all technologies based on the linear growth model. The values are averaged over 51 years and show the amounts of MW_e per year that must be constructed. The major difference between the two scenarios is evidently the

Table 1.3. Capacity Additions for Power Plants

	Total world 1989 power plants (MW_e)	Average new constructed electrical capacity (in MW_e/yr)[a] over period 1989–2040 if linear growth to coal–oil scenario supply potential			Average new constructed electrical capacity (in MW_e/yr)[a] over period 1989–2040 if linear growth to this studies supply potential		
		Cap. add.	Replacement (5%/yr)	Total	Cap. add.	Replacement (5%/yr)	Total
Gas	333,000	4,800	14,280	19,080	23,560	38,000	61,600
Oil	260,000	9,650	18,560	28,210	620	72,80	7,900
Oil tar–shale	—	6,900	8,750	15,650	22,000	28,000	50,000
Coal	1,204,000	59,240	104,150	163,390	0	9,000	9,000
Nuclear	330,000	?	?	?	0	<16,500	<16,500
Renew							
solar therm./electr.[b]	~500	2,000	2,500	4,500	2,000	2,500	4,500
photovoltaics	~500	6,200	7,900	14,100	28,000	35,700	63,700
wind	>2,000	15,600	19,800	35,400	23,000	29,700	52,700
water	633,600	44,000	35,600 (2.5%/yr)	79,600	25,600	31,300 (2.5%/yr)	56,900
oceans & tidal	} 8,000	} 12,400	15,900	28,300	} 9,300	11,900	21,200
geothermal							
biomass[c]	<100	3,900	10,000 (10%/yr)	13,900	9,300	23,800 (10%/yr)	33,100
Totals (MW_e and MW_e/yr)	~2,771,700	164,690	237,440	402,130	143,380	233,680	377,100

[a]Note that the data are based on linear growth (assumed is a constant growth rate per year over the total 50-year period); with compound growth rate the numbers are higher, e.g., for natural gas "total" figure in the last column, average capacity installation is 94,000 MW/y.
[b]Assume: 10% conversion to electricity of all solar thermal energy.
[c]Assume: 10% biomass energy used for electric power, at 25% generating efficiency; 90% biomass energy for heating.

number of coal and gas plants. In the coal–oil scenario the construction of new power plants (addition plus replacement) contributes to an average of 160,000 MW_e per year, whereas in the gas scenario an average of 61.600 MW_e per year of new natural-gas-based installed capacity will be necessary. The major difference is the scale of the plants involved. Whereas for cost reasons coal plants tend to be large, natural-gas-based plants, even those using existing technologies, can be built in small capacities. This argument, plus the exhaustion of the fuels as calculated in Table I.2, makes the natural gas scenario more likely. As can be derived from Table I.3, all technologies other than coal- and gas-based do not require unrealistically high and extensive capacity installations per year. In fact, given the indications about the potential market, the electric utility industry could have a very healthy future.

With respect to the role of fuel cells, three major market segments are expected:

1. The natural gas market, in which fuel cells are expected to gain a large market share. As an illustration, the Japanese Ministry of International Trade and Industry announced that in the year 2000, 1100 MW_e of PAFC should be installed in Japan, and in the year 2010, 5800 MW_e. These are impressive targets that Japanese fuel cell manufacturers (and perhaps some others) will have to make a strong effort to comply with. In the scenario of increased gas usage, these Japanese numbers may represent just the tip of the iceberg.
2. Biomass energy, for which we have assumed 10% to be used for electric power generation, at 25% efficiency, provides a potential market of 30,000 MW_e per year. Because of the enhancing effect of CO_2 on hydrogen production via steam reforming, the fuel cell market share for these applications will become very large once mature technology at low maintenance levels is available.
3. Photovoltaics, other solar electricity generators, and wind energy will, for many applications, need to be complemented with intermediate storage of energy and fuel cells to produce at the time (and perhaps at the location) where the power is needed.

It can be concluded that for most of the future power markets the fuel cell will be an excellent competitor, especially if all its additional advantages (low emissions, siting flexibility, power quality, flexibility for the planner, etc.) are taken into account.

Even in the coal scenario, the potential for fuel cells to penetrate in the sectors mentioned show average potential markets of perhaps over 33,000 MW_e per year. This market size would be comparable to the current yearly production of gas turbines for power electricity generation.

The trend from centralized to decentralized power production seems to be irreversible. The current worldwide electric power production is largely based on centralized networks and has been shown to have a number of disadvantages (due to emissions, transmission losses, visual pollution, distances between production site and end user, long lead times for plant construction), and, perhaps most significant, involves large and long-term financing requirements. Electric power planners plan 10 years in advance. Special deals are made with governments, tax benefits granted, monopoly positions created, and long-term feedstock prices guaranteed, all just to make it possible to build large power plants. The rich part of the world has apparently succeeded in doing so because of its financial strength. The most populated part of the world, where most growth will be, will not likely be building central power plants in essentially the same manner. Low investment costs, phased installations, and shorter term planning represent the only realistic possibility for these countries, and this will lead to decentral power generation systems. Logistically this also makes more sense from other points of view, such as infrastructure requirements, distances to population centers, and political instability. Modern technologies, and especially fuel cells, will play an increasingly important role here, since for the first time these technologies enable decentral units to be highly efficient and cleaner. Furthermore, the central power producers themselves are becoming increasingly aware of the advantages of such small dispersed systems. The obvious advantages for planning, the trend of power distributors to become service-oriented rather than production-oriented, and the possibility for phased investment and short delivery times are important driving forces that will ultimately make fuel cell systems commercially successful.

1.7. CONTENTS OF THIS BOOK

The editors were most fortunate to find a distinguished group of internationally renowned experts prepared to make contributions to this book. As a consequence, this book contains the expertise not only of fuel cell R&D but also of the system technology involved and its economics. All chapters give concise overviews of the subject involved, with each author giving his or her logical explanations. Due to the self-contained nature of each chapter, a number of issues have been repeated, sometimes from a different point of view.

The book begins with a description of the early history of fuel cells (Chapter 1), describing the pioneering efforts, failures, and successes over the last 150 years.

Chapter 2 reviews the principal aspects of fuel cell technology and is an overall introduction to how fuel cells work, what they are, and their respective progress in R&D.

In Chapter 3, a treatment of the electrochemical (thermodynamic and kinetic) aspects of fuel cell operation of all major types is given. Chapter 4 discusses the conversion of fossil fuels to the fuel cell's major fuel, hydrogen. This "fuel processing" explanation is followed by Chapter 5, which extensively describes the general characteristics of fuel cell systems, including a coverage of efficiency measures, environmental aspects, and the advantages of fuel cells for decentral power production and other applications.

Chapter 6 contains a systematic treatment of design and optimization of fuel cell power plants, based mainly on external processing of fossil fuel and integrated power production.

Chapters 7–11 treat the research, development, and demonstration (R, D&D) of the most important fuel cells types: alkaline fuel cell (AFC), phosphoric Acid Fuel cell (PAFC), molten carbonate fuel cell (MCFC), solid oxide fuel cell (SOFC), solid polymer (electrolyte) or proton conducting fuel cell [SP(E)FC].

These chapters are built on the same principal structure, but since some of the cell types have almost reached commercial application and others are still in R&D stage, different aspects of fuel cell R, D&D have been emphasized in each of Chapters 7–11.

Chapter 12 reviews the economics of power plants at different capacities, with major emphasis on natural-gas-fired PAFC and MCFC power plants with external reforming, and shows the effect of cogeneration, mass production, and economy of scale on system cost and cost of electricity (COE). It also includes a comparison with the COE of other technologies.

Chapter 13 reviews marketing aspects, identifies most likely sectors in which fuel cells will be introduced, and highlights the considerable market potential of the technology.

1

History

J. A. A. Ketelaar

1.1. INTRODUCTION

The development of the fuel cell (FC) covers 150 years, since invention in 1839. This long period has been characterized by ups and downs.

Such a very long term technological development time, without commercialization of any large-scale applications, has to be considered as an exceptional case for an important invention. Just a few inventions from the 19th century in the area of energy transformation devices include the electric motor, the dynamo, the steam turbine, the internal combustion engine and the Hall–Héroult molten salt electrolysis process for the production of aluminum.

The great promise of fuel cells as a means for efficient production of electric energy from the oxidation of a fuel was recognized nearly from the start, just as the advantages of the other inventions mentioned were soon recognized.

In 1896, W. W. Jacques thought that the fuel cell, in this case the direct coal fuel cell, would make the dynamo (1867) an antique device, and he described a large ship with propulsion from fuel cells.

Each of the main fuel cell types studied today has a long history. Only one type, the direct coal FC, has been abandoned, but it was not until 1939.

Instead of comprehensively reviewing past fuel cell research, this chapter centers on past investigations which are related to present-day fuel cell research and technology.

1.2. THE FIRST 100 YEARS, 1839–1939

1.2.1. William R. Grove (1811–1896) Invents the Fuel Cell

William Grove, a barrister in Swansea, Great Britain, was the true inventor of the fuel cell. Later he became a professor of experimental philosophy at the Royal Institution in London and then its vice-president. Still later he was knighted and became a Lord Justice.

J. A. A. Ketelaar • University of Amsterdam, Park de Eschhorst 6, 7461 BN Rijssen, The Netherlands.

Fuel Cell Systems, edited by Leo J. M. J. Blomen and Michael N. Mugerwa. Plenum Press, New York, 1993.

The first hydrogen–oxygen, very dilute sulfuric acid FC was described by Grove in a postscript from January 1839 to a letter he wrote in December 1838 to the *Philosophical Magazine.*[1] The new cell consisted of two platinum strips surrounded by closed tubes containing hydrogen and oxygen, respectively, formed by preliminary electrolysis of the electrolyte.

Three years later in a remarkable paper,[2] Grove showed that he was already well aware of the fact that the three-phase contact is essential: "As the action could only be supposed to take place with ordinary platina foil, at the line or water-mark where the liquid, gas and platina met." He sought to extend this contact by coating the electrodes with "spongy platina."

Grove also reported the electrolysis of potassium iodide solution and of water by 4-cell and 26-cell series batteries, respectively (Fig. 1.1). He observed that the hydrogen was used up twice as fast as the oxygen in his FC. He now used stronger sulfuric acid, hydrogen obtained from zinc and sulfuric acid, and substituted air for oxygen.

Grove was also troubled by the lack of invariance of cell performance, which is still the major problem of the present-day fuel cell.

Shortly after,[3] Grove reported on a hydrogen–chlorine fuel cell, and he found that "other volatile bodies such as camphor, essential oils, ether and alcohol associated with oxygen gave a continuous current." He clearly stated "that every chemical synthetic action may by a proper disposition of the constituents, be made to produce a voltaic current." Grove also foresaw the fuel cell as a possible source of commercial electricity,[4] especially if hydrogen as a fuel could be replaced by coal, wood, or other combustibles.

It can be said that the very first fuel cell invented by Grove foreshadowed the phosphoric acid fuel cell (PAFC) of more recent date.

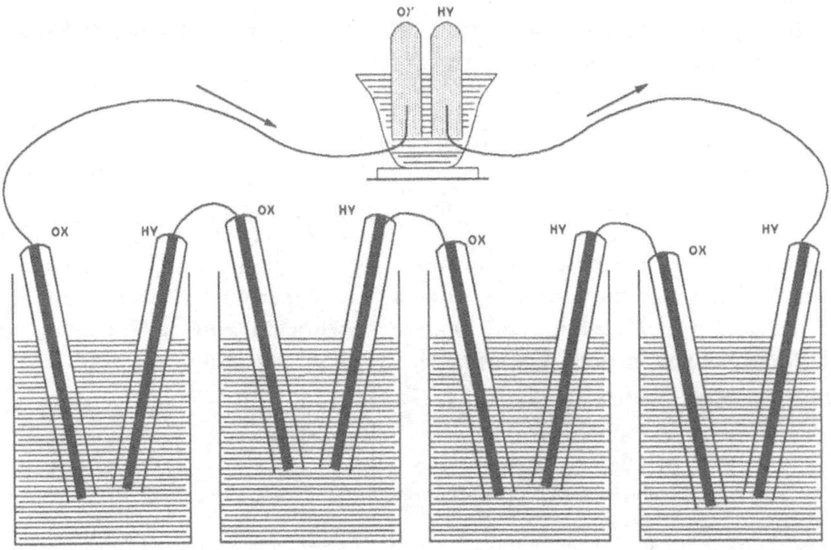

Figure 1.1. Four cells of Grove's battery to drive an electrolytic cell, 1842.[2]

1.2.2. Successors to Grove

An improvement of the Grove cell made by using porous platinized coke as electrodes was the subject of the first U.S. patent on the fuel cell, assigned to Vergnes in 1860.[5] In 1913 the same electrode was described by Siegl.[6] For the first time he announced that his gas element (hydrogen or town gas with air or oxygen) was commercially available from an industrial producer. This again foreshadowed today's practice.

An important step forward in the development of the H_2–O_2 fuel cell is due to Mond and Langer.[7] They found, as Grove had observed, that the efficiency of the platinum black coating of the electrodes was soon lost if it became wet. Thus they constructed a FC with a diaphragm from plaster of Paris, earthenware, asbestos, or pasteboard, impregnated with dilute sulfuric acid. The electrodes of thin perforated platinum foil, coated with platinum black, were placed on the dry backside.

Mond and Langer obtained a current density (CD) of 2.8–3.5 mA/cm^2 at 0.73 V with cells of 700 cm^3 surface area, covered with 0.35 g of Pt foil and 1 g of Pt black! The efficiency was calculated by them to be 50% of the heat of combustion. They recognized the strong polarization (overvoltage) of the oxygen electrodes, and the roles of internal resistance and concentration polarization. Also, stacks of cells in parallel were constructed. As cheap fuel impure hydrogen from the reaction of steam with coke, coal, or iron was considered, and the need to remove CO was recognized, but they presented no method for doing so.

The fuel cell concept obtained a boast from W. Ostwald,[8] one of the godfathers of physical chemistry. He clearly stated the advantages of producing electricity from the direct conversion of the (free) energy of the fuel combustion reaction in a galvanic cell, over the usual way via steam engine and dynamo; the latter process at that time had an efficiency of only 10%! Ostwald also mentions the possible environmental advantages: "no smoke, no soot, no fire."

However, the same fundamental points had already been made 14 years earlier by C. Westphal in a patent.[9]

It seems curious that, starting with Grove, for a long time only sulfuric acid was used as an electrolyte. It was in 1902 that J. H. Reid first described fuel cells with aqueous KOH;[10] P. G. L. Noel also described them in 1904.[11]

The present alkaline fuel cell thus made its entrance 63 years after Grove's invention of the acid FC.

1.2.3. The Direct Coal Fuel Cell

Since during the nineteenth century coal was the only primary fuel in large-scale use, much attention was given to fuel cells based on coal as fuel, working at higher temperatures with molten salts as electrolyte. A. C. and A. E. Becquerel in 1855 and later, P. Jablochkoff used a carbon rod in molten sodium nitrate in a platinum or iron crucible serving also as counterelectrode.[12,13]

Molten potassium or sodium hydroxide, mostly containing some water, were used as electrolyte by several fuel cell inventors of direct coal cells, especially by the American engineer J. J. Jacques, who performed large-scale experiments (1896).[14] He constructed by far the largest batteries ever built and tested during this 100-year period. Rated 1.5 kW, they consisted of large cells with iron vessels, serving as cathodes and with molten KOH at 400–500°C. A carbon rod served as anode, and air was blown in at the bottom near the wall. There were 100 cells in series, delivering a current density of $100\ mA/cm^2$ at 1 V. Jacques was well aware of the difficulty with the formation of carbonate from carbon and the supply air. He used pure carbon instead with a small ash content.

The second unique feature was the reported production of electric power for as long as six months, instead of only a few hours as in all other tests in this period.

Though the fuel cells lacked invariance and since few reliable data are available, the work of Jacques drew much publicity and is of great historical interest.

Later it was clearly shown that in all these direct coal FCs the galvanic, electrochemical reaction (*E*) was not the oxidation of carbon, but the carbon first reacted chemically (*C*) with the electrolyte.

It was W. Borchers who, in 1897, concludes that a direct carbon FC could never be expected to perform satisfactorily for several reasons (e.g., the ash content soils the electrolyte), but that the major difficulty was the chemical reaction between carbon and electrolyte.[15]

Liebenow and Stresser published many measurements on the potential of carbon and of active and passive iron in molten KOH and on the open-circuit voltage (OCV) of the cell: C–KOH–Fe.[16] They also observed slow dissolution of carbon in molten alkali. This was the first fuel cell publication from a laboratory of an industrial producer, namely Accumulatorenfabrik AG, later Varta AG (Germany).

1.2.4. F. Haber and His Pupils

Haber and Brunner[17] fully elucidated the mechanism of the so-called direct carbon fuel cell. The passive iron, just as platinum, forms a well-defined oxygen electrode, especially in the presence of small amounts of manganate, as an impurity present in KOH.

Instead of the expected (*E*) anodic oxidation of carbon,

$$\text{(E)} \qquad C + 6OH^- \rightarrow CO_3^{2-} + 3H_2O + 4e,$$

in reality with some moisture present, the *CE* (chemical + electrochemical) mechanism proved to be

$$\text{(C)} \qquad C + 2OH^- + H_2O \rightarrow CO_3^{2-} + 2H_2$$

followed by

$$\text{(E)} \qquad 2H_2 + 4OH^{2-} \rightarrow 4H_2O + 4e$$

Thus the carbon fuel cell came to be a H_2–O_2 fuel cell. Analogous reactions occur with CO and generator gas.

Related to the (*CE*) mechanism of the direct carbon fuel cell is that of the CO fuel cell extensively studied by K. A. Hofmann.[18] Here the fuel electrode is a copper strip in alkaline solution, combined with a platinum–air electrode. Thus, in reality the former electrode is a H_2 electrode. Moreover, the electrolyte is not invariant because of the formation of carbonate.[19]

Haber and his pupils investigated the thermodynamical aspects of the reversible EMF of the H_2–O_2 fuel cell.[20,21] They measured the OCV over the temperature range 340–580°C for different pressures and concentrations of both gases. Haber used as electrolyte a thin glass disk covered on both sides with platinum or gold and joined on both sides to tubes containing the gases. The correspondence between the precise experimental and theoretical values of the OCV was excellent. These cells can be characterized as the first fuel cell type with an ion exchange membrane (solid polymer fuel cell, or SPFC), the polymer being inorganic rather than organic in nature as in more recent fuel cells of this type.

Haber's pupil Beutner, however, was aiming at applications with a high-temperature fuel cell.[22] He used palladium foil as a diffusion H_2 electrode and different molten salts (e.g., KF + NaCl and Li, NaCl) at 600–800°C as electrolytes. Current densities up to 14 mA/cm^2 were realized.

1.2.5. W. Nernst

Well known for his many contributions to physical chemistry, Nernst was also the inventor of the Nernst glower or lamp in 1897. It consisted of a thin rod or tube of a high-temperature anion-conducting solid material from zirconium oxide with 15% yttrium oxide. This is the prototype of the electrolyte in the present SOFC (solid oxide fuel cell). It is reported that Nernst had already made such a fuel cell in 1900.

He also invented the redox cell as an indirect H_2–O_2 fuel cell.[23] The electrolyte is an acid solution with multivalent ions of, e.g., titanium, thallium, or cerium, which can be oxidized and reduced by oxygen and hydrogen, respectively. The electrolyte solution flows through both halves of the galvanic cell, separated by a diaphragm. The chemical regeneration (*C*) takes place in the outside parts of both cycles.

$$\text{Anode (E)} \quad X^{2+} \rightarrow X^{4+} + 2e,$$

$$\text{(C)} \quad H_2 + X^{4+} \rightarrow X^{2+} + 2H^+;$$

$$\text{Cathode (E)} \quad X^{4+} + 2e \rightarrow X^{2+},$$

$$\text{(C)} \quad X^{2+} + \tfrac{1}{2}O_2 + H_2O \rightarrow X^{4+} + 2OH^-.$$

Overall reaction: $H_2 + \frac{1}{2}O_2 \rightarrow H_2O$, where X could be cerium. The advantage is that with slow regeneration reactions compensation could be found by larger outside volumes.

E. K. Rideal and his school worked for a long time on redox cells. However, with all redox cells proposed, reactions were too slow to be usable for a fuel cell.[24]

Nernst collaborated for some time on fuel cells with researchers at Accumulatorenfabrik AG, but no product was ultimately realized. Of his many pupils only W. Schottky published a theoretical paper in which he proposed a SOFC based on the Nernst mass as solid electrolyte.[25]

Many suggestions for fuel cells were made in patents, however, in general without performance data and in many cases the patent was not based on experimental results. Nevertheless hydrophobing by paraffin of porous electrodes to prevent drowning was described by Jungner in 1920.[26] Now this is generally done with the help of teflon (polytetrafluoroethylene). To overcome the same defect, double-porosity gas electrodes were used by A. Schmid to obtain current densities of 14.5 mA/cm^2.[27] He perfected the gas diffusion electrodes very much indeed.

A patent from Niederreither[5,9] covered a H_2–O_2 fuel cell with gas pressures up to 600 atm, though very probably not based on experiments; much later the idea found application in the work of Bacon (see Section 1.3.1).

1.2.6. E. Baur and His School

In the last part of the first 100-year period of fuel cell development, the field was dominated by E. Baur (1873–1944) and his pupils, first in Braunschweig, but mostly in Zürich, where he resided as a professor of physical chemistry at both (technical) universities.

One of his first pupils, I. Taitelbaum,[28] made a study of the fuel cell with molten NaOH (380°C) with the addition of manganate or vanadate as catalysts in a passivated iron crucible, comparable to the Jacques cell. Instead of slowly reacting carbon, he added many kinds of reactive materials to the melt, such as sugar, CO, lignite, town gas, sawdust, and heavy oil. He operated his cells, which as Haber and Brunner[17] had shown were really H_2–O_2 cells, for up to 6 h and used voltage–current diagrams. It is remarkable that here for the first time a diaphragm of porous MgO was used, both as a porous mass and as a powder. It served as a prototype for the molten carbonate cell (MCFC) of much later date with a matrix electrolyte (Davtyan, Ketelaar, and Broers etc.)

In a subsequent publication by Baur and Ehrenberg,[29] measurements were reported at the much higher temperature of 1000°C with molten silver as oxygen cathode and a carbon rod, platinum–CO, or platinum–H_2 as anodes.

As electrolyte they used not NaOH but molten salts, such as K, Na carbonate, borax, and cryolite. Now the true reversible C–O_2 potential could be measured as 0.97–1.01 V (calculated 0.994 V) for the Boudouard equilibrium of carbon with 99.2% CO and 0.8% CO_2. Also the H_2–O_2 fuel cell was studied with iron as anode. At a cathodic current density of 40 mA/cm^2 and anodic of 22 mA/cm^2, the voltage was still 50% of the open-circuit voltage. The experiments lasted up to 1.5 h.

Later Baur, Treadwell, and Trumpler stated that they had shown it was technically possible to construct stable, powerful cells.[30] However, it seemed doubtful whether their cells would be economically useful too. In these different constructions the matrix electrolyte of molten Na and K carbonate in a porous MgO ceramic was used at 750–800°C. The H_2 or CO anode was small pieces of thin iron wire, and the air cathode was magnetite or oxidized iron. The fuel cells were in use for 6 h at most. The authors were aware that further oxidation of the air cathode would result in increasing internal resistance and loss of performance.

Baur and his pupils tried many more modifications of fuel cells, especially with different forms of the air electrode, in the form, for example, of small carbon grains.[30]

Accidentally, when blowing into the air electrode, they observed the beneficial effect of adding CO_2 to the airstream at the cathode to prevent concentration polarization in the molten carbonate electrolyte.[31] Much later the addition of CO_2 to the air was the subject of patents by Gregor in 1939.[5]

In an extensive study of the H_2–O_2 fuel cell, Tobler used platinized graphite with hydrophobing by paraffin as the hydrogen electrode.[32]

In two papers Baur and Preis reported on fuel cells with solid electrolytes.[33] The interest was due to a theoretical paper on the subject by W. Schottky.[25] They tried several ceramics from clay and kaolin with additives, such as CeO_2 and lithium silicate. However, only Nernst mass, ZrO_2 with 15% Y_2O_3, had a sufficient high conductance.

Figure 1.2 shows a cell where the solid electrolyte is a thin-walled tube from Nernst mass, closed at one end and filled with coarse coke or iron powder as the oxygen electrode. The tube is placed in a mass of coarse magnetite powder as the anode through which the fuel gas, H_2, CO, or town gas, circulates. A battery of eight tubes in parallel in a common magnetite bath gave an OCV of 0.83 V (0.2 V below the theoretical value). With a current of 70 mA (equal to 0.35 mA/cm^2 of active surface) the voltage dropped to 0.65 V. The internal resistance of a single cell was about 2 Ω.

In the second publication the authors described a single cell with once again a thin tube filled with coke powder, now placed in a wider earthenware or porcelain tube filled with magnetite.[33]

They tested many other conducting materials, perhaps for economic reasons, as none performed better than Nernst mass. The conducting material consisted of three parts of tungsten oxide (WO_3), one part of CeO_2 (as burned monazite ore), and two parts of clay, the latter being added to give sufficient plasticity to the green mixture.

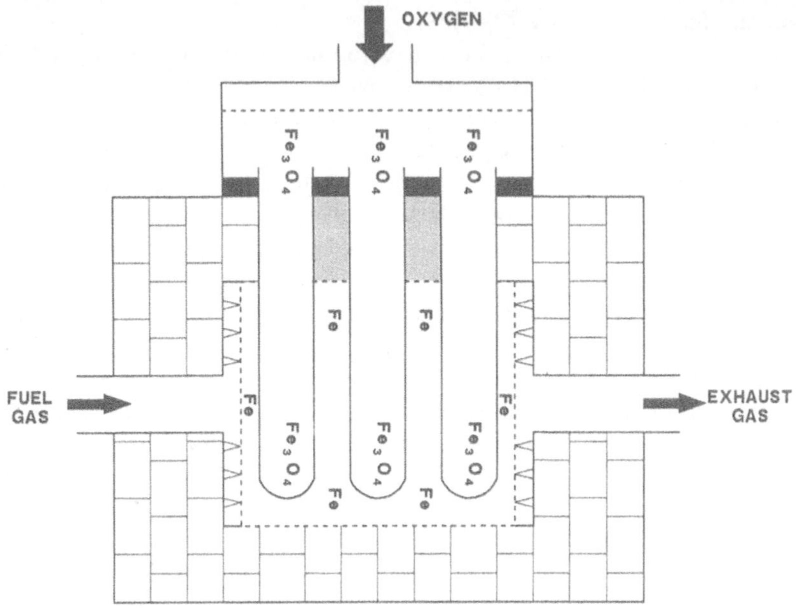

Figure 1.2. Baur and Preis, Solid electrolyte fuel cell, 1937.[33]

They made a battery of 18 cells in six units in series, each with 3 cells in parallel. The stack gave 55 mA at 90% of the OCV of 0.7 V per cell. For the stack a current–voltage relation as a function of time was given.

In 1939, in his last publication Baur provided a review.[34] He stated that in the whole period since the discourse of Ostwald in 1894,[8] the target that had been proclaimed to be a galvanic cell which would directly deliver electric energy with a high efficiency from the heat of combustion of a fuel had not been reached (in nearly 100 years!). He also said that an unsurpassable barrier to further development might well be the difficulty of building large blocks from very thin and fragile elements at high temperatures.

For future research he felt that, in the first place, molten NaOH or sulfuric acid should be tried as electrolytes in a temperature range of 200–300°C. If electric polarization would be too large because of too slow reactions, then one would have to go to even higher temperatures. The electrolyte would have to be a molten salt (e.g., Na, K carbonate). The molten salt should be contained in porous magnesia to prevent wetting of the likewise porous electrodes.

The only alternative would be the solid electrolyte cell at 1000–1100°C where polarization is absent. However, conductivity of ionic conductors is too low.

Baur thus predicted that development today indeed centers on PAFC, MCFC, and SOFC. He reasoned that in the future a volume specific power of $3\,kW/m^3$ could be reached on improving from the present 1 W/l at 80% OCV. He compared this with diesel engines with values of $30–70\,kW/m^3$.

Imagining a cube of 10 m edge as a unit, there was a great fear that in the case of a small defect inside, the whole block would have to be demolished, just as a partially solidified blast furnace.

Baur's last sentence was: "even when in the end only 50% of the combustion energy of the fuels could be delivered as electric energy at the switchboard of the fuel cell power plant it would be a revolution in the energy economy of the world." This is especially understandable, since at that time a new conventional plant had an efficiency of only 20–30%.

In summary, around 1940 fuel cell development had apparently reached only a relatively low level after 100 years of intensive research.

To characterize this era it can be said that through investigations of many people, including outstanding scientists and engineers such as Grove, Jacques, Mond, Haber, Nernst, and Baur and his pupils, work had been done on different types of fuel cells, all precursors of the present fuel cell types. Most of the problems, but also most of the measures taken nowadays to overcome such problems, were discovered.

Much insight was obtained. Advantages and disadvantages, and the great promises and the great problems associated with this source of electric energy became well understood. However, all experimental work, with the possible exception of that of Jacques,[14] was restricted to small size cells, at best arranged in stacks of some 20 cells delivering small currents for only a short time.

1.3. THE NEXT 50 YEARS, 1939–1989

Three men, F. T. Bacon, O. K. Davtyan, and E. W. Justi, have taken fuel cells from research into technological development at the very beginning of this modern period.

1.3.1. F. T. Bacon (1904–1992)

Bacon was engaged as an engineer by C. A. Parsons and Co. Ltd. from 1925 to 1940. In 1937, he proposed to combine pressurized water electrolysis with intermediate hydrogen storage, combined with a pressurized hydrogen–oxygen fuel cell. In order to eliminate the use of precious metals and to increase current density, Bacon, in 1939, built a single (high-) pressure reversible cell with 27% KOH at 100°C, with gas pressures of 220 bar with a current density of 13 mA/cm^2 at 0.89 V as best performance.

In 1946, Bacon returned to Cambridge University to work first with E. K. Rideal, sponsored by the semigovernmental Electrical Research Association.[35,37] During that period a new fuel cell was built, based on feed gases from cylinders, with double-porosity sintered nickel electrodes. The oxygen electrodes were coated with lithiated nickel oxide by impregnation with a LiOH solution and heating to 700°C; this resulted in a corrosion-resistant electrode. In 1954, a six-cell battery was built with 12.5-cm-diameter electrodes which could generate 150 W at 41 bar and 200°C. It was shown at an exhibition and the first publication appeared.[35] Once again no interest was shown by British industry.

However, the newly formed National Research Development Corp. acted as a sponsor, and in 1956 Bacon could start further development work. In 1959, a 40-cell, filter-press-type battery with 25-cm-diameter electrodes was built. This battery could deliver 6 kW at a current density of 700 mA/cm^2 at 200°C and a pressure of 38 bar. It could be used for a forklift truck and for welding, and a demonstration was given in August 1959.

In the meantime, Pratt & Whitney Aircraft in the United States had obtained licenses to Bacon's U.S. patents. In September 1959, Allis-Chalmers Manufacturing Co. demonstrated a tractor driven by 15 kW of low-temperature, low-pressure hydrogen–oxygen fuel cells, built up from a much larger number of 1008 cells than the Bacon 40-cell battery.

From 1961 a new industrial firm, Energy Conversion, was formed in Great Britain to develop fuel cells commercially with Bacon as a consultant. This firm joined forces with Pratt & Whitney in developing a battery for the Apollo space missions.

For weight reasons the cell pressure was lowered, but the temperature was increased to 200°C, with 75% KOH (thus lowering the vapor pressure to 0.5 bar), whereas the total pressure was 3.5 bar. The battery consisted of three units of 100 kg each with 31 cells in series and a 57-cm diameter. Each unit could deliver 1.4 kW at 27–31 V with an average of 0.6 kW. Module life was limited by the slow corrosion of the cathode coated with lithiated nickel oxide, and design life was only 400 h.[36,37]

Commercialization for other applications was seriously hampered by the heavy weight, the sensitivity to impurities in the gases, and the complicated controls necessary for the regulation of water removal by evaporation from the H_2 side. After the Apollo missions no further high-pressure cells were built.

With the space missions the fuel cell found its first real application. For the same performance primary and secondary batteries would have had a much greater weight. Moreover, in this application of the fuel cells, cost and a long lifetime were not issues.

1.3.2. O. K. Davtyan (USSR, Moscow, Tiflis, Odessa)

Davtyan began fuel cell research by investigating the possibility of converting the "poor" gas from underground coal gasification into electric energy. In the USSR extensive experiments were made at that time with this method of coal exploitation.

He probably started his research in 1938, but the first two publications appeared in 1946.[38] His choice was a high-temperature cell at about 700°C with a solid ionic conductor as electrolyte. Davtyan, who was certainly familiar with the publications of Baur and Preis,[33] chose a melted mixture of 27% prebaked monazite sand, 20% WO_3, 10% sodium glass, some clay, and, as a new ingredient, 43% Na_2CO_3. He observed a very low specific resistivity of this preparation of 13.6 Ω · cm at 700°C and 1.3 Ω · cm at 900°C. These values are much lower than those of Baur and Preis, who had observed 100–150 Ω·cm at about 1000°C. Davtyan thought that the tungsten and cerium present would have catalytic action on the electrode reactions, because of the presence of different valencies of these elements.

In the second publication he described the construction and performance of a fuel cell with this electrolyte in the form of a 4-mm-thick disk. The porous electrodes were made of baked slurries with 60% Fe_2O_3 and 20% clay, with 20% iron powder for the fuel gas anode and with 20% magnetite, Fe_3O_4, for the air electrode, respectively. The cell was operated on generator gas and on town gas. At 700°C with generator gas at a current density of 20 mA/cm^2, a voltage of 0.79 V was observed, compared to an OCV of 0.85 V (which is rather low).

The experiments were conducted over 10–20-h time periods. Reproductions of Davtyan's cell are given by Euler[9] and Broers.[39]

Davtyan and co-workers, from 1946 to 1971, published many papers on subjects related to fuel cells,[40] almost all in Russian. A bibliography is given by Euler[9] and Broers.[39]

Later V. S. Daniel-Bek followed up the work of Davtyan.

1.3.3. E. W. Justi (Technical University at Braunschweig, Germany)

Justi started research on fuel cells around 1948. Soon he shifted to the low-temperature alkaline H_2–O_2-type fuel cell. The innovation here was the circumvention of the need for noble metals by his invention, together with A. Winsel, of the DSK (double skeleton) fuel cell electrodes.[41] Raney nickel (1925) is well known as a hydrogenation catalyst in organic chemistry; it is obtained from 50% aluminium–nickel alloy after the aluminium has been dissolved with 30–50% KOH at 80–100°C. For the DSK electrode this alloy was intimately mixed with pure nickel, both as powders. The mixture was compressed at high pressures, sintered, and the Al was then dissolved. The DSK electrode thus consisted of very fine, very active nickel in a porous, ductile skeleton of nickel. The electrode has a uniform pore diameter, thus in principle eliminating the need for hydrophobation.

The H_2–O_2 fuel cell with DSK electrodes in 6 N KOH at 67°C could deliver a current density of 250 mA/cm^2 at 0.615 V, half of the OCV. The power volume density of 0.256 W/cm^3 was somewhat lower than the value of 0.355 W/cm^3, which Bacon had realized at 200°C and 41 bar.

Long-term continuous testing of a three-cell stack at 31–35°C did not show any differences in the current–voltage curves over a period of 18 months. The water formed was removed by replenishing the electrolyte. An electrode lifetime of 1000 A-h/cm^2 was observed, equivalent to 10,000 h, at a current density of 100 mA/cm^2. At the optimal working temperature of 80°C results might even be considerably better.

Varta AG further developed these cells. The electrodes were improved by adding layers with small pores DSK electrode to both sides of a layer with larger pores which permits entrance of the gas (Janus electrode). Larger stacks were made, e.g., 20 cells in filter press and in staple form. Siemens AG also produced fuel cell batteries of about the same type.

Both firms found applications in television repeater stations, with 100-W, 12-V and 25-W, 28-V batteries, respectively (Ref. 9, p. 26). The hydrogen and oxygen were supplied from pressure cylinders. Installations of both types have been operational for several years without any defects. For removal of the water formed, the circulating electrolyte had to be replaced every three months, depending on the volume of the electrolyte storage vessel.

Siemens has continued the development of a 7-kw alkaline fuel cell (AFC) battery.[42,43]

1.3.4. Molten Carbonate Fuel Cell

Early in 1950, J. A. A. Ketelaar and G. H. J. Broers began FC research at the University of Amsterdam under contract of TNO, the Dutch semigovernmental organization for applied research. While repeating Davtyan's experiments, Ketelaar and Broers rapidly saw that neither the electrolyte nor the electrodes were invariant over a period of a few days. From the temperature dependence of the resistivity of Davtyan's complex electrolyte (Section 1.3.2) it could be shown that this electrolyte was not a real solid at the working temperature of 650–750°C. It was indeed a matrix or paste electrolyte,[44] the matrix consisting of solid La–Ce–Th oxide resulting from the transformation of the phosphates of these elements constituting the monazite ore. The liquid phase consisted of a mixture of the phosphates, wolframates, and silicates of sodium, together with excess sodium carbonate. This was confirmed by chemical analysis.

The next step was to take a simpler and cheaper mixture by using porous sintered MgO disks impregnated for 40% with a low melting mixture of Li_2–K_2CO_3 or the ternary Li_2–Na_2–K_2CO_3 with compositions approximately corresponding to eutectic.[39] Later, MgO powder was used as matrix. Ketelaar and Broers used thin layers of metal powder as electrodes, covered by metal gauze, silver being used for the O_2, CO_2 cathode, and nickel for the fuel electrode with perforated stainless steel disks as supports (see Fig. 1.3).[39,45–47]

The cells were operated up to 4500 h on town gas, hydrogen, carbon monoxide, and natural gas (with steam added). Polarization was absent. However, the internal resistance rose slowly with time, and this was found to be due to the loss of electrolyte, especially Li_2CO_3, by evaporation and from leakage to the asbestos spacers. A current density of 50 mA/cm^2 at 0.7–0.8 V could be reached.

In 1963, some electrolyte samples were made with MgO replaced by Al_2O_3 powder. Experiments with a premelted paste with the ternary carbonate mixture

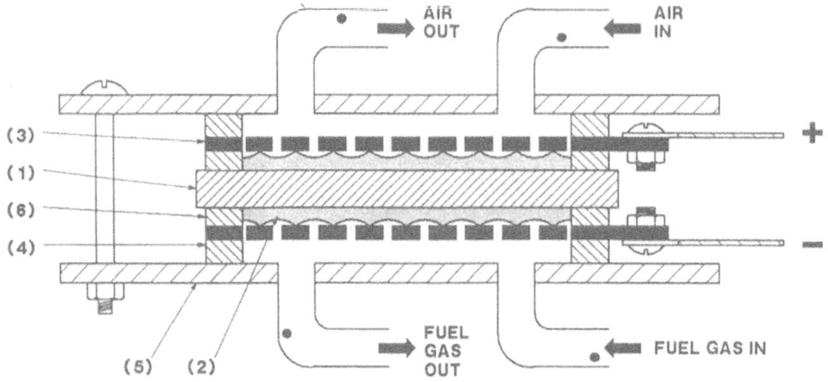

Figure 1.3. Ketelaar and Broers, Molten carbonate fuel cell, 1958.[39] (1) electrolyte disk, (2) fine metal powder and metal gauze support, (3) perforated stainless steel plate, (4) asbestos gasket, (5) steel cover plates, (6) mica ring gaskets; the bolts clamping the cell together are also isolated by mica rings.

gave a surprisingly good performance, with a current density of 100 mA/cm^2 at 0.7 V, which could be maintained for six months. This appeared to be due to the formation *in situ* of a solid phase of $LiAlO_2$, lithium aluminate. It is in use worldwide as matrix material.

In the presence of steam and nickel, methane was shown to be a satisfactory fuel. This led to the concept of internal, instead of external, reforming of natural gas by adding the appropriate catalyst to the gas chamber of the anode, thus using directly the heat of the fuel cell.

Postmortem analysis of electrolyte disks showed that after long-term use some silver from the cathode had been dissolved in the electrolyte and precipitated at two thirds of the distance from the cathode. Notwithstanding the promising results, the work was stopped by TNO in 1969 as industry lacked interest, and the economics in competition with conventional and nuclear power looked poor at the then prevailing low prices of oil and gas. Also, worldwide the great interest for fuel cells in the 1960s was waning around 1970. However, the oil crisis in 1973, with its higher oil and gas prices, led to renewed interest.

Many others were for some time engaged in MCFC research, including Gorin at Consolidated Coal, Texas Instruments, Weiniger and Douglas at General Electric Research (U.S.), Gaz de France, Chambers and Tantram (G.B.), but only General Electric continued the effort until now. See the extensive literature and the relevant chapters in this book.[9,14]

However, special mention will be made of the efforts for the MCFC at the Institute of Gas Technology, Chicago, in 1960 by B. S. Baker and his colleagues, which work continues today.

The point of departure for the program was the work of Broers and Ketelaar. Since then very great progress has been made in the development of the MCFC in all aspects. Advances in performance were impressive, e.g., at 0.7 V current densities rose from 35 to 350 mA/cm^2 in a few years. Important improvements were made on the anode interface and on the matrix electrolyte, accompanied by considerable increase in size of the cell surface area and the construction of cells and stacks; 1–2-kW stacks with 1000-cm^2 cell area have been operated for over 5000 h. This development is in large part due to efforts from United Technologies and Energy Research Corp. At a theoretical fuel utilization of 100%, with internal reforming and at a voltage of 0.73 V, the efficiency may reach 64%. In reality, the fuel utilization will be around 85% and the stack efficiency is thus limited to 54%, based on the HHV of the fuel.

In Japan, now second to the United States in fuel cell efforts, R&D was closed down in 1970 but taken up again in 1981 in the government-sponsored Moonlight project, particularly by the electrical manufacturing companies Mitsubishi Electric, Fuji, Hitachi, and Toshiba. Here an MCFC with a current density of 150 mA/cm^2 and a voltage of 0.74 V per cell, falling to 0.69 V after 10,000 h operation, was realized.[42,43,48]

Fuji Electric has reported on a 24-cell MCFC stack of 7 kW with 2500 cm^2 of cell area and a current density of 150 mA/cm^2 at 0.79 V.

Relatively new in America, Japan, and The Netherlands is the use of tape casting for the $LiAlO_2$ matrix instead of hot pressing of the electrolyte. At the same current density of 150 mA/cm^2 an increase in cell voltage of 0.1 V has thus been obtained.

1.3.5. Solid Oxide Fuel Cell

Schottky's proposal for a true solid electrochemical cell was followed up by Baur and Preis,[33] e.g., by using Nernst mass, ZrO_2, with 10–15% Y_2O_3. In 1951, it was shown that the addition of 15% CaO produces an electrolyte with a slightly higher conductivity. As the conductivity at 1000°C is still only 10% of that of molten carbonate at 650°C, the SOFC electrolyte must be very thin, only 50 microns. Apart from the restriction on thickness, the construction of a single SOFC is in principle simple. For the air electrode La–Sr manganite is used instead of platinum, and the fuel electrode is made of a nickel–zirconia cermet. Because of the high temperature, not only H_2 but also CO and hydrocarbons, together with steam, can be used directly as fuels. However, the OCV is lower than for other FCs because of the temperature effect.

The most advanced program since 1958 has been undertaken by Westinghouse Electric Corp.[42,43,48] Since 1980, the effort was expanded with new fabrication techniques such as flame spraying the subsequent layers. Ingenious solutions were found for interconnecting the cells. The single cells are short rings on a zirconia support tube.

A 5-kW, 24-cell stack was built in 1986, which after preheating could maintain its temperature of 1000°C. The fuel consumption was 85%; burning the 15% cell output inside the system helped here.

Argonne National Laboratory (United States) has built a pressurized monolithic battery with a construction much like corrugated pasteboard.

In Europe, SOFC research is mainly done by Dornier Systems (Germany), starting from their experience in the Hot-Elly program for water vapor electrolysis with solid electrolytes. A multinational project sponsored by the CEC (Commission of the European Communities) was started in 1987.[42]

1.3.6. Phosphoric Acid Fuel Cell

The PAFC as a low-temperature (150–200°C) acid FC is directly related to the 150-year-old first FC from Grove. On the other hand, it also was the last type of FC to be introduced, as late as 1967, as part of the Target program by the American Gas Association and United Technology Corp. (UTC), later International Fuel Cells Corp., a joint venture of UTC with Toshiba Corp. Also, Westinghouse and Engelhard Industries in America and several firms in Japan have developed extensive programs. U.S. efforts were mainly sponsored by the Department of Energy and the Gas Research Institute.

In origin, the PAFC goes back to the fruitless experiments to develop a direct hydrocarbon fuel cell. As sulfuric acid at temperatures of 80–100°C is reduced by the fuel, this acid was replaced by phosphoric acid, with which temperatures up to 200°C could be realized.[14] Though the current density for hydrocarbons as a fuel remained too low, this was not the case for hydrogen as a fuel, even with "dirty" gas containing CO and some H_2S.

Rapid progress was made after graphite had been shown to be a satisfactory construction material. From 1967 to 1983, UTCs unit size grew from 12.5 kW, via 40 kW, to 1 MW, and finally in 1983 a pilot plant of 4.8 MW was built in New York City.[43] Also the necessary platinum amount could be decreased from 20 to 0.5 mg/cm^2 on a high-surface carbon carrier. Performance also rose by operating under pressure and by using a crossflow configuration of ribbed substrate stacks. The 4.8-MW FC power plant was designed for operation at 3.4 bar and consisted of 240-kW, 400-cell units with a current density of 270 mA/cm^2 at 0.65 V.

In Tokyo a similar UTC pilot plant of 4.8 MW operated very successfully. For Japanese industrial development of fuel cells see Kishida[42,43,48] and the relevant chapters in this book.

Westinghouse's PAFC differs from that of UTC by using air instead of water cooling.

In both the United States and Japan, and recently in Europe, a number of small dispersed PAFC power units of different capacities are operating. The overall efficiency for the large units is now 40–45%.

1.3.7. Solid Polymer Fuel Cell

For the NASA Gemini space missions from 1962 to 1966, SPFC technology, developed by General Electric Company, was used. The electrolyte was a thin (~0.25 mm) membrane of a H^+ ion exchanger, at first a sulfonated polystyrene and later a very stable sulfonated polytetrafluoroethylene (Nafiion).

In the NASA Gemini space program there were two pressure vessels on board with a 1-kW battery of 96 cells in three parallel groups of 32 cells. Power density was 38 mW/cm^2 at a voltage of 0.83 V per cell. Water was removed with a wick in each cell. This was also the weak spot, as it was not very reliable and a dry membrane is known to develop a very high resistance. No other application could as yet be found for this type of FC, because of the heavy platinum load necessary and this sensitivity to drying out.

1.3.8. Alkaline and Direct Methanol Fuel Cells

In Section 1.3.3, AFCs with Justi's DSK electrodes were mentioned.

1.3.8.1. AFC, Union Carbide

At Union Carbide, Kordesch began in 1963 to develop this cell as a power source for vehicles. The electrodes consist of active carbon on a sintered nickel support. A 32-cell battery delivered 1 kW at 24 V, at a current density of 50 mA/cm^2, and with a lifetime of many thousands of hours. He succeeded in reducing electrode thickness. The work stopped in 1969.

In Japan, Fuji Electric in 1972 developed a 10-kW power unit with an AFC at a current density of 100 mA/cm^2 and a hydrogen–air fuel cell for the Moonlight project.

1.3.8.2. AFC, Elenco

Elenco, a Belgian–Dutch firm, started in 1976 to develop an AFC battery for traction. Modules with 24 cells for hydrogen–air operation can deliver 0.5 kW with a current density of 100 mA/cm^2 at 0.7 V per cell. The electrolyte is 30% KOH, and a pair of electrodes contains only 0.7 mg/cm^2 platinum. The lifetime is 5000 h.[42,43,48]

The major application of the 10–15-kW batteries is power sources for city buses, garbage trucks, etc., which would cause less pollution and noise in cities. For starting and sufficient acceleration a secondary battery is also needed (for high power over short time), which is reloaded automatically.

1.3.9. Methanol Fuel Cell

For the application of fuel cells in electric vehicles the ideal solution would be the direct methanol acid FC (DMFC). Methanol is a relatively cheap liquid and is thus more attractive than hydrogen as a fuel for transportation purposes.

Since 1960–1970, Shell and ESSO–Exxon have studied the DMFC extensively. However, the current density obtained remained much too low because of the high overpotential of the CH_3OH anode. Also, the activity of the Pt–Ru catalyst soon drops due to self-poisoning. The achieved current density of only 15 mA/cm^2 at 0.5 V at 60°C is far removed from the required performance.[49] The low power density of 7 W/kg and the high cost of noble metals were considered unacceptable. Recently, Hitachi in Japan and the CEC in Europe have begun further investigations on DMFC.[42,48]

REFERENCES

1. W. R. Grove, *Phil. Mag.* **14,** 127–130 (1839).
2. W. R. Grove, *Phil. Mag.* **21,** 417–420 (1842).
3. W. R. Grove, *Proc. R. Soc. London* **4,** 463–465 (1833); **5,** 557–559 (1845).
4. W. R. Grove, *Phil. Mag.* **8,** 404–405 (1854); added to M. Matteucci, *Phil. Mag.* **8,** 399–403 (1854).
5. U.S. patent 28317, May 15, 1860 to M. Vergnes quoted from S. G. Meibuhr, *Electrochim. Acta* **11,** 1301–1308 (1966).
6. K. Siegl, *Elektrotechn. Z.* **34,** 1317–1318, Heft 46 (1913).
7. L. Mond and C. Langer, *Proc. R. Soc. London* **46,** 296–303, 365–366 (1989); see Ref. 5.
8. W. Ostwald, *Z. Elektrotechn. Elektrochem.* **1,** 81–84, 122–125 (1894).
9. K.-J. Euler, *Entwicklung der electrochemischen Brennstoffzellen* (Verlag Karl Thieme, München, 1974), with extensive bibliography; C. Westphal, pp. 61–62.

10. J. H. Reid, U.S. patent 736016,017 (1902) and USP 757,637 (1904), Ref. 9, p. 72 and Ref. 5; *Electric. World Eng.* **42,** 981–982 (1903).
11. P. C. L. Noel, Fr. patent 350,111 (1904); Ref. 9, p. 72).
12. A. C. Becquerel and A. E. Becquerel, *Traité d'Electricité,* I (Libr. Fermin, Didot, Paris, 1855), in Ref. 9, pp. 55, 211.
13. P. Jablochkoff, *C. R. Acad. Sci. Paris* **85,** 1052–1053 (1877).
14. H. A. Liebhafsky and E. J. Cairns, *Fuel Cells and Fuel Batteries* (Wiley, New York, 1968), ch. 2; W. W. Jacques, *Harpers Mag.* **26**(559), 144–150 (1896); see Ref. 9 for details and figures, and Ref. 15, pp. 129–130 for criticism.
15. W. Borchers, *Z. Elektrochem.* **4,** 129–136, 165–171 (1897), many references to older FC publications.
16. C. Liebenow and L. Strasser, *Z. Elektrochem.* **3,** 353–362 (1897).
17. F. Haber and L. Brunner, *Z. Elektrochem.* **10,** 697–713 (1904); **12,** 78–79 (1906).
18. K. A. Hofmann, *Ber. Dtsch. Chem. Ges.* **51,** 1526–1537 (1918); **52,** 1185–1194 (1919); **53,** 914–921 (1920).
19. F. Haber and A. Moser, *Z. Elektrochem.* **11,** 593–609 (1904).
20. F. Haber, *Z. Anorg. Allg. Chem.* **51,** 245–288, 289–314, 356–368 (1906).
21. F. Haber, Energie a. Kohle u. gasform. Brennstoffe, Österr. Pat. 27,743 (1907); from Ref. 9.
22. R. Beutner, *Z. Elektrochem.* **17,** 91–93 (1911).
23. W. Nernst, DRP 264,026, 264,424 (1912); Ref. 9, p. 216.
24. E. K. Rideal, *Z. Elektrochem.* **62,** 325–327 (1958); A. M. Posner, *Fuel* **34,** 330–338 (1955).
25. W. Schottky, *Wiss. Veroff. Siemens Werke,* **14**(2), 1–19 (1935); *Chem. Abstr.* **20,** 5358 (1935).
26. E. W. Jungner, DRP 348, 393 (1920); Ref. 9, pp. 127, 158.
27. Alfred Schmid, *Die Diffusionsgaselektrode* (Verlag Ferd. Enke, Stuttgart, 1923; *Helv. Chim. Acta* **7,** 370–373 (1924). This author obtained many patents on the FC; see Ref. 9, pp. 217, 218 and Ref. 14.
28. I. Taitelbaum, *Z. Elektrochem.* **16,** 286–300 (1910), with additional remarks from E. Baur, 300–302.
29. E. Baur and H. E. Ehrenberg, *Z. Elektrochem.* **18,** 1002–1011 (1912).
30. E. Baur, W. D. Treadwell, and G. Trumpler, *Z. Elektrochem.* **27,** 199–208 (1921).
31. E. Baur, *Z. Elektrochem.* **39,** 168–169 (1933), **40,** 249–252 (1934), E. Baur and R. Brunner, *Z. Elektrochem.* **41,** 794–796 (1935) (here the effect of CO_2 is reported); **43,** 725–726 (1937). See also Ref. 14.
32. J. Tobler, *Z. Elektrochem.* **39,** 148–167 (1933), E. Baur and J. Tobler, *Z. Elektrochem.* **39,** 169–180 (1933), an excellent review.
33. E. Baur and H. Preis, *Z. Elektrochem.* **43,** 727–732 (1937); **44,** 695–698 (1938).
34. E. Baur, *Bull. Schweiz. Elektrochem. Verein* **30,** 478–481 (1939).
35. F. T. Bacon, *BEAMA J.* **6,** 61–67 (1954); *Electrochim. Acta* **14,** 569–585 (1969); also Refs. 9, 14.
36. F. T. Bacon, in *Fuel Cells* (G. T. Young, ed.) (Reinhold, New York, 1960, pp. 51–77.
37. A. M. Adams, F. T. Bacon, and R. G H. Watson, in ed. *Fuel Cells* (W. Mitchell, Jr., ed.) (Academic Press, New York, 1963), pp. 130–192.
38. O. K. Davtyan, *Bull. Acad. Sci. USSR Class Sci. Technol.* **1,** 107–114 (1946); **2,** 215–218 (1946) (Russian), also Refs. 9, 14.
39. G. H. J. Broers, *High Temperature Galvanic Fuel Cells,* Ph.D. thesis, University of Amsterdam (1958).
40. O. K. Davtyan, G. A. Teterin, and M. V. Diminsky, *Sov. Electrochem.* **6,** 773–776 (1970).
41. E. W. Justi and A. W. Winsel, *Kalte Verbrennung-Fuel cells* (Steiner Verlag, Wiesbaden, 1962); English transl., Pergamon Press, New York (1965).
42. CEC, *Italian Fuel Cell Workshop Proceedings*, Taormina, Italy (P. Zegers, ed.) (CEC, Brussels, 1987).
43. J. Appleby, ed., *Fuel Cells. Trends in Research and Application* (Hemisphere, Washington, DC, 1987).
44. J. A. A. Ketelaar, in *Third Congress International d'Electrothermie* (1953), Sect. 5, pp. 1065–1068.
45. G. H. J. Broers and J. A. A. Ketelaar, *Ind. Eng. Chem.* **52,** 303–306 (1960).
46. G. H. J. Broers, M. Schenke, and G. G. Piepers, *Adv. Energy Conv.* **4,** 131–147 (1964).
47. G. H. J. Broers and M. Schenke, in *Symp. Am. Chem. Soc.* (B. S. Baker, ed.) (American Chemical Society, Washington DC, 1965) pp. 225–250.

48. *Molten Carbonate Fuel Cell Technology*, Vol. of *Proceedings of the Electrochemical Society*, (J. R. Selman and T. D. Clear, eds.) (Pennington, NJ, 1984), pp. 84–13; *Fuel Cell Seminar Abstracts*, Tucson, AZ (Courtesy Ass., Washington DC, 1985).
49. K. R. Williams, O. P. Gregory, and P. Jones, in *Hydrocarbon Fuel Cell Technology*, *Symp. Am. Chem. Soc.* (B. S. Baker, ed.) (American Chemical Society, Washington, DC, 1965), pp. 143–149.

2

Overview of Fuel Cell Technology

S. Srinivasan, B. B. Davé, K. A. Murugesamoorthi, A. Parthasarathy, and A. J. Appleby

2.1. FUNDAMENTAL ASPECTS OF FUEL CELL SYSTEMS

2.1.1. Thermodynamic Aspects

The fuel cell is an electrochemical device which converts the free-energy change of an electrochemical reaction (AGO) into electrical energy. One may thus write the expression

$$\Delta G^\circ = -nFE_r^\circ, \tag{2.1}$$

where E_r° is the reversible potential of the cell. The simplest and most commonly encountered fuel cell reaction is

$$H_2 + \tfrac{1}{2}O_2 = H_2O. \tag{2.2}$$

The free-energy change of this reaction under standard conditions of temperature and pressure ($T = 25°C$, $P_{H_2} = P_{O_2} = 1$ atm, H_2O in liquid state) is 56.32 kcal mole^{-1}. The number of electrons transferred in this reaction is 2. Thus, the reversible potential is 1.229 V.

The variations of the reversible potential (E_r) with temperature and pressure are expressed by the equations

$$E_r = E_r^\circ + \left(\frac{\partial E}{\partial T}\right)_p (T - 298) + E_r^\circ + \frac{\Delta S}{nF}(T - 298), \tag{2.3}$$

$$E_r = E_r^\circ - \frac{(\Delta n)RT}{nF} \ln P, \tag{2.4}$$

S. Srinivasan, B. B. Davé, K. A. Murugesamoorthi, A. Parthasarathy, and A. J. Appleby • Center for Electrochemical Systems and Hydrogen Research, Texas Engineering Experiment Station, The Texas A&M University System, College Station, Texas 77843-3402. *Present address* of B. B. Davé: Nalco Chemical Company, Naperville, Illinois 60566. *Present address* of K. A. Murugesamoorthi: AT&T Bell Laboratories, Energy Systems, Mesquite, Texas 75149. *Present address* of A. Parthasarathy: Department of Chemistry, Colorado State University, Fort Collins, Colorado 80522.

Fuel Cell Systems, edited by Leo J. M. J. Blomen and Michael N. Mugerwa. Plenum Press, New York, 1993.

where Δn is the change in number of gas molecules during the reaction. The entropy change, ΔS, for reaction (2.1) is -39 entropy units, while Δn is $-\frac{3}{2}$. Thus, Eqs. (2.3) and (2.4) show that the reversible potential decreases with an increase of temperature, while the behavior is opposite with an increase of pressure.

At temperatures above 100°C, water is produced as a vapor in the cell. The value of ΔS is considerably less when water is produced in this state than as a liquid. Thus $\partial E/\partial T$ is -0.25 mV/°C in the former case and is -0.54 mV/°C in the latter. The effect of pressure on E_r is also less when water is produced as a vapor than as a liquid ($\Delta n = -\frac{1}{2}$ in the former and $-\frac{3}{2}$ in the latter case).

Hydrogen is not a primary fuel. Thus, attempts were made in the 1960s to use primary fuels such as hydrocarbons (CH_4 to $C_{10}H_{22}$) and coal in fuel cells. However, due to the high degree of irreversibilities of the anodic oxidation reactions of the hydrocarbons, these attempts proved futile. Thus, these fuels were processed to produce hydrogen for low-temperature fuel cells and H_2 and CO for the higher temperature fuel cells by steam-reforming reactions of the hydrocarbons and by gasification of coal. Several other types of fuels (methanol, ethanol, ammonia, hydrazine) were also researched in the 1960s. Practically all of these fuel cell reactions have a thermodynamic reversible potential from 1.0 to 1.2 V.

Even if there were no efficiency losses in H_2–O_2 fuel cells (due to activation, mass transport, and ohmic overpotentials), heat would still have to be rejected from a fuel cell because ΔS is negative for reaction (2.2).[1] Thus, the theoretical efficiency of H_2–O_2 fuel cells at 25°C, based on the enthalpy change of the reaction (commonly referred to as the higher heating value by mechanical engineers), is 83%. There is only one fuel cell reaction where ΔS is positive, namely,

$$C + \tfrac{1}{2}O_2 \rightarrow CO. \tag{2.5}$$

For this reaction, the theoretical efficiency is 137% at 150°C. The reaction

$$C + O_2 \rightarrow CO_2 \tag{2.6}$$

has entropy change of 0 e.u. Thus, for this reaction, the theoretical efficiency is close to unity. As a rough rule of thumb, if Δn is positive, the entropy change is positive (due to increasing disorder), while if Δn is negative (increasing order), ΔS is negative, and if Δn is zero ΔS is also zero.

2.1.2. *Electrode Kinetics Aspects*

The vitally important role of electrode kinetics on the performance of fuel cells (particularly those operating at low and intermediate temperatures, 25–200°C) is best illustrated by a typical cell potential versus current density plot (Fig. 2.1). Three distinct regions are illustrated in this plot. The predominant cause of the difficulties in attaining high energy efficiencies and high power densities in low- to medium-temperature fuel cells is the low electrocatalytic activity of most electrode materials for the oxygen electrode reaction. The hydrogen electrode shows a linear relationship of its half-cell potential versus

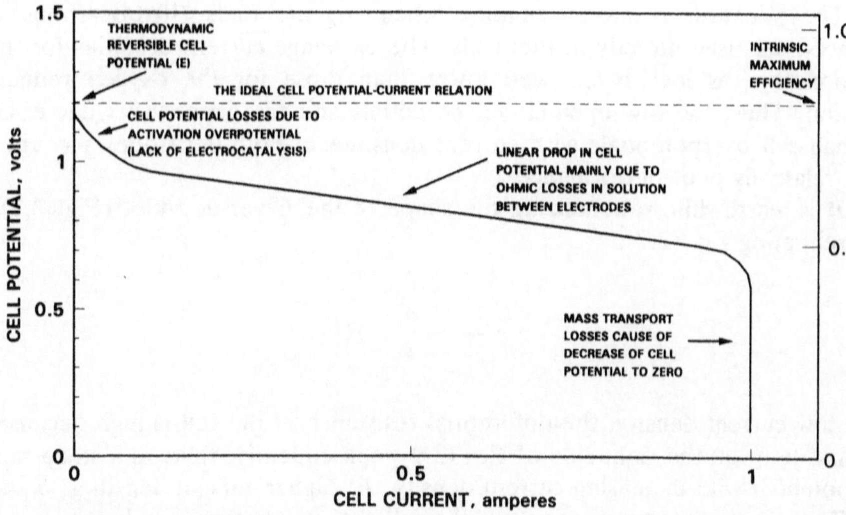

Figure 2.1. Typical plot of cell potential vs. current for fuel cells illustrating regions of control by various types of overpotentials.

current density plot from zero to the highest value of current density in the fuel cells using phosphoric acid ($T \approx 200°C$), potassium hydroxide ($T = 80°C$), or a proton-conducting electrolyte (Nafion or Dow membrane, $T = 85°C$). This is not the case with the oxygen electrode, where a semiexponential relation between its half-cell potential and current density is observed. Thus, at low current densities (say $<100\ mA/cm^2$), the entire loss in the fuel cell potential from the reversible value is due to activation overpotential at the oxygen electrode.

Another problem encountered is that the reversible potential is not attained even at zero current density. This problem is again due to the oxygen electrode. The exchange current density for this reaction is so low that competing anodic reactions (for example, oxidation of the platinum electrocatalyst, corrosion of carbon, oxidation of organic impurities) play a significant role. The net result is that the open-circuit potential is a mixed potential, which is lower than the reversible potential for the H_2–O_2 fuel cell reaction by about 0.1 to 0.2 V. Thus, even at close to zero current densities, the efficiency of a fuel cell is lower than its theoretical value by 8–16%. The cell potential (E) versus current density (i), from a current density of 0 to the value at the end of the linear region, may be expressed by the relation

$$E = E_o - b \log i - Ri, \tag{2.7}$$

where

$$E_o = E_r + b \log i_o, \tag{2.8}$$

the parameters b and i_o are the Tafel slope and exchange current density for the oxygen reduction reaction, and R accounts for the linear variation of overpotential (predominantly ohmic) with current density, which is observed in the intermediate range.

The situation is more complex when organic fuels (hydrocarbons and alcohols) are used directly in fuel cells. The exchange current densities for these reactions are as low as or even lower than those for the oxygen reduction reaction. Thus, the low open-circuit potentials and the exponential decrease of the half-cell overpotentials with current densities at both electrodes account for their relatively poor performance.

It is worthwhile rationalizing the shape of the E versus i plot (Fig. 2.1) by differentiating Eq. (2.7):

$$\frac{\partial E}{\partial i} = -\frac{b}{i} - R. \tag{2.9}$$

At a low current density, the differential resistance of the cell is high because of the first term on the right side of Eq. (2.9); consequently, there is a steep fall of cell potential with increasing current density. At higher current densities, $b \ll R$.

Thus, in the intermediate range, the cell potential varies linearly with current density. At high current densities, mass transport limitations, due to the low rates of supplies of reactants to the electrocatalytic sites or of products away from these sites, predominate and cause the sudden drop of cell potential to near-zero values. Advances have been made in the fabrication of fuel cell electrodes with optimized structures; thus, mass transport limitations are rarely encountered at current densities up to a few A/cm^2. In high-temperature fuel cells using molten carbonate ($T \approx 650°C$) and solid oxide ($T \approx 1000°C$) electrolytes, the exchange current densities of the fuel cell reaction are quite high ($>1\ mA/cm^2$), and thus the cell potential versus current density plot is linear throughout the entire current density range (0 to a few hundred mA/cm^2).

Theoretical and experimental electrode kinetic studies of fuel cell reactions (2.1)–(2.6) have led to the engineering design, development, and demonstrations of fuel cell power plants exhibiting high levels of performances (high energy efficiency, high power density, and reduction in noble metal loading). The major accomplishments are (i) the design and fabrication of porous gas diffusion electrodes with optimized structures to enhance (a) diffusion of dissolved gases (H_2, O_2) to reaction sites, (b) electrochemically active surface sites, and (c) ionic transport through porous electrodes; (ii) use of supported electrocatalysts (Pt crystallites on high-surface-area carbon) to significantly reduce noble metal loadings, as compared with loadings when unsupported platinum electrocatalysts were used; (iii) inhibition of CO poisoning in phosphoric acid fuel cells by elevation of operating temperature; (iv) utilization of alloys and heat-treated metal-organic macrocyclics as electrocatalysts which exhibit higher exchange current densities than platinum for oxygen reduction; (v) development of non-noble metal electrocatalysts for high-temperature fuel cells with molten carbonate or solid oxide electrolytes; and (vi) use of thin electrolyte layers to minimize ohmic overpotentials.

The efficiency (ε) of a fuel cell varies with current density in the same manner as that of the cell potential with current density (cf. Fig. 2.1) because

$$\varepsilon = \frac{nFE}{\Delta H}. \tag{2.10}$$

The power density of a fuel cell is expressed by the relation

$$P = Ei. \tag{2.11}$$

The shape of P versus i or P versus E plot is a parabola if the E versus i relation is linear. The parabola is distorted for low- and intermediate-temperature fuel cells because of the semilogarithmic relation between cell potential and current density at low values of i and sudden drops of cell potential at high current densities.

The irreversible losses in fuel cells lead to waste heat generation ($\dot{Q}$), which may be expressed by

$$\dot{Q} = -\frac{4.18T(\Delta S)i}{nF} + i\sum \eta + i^2R. \tag{2.12}$$

In this equation the first term represents the entropic loss, which cannot be overcome due to thermodynamic considerations (Section 2.1.1), the second term is due to activation and mass transport overpotentials, and the third is due to ohmic heating effects.

Due to the heat generation in fuel cells, it is necessary to incorporate cooling subsystems in the fuel cell system. A considerable portion of this heat is efficiently used in one or more of the following ways: fuel processing, which usually involves an endothermic reaction; space heating; hot water; and, occasionally in high-temperature fuel cells, for chemical processes and for enhancing electricity generation using gas turbines.

2.1.3. Classification of Fuel Cells and Scope of This Chapter

Since the invention of fuel cells by Sir William Grove in 1839, several types have been developed in the nineteenth and the twentieth centuries (see Chapter 1). Nearly as many classifications have appeared in the literature[1–5] because there are a vast number of variables among the fuel cell systems, such as types of fuel and electrolyte, operating temperature, primary and regenerative systems, and direct or indirect systems. Over the years, particularly since the energy crisis of 1973, which was instrumental in the renaissance of fuel cells, there has been some streamlining, and it is customary to classify fuel cells by the type of electrolyte—alkaline (normally KOH), acid (mainly phosphoric, acids such as the "super-acids" are still in the research stage), molten carbonate (the presently used electrolyte is 62% Li_2CO_3 and 38% K_2CO_3), solid oxide (the most advanced is yttria-stabilized zirconia), and solid polymer (proton-conducting membranes, now mainly Nafion from DuPont and Dow membrane).

This chapter presents an overview of fuel cell technology, which is based on the type of electrolyte, and focuses on the electrochemical cell stack, the heart of the fuel cell power plant. The topics covered are (i) principles of operation (ii) operating conditions, (iii) typical cell materials and configurations, (iv) methods for cooling electrochemical cell stacks, (v) acceptable contaminant levels, and (vi) applications and economics. The two major subsystems, other than the electrochemical cell stack, in a fuel cell power plant for electric power generation are the fuel processor and power conditioner. The fuel processor converts the primary and/or portable fuel (natural gas, petroleum-derived hydrocarbons, methanol coal) into H_2 or H_2 and CO. These secondary fuels (H_2, CO) are considerably more electroactive in the electrochemical cell stack than the primary ones. Fuel processing technology is highly advanced and efficient (see Chapters 4 and 6). Power conditioning is required to convert direct current, generated in the electrochemical cell stack, to alternating current, which is generated in conventional power plants (thermal or nuclear). Great strides have been made in power conditioning technology due to the incorporation of semiconductor and integrated circuit technology. The major challenge in interfacing the electrochemical cell stack with the power conditioner is to maintain a constant ac voltage which should be independent of the dc stack voltage variation and is a function of power outputs from the electrochemical cell stack. The efficiency of dc to ac conversion in megawatt-size power plants is over 90% (in some cases over 96%), and the projected capital cost of power conditioners is about $75/per kilowatt.

2.2. ALKALINE FUEL CELLS

2.2.1. Background and Principles of Operation

The pioneer of alkaline fuel cell research, development, and demonstration is Dr. Francis T. Bacon.[6] Bacon commenced his research in this area in 1932 and had completed the construction of a 5-kW hydrogen–oxygen fuel cell power plant and its performance evaluation in 1952. One of the main reasons why Bacon chose an alkaline electrolyte was to use non-noble metal electrocatalysts. In this fuel cell, the electrocatalysts were nickel for the anode and lithiated nickel oxide (lithium improves the electronic conductivity and the corrosion resistance of nickel oxide) for the cathode. In order to obtain a stable three-phase zone, he designed the electrodes for dual porosity, the larger pores on the gas side and the smaller ones on the solution side. The electrolyte in the Bacon cell was 30% KOH, and the cell operating temperature and pressure were 200°C and 50 atm, respectively.

A schematic illustrating the operating principles of an alkaline fuel cell is illustrated in Fig. 2.2. The half-cell reactions are

$$\text{Anode} \qquad H_2 + 2OH^- \rightarrow 2H_2O + 2e^- \tag{2.13}$$

$$\text{Cathode} \qquad O_2 + 2H_2O + 4e^- \rightarrow 4OH^- \tag{2.14}$$

Potassium hydroxide, which is the most conducting of all alkaline hydroxides, has always been the electrolyte of choice. The hydroxyl ions are the conducting species in the electrolyte. Though water is produced at the anode in this cell, there is some migration of water to the cathode, and, thus, product water exits the cell from the anode (about $\frac{2}{3}$) and the cathode (about $\frac{1}{3}$). In one of the original versions of the alkaline fuel cells, molten KOH (80%) was used as the electrolyte. In practically all other fuel cell systems, its concentration is around 30%, which is approximately the optimal value from the point of view of conductivity.

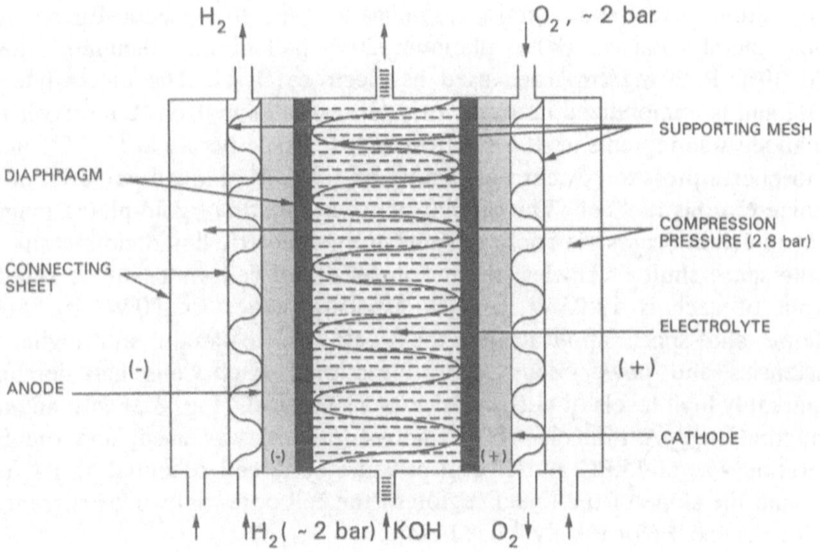

Figure 2.2. Principles of operation of alkaline fuel cells (Siemens).

2.2.2. Operating Conditions

The unique advantage of an alkaline electrolyte over an acid one for a fuel cell is that noble metal electrocatalysts are not necessary for the former. In addition, even with the non-noble metal or oxide electrocatalysts, the oxygen electrode performance is extremely good. Fuel cells have been developed with platinum electrocatalysts as well. In recent times, it was shown that a heat-treated cobalt tetraphenyl porphyrin deposited on high-surface-area carbon exhibits the highest activity ever reported for oxygen reduction.[7,8] The reason for this is that anion adsorption is minimal from alkaline electrolytes. The Tafel slope for oxygen reduction from alkaline electrolytes is <60 mV/decade throughout the entire operating range. Alkaline fuel cells generally operate at around 60–80°C. In some cases, the operating pressure is a few atmospheres, but most often it has been at atmospheric pressure. From a system point of view, the methods for product water and heat removal are critical, particularly because the alkaline fuel cell operates at less than 100°C. However, these problems have been resolved by several manufacturers of alkaline fuels, as described next.

2.2.3. Typical Cell Materials and Configurations

The Pratt and Whitney Division of United Technologies Corporation inherited the Bacon fuel cell technology for the development of 1.5-kW power plants for NASA's Apollo missions.[9] Unlike the Bacon fuel cells, the Pratt and Whitney fuel cell used KOH at a concentration of 80–85% (molten KOH) and operated the fuel cell at 250°C and close to atmospheric pressure. Heat and water vapor removal were effected by using closed-loop hydrogen recirculation. These fuel cell systems were used by NASA in the Apollo missions (1960s and early 1970s) lasting two to four weeks. For the power systems for the space shuttle flights of the 1970s to the 1990s, International Fuel Cells–United Technologies

Corporation made some major modifications to the pseudo-Bacon system.[10] Noble metal catalysts (80% platinum–20% palladium: 10 mg/cm^2, and 90% gold–10% Pt:20 mg/cm^2) are used as electrocatalysts. The electrolyte is 35% KOH and is immobilized in a reconstituted asbestos matrix. A reservoir plate on the anode side replenishes the electrolyte. The cells operate at 60–70°C under 4.0 atmospheres pressure. A current density of 1 A/cm^2 at a cell potential of 0.8 V is obtained in this fuel cell. The bipolar plates are made of gold-plated magnesium. Heat rejection occurs via cooling plates using dielectric liquid circulation. In each of the space shuttle vehicles, there are three fuel cell systems (the rated power output of each is 12.0 kW, and the specific power is 100 W/kg). For some defense and space applications, the goals are to attain still higher energy efficiencies and power densities. International Fuel Cells has demonstrated remarkably high levels of performance in a single cell (Fig. 2.3) with an advanced configuration.[11] A thin electrolyte matrix (50 μm) was used, and the fuel cell operation was at 150°C and 15 atm pressure. The cell potential at 1 A/cm^2 was 1 V, and the slope of the linear region of the cell potential versus current density (up to 8 A/cm^2) plot is only 0.05 Ω·cm^2.

Another advanced alkaline fuel cell system was developed by Siemens for military (underwater) applications.[12] High-surface-area Raney nickel, containing 1–2% Ti (nickel loading, 120 mg/cm^3), and Raney silver (60 mg/cm^2 Ag with small amounts of Ni, Bi, and Ti as additives to prevent sintering) are the electrocatalysts for hydrogen oxidation and oxygen reduction. A problem with the nickel electrocatalyst is its possible irreversible oxidation during potential excursions to moderately high values (>0.2 V per reversible hydrogen electrode). Nickel oxide, thus formed, inhibits the hydrogen oxidation reaction. Siemens has

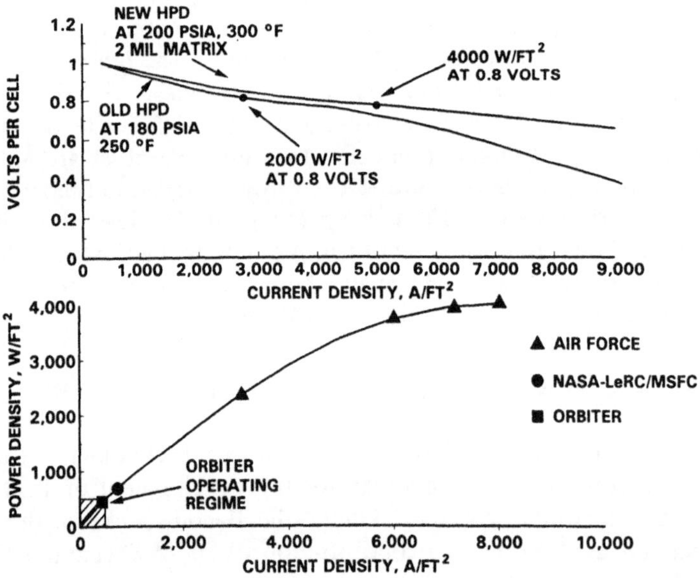

Figure 2.3. Current and projected performance of International Fuel Cells–United Technologies alkaline fuel cells for space applications; (a) cell potential vs. current density; (b) power density vs. current density.

built several 7-kW power plants. The power plant contains 70 monopolar cells (340 cm^2 active area) electrically connected in series and hydrolytically in parallel for electrolyte circulation. The electrolyte circulates through a seperate water evaporator unit. Hydrogen is circulated through the evaporator spaces, and the steam carried out is subsequently condensed. The rated performance of the cell is a current density of 420 mA/cm^2 at a cell potential of 0.77 V. The lifetime is about 3000 h. There is some performance deterioration (5% in 1000 h), probably because of the slow dissolution of the silver electrocatalyst.

A novel approach for the construction of low-cost, lightweight bipolar alkaline fuel cell power plants was demonstrated by Exxon, Alsthom, and Occidental Chemical.[13] In this system the anodes and cathodes were identical and had a platinum loading of 0.3 mg/cm^2. Goffered bipolar plates, made with 30% conductive carbon black and 70% propylene, were used. The goffering in the plates was approximately 0.4 mm wide and 0.5 mm deep, with a web thickness of 0.35 mm. The active area of the cell was 400 cm^2. The bipolar plate was incorporated in a 40% talc-filled polypropylene frame, which had an external area of 900 cm^2 with internal manifolding for gas and electrolyte (6 N KOH) distribution. An injection-molding operation was used for the construction of the bipolar plate–frame fixture. The stacks, assembled in subgroups of 100 cells to minimize shunt currents during fuel cell operation, were process-air-cooled. An assembly line was set up for pilot plant production of 400 cells/day, equivalent to 16–20 kW/day. An illustration of the components and layout of a two-cell stack is shown in Fig. 2.4. Cell performance was reasonably good—a cell potential of 0.72 V (the goal was 0.8 V) at a current density of 150 mA/cm^2 with hydrogen–scrubbed air as reactants. A component lifetime of 8000–10,000 h was demonstrated. For a power plant of similar construction, sulfuric acid was used as the electrolyte.

Elenco, a consortium of the Belgium Atomic Energy Commission, Bekaert, a Belgium company and Dutch State Mines, embarked on an alkaline fuel cell research, development, and demonstration program in the early 1970s.[14] The power plant was designed for transportation applications, particularly city buses. The electrodes contained one or two layers of catalyzed (0.3–0.4 mg/cm^2 Pt) carbon on a nickel screen. The multicell stack, with a monopolar configuration and edge collection, has 24 cells in series (electrode area 300 cm^2). The electrolyte is 6 M KOH and circulates through the multicell stack. The operating temperature is 65°C. The multicell stack is rated for a power output of 400–500 W and weighs 4 kg. The goal for the power density is 150 W/kg, which means that a 50% improvement in cell performance (presently, 70 mW/cm^2) is necessary to meet this goal. Performance evaluation, including life testing, has been conducted on 0.5- and 5-kW modules. These modules have been in operation for over 5000 h, and a 4%/1000 h degradation rate has been reported. The power plant for electric vehicles (rated power output 10 kW) incorporates a series–parallel connection of the modules to provide the desired voltage for electric vehicle traction systems: 96, 144, or 216 V. A 10-kW system has been installed and tested in a delivery-type vehicle (Volkswagen minivan). Elenco's view is that hydrogen is a strong contender for a transportation fuel in Europe, unlike the United States, because of the relatively small resources of fossil fuels in several European countries. Furthermore, the prospects of replacing the platinum electrocatalyst

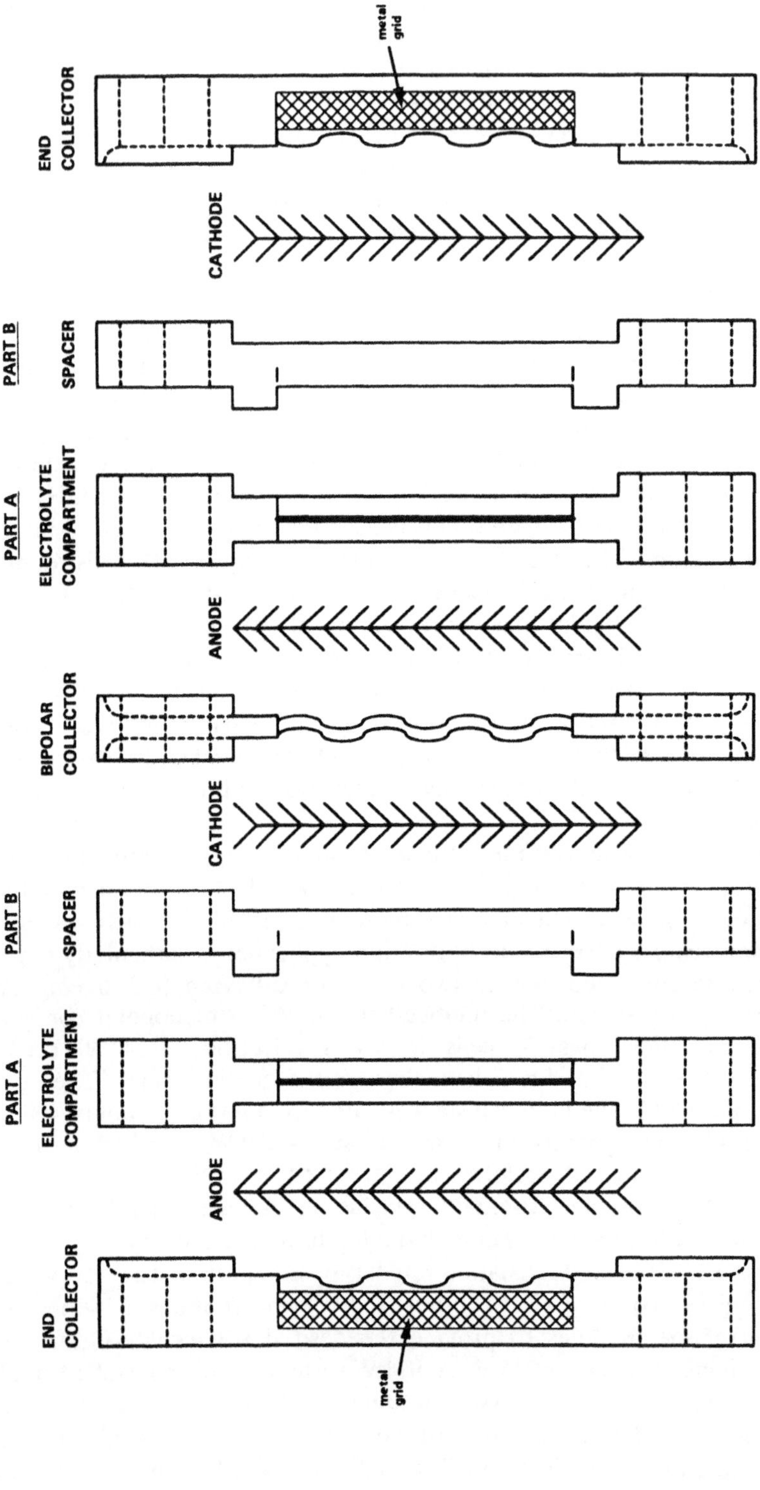

Figure 2.4. Components and layout of Exxon–Alsthom–Occidental two-cell stack alkaline fuel cell.

with non-noble metal ones are being studied and look promising. The goals at Elenco are to mass-produce hydrogen-fueled power plants at a cost of $200/per kilowatt. Even though this value may sound high for the transportation application, by use of fuel cell–battery hybrid power sources or by further improvement of power densities of the alkaline system, it should be possible to reduce the power plant costs by at least a factor of 2.

2.2.4. *Methods for Cooling Electrochemical Cell Stacks*

State-of-the-art alkaline fuel cell systems are designed to generate electric power at relatively low temperatures (60–80°C). Except for the space power system, the liquid electrolyte is circulated through the cell stack in other fuel cell systems; a solution to the heat rejection problem is thus provided (i.e., external heat exchange for cooling the electrolyte). Some terrestrial fuel cells are process-air-cooled. As stated in Section 2.2.3, heat and water removal in the apollo space fuel cell were carried out by using closed-loop hydrogen circulation (glycol cooling in the heat exchanger), while in the fuel cell for space shuttle flights cooling plates, with dielectric liquid circulation, were incorporated in the cell stack.

2.2.5. *Acceptable Contaminant Levels*

The alkaline fuel cell system is the most efficient of all fuel cells, but the major challenge is the complete removal of CO_2 from the anodic and cathodic gas streams before their entry into the electrochemical cell stack. Even the small level of CO_2 in the air (about 350 ppm) is sufficient to carbonate the electrolyte and form solid deposits in the porous electrode. Several schemes (carbonate or ethanol amine scrubbers, physical adsorption–Selexol process, Fluor-solvent process, and pressure-swing absorption) have been proposed for reduction of the CO_2 level from 25% to 500 ppm in reformed fuels. Removal of small amounts of CO_2 by diffusion through membranes and use of electrochemical concentration cells (i.e., oxidation of impure hydrogen at anode and production of pure hydrogen at cathode) have been proposed.

2.2.6. *Applications and Economics*

The alkaline fuel cell system, using pure H_2 and O_2 as reactants, has established its space application. This system, coupled with an alkaline or a solid polymer electrolyte water electrolyzer, is a strong contender for NASA's lunar and Mars missions, as well as for the space stations. For terrestrial applications, this system will have to use pure hydrogen (purity >99.99%), as in space applications, and air as the cathodic reactant. The challenges are to economically remove CO_2 from these reactants and to develop safe, lightweight methods of storing gaseous hydrogen. Though storage of hydrogen as a cryogenic liquid or as a metal hydride has been proposed, these methods are not sufficiently efficient (from energy, weight, and/or volume considerations), particularly for the transportation application. The alkaline fuel cell system is one of the most attractive systems for the transportation application because it is capable of achieving high energy efficiencies and high power densities without using noble metal electrocatalysts. At present, it appears best to store hydrogen as a compressed gas in lightweight cylinders (fiberglass–reinforced aluminum rather than steel) to increase the amount of hydrogen stored from 1% to 4%. Another application of alkaline fuel cell systems is for standby power, for example, for telephone companies. Estimating the capital costs of alkaline fuel cell systems for terrestrial applications is premature without having a fairly detailed engineering design of the power plant. However, using relatively inexpensive materials for components, as in the Exxon, Alsthom, and Occidental fuel cell system and if Pt can be successfully replaced, it should in the future be possible to attain capital costs of the alkaline fuel cell system that are considerably lower than that of the acid electrolyte systems, which use noble metal electrocatalysts and, in one of the latter cases, use expensive proton-conducting membranes (Section 2.6). For the transportation application, the goal is to reduce capital costs to $50–$100/per kilowatt. For other applications, such as standby power, capital costs of $500/per kilowatt are affordable. Note that the early markets, space and defense, allowed for $5000–$15,000 per kilowatt or more.

2.3. PHOSPHORIC AND OTHER ACID ELECTROLYTE FUEL CELLS

2.3.1. Background and Principles of Operation

Apart from alkaline fuel cell systems, which were developed for space applications, the phosphoric acid fuel cell (PAFC) system in the only other one ready for applications. Since the early 1960s, over $300 million has been spent on the development of PAFC power plants, leading to the demonstrations of power plants rated fro 1 kW to 5 MW.

Phosphoric acid is the electrolyte of choice for acid electrolyte fuel cells, particularly with hydrogen produced by steam reforming of organic fuels, such as hydrocarbons (natural gas), and alcohols (methanol or ethanol) as the anodic reactant. The PAFC is (i) CO_2-rejecting, (ii) can tolerate 1–2% CO at the current operating temperature of 200°C, and (iii) can utilize the waste heat from the electrochemical cell stack efficiently for the endothermic steam-reforming reaction, as well as for providing space heat or hot water. The principles of operation of the cell are schematically represented in Fig. 2.5. The reformed fuel and oxidant enter the anode and cathode gas chambers, dissolve in the electrolyte, and diffuse to the electrocatalyst sites in the porous gas-diffusion electrodes. The anodic and cathodic reactions and the transport of hydrogen ions from the anode to the cathode are indicated in Fig. 2.5. In the mid-1960s, the platinum loading of the electrodes was about 10 mg/cm^2 (unsupported platinum back); the major accomplishment in the late 1960s was the development of supported platinum electrocatalysts, which made it possible to reduce the platinum loading to less than 5% of this value. The attractive features of phosphoric acid are (i) stability in the electrochemical environment at temperatures up to at least 225°C, (ii) reasonably good electrolyte conductivity at temperatures above 150°C, and (iii) efficient rejection of product water and waste heat at the operating temperature. The initial problem encountered with PAFCs

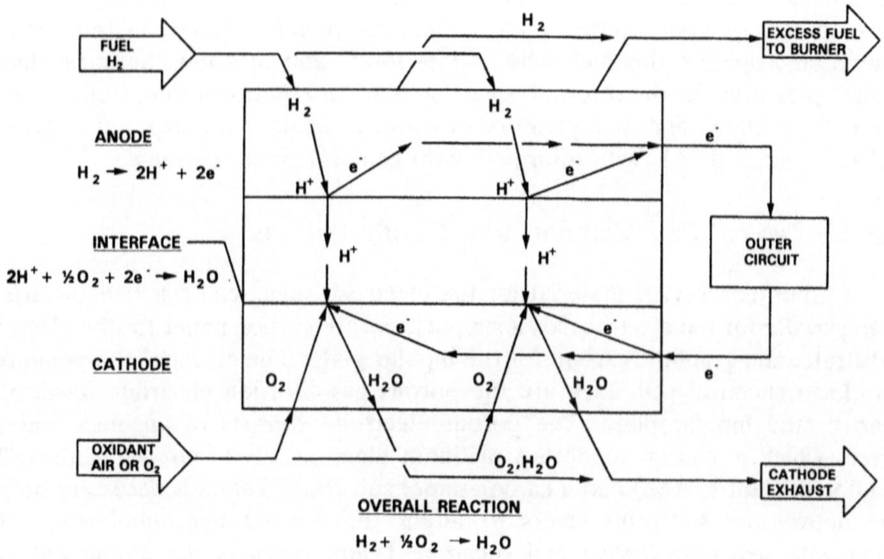

Figure 2.5. Principles of operation of phosphoric acid fuel cell schematic.

was that the kinetics of oxygen reduction was considerably slower than in other acids, such as sulfuric or perchloric.[2–4] However, increasing the operating temperature to above 150°C allowed reasonably high rates of oxygen reduction to be achieved. The reason for the slow kinetics below 150°C is that the adsorption of the phosphoric acid molecule and/or anions of this acid inhibits oxygen adsorption, an intermediate step in oxygen reduction. Above 150°C, phosphoric acid is predominantly in the polymeric state, as pyrophosphoric acid. This acid is strongly ionized, and probably because of the relatively large size of the anions ($H_3P_2O_7^-$) of this acid, anion adsorption is minimal. Several publications have appeared[2–4] on the kinetics of hydrogen oxidation and of oxygen reduction on platinum in phosphoric acid. The rate of the former reaction is controlled by the dissociative adsorption of hydrogen on the platinum surface and is susceptible to CO poisoning, a site-blocking mechanism. As stated earlier, the effect of CO poisoning is strongly decreased above 150°C. The rate-determining step for oxygen reduction on platinum in phosphoric acid is generally accepted as

$$M + O_2 + H^+ + e^- \rightarrow MHO_2. \tag{2.15}$$

One may expect a Tafel slope of $2.303 \times 2RT/F$ for this reaction mechanism and for it to depend on temperature, but as for hydrogen evolution, anomalous behavior—independence of Tafel slope on temperature—is observed.[15]

2.3.2. Operating Conditions

The most advanced PAFC system was developed by International Fuel Cells–United Technologies Corporation.[16] Within the last 20 years, significant progress has been made by Westinghouse/Energy Research Corporation,[17]

Engelhard,[18] Toshiba,[19] Mitsubishi,[20] Fuji,[21] Sanyo,[22] and Hitachi[23] in the development of PAFC power plants. Most companies, other than International Fuel Cells, operate the fuel cells at 150–190°C and at atmospheric or slightly higher pressures. In the most advanced system, International Fuel Cells operates the cells at 200°C and at a pressure of 8 atm to achieve a performance level of 325 mA/cm^2 at 0.73 V with reformed hydrogen and air as reactants.

2.3.3. *Typical Cell Materials and Configurations*

Carbon is the vital material for the electrochemical cell stack: high-surface-area powder for the electrocatalyst support, porous carbon paper for the electrode substrate, and graphitic carbon for the bipolar plate. The essential components of an electrochemical cell stack are the porous gas-diffusion electrode, electrolyte matrix, and bipolar plate. The porous electrode consists of an electrocatalyst layer, which is carbon-supported platinum electrocatalysts bonded with Teflon (Teflon content is 30%) and a carbon paper substrate. Teflon is necessary both in the active and substrate layers to attain the desired hydrophobicities. The composite structure—active and substrate layers—permits the attainment of a stable three-phase zone necessary for the efficient electrochemical processes, namely, mass transport of reactants to the electrocatalytic sites, electrochemical reaction at the electrocatalyst–electrolyte interface, transport of protons from the anode to the cathode, and transport of the product water from the electrocatalyst layer via the substrate layer of the oxygen electrode and the gas channels to the external environment.

The state-of-the-art electrolyte matrix is Teflon-bonded silicon carbide. The matrix is microporous and is generally deposited on the cathode. Since there is some loss of electrolyte, electrolyte replenishment means are incorporated in the cell. Several designs for the bipolar plate have been evaluted. International Fuel Cells[16] uses a ribbed substrate configuration (Fig. 2.6). Thin impervious graphite sheets between the ribbed substrates serve as plates in adjacent cells in the multicell stack. The ribbed substrates are porous and provide channels for the gases to enter the porous electrode. In addition, the ribbed substrate on the anode side also serves as an electrolyte reservoir. The ribbed substrate can be manufactured and cut to the desired size in a continuous operation in order to significantly reduce the production cost. Westinghouse/Energy Research

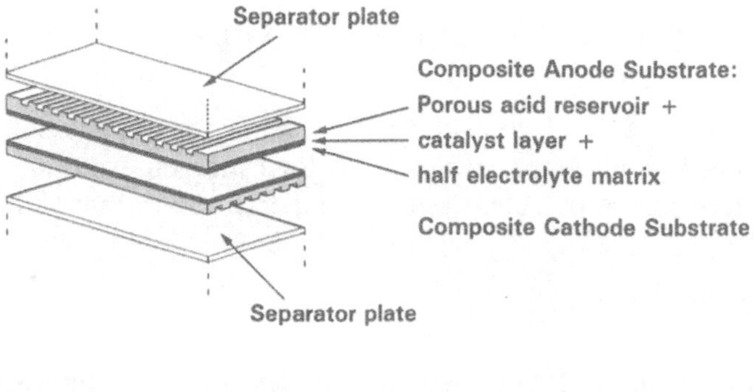

Figure 2.6. International Fuel Cells phosphoric acid fuel cell ribbed substrate cell configuration.

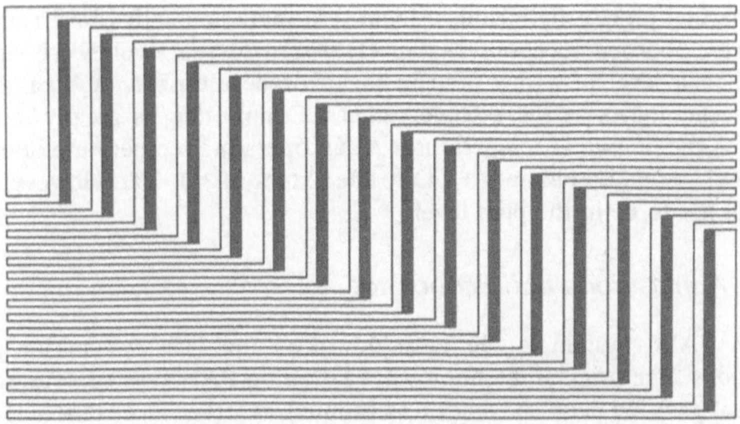

Figure 2.7. Westinghouse–ERC Z pattern bipolar plate for phosphoric acid fuel cell.

Corporation[17] utilizes a Z pattern bipolar plate molded with graphite and a binder (Fig. 2.7). This plate is designed for the reactant gas (fuel or air) to enter half of one side of the cell and leave on the other half of the opposite side. Several other designs of the bipolar plate have also been evaluated, and some are being utilized.

2.3.4. Methods for Cooling of Electrochemical Cell Stacks

State-of-the-art PAFC systems operate at 190–210°C. Because of the relatively high operating temperature, compared with low-temperature fuel cell systems, such as those with alkaline (Section 2.2) and solid polymer (Section 2.6) electrolytes, water and heat rejection are relatively simpler. Passive cooling methods (radiation of heat from the electrochemical cell stack to the surroundings) are inadequate for large systems. Active cooling methods involve liquids or gases. The IFC PAFC system incorporates a number of parallel thin-walled two-pass copper U-tubes passing through the graphite separators in the stacks. An alternative design consists of coolant tubes connected to the stack coolant supply conduits by dielectric hoses with high length-to-diameter ratios to prevent short-circuiting within the stack in case of flaws in the protective coatings. Several Japanese companies, especially Toshiba, Fuji, and Mitsubishi, employ liquid cooling methods. Gas cooling can be effected by pumping gases through cooling channels located, say, after every fifth cell. The alternative is to process-gas-cool, i.e., supply air at several times the stoichiometric flow rate. Liquid cooling is more efficient than gas cooling, but requires complex manifolding. Gas cooling is reliable and economical.

2.3.5. Acceptable Contaminant Levels

A problem encountered in the 1960s and early 1980s was the poisoning of the hydrogen electrode by CO and H_2S which were present in the reformed fuels.

Since the fuel processing includes steam reforming and shift conversion, the CO levels were reduced to about 1–2%. Hydrogen sulfide levels were significantly reduced to a level of a few ppm in the hydrodesulfurizer. A breakthrough in PAFC technology was the discovery that CO poisoning is greatly inhibited by increasing the operating temperature. At an operating temperature above 180°C, the PAFC system is tolerant to CO to the extent of <1–2%. However, the H_2S level still has to be in the ppm level.

2.3.6. Applications and Economics

The PAFC system is the most advanced fuel cell system for terrestrial applications. The original intention was to develop PAFC systems (using naphtha or natural gas) for peak shaving or as spinning reserves in electric utilities. This application is still being stressed in Japan and Italy. However, in the United States, the PAFC system still has a hard competitor, namely gas turbines, for this application. The most recent intended application of the PAFC system is for on-site integrated energy systems to provide electrical and heat (space heating, hot water) energy for apartment buildings, shopping centers, hospitals, hotels, etc. These fuel cell systems will use natural gas. International Fuel Cells has initiated a program to build one hundred 200-kW fuel cell systems for these applications. These systems are being sold for $575,000, or $2875/per kilowatt. Japan is engaged in building a few megawatt-size PAFC systems for electric utility applications, and Fuji produces relatively high-priced 50-kW systems in series, mainly for the home market.

2.3.7. Research on Other Acid Electrolytes for Fuel Cells

Historically, sulfuric, perchloric, and hydrofluoric acids, which are frequently used in chemistry laboratories, were investigated as electrolytes in the first active era of fuel cell research and development. Sulfuric acid was found to be unstable at the operating potentials of the anode (sulfuric acid is reduced to sulfurous acid and to some extent even to H_2S or S). Perchloric acid, being a strong oxidizing agent, can cause the fuel to explode. Furthermore, these aqueous electrolytes lose water at temperatures above 100°C. In addition, these acids also have relatively high vapor pressure. For all the above reasons, these acids cannot compete with phosphoric acid, particularly for fuel cell operation at over 150°C, which is the minimum temperature essential to minimize carbon monoxide poisoning. Hydrochloric and hydrobromic acids are the ideal electrolytes for the regenerative H_2–Cl_2 and H_2–Br_2 fuel cells. The chlorine and bromine electrode reactions are considerably faster than the oxygen electrode reaction (exchange current densities for the former reactions about 10^{-3} A/cm^2, while that for the latter is 10^{-9} A/cm^2 on smooth electrode surfaces). Moreover, there are no complicated changes in the chemistry of the surfaces (say oxide formation), nor are different electrocatalysts required for forward and reverse reactions. Thus, the performances of regenerative H_2–Cl_2 and H_2–Br_2 fuel cells are excellent in terms of efficiency and power density. The main challenge is the selection of stable materials for fuel cell components in the highly corrosive environment.

Great interest in finding suitable fluorinated sulfuric acids as fuel cell electrolytes stemmed from investigations with trifluoromethane sulfonic acid

(TFMSA).[24–25] The attractive features of TFMSA are (i) its high ionic conductivity (fluorinated sulfonic acids are often, though incorrectly classified as "superacids"), (ii) good thermal stability, (iii) higher solubility of oxygen than in nonfluorinated acids, and (iv) low degree of adsorption of the acid anion on the electrode surface. Due to the third and fourth characteristics, the oxygen reduction kinetics on platinum in this acid is more than one order of magnitude higher than in other aqueous acids. The difficulties encountered when using this acid as a fuel cell electrolyte are (i) high wettability of Teflon, which causes a flooding problem of the fuel cell electrode, and (ii) acid concentration management at the possible operating temperatures of less than 100°C.

To overcome the problems encountered with TFMSA, but taking into account the attractive features of fluorinated acids (superacids), there was a considerable effort to develop similar acids with higher molecular weights (for use in fuel cells at above 100°C) and with two or more acid groups, say disulfonic acid.[3,5] Instead of fluorinated sulfonic acids, fluorinated carboxylic, phosphoric, and antimonic acids were also investigated. Another type of acid electrolyte which has been investigated is fluorinated disulfone imide acid. Until now, attempts to find an alternative acid to phosphoric acid as a fuel cell electrolyte have not been successful. However, the positive aspect is that significant progress has been made in developing the solid polymer electrolyte fuel cell, which uses a perfluorinated sulfonic acid polymer as the electrolyte (see Section 2.6).

2.4. MOLTEN CARBONATE FUEL CELLS

2.4.1. Principles of Operation

A schematic of a single cell in a molten carbonate fuel cell (MCFC) is shown in Fig. 2.8. Molten alkali carbonate mixture, retained in a porous lithium aluminate matrix, is used as the electrolyte. At the cathode, made of lithiated NiO, oxygen reacts with carbon dioxide and electrons to form carbonate ions:

$$\tfrac{1}{2}O_2 + CO_2 + 2e^- \rightarrow CO_3^{2-}. \tag{2.16}$$

The ionic current through the electrolyte matrix is carried by the carbonate ions from cathode to anode. At the anode (nickel with 10% Cr), oxidation of hydrogen consumes the carbonate ions and forms water vapor and carbon dioxide:

$$H_2 + CO_3^{2-} \rightarrow H_2O + CO_2 + 2e^-, \tag{2.17}$$

releasing electrons to the external circuit. In a practical MCFC, carbon dioxide produced at the anode must be transferred to the cathode where it is consumed. The transfer of CO_2 from the anode exhaust to the cathode inlet can be carried out in two ways: (i) burn the spent anode stream with excess air and mixing it with the cathode inlet gas after removing the water vapor; (ii) use a "product exchange device" to separate CO_2 from the anode exhaust.[26] The latter method would provide a richer oxidant because it is not diluted by nitrogen, thereby resulting in a higher cell voltage.

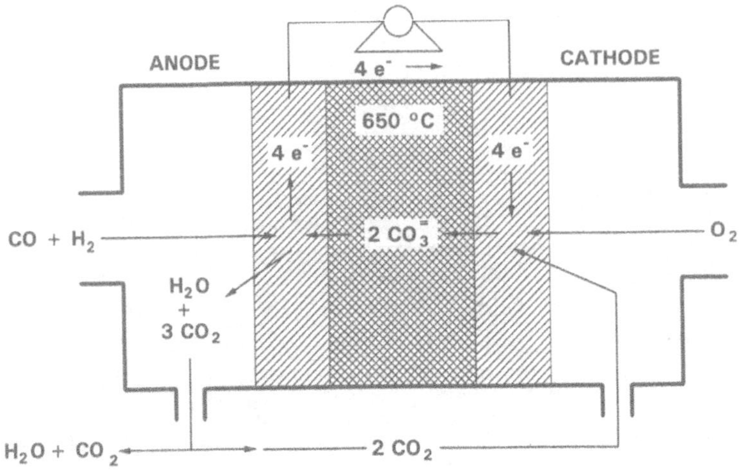

Figure 2.8. Schematic of a molten carbonate fuel cell.

2.4.2. Operating Conditions

Since the MCFC operates at about 650°C,[3,5,26] a power plant with cell stacks of this type has several favorable characteristics for utility power generation: (i) polarization losses are reduced to such an extent that it does not require expensive catalysts as do low-temperature fuel cells, such as PAFC; (ii) the operating temperature is high enough to produce high-quality waste heat, yet it is sufficiently low that the decrease in the free energy is not very large (e.g., though the thermodynamic reversible potential is about 200 mV lower than at room temperature, the OCV is nearly the same as in the PAFC), and (iii) the high-temperature waste heat can be used for fuel processing and cogeneration, a bottoming cycle, and internal reforming of methane. The MCFC nominally operates at 160 mA/cm^2 to produce 0.75 V per cell at atmospheric pressure and 75% fuel (hydrogen) utilization. Under pressurized conditions, the performance of MCFC is better.

2.4.3. Typical Cell Materials and Configurations

Characteristics of state-of-the-art components of MCFC are given in Table 2.1. All three components of the MCFC are now manufactured by tape-casting processes. The anode is made of nickel–chromium alloy (2–10 wt% Cr). The addition of Cr prevents the sintering of the porous anode, since it forms $LiCrO_2$ at the grain boundaries and prevents metal diffusion. Recently, small metal oxide (Al_2O_3, $LiAlO_2$, etc.) particles have been incorporated in the anode to prevent mechanical creep.

In addition to its electrocatalytic function, the anode is also required to act as an electrolyte reservoir barrier for gas crossover and to provide structural support for the cathode and the thin electrolyte matrix in the cell stack. The gas crossover is prevented by a thin structure of fine porosity which is filled with molten

Table 2.1. Characteristics of State-of-the-Art Cell Components for Molten Carbonate Fuel Cells

Component	Current status
Anode	Ni–10 wt% Cr 3–6 μm pore size 50–70% porosity 0.5–1.5 mm thickness 0.1–1 m^2/g
Cathode	Lithiated NiO 7–15 μm pore size 70–80% porosity 0.5–0.75 mm thickness 0.5 m^2/g
Electrolyte support	γ-$LiAlO_2$ 0.1–12 m^2/g
Electrolyte[a]	62 Li–38 K 50 Li–50 Na 70 Li–30 K ~50 wt%
Fabrication process	Tape cast 0.5 mm thickness

[a]Mole percent of alkali carbonate salt. From Kinoshita *et al.*[3]

carbonate electrolyte at the anode–electrolyte interface. This layer is called the *bubble pressure barrier* (BPB).

The pore-size distributions for the electrodes, electrolyte matrix, and BPB play very important roles in establishing a stable electrolyte–gas interface in the porous electrodes because the balance in capillary pressures is used for the electrolyte distribution in the cell components. The electrolyte tile in MCFC consists of a eutectic mixture of 68% Li_2CO_3 and 32% K_2CO_3, retained in a porous γ-$LiAlO_2$ matrix. The tile not only conducts the carbonate ions from cathode to anode, but it also separates the fuel and oxidant gases. The electrolyte composition is favorable, but by no means is it optimized for MCFC performance and stability. The cations, present in the molten alkali carbonate electrolyte, have a strong effect on the performance and endurance of MCFC. High Li and Na content in the electrolyte increases ionic conductivity, while high K promotes gas solubility.

The cathode consists of porous lithiated NiO, generally formed by *in situ* oxidation and lithiation of sintered nickel. Since lithiated NiO is a non-stoichiometric compound ($Li_xNi_{1-x}O$, where $0.022 \le x \le 0.04$), the electronic conductivity is several orders of magnitude higher than NiO. As shown in Table 2.1, the typical thickness of the cathode is lower than that of the anode in order to lower total electronic resistance and to obtain better performance.

The oxygen reduction reaction at the cathode is complex and depends on the cations present in the electrolyte. The oxygen reacts with the carbonate ions and forms peroxide and/or superoxide ions. In a pure Li_2CO_3 or a Li-rich melt, the peroxide species is dominant, whereas in a K-rich melt the superoxide is dominant.[27–30] In the eutectic mixture of 68% Li_2CO_3 and 32% K_2CO_3, both peroxide and superoxide species are present.[31] The electrolyte distribution is critical to cathode performance because it is very sensitive to the degree of pure filling volume by electrolyte. The optimal pore filling is about 20%, and even then the overpotential losses at the cathode are considerably greater than those occurring at the anode.[32]

The dissolution of NiO causes a serious problem concerning the cell lifetime. Although the solubility of nickel is far less than that of silver and copper (≈10 ppm), dissolved NiO may diffuse in the electrolyte toward the anode, where metallic Ni may electrodeposit because of the relatively negative potential (i.e., the hydrogen partial pressure is sufficiently high). The deposition of metallic Ni grains in the electrolyte matrix can eventually cause the electrolyte to conduct electronically and short-circuit the cell. This phenomenon is proportional to the partial pressure of CO_2 and may involve the reaction

$$NiO + CO_2 \rightarrow Ni^{2+} + CO_3^{2-}. \tag{2.18}$$

The dissolution of NiO is also a function of electrolyte basicity; hence, it depends on the cations in the metal. Research is in progress to develop alternative cathode materials and to optimize the electrolyte composition[33] to solve the problem of NiO dissolution.

The bipolar plates in the MCFC stacks are usually fabricated from thin (~15 mil) sheets of an alloy (e.g., type 310S or 316L stainless steel). The anode side of the bipolar plate is coated with a Ni layer to prevent corrosion.

2.4.4. *Methods for Cooling of Electrochemical Cell Stacks*

The advanced-form MCFC, referred to as the internal reforming molten carbonate fuel cell (IRMCFC), may consume the lower hydrocarbons (e.g., CH_4)

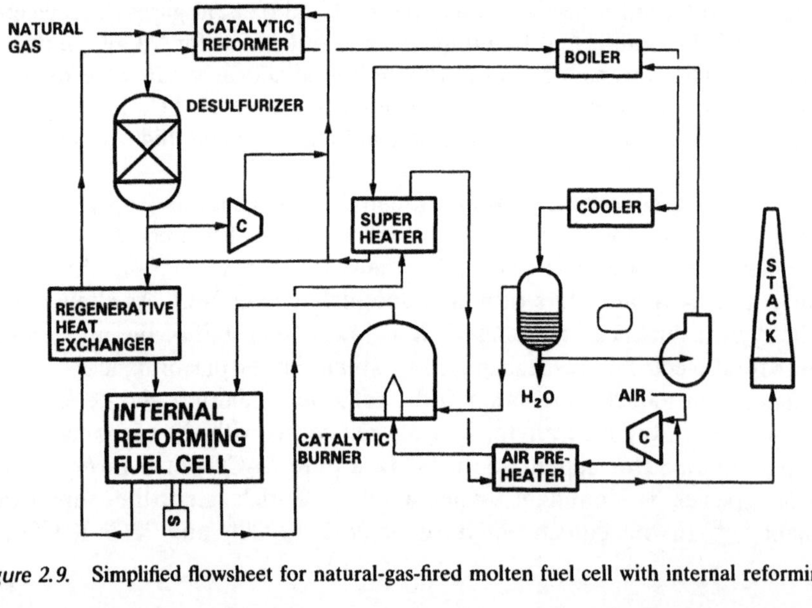

Figure 2.9. Simplified flowsheet for natural-gas-fired molten fuel cell with internal reforming.

Table 2.2. Contaminants and Their Effects on the Molten Carbonate Fuel Cells

Class	Contaminant	Potential effect
Particulates	Coal fines, ash	Plugging of gas passages
Sulfur-containing compounds	H_2S, COS, CS_2, C_4H_4S	Voltage losses Reaction with electrolyte via SO_2
Halogen-containing compounds	HCl, HF, HBr, $SnCl_2$	Corrosion Reaction with electrolyte
Nitrogen-containing compounds	NH_3, HCN, N_2	Reaction with electrolyte via NO_x
Trace metals	As, Pb, Hg, Cd, Sn, Zn, H_2Se, H_2Te, AsH_3	Deposits on electrode Reaction with electrolyte
Hydrocarbons	C_6H_6, $C_{10}H_8$, $C_{14}H_{10}$	Carbon deposition

G. L. Anderson and P. C. Garrigan, in Proceedings of the Symposium on Molten Carbonate Fuel Cell Technology (R. J. Selman and T. D. Claar, eds) (The Electrochemical Society, Inc., Pennington, NJ, 1984, Table 1, p. 299.

directly. The IRMCFC is intrinsically efficient because the water vapor and the heat produced by the exothermic reaction occurring at the anode (Eq. (2.17)) provide the thermal energy and the reactant required for steam reforming methane. Since the standard nickel anode does not have sufficient activity to reform the required amount of methane to hydrogen, a supported Ni catalyst may be required in IRMCFC. A simplified flowsheet for natural-gas-fired IRMCFC is shown in Fig. 2.9. A recent study showed that an IRMCFC system could operate on natural gas with an overall efficiency greater than 50% and perhaps over 60% in the future.

2.4.5. *Acceptable Contaminant Levels*

The MCFC does not have a problem of anode CO poisoning as in low-temperature fuel cells (PAFC, SPFC), in fact, CO in the anode gas is used as a fuel. The fuel gas derived from fossil fuels, such as coal and natural gas, may contain several contaminants. Some of the contaminants and their effects on performance of MCFC are given in Table 2.2. The performance of the anode degrades even in the presence of small amounts of sulfur (~1 ppm) in the fuel or oxidant gas stream, although the degradation is reversible. The adverse effect of sulfur compounds on the performance of MCFC requires prior removal of the sulfur compounds from the fuel gas or periodic purging of sulfur from the system.

2.4.6. *Applications and Economics*

When used with a coal gasifier, the MCFC with its higher fuel efficiency and high-grade waste heat will allow a heat rate better than 6800 Btu/kWh or 50% efficiency for HHV coal to ac power[34] by using the MCFC in a combined cycle.[35] The PAFC could produce only about 9600 Btu/kW-h or 36% efficiency.[36] Plans are under way to design and construct MCFC power plants of megawatt-size capacity for base load and intermediate load power generation and cogeneration applications. The projected target cost estimate of an MCFC power plant is $1000 to $1500/per kilowatt.

2.5. SOLID OXIDE FUEL CELLS

2.5.1. Principles of Operation

The solid oxide fuel cell (SOFC) is an all-solid-state power system which uses yttria-stabilized zirconia as the electrolyte layer. This system can provide high-quality waste heat for (i) cogeneration applications and (ii) bottoming cycles, utilizing conventional steam turbines for additional electricity generation. Chemical to electrical energy efficiencies greater than 50% can be readily obtained. Yttria-stabilized zirconia is sufficiently ionically conducting above 1000°C for use as an electrolyte layer in fuel cells; the oxide ion transference number is close to unity. The search for intermediate-temperature solid electrolytes has not been too successful to date. Strontium-doped lanthanum manganite and nickel–zirconia cermet are used as cathode and anode materials, respectively. Mg- or Sr-doped lanthanum chromite is used as the interconnect layer for the multicells.

The operating principle of a SOFC is shown in Fig. 2.10. At the anode, hydrogen reacts with oxide ions transported through the electrolyte to form water. This is accompanied by the release of electrons to the external circuit. The electrons from the external circuit react with oxygen at the cathode and produce oxide ions. The overall process is the reaction of oxygen with hydrogen to produce water. Carbon monoxide can also be used, instead of hydrogen, and the corresponding reaction product will be carbon dioxide.

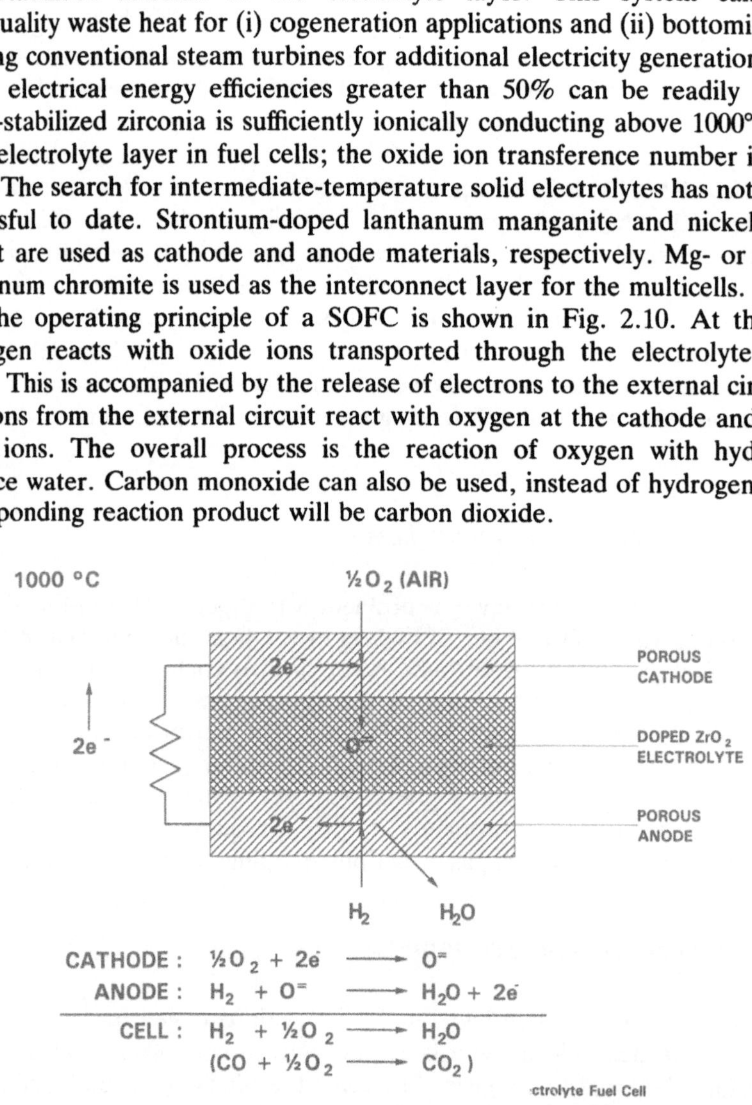

Figure 2.10. Principle of operation of a high-temperature solid oxide electrolyte fuel cell.

2.5.2. Typical Cell Materials, Configurations, and Operational Conditions

A major technical challenge in SOFC technology is the fabrication of the component layers of the electrochemical cell stack. Many researchers are working on (i) the materials aspects of SOFCs, (ii) production of intermediate temperature SOFCs, and (iii) fabrication of monolithic solid oxide fuel cells (MSOFC) by techniques other than tape calendaring.[36–39] Since there are four ceramic layers involved, matching their thermal expansion coefficients and sinterabilities is difficult.

The cathode material must (i) be stable in an oxidizing environment, (ii) have an electronic conductivity of at least $50\ \Omega^{-1}\ cm^{-1}$, (iii) have a porosity level of about 30%, and (iv) exhibit a good activity for oxygen reduction under the operating conditions. Its firing shrinkage profile and thermal expansion coefficient should also match with those of other cell components. Strontium-doped lanthanum manganite, which is currently used as the cathode material for SOFCs, has an electrical conductivity of $80\ \Omega^{-1}\ cm^{-1}$ at 1000°C and is stable in an oxidizing environment. By careful control of the stoichiometry and powder characteristics, other properties such as firing shrinkage profile, thermal expansion coefficient, and porosity can be tailored to match those of the other cell components.

The anode material must be stable in the reducing environment, and the conductivity should be more than $120\ \Omega^{-1}\ cm^{-1}$. Nickel–zirconia cermet is being used as the anode material, for which yttria-stabilized zirconia serves as a porous substrate for the nickel phase. Nickel conducts electrons and also serves as a catalyst for the electrochemical oxidation. Anode layers with an electrical conductivity of $615\ \Omega^{-1}\ cm^{-1}$ have been fabricated.[40] However, controlling the thermal expansion coefficient has been the major challenge with respect to the anode layer. A nickel content below 35% will drastically reduce the electronic conductivity of the material, under which conditions the ionic conductivity in the zirconia will dominate and affect cell performance. This minimum amount of nickel is incorporated in the anode to reduce the mismatch of the thermal expansion coefficient with other cell components.[40]

The electrolyte layer for the SOFCs must have high ionic and negligible electronic conductivities. The ionic transference number should be close to unity. This layer must also be very dense to avoid the crossover of reactants. Gas permeabilities of less than $10^{-7}\ cm^2/s$ reduce the cross-leakage current density to a level of $1\ mA/cm^2$. It was found that yttria-stabilized zirconia samples with densities less than 90% of the theoretical value have permeabilities greater than $10^{-7}\ cm^2/s$ and those with 92–93% of the theoretical density have negligible permeabilities (too small to be measured). Hence, it can be concluded that yttria-stabilized zirconia samples having densities of more than 94% of the theoretical density are suitable for SOFCs. Compared to other cell components, the electrolyte layer has an extremely low electronic conductivity. To reduce ohmic overpotential losses, the layer should be as thin as possible. The yttria-stabilized zirconia electrolyte layer is the most brittle among all SOFC components.

The material currently used for the SOFCs interconnect layer is Mg or Sr-doped $LaCrO_3$. Like all chromium oxides this material has a sintering problem, which makes it difficult to prepare a high-density material in air. It is necessary to sinter it at 1650°C in a hydrogen atmosphere to obtain a dense material. However, this high temperature is not compatible with other SOFC components because of interdiffusion. Also, the cathode material, Sr-doped lanthanum manganite, decomposes at this temperature. Hence, it is necessary to develop suitable interconnect materials which can be sintered at about 1400°C. Suitable sintering aids can also be developed to sinter lanthanum chromite.

Three major designs for SOFCs are being developed, namely tubular, monolithic, and planar. The tubular design was designed and developed by Westinghouse Electric Corporation.[37,41] Figure 2.11 shows the cross section of a bundle of tubular SOFCs. It is the most developed of the three designs and is expected to be commercialized in the 1990s. This design uses a porous calcia-stabilized zirconia ceramic tube. It is prepared by extrusion followed by sintering and is closed at one end. It is the support tube for depositing the fuel cell component layers and also delivers the oxidant to the reaction sites in the cathode layer above it.

The support tube is about 2 mm thick and 13 mm in diameter. Tube length has been increased from 30 to 78 cm, and future cells for larger power plants will have lengths of 2 m or more. The porosity of the support tube is about 30%, and the pore size varies from 2 to 10 μm. The cathode layer is deposited on the support tube by coating a slurry of La(Sr)MnO_3, evaporating the solvent, and sintering. The porosity of this layer is designed to be about 30% and its thickness about 100 μm. The dense electrolyte layer is formed by chemical vapor deposition (CVD) and electrochemical vapor deposition (EVD) at 1150°C using $ZrCl_4$, YCl_3, H_2, and H_2O with O_2 in the vapor state. This is a good technique for depositing dense films on a porous substrate, and it is a key element in the Westinghouse technology. The porous Mg-doped $LaCrO_2$ interconnect layer is

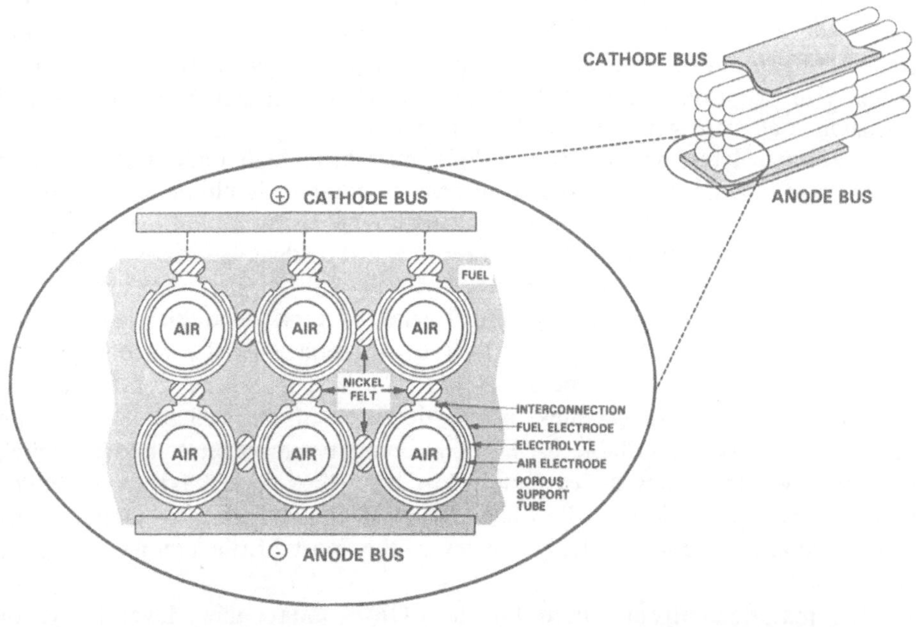

Figure 2.11. Westinghouse tubular solid oxide fuel cells in its bundle configuration.

also deposited by EVD from appropriate chloride vapors. The Ni–$ZrO_2(Y_2O_3)$ cermet anode is prepared by coating a slurry of NiO and yttria-stabilized zirconia, followed by EVD. The porosity and thickness of the anode layer are 30% and 20 μm, respectively. Air enters the tubular cell via a concentric alumina tube, and the fuel flows by outside the electrochemical cell.

A 400-W experimental unit, constructed by Westinghouse, was installed and tested for 1760 h at the Tennessee Valley Authority (TVA) in 1986. Two 3-kW SOFC systems were built by Westinghouse and tested successfully by Tokyo Gas Company and Osaka Gas Company in Japan during 1987. Easy installation, operation, and maintenance were demonstrated. The cell showed less than 2% cell potential degradation for 1000 h and a lifetime of greater than 15 h.[41]

The test data of the Westinghouse fuel cell (Fig. 2.12) shows improvement in cell performance and life. These systems also exhibit low emission of pollutants and low noise level. Two 3-kW systems were successfully operated over 5000 h with direct fueling of natural gas. In this system, air is preheated to 700°C for start-up purposes and then cooled to 500°C during cell operation, by a subsystem which can deliver up to 600 L of electrically heated air per minute. The exhaust gas mixture contains about 77% N_2, 17 O_2, 4% H_2O, and 2% CO_2 and is 700°C.

The support tube in the Westinghouse design accounts for 50% of the cell materials. The weight of this tube is much higher than that of the cell itself, and thus it has a significant effect on the specific power density. The EVD process for depositing the cell components is quite expensive. On the other hand, the MSOFC design, proposed by Argonne National Laboratory[41] in 1983, has no support tube; the ceramic cell components support themselves, so this design makes it possible to increase specific power and power density. It is projected that power densities of 4×10^3 kW/m^3 and 8.08×10^3 W/kg, compared with 140 kW/m^3 and 100 W/kg, respectively, for the Westinghouse fuel cell system, can be attained. MSOFCs[38–41] are still in the early stage of development, and more research and development is needed before they become practical.

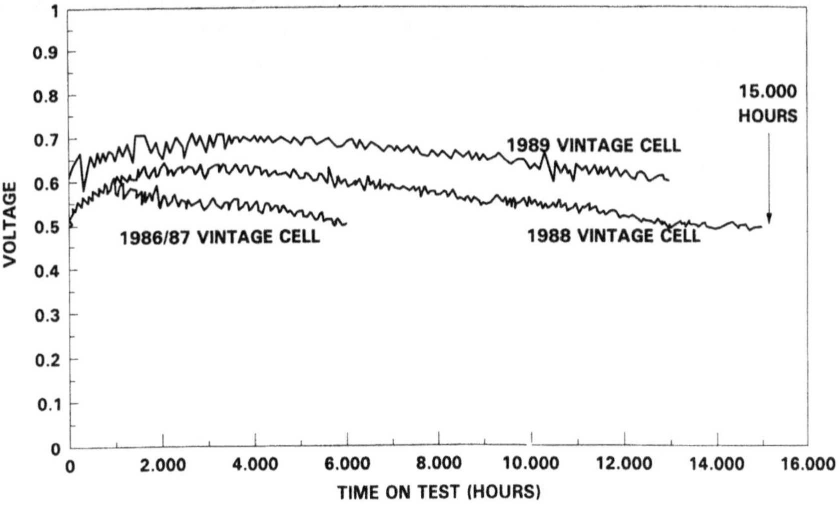

Figure 2.12. Cell test data verifying improved cell performance and life of tubular solid oxide fuel cell.

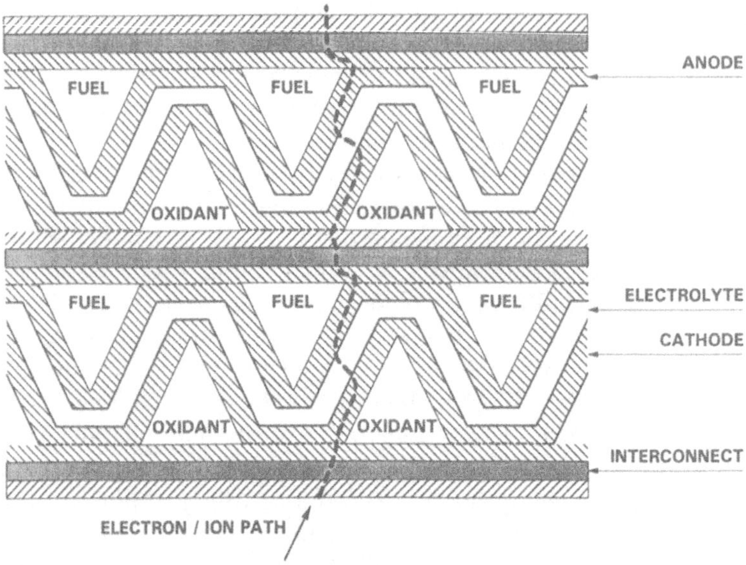

Figure 2.13. Coflow monolithic solid oxide fuel cell.

Figure 2.13 is a schematic of an MSOFC and indicates the component materials for a coflow configuration. Currently MSOFCs are fabricated by the following steps: (i) slurries of electrodes, electrolyte, and interconnect are initially made; (ii) a tape-casting process is utilized for the production of thin ceramic layers; (iii) these layers are then laminated, corrugated, and stacked; and (iv) the electrochemical cell stack is finally sintered. During the corrugating process, the laminated layers are compressed into a honeycomb structure. These honeycomb structured layers are then stacked atop each other. The MSOFC electrochemical cell stack is then sintered at high temperatures. In this type of design, fuel and air flow in the same direction (i.e., coflow). Another design of MSOFC utilize a crossflow of fuel and air. Crossflow stacks, consisting of four cells, have been successfully fabricated and tested. Several coflow stacks have also been fabricated for demonstration purposes. A single cell with a surface area of 14.5 cm^2 has exhibited a power output of 4.4 W and has been operated for over 100 h.

Planar SOFCs may be advantageous over tubular SOFCs for a large power plant.[42] With the tubular design, the electrolyte layer is deposited on the cathode layer by EVD. During this process, the electrolyte vapor partially enters the cathode layer and blocks the pores, which is then followed by the deposition of a uniform dense layer of the electrolyte on top of the cathode. The ohmic resistance of this layer could reach relatively high values. This is not true for planar cells, which are fabricated by tape-casting.[43] Consequently, the ohmic resistance of the planar cells is much less than that of the tubular cells. The power density and the specific energy could be higher than those of tubular cells. The planar configurations are easier to fabricate than the MSOFCs. However, this design is in the research and development stage. Solutions to sealing problems may be a challenge. Several research groups are working on this design. Fuji Electric Corporate Research and Development Ltd.[44] has recently tested a planar SOFC with a cell area of 7 cm^2 that uses dry air and hydrogen as reactants. The power output was 0.22 W/cm^2, the open-circuit voltage was 1.07 V, and the short-circuit current at 33% fuel utilization was 1 A/cm^2.

2.5.3. Methods for Cooling of Electrochemical Cell Stacks

Cell cooling is normally carried out by using an excess flow of air, and the sensible heat contained in the effluent gases combined with combustion of the exhaust fuel is used to preheat the air entering the cell. In effect, this configuration integrates a high-temperature heat exchanger and the fuel cell stack without the need for any gastight seals.

2.5.4. Acceptable Contaminant Levels

SOFCs operate equally well on dry or humidified hydrogen or carbon monoxide fuel or on their mixtures. Moreover, the high operating temperature considerably minimizes catalyst poisoning. A concentration of 50 ppm of hydrogen sulfide in the fuel lowers the operating cell potential by only about 5%; the cell potential reverts to the original voltage when it is removed from the fuel mixture. In a typical fuel cell consisting of 25% H_2–H_2O and 75% CO–CO_2 at 700°C, the tolerance of the nickel anode to H_2S is about 5 ppm, whereas at 1000°C it is about 90 ppm. Cobalt cermet tolerance to sulfur at 1000°C is about 200 ppm. These sulfur tolerances are about one to two orders of magnitude higher than for other types of fuel cells.

2.5.5. Applications and Economics

SOFCs are very attractive for electric utility and industrial applications. The high operating temperature and tolerance to impure fuel streams make SOFC systems especially attractive for utilizing H_2 and CO from natural gas steam-reforming and coal gasification plants. The target is to ultimately reduce the capital cost of the fuel cell power plant to about $1000 to $1500/per kilowatt.

2.6. SOLID POLYMER ELECTROLYTE FUEL CELLS

2.6.1. Principles of Operation

The solid polymer electrolyte fuel cell (SP(E)FC)[1–5,45] is perhaps the most elegant of all fuel cell systems in terms of design and mode of operation. It consists of a solid polymeric membrane which acts as an electrolyte. The membrane is sandwiched between two platinum-catalyzed porous electrodes. A single cell assembly, shown in Fig. 2.14, can be mechanically compressed by screws or pneumatic pressure. The fuel cell requires humidified gases, hydrogen and oxygen. The electrochemical reactions that occur at the platinum electrocatalyst sites are as follows;

$$\text{Anode} \quad H_2 \rightarrow 2H^+ + 2e^-, \tag{2.19}$$

$$\text{Cathode} \quad O_2 + 4H^+ + 4e^- \rightarrow 2H_2O. \tag{2.20}$$

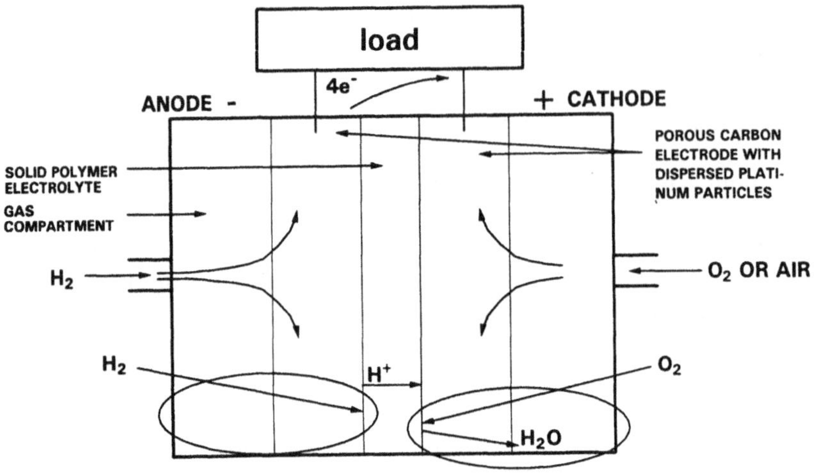

Figure 2.14. Schematic of solid polymer electrolyte fuel cell.

The overall fuel cell reaction is thus as expressed by Eq. (2.2). The oxygen reduction reaction is the slower reaction (the exchange current for this reaction is at least three orders of magnitude lower than that of hydrogen oxidation), and thus the challenge is the enhancement of the electrocatalytic activity of this reaction. Several fundamental investigations on oxygen reduction kinetics in acid electrolytes have been carried out.[46–48] However, it is only recently that detailed studies of the electrode kinetic parameters at the platinum–isomer interface were carried out.[49–52]

The functioning of a fuel cell relies on the formation of a stable three-phase boundary in the immediate vicinity of the electrocatalyst site. This boundary, for fuel cells with liquid electrolytes, is formed between the gas within the pores of the electrode, the liquid electrolyte which has worked its way into the pores to the electrocatalytic. Alternatively, a thin film of the electrolyte layer may form on the electrocatalyst, and the reactant gas can dissolve in the electrolyte interface and then diffuse through the electrolyte film to the electrocatalyst.

In the SP(E)FC there is no liquid electrolyte and it is very probable that a thin film forms on the electrocatalyst. The presence of excess liquid water hampers the easy access of gas into the porous structure of the electrode, and the fuel cell undergoes a decrease in performance due to mass transport limitations of the oxidant gas. This flooding situation is exacerbated by the ingress of water from the hydrogen side. The proton formed during hydrogen oxidation is usually strongly hydrated and causes transport of water from the anode to the cathode. This phenomenon has further repercussions in that the loss of water causes drying out of the membrane and an increase in ohmic behavior at the anode–membrane interface. It is therefore of paramount importance that methods for efficient addition of water to the hydrogen electrode and removal of water from the oxygen electrode be divided.

2.6.2. *Operating Conditions*

The solid polymer electrolyte fuel cell operates at a lower temperature than the PAFC, MCFC, and SOFC, which are in a more advanced state of

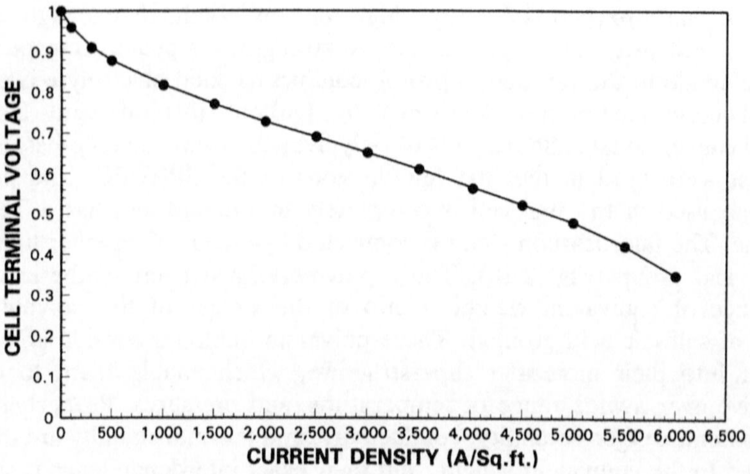

Figure 2.15. Terminal voltage vs. current density of Ballard Technologies Corporation's solid polymer electrolyte fuel cell showing the best performance of its kind.

development. The limit on the temperature at which the fuel cell operates is set by the thermal stability and conductivity characteristics of the polymeric membrane that is used as its electrolyte. With Nafion, it is best not to exceed an operating temperature of 85°C. With the new class of ionomers developed by the Dow Chemical Company, this temperature can be increased by another 10–20°C. Operating pressures for oxygen can be from atmospheric to 8 atm. Air pressures up to 8 atm have been used. Operation at higher pressure is necessary for attaining higher power densities, particularly with air as the cathodic reactant. The pressures, in general, are maintained equal on either side of the membrane. This minimizes the problem of gas crossover through the membrane. The crossover reduces the cell potential and also increase the risk of forming an explosive mixture of hydrogen and oxygen. The best performance so far (Fig. 2.15) in a SP(E)FC (Pt loading, 4 mg/cm^2) has been reported by Ballard technologies Corporation.[53]

As mentioned in Section 2.6.1, the water management in the membrane and electrode assembly of the SP(E)FC is fairly complex and requires dynamic control to match the varying operating conditions of the fuel cell. A simple humidification scheme, used at Texas A&M University and Los Alamos National Laboratory, is to disperse the gas through a ceramic frit immersed in a column of water. The disadvantage of this method is that the humidification depends greatly on the flow rate of the gases introduced into the fuel cell. High flow rates cause entrainment of the water as droplets in the gas stream. The flow rate is usually determined by stoichiometric flow requirements. Steam generation and atomization techniques have also been tried for humidifying the reactant gases.

2.6.3. Typical Cell Materials

The solid polymer electrolyte fuel cell has made a great revival, mainly due to the success at DuPont and Dow Chemical Company in developing perfluorinated sulfonic acid membranes. These polymers have the following desirable properties: high oxygen solubility, high proton conductivity, high chemical stability, low density, and high mechanical strength. As proton transport occurs from the anode to the cathode, a proton-conducting solid electrolyte is essential. The polymeric membranes developed by DuPont (Nafion) and Dow have replaced conventional sulfonic acids of polydivinylbenzene–styrene-based copolymers that were used in the first few versions of the SP(E)FC. The polymeric electrolyte used in the fuel cell is completely fluorinated and has a Teflon-like backbone. The fluorocarbon chain is connected by means of an ether linkage to a sulfonic acid group (Fig. 2.16). These polymers have been synthesized over a wide range of equivalent weights (ratio of the weight of the polymer to the number of sulfonic acid groups). These polymers imbibe a considerable amount of water into their molecular superstructure, which enable them to be fairly conductive over a wide range of temperatures and pressures. Properties such as water content, oxygen solubility, conductivity, and thermal stability are intimately connected to the equivalent weight, and their exact interdependence is still being investigated.

The gas-diffusion electrodes that have been used so far contain unsupported ($2–10\,mg/cm^2$) or supported platinum (platinum supported on carbon; $0.4\,mg/cm^2$, 10% or 20% Pt on C) electrocatalysts. The latter type of electrode is essentially the same as that developed for the PAFC and has not yet been optimized for SP(E)FC. The electrodes have a carbon backing (cloth or paper). A Teflon emulsion is used to bond the platinum particles (nm diameter) to the carbon layer. The active layer is deposited a few microns thick on the substrate layer (about $10\,\mu m$ for unsupported and $50\,\mu m$ for supported electrocatalysts). Recent studies at Los Alamos National Laboratory[54] and at Texas A&M University[55] have shown that the performance of electrodes with low platinum loading ($0.4\,mg/cm^2$) is comparable to those with high platinum loading (compare results in Figs. 2.15 and 2.17). This was achieved by impregnating the active layer of the former type of electrode with a proton conductor.[54,56]

The graphite support structures have gas flow paths to allow for laminar flow of gases across the electrode. Rectangular and circular flow paths in axial and concentric directions have been attempted at General Electrics, United Technologies, and Delco Remy Division of Genera Motors Corporation. Flow-modeling studies on these electrodes are in progress in several laboratories.

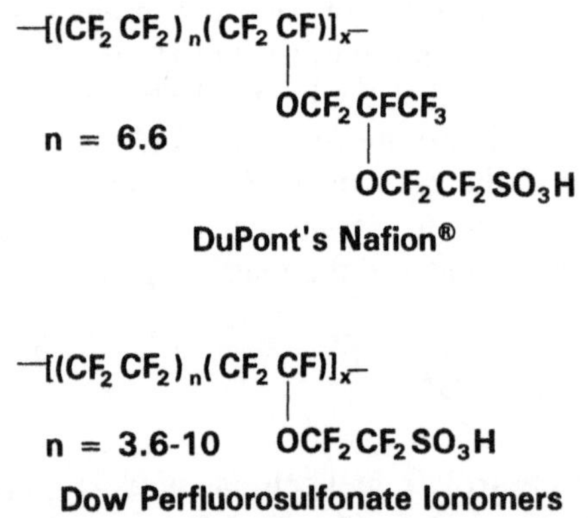

Figure 2.16. Comparison of DuPont's Nafion and Dow perfluorosulfonate ionomer membranes.

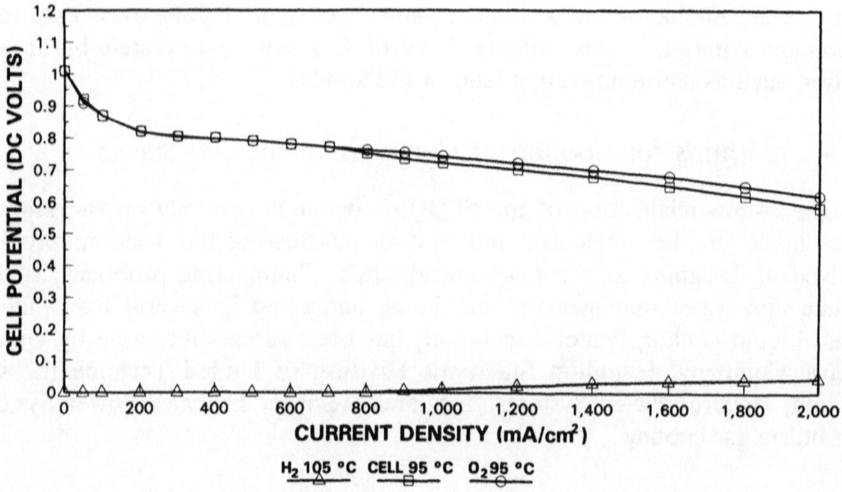

Figure 2.17. Cell and half-cell potentials vs. current density plots for a single cell with Dow membrane (thickness: 125 μm) and Pt-sputtered Prototech electrodes (Pt loading 0.45 mg cm^{-2}) operating at 95°C with H_2–O_2 at 4–5 atm.

2.6.4. Acceptable Contamination Levels

In contrast to the alkaline fuel cell, which is very sensitive to CO_2, the SP(E)FC is insensitive to CO_2 in the oxidant. This affords the possibility of using reformed gas directly as the oxidizer. The major contaminant, however, is carbon monoxide. Hamilton Standards Division of United Technologies Corporation has shown a dramatic decrease in performance (Fig. 2.18) when there is even 0.17% carbon monoxide in the oxygen gas.[57] Ballard Technologies has demonstrated that the CO level can be significantly reduced (from 1–2% to 100 ppm) by passing

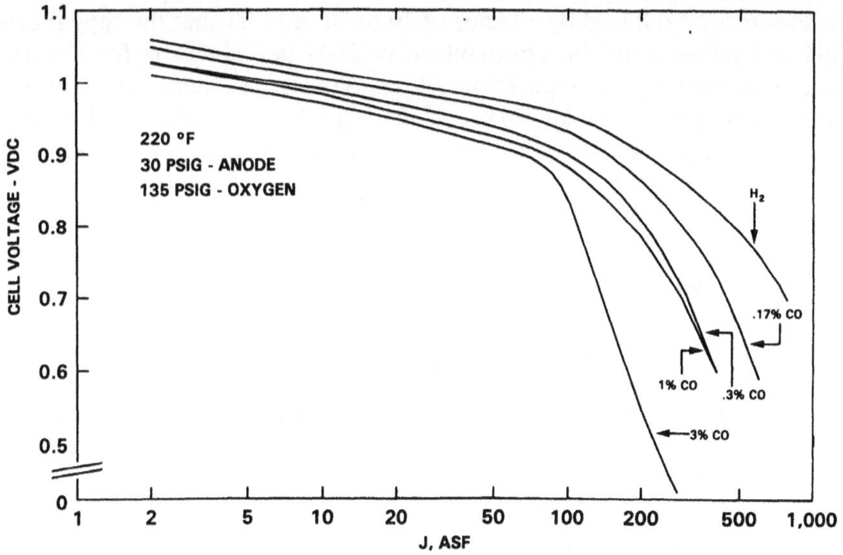

Figure 2.18. Effect of CO poisoning on the H_2 electrode in a H_2–O_2 solid polymer electrolyte fuel cell (Hamilton Standard).

the reformed methanol and a small amount of oxygen (1–2%) over a platinum-on-alumina catalyst.[53] This reduced level of CO can be tolerated by an alloy catalyst, such as platinum–ruthenium, in the anode.

2.6.5. Methods for Cooling of Electrochemical Cell Stacks

The commercialization of the SP(E)FC depends not only on the improvements made on the single cell but also on addressing the scale-up problems involved in designing an electrochemical stack. Engineering problems, such as thermal and water management, are being addressed in several R&D laboratories. Liquid cooling (water circulation) has been successfully used by General Electric Company, Hamilton Standards Division of United Technologies Corporation, Ballard, Pavco Systems, Inc, and Siemens. Ergenics Power Systems, Inc., utilize gas cooling.

2.6.6. Application and Economics

As mentioned earlier, the high performance (efficiency–power density) of the SP(E)FC has made it an attractive candidate for terrestrial and extraterrestrial applications. Its performance evaluation in space and in deep ocean is already in progress. It shows promise for transportation applications because of the high power densities reported in cells with low-platinum-loading electrodes, hydrogen and air as reactants, and the Dow membrane as the solid polymer electrolyte.[56]

At present, the major challenge is economic, namely significantly reducing the costs of the proton-conducting membranes and the platinum-catalyzed electrodes. With the progress made in achieving high power densities (>0.5 W/cm^2) with low-platinum-loading electrodes, the noble metal costs can be reduced to about \$30–\$40 per kilowatt. DuPont's Nafion and Dow's state-of-the-art membranes cost about \$60–\$200 per kilowatt. It is imperative that the membrane costs be reduced by a factor of 10 to 20 in order that the capital cost of the fuel cell power plant be affordable (say \$200 per kilowatt) for terrestrial, particularly transportation, applications. Technological problems, such as thermal and water management of high-power-density fuel cells, need novel solutions. There is increasing interest in developing regenerative solid polymer electrolyte fuel cells for NASA's lunar and Mars missions as well as for the space station.[57]

2.7. DIRECT METHANOL FUEL CELLS

The direct methanol fuel cell (DMFC) has been for the last 30 years, and still is, the Fuel Cell researcher's dream. Though methanol is the simplest organic liquid fuel (which can most economically and efficiently be produced on a large scale from the relatively abundant fossil fuels—coal and natural gas—and is the most electroactive organic fuel), its electrochemical activity is several orders of magnitude (at least three) less than that of hydrogen. Moreover, the steam reforming of methanol occurs at a considerably lower temperature (200° versus 700°C) than that of other organic fuels (say, natural gas, ethanol). Thus, in the near to intermediate term, from the technical and economic points of view, it may be more advantageous to interface methanol reformers with the electrochemical cell stack in fuel cell power plants. However, in the long term, the ideal fuel cell power plant for transportation, remote power, and other portable power applications will be the one using methanol directly as the anodic reactant. The challenges to meet this goal are as follows:

- The discovery of electrocatalysts with exchange current densities (based on real surface area) of at least 10^{-5} A/cm^2 for the anodic oxidation of methanol to CO_2.
- Inhibition of poisoning of the electrocatalysts by methanol or by intermediates (such as CO, —COH) produced during the oxidation, to an extent such that the time variation of overpotential is less than 10 mV/1000 h.
- Finding methods to prevent transport of methanol from the anode to the cathode.
- Discovering electrocatalysts for oxygen reduction which are inert toward methanol.

2.8. RELATIVE ADVANTAGES AND DISADVANTAGES OF DIFFERENT TYPES OF FUEL CELLS

Since the late 1950s there has been a great incentive to develop fuel cell power plants for space and terrestrial applications. The lag time for this development was surprisingly short for space applications, about five years. The main reason for this short lag time is that for space applications cost is of little consideration, while weight, volume, and reliability of power plants are critical. Moreover, for space applications, cryogenic hydrogen and oxygen are the ideal fuels, and from an electrochemical point of view, hydrogen and oxygen are the best fuels. However, for terrestrial applications, fuel cells have to compete, technically and economically, with conventional power plants which have been in an advanced state of development for more than 50 years. The enthusiasm and stimulation for the development of fuel cells evolved in the space era, which began in the 1960s and was, and is, reinforced by the energy crisis in 1973, the environental problems in the 1980s, and the Gulf crisis of 1990.

The relative advantages and disadvantages of different types of fuel cells are best assessed on a quantitative basis. For this purpose, the factors taken into consideration are (i) fuel efficiency; (ii) power density, which is important from the points of view of minimizing the weight, volume, and capital costs of the power plant; (ii) projected rated power level; (iv) projected lifetime; and (v) projected capital cost. The best estimates at the present time for these parameters are provided in Table 2.3. This table also cites intended applications and projected time frame and the fuels which may be used in these types of fuel cells. With the growing interest in developing fuel cells in several countries (United States, Japan, Netherlands, Italy, Belgium, Germany, India, Taiwan, Korea, etc.), particularly in countries which have to import fossil fuels (such as Japan, Italy, Belgium, Korea), the emphasis now appears to be on the engineering design, construction, and performance evaluation of fuel cell power plants. Thus, after the 30-year investment in research and development, fuel cell power plant

Table 2.3. Relative Advantages and Disadvantages of Different Types of Fuel Cell Power Plants

Type of fuel cell and of fuel	Fuel efficiency (%)		Power density (mW cm^{-2})		Rated power level projected (kW)	Lifetime projected (h)	Capital costs projected ($/kW)	Applications, time frame
	Present	Projected	Present	Projected				
Alkaline H_2	40	50	100–200	>300	10–100	>10,000	>200	Space 1960– Transportation 1996– Standby power 1966–
Phosphoric Acid; CH_4, CH_3OH	40	45	200	250	100–5000	>40,000	$1000–	Onsite integrated energy systems peak sharing 1992–
Molten carbonate; CH_4; coal	45	50–60	100	200	1000–100,000	>40,000	$1000	Base load and intermediate load power generation, cogeneration 1996–
Solid oxide fuel cell CH_4; coal	45	50–60	240	300	100–100,000	>40,000	$1500	Base load and intermediate load power generation, cogeneration 2000– Regenerative 2010– Space and terrestrial Space
Solid polymer H_2; CH_3OH	45	50	350	>600	1–1000	>40,000	>200	Space 1960– Transportation 1996– Standby power 1992– Underwater 1996–
Direct methanol CH_3 O + 1	30	40	40	>100	1–100	>10,000	>200	Transportation 2010– Remote power 2000–

manufacture appears to be an emerging technology in the last decade of this century and will find many applications in the twenty-first century.

ACKNOWLEDGMENTS

The authors wish to express their appreciation to the following sponsors of the fuel cell projects in the Center for Electrochemical Systems and Hydrogen Research at Texas A&M University (TAMU); DARPA-ONR; Energy Research in Applications Program–Texas Higher Education Coordinating Board; Electrochem, Inc./Center for Space Power, TAMU; NASA–Johnson Space Center for solid polymer electrolyte fuel cells; U.S. Department of Energy for molten carbonate and superacid electrolyte fuel cells; and Electric Power Research Institute/Center for Space Power, TAMU for solid oxide electrolyte fuel cells.

REFERENCES

1. J. O'M. Bockris and S. Srinivasan, *Fuel Cells: Their Electrochemistry* (McGraw-Hill, New York, 1969).
2. S. S. Penner, ed., *Assessment of Energy Needs for Advanced Fuel Cells,* (Elsevier, New York, 1986).
3. K. Kinoshita, F. R. McLarnon, and E. J. Cairns, *Fuel Cells: A Handbook* (U.S. Department of Energy, Morgantown, WV, 1988), DOE/METC88/6096.
4. S. Srinivasan, *J. Electrochem. Soc.* **136,** 41C (1989).
5. A. J. Appleby and F. R. Foulkes, *Fuel Cell Handbook* (Van Nostrand Reinhold, New York, 1989).
6. A. M. Adams, F. T. Bacon, and R. G. H. Watson, in *Fuel Cells* (W. Mitchell, ed.) (Academic Press, New York, 1963).
7. E. Yeager, Presentation at DOE Contractors Meeting (Technology Base Research Project), Cleveland, OH, April 14–15, 1986.
8. F. Solomon, Ext. Abstracts, Electrochemical Society Spring Meeting, Toronto, Canada, 1985.
9. C. C. Morrill, *Proceedings of the Annual Power Sources Conference,* vol. 19, No 32, 1965.
10. J. K. Stedman, *Fuel Cell Seminar Abstracts* (Courtesey Associates, Inc., Washington, DC, 1985), p. 138.
11. R. Warnock, Wright Patterson Air Force Base, Dayton, OH, personal communication, 1988.
12. K. Strasser, *J. Power Sources* **29,** 149 (1990).
13. A. J. Emery, *National Fuel Cell Seminar Abstracts* (Courtesey Associates, Inc., Washington DC, 1983), p. 98.
14. H. van den Broek, *J. Power Sources* **29,** 210 (1990).
15. E. Yeager, D. Scherson, and B. Simic-Glavski, Abstract 706, The Electrochem. Soc., Ext. Abstracts, vol. 83–1, p. 1043, San Francisco, CA, May 8–13, 1983.
16. L. M. Handley and R. Cohen, EM-2123, EPRI, Palo Alto, CA, 1981.
17. E. V. Somers, R. E. Grimble, and E. J. Vidt, EPRI-EM-1365, EPRI, Palo Alto, CA, 1980.
18. A. Kaufman, Abstracts, Fuel Cell Seminar, Tucson, AZ, Oct. 28, 1986, p. 194.
19. M. Ueno, Presentation at Seminar on New Japanese Technology, Cleveland, OH, October 25, 1982.
20. Presentation by MELCO to Netherlands Fuel Cell Group, Amsterdam, October, 1984.
21. R. Anahara, *J. Power Sources* **29,** 109 (1990).
22. S. Sumi, Abstracts, Fuel Cell Seminar, Tucson, AZ, October 28, 1986, p. 195.
23. N. Itoh, *J. Power Sources* **29,** 29 (1990).
24. A. Adams, R. Foley and H. Barger, *J. Electrochem. Soc.* **124,** 1228 (1977).
25. A. J. Appleby and B. S. Baker, *J. Electrochem. Soc.* **125,** 404 (1978).
26. J. R. Selman, *Energy* **11,** 153 (1986).
27. A. J. Appleby and S. B. Nicholson, *J. Electroanal. Chem.* **38,** App. 14 (1972).
28. A. J. Appleby and S. B. Nicholson, *J. Electroanal. Chem.* **53,** 105 (1974).

29. A. J. Appleby and S. B. Nicholson, *J. Electroanal. Chem.* **83,** 309 (1977).
30. A. J. Appleby and S. B. Nicholson, *J. Electroanal. Chem.* **112,** 71 (1980).
31. S. H. Lu, Ph.D. dissertation, Illinois Institute of Technology, Chicago, IL, 1985.
32. R. Bernard, 15th Workshop on Molten Carbonate Fuel Cells, EPRI/DOE Workshop, Morgantown, WV, 1987.
33. A. Pigeaud, C-Y. Yuh and H. C. Maru, "Optimization of Molten Carbonate Fuel Cell Electrolyte", National Cell Fuel Seminar Abstracts, 1988, p. 193.
34. T. L. Bonds, M. H. Dawes, A. W. Schnake, and L. W. Spradin, *Fuel Cell Power Plant Integrated System Evaluation* (Electric Power Research Institute, Palo Alto, CA, 1981), EM-1670.
35. J. Brown, B. Nazar, and V. Varwa, *Site-Specific Assessment of a 50 MW Coal Gasification Air-Cooled Fuel Cell Power Plant* (Electric Power Research Institute, Palo Alto, CA, 1983), RP1091-11.
36. A. O. Isenberg, in *Proceedings of the Symposium on Electrode Materials and Processes for Energy Conversion and Storage* (J. D. E. McIntyre, S. Srinivasan, and F. G. Will, eds.) (The Electrochemical Society, Inc., Pennington, NJ, 1977), p. 682.
37. S. C. Singhal, ed., Proceedings of the First Symposium on Solid Oxide Fuel Cells, Hollywood, FL, 1989.
38. N. Q. Minh, C. R. Home, F. S. Liu, D. M. Moffatt, P. R. Staszak, T. L. Stillwagon, and J. J. Van Ackeren, in *Proceedings of the 25th Intersociety Energy Conversion Engineering Conference* (P. A. Nelson, W. W. Schertz, and R. H. Till, eds.) American Institute of Chemical Engineers, New York, 1990), vol. 3, p.230.
39. K. A. Murugesamoorthi, S. Srinivasan, D. L. Cocke, and A. J. Appleby, in *Proceedings of the 25th Intersociety Energy Conversion Engineering Conference* (P. A. Nelson, W. W. Schertz, and R. H. Till, eds.) (American Institute of Chemical Engineers, New York City, 1990), vol. 3, p. 246.
40. U. Balachandran, S. E. Dorris, J. J. Picciolo, R. B. Poeppel, C. C. McPheeters, and N. Q. Minh, in *Proceedings of the 24th Intersociety Energy Conversion Engineering Conference* (P. A. Nelson, W. W. Schertz, and R. H. Till, eds.) (American Institute of Chemical Engineers, New York, 1989), p. 1541.
41. W. G. Parker, in *Proceedings of the 25th Intersociety Energy Conversion Engineering Conference,* (P. A. Nelson, W. W. Schertz, and R. H. Till, eds.) (American Institute of Chemical Engineers, New York, 1990), vol. 3, p. 213.
42. D. Fee, Presented at a meeting of the DOE/AFCWG, Argonne National Laboratory, Argonne, IL, April 19, 1984.
43. M. Hsu and H. Tai, Abstracts, Fuel Cell Seminar, Phoenix, AZ, p. 115, 1990.
44. K. Koseki, Abstracts, Fuel Cell Seminar, Phoenix, AZ, p. 107, 1990.
45. H. J. R. Maget, in *Handbook of Fuel Cell Technology* (C. Berger, ed.) (Prentice Hall, Englewood Cliffs, NJ, 1967).
46. A. Damjanovic, A. Dey, and J. O'M. Bockris, *Electrochim. Acta* **11,** 791 (1966).
47. A. Damjanovic and V. Brusic, *Electrochim. Acta* **15,** 1281 (1970).
48. A. J. Appleby, in *Modern Aspects of Electrochemistry, Vol. 7* (J. O'M. Bockris and B. E. Conway, eds.) (Plenum, New York, 1974), p. 369.
49. S. Gottesfeld, I. D. Raistrick, and S. Srinivasan, *J. Electrochem. Soc.* **134,** 1455 (1987).
50. D. R. Lawson, L. D. Whiteley, C. R. Martin, M. N. Szentirmay, and J. I. Song, *J. Electrochem. Soc.* **135,** 2247 (1988).
51. W. Paik, T. E. Springer, and S. Srinivasan, *J. Electrochem. Soc.* **136,** 644 (1989).
52. A. Parthasarathy, C. R. Martin, and S. Srinivasan, *J. Electrochem. Soc.*, **138,** 916 (1991).
53. K. Prater, *J. Power Sources* **29,** 239 (1990).
54. E. A. Ticianelli, C. R. Derouin, and S. Srinivasan, *J. Electroanal. Chem.* **251,** 275 (1988).
55. S. Srinivasan, D. J. Manko, H. Koch, M. A. Enayetullah, and A. J. Appleby, *J. Power Sources* **29,** 367 (1990).
56. I. D. Raistrick, U.S. patent #4,876,115, 1990.
57. Final Report, United Technology, Hamilton Standard for LANL #9-X53-D6272-1.

3

Electrochemistry of Fuel Cells

Embrecht Barendrecht

3.1. INTRODUCTION

As pointed out in Section 3.2, electrochemical energy converters principally utilize the chemical energy of fuels with a higher efficiency than heat engines. Most of the fuels used are hydrocarbons, so, directly or indirectly, natural gas-, oil-, or coal-based. The consequence of this statement is that fuel cells can contribute considerably to mitigate the greenhouse effect. Other advantages will be discussed thoroughly in the next chapters. We here confine ourselves to electrochemical aspects, though some overlap with other approaches is unavoidable. In Fig. 3.1, the working principle of an aqueous acid fuel cell is given.[1–3] This simplified picture of how a fuel cell works is sufficient to explain the principle. It shows that at the anode hydrogen is oxidized to H^+ ions and that at the cathode dioxygen is reduced, with the formation of water. Moreover, it also shows that gas-diffusion electrodes are essential for functioning.

Aspects and/or characteristics that will be common to all types of fuel cells, such as the thermodynamics, the electrocatalysis, and the gas-diffusion electrode, will be dealt with in Section 3.2. In Sections 3.3, 3.4, 3.5, and 3.6, the alkaline, the phosphoric acid, the molten carbonate, and the solid oxide fuel cell, respectively, will be treated. In Section 3.7 other types of fuel cells will be treated briefly. Some problem areas in R&D are the subject of Section 3.8.

3.2. GENERAL

3.2.1. General Characteristics of the Principal Types of Fuel Cells

There are a number of features common to all fuel cells. The first principles of thermodynamics are the same for all, but kinetics and catalysis, though governed by similar laws, are strongly dependent on the chemical systems

Embrecht Barendrecht • Eindhoven University of Technology, De Naaldenmaker 2, 5506 CD Veldhoven, The Netherlands.

Fuel Cell Systems, edited by Leo J. M. J. Blomen and Michael N. Mugerwa. Plenum Press, New York, 1993.

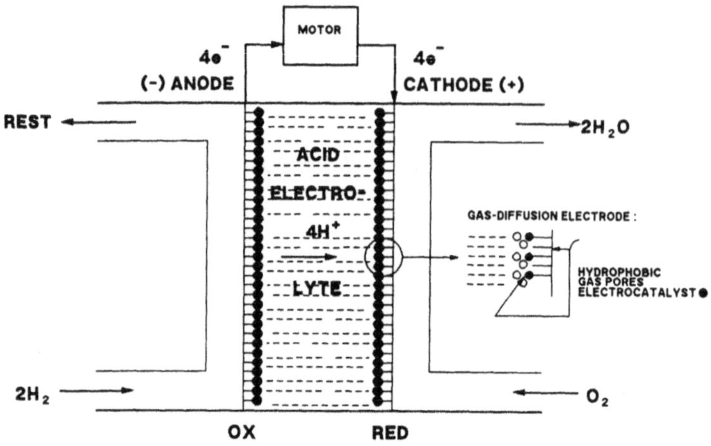

Figure 3.1. Operating principle of an aqueous acid fuel cell.

involved. The gas-diffusion electrode for the several fuel cells, however, has a lot of features in common, and will be treated in Section 3.2.3.2.

3.2.2. The Thermodynamics and Electrode Kinetics of Fuel Cells

3.2.2.1. Thermodynamics

A fuel cell is, as already said, an electrochemical system in which the chemical energy of a fuel is converted directly into electrical energy with the aid of an oxidant (e.g., oxygen). The conversion takes place by charge separation, in principle without the intermediate heat: i.e., heat engines are avoided in the production of electrical energy, and so the Carnot restriction. This in no way means that the stored chemical energy can thus be converted for 100%.

Under the assumption that the cell is working isothermally, which is usually true, then, during the cell reaction and depending on magnitude and direction of the entropy change (see Table 3.1), heat flows will occur. This change will be negative if the heat flow is to the surroundings. According to the second law of thermodynamics, one can write for a reversible process:

$$\Delta Q_{\text{rev}} = T\,\Delta S = \Delta H - \Delta G \quad \text{(in J)}, \tag{3.1}$$

with

$\Delta Q_{\text{rev}}\ (\Delta q_{\text{rev}})$ = reversible change in thermal energy (per mole)
T = absolute temperature in K
$\Delta S\ (\Delta s)$ = entropy change (per mole)
$\Delta H\ (\Delta h)$ = enthalpy change (per mole)
$\Delta G\ (\Delta g)$ = Gibbs free energy change (per mole)

For all energy-producing systems, ΔH is negative.

In fact, the cell reaction is irreversible and so both the anode and the cathode reaction run with overpotentials, in some cases even to a substantial extent; moreover, the conductivity of the electrolyte is finite and is the cause of heat

Table 3.1. Thermodynamic Data of the Most Important Fuel Cell Reactions (per mole)

Reaction	T (K; °C)	$\Delta g°$ (kJ)	$\Delta h°$ (kJ)	$\Delta s°$ J/K	n	ν^a (°)	$U°$ (V)	η_{rev}
$H_2 + \frac{1}{2}O_2 \rightarrow H_2O(l)$	298; 25	−237	−285	−162	2	−1.5	1.23	0.83
$H_2 + \frac{1}{2}O_2 \rightarrow H_2O(g)$	298; 25	−229	−242	−44	2	−0.5	1.18	0.95
$H_2 + \frac{1}{2}O_2 \rightarrow H_2O(g)$	378; 80				2	−0.5		
$H_2 + \frac{1}{2}O_2 \rightarrow H_2O(g)$	1,273; 1,000				2	−0.5		
$NH_3 + \frac{3}{4}O_2 \rightarrow \frac{1}{2}N_2 + \frac{3}{2}H_2O(l)$	298; 25	−226	−225	−97	3	−1.25	1.17	0.89
$N_2H_4 + O_2 \rightarrow N_2 + 2H_2O(l)$	298; 25	−312	−311	+2	4	−1.0	1.61	1.00
$C + \frac{1}{2}O_2 \rightarrow CO$	298; 25	−137	−110	+89	2	+0.5	0.71	1.24
$C + \frac{1}{2}O_2 \rightarrow CO$	773; 500	−180	−110	+90	2	+0.5	0.93	1.63
$C + \frac{1}{2}O_2 \rightarrow CO$	1,273; 1,000	−224	−113	+87	2	+0.5	1.16	1.97
$C + O_2 \rightarrow CO_2$	298; 25	−394	−384	+33	4	0	1.02	1.00
$C + O_2 \rightarrow CO_2$	773; 500	−396	−394	+2	4	0	1.03	1.00
$C + O_2 \rightarrow CO_2$	1,273; 1,000	−396	−395	+1	4	0	1.03	1.00
$CO + \frac{1}{2}O_2 \rightarrow CO_2$	289; 25	−257	−283	−86	2	−0.5	1.33	0.91
$CO + \frac{1}{2}O_2 \rightarrow CO_2$	923; 650				2	−0.5	1.00	
$CO + \frac{1}{2}O_2 \rightarrow CO_2$	1,273; 1,000	−172	−281	−86	2	−0.5	0.89	0.61
$CH_3OH(l) + \frac{3}{2}O_2 \rightarrow CO_2 + 2H_2O(l)$	298; 25	−704	−727	−77	6	−0.5	1.21	0.97
$CH_4 + 2O_2 \rightarrow CO_2 + 2H_2O(l)$	298; 25	−802	−804	−6	8	−2.0	1.04	1.00
$CH_4 + 2O_2 \rightarrow CO_2 + 2H_2O$	923; 650				8	0		
$CH_4 + 2O_2 \rightarrow CO_2 + 2H_2O$	1,273; 1,000				8	0		

[a] ν = molar change (volume).

production if an electrical current is generated. So, the total heat flux, $\Delta\dot{Q}_{tot}$ (in W or $J \cdot s^{-1}$), is

$$\Delta\dot{Q}_{tot} = \Delta\dot{Q}_{rev} + \Delta\dot{Q}_{irrev} = \left(\Delta Q_{rev} \cdot \frac{I}{n\mathbf{F}}\right) + (-I^2R_{elt} - I\eta_{tot}). \tag{3.2}$$

Because $\Delta G/n\mathbf{F} = -U$, and $-U_i = -U + IR_{elt} + \eta_{tot}$, the heat flux is

$$\Delta\dot{Q}_{tot} = \left(\Delta H \cdot \frac{I}{n\mathbf{F}}\right) + IU_i, \tag{3.3}$$

with

I = current (A) produced by the cell
R_{elt} = electrolyte resistance in the cell (Ω)
η_{tot} = total overvoltage (V) in the cell for a given I-value
U = electromotive force (V)
U_i = cell voltage (V) for a given I-value

As can be seen from Table 3.1, $\Delta h°$ is highly independent of changes in temperature for constant aggregation states (as long as no phase transition occurs, denoted by v). So, for a reversibly operating cell the voltage temperature coefficient is

$$\left(\frac{\partial U}{\partial T}\right)_p = \frac{\Delta S}{n\mathbf{F}}. \tag{3.4}$$

For most fuel cell systems, ΔS is negative and this means that U will drop with a temperature increase. This effect is important in considering high-temperature fuel cells, namely the molten carbonate fuel cell and (even more) the solid oxide fuel cell. On the other hand, however, the overvoltage will decrease drastically at higher temperatures, so the power output can be greater. For this reason it is better to compare the performance of a fuel cell on the basis of power density ($W \cdot m^{-2}$) than current density ($A \cdot m^{-2}$), as is mostly done (see also Chapter 12 for yet another reason to use power density for performance).

3.2.2.2. *Efficiency of a Fuel Cell*

The *electrical conversion efficiency* of a fuel cell, η, is

$$\eta = \frac{\text{(produced electrical energy)}}{\text{(total thermal energy produced during a chemical reaction)}}. \tag{3.5}$$

If the cell operates reversibly, then

$$\eta_{\mathrm{rev}} = \frac{\Delta G}{\Delta H} = \frac{-U}{-U + T\,\Delta S} = \frac{U}{U - T(\partial U/\partial T)_p}. \tag{3.6}$$

Because the temperature coefficient is for most cases between -0.1 and -1.0 mV/°C, η_{rev} will nearly always be larger than 0.9. For the fuel cell reaction: $H_2 + \frac{1}{2}O_2 \rightarrow H_2O(l)$, with $(\partial U/\partial T)_p = -0.84$ mV/°C at 298 K, η_{rev} nearly always amounts to 0.83; thus a fuel cell operating at room temperature is in this respect even a rather unfavorable one.

In reality, a fuel cell is not operating reversibly, so the real electrical conversion efficiency is

$$\eta = \frac{-n\mathbf{F}U_i}{\Delta H} = \frac{-U_i}{\Delta H/n\mathbf{F}} = \frac{-(\Delta G/\Delta H)U_i}{(\Delta G/\Delta H)(\Delta H/n\mathbf{F})} = \eta_{\mathrm{rev}}\left(\frac{U_i}{U}\right) = \eta_{\mathrm{rev}}\eta_u. \tag{3.7}$$

The quotient U_i/U is always <1 and is called the *voltage efficiency*, η_U. From electrode kinetics considerations it is well known that U_i is also dependent on $\log I$: the higher I is the lower U_i, and thus the efficiency is. In what has been stated here, the implicit assumption was that the *current efficiency*, η_i, is 1. As this is not always the case, the expression for η becomes

$$\eta = \eta_{\mathrm{rev}}\eta_U\eta_i. \tag{3.8}$$

For the whole fuel cell system, including the pretreatment of fuel gas, cooling of the system, recycling or reprocessing of the reaction gases, and so on, this η

has to be mutliplied by the *system efficiency*, η_s, so that the ultimate *electrical conversion efficiency*, of the fuel cell system is

$$\eta = \eta_{rev}\eta_U\eta_i\eta_s. \tag{3.9}$$

It is therefore clear that R&D must be concentrated on increasing η_U (electrocatalysis, electrolyte, and cell design), η_i (electrocatalysis, in this case selectivity), and η_s (total system and stack design). Further, it is clear that the fuel cell system produces both electrical energy and heat. Generally, it is important to strive for a low ratio of heat versus electrical energy. Note that improving η is part of fuel cell R&D by many companies (see Chapters 7–9). Improving η_s (which also has substantial emphasis in this book, since η_s can be larger than 1, and thus improving the ultimate η over the fuel cell efficiency itself) will be more extensively dealt with in Chapter 6.

3.2.2.3. *Reversible Cell Voltage*

For the most common fuel cell types (alkaline fuel cell (AFC) at 25°C and 80°C, phosphoric acid fuel cell (PAFC) at 200°C, molten carbonate fuel cell (MCFC) at 650°C, and solid oxide fuel cell (SOFC) at 1000°C) the formulas for the reversible voltages U, together with the values at standard concentrations and pressures, are given. As oxidant; dioxygen in air is used.

If we use *hydrogen* as a fuel, the anode (negative electrode, N.E.) reaction is

$$H_2 \rightarrow 2H^+ + 2e^- \quad \text{(acid, in our case phosphoric acid, PA),} \tag{3.10a}$$

$$H_2 + 2OH^- \rightarrow 2H_2O + 2e^- \quad \text{(alkaline, A),} \tag{3.10b}$$

$$H_2 + CO_3^{2-} \rightarrow H_2O + CO_2 + 2e^- \quad \text{(molten carbonate, MC),} \tag{3.10c}$$

$$H_2 + O^{2-} \rightarrow H_2O + 2e^- \quad \text{(solid oxide, SO).} \tag{3.10d}$$

The cathode (positive electrode, P.E.) reaction is

$$\tfrac{1}{2}O_2 + 2H^+ + 2e^- \rightarrow H_2O \quad \text{(acid, PA),} \tag{3.11a}$$

$$\tfrac{1}{2}O_2 + H_2O + 2e^- \rightarrow 2OH^- \quad \text{(alkaline, A),} \tag{3.11b}$$

$$\tfrac{1}{2}O_2 + CO_2 + 2e^- \rightarrow CO_3^{2-} \quad \text{(molten carbonate, MC),} \tag{3.11c}$$

$$\tfrac{1}{2}O_2 + 2e^- \rightarrow O^{2-} \quad \text{(solid oxide, SO).} \tag{3.11d}$$

a. *Acid-Type Fuel Cell.* For acid-type fuel cell reactions the equilibrium potentials and cell voltage can be written as follows:

$$\text{N.E.} \quad E\left(\frac{H^+}{H_2}\right) = E^\circ\left(\frac{H^+}{H_2}\right) + \frac{\mathbf{R}T}{2\mathbf{F}}\ln\left(\frac{[H^+]^2}{[H_2]}\right), \tag{3.12a.1}$$

$$\text{P.E.} \quad E\left(\frac{O_2}{H_2O}\right) = E^\circ\left(\frac{O_2}{H_2O}\right) + \frac{\mathbf{R}T}{2\mathbf{F}}\ln\left(\frac{[O_2]^{1/2}[H^+]^2}{[H_2O]}\right), \tag{3.12a.2}$$

$$\text{Cell emf} \quad U\left(\frac{O_2}{H_2}\right) = -E\left(\frac{H^+}{H_2}\right) + E\left(\frac{O_2}{H_2O}\right),$$

resulting in

$$= U^{\circ}\left(\frac{O_2}{H_2}\right) + \frac{\mathbf{R}T}{2\mathbf{F}}\ln\left(\frac{[H_2][O_2]^{1/2}}{[H_2O]}\right), \qquad (3.12a)$$

i.e., a simplified equation, for $[H^+]_{P.E.} = [H^+]_{N.E.}$ and $[H_2O]_{P.E.} = [H_2O]_{N.E.} = [H_2O]$.

b. Alkaline-Type Fuel Cell. Equilibrium potentials and cell voltage are as follows:

$$\text{N.E.} \quad E\left(\frac{H_2O}{H_2}\right) = E^{\circ}\left(\frac{H_2O}{H_2}\right) + \frac{\mathbf{R}T}{2\mathbf{F}}\ln\left(\frac{[H_2O]^2}{[H_2][OH^-]^2}\right) \qquad (3.12b.1)$$

$$= E^{\circ\prime}\left(\frac{H_2O}{H_2}\right) + \frac{\mathbf{R}T}{2\mathbf{F}}\ln\left(\frac{[H^+]^2}{[H_2]}\right),$$

$$\text{P.E.} \quad E\left(\frac{O_2}{OH^-}\right) = E^{\circ}\left(\frac{O_2}{OH^-}\right) + \frac{\mathbf{R}T}{2\mathbf{F}}\ln\left(\frac{[O_2]^{1/2}[H_2O]}{[OH^-]^2}\right)$$

$$= E^{\circ\prime}\left(\frac{O_2}{OH^-}\right) + \frac{\mathbf{R}T}{2\mathbf{F}}\ln\left(\frac{[O_2]^{1/2}[H^+]^2}{[H_2O]}\right), \qquad (3.12b.2)$$

$$\text{Cell emf} \quad U\left(\frac{O_2}{H_2}\right) = -E\left(\frac{H_2O}{H_2}\right) + E\left(\frac{O_2}{OH^-}\right),$$

resulting in

$$= U^{\circ}\left(\frac{O_2}{H_2}\right) + \frac{\mathbf{R}T}{2\mathbf{F}}\ln\left(\frac{[H_2][O_2]^{1/2}}{[H_2O]}\right), \qquad (3.12b)$$

i.e., a simplified equation, for $[H^+]_{P.E.} = [H^+]_{N.E.}$ and $[H_2O]_{P.E.} = [H_2O]_{N.E.} = [H_2O]$.

c. Molten Carbonate Fuel Cell. Equilibrium potentials and cell voltage (for *hydrogen* as a fuel) are

$$\text{N.E.} \quad E\left(\frac{H_2O}{H_2}\right) = E^{\circ}\left(\frac{H_2O}{H_2}\right) + \frac{\mathbf{R}T}{2\mathbf{F}}\ln\left(\frac{[H_2O][CO_2]}{[H_2][CO_3^{2-}]}\right), \qquad (3.12c.1)$$

$$\text{P.E.} \quad E\left(\frac{O_2}{CO_3^{2-}}\right) = E^{\circ}\left(\frac{O_2}{CO_3^{2-}}\right) + \frac{\mathbf{R}T}{2\mathbf{F}}\ln\left(\frac{[O_2]^{1/2}[CO_2]}{[CO_3^{2-}]}\right), \qquad (3.12c.2)$$

$$\text{Cell emf} \quad U\left(\frac{O_2}{H_2}\right) = U^{\circ}\left(\frac{O_2}{H_2}\right) + \frac{\mathbf{R}T}{2\mathbf{F}}\ln\left(\frac{[H_2][O_2]^{1/2}}{[H_2O]}\right), \qquad (3.12c)$$

i.e., a simplified equation for $[CO_3^{2-}]_{P.E.} = [CO_3^{2-}]_{N.E.}$, $[H_2O]_{P.E.} = [H_2O]_{N.E.} = [H_2O]$, and $[CO_2]_{P.E.} = [CO_2]_{N.E.}$

For *carbon monoxide* as a fuel we have for the anode (N.E.) reaction

$$CO + CO_3^{2-} \rightarrow 2CO_2 + 2e^-, \qquad (3.10c')$$

if the shift reaction

$$CO + H_2O \rightarrow CO_2 + H_2$$

is slow; the cathode (P.E.) reaction is (3.11c). So

$$\text{N.E.} \qquad E\left(\frac{CO_2}{CO}\right) = E°\left(\frac{CO_2}{CO}\right) + \frac{\mathbf{R}T}{2\mathbf{F}}\ln\left(\frac{[CO_2]^2}{[CO][CO_3^{2-}]}\right), \qquad (3.12c'.1)$$

$$\text{P.E.} \qquad \text{see (3.12c.2)},$$

$$\text{Cell emf} \qquad U\left(\frac{O_2}{CO}\right) = U°\left(\frac{O_2}{CO}\right) + \frac{\mathbf{R}T}{2\mathbf{F}}\ln\left(\frac{[CO][O_2]^{1/2}}{[CO_2]}\right), \qquad (3.12c')$$

i.e., a simplified equation for $[CO_3^{2-}]_{P.E.} = [CO_3^{2-}]_{N.E.}$ and $[CO_2]_{P.E.} = [CO_2]_{N.E.} = [CO_2]$.

d. Solid Oxide Fuel Cell. Equilibrium potentials and cell voltage with hydrogen as a fuel (the shift reaction is mostly very fast, so the carbon monoxide oxidation is not treated here) are

$$\text{N.E.} \qquad E\left(\frac{H_2O}{H_2}\right) = E°\left(\frac{H_2O}{H_2}\right) + \frac{\mathbf{R}T}{2\mathbf{F}}\ln\left(\frac{[H_2O]}{[H_2][O^{2-}]}\right), \qquad (3.12d.1)$$

$$\text{P.E.} \qquad E\left(\frac{O_2}{O^{2-}}\right) = E°\left(\frac{O_2}{O^{2-}}\right) + \frac{\mathbf{R}T}{2\mathbf{F}}\ln\left(\frac{[O_2]^{1/2}}{[O^{2-}]}\right), \qquad (3.12d.2)$$

$$\text{Cell emf} \qquad U\left(\frac{O_2}{H_2}\right) = U°\left(\frac{O_2}{H_2}\right) + \frac{\mathbf{R}T}{2\mathbf{F}}\ln\left(\frac{[H_2][O_2]^{1/2}}{[H_2O]}\right), \qquad (3.12d)$$

i.e., a simplified equation for $[O^{2-}]_{P.E.} = [O^{2-}]_{N.E.}$ and $[H_2O]_{P.E.} = [H_2O]_{N.E.} = [H_2O]$.

e. General Remarks, Temperature, and Pressure Dependence. Comparing (3.12d) with (3.12a), (3.12b), and (3.12c), it follows that these formulas are identical under the given conditions. What differs are the respective $U°$-values and $[H_2O]$-values. As to the latter, one must be aware that for the MCFC and the SOFC, initially no water at all is present so that, according to (3.12c) and (3.12d), the initial driving emf is infinite, at least theoretically. The current distribution must then be temporarily nonuniform, and since the degree of conversion of hydrogen to water is locally dependent so are the other relevant concentrations, such as those of H^+, CO_3^{2-} and O^{2-}, both at N.E. and P.E. It is for this reason that the practical values of $U(O_2/H_2)$ for the AFC, the PAFC, the MCFC, and the SOFC as denoted, respectively, in (3.12a)–(3.12d), are only approximate.

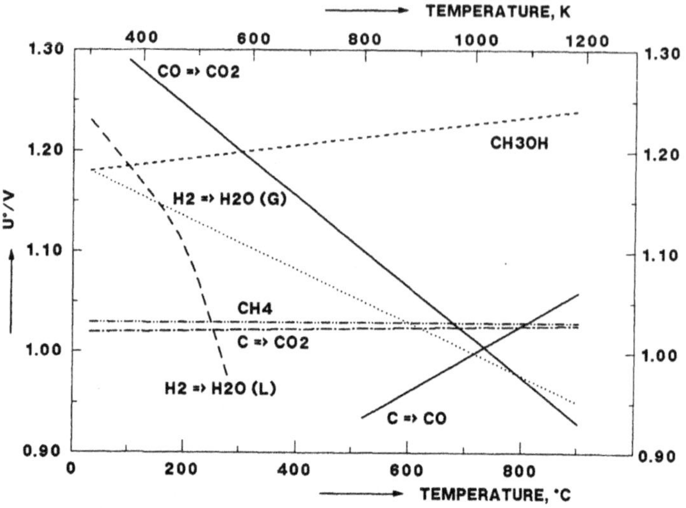

Figure 3.2. Standard reversible voltage, $U°$, as a function of temperature for the most important fuel cell reactions.[4]

In Fig. 3.2 the *temperature* dependence of the standard reversible voltages, $U°$, for some selected fuel cell reactions are given.[4] As shown in Table 3.1 and Fig. 3.2, Δs is negative if v (the change in number of gas molecules when going from reactants to products) is negative; even so $\Delta s = 0$ if $v = 0$, and $\Delta s > 0$ if $v > 0$.

The reversible voltage is also a function of *pressure, p*:

$$U = U° - \frac{v\mathbf{R}T}{n\mathbf{F}} \ln p = U° - \frac{1}{n\mathbf{F}} \int_1^p \Delta V \, dp, \tag{3.13}$$

with ΔV, the volume change for the reaction, going from 1×10^5 to $p \times 10^5$ Pa.

Another important variable is the water activity and its effect on the emf. For an alkaline hydrogen–oxygen fuel cell, the variation of $U°$ as a function of the KOH concentration is given in Table 3.2.

From Fig. 3.2, it is evident that if H_2 or CO are used as a fuel, the standard reversible potential, $U°$, decreases with temperature; this potential remains constant for methane and increases with temperature only for methanol.

Table 3.2. $U°$-Values as a Function of the KOH Concentration at 25°C

KOH concentration ($mol \cdot kg^{-1}$)	$U°$ (V)
0.18	1.229
1.8	1.230
3.6	1.232
5.4	1.235
7.2	1.243
8.9	1.251

From Ref. 2.

3.2.2.4. Electrode Kinetics

There are outstanding textbooks, covering the basics of electrode kinetics.[5-8] In this chapter we confine ourselves to the first principles. The fuel cell reactions are such that reduction of the oxidant, OX (in nearly all cases oxygen), takes place at the positive cathode, and oxidation of the reductant, RED (fuel), at the negative anode. Note that the signs of both electrodes are contrary to the signs in the case of electrolysis, because in fuel cells $\Delta G < 0$. At open-circuit voltage, U_{ocv}, i.e., when the fuel cell is not producing electric energy, the measured voltage is as given by (3.12a)–(3.12d). However, if the fuel cell is producing electric energy, and so a current I is passing through the cell, U_{ocv} decreases to some value U_I for which are responsible

The *ohmic drop,* caused by the electric resistance of the electrolyte, R_{elt}, or the membrane, R_m, if present, etc.; in general, the ohmic drop is proportional to I.
Overpotentials at both the anode and cathode; in general, these overpotentials are proportional to $\log I$, if diffusion of oxygen and/or fuel is not rate-determining.

Both effects cause U to decrease (see also Fig. 3.3):

$$U_I = U_{ocv} - I(R_{elt} + R_m + \cdots) - |\eta_a| - |\eta_c|, \tag{3.14}$$

with η_a and η_c, respectively, the overpotentials at the anode and the cathode.

Once again, this picture is only approximate. During operation, the electrolyte resistance can change as a result of temperature changes or changes in electrolyte composition and concentration, especially near the electrode. On the other hand, η_a and η_c are logarithmic functions of the current. For a simple electrode reaction,

$$\mathrm{OX} + ne^- \rightleftharpoons \mathrm{RED}. \tag{3.15}$$

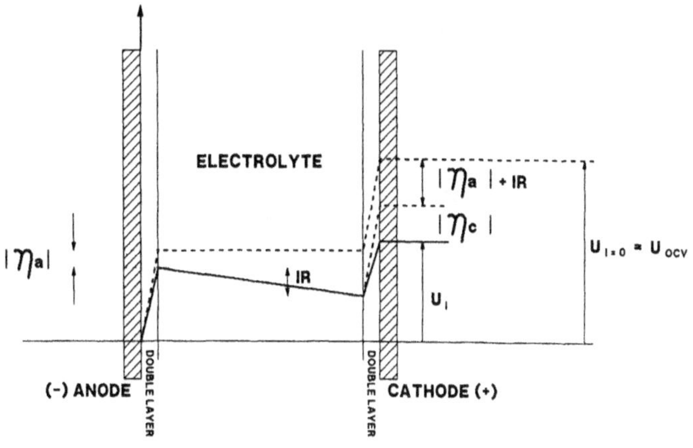

Figure 3.3. Overpotentials, η_a and η_c and ohmic drop in a fuel cell (.....): $I = 0$; (——) $I \neq 0$.

the following (again simplified) relations apply:

$$I = n\mathbf{F}k^{\circ\prime}[c^*_{\mathrm{RED}} \exp\{\alpha_a nf(E - E^\circ)\} - c^*_{\mathrm{OX}} \exp\{-\alpha_c nf(E - E')\}], \quad (3.16a)$$

or

$$I = I_0[\exp(\alpha_a nf\eta) - \exp(-\alpha_c nf\eta)], \quad (3.16b)$$

with

$$I_0 = n\mathbf{F}k^{\circ\prime}(c^*_{\mathrm{OX}})^{\alpha_a}(c^*_{\mathrm{RED}})^{\alpha_c} = I^\circ_0(c^*_{\mathrm{OX}})^{\alpha_a}(c^*_{\mathrm{RED}})^{\alpha_c}, \quad (3.16c)$$

and

$$\eta = a + b \log|I|, \quad (3.17a)$$

with the Tafel slope

$$b = \frac{d\eta}{d \log|I|}. \quad (3.17b)$$

In these equations, n, $\mathbf{F}$, E, and E° have their usual significance. $k^{\circ\prime}$ is the standard heterogeneous rate constant ($m^3 \cdot s^{-2}$); c_{RED}, c_{OX} represent the concentrations ($mol \cdot dm^{-3}$) of the reductant (fuel) and the oxidant (oxygen), with the degree symbol representing the concentration at the phase boundary, and the asterisk the concentration in the bulk of the solution, respectively;

$$f = \frac{\mathbf{F}}{\mathbf{R}T};$$

I_0 (i_0) is the exchange current (density) in $A \cdot m^{-2}$; I°_0 is standard I_0; α_a, α_c are the transition coefficients for the anodic and cathodic processes, respectively; $\alpha_a = 1 - \alpha$ and $\alpha_c = \alpha$, because $\alpha_a + \alpha_c = 1$; and $\eta = E - E_{eq}$, i.e., the difference between the actual and the equilibrium E values ($E_{eq} = E$ at $I = 0$); see Fig. 3.4.

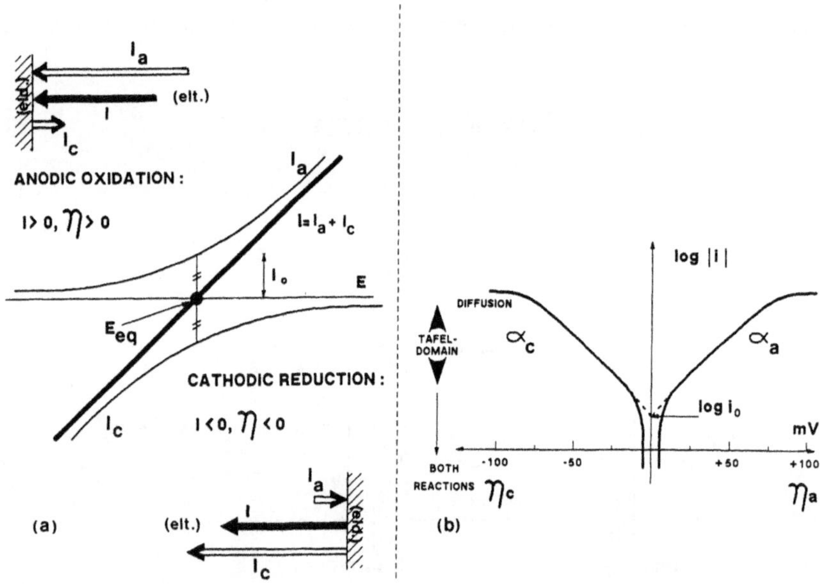

Figure 3.4. Explanation of the Butler–Volmer equation:[8] (a) $I = f(E)$ curve at E_{eq}, arrows give direction-flow electrons; (b) illustration of the Tafel equation, $\eta = f(\log|i|)$, see text.

As soon as the fuel cell reaction starts, the concentrations of reductant and oxidant in the immediate vicinity of the anode, $c^{\circ}_{\rm RED}$, and the cathode, $c^{\circ}_{\rm OX}$, respectively, decrease and new reactants are supplied by diffusion. In general, diffusion becomes the rate-determining step (r.d.s.), as can be seen from Fig. 3.5; moreover, the diffusion (and so the current) is governed by Fick's law:

$$I_i = n\mathbf{F} \cdot D_i(c_i^* - c_i^{\circ})/\delta_{\rm N}, \tag{3.18}$$

with the subscript i representing species i, D_i the diffusion coefficient ($\rm m^2 \cdot s^{-1}$), and $\delta_{\rm N}$ the so-called Nernstian diffusion layer (m).

As illustrated in Fig. 3.5, the concentration gradient $(c_i^* - c_i^{\circ})/\delta_{\rm N}$ approaches a maximum value and so does the current for $c_i^{\circ} \to 0$. In that case $I \to I_{\rm d}$, $I_{\rm d}$ being the diffusion current. From (3.18), it can easily be derived that $I \sim c^* - c^{\circ}$, and so $I_{\rm d} \sim c^*$.

Equations (3.16a)–(3.16c) are valid when the diffusion rate is infinite, which is only rarely true. If diffusion is also rate-determining, then the aforementioned equations must be transformed to

$$I = n\mathbf{F}k^{\circ\prime}[c^{\circ}_{\rm RED}\exp\{\alpha_{\rm a}nf(E - E^{\circ})\} - c^{\circ}_{\rm OX}\exp\{-\alpha_{\rm c}nf(E - E')\}], \tag{3.19a}$$

$$I = I_0\left[\frac{c^{\circ}_{\rm RED}}{c^*_{\rm RED}}\exp\{\alpha_{\rm a}nf\eta\} - \frac{c^{\circ}_{\rm OX}\exp\{-\alpha_{\rm c}nf\eta\}}{c^*_{\rm OX}}\right]. \tag{3.19b}$$

From a didactic point of view, (3.16b) and (3.19b) are easier to handle: the exchange current (density), I_0 (i_0), is a direct measure of the electrode reaction rate; a high value means that the reaction proceeds reversibly. Moreover, the overpotential, η, the distance to the equilibrium potential, $E_{\rm eq}$, gives us a better insight into what happens, as shown in Figs 3.4a, b.

For a highly irreversible reaction, the interaction region is negligible because I_0 is very small.

In principle, the measured overpotential η ((3.19b)) is the sum of the activation overpotential, $\eta_{\rm a}$, as described by (3.16b) and the diffusion overpotential, $\eta_{\rm d}$. However, an electrochemical reaction is rarely a pure electron transfer process. Even for such a seemingly simple process as hydrogen evolution, one of

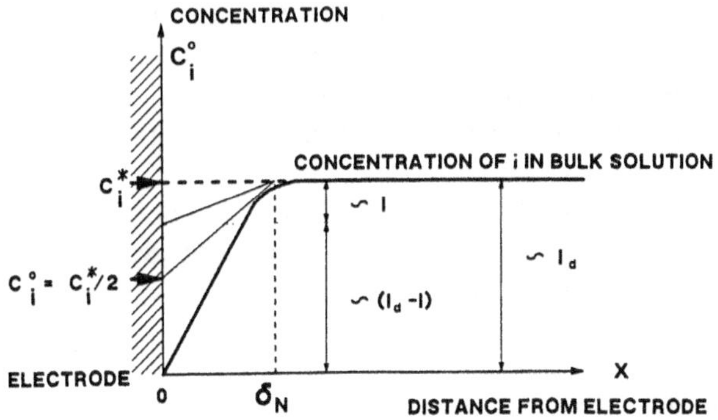

Figure 3.5. Concentration profiles at an electrode: c_i^* = concentration of species i in bulk solution; c_i^0 = concentration of species i at the electrode; $\delta_{\rm N}$ = Nernst diffusion layer thickness.

the possible mechanisms is

$$2H^+ + 2e^- \rightarrow 2H_{ad}, \tag{3.20a}$$

$$2H_{ad} \rightarrow H_{2,ad} \rightarrow H_2\uparrow. \tag{3.20b}$$

Reaction (3.20a) is by evidence an electron transfer reaction, but the second is a purely chemical reaction. Most electrochemical reactions are a combination of electron transfer and chemical reactions. It is the purpose of research in electrode kinetics to unravel the substeps, composing the net reaction, with their sequence, and to determine the pertinent rate constants. It is beyond the scope of this chapter to go into details as to the methods used. In general, we distinguish between *in situ* and *ex situ* methods. All electrochemical methods, such as linear potential sweep voltammetry, cyclovoltammetry, potential and current pulses, impedance and hydrodynamic methods are *in situ* methods, as are some spectrometric methods such as ellipsometry. Ultrahigh vacuum spectroscopic methods are *ex situ* methods. As to their use and potentialities see Refs. 5, 6, 7.

3.2.3. *Elements of a Fuel Cell*

In constructing a fuel cell, the main elements are the electrode material and the electrolyte (electrocatalysis), but their geometrical configuration (e.g., a porous electrode) and how they are assembled to make the cell and, ultimately, the stack (electrochemical engineering) are also important. The latter topic will be treated elsewhere (see Chapters 7–11).

3.2.3.1. *The Electrolyte in Relation to Fuel Gas and Oxidant Gas Composition; Effect of Temperature*

From the next sections it will become clear that every type of fuel cell has its own requirements as to, e.g., the optimal electrolyte composition. First, the fuel gas must not contaminate and thus deterioriate the electrolyte; e.g., an alkaline electrolyte cannot endure CO_2 or SO_2 or it will change in composition and, in combination with the electrode material, the electrocatalytic properties will change as well; the same applies to the oxidant gas. Second, the electrolyte composition and concentration must be such that the fuel gas and oxidant gas solubility does not become too low, since otherwise the kinetics will be hampered. Moreover, other conditions, such as maximum conductivity, must be fulfilled. Surface tension will be treated separately for the various types of fuel cells.

Particularly for the MCFC, where a partially flooded condition is regulated not by adding PTFE (Teflon) to the electrode but by changing the porous structure and the carbonate composition, the surface tension of the melt is of great importance.

Even so, the temperature will have an effect on all of the aforementioned properties: generally, if the temperature rises, the conductivity increases; the solubility of the gases, however, decreases. Where relevant, these effects will be treated in more detail for various fuel cells in the next sections.

3.2.3.2. The Gas-Diffusion Electrode

In order to have an acceptable gas-mass transport and, as a consequence, an acceptable current per unit of geometric area, it is clear that a special geometric structure of the electrode is required. From the first law of Fick,

$$i_l = \frac{n\mathbf{F} \cdot D_g A c_g}{W} \quad (\mathrm{A \cdot m^{-2}}) \tag{3.21}$$

with i_l the limiting current density ($\mathrm{A \cdot m^{-2}}$), D_g and c_g the diffusion coefficient ($\mathrm{m^2 \cdot s^{-1}}$) and concentration of the dissolved gas ($\mathrm{kg \cdot mol \cdot m^{-3}}$), respectively, W the thickness of the electrolyte film through which the gas had to diffuse (m), and A the real surface area of the film per unit geometric area. It is clear that, whichever the geometric structure may be, A must be as large, and W as small, as possible. So a three-phase boundary of solid (electrode), gas (fuel, oxidant), and liquid (electrolyte) is necessary to obtain current densities up to $5 \times 10^3\ \mathrm{A \cdot m^{-2}}$ (thereby keeping in mind that gas dissolves in the electrolyte to an extent of $10^{-6}\ \mathrm{kg \cdot mol \cdot m^{-3}}$ and that D_g is about $10^{-9}\ \mathrm{m^2 \cdot s^{-1}}$). If we suppose W to be 10^{-6} m, then for a current density of, say, $5 \times 10^3\ \mathrm{A \cdot m^{-2}}$, the specific area has to be (for $n = 2$) about $2.5 \times 10^4\ \mathrm{m^2 \cdot m^{-2}}$. So, to achieve this, the electrode must have a three-dimensional structure. In this respect, the experiment of Will[9] is revealing. For a thin platinum wire partially submerged in an electrolyte, he found that the electrolyte spread up the wire a considerable distance from the edge of the bulk meniscus. For a reversed wire, i.e., a pore, the same applies: a thin film of electrolyte covers the interior surface of the pores of a porous electrode, see Fig. 3.6, Refs. 10–14, and Chapter 9 of this volume.

Before dealing with more realistic models, we mention the *simple-pore* model (see Fig. 3.6a, the pore surface is not, or hardly, wetted). For this model, in nearly all cases the current densities predicted are one or more orders of magnitude lower than those experimentally determined, due to the relatively small gas–electrolyte surface area accessible; so, the length of the pores hardly influences the current density. It is generally accepted that the gaseous reactants dissolve first into the electrolyte at the gas–electrolyte interface (this is not only assumed for this model but also for each other model). And, of course, the extent of this three-phase-contact region is dependent on the wetting properties of the electrode material–electrolyte composition combination; for the simple-pore model this region is very small. Furthermore, the position of the three-phase-contact domain is dependent on the differential pressure Δp; at equilibrium (see Fig. 3.7)

$$\Delta p = \frac{2\gamma \cos \Theta}{r_e}. \tag{3.22}$$

The equilibrium pore radius, r_e, at which both gas and electrolyte are present in the pore, is proportional to γ, the surface energy, and $\cos \Theta$, where Θ is the contact angle between the gas–electrolyte and the electrode–electrolyte interfaces. Because γ and Θ are material-dependent and thus known, the variable Δp

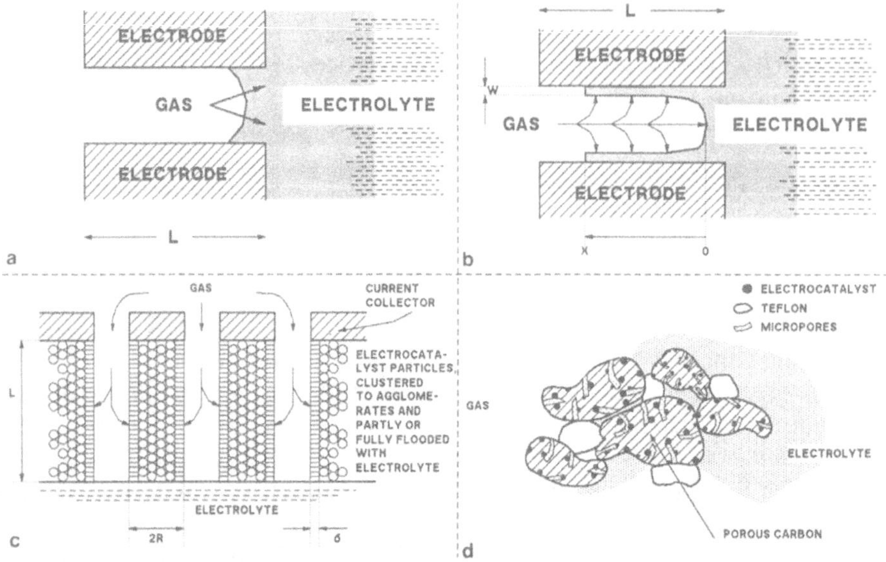

Figure 3.6. Models of the gas-diffusion electrode: (a) simple-pore model; (b) thin-film model; (c) agglomerate model; (d) enlarged part of a partly flooded porous electrode.

determines if the pore is gas-filled (for $r > r_c$) or electrolyte-filled (for $r > r_c$); i.e., Δp can significantly change the behavior of the system.

The models of porous electrodes presented here are based on the following assumptions:

1. Diffusion is the only transport mechanism, so the influence of convection and migration is neglected.
2. Only rate processes within the pores are considered, and the gas and fluid phases outside the pore domain are assumed to have a constant composition.
3. Physical parameters such as diffusion coefficient, viscosity, temperature, pressure, have a constant value.
4. The current density (i)–overpotential (η) relation for the generalized electrode reaction, which can be written as

$$gG + lL \pm ne^- \rightarrow pP, \tag{3.23}$$

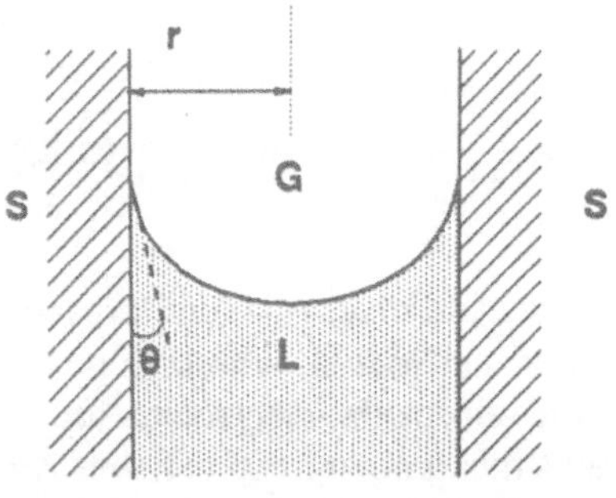

Figure 3.7. Three-phase contact within a pore with radius r, contact angle Θ; S, solid electrode; L, liquid electrolyte; G, gas.

where G is the gaseous reactant (O_2 or fuel), L is the liquid reactant (H_2O, H^+, etc.), and P is the reaction product, g, l, and p are stoichiometric numbers, and ne^- is positive for a reduction and negative for an oxidation, is

$$i = i_0[\chi_a \exp \alpha_a n' f\eta - \chi_c \exp -\alpha_c n' f\eta], \tag{3.24}$$

$f = \mathbf{F}/\mathbf{R}T$, $\chi_a = (c_g^\circ/c_g^*)^{g'}(c_l^\circ/c_l^*)^{l'}$, and $\chi_c = (c_p^\circ/c_p^*)^{p'}$ for oxidation of the fuel gas, and $\chi_a = (c_p^\circ/c_p^*)^{p'}$ and $\chi_c = (c_g^\circ/c_g^*)^{g'}(c_l^\circ/c_l^*)^{l'}$ for reduction of the oxidant, O_2; g', l', p' are the stoichiometric numbers in the rate-determining step, and n' is the number of electrons passing in the r.d.s. The other symbols have their usual significance. At high overpotentials, the rate of the reverse reaction is negligible, so for the reduction of oxygen only the second term of (3.24) is of interest.

More realistic is the *thin-film* model, especially suitable for the lower temperature fuel cells where the wettability can be adjusted, among other things, with PTFE. In this model the three phases are not only rather close to each other, but they are also extended over a large surface area per unit geometric area (Fig. 3.6b).

A thin-film model,[10,12] has additional assumptions:

5. A significant part of the inner surface of the cylindrical pore (radius R) is covered with a thin film of electrolyte (length $L = 2$–5 mm; thickness $W = 0.5$–$1\ \mu$m).
6. The supply of the reactant gas to the inner part of the pore is not rate-determining.
7. The gaseous reactant dissolves so fast into the electrolyte film that the equilibrium concentration is easily reached at the gas–electrolyte interface. From there it moves by diffusion to the active sites of the pore surface.

Even this thin-film model is not easy to describe. The reaction products mostly cause a nonuniform concentration pattern over the length of the film and so an overpotential as a function of x.

This thin-film model has yielded valid predictions in a number of cases.

Another model, the *agglomerate* model, (Fig. 3.6c and d), is still more sophisticated but also more realistic, albeit at the price of greater mathematical complexity.[10,11,14] The electrode is supposed to consist of small catalyst particles forming the agglomerates together with the interstitial electrolyte. Macropores between those agglomerates are gas-filled. The agglomerate may be covered by an electrolyte film, but this covering is not essential because the liquid-filled micropores are supposed to be the main sites where the electrode reaction goes on. This agglomerate model is very suitable for describing the electrode processes in the MCFC, which has a different porous structure. There are no known materials that can serve to wet-proof a porous structure, such as PTFE in low-temperature fuel cells, to prevent ingress by molten carbonates. To obtain a stable three-phase contact, it is necessary to carefully tailor the porous structures

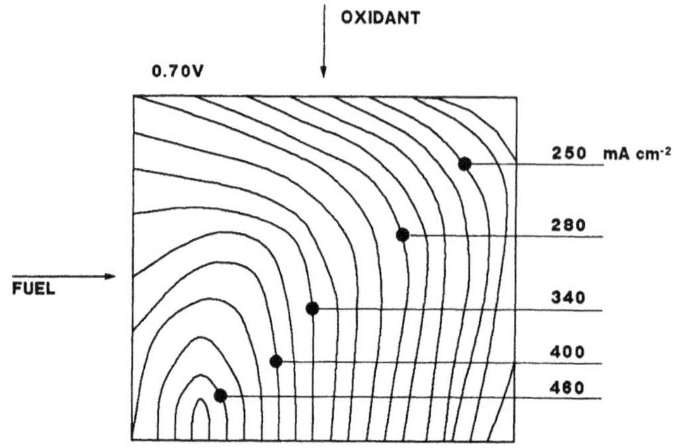

Figure 3.8. Example of current distribution for a flat plate at a constant potential.

of both the electrodes and the electrolyte matrix ($LiAlO_2$). This "pores-spectrum matching technique" is in principle based on Eq. (3.22).[15] (Ref. 12, p. 161).

For SOFC, where no liquid electrolyte is present, electrode flooding does not occur. The only requirement is that the porous structure is sufficiently thin (conductivity!) and yet porous enough to offer an extensive electrode–solid electrolyte interfacial region for the electrode reaction.

3.2.3.3. *Current Distribution*

A good current distribution over the whole porous electrode surface is essential for good functioning. One of the reasons why current distribution is often so inhomogeneous is the exhaustion of the fuel gas and the oxidant gas in passing the electrode. And, of course, the electronic conductivity of the porous electrocatalytic system and the ionic conductivity of the electrolyte within the micropores also play, and often, a decisive role, causing an uneven potential distribution both over and within the electrode. The problem is mathematically complicated because, as argued before, a model for this current distribution requires integration of both macroscopic (the two-dimensional electrode surface) and microscopic (pores, i.e., in-depth distribution) aspects. Moreover, a locally higher current density causes a higher temperature at that spot, changing catalysis and kinetics.

To counteract that process, the two gas streams flow countercurrent; nevertheless, current density differences can be considerable. A typical example, to give an idea about the extent, is shown in Fig. 3.8.

3.3. *THE ALKALINE FUEL CELL*

3.3.1. *General Characteristics*

As highlighted in Chapter 1, the history of fuel cells which can produce power in the kilowatt range with a certain reliability began in the 1930s, culminating in a 5-kW system in the mid-1950s engineered by Bacon.[16] His fuel cell was of the alkaline type, used 30% KOH as electrolyte, and operated at 200°C and at a pressure up to 50 atm. The electrodes were Ni-based: the anode was a sintered nickel electrode of dual porosity, the cathode a lithiated nickel oxide electrode. At such relatively high operating temperature and pressure, the performance of the Bacon fuel cell was excellent. At 200°C and 45 atm, a cell voltage of 0.78 V could be attained at 800 mA/cm^2. To be more reliable, especially for space flights, lower operating pressures were used in practice. The success stimulated also terrestrial applications and in the 1960s some AFCs were used in tractors and automobiles, mostly in a FC–battery hybrid fashion.[15,17] For that application, temperature and pressure were lowered to ambient values 50–80°C and 1 atm). To have acceptable catalysis, Ni was replaced by Pt supported on porous carbon, made more hydrophobic with a Teflon binder. Such an electrode consisted of a layer of catalyzed porous carbon on a Ni-screen collector, backed with (hydrophobic) porous Teflon on the gas side, see Fig. 3.6d. The multicell arrangement was monopolar because of the circulating electrolyte (in contrast to bipolar arrangements where the electrolyte is fixed, as in the PAFC and the MCFC). The electrolyte was circulated so that heat could be removed and water eliminated by evaporation.[15]

At the anode pure (CO_2 and CO-free) hydrogen is oxidized:

$$2H_2 + 4OH^- \rightarrow 4H_2O + 4e^-, \tag{3.10b}$$

and at the cathode oxygen is reduced (see Fig. 3.9):

$$O_2 + 2H_2O + 4e^- \rightarrow 4OH^- \tag{3.11b}$$

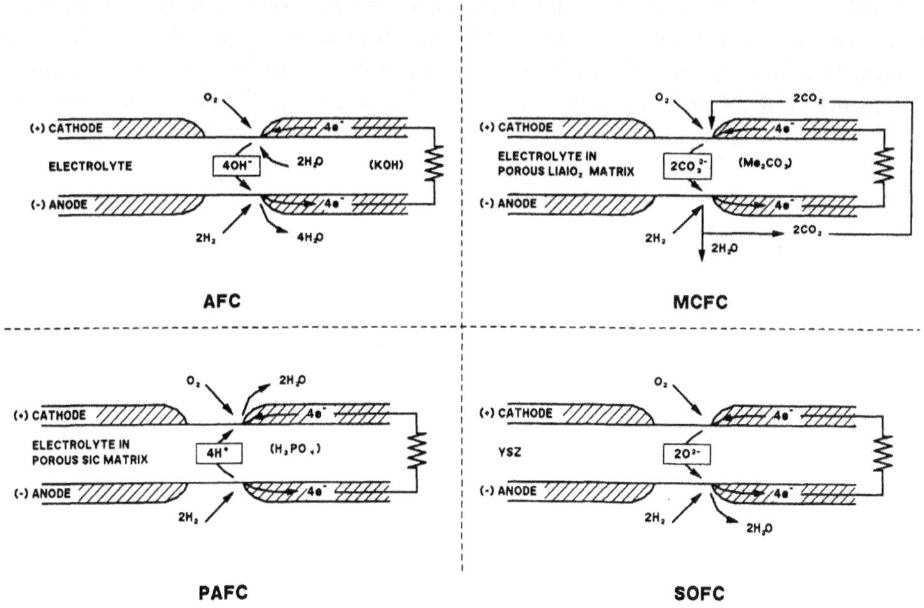

Figure 3.9. Electrode reactions for AFC, PAFC, MCFC, and SOFC.

so the net cell reaction is

$$2H_2 + O_2 \rightarrow 2H_2O + \text{electric energy} + \text{heat}, \tag{3.25}$$

and thus water and heat have to be removed during operation.

This can be easily accomplished because the KOH electrolyte is circulated and water can be removed by evaporation.

Fuel pretreatment must be such that the initial CO_2 concentration in (reformed) carbonaceous fuels (which can amount to 25%) must be reduced to a few ppm before introduction into the AFC. Mostly, CO_2-containing gases are sent in three passes through a solvent which is thermally regenerable, such as ethanolamine or similar compounds:

$$2RNH_2 + CO_2 \underset{115°C}{\overset{27°C}{\rightleftharpoons}} RNH \cdot CO \cdot HNR + H_2O. \tag{3.26}$$

With this system, in the first pass the CO_2 level can be reduced to 50 ppm. Other known methods are pressure cycling, temperature cycling with molecular sieves, or the use of a Pd–Ag membrane; they will not be explained here further.

Storage of hydrogen in fiber-wound aluminum cylinders or in certain metal alloys (Ti–Ni), which can form hydrides, is a real option for transportation.

3.3.2. *The Electrode Reactions*

3.3.2.1. *The Anode Reaction; Hydrogen Oxidation Catalysis*

The net anode reaction has been given by (3.10b), and its corresponding equilibrium potential by (3.12b.1). In the patent and open literature[15,17] (because of cost reduction) there is a tendency to lower Pt (Rh) contents from 10 to 0.5 mg/cm^2 and even lower. The mechanism of H_2 oxidation is well-studied. Nevertheless, the polarization due to H_2 oxidation is still some 20% of the total polarization at high current densities, so that the AFC's anodic reaction remains a significant field for further study. The contribution to polarization is, however, the largest for the cathodic oxygen reduction (see Section 3.3.2.2 and Fig. 3.10).

In earlier AFCs Raney-Ni catalysts (120 mg Ni $\cdot$ cm^{-2}) were used; for control of sintering, they contained Ti as well as Al. Also Ag supported by carbon was applied (1.5–2 mg Ag $\cdot$ cm^{-2}).

3.3.2.2. *The Cathode Reaction; Oxygen Reduction Catalysis*

The net cathode reaction is given by (3.11b), and the equilibrium potential by (3.12b.2). Till now, Pt is considered to be one of the best electrocatalysts; nevertheless, the oxygen reduction reaction is rather irreversible, even in alkaline solutions: for a 0.25 mg/cm^2 Pt cathode, operating in 30% KOH at 70°C and 100 mA/cm^2, the potential is 0.868 V (reversible hydrogen electrode, RHE), while in 96% H_3PO_4 at 165°C and 100 mA/cm^2 the potential is 0.730 V (versus RHE). For a possible explanation of this remarkable behavior, see Ref. 18. In any case, this results in efficiencies of some 10–15% higher for AFCs compared to PAFCs. This effect is also real for other applied electrocatalysts. The polarization

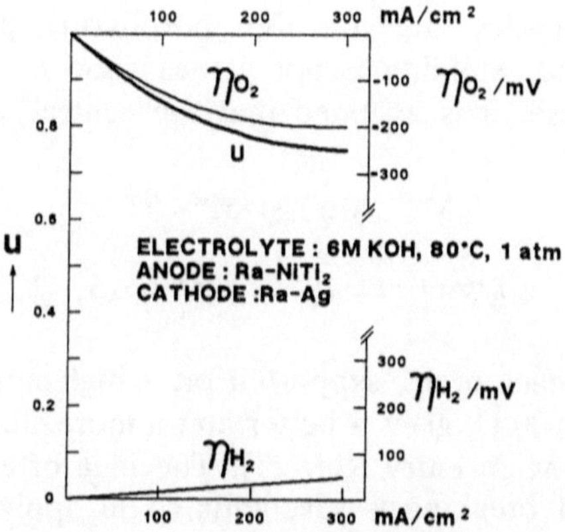

Figure 3.10. Polarization characteristics of an AFC.

of the cathode is higher for air than for pure oxygen and the difference can amount to some 60 mV at 100 mA/cm^2 (the same applies to the anodic oxidation of hydrogen if the partial pressure is lowered). At the present state-of-the-art, the IR-free cell voltage in a 12 M KOH solution at 65°C and at 100 mA/cm^2 is about 0.87 V at the start; current densities up to 300 mA/cm^2 can be maintained over longer periods.[8] Nevertheless, besides its too high polarizability, strong reasons to look for other electrocatalysts are the high cost for Pt and its tendency to dissolve slowly.

As to the mechanism and rate-determining steps, much has been published and many studies[19–22] have been carried out with the rotating ring disk electrode (Section 3.8.2). The formation of the hydrogen peroxide anion (HO_2^-) must be avoided because of its deteriorating effect on the carbon support and especially of its lowering effect on the cell voltage (compare Fig. 3.11: −0.08 V instead of 0.40 V). There is agreement about the first step,

$$O_2 + e^- \rightarrow O_2^-, \qquad (3.27)$$

being a rate-determining step. A good electrocatalyst for the cathodic reduction process of O_2 causes O—O bond breaking in an early stage of the process so that the reaction can proceed to OH^- ((3.11b), with $n = 4$ per O_2).

A very promising class of electrocatalysts is the transition metal macrocyclic catalysts with Co or Fe as a metal (see Refs. 23 and 24 for an extensive literature

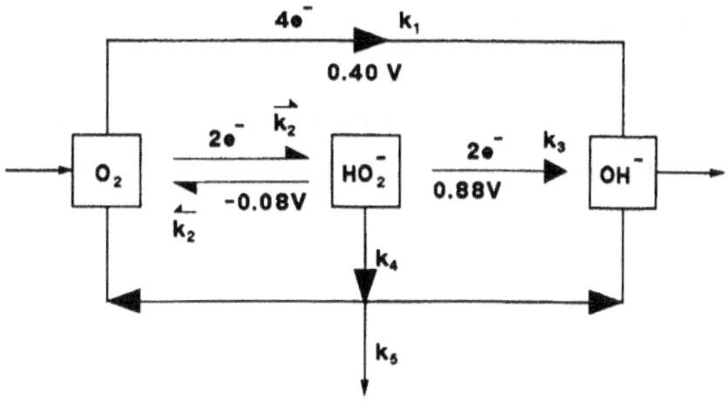

Figure 3.11. Scheme of the cathodic oxygen reduction in an alkaline medium.

review). The macrocycles are of the porphyrin, phthalocyanine, tetra-azaannulene, etc., type. Stability is not yet as good as required for long life application. In any case, it is assumed that the central ion undergoes a redox process:

$$M^{n+} + e^- \rightarrow M^{(n-1)+}, \tag{3.28}$$

followed by

$$M^{(n-1)+} + O_2 \rightarrow M^{n+} + O_2^-. \tag{3.29}$$

Pyrolysis of the macrocycle, supported on a high-surface-area carbon, at a temperature of about 800°C, gives a better and a more stable electrocatalyst in a number of cases (see as an entry Ref. 23). The idea of end-on adsorption as a favorite way for bond breaking is substantiated in applying dicofacial dimetal porphyrins as an electrocatalyst.[26] Other electrocatalysts are Ag with Hg added, mixed inorganic oxides of the spinel and perovskite type, such as $LaNiO_3$,[15] etc.

In this context, the properties of the carbon support (and the same applies for the anode), its surface lattice, its porosity, its hydrophobicity (PTFE!), etc., play a very important role.[15] A scheme of the oxygen reduction is given in Fig. 3.11a.

3.3.2.3. *The Electrolyte Composition*

All AFCs work with potassium hydroxide, not with sodium hydroxide, since, if by some accident CO_2 enters the system in the case of sodium hydroxide, sodium bicarbonate is formed, which is not very soluble in concentrated (necessary for a high conductivity) sodium hydroxide solutions and therefore can block pores of the electrode system and reduce their efficiency. Moreover, highly concentrated potassium hydroxide in water has an extremely low vapor pressure. Concentrations are up to 50%, (i.e., 13.6 M). At higher temperatures, even higher concentrations can be used, leading to solidification in thin matrices, such that high current densities (for short times even up to 9 A/cm^2) can be achieved.[27]

3.3.2.4. *Miscellaneous Remarks*

a. Temperature Dependence. The AFC normally starts at ambient temperature. At KOH concentrations from 20% to 50% it operates best at 60–80°C; at higher temperatures evaporation, especially at low conversion rates, requires supplementing with water. Operation at higher temperature is favorable for catalysis, especially for the oxygen reduction.

b. Pressure Dependence. A temperature higher than 100°C needs a higher pressure. Kinetics is then, of course, faster. In spacecraft, pressures up to 50 atm at 200°C have been used together with pristine H_2. Efficiencies higher than 60% (higher heating value, HHV) have easily been obtained successfully even when operating 5000 h or more.[15,28]

c. Contaminants. Carbon dioxide must be removed, both in hydrogen and in air (380 ppm CO_2) down to about 10 ppm by volume, to prevent the hydroxide from becoming carbonate, with a resulting decline in catalytic power. Also, sulfur dioxide and hydrogen sulfide have deleterious effects on catalysis and must be removed.

d. Stability. It must be said that, generally, the AFC is a stable and reliable fuel cell for several thousand operating hours. For that reason the cell is very suitable for space application. Because the cell operates from ambient to 80°C, transportation applications (city buses, military) are obvious. The price (Pt) is, however, the limiting factor for several markets and applications.

3.3.3. Alternative Types

Since catalysis works better in alkaline (as well as in bicarbonate) than in acid solution, carbonation by the presence of CO_2 in fuel gas and air must be avoided. Therefore, an alkaline *methanol* fuel cell (even though kinetics and catalysis in an alkaline electrolyte were indeed favorable) was not realized (methanol is readily available and a good fuel alternative). CO_2-rejecting alkaline electrolytes such as cesium carbonate–bicarbonate solutions offer a possibility (see Ref. 28, p. 281), but at the penalty of higher operating temperatures, which would be a disadvantage in certain applications.

Another AFC is the *hydrazine* fuel cell. The anode reaction suggest a higher cell voltage; in reality it is near that of a hydrogen-fueled AFC under the same operating condition.

$$N_2H_4 + 4OH^- \rightarrow N_2 + 4H_2O + 4e^-, \qquad E^\circ = 1.56\text{ V}. \qquad (3.30)$$

This fact suggests that hydrazine is only acting as a source of hydrogen. The anode is a Pd-activated porous Ni electrode, and the electrolyte is a 25% KOH solution with 3% that of hydrazine. The cell operates at 70°C.[2] Because hydrazine is both expensive and poisonous, its future is doubtful.

3.3.4. Future Research and Development

Current AFCs normally operate with monopolar electrodes and nonsolidified electrolytes (contrary to the PAFC and MCFC). Solidifying the KOH electrolyte in an inert matrix offers, moreover, the possibility of reducing ohmic drop (and thus operating at higher current densities) and facilitating introduction of bipolar plate electrodes (and thus simpler stack construction).[27]

Improving the electrocatalysis, especially of oxygen reduction, is still a matter of necessity. New developmets are described in Chapter 7.

3.4. THE PHOSPHORIC ACID FUEL CELL

3.4.1. General Characteristics

The PAFC, now close to commercialization, is mainly intended for stationary dispersed power plants and/or stationary base-load power plants. The materials and technology required have identified acid fuel cell research needs and goals such that, as a result of R&D executed, demonstration and field tests of several capacities, such as 40-kW and 4.8-MW power plants, have been completed fairly successfully. Nevertheless, further R&D is necessary to make this first-generation fuel cell cost-competitive and reliable (see Chapter 8).

Compared with the AFC, the PAFC has some drawbacks, the most obvious being its lower efficiency. If under defined conditions an AFC operates at 0.9 V, then a PAFC operates at 0.7 V, notwithstanding its higher operation temperature (200°C versus 80°C), at which catalysis is supposed to run faster. A convincing advantage, however, is that the PAFC tolerates CO_2 and a CO level up to 1–3%. Therefore, carbonaceous fuels (i.e., natural gas, coal gas) are, after converting to a hydrogen-rich gas, welcome as a fuel, without extensive purification. When using hydrocarbons, it must be kept in mind that at higher temperatures direct cracking of fuel and the accompanying irreversible deposition of carbon will occur. Adding steam can eliminate this last possibility. Steam reforming outside the cell to produce H_2 is a prerequisite in using hydrocarbons for the PAFC, since direct oxidation at 200°C would be a very slow process. Steam reforming is an endothermic process and will be dealt with in Chapter 4; see Chapter 2.[15]

The PAFC can operate at relatively high current densities. About 300 mA/cm^2 can be easily achieved at proven lifetimes of more than 25,000 h. A synthesis-gas-fueled PAFC shows an (HHV-based) overall system efficiency (based on natural gas) of about 40% at a current density of 150 mA/cm^2, while under similar conditions an AFC only achieves some 36% (and is therefore not widely economically accepted for application by electric utilities). This must be compared with the high AFC performance on *pure* H_2; therefore the use of alkaline fuel cells is mostly justified when pure H_2 is available from nonfossil sources (e.g., in chlorine plants). Note, however, that the PAFC performance on pure H_2 is also much higher than on synthesis gas.

The PAFC makes use of highly concentrated phosphoric acid (95%) as the electrolyte (i.e., pyrophosphoric acid, $H_4P_2O_7$), with a high ionic conductivity (at 200°C, $>0.6\,\Omega^{-1}\cdot cm^{-1}$). The electrolyte is held in a porous SiC matrix by capillary action and so immobilized. Thus, stack building with bipolar plates is possible. Volatility at that temperature is low, and replenishment can be postponed for longer periods. Sulfuric acid is unsuitable, because of its volatility and decomposition. Pt on a porous carbon support is used as anode and as cathode material (see Figs 3.6d and 3.9).

At the anode hydrogen is oxidized:

$$2H_2 \rightarrow 4H^+ + 4e^- \tag{3.10a}$$

and at the cathode oxygen is reduced:

$$O_2 + 4H^+ + 4e^- \rightarrow 2H_2O \tag{3.11a}$$

3.4.2. *The Electrode Reactions*

3.4.2.1. *The Anode Reaction; Hydrogen Oxidation Catalysis*

The net anode reaction is given by (3.10a), its potential by (3.12a.1). The catalysis with Pt is good (usually 0.25 mg/m^2 of colloidal Pt in 10% loading on porous carbon—Vulcan XC-72, e.g.—combined with about 40 wt% colloidal Teflon).[15] Corrosion of Pt and carbon is of minor importance.

The anode substrate can act as an effective reservoir because hydrogen-diffusion polarization is not much affected for a reasonable acid fill.

3.4.2.2. The Cathode Reaction; Oxygen Reduction Catalysis

The net cathode reaction is given by (3.11a), its potential by (3.12a.2). Together with (3.12a.1), the cell voltage can be given by (3.12a) if we assume that $[H^+]_{P.E.} = [H^+]_{N.E.}$ and $[H_2O]_{P.E.} = [H_2O]_{N.E.}$

As for all other fuel cells, oxygen reduction is most strongly impeded, notwithstanding the rather high operating temperature and the use of the currently most favored catalyst material, platinum. It is essential that the smallest possible stable Pt particles are used in order to give the maximum number of Pt catalytic sites. This not only serves system cost-effectiveness, but also reduces overpotential[29] because of the greater number of accessible catalytic sites. Of utmost importance is the support used: porous carbon. It is well known that generally carbon corrodes at 200°C in a pyrophosphoric acid medium and that this corrosion is enhanced drastically if the water–vapor pressure is above 100 mm Hg (Eq. (3.11a) shows that water is formed during cathodic reduction). Pretreatment of the porous carbon (graphitization at 2700°C) while maintaining a high surface area (200 m^2/g or more is desirable) is essential.[15,30]

As an extra complication, Pt is fairly soluble under PAFC conditions. As the Pt crystallites are very small, one may expect even higher solubilities than for bulk Pt (Kelvin effect). Fortunately, this process does not happen, though the reason is unknown. In any case there must be strong catalyst-support interactions with, as a result, a stabilizing effect on the 2–5 nm Pt agglomerates. In the long run, however, these agglomerates tend to grow, thus causing the overpotential to increase[29] and cell efficiency to decrease.

In addition, binary alloys, such as Pt–V or Pt–Cr, were examined; they show higher stability and corrosion resistance than Pt,[31] where a volcano plot, i.e., activity versus nearest-neighbor distance, shows higher activities for these alloys than for Pt or Ru. It is in this domain of improved cathode materials that cost reduction may be achieved with some probability of success.

The mechanism of oxygen reduction has been extensively studied.[23,24] It is now generally accepted that the first step is the uptake of an electron by adsorbed oxygen:

$$O_2(\text{ad}) + e^- \rightarrow O_2^-(\text{ad}), \qquad E^\circ = -0.33\text{ V}. \tag{3.27}$$

It is also clear that the Pt-lattice structure plays a role in the type of adsorption (bridge, end-on). This lattice structure changes if the Pt surface is oxidized. As a consequence, the type of adsorption also changes now.

In the case of (pyro)phosphoric acid, it has been proven that the anion is specifically adsorbed, as is also the neutral molecule, which therefore inhibits the reduction process. The (possible) mechanism has been extensively discussed[23,24,32] and will not be treated further here. In any case, oxygen is reduced in a four-electron process to water, without formation of hydrogen peroxide (see Fig. 3.12).

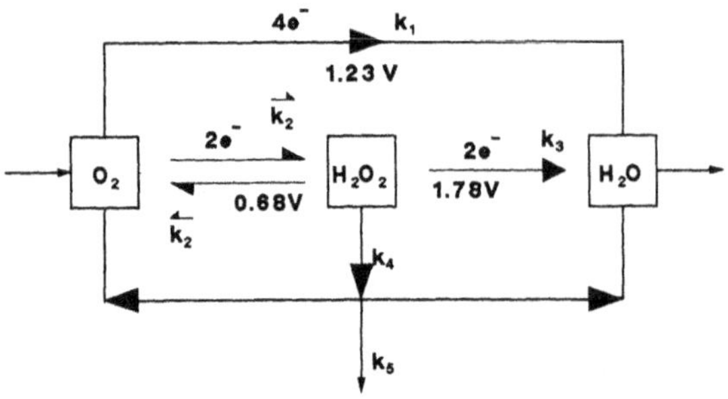

Figure 3.12. Scheme of the cathodic oxygen reduction in an acid medium.

3.4.2.3. The Electrolyte

The electrolyte accepted up to now is phosphoric acid in a concentration of about 95–97 wt%, so at the operating temperature pyrophosphoric acid is formed.

$$2H_3PO_4 \rightarrow H_4P_2O_7 + H_2O. \tag{3.31}$$

As pointed out, with this acid—and this is its only disadvantage—catalysis is not as desired. In that respect fluorinated sulfonic acids such as trifluoromethanesulfonic acid, CF_3SO_3H, perform much better, even at lower temperatures.[33] Not only the extent of, and blocking by, adsorption plays a role (see Section 3.4.2.2), but more deleterious is the much lower solubility of O_2 in phosphoric acid. The reason why the fluorosulfonic acids did not cause a breakthrough as acid electrolytes is that they lose water at $T > 110°C$ and then exist as hydrates: with increasing temperature their conductivity also drops, becoming less than $0.1\ \Omega^{-1} \cdot cm^{-1}$.

3.4.2.4. Effects of Temperature and Pressure

a. Temperature. Temperature increase has a beneficial effect on the kinetics and surpasses the thermodynamically ordered penalty. So, it makes sense to increase the temperature from, say, 190°C to 210°C. At a pressure of 5 bar and operating at 200 mA/cm^2 a 10°C increase in temperature gives some 15 mV gain in cell voltage. However, a penalty is the increased corrosion of both the carbon support and the electrode material. A good compromise is therefore an operating temperature of about 200°C.

b. Pressure. A pressure increase (see Eqs. (3.12a) and (3.13)) has a beneficial effect on cell voltage. It is not quite clear if there is some penalty for corrosion. In any case, an increase of pressure from 1 to 5 bar, at 205°C and 200 mA/cm^2, causes the cell voltage to increase with some 120 mV. The gain in cell voltage must, of course, compete with the accompanying higher costs of pressurized equipment.

3.4.2.5. Effects of Contaminants; Stability

Carbon dioxide does not affect the performance of the PAFC. Carbon monoxide, on the contrary, can only be tolerated up to about 1–3%, and sulfur compounds, as hydrogen sulfide or sulfur dioxide, up to 50 ppm. So, in most cases, some fuel cleanup is required.

3.4.3. Alternative Types of Acid Fuel Cells

In Section 3.4.2.3 the use of alternative acid electrolytes was mentioned. Progress in that area has not been convincing until now. It is, however, of interest, to consider a direct methanol fuel cell (i.e., the use of methanol in the cell without reforming). Such a cell, suitable for automotive use, would allow methanol to be used in the vehicle along with the existing fuel distribution infrastructure. Though initially an alkaline medium with methanol operates better, in the long run (because of necessary electrolyte replenishment) an acid electrolyte must be preferred.

The cell reaction

$$2CH_3OH + 3O_2 \rightarrow 2CO_2 + 4H_2O \qquad (U° = 1.21\ V) \tag{3.32}$$

has a sufficiently high voltage to justify exploration of its potentialities. At the anode, the oxidation can be formulated as

$$CH_3OH + H_2O \rightarrow CO_2 + 6H^+ + 6e^-. \tag{3.33}$$

This reaction proceeds rather slowly with Pt as a catalyst.

Other catalysts are being studied extensively. In most cases, they are based on noble metals or their alloys, e.g., Pt–Ru.[34] Of course, reaction (3.33) does not proceed without rate-determining steps.

The very high initially observed activity rapidly decreases owing to adsorption of poisons formed from dehydrogenated methanol. In fact, catalysts still need to be developed that are capable of adsorbing CH_3OH and its oxidized intermediates ([COH]?), together with H_2O in the same low-potential region (see Ref. 35, p. 93).

3.4.4. Future Research and Development

Special points of interest concern

- Optimization of the porous electrode structure, with regard to wettability (degree of flooding, pore-size distribution) and corrosion resistance
- Optimization of the cathodic oxygen reduction reaction

A large effort may be required to find an effective substitute for Pt, if necessary. Perhaps, metal macrocycles (with Fe?) might prove to be more active than Pt, even at high current densities. Stable compounds can be expected if the metal macrocycle is pyrolyzed, similar to the description in the alkaline case; see Section 3.2.2. New developments are described in Chapter 8.

3.5. THE MOLTEN CARBONATE FUEL CELL

3.5.1. General Characteristics

The MCFC is generally considered a second-generation fuel cell. It can be used with coal gas, and even more so with natural gas, as a fuel. Nevertheless, it needs substantial R&D before the commercialization stage is reached. Its behavior is closely connected with its unique electrochemistry and melt chemistry. The cell operates at about 650°C with a binary alkaline carbonate mixture as the electrolyte. Assuming that coal gas or natural gas, after reforming and shifting, are ultimately presented as hydrogen at the fuel electrode (the anode), the net anode reaction is

$$2H_2 + 2CO_3^{2-} \rightarrow 2H_2O + 2CO_2 + 4e^-. \tag{3.10c}$$

At the cathode the net reaction is

$$O_2 + 2CO_2 + 4e^- \rightarrow 2CO_3^{2-}. \tag{3.11c}$$

The charge transport in the electrolyte is served by carbonate ions (going from cathode to anode). Because the anode reaction will deplete the carbonate present, it is clear that stoichiometric supply becomes imperative. CO_2 recycling, whereby CO_2 functions as a Lewis acid, is therefore an essential part of the process which, in a simplified form, is given in Figs 3.9 and 3.13.

Because of the high operating temperature, the ideal efficiency (i.e., $\Delta G/\Delta H$) of the overall cell reaction decreases (since $T\,\Delta S$ increases). However, a higher operating temperature also reduces polarization losses, such that the actual efficiency is even increased. Moreover, the MCFC operating temperature is high enough to produce valuable waste heat for fuel (pre)processing, for bottoming cycles, and/or cogeneration processes, or even to provide compression work to let the MCFC operate at higher pressures and so improve the cell voltage. The most important use of this heat, however, is for the reforming of methane or coal gas externally or internally (in the latter case in an even more efficient way). The idea for the present MCFC stems from Davtyan and was further pioneered by Broers[36] and Ketelaar; they perfected the idea so that present research is largely based on their work.

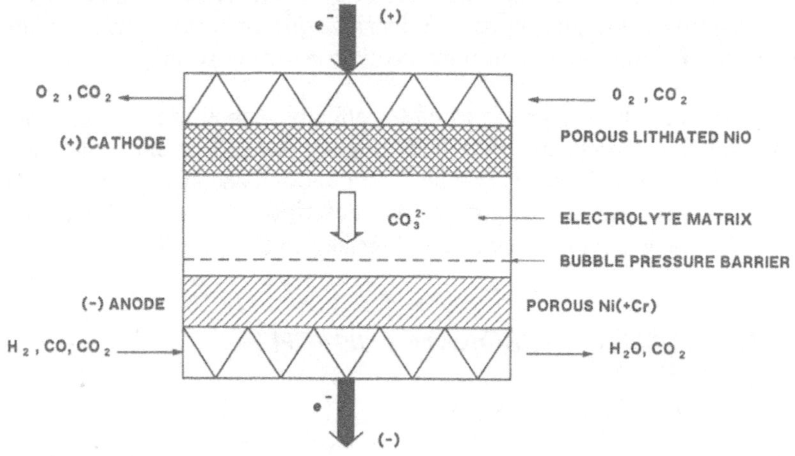

Figure 3.13. Electrode reactions in a molten carbonate fuel cell.

Some design goals for MCFC are[15]

a Lifetime up to 40,000 operating hours (about 4.5 years)
b. Performance per cell at 0.85 V to be 160 mA · cm^{-2} (for low Btu gas)
c. Scaling up to stacks with 100 cells, each up to 1 m^2/cell

Nowadays, problems such as creep, wetting, anode and cathode material, corrosion, and dissolving of cathode material, have been solved to such an extent that commercialization comes within sight. Electrolyte management over longer periods still remains problematic.

If internal reforming (because waste heat and water are directly available within the cell for the direct conversion of desulfurized methane to hydrogen; the so-driven reforming reaction takes place in the anode chamber) can be accomplished successfully, efficiencies can be increased to some 60%. A thorough review is given by Selman in Chapter 9 (Ref. 15, pp. 153–207).

3.5.2. The Electrode Reactions; Electrolyte

Figure 3.13 shows a single cell in an MCFC cell stack, together with material and charge flow. Since CO is shifted, with the produced H_2O [Eq. (3.10c)] according to

$$CO + H_2O \Leftrightarrow H_2 + CO_2, \tag{3.34}$$

the anode reaction

$$CO + CO_3^{2-} \rightarrow 2CO_2 + 2e^- \tag{3.10c'}$$

will be of minor importance.

The anode and cathode are porous and so facilitate the extent of the electrode reaction. For both, in principle, the agglomerate model is applicable. The pores, containing electrochemically active sites, are partially flooded with the molten carbonate and elsewhere covered with a thin film of the carbonate melt. At the cathode, where reduction of oxygen occurs, electrons are supplied to the reaction sites; at the anode, where the fuel is oxidized, electrons are removed. Carbonate ions travel from the cathode to the anode through a porous tile of the insulating and chemically inert ceramic material, γ-$LiAlO_2$, and CO_2 has to be supplied at the anode [Eq. (3.11c)].

Both the fuel gas (H_2) and oxidant (O_2) are supplied to the electroactive sites as dissolved species. The products from the anode reaction (CO_2 and H_2O) must be transported away from the electrode sites. Though the net electrode reactions are known, the reaction mechanisms of oxidation and reduction, together with the role of the molten carbonate, are not fully understood. More insight into these mechanisms certainly will help to optimize the process.

3.5.2.1. The Anode Reaction; Fuel Oxidation Catalysis

Soon after the first pioneering experiments, pure nickel was used as cathode material. Because of sintering, alloying with chromium up to 10 wt% became common practice around 1976. Pore size is 3–6 μm, porosity is 50–70%, and thickness is 0.5–1.5 mm; fabrication is now mostly by tape casting. Since under compression (stack construction) the anode still tends to creep, there is a continuing search for alternative anode materials. $LiFeO_2$ (also used as a cathode material (Ref. 28, p. 188)). Li_2TiO_3, and others are being studied. At the moment, much attention is being paid to the Ni–5% Al alloy.

The Al phase oxidizes, so an internal dispersion of oxide particles within the Ni phase is formed, thus impeding Ni deformation and Ni sintering (Ref. 28, p. 390). Also Ti–Ni hydrides (Ref. 28, p. 305) and Cu-based materials are being investigated (Ref. 17, p. 64). Contrary to the cathode, the anode also functions as an electrolyte reservoir for the electrolyte tile, since polarization of the anode is relatively insensitive to its degree of filling by the electrolyte. Too rapid an electrolyte flow from the anode to the tile (to replenish electrolyte loss) must, however, be prevented, and this is ensured by a thin layer of small (ca. 1 μm) but uniform porosity at the anode–electrolyte interface (and fabricated as an integral part of the anode), the "bubble-pressure barrier." In this way, gas crossover and thus mixing of the reactant gases are prevented.

Since the anode material operates in a reducing environment, the original pore structure may be preserved to a considerable extent during prolonged operation. The hydrogen oxidation reaction is the dominant one, since it is generally assumed that CO oxidation kinetics are rather slow. Nevertheless, more insight into the reaction mechanism, together with the related melt chemistry (see Section 3.5.2.3), may lead to a better understanding of the internal reforming process and so help to further optimize the MCFC.

Two types of mechanisms are proposed for the anode reaction. The first[37] is

$$2H_2 + 4M \rightleftharpoons 4MH \qquad (M = Ni), \tag{3.35}$$

$$2MH + 2CO_3^{2-} \rightarrow 2OH^- + 2CO_2 + 2M + 2e^- \qquad (\text{r.d.s.}), \tag{3.36}$$

$$2MH + 2OH^- \rightleftharpoons 2H_2O + 2M + 2e^-. \tag{3.37a}$$

The second is reaction (3.35) with (3.36):

$$4MH + 4CO_3^{2-} \rightarrow 4OH^- + 4CO_2 + 4M + 4e^- \qquad (\text{r.d.s.}),$$

and

$$4OH^- + 2CO_2 \rightleftharpoons 2H_2O + 2CO_3^{2-}. \tag{3.37b}$$

Small amounts of S-containing contaminants degrade the performance and these contaminants must therefore be removed.

3.5.2.2. The Cathode Reaction; Oxygen Reduction Catalysis

Lithiated NiO is the current cathode material, and is always applied in porous form; its pore size is between 5 and 15 μm, and the porosity is rather high

(70–80%). Its thickness is 0.50 to 0.75 mm. It starts as a porous Ni mass, which in the first hours of operation is oxidized to NiO and also becomes lithiated. The original structure then changes drastically; in addition to the original pores, very small new pores arise which cause a characteristic wetting pattern, with the very small pores preferentially filled by the melt and the larger, gas-filled, pores wetted by a thin film. Notwithstanding the rather optimum porous structure, oxygen reduction is slow (complex kinetics) even at 650°C, and therefore it is out of the question with this material to lower the temperature because of corrosion.

Therefore, alternative electrode materials are being searched for, the more so because NiO tends to dissolve, especially at higher CO_2 pressures and at higher temperatures. As said, the reduction mechanism is rather complex and governed by the melt chemistry, especially the acidity of the melt (see Section 3.5.2.3), which in turn is strongly dependent on the cationic composition. In less acidic melts (Li rich and/or if K is replaced by Na) the peroxide mechanism should prevail:[38,39]

$$O_2 + 2CO_3^{2-} \rightleftharpoons 2O_2^{2-} + 2CO_2, \tag{3.38a}$$

$$2O_2^{2-} + 4e^- \rightarrow 4O^{2-} \quad \text{(supposed to be the r.d.s.)}, \tag{3.38b}$$

$$4O^{2-} + 4CO_2 \rightleftharpoons 4CO_3^{2-}, \tag{3.38c}$$

(where O_2^- is the superoxide ion, O_2^{2-} is the peroxide ion, and O^{2-} is the oxide ion). In acidic melts, however, the superoxide mechanism is dominant:[38,39]

$$\tfrac{1}{3}(3O_2 + 2CO_3^{2-} \rightleftharpoons 4O_2^- + 2CO_2). \tag{3.39a}$$

$$\tfrac{2}{3}(2O_2^- + 2e^- \rightarrow 2O_2^{2-}) \quad \text{(supposed to be the r.d.s.)}, \tag{3.39b}$$

$$\tfrac{2}{3}(2O_2^{2-} + 4e^- \rightleftharpoons 4O^{2-}), \tag{3.39c}$$

$$\tfrac{2}{3}(4O^{2-} + 4CO_2 \rightleftharpoons 4CO_3^{2-}), \tag{3.39d}$$

These results have been achieved at fully immersed gold electrodes; rather low exchange current densities were found. With other techniques and by other researchers, higher exchange current densities were calculated.[40–42] There is still a considerable lack of understanding of the precise cathode mechanism, as governed by electrode material and melt chemistry. Alternative cathode materials mentioned are $LiFeO_2$ (one of the most prominent materials), followed by Li_2MnO_3. However, conductivity must be improved; doping with Co, Cu, or Mg is under study (Ref. 28, p. 188).

A special aspect of cathode behavior is its long-term solubility. NiO is known to dissolve (see the next section).

3.5.2.3. Composition of the Electrolyte; Matrix Effects

The melt chemistry is of utmost importance for the well functioning of the MCFC: it influences the electrocatalytic behavior of the anode and cathode, and the solubility of the cathode. Moreover, the composition is of importance for overall conductivity, wetting, corrosion (separator plate!), and volatility. Therefore, it is startling that until recently every manufacturer used a mixture of lithium/potassium carbonate in a 62:38 mol% ratio. This mixture was suspended in a porous, insulating, and chemically inert ceramic matrix of predominantly γ-$LiAlO_2$, thereby forming an "electrolyte tile." The matrix design is important for porosity and tortuosity, and hence for strength and achievable conductivity for a certain melt composition. Its fabrication is by tape casting, with a thickness of about 0.5 mm; there is a strong tendency to go thinner in order to avoid too large an ohmic drop.

In general, it turns out that electrocatalysis, especially for the cathodic reduction of oxygen, is enhanced in a less acidic melt (as in aqueous solutions) by substituting sodium for potassium. Its advantages are (Ref. 28, p. 193, where a 60 mol% Li–40 mol% Na carbonate melt is studied):

The NiO solubility is decreased by a factor of 3;
Sodium carbonate is less volatile than potassium carbonate, since the reduction of potassium carbonate by hydrogen ($K_2CO_3 + H_2 \rightarrow 2K\uparrow + CO_2 + H_2O$) does occur to a lesser extent with sodium carbonate.
The dissolution rate of NiO is decreased.

In particular, the latter aspect is of utmost importance, since with the classic melt composition Ni deposition can occur near the anode by reduction with hydrogen:

$$Ni^{2+} + CO_3^{2-} + H_2 \rightarrow Ni\downarrow + CO_2 + H_2O. \qquad (3.40)$$

The dissolution process is enhanced by higher pressures of CO_2, because this functions as a Lewis acid:

$$NiO + CO_2 \rightarrow Ni^{2+} + CO_3^{2-} \qquad (3.41)$$

In this respect, it is of interest to note that addition of earth alkali metals (e.g., 5 mol% $BaCO_3$) reduces dissolution even more (Ref. 28, p. 291). It is not quite clear why.

Other aspects, like wetting and corrosion will be treated in other chapters.

3.5.2.4. *Effects of Temperature and Pressure*

a. *Effect of Temperature on MCFC Voltage.* In principle, the temperature influence on the emf of the MCFC is given by (3.12c), which shows that the equilibrium composition of the fuel and oxidant gas is of interest. For the fuel gas, the equilibrium constant, K, of the water–gas shift reaction

$$CO + H_2O \rightleftharpoons CO_2 + H_2 \qquad (3.34a)$$

(which achieves rapid equilibrium at the anode at 923 K), is given by

$$K = \frac{p[CO]p[H_2O]}{p[CO_2]p[H_2]} \qquad (3.34b)$$

and increases with temperature (p is partial pressure). As an example (Ref. 17, pp. 72–74) for a fuel gas composition (in atm) of 77.5% H_2–19.4% CO_2–3.1% H_2O (saturated at 25°C) at 1 atm, the equilibrium composition of the gas at 990 K is $p(H_2) = 0.649$, $p(CO_2) = 0.068$, $p(CO) = 0.126$, and $p(H_2O) = 0.157$, resulting in a theoretical emf of 1.143 V ($K = 0.454$) for an oxidant gas composition of 30% O_2–60% CO_2–10% N_2. At 800 and 1000 K, the respective K-values are 0.247 and 0.727, and the emf's are 1.155 and 1.133 V. One must be aware of the exothermic carbon deposition reaction (Boudouard reaction $2CO \rightleftharpoons C + CO_2$), which, for a certain gas composition, can occur if the temperature is below a certain value. In addition, and which is of interest in internal reforming, too high a temperature must be avoided because of the endothermic methane decomposition reaction ($CH_4 \rightleftharpoons C + 2H_2$).

However, although thermodynamics predict that the emf decreases with an increase in temperature, polarization is lower at higher temperatures, and the net result is a higher "dynamic" cell voltage at higher temperature; this effect is more significant at the cathode. At 160 mA/cm^2, cathode polarization shows a decrease of 160 mV, with a change in temperature from 550 to 650°C.[43] An extra positive impact of temperature rise is the remarkable decrease in ohmic drop.

b. Effect of Pressure on MCFC Voltage. In the foregoing sections, pressure influence as a result of the water–gas shift reaction has been mentioned. This influence will be stated more explicitly here. From the Nernst law and gas solubilities, it follows that increased pressure (from p_1 to p_2) results in an increased cell voltage, ΔU_p. Exact calculation is impossible, for the same reasons as for the temperature influence: the observed value may be more than a factor of 2 higher than predicted thermodynamically. For 160 mA/cm^2 at 650°C and for a wide range of fuel gas compositions, the following relationship was reported:[44]

$$\Delta U_p \quad (\text{mV}) = 76.5 \log\left(\frac{p_2}{p_1}\right). \tag{3.42}$$

In changing pressure, care must be taken that gas composition is such that undesirable side reactions, such as carbon deposition, do not occur.

3.5.2.5. *Effects of Contaminants; Stability*

The performance of an MCFC can decrease dramatically if certain impurities in the fuel gas or oxidant gas are present; these contaminants, if exceeding a certain concentration level, can also lead to reduction in cell life.

Sulfur compounds, even if present in low-ppm concentrations in the fuel gas, are detrimental. Gasified coal and natural gas, both major sources of fuel gas for the MCFC, can contain considerable amounts of sulfur compounds. Less than 10 ppm H_2S in the fuel gas and less than 1 ppm SO_2 in the oxidant can be tolerated.[45] There is abundant literature available (Ref. 17, pp. 76–81). At a Ni anode H_2S absorbs and, consequently, blocks electrochemical and water–gas shift reaction sites: NiS_x is formed, according to

$$x(H_2S + CO_3^{2-} \rightarrow H_2O + CO_2 + S^{2-}), \tag{3.43}$$

$$xS^{2-} + Ni \rightarrow NiS_x + 2xe^-. \tag{3.44}$$

At low concentration levels of H_2S, permanent damage can be avoided, because of

$$NiS_x + xH_2 \rightarrow Ni + xH_2S. \qquad (3.45)$$

Halogen-containing compounds are also detrimental, because of cathode corrosion and because the LiCl and KCl formed will enhance electrolyte losses (higher vapor pressure).

In conclusion, contaminants harm stability, and so sintering and creep of the anode, and dissolution of the cathode as a function of cathode material, electrolyte composition with high $p(CO_2)$, corrosion of separator plates, volatility of the electrolyte, etc. Progress in all these areas will help to attain the 40,000-h cell operating time.

3.5.3. *The Internal Reforming MCFC (IR MCFC)*

Internal reforming (i.e., performing the chemical reforming process in the fuel cell itself, near the electrochemically active sites) eliminates the need for an externally sited, separate fuel processor. This cost-effective alternative also has the advantage of enhancing the efficiency of the system because of better heat and water economy. Now the endothermic steam-reforming reaction, which for methane is

$$CH_4 + H_2O \rightarrow CO + 3H_2, \qquad \text{with } \Delta H(650°C) = 53.87\ \text{kcal/mol}, \qquad (3.46)$$

takes place nearly simultaneously with, and in the immediate vicinity of, the electrochemical oxidation of hydrogen. Both the heat and the water necessary are produced by the electrochemical oxidation reaction. Internal reforming eliminates the need for external heat exchange, and the water produced during the electrochemical oxidation of hydrogen also suffices for the water–gas shift reaction. The reforming reaction is favored by high temperature and low pressure, and thus an IRMCFC operates best at atmospheric pressure (Ref. 29, p. 403). Ni, supported on MgO or $LiAlO_2$, has sufficient catalytic activity for sustaining the reforming process at 650°C; methane conversion approaches 100% at fuel utilizations greater than 50%. R&D is still in progress.

3.5.4. *Future Research and Development*

From the foregoing, it should be clear that it is impossible to tackle only a single problem: an alternative carbonate electrolyte to avoid cathode dissolution has its implications for the anode process and vice versa. In this section, a short summary of the described problem areas is given:

1. *Electrodes*
   a. Anode: electrocatalysis; creep; internal reforming
   b. Cathode: electrocatalysis; dissolution as a function of melt composition; cathode material and gas composition
2. *Electrolyte*
   a. Matrix: elimination of cracking to avoid gas crossover; reducing thickness versus enhancing strength; improving porous structure (conductivity)

b. Electrolyte: optimizing melt chemistry in relation to wetting and electrocatalysis at both anode and cathode: alternative electrolyte compositions; corrosion and dissolution of separator plates and cathode, respectively; electrolyte inventory and distribution: how to avoid segregation of cations, electrolyte displacement during operation; evaporation
3. *Contaminants*: sulfur tolerance
4. *CO_2 management*: study of various electrochemical or diffusion concepts: e.g., ceramic type membranes with different permeabilities for CO_2 and H_2
5. *Internal Reforming*: catalyst degradation and wetting; effect of gas composition.

New developments are discussed in Chapter 9.

3.6. THE SOLID OXIDE FUEL CELL

3.6.1. General Characteristics

The history of the SOFC has its roots in developing steam electrolysis for producing hydrogen. This technology offers an all-solid-state electrolysis system operating at a sufficiently high temperature (1000°C) to provide high-quality hydrogen at rather low cell voltage ($U° \approx 0.9$ V). At those conditions, electrocatalysis is excellent.

Just like the electrolysis system, the fuel cell system is based on ZrO_2, doped with 8–10 mol% Y_2O_3 (i.e., yttria-stabilized zirconica, YSZ). At 1000°C, conduction in the solid oxide is by oxide ions, O^{2-}. The electrochemical reactions occurring in the SOFC, utilizing H_2 and O_2, are

$$2H_2 + 2O^{2-} \rightarrow 2H_2O + 4e^-, \tag{3.47}$$

at the anode and

$$O_2 + 4e^- \rightarrow 2O^{2-} \tag{3.48}$$

at the cathode, as illustrated in Fig. 3.9. If the fuel gas consists of CO and no water is present for shifting, then the anode reaction is

$$2CO + 2O^{2-} \rightarrow 2CO_2 + 4e^-. \tag{3.49}$$

The principle of the SOFC can also be applied in a *sensor* for measuring oxygen partial pressure. The open-circuit potential can be expressed in terms of the partial pressures of O_2 in the anode (subscript a) and cathode (subscript c) compartments [cf. Eq. (3.12d)]; then

$$U = \frac{\mathbf{R}T}{4\mathbf{F}} \ln\left\{\frac{p(O_2)_c}{p(O_2)_a}\right\}. \tag{3.50}$$

The SOFC now acts as a concentration cell, and there is a direct logarithmic relationship between the partial pressure in the anode compartment and the cell voltage. (In the cathode compartment air is used as a reference, and hydrogen is not present in the anode compartment.)

As in steam electrolysis, electrocatalysis and, thus, kinetics are fast, and it seems that the thermodynamically prescribed penalty of a lower cell voltage at higher temperatures is more or less compensated by faster kinetics. If natural gas is being used, the CH_4 steam-reforming reaction can occur rapidly at the high operating temperature so that only minor precautions or measures have to be taken. Even so, the water–gas shift reaction, involving CO (e.g., from coal gas), is ensured. As an extra advantage CO_2 recycling, necessary for the MCFC, is not necessary. Up-to-date reviews are available (Refs. 15, 17, and Chapter 10 of this book).

At the moment *three designs* are operative:

1. The planar geometry design, common for the PAFC and the MCFC; gas sealing at high temperatures still constitutes a problem (Sohio).
2. The tubular design, advocated by Westinghouse.
3. The monolithic design, proposed by Argonne National Laboratory, is the most compact construction.

Fabrication and (inter)connection technology, however, present the most problems at the moment.

The SOFC can now operate at $0.2\,\mathrm{W\cdot cm^{-2}}$ with a cell resistance of $0.37\,\Omega\cdot\mathrm{cm^{-2}}$.

3.6.2. *The Electrode Reactions; Electrolyte*

It is important to note that in an SOFC a three-phase interface (necessary in other types of fuel cells) is not essential because oxidation and reduction of the respective gases can occur over the entire electrode surfaces of the anode and cathode, respectively. See Refs. 15 and 17.

On the other hand, it is possible to manufacture the components for most fuel cell types separately (at least partly). For an SOFC this is not easy, if not impossible, because the cell is all solid state, and conductivity requires a thin-layer technique with compact buildup; for the same reason, cell interconnections play an important role. *Chemical* and *electrochemical vapor deposition* (CVD and EVD, respectively) techniques have been developed to produce thin layers of refractory metal (Me) oxides suitable for both the electrolyte and the aforementioned interconnections.[17] Pore closure takes place with CVD; then, with the EVD technique, the appropriate $MeCl_x$ vapor is introduced on one side of the porous ZrO_2 support and a H_2–H_2O Mixture at the other side.

Thus, two galvanic systems are formed on both sides of the support tube (see Fig. 3.14):

$$MeCl_x + yO^{2-} \rightarrow MeO_y + \tfrac{1}{2}xCl_2 + 2ye^-, \tag{3.51}$$

and

$$yH_2O + 2ye^- \rightarrow yH_2 + yO^{2-}, \tag{3.52}$$

so that the net result (if $y > x$) is

$$MeCl_x + yH_2O \rightarrow MeO_y + xHCl + (y - \tfrac{1}{2}x)H_2; \tag{3.53}$$

That is, a dense and uniform metal oxide layer, MeO_y, is formed, of which the

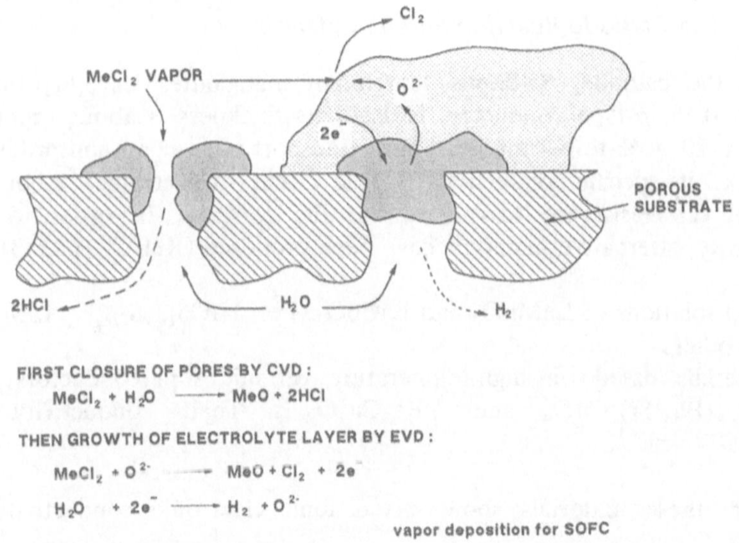

Figure 3.14. Principle of chemical (CVD) and electrochemical vapor deposition (EVD).

deposition rate is dependent on the diffusion rate of the ionic species and/or the concentration and diffusion rate of electronic charge carriers (Ref. 28, pp. 157, 179, 183). In this way gastight thin films of zirconia electrolytes and other types of highly conductive, thin, bipolar plates can be formed; codoping is also under investigation (Ref. 28, p. 157).

Because of the high operating temperature of the SOFC, the materials used must be chemically stable in both oxidizing and reducing environments, have a high conductivity, and be thermochemically compatible, all at the high temperature involved!

Very important are the cell interconnections: the material must be only electronically conductive, and its thermal expansion coefficients (about 10^{-5}/°C from room temperature to 1000°C) must match the values for the electrolyte and electrode materials. All cell materials must be capable of withstanding thermal cycling. Mg-doped lanthanum chromite, $LaCrO_3$ (specific conductivity $20\ \Omega^{-1} \cdot cm^{-1}$), meets the requirements for an interconnection material.

3.6.2.1. *The Anode Reaction and Its Catalysis*

A suitable anode material is a porous Ni–ZrO_2 cermet (YSZ) about 150 μm thick and 20–40% in porosity: mass transport of reactant and product gases is rapid. At 1000°C the specific conductivity is ca. $1000\ \Omega^{-1} \cdot cm^{-1}$. This material is satisfactory if hydrogen is used as a fuel. However, for the direct electrochemical oxidation of methane, only Pt has a catalytic behavior such that the major products are CO_2 and H_2O, although with low conversion values. Further investigations are necessary. With metal oxides of the general formula $LaSr_{1-x}M_xO_3$ a significant increase in catalytic behavior is shown (Ref. 28, pp. 179); its behavior in a reducing environment is, however, as yet unknown.

Research into mechanism and kinetics of methane oxidation is still only in infancy.

3.6.2.2. *The Cathode Reaction and Its Catalysis*

For the cathode, Sr-doped lanthanum manganite ($La_{1-x}Sr_xMnO_3$, with $x = 0.10$–0.15, p-type conductor) is used. Its thickness is about 1 mm and the porosity is 20–40% to permit rapid mass transport of reactant and product gases. At 1000°C, its specific conductivity is $77\,\Omega^{-1}\cdot cm^{-1}$ (electronic). About 50% of the total cell resistance loss occurs at the cathode. In order to improve conductivity, alternative materials have been proposed (Ref. 28, p. 183):

Solid solutions of $LaMnO_3$ and $LaCoO_3$, i.e., $LaCo_{1-x}Mn_xO_3$ with $x \approx 0.2$, and others.
Materials based on high-temperature ceramic superconductors, such as $La_{2-x}(Ba, Sr)_xCuO_4$ and $YBa_2Cu_3O_{7-x}$. Their conductivity exceeds $200\,\Omega^{-1}\cdot cm^{-1}$.

However, these materials show mixed ionic–electronic conductivity in the 800°–1000°C range.

As to the mechanism of the oxygen reduction, it is supposed that oxygen is first adsorbed on the electrocatalyst in a rate-determined step, followed by rapid electronation (3.48). There also is a view (Ref. 15, p. 223) that surface and bulk diffusion are not rate-determining steps; see also Section 3.6.3.

3.6.2.3. *The Electrolyte*

Traditionally, YSZ has been used as the electrolyte. Because the ionic specific conductivity at 1000°C is rather low ($0.1\,\Omega^{-1}\cdot cm^{-1}$), the thickness of the layer must be kept small (40 μm). The electrolyte is applied to the anode and cathode by EVD. The solid oxide must be free of porosity, and the transport number for O^{2-} must be close to unity. Moreover, its electronic conductivity must be as close to zero as possible.

The development of alternative solid electrolytes is hampered by the fact that nearly all other oxides are chemically less stable in the extreme oxidizing and (especially) reducing environments encountered in an operating SOFC. More research is required because of the moderate conductivity of YSZ. A further problem with these alternative materials is that they are susceptible to significant electronic conduction.

3.6.2.4. *Effect of Temperature and Pressure*

a. Temperature. Though higher temperatures cause a penalty (prescribed by thermodynamics) in the temperature domain of 900–1000°C, the penalty is compensated such that at $300\,mA\cdot cm^{-2}$ a rise of more than 70 mV in cell voltage/100°C is observed (Ref. 17, p. 98). From these data it is evident that the dramatic increase in cell voltage is mainly due to a sharp decrease in ohmic polarization if the temperature rises.

b. Pressure. Till now, data are not available, though one may expect enhanced performance with an increase in pressure, according to (3.12d).

3.6.2.5. Miscellaneous

a. Contaminants. In general, sulfur-containing compounds harm the performance of the SOFC, though one or two orders of magnitude less than for other fuel cell systems.[28] However, up to 50 ppm H_2S, there is reversibility of sulfur chemisorption at 1000°C. Other contaminants, such as heavy metals, have, until now, not been covered in performance studies.

b. Stability. For long-term stability under different gas compositions, thermocycling, etc., reliable data are not available.

3.6.3. Future Research and Development

To further optimize the SOFC, material research is necessary in both the (electro)chemical and thermomechanical (thermocycling!) fields.

In the more fundamentally oriented area of electrochemistry, it is of interest to study the electron transfer and mass transport control of the electrode reactions, aiming at defining solid-state and surface diffusion of oxygen to the reaction sites, at a distance from the three-phase interface, in addition to the molecular and Knudsen diffusion in the small pores of the electrodes. In analogy with three-phase electrode processes occurring in liquid-electrolyte fuel cells, it is important to study the direct reduction of oxygen at the air electrode–YSZ electrolyte interface. The same applies to the fuel electrode kinetics (hydrogen and methane), with the effects caused by contaminants (e.g., sulfur-containing gases).

Concerning the electrocatalysts, the effects of mixed ionic–electronic conductivity should be understood. Aging effects (sintering of nickel particles, porosity (structure) change, interdiffusion between YSZ and Sr-doped $LaMnO_3$) need further study, since they can influence long-term stability in a negative way.

It is clear that at the high operating temperature involved the grain-boundary morphology, connected with segregation, sintering, etc., can change and this affects long-term stability. It is also well known that thermochemical and electromechanical stresses influence this morphology and, thus, the performance.

3.7. OTHER TYPES OF FUEL CELLS

3.7.1. Introduction

The main types of fuel cells have been treated in Sections 3.3–3.6. The hydrazine fuel cell was also reviewed in Section 3.3, and the methanol fuel cell was treated in Section 3.4.3. Besides, in the area of acid-type fuel cells, the solid polymer (or proton) electrolyte (SPE) or polymer exchange membrane (PEM) fuel cell is a reliable and interesting alternative fuel cell because it operates in the low-temperature region, and becomes more and more of interest for transport applications. An indirectly operating fuel cell, because it works like a redox cell, is also treated here. Finally, brief attention is given to a combination of a fuel cell and an electrosynthesis device.

3.7.2. The Solid Polymer Electrolyte Fuel Cell

3.7.2.1. General Characteristics

As the SOFC, the SPEFC operates in reverse of water electrolysis and has the same advantages: no corrosive liquid in the cell and, consequently, minimal corrosion problems and long life; the cell can also withstand relatively large pressure differences. Nevertheless, there are disadvantages, otherwise the SPEFC would have ousted other fuel cells. The polymer electrolyte is expensive; there are still problems with water management in the membrane and with the adherence of the porous electrocatalyst onto the membrane.[17]

A good overview of this cell type is given in (Ref. 15, p. 137) and Chapter 11 of this book. Because the electrolyte is solidified in the polymer matrix (a proton-exchange polymer), and since water is formed during operation ($2H_2 + O_2 \rightarrow 2H_2O$), this water has to be removed. The principle of its design is shown in Fig. 3.15.

3.7.2.2. The Electrolyte Matrix

The electrolyte is a polymer ion-exchange membrane of acid type and is, in principle, a perfluorosulfonic acid membrane, of which Nafion (registered trademark of duPont de Nemours) is the best known; it is electrochemically stable at temperatures up to about 100°C. These membranes have a PTFE polymer chain as a backbone, several units (n = 6–10) in length, with a flexible branch pendant to this chain, a perfluorinated vinyl polyether ($m \geq 1$) with a terminal acidic (sulfonic) group to provide the cation- (proton-) exchange capability. As an example, such an ionomer unit may have the following structure (equivalent weight about 1200):

$$\begin{array}{l} —[(CF_2—CF_2)_n—CF_2—CF]— \\ \qquad\qquad\qquad\qquad\quad | \\ \qquad\qquad\qquad\quad [OCF_2—CF(CF_3)]_m—OCF_2—CF_2—SO_3H \end{array} \qquad (3.54)$$

Such membranes (thickness ~0.3 mm) have a high proton conductivity ($>2\,\Omega^{-1} \cdot cm^{-2}$); the proton transport number is unity with a low electro-osmotic water transport rate, though the water content may be about 30%. The hydrogen and oxygen permeabilities are small: 3–5 × $10^{-4}\,cm^3 \cdot cm/cm^2 \cdot h \cdot atm$ at 25°C.

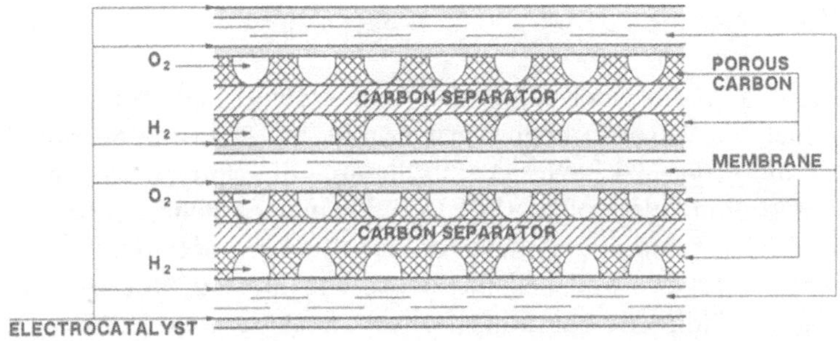

Figure 3.15. Design of a solid polymer electrolyte fuel cell.

The membrane is stable against chemical attack in strong bases, strong oxidizing and reducing acids, hydrogen peroxide, chlorine, etc., up to temperatures of 125°C. The proton conductivity is good because its concentration is up to 4 M and more. A requirement is that the porous electrocatalyst must have an intimate and durable contact with the membrane surface.

There is also intensive research to apply solid-state proton conductors.[17,46] One of the most interesting conductors is hydrogen uranyl phosphate. The operating temperature range is from 100 to 400°C.

3.7.2.3. *The Electrodes*

Because of the membrane, the cell usually works in the temperature range of 80–100°C, with hydrogen as a fuel. At the anode, hydrogen is oxidized:

$$2H_2 \rightarrow 4H^+ + 4e^-, \tag{3.10a}$$

and at the cathode oxygen is reduced:

$$O_2 + 4H^+ + 4e^- \rightarrow 2H_2O, \tag{3.11a}$$

while the proton transport is via the membrane. Both reactions use Pt supported on carbon as an electrocatalyst, with a Pt loading of $0.4\ \text{mg} \cdot \text{cm}^{-2}$. At 80°C and 3 atm hydrogen/5 atm air, the cell has a performance of 0.78 V at about $200\ \text{mA} \cdot \text{cm}^{-2}$. There is a slightly better CO tolerance than with AFC, similar to PAFC.

Increasing pressure and temperature has a beneficial effect. Though polarization increases, higher current densities up to $600\ \text{mA} \cdot \text{cm}^{-2}$ and even higher can be tolerated. The electrocatalysis for oxygen reduction is much better than that in concentrated phosphoric acid.

3.7.2.4. *Future Prospects for SPEFC*

It is clear that immobilized liquid electrolytes in the low-temperature domain have striking advantages because of the total nonvolatility of the electrolyte system and of a drastic reduction of corrosion and materials problems, ease in edge sealing of cells, and the like. Moreover, the cell produces pure water. However, much polymer research will be required to develop satisfactory, low-cost systems.

If it is possible to develop an effective anion-exchange (alkaline!) polymer, then a superior (alkaline) fuel cell could be produced. For the oxygen electrode perhaps pyrolyzed transition metal complexes might be applied. New developments are described in Chapter 11.

3.7.3. *The Redox Fuel Cell*

In the classical form, the redox cell operates as an on-site storage galvanic cell with soluble reactants, e.g., the combination

$$Cr^{2+}/Cr^{3+} \quad (E^\circ = -0.41\ \text{V versus the normal hydrogen electrode, NHE}) \tag{3.55}$$

and

$$Fe^{3+}/Fe^{2+} \quad (E^\circ = 0.77 \text{ V versus NHE}).$$

In the charged state mainly Cr^{2+} (in the "negative" compartment) and Fe^{3+} (in the "positive" compartment) are present. During discharge Cr^{3+} and Fe^{2+} are formed.

Charging occurs by input of electric energy, as with secondary batteries. However, it is also possible to make use of redox systems such that charging can occur with hydrogen and oxygen.[47] The advantage of this redox-type fuel cell is that it is possible to use simple and inexpensive electrode materials and a much simpler electrode design (no porous electrodes); perhaps, fuels other than hydrogen, say methanol, are possible. In Fig. 3.16, the working principle is sketched, according to the literature (Ref. 28, p. 313). The redox systems are

$$Ti(OH)^{3+} + H^+ + e^- \rightleftharpoons Ti^{3+}H_2O \tag{3.56a}$$

for the negative compartment with $E^\circ = 0.06$ V versus NHE, and

$$VO_2^+ + 2H^+ + e^- \rightleftharpoons VO^{2+} + H_2O \tag{3.56b}$$

for the positive compartment with $E^\circ = 1.00$ V versus NHE.

The titanium redox couple can be regenerated with hydrogen at 60°C with, as a heterogeneous catalytic system, over $Pt-Al_2O_3$:

$$2Ti(OH)^{3+} + H_2 \rightarrow 2Ti^{3+} + 2H_2O. \tag{3.57a}$$

The vanadium redox couple can be regenerated with oxygen at 75°C in concentrated nitric acid:

$$2VO^{2+} + \tfrac{1}{2}O_2 + H_2O \rightarrow 2VO_2^+ + 2H^+. \tag{3.57b}$$

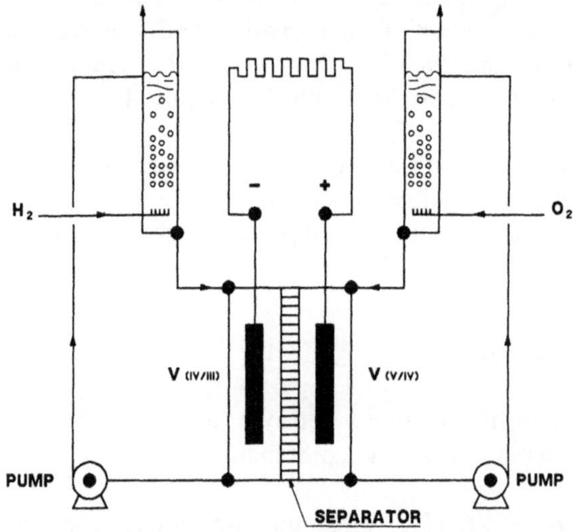

Figure 3.16. The redox fuel cell.

The ionic conductance in the cell is maintained with a proton-exchange membrane like Nafion. If the redox systems are made up of the same element, i.e., if for the Ti couple a V couple is also used [e.g., V(IV)/V(III)], then it is possible to make use of less expensive separators, as in batteries. R&D is in progress.

3.7.4. The Fuel Cell as a Chemical Reactor

It has been tempting to combine electric power generation with chemical synthesis. Several interesting systems have been developed, with existing electrochemical synthesis systems.

As an example, the production of chlorate has been chosen.[48] In Fig. 3.17a is shown how hypochlorite, produced in an electrolytic cell, C, is converted into chlorate in a special reactor, R. The main reactions in cell C are

$$6Cl^- \rightleftharpoons 3Cl_2 + 6e^-, \qquad E° = 1.36 \text{ V versus NHE,} \tag{3.58a}$$

$$3Cl_2 + 3H_2O \rightarrow 3ClO^- + 3Cl^- + 6H^+, \tag{3.58b}$$

and in reactor R they are

$$3ClO^- \rightarrow ClO_3^- + 2Cl^-. \tag{3.58c}$$

The hydrogen produced in this way is primarily burned in most cases. However, it can also be "burned" electrochemically in a fuel cell to produce electric energy which can be fed to the cell again, thus saving energy. What happens is that in this way the original $E°$-value for

$$Cl^- + 3H_2O \rightleftharpoons ClO_3^- + 6H^+ + 6e^-, \qquad E° = 1.45 \text{ V}, \tag{3.58d}$$

can be decreased by 1.23 V (for the fuel cell) to 0.22 V; the overall reaction is then

$$Cl^- + 1.5O_2 \rightleftharpoons ClO_3^-, \qquad E° = 0.22 \text{ V}. \tag{3.58e}$$

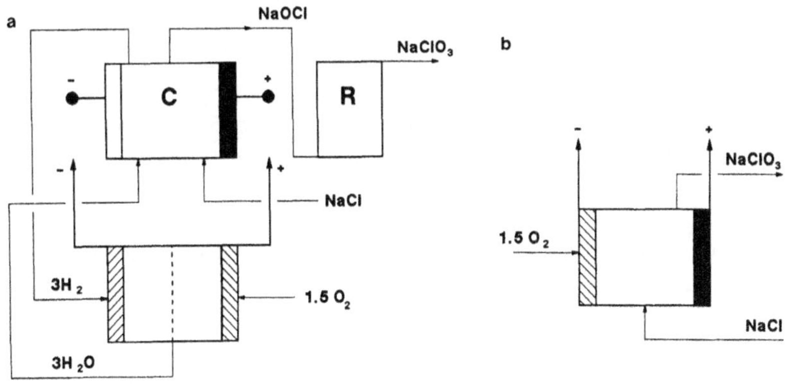

Figure 3.17. Combination of a fuel cell and an electrochemical reactor for the production of chlorate: (a) the two units separated; (b) the combined unit with an oxygen-consuming electrode.

This approach is costly, since two electrochemical cells are necessary. Since oxygen has to be reduced and the chloride ion has to be oxidized, it must be possible to make use of an oxygen-consuming electrode, as in a fuel cell (Fig. 3.17b). In fact, reaction (3.58d) is a combination of reactions (3.58a), (3.58b), and (3.58c). At the oxygen-consuming electrode the cathode reaction is

$$1.5O_2 + 6H^+ + 6e^- \rightarrow 3H_2O, \qquad E^\circ = 1.23\ \text{V}. \tag{3.58f}$$

This combination of processes in one cell offers a challenge for both fuel cell and electrosynthesis technology. Momentarily, realizations in a technical process are not known. However, one can imagine that when the gas-diffusion electrode has grown into a more mature concept, such challenges will be realized. Even so, it is clear that there are also possibilities for a hydrogen-consuming electrode. The corresponding electrochemical reduction reaction must then meet thermodynamic criteria as described above for the chlorate process.

3.8. PROBLEM AREAS IN RESEARCH AND DEVELOPMENT

Fuel cell research and development is at a critical point in time. A breakthrough to commercialization is at hand. Nevertheless, there is a need for further research as to materials (better electrocatalysts, membranes, corrosion resistivity, a.s.o.), transport properties (gas-diffusion electrodes; in general, electrolyte matrix and electrode design), diagnostic tools (electrochemical, spectrochemical, and other methods), and long-term stability. All these apply to existing fuel cells, but novel concepts for electrode, electrolyte, and cell are being developed. Because most problems have already been described, we confine ourselves to a summary here.

3.8.1. New Materials

In the foregoing sections, the material problems for the AFC, PAFC, MCFC, SOFC, and SPEFC have been dealt with in some detail. Here, only possible future developments are mentioned.

a. AFC. Noble metals may be replaced by non-noble metals and their alloys, metal macrocycles (eventually pyrolyzed or built into polymeric structures to enable the four-electron pathway). Systems which reject CO_2 (anion-exchange membranes, the remainder built up like the SPEFC) or which can operate at higher temperatures may be produced. In any case lifetimes extending to 40,000 h are required. Further research on the applicability of alkaline hydroxide melts is advised.

b. PAFC. The poor electrocatalysis of the cathodic oxygen reduction needs to be improved by studying novel catalyst systems, including (pyrolyzed) metal macrocycles. For the anode, catalysts with an improved tolerance to carbon monoxide and sulfur compounds are needed. Since the electrode kinetics' properties of phosphoric acid are poor, alternative electrolytes have to be found that yet retain the other requisite properties of phosphoric acid (cheaper polymer electrolytes?).

c. MCFC. Although some key issues with regard to the electrocatalysts for anode and cathode have been addressed, fundamental metal, ceramic, corrosion, and solid-state science is necessary to develop better and more stable electrodes, separator plates, and an electrolyte matrix. This type of research has to be carried out as a function of the melt chemistry (not limited to alkali carbonates) with sufficient emphasis on the internal reforming mode, stability, and tolerance for both sulfur compounds and hydrogen chloride.

d. SOFC. Material (ceramic) research for cell components is a must to enhance reliability and ease of manufacturing. However, even more than for the aforementioned fuel cell types, the materials can only be used in compatible sets: a change in one component may have negative consequences for other parts of the cell or for the entire fuel cell concept. The detailed electrochemical and thermochemical (diffusion of electrode material into the electrolyte with its influence on the electronic and ionic properties of the oxide ion-conducting electrolyte) processes are not well understood and need elucidation.

e. SPEFC. Polymer electrolyte membrane research to arrive at cheaper modified cation-exchange membranes (proton conductors), operating at temperatures ≥150°C, should be enhanced. At the same time, as pointed out for the PAFC, other electrode materials should be considered, if possible, for the direct oxidation of methanol.

f. Miscellaneous. For transport applications, further research into the direct oxidation of methanol at moderate temperatues (say 100–150°C) should be recommended. The low-temperature fuel cells as a chemical reactor need further attention as well.

3.8.2. Diagnostic Tools

A better insight into the electrode kinetics and electrocatalysis of the reactions in fuel cells is a primary requirement for making them function better. The different electrochemical methods cannot be applied to all types of fuel cell reactions. For instance, the hydrodynamic method with rotating electrodes is not applicable for studying reactions in the SOFC. Moreover, one must be fully aware of the fact that results obtained at "bare" solid electrodes are not simply transferable to porous systems, because of a different site distribution and because of a difficult correction for diffusion polarization.

Some of the possible electrochemical and other methods will be reviewed shortly; see the literature for a more extensive explanation.[5–7]

a. Linear-Sweep Voltammetry (LSV). Measuring the current as a function of the potential, varying linearly with time, is the most important electrochemical method and can be regarded as yielding a "fingerprint" of the electrochemical system. This method can be employed for studying the electrode reactions of all fuel cell systems. The same applies to the next method.

b. Cyclovoltammetry Method. The difference is that after the first linear potential sweep, this sweep is reversed. In most cases, cycling is repeated three times. It has become a very popular technique to obtain valuable information about fairly complicated electrode reactions (which most fuel cell electrode reactions are). Potential range and potential sweep rate can be varied. The system to be studied must be free of convection as far as possible. The method is very revealing, especially for studying the redox states of the electrocatalyst with and without reductant or oxidant.

c. Hydrodynamic Methods. Hydrodynamic methods operating with rotating electrodes are only applicable in liquid electrolytes, with viscosities which are not too high. A well-known example is the study of the cathodic reduction of oxygen for several types of electrocatalytic materials in an alkaline hydroxide electrolyte, in order to establish the production of the nondesired hydrogen peroxide.

If temperature- and corrosion-resistant materials can be found for applying the rotating ring-disk electrode in molten carbonate, this would ease the study of the cathodic reduction of oxygen considerably (even so the anodic oxidation of hydrogen and carbon monoxide).

d. Pulse Methods. Pulse methods can be successfully applied in studying all types of electrode reactions. Both the potential step method (*chronoamperometry* $I = f(t)$) and the current step method (*chronopotentiometry* $E = f(t)$) are useful, each giving complementary information. The first method gives current transients and is suitable for separation of charge (faradaic and double layer) and mass transfer effects if the system is under mixed control (as is usually the case). Sampling time must be at least $10\,\mu s$. The potential transient technique (chronopotentiometry) offers the possibility of discriminating between parallel and consecutive reaction paths and determining kinetic parameters from the so-called transition time (time between two potential plateaux).

e. AC Impedance Spectroscopy. AC impedance spectroscopy is a method which is impossible to put aside. Nevertheless, one should be very cautious in interpreting the complex plane plots. As for the aforementioned methods, the method is not self-sufficient, although a lot of valuable information can be obtained: double-layer capacitance, kinetic parameters, the occurrence of parallel and consecutive reactions, mass transport parameters, etc. The resulting signal has the same frequency as the signal applied; only in the case of a pure resistance there is no phase angle difference. This feature facilitates data processing.

In addition to these electrochemical methods, *spectrochemical* and other physicochemical methods can give valuable information, such as Auger, XPS, LEED, SEM, photoacoustic spectroscopy, and ESR. The *ex situ* methods do not provide information on what happens during charge transport. However, postmortem analysis can supply rather interesting information about structure and surface composition change.

The study of *porous* systems is a separate issue.[49] Pulse methods and AC impedance methods are indispensable tools. For liquid electrolytes, *micro*electrodes (to simplify ohmic drop corrections) and *thin-layer* cells (to avoid mass transport corrections) are worth considering.[5–7]

LIST OF SYMBOLS

a	activity ($mol \cdot dm^{-3}$); term in Tafel equation
A	electrode surface (m^2)
b	Tafel slope ($= \partial E/\partial \log \lvert i \rvert$)
c	concentration ($c°$ at electrode surface, c^* in the bulk of the solution) ($mol \cdot dm^{-3}$)
C	capacity (F), coulomb ($A \cdot s$)
D	diffusion coefficient ($m^2 \cdot s^{-1}$)
e	charge of an electron (1.602×10^{-19} C)
e^-	electron
E	electrode potential ($E°$, standard E versus NHE; E_i, at c.d. i; E_{eq}, equilibrium potential, at $i = 0$) (V)
f	$\mathbf{F}/\mathbf{R}T$
F	farad
$\mathbf{F}$	faraday (96,495 $C \cdot mol^{-1}$)
F	field force ($V \cdot m^{-1}$)
G	conductivity (Ω^{-1}); free enthalpy (J; per mole g); gas state
H	enthalpy (J; per mole h)
i	current density (I/A), c.d. ($A \cdot m^{-2}$)
i_a	anodic c.d. (positive sign)
i_c	cathodic c.d. (negative sign)
i_0	exchange c.d.
i_l	limiting c.d.
I	current (A)
$k°$	standard rate constant
k'	heterogeneous rate constant
K	equilibrium constant
K	kelvins
L	liquid state
n	number of electrons in an overall electrode reaction (n', in a rate-determining step)
ox	oxidation
OX	oxidator
p	pressure (atm)
Q	thermal energy (J; per mole q); charge (C)
r	radius (m)
R	resistance (Ω)
$\mathbf{R}$	gas constant (8.314 $J \cdot K^{-1} \cdot mol^{-1}$)
red	reduction
RED	reductor
S	entropy ($J \cdot K^{-1}$; per mole s); solid state
t	time (s)
T	temperature (K)
U	cell voltage ($U°$, at standard conditions) (V)
V	volume (m^3)
x	distance to electrode (m)
z	charge of an ion
Z	impedance (Ω)

Greek symbols

α	transfer coefficient (α_a anodic; α_c cathodic)
γ	surface energy
δ	thickness diffusion layer (m)
η	overpotential ($= E_i - E_{eq}$) (V), efficiency
θ	contact angle (rad)
κ	specific conductivity ($\Omega^{-1} \cdot m^{-1}$)
ν	stoichiometric coefficient; difference of stoichiometric coefficient: right side minus left side of the reaction equation (see Table 3.1.)

Subscripts, superscripts, other symbols

a	anode
c	cathode
g	gas
l	liquid
s	solid
o	standard
[ ]	concentration
N.E.	negative electrode
P.E.	positive electrode
YSZ	yttria-stabilized zirconia
NHE	normal hydrogen electrode
RHE	reversible hydrogen electrode
r.d.s.	rate-determining step

REFERENCES

1. C. Berger, ed., *Handbook of Fuel Cell Technology* (Prentice Hall, Englewood Cliffs, NJ, 1968).
2. A. McDougall, *Fuel Cells* (MacMillan, London, 1976).
3. B. V. Tilak, R. S. Yeo, and S. Srinivasan, in *Comprehensive Treatise of Electrochemistry*, Vol. 3 (J. O'M. Bockris *et al.*, ed.) (Plenum, New York, 1981), Ch. 2.
4. G. H. J. Broers, TNO-Netherlands, File No. DA-91-591-EUC-1023 (1959).
5. A. J. Bard and L. R. Faulkner, *Electrochemical Methods: Fundamentals and Applications* (Wiley, New York, 1980).
6. R. E. White, J. O'M. Bockris, B. E. Conway, and E. Yeager, eds., *Comprehensive Treatise of Electrochemistry*, Vol. 8 (Plenum, New York, 1984).
7. E. Yeager, J. O'M. Bockris, B. E. Conway, and S. Sarangapani, eds., *Comprehensive Treatise of Electrochemistry*, Vol. 9 (Plenum, New York, London, 1984).
8. K. Tomantschger, F. McClusky, L. Oporto, A. Reid, and K. Kordesch, *J. Power Sources* **18,** 317 (1986).
9. F. G. Will, *J. Electrochem. Soc.* **110,** 145 (1963).
10. D. T. Wasan, T. Schmidt, and B. S. Baker, *Mass Transfer in Fuel Cells: I. Models for Porous Electrodes;* Chem. Eng. Progr. Symp. Ser. **77,** Vol. 63 (1967).
11. E. H. Camara and G. G. Wilson, eds., *Fuel Cells: Technology Status and Applications*, (Institute of Gas Technology, Chicago, 1982).
12. S. Srinivasan and H. D. Hurwitz, *Electrochim. Acta* **12,** 495 (1967).
13. J. Giner and C. Hunter, *J. Electrochem. Soc.* **116,** 1124 (1969).
14. C. Y. Yuh, *Potential Relaxation and AC Impedance of Porous Electrodes*, PhD. thesis (Illinois Institute of Technology, Chicago, 1985).
15. S. S. Penner, ed., *Assessment of Research Needs for Advanced Fuel Cells* (U.S. Department of Energy, 1985) DOE/ER/300.60-T1, with extensive literature citations.

16. F. T. Bacon, *Electrochim. Acta* **14,** 569 (1969).
17. K. Kinoshita, F. R. McLarnon, and E. J. Cairns, *Fuel Cells: A Handbook* (U.S. Department of Energy, 1988), DOE/METC-88/6096; DE 88.01.0252, with extensive literature citations.
18. A. van der Putten, A. Elzing, W. Visscher, and E. Barendrecht, *J. Electroanal. Chem.* **221,** 95 (1987).
19. A. Damjanovic, M. A. Genshaw, and J. O'M. Bockris, *J. Electrochem. Soc.* **114,** 1108 (1967).
20. H. S. Wroblowa, Y. C. Pan, and G. Razumney, *J. Electroanal. Chem.* **69,** 195 (1976).
21. A. J. Appleby and M. Savy, *J. Electroanal. Chem.* **92,** 15 (1978).
22. A. van der Putten, W. Visscher, and E. Barendrecht, *J. Electroanal. Chem.* **195,** 63 (1985).
23. M. R. Tarasevich, A. Sadkowski, and E. Yeager, *Comprehensive Treatise of Electrochemistry,* vol. 7 (B. E. Conway, ed.) (Plenum, New York, 1983), Ch. 6.
24. E. Yeager, *Electrochim. Acta.* **29,** 1527 (1984).
25. A. van der Putten, A. Elzing, W. Visscher, and E. Barendrecht, *J. Electroanal. Chem.* **205,** 233 (1986).
26. J. P. Collman, M. Marocco, P. Denisevich, C. Koval, and F. C. Anson, *J. Electroanal. Chem.* **101,** 117 (1979).
27. Extended Abstracts 88-2, Fall Meeting of the Electrochemical Society, Chicago, October 9–14, 1988, p. 73.
28. *Fuel Cell Seminar Abstracts,* October 23–26, 1988, Long Beach, CA; sponsored by the National Fuel Cell Coordinating Group.
29. A. Elzing, A. van der Putten, W. Visscher, and E. Barendrecht, *J. Electroanal. Chem.* **200,** 313 (1986).
30. S. Sarangapani, J. R. Akridge, and B. Schumm, eds., *The Electrochemistry of Carbon* (The Electrochemical Society, Princeton, NJ, 1984).
31. V. Jalan and E. J. Taylor, *J. Electrochem. Soc.* **130,** 2299 (1983).
32. T. Maoka, *Electrochim. Acta.* **33,** 371, 379 (1988).
33. P. N. Ross, *J. Electrochem. Soc.* **130,** 882 (1983).
34. M. M. P. Janssen and J. Moolhuysen, *Electrochimica Acta* **21,** 869 (1976).
35. W. E. O'Grady, ed., *The Electrocatalysis of Fuel Cell Reactions,* Proceedings Vol. 79-2 (The Electrochemical Society, Princeton, NJ, 1979).
36. G. H. J. Broers, *High Temperature Galvanic Fuel Cells,* thesis (University of Amsterdam, The Netherlands, 1958).
37. P. G. P. Ang and A. F. Sammells, *J. Electrochem. Soc.* **127,** 1287 (1980).
38. J. Jewulski and L. Suski, *J. Appl. Electrochem.* **14,** 135 (1984).
39. A. J. Appleby and S. B. Nicholson, *J. Electroanal. Chem.* **53,** 105 (1974), **83,** 309 (1977); **112,** 71 (1980).
40. S. H. Lu, Ph.D. thesis (Illinois Institute of Technology, Chicago, 1985).
41. I. Uchida, T. Nishina, Y. Mugikura, and K. Itaya, *J. Electroanal. Chem.* **206,** 229 (1986).
42. I. Uchida, Y. Mugikura, T. Nishina, and K. Itaya, *J. Electroanal. Chem.* **206,** 241 (1986).
43. R. J. Selman, *Energy* **11,** 153 (1986).
44. T. G. Benjamin, E. H. Camara, and L. G. Marianowski, *Handbook of Fuel Cell Performance* (Institute of Gas Technology, U.S. Department of Energy, 1980).
45. L. G. Marianowski, *Prog. Batt. Solar Cells* **5,** 283 (1984).
46. J. Jensen, ed., *Solid State Protonic Conductors I for Fuel Cells and Sensors* (Odense University Press, Odense, Denmark, 1982).
47. J. T. Kummer and D. G. Oei, *J. Appl. Electrochem.* **15,** 619 (1985).
48. R. W. Spillman, R. M. Spotnitz, and J. T. Lundquist, Jr., *Chemtech* 176 (1984).
49. E. Yeager, J. O'M. Bockris, B. E. Conway, and S. Sarangapani, eds., *Comprehensive Treatise of Electrochemistry,* Vol. 6, (Plenum, New York, 1983), Ch. 5 especially.

4

Fuel Processing

P. Pietrogrande and Maurizio Bezzeccheri

4.1. INTRODUCTION

Fuel cells do require fuel to operate. As shown in the previous chapter, each type of electrolyte has its own fuel cell reaction chemistry: without losing generalization we can focus on the typical phosphoric acid electrolyte cell (PAFC) to illustrate what a fuel processor is.

PAFC fuel cells convert the free energy available from hydrogen molecules into electric power by the electrolytic reaction of hydrogen and oxygen. This oxidation reaction is carried out in an isothermal environment, resulting in a very favorable conversion efficiency of fuel energy to electric power, substantially higher than the efficiency obtainable with conventional thermal cycles. However, since molecular hydrogen is not normally available at prices comparable with conventional fossil fuels, a fuel processing step is required to convert fuels such as coal, oil, or natural gas into hydrogen-rich gas. The fuel processor is delegated to perform such a conversion.

Furthermore, each electrolyte type imposes its own requirements: PAFCs, for instance, are quite insensitive to acid components such as CO_2 in the feed gas, do not require complete carbon monoxide removal, and tolerate traces of other contaminants. On the contrary, alkaline fuel cells (AFC) are very sensitive to acid gas components; solid polymer fuel cells (SPE) are sensistive to carbon monoxide; molten carbonate fuel cells (MCFC) can accept high levels of CO, which is internally converted to hydrogen, but do require substantial amounts of CO_2 in the oxidizing gas to compensate for the carbonate transfer to the other electrode. All the unit operations that deal with the preparation of oxidant and reducing gases to be fed to fuel cells can be enclosed in the fuel processing section.

This chapter is designed to provide sufficient background on those aspects of chemical engineering required to understand fuel processing for fuel cells. The chapter describes the most common pathways to the production of hydrogen and

P. Pietrogrande • Bain, Cunio & Associates, Via Lutezia 8, 00198, Rome, Italy. Maurizio Bezzeccheri • ICROT S.p.A., Via S. Giovanni d'Acri 6, 16152, Genova, Italy.

Fuel Cell Systems, edited by Leo J. M. J. Blomen and Michael N. Mugerwa. Plenum Press, New York, 1993.

reducing gases. It then focuses on the most common processes, describing current industrial practices and possible future developments. Finally, it addresses those purification technologies which can contribute to the overall system engineering of fuel cell power plants.

Note that terms like "reducing gas," "synthesis gas", or "syngas," "hydrogen-rich gas" or sometimes "reformate" are utilized by different specialists to indicate gas mixtures rich in hydrogen and carbon monoxide. The latter, for many practical purposes, may be regarded as equivalent to hydrogen, especially in the presence of water vapor, as will be explained later. Most fuel processing thus deals with hydrogen and its production.

4.2. RAW MATERIAL OPTIONS FOR HYDROGEN

Fuel cells are an energy conversion device: they utilize the chemical potential of hydrogen molecules to produce oxidized molecules plus electric power. Within the scope of utilization of fuel cells there are applications such as aerospace power generators, where the energy source is selected on the basis of its specific energy content (kJ/kg) rather than its economic availability. When, however, fuel cells are considered as substitutes of conventional electric power generators, availability, cost, and logistics of the energy source are more important.

Fuel cells can utilize primary energy sources (those available, although in a limited amount, in nature), secondary energy sources (those made available by engines or other conversion devices which utilize primary energy sources), and renewable sources (those that can be continuously renewed by natural conversion processes occurring without the intervention of humans). In any case, hydrogen is not available as such in nature: it is a widely utilized chemical whose industrial production has been considered common technology for over a century. The association of hydrogen to energy is, however, much less common, and to date, beside fuel cells, hydrogen is commonly utilized as a fuel only for space vehicles.

Scientists and technologists consider hydrogen a very interesting energy vector because of its low environmental impact (combustion of hydrogen results in just water production), its high energy content (kJ/kg), and the possibility of transporting and using it in a variety of applications; in the near future we may drive hydrogen-powered cars, fly in hydrogen-powered airplanes, and cook and heat our homes with appliances directly utilizing hydrogen combustion.

Hydrogen is obtained in laboratory experiments by chemical reactions such as those produced by immersing zinc in strong acids or strong basic solutions. This technique is, however, not practical for industrial applications due to cost considerations.

Hydrogen can be produced in large amounts from primary energy sources, such as fossil fuels (coal, oil, or natural gas), from a variety of chemical intermediates (refinery products, ammonia, methanol) and from alternative resources such as biomass, biogas, and waste materials. Hydrogen can also be produced by water electrolysis, which uses electricity to split hydrogen and oxygen elements, and can be regarded as a secondary energy source.

Figure 4.1 illustrates the most obvious pathways for the production of hydrogen from different energy sources. From the figure the close link which is established between energy resources (nuclear, fossil, alternative, or renewable)

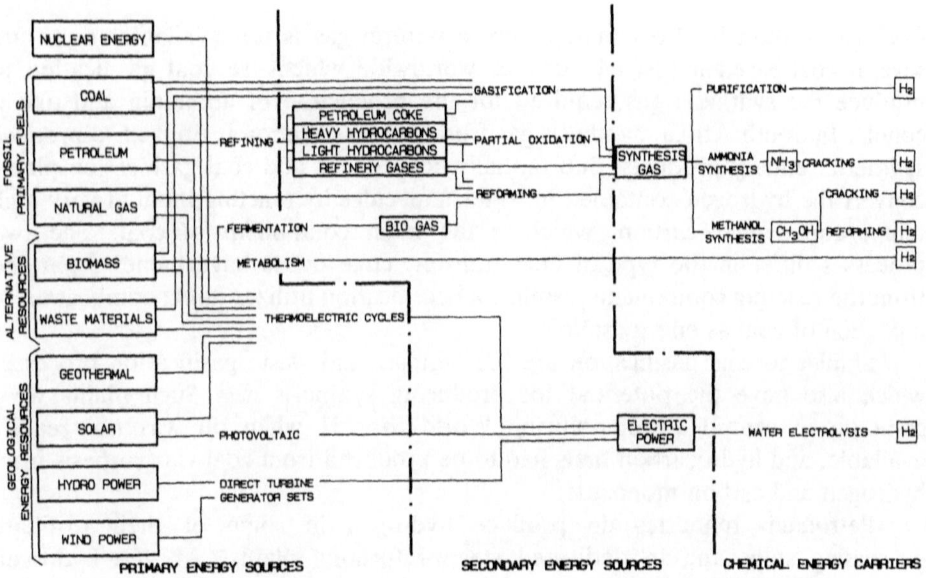

Figure 4.1. Hydrogen production pathways from different energy sources.

and hydrogen is evident: hydrogen can, to some extent, be considered as an intermediate energy vector, whose potential applications span from substituting many fuel and electric duties to upgrading raw materials. Fuel cells constitute just one of many hydrogen utilization possibilities in future energy applications.

Each box in Fig. 4.1 represents either a primary or intermediate source of energy, while the lines represent processing options. Not all the pathways shown are technically and economically feasible: for instance, bacteria can produce hydrogen,[1] but it would be completely unrealistic at this point to assume that appreciable amounts of industrial-quality hydrogen can be produced from biomass by the utilization of microorganisms.

Nuclear energy is characterized by very high energy release in small volumes: efficient, high-temperature heat removal technologies would reduce the complexity of reaction core cooling by the primary circuit. Coupling nuclear reactors with methane steam reforming would result in intense heat removal at temperatures around 900°C. The hydrogen produced would then be cooled by secondary loops connected to the thermal cycle for electric power generation; by-product hydrogen could be distributed as an energy vector or utilized in fuel cells.

Closer to practical utilization in the near term, base-load conventional nuclear power plants could be connected, during off-peak electric demand, to water electrolyzers which would produce hydrogen and by-product oxygen, making it possible to maintain large power plants at their optimal operating load while adapting the grid power requirement by this secondary electric power utilization option. Obviously this solution could also be associated with other power plant technologies, including hydroelectric and solar systems: in remote locations hydrogen produced could be stored in bottles or as liquid and transported in a manner which is often cheaper and more practical than electric power lines.

The use of coal for hydrogen production has been very common for at least a century: coke oven and town gas contain mostly hydrogen and carbon monoxide.

Coal gasification is also common where natural gas is not available: there are several coal-based industrial facilities worldwide which use coal gasification to produce the synthesis gas required for the production of ammonia and oxoalcohols. In South Africa, synthesis gas from coal gasification is utilized to produce synthetic refinery products. Substantially coke ovens and coal gasification plants convert the hydrogen contained in water molecules by reacting them at very high temperatures over carbon, which is the main component of coal. The two processes differ in the type of heat transfer: coke ovens have flames separated from the reacting components, while coal gasification utilizes direct combustion of a portion of coal as energy source.

Similar to coal gasification are the biomass and waste gasification processes which also have the potential for producing synthesis gas. Such plants were common in central Europe during World War II when oil was not readily available, and hydrocarbon fuels had to be produced from coal via synthesis from hydrogen and carbon monoxide.

Petroleum refineries do produce hydrogen in some of their distillate processing steps and in dedicated steam-reforming plants. All the hydrogen produced is normally utilized internally to upgrade refinery products. In some cases, however, refinery gas containing appreciable amounts of hydrogen is utilized internally as fuel gas.

Petroleum coke gasification has also been considered by refineries to produce hydrogen for their internal use (such as hydrotreating and desulfurization): on a purely technical basis this would make sense because the heavier the crude processed (which is the current trend), the more hydrogen is required to process the products and the more coke that is generated. Normally such refineries are even forced to import extra natural gas to produce the hydrogen required to upgrade their products and to provide auxiliary firing. Petroleum coke, on the other hand, poses logistic and environmental problems and needs to be disposed of. It might instead be utilized as raw material in gasification plants, producing the required hydrogen, reducing the import of natural gas, while avoiding the need for its disposal.

Natural gas steam reforming is responsible for over three quarters of the total hydrogen and synthesis gas production. The process is discussed more extensively later in this chapter. It is based on the reaction of water vapor and methane at high temperature over a catalyst. Other hydrocarbon-containing gases are suitable for hydrogen production. This is the case for landfill gas, digester gas, and, in general, all biogases produced by anaerobic fermentation of biomass and wastes.

Secondary sources of hydrogen include coke oven gas, by-product hydrogen from chloroalkaline production, and from manufacturing of styrene, methanol, and ammonia. The last two are the result of a synthesis reaction, which can be reversed under appropriate circumstances to yield the hydrogen-rich gas mixture they were prepared from. Local market and production situations often make such secondary sources very interesting for pure hydrogen or synthesis gas production.

The availability of by-product hydrogen is obviously a site-specific issue, which needs to be analyzed case by case; typical aspects to be addressed include considerations of purity, transportation, reliability of supply, and delivery conditions.

Electricity is also a secondary energy source. It can be produced from combustion of fossil or biological fuels via thermal cycles, from solar energy via photovoltaic conversion or by direct mechanical conversion of kinetic energy. Water electrolysis is indeed a very common process, utilized for small applications of hydrogen. Since electricity is widely available, it is indeed a very convenient secondary source of hydrogen. However, electric power available to the user implies significant energy losses for its production and distribution. If hydrogen is to be used for energy applications, then the electric conversion and transportation efficiencies, have to be added to the conversion efficiency of water electrolysis, and they often contribute well below 30% of the energy content of the primary energy source from which they have been extracted.

4.3. INDUSTRIAL HYDROGEN PRODUCTION

Hydrogen is a commodity chemical in today's industrial base, although commercial hydrogen sales represent less than 10% of world's hydrogen production, estimated to be 20 million tons per year. World production and utilization of hydrogen are provided in Fig. 4.2 and refer to a 1983 situation.[2] It is evident from the figure that most of the hydrogen is produced by hydrocarbon (natural gas or light distillates) steam reforming, and the largest consumption is in the refining and petrochemical industries. In such cases the hydrogen plant is one of the main processing units and is closely integrated with the rest of the facility. The hydrogen plant in Fig. 4.3a, for instance, is the world's largest, with a capacity in excess of 140,000 N · m^3/h of pure hydrogen. The large steel building in the center is a 640-tube reformer. Smaller-capacity plants often utilize very similar technology, although to an experienced eye they may look quite different: for example, the unit in Fig. 4.3b which has a capacity of 250 N · m^3/h. In this case the reformer heater, containing just four catalyst-filled tubes is the cylindrical structure on the left side of the picture. The main differences between the two are limited to the mechanical arrangements of the major equipment and the level of thermal integration with the surrounding units. Although economy of scale is significant, steam reforming is a viable industrial solution from 100 to 200,000 N · m^3/h capacity.

Coal gasification plants look much different from steam-reforming systems. The difference is not just in the gasification reactor itself, one of which is shown in Fig. 4.4, but also in the ancillary equipment, such as for coal handling, sulfur

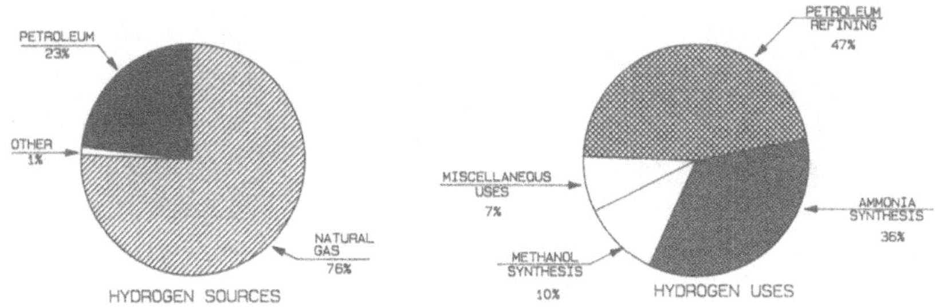

Figure 4.2. World production and utilization distribution.

Figure 4.3. Hydrogen production facilities based on natural gas steam reforming: (a) one of the world's largest reformers, Al Jubail refinery, (b) a small reformer in Annunziata's styrene manufacturing plant, Frosinone, Italy.

Figure 4.4. Coal gasification pilot reactor.

removal, water treatment, and air separation. The plant complexity is such that only large facilities are economically justifiable. A small pilot plant, producing almost $1500\,N \cdot m^3/h$ of equivalent pure hydrogen, has been built.

The other common hydrogen production technology is water electrolysis. Although the high electric power consumption makes this alternative not very attractive for most industrial users, some specific advantages of water electrolysis make it a good candidate for remote locations where electric power is available at low cost, for very small capacities, or as a backup system. Figure 4.5 shows one such unit. Such units are built in standard modules and can be designed from very small capacities to over $100{,}000\,N \cdot m^3/h$.

Very small hydrogen users utilize methanol converters and ammonia crackers to produce a reducing gas which can be purified further. At the moment such systems do not represent a significant alternative to steam reformers for industrial hydrogen production because of the high cost of feedstock. However, methanol reformers or cracking units may find very interesting applications in fuel cells, and

Figure 4.5. Water electrolysis unit for hydrogen production.

are discussed in some detail in this chapter. A portable methanol converter with 4.5-N · m^3/h capacity is shown in Fig. 4.6.

Hydrogen can also be effectively recovered from many industrial gases, such as coke oven, electrolytic, or petrochemical off-gas. In some cases the recovery consists only of impurities removal; in other cases the recovery consists of reacting CO-containing gases with water vapor to yield hydrogen and carbon dioxide.

Purification of hydrogen can be achieved, depending on conditions, by scrubbing it with proper solutions (absorption), by passing it through semipermeable membranes or molecular sieves (adsorption), by cryogenic distillation, or by reacting hydrogen with special metallic compounds to form hydrides; through purification, gas streams which are valued as fuels can be upgraded to much higher priced products: it should be considered that commercial hydrogen is often priced to 20 times its fuel value. A short evaluation of hydrogen recovery is also made in this chapter to help assess possible future H_2 sources for fuel cells.

Figure 4.6. Portable methanol reformer. Courtesy of Tecnars.

4.4. STEAM REFORMING OF HYDROCARBONS

Steam reforming is the most commonly used method to produce hydrogen-rich gases. Methane (the major component of natural gas) reacts with steam according to the equilibrium reaction

$$CH_4 + H_2O \Leftrightarrow CO + 3H_2,$$

which is endothermic when proceeding from left to right.

High temperatures and low pressures favor the production of carbon monoxide and hydrogen. While the pressure is usually fixed by the end-use requirements of the product gas, high temperatures are always required to achieve acceptable conversion. To obtain such temperatures on the process gas side, typically 800–900°C, the catalyst-filled tubes in which the reaction takes place are immersed inside the radiant section of a furnace (the primary reformer) where heat is exchanged directly between the flame and the tube skin. A schematic of a reforming furnace is shown in Fig. 4.7.

The construction material available for the catalyst-filled tubes presently limits the tube skin temperature to approximately 1000°C, corresponding to a process gas temperature (inside the tube) of about 920°C. At this operating temperature, with a typical steam-to-carbon molar feed ratio of 3.5 and at moderate pressure, about 98% of the methane is converted. If it is necessary to

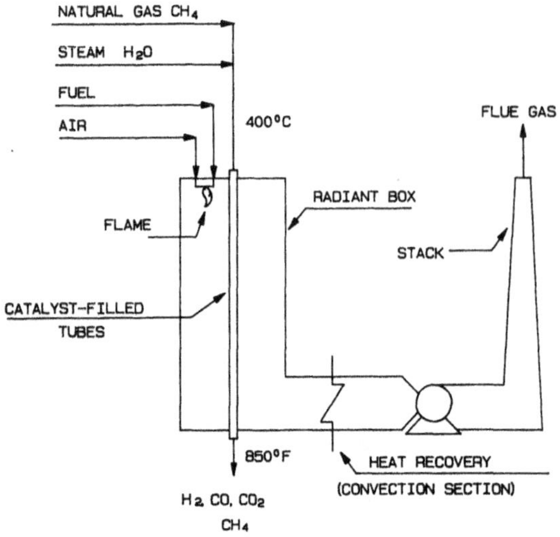

Figure 4.7. Steam reforming furnace.

obtain a higher methane conversion, an additional direct combustion step (secondary reforming) is required. Controlled quantities of air or oxygen are mixed with the hot primary reformer effluent gas in a refractory-lined vessel filled with catalyst. The resulting partial combustion increases the syngas temperature, promoting additional methane conversion. When secondary reformers are used, methane conversions above 99.6% are obtained even at relatively high gas pressures. Outlet temperatures up to 1050°C are common for secondary reformers, and special syngas coolers are utilized for most applications.

Figure 4.8 shows a schematic of a combined primary and secondary reformer system. Shifting part of the hydrogen production from the primary to the secondary reformer has some advantages over using a primary reformer alone,

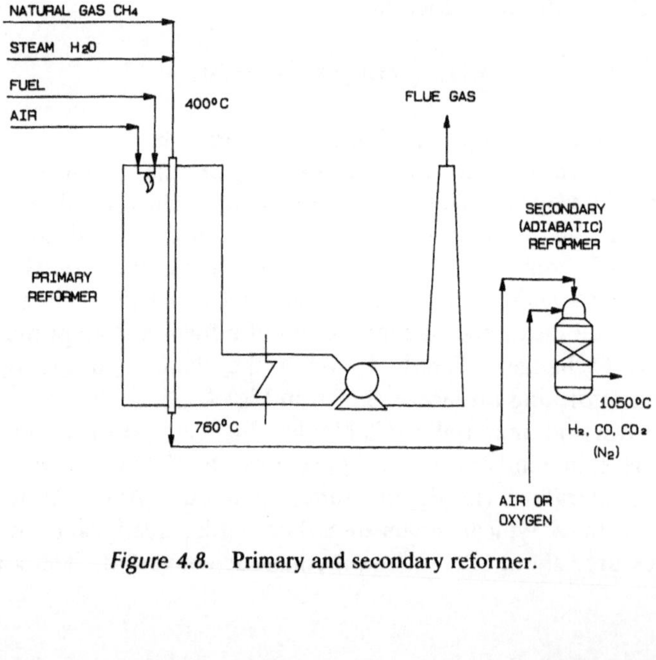

Figure 4.8. Primary and secondary reformer.

such as the possibility of operating the primary reformer furnace at less severe conditions, the better overall conversion efficiency obtainable, and the reduced equipment cost in large plants. The latter effect is due to the fact that the primary reformer cost is mostly based on the required heat transfer surface, which is directly related to a fractional exponent of the capacity ratio, usually to the power 0.8.

The major disadvantages of a two-stage reformer system are the need for additional equipment and the diluting effect of nitrogen if air rather than oxygen is used in the secondary reformer. However, many ammonia plants make use of the secondary reformer with air to introduce nitrogen in the syngas, since ammonia production requires a mixture of hydrogen and nitrogen in a ratio (3:1) which is easily obtainable by proper operation of the secondary reformer. This type of configuration appears to also be suitable for "intermediate-scale" (about 100 MW) phosphoric acid fuel cell applications.

Another widely adopted improvement to the natural gas processing system is the use of gas turbines to provide the combustion oxygen for the primary reformer furnace. Gas turbines operate with large excess of air, producing hot exhaust gas at 450–550°C, which still contains 14–16% oxygen. Such gas is used as preheated combustion air in the furnace to reduce the firing duty. This configuration has been successfully applied in the ammonia plant, whose reformer is pictured in Fig. 4.9 and in several other recently designed petrochemical plants; a conceptual sketch is given in Fig. 4.10. The major reason for the gas turbine's

Figure 4.9. Large steam-reforming furnace. Courtesy of Exxon Chemical Company and KTI Corp.

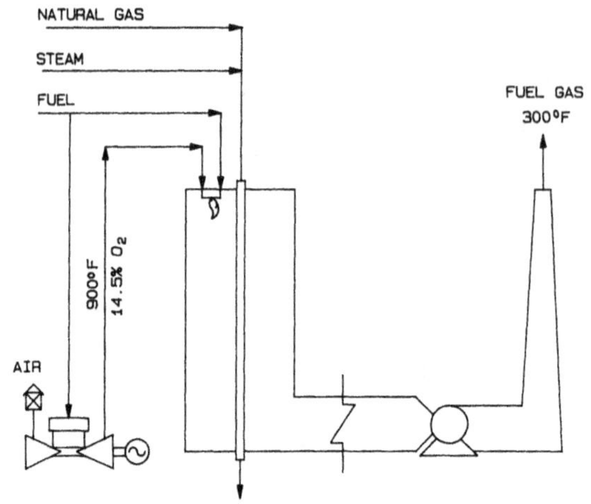

Figure 4.10. Gas turbine and reformer furnace integration.

integration with fired heaters is a net overall energy savings due to extensive recovery of the exhaust gas enthalpy as process heat.

Side aspects of such integration, however, are as important as the improved energy efficiency; the combustion of fuel with the oxygen-depleted exhaust gas results in reduced adiabatic flame temperature in the furnace burners, thus reducing thermal NO_x formation within the furnace.[3] It has been demonstrated in a utility boiler[4] that up to 50% of the NO_x produced in the gas turbine is destroyed in the furnace's flame. This makes the combination of gas turbine and reformer furnace a very efficient combustion system with low atmospheric emissions.

The technology for steam reforming, the basis of the above description, is well established. There are, however, six main areas for possible improvements: metallurgy, heat transfer, geometry, process integration, combustion, and catalysis.

The reaction conversion rate is very dependent upon exit process gas temperature, which is in turn related to maximum allowable skin temperature on the catalyst-filled tubes exposed to the direct flame radiation. Normally the limiting parameter in the design of a reforming furnace is given by the skin tube temperature. Materials such as manaurite do offer acceptable mechanical performance up to 1100°C. It should be noted, however, that reactors often have only a small portion of their heat transfer surface exposed to the maximum temperature, while the rest is at less extreme condition. In such a case, a large area of the material utilized is overdesigned for its actual service.

Table 4.1 presents the characteristics and performances of different materials utilized for steam-reforming services. Fuel cell power plants are and will be normally designed at relatively low pressures. This means that the mechanical performance of the reformer tube wall is not as critical as in ammonia plants, where the reformer tube operates at 45 bar, and thus tube wall thickness becomes the limiting design parameter.

It should also be clear that fuel cell systems are highly integrated. Improving the performance of the steam-reforming reaction does not automatically contribute to a sensible overall performance improvement; lower internal reformer

Table 4.1. A Selection of Heat-Resistant Cast Alloys for Service at 800–1200°C. Comparison table, Cast and Wrought Alloys[7]

Alloy designations		Chemical composition (approx. values, %)							Other	Stress to produce rupture in 100,000 h (note values according to lower scatterband)			
		C	Si	Mn	Cr	Ni	Nb	W		750°C psi	870°C psi	980°C psi	1090°C psi
ACI_x others Cr/Ni, others													
HK-40	25/20 Si	0.40	1.50	1.50	25	20	—	—	—	4,500	2,250	875	—
HP Mod, 36X	25/35 Nb	0.40	1.50	1.50	25	35	1.5	—	—	6,550	3,660	1,450	360
HP Mod, 36XS	25/35 WNb	0.45	1.50	0.70	25	35	Present	Present	—	—	3,500	1,310	350
G4868	30/30 Si	0.55	2.00	1.30	30	30	—	—	—	5,250	2,500	975	320
NA 22 H	28/48 W	0.45	1.50	1.50	28	48	—	5.0	—	—	3,200	1,450	440
Supertherm	28/35 WCo	0.50	1.20	1.20	28	35	—	5.5	Co = 15	—	3,650	1,600	610
H110	30/33 W	0.55	1.20	1.20	30	33	—	4.5	+ MT	—	2,700	1,300	570
Transfer lines & manifolds													
800 H cast	20/32 Nb	0.10	1.50	1.50	20	32	1.0	—	—	5,100	2,200	660	—
800 H forged	20/32	0.10	1.00	1.00	20	32	—	—	Al + Ti ⩽ 0.7	4,200	1,600	490	—
AISI, others Cr/Ni, others													
310 SS	25/20	0.15	0.75	2.00	25	20	—	—	—	—	—	—	—
800	20/32	0.05	0.35	0.75	20	32	—	—	Al + Ti ⩽ 0.7	4,000	1,600	620	—
800 H	20/32	0.10	0.35	0.75	20	32	—	—	Al + Ti ⩾ 0.85	5,200	2,100	765	—
802	20/32	0.35	0.75	0.75	20	32	—	—	Al = 0.5 Ti = 0.75	5,300	3,200	950	
800 DS	18/37	0.15	2.20	0.75	20	37	—	—	Al = 0.3, Ti = 0.3				

Monaurite 36X and 38XS (FAM, Pompay)
Maerker G4868 and H110 (Schmidt and Clemens)
NA22H and Supertherm (ABEX)
INCOLOY alloy 800, 800H, 802, 800DS (INCO)

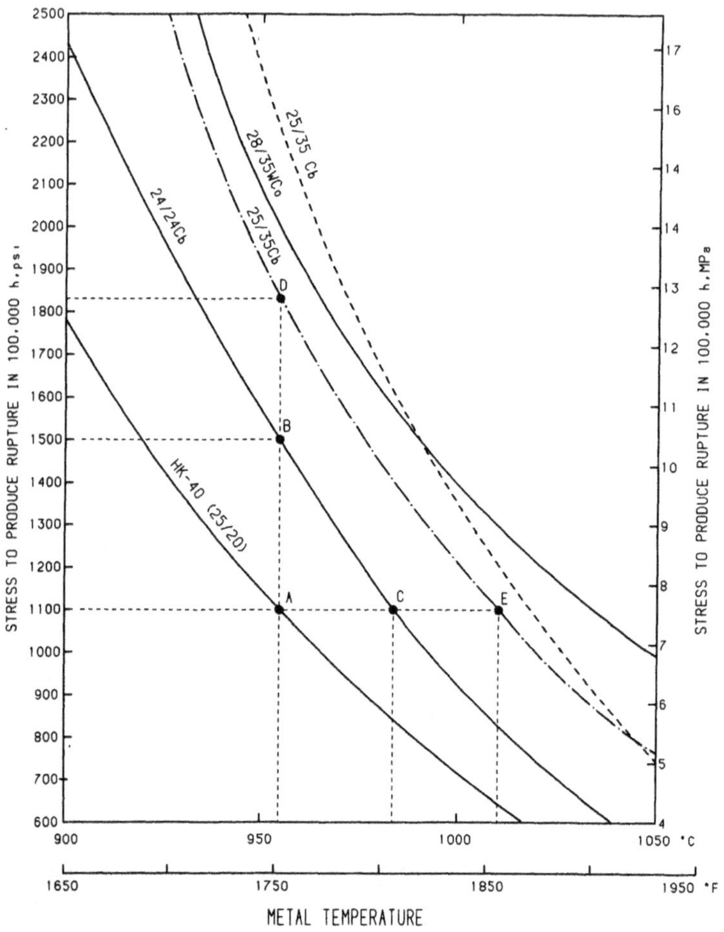

Figure 4.11. Stress and tube failures—sensibility to temperature.

temperatures may result in slightly higher methane slip, but methane is required for firing the combustion side of the reformer, and it is not much of a problem if a small quantity of it flows through the entire process unreacted.

Much more important is the performance of the tube materials to improve equipment reliability. The curve in Fig. 4.11 is an example of how stress and tube failures are sensible to temperature. A maldistribution on the flue gas side (due to improper flame shape) or the process gas side (due to uneven catalyst distribution, or carbon formation over a portion of the tube), has a direct effect of overheating the tube skin, reducing its mechanical performance and initiating a stress-induced rupture. Safe material selection is thus based on considering possible maldistribution and using a conservative approach.

To give an indication on how stress rupture is important for plant reliability, it is sufficient to indicate that the plant in Fig. 4.12, utilizing a regenerative tube and exposed to relatively moderate temperatures, had a visible torsion effect after about 6000 h of operation, presumably due to catalyst maldistribution. The tubes of the same material, and operating at similar design conditions mounted on a hydrogen plant built for Rivoira in 1973, have, since their start-up, operated at full capacity for over 135,000 h with no mechanical damage and have afterwards

Figure 4.12. Regenerative reformer tube configuration. UTC Design, Barcelona.

passed the x-ray inspections satisfactorily. In this case high-pressure drop in each catalyst bed has enabled a very even flow with little temperature maldistribution.

Developments in metallurgy are thus useful only after all aspects connected to heat transfer and geometry have been thoroughly explored. The temperature profile of the catalyst-filled tube is a very critical point in defining the heat transfer characteristics. Figure 4.13 presents the three types of reforming furnaces most commonly used in industry. They differ with respect to the position of the burners and the type of flow pattern. The first one, the top-fired furnace design, is characterized by cocurrent flow of process and synthesis gases. This ensures that the maximum cooling from the endothermic reaction is achieved in the area where the tube is exposed to maximum radiation from the flame. At the exit of the tube, where most of the methane has already reacted and the process gas is hotter, flue gas is cooler and skin tube temperature is still acceptable. The skin temperature of the catalyst tube is thus almost constant, while heat flux varies significantly.

The second type of furnace, the bottom-fired furnace, offers countercurrent flue gas and process gas flow. Maximum heat fluxes are thus obtained close to the bottom of the furnace, in correspondence to maximum process gas temperatures, when most of the methane conversion has already occurred. With this solution, if high methane conversions are to be achieved, there is the risk of tube skin overheating caused by insufficient heat absorption by the endothermic reaction

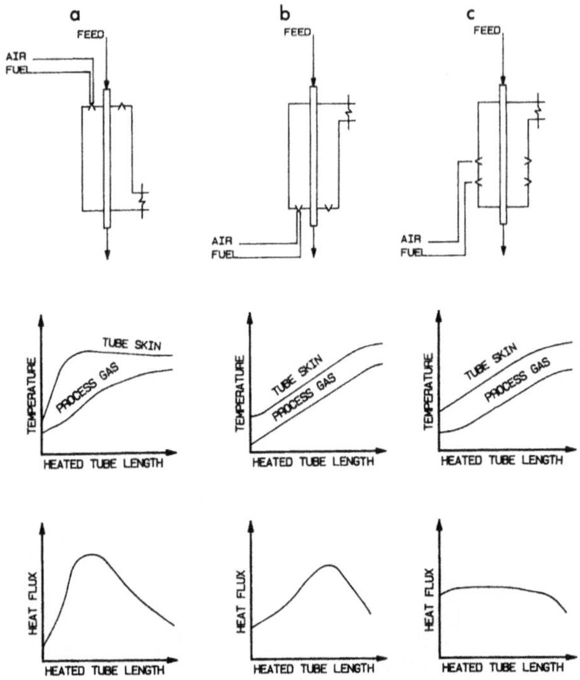

Figure 4.13. Three typical configurations for steam-reforming furnace: (a) top-fired, cocurrent; (b) bottom-fired, countercurrent; (c) side-fired.

which has already taken place. Such reformers are thus utilized only in those processes where the process gas leaving the reactor tubes is still allowed to contain appreciable levels of unreacted methane, as primary reformers of ammonia plants.

The third type of furnace, the side-fired design, offers a very homogeneous heat flux along the entire length of the tube. However, the skin tube temperature has a maximum in correspondence to the lower part of the reaction tubes, forcing the designer to select the tube material on the basis of this specific hot portion. Side-fired reformers require a larger number of burners compared to top- or bottom-fired ones. This is perceived as a disadvantage for all applications when forced air is used (as in preheated air or gas turbine exhaust) and it also requires more careful adjustment of fuel firing. As a result, the trend has been toward top-fired furnaces, which combine the cheapest design with multipurpose application and smaller furnace dimension.

In large chemical and petrochemical plants the steam reformer is well integrated with all other equipment and with the utility systems. Energy optimization is thus achieved on a broader range of scope than just hydrogen production. Typical radiant efficiencies of top-fired steam reformers, for instance, are in the range of 55–60% when nonpreheated combustion air is employed. This means that only about half of the energy released by the combustion is actually used to provide endothermic and sensible heat for the reaction to proceed. The remaining half of the energy is available through the flue gases. Large steam reformers are thus equipped with a convection section which recovers the sensible heat from flue gas, bringing the overall thermal efficiency of the reformer to over

85–88%. The most obvious service for the convection section is the preheating of feed and combustion air. The energy available in the flue gas, however, far exceeds the requirements of such services. In many cases, therefore, the heat is used to produce high- or medium-pressure steam, part of which is then used as process steam in the reformer, the rest being exported. A typical refinery hydrogen plant of 60,000 $N \cdot m^3/h$ hydrogen capacity would produce over 50 tons/h of high-pressure (HP) steam, which can be sent to the main refinery HP steam network system. The same plant would be utilizing approximately 30 tons of medium-pressure (MP) steam from the refinery MP steam network.

In addition to the heat available in the flue gases, conventional hydrogen plants have sensible energy available from the reformer effluent gas, which is normally at temperatures in excess of 800°C. Specially designed process gas boilers are used to reduce the temperature to 300–350°C before the gas enters the shift conversion reactors. A process gas boiler is a sort of fire-tube boiler with internal refractory materials to withstand the severe conditions at the inlet. They can be designed to produce steam at any level of pressure up to 60 bars.

Small hydrogen plants are often located in industrial areas where there is little integration among the units. In such case there is little or no use for the extra steam available from the reformer convection section and process gas boiler.

This situation is similar for fuel cell power plants, where there is often no use for export steam. In some cases, actually, fuel cells themselves produce saturated steam which may be utilized as process steam in the fuel processing section. This is the case for most currently available phosphoric acid stacks. To reduce energy rejection from the steam-reforming process, in such cases regenerative tube designs and pressurized combustion reformers have been employed.

These last-mentioned concepts have been applied for some time to steam reformers in commercial hydrogen plant use, with limited success. The solution is, on the other hand, very common for fuel cell power plants, where it has proved to be a reliable and effective means of producing synthesis gas. The idea consists of a reactor as in Fig. 4.14, where the catalyst-filled annulus is exposed on one side to the radiant heat transfer from flue gas, and on the other side to mostly convective heat transfer from process gas. With this geometry the reformer effluent, before leaving the reactor, is cooled to 500–600°C, reducing complexity and the duty of downstream process gas cooling equipment, and in the meantime supplying process heat to the reforming reaction. The net effect is a saving in firing heat duty (part of the reaction heat is recovered by the regenerative tube exchanger), resulting also in a smaller flue gas rate, and thus reduced convection section duty. This design has one other major advantage: the tube is free to expand linearly, and all connections are positioned in the cold end section. This is quite important for those services which require frequent thermal cycles, and it reduces the cost of conventional reformers' pigtails. In effect, design practice on pigtails, at a location where the reformer effluent is at temperatures in excess of 700°C, constitutes a complex technology, for which the knowhow is specific for each major equipment supplier. With regenerative-type tubes this problem is significantly reduced, improving the overall availability expectation of hydrogen production. The equipment shown in Fig. 4.15 is representative of a regenerative-tube reformer like that developed by United Technology Corporation (now IFC) for fuel cell application.

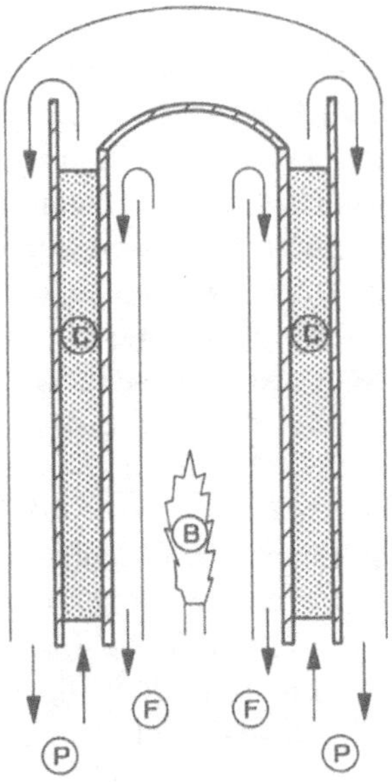

Figure 4.14. Regenerative-type reformer tube.

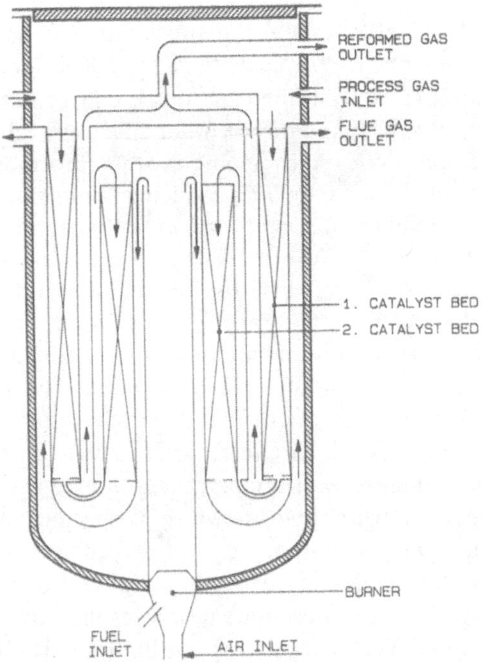

Figure 4.15. Schematic of Haldor Topsøe's reformer featuring a combination of countercurrent and cocurrent arrangements.

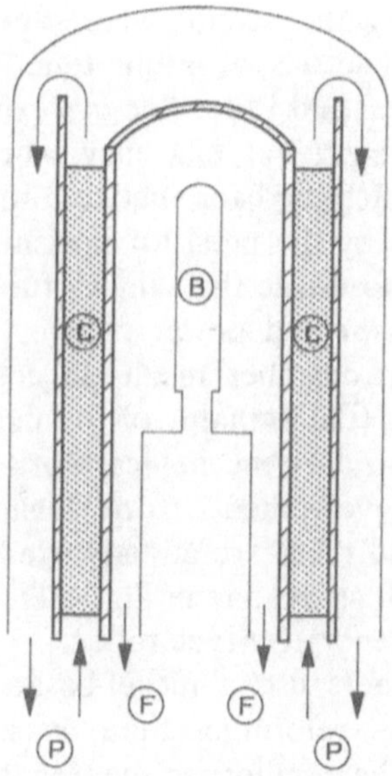

Figure 4.16. KTI compact regenerative reformer.

Similar reformers have been utilized in commercial hydrogen plants with some success. The schematic of Fig. 4.14 shows a countercurrent flow between flue gas and process gas. With reactors which have such arrangements and have conversions close to 100% one critical point is the heat transfer on the hottest section of the tube, as discussed above. To reduce the risk of unacceptably high tube skin temperatures, this type of reformer tube is often protected at the hottest section by a ceramic cup. One way to avoid excessive skin tube temperatures uses a different geometry, like that utilized by HALDOR TOPSOE for the pilot plant built in Houston, and is shown in Fig. 4.15. In this configuration flue gas and process gas are exchanging heat cocurrently in the hottest section and countercurrently in the section where skin tube temperatures are no longer critical.

The use of innovative combustion technology is also shown in the reformer concept whose schematic is given in Fig. 4.16, where a radiant burner is utilized. In this case the reformer is of the type developed by KTI and resembles in its heat profile a side-fired reactor, where heat is homogeneously supplied along the entire length of the radiant section. Tube impingement is here prevented because part of the radiant heat is transferred by graybody radiation between the burner body and the tube, which is not directly exposed to the flame; flue gas temperatures leaving the burner surface are thus much lower and cannot damage the tube. As a consequence of the extremely constant temperature profile, the burner and reformer tube can be extremely close to each other, thereby reducing overall external dimensions.

Such solutions make it possible to recover the heat at high temperature for process duties, thus reducing fuel firing rates and flue gas flow rates. The systems shown are normally coupled with a section where flue gas convective heat transfer is enhanced by special geometries or promoters. This results in further heat recovery, reduced firing rates, and lower flue gas exit temperatures.

To further improve system net efficiency when no export steam can be produced, many reformers designed for fuel cell applications utilize pressurized combustion. This is favored by the need for pressurized airflow for the fuel cell stacks and the requirement to utilize the exhaust fuel gas from the stacks as fuel: The integration of compressor and power recovery turbine is such that pressurization of the combustion chamber results in cost reduction (more compact design) and energy recovery (the enthalpy of the flue gas leaving the reformer is utilized in the expander to produce mechanical work instead of heat). Pressurized combustion is a possible development in the technology for high-temperature heat transfer equipment, and there are already examples of commercial application in hydrogen plants, such as the one in Fig. 4.17, which has been in operation since 1983 and utilizes a regenerative-type tube.

Industrial steam reformers use a nickel-based reforming catalyst. At the operating conditions of the reforming tube, such catalyst gives quite good performance, and typically the equilibrium theoretical composition and the actual composition of the effluent gas differ only by very small amounts. Many suppliers of reformers do use "approach to equilibrium" to measure the effectiveness of the catalyst bed: this is defined as the difference between the actual gas temperature and the temperature at which the actual effluent gas composition would be at equilibrium. Typically, approaches to equilibrium are limited to 10–25°C, indicating quite good performance for the catalyst. This implies that in conventional steam reformers, catalyst activity is a much less critical factor than heat transfer, which is the limiting factor engineers utilize to design equipment. With

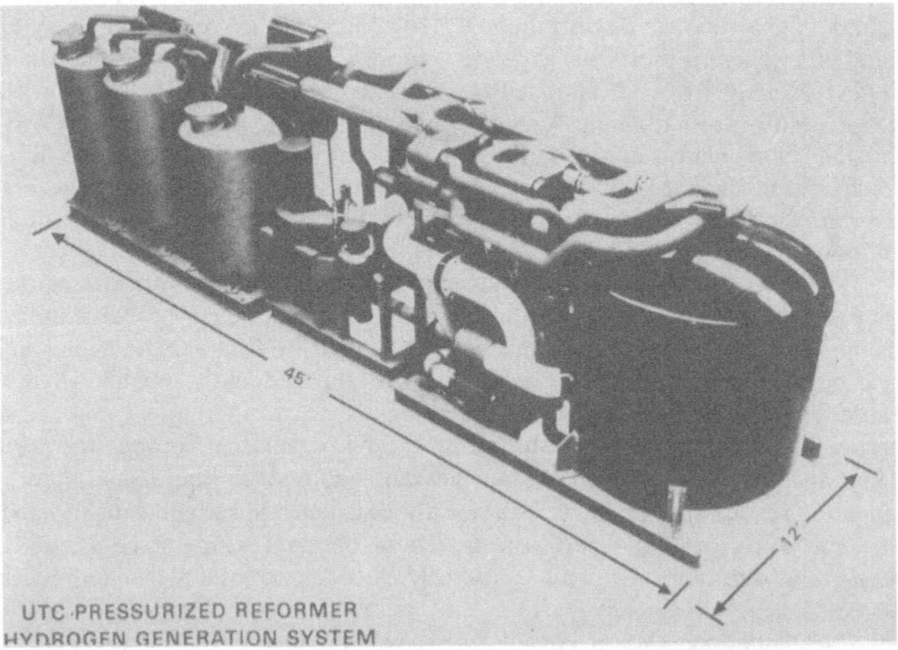

Figure 4.17. Pressurized combustion reformer. UTC Design, Barcelona.

Figure 4.18. **Experimental setup for catalyst demonstration to perform with untreated diesel fuel oil.**

regenerative tube designs, catalyst activity becomes more important, and improved catalysts are often used. The reforming catalyst is very sensitive to the presence of some impurities, including sulfur compounds, chlorine, and heavy metals. Feed gas is thus normally treated prior to entering the reformer in order to remove all contaminants. There are, however, certain catalysts suitable for steam reforming and which can tolerate certain impurities in the feed gas. Figure 4.18 shows an experimental setup developed to demonstrate the capability of a special type of catalyst to perform satisfactorily with untreated diesel fuel oil. The plant was designed for integration with phosphoric acid fuel cell stacks, for a U.S. Department of Defense project.

Another type of catalyst has been developed recently by the National Research Board in Italy for application in molten fuel cell power plants. It is a high-activity nickel-type catalyst which comprises a carrier composed of high-purity ultrafine single-crystalline magnesia. All such developments will contribute, even if marginally, to the overall improvement of fuel cell system technology.

Steam reforming of methane is just one case of the steam-reforming reaction. Liquefied petroleum gases (LPG) and naphtha are often used as feedstock where natural gas is not available; other gases, such as refinery gas or even biogas, may be utilized as well whenever distance from the source and other circumstances make it possible. The basic process is the same, although some extra considerations need to be made. First of all, the ratio of steam to carbon, which is a measure for the amount of dilution steam required, needs to be optimized; secondly, the sulfur removal from feedstock needs to be considered.

The main role of steam in the reforming reaction is to push the equilibrium toward H_2 and CO formation. Steam, however, is also reducing partial pressures, which also helps the equilibrium reaction, due to the fact that for each 2 moles which react, 4 are formed, with a net increase in volume. In addition, steam prevents carbon formation in the reforming tubes and in downstream cooling equipment.

Under certain unfavorable conditions, carbon formation may occur by virtue of the following two principal reactions:

$$2CO \rightarrow CO_2 + C \quad \text{(Boudouard)},$$

$$CO + H_2 \rightarrow H_2O + C \quad \text{(CO hydrogenation)}.$$

Carbon formation may occur only if the gas composition exceeds the equilibrium constant for the above reactions. Carbon may form on the catalyst itself during reforming or on downstream heat transfer equipment as the gas is cooled. Increasing levels of steam generally diminish the tendency for carbon formation. To establish a detailed temperature and composition profile along the reformer tube is a complex matter which requires kinetic simulation models coupled with heat transfer simulation models.

When considering the carbon formation tendencies, however, an equilibrium assumption for each location in the reformer provides an approximation which increases in validity toward the hotter portions of the tube. Using this assumption, the carbon formation tendency of any reformer feedstock may be predicted throughout the tube length. Once the gas exits the catalyst bed, the composition is fixed. As the gas is cooled, the thermodynamic driving force for carbon formation increases. In fact, reformer effluent gas from any process will eventually enter the carbon formation regime at some point during the cooling process. However, also as the gas is cooled, the kinetic rate of carbon formation decreases. Data collected from years of design and operating experience on this subject indicate that if the equilibrium carbon formation temperature is below about 750°C, no carbon will form in a process gas boiler due to kinetic rate limitations. The equilibrium constant for carbon formation by the Boudouard reaction at this limiting temperature is about 0.28.

In a conventional hydrogen plant, the reformer effluent is generally cooled by using a process gas boiler. In this type of device, gas quenching occurs quite rapidly. Furthermore, relatively low tube wall temperatures are maintained. These conditions serve to diminish the tendency for carbon formation, which may only occur on the heterogeneous tube wall surface. In the regenerative-type reformer tubes such as those utilized in fuel cell systems, the tube wall temperature is maintained at a relatively higher value as the reformer effluent is cooled into the carbon formation regime. Due to this fact, the critical values generally applied in conventional hydrogen plants may not be directly applicable in the case of the fuel cell plant utilizing regenerative reformers. More stringent criteria are necessary for such a case. Normally steam reformers operate with steam–carbon ratios in excess of 2–3 (mole/mole) to have sufficient safety margin. To protect the reformer catalyst and the downstream equipment from carbon formation due to misoperation or accidents, most plants are equipped with trip and alarm systems on the ratio controller between the steam and hydrocarbon feed. Such systems allow for certain variation of operating parameters, but are designed to shut off the feedstock valve well before a critical steam–carbon ratio is reached.

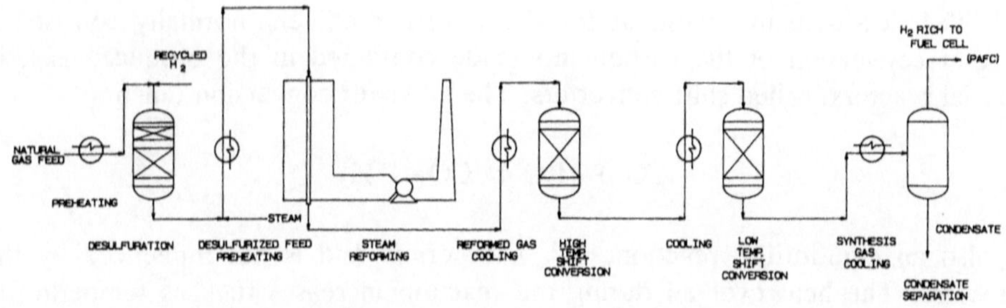

Figure 4.19. Simplified flow diagram of PAFC system fuel processor.

Although most important, the steam-reforming reaction is just one of the process steps required to produce fuel cell feed from hydrocarbon feedstocks. Fig. 4.19 shows a conceptual flow diagram of a possible process route. The main unit operations are the following:

- Feedstock heating
- Feedstock desulfurization
- Desulfurized feedstock heating
- Steam reforming
- Reformed gas cooling
- Shift conversion
- Synthesis gas cooling

As already discussed, the most common reforming catalyst is supported nickel oxide. This catalyst offers good performance up to about 1000°C and is available from different manufacturers, normally as Rashig rings. The catalyst is deactivated by sulfur, chlorine, and arsenic; the latter produces an irreversible loss of activity, while sulfur and chlorine causes reversible effects.

Natural gas and other light hydrocarbons may contain sulfur contaminants which must be removed before the reformer. The type of contaminants and their quantity determine the most appropriate removal process: low quantities of hydrogen sulfide may be removed at ambient temperature by activated carbon beds and by iron sponges or at higher temperature by zinc oxides. Sulfur-containing contaminants such as carbonyl disulfide and heavy mercaptans may require a hydrogenation step which converts these to easily removable H_2S. Typical hydrotreating catalysts are cobalt–molybdenum or nickel–molybdenum oxides supported on alumina, which are active in the range of 250–450°C. In typical installations where the feedstock contains heavier species of sulfur than H_2S, a combination of a hydrodesulfurization catalyst and a zinc oxide absorbtion bed is applied in the same vessel.

Chlorine may be accidently introduced into the reformer through the process steam cycle, since stream is not passed through any type of guard bed. To prevent this, the control of boiler feedwater quality needs to be performed regularly or continuously. Arsenic and other trace metals eventually present in vaporized feedstocks will be removed by the nickel–molybdenum oxide catalyst and do not present commonly encountered problems even for hydrogen plants operating on naphtha feedstocks.

The downstream treatment for the reformer effluent normally consists of further conversion of the carbon monoxide contained in the produced gas, in special reactors, called shift converters. The CO shift conversion reaction

$$CO + H_2O \Leftrightarrow CO_2 + H_2$$

is also an equilibrium reaction; it is exothermic and is not influenced by the pressure. The heat evolved during the reaction increases the gas temperature, which in turn depresses further conversion. To increase the overall hydrogen yields, thus reducing CO, two reactors are often used in series, separated by a gas-cooling step. The catalyst used for the CO shift reaction differs depending on the service. The most common catalyst is a mixture of iron and chromium oxide normally available in tablets. The catalyst is active in the range 330–530°C, and it is customary to introduce the feed gas at the lowest possible temperature (350°C) in order to maximize CO conversion. The temperature rise in the catalyst bed depends on the feed gas CO content and is, in typical H_2 plants, about 60°C. Typical CO content of effluent gas from high-temperature shift converters is 1–2 vol%; if higher CO conversion is required, a second stage of shift conversion, after cooling, is performed.

When CO in product gas needs to be reduced to the maximum extent, a low-temperature shift conversion catalyst needs to be used. This catalyst is based on copper and zinc oxide supported by alumina. Due to the presence of copper, which is very sensitive to chlorine and sulfur, the low-temperature CO shift catalyst is normally protected by a guard bed; the catalyst requires special care during its reduction, start-up, and shutdown procedures.

To minimize carbon monoxide content in the effluent gas, designers tend to reduce inlet temperature as much as possible (the catalyst is active in the 200–250°C range) while trying to avoid operation close to the gas dew point. The temperature increase in low-temperature CO shift converters is normally limited to 10–20°C because of the already low level of carbon monoxide present in the feed.

As mentioned, in many cases hydrogen and carbon monoxide can both be considered as feed for fuel cells. This is because CO shift conversion is a relatively simple reaction which can occur either on a separate, catalyst-filled vessel (as when feeding PAFC, AFC, SPECF) or directly over the stack electrode, which acts as catalyst and immediately consumes the H_2 molecules formed. Sources of by-product CO could thus be considered almost as an equivalent to sources of by-product hydrogen.

4.5. *PARTIAL OXIDATION OF HEAVY HYDROCARBONS*

The partial oxidation of hydrocarbon liquids is a process used to produce hydrogen where natural gas is not economically available or excess heavy oil is available at low cost. The hydrocarbon liquid reacts with steam to form synthesis gas according to the reaction

$$C_nH_m + nH_2O \Leftrightarrow nCO + (n + \tfrac{1}{2}m)H_2.$$

At high temperatures this endothermic reaction produces appreciable quantities of hydrogen. The heat of reaction, rather than coming through the catalyst tube wall (as in hydrocarbon steam reformers), is produced by the combustion of a portion of the feed with controlled quantities of oxygen (or air) in a high-temperature refractory-lined reactor:

$$C_nH_m + \left(n + \frac{m}{4}\right)O_2 \rightarrow nCO_2 + \tfrac{1}{2}mH_2O.$$

Structurally, the partial oxidation reactor is very similar to the secondary reformer, but it does not contain any catalyst.

The two processes applied commercially worldwide are the Texaco and Shell processes. They are similar in some respects, but use different mechanical designs for the reactor and the cooling systems. Schematics of the Shell reactor design and of the overall Shell system are shown in Figs. 4.20 and 4.21.

Liquid hydrocarbons, and specifically oil residue, can contain sulfur compounds which react in the reducing–cracking atmosphere of the reactor to produce COS and H_2S. The presence of sulfur compounds in the syngas requires additional downstream equipment to remove the sulfur before final use of the gas and impose careful selection of materials and catalysts.

The design of systems for heat recovery from partial oxidation reactors is critical due to the high temperature of the effluent and the presence of uncombusted carbon and other solid particles which can foul the cold surfaces of the coolers. The schematic design of the Shell partial oxidation syngas cooler is shown in Fig. 4.22.

The effluent from partial oxidation systems contains several contaminants as well as carbon and ash. The treatment of such effluent utilizes well-established techniques which require substantial capital investment.

4.6. *COAL GASIFICATION*

Coal gasification is the endothermic reaction of carbon atoms contained in coal with water vapor, according to the equation

$$C + H_2O \Leftrightarrow CO + H_2.$$

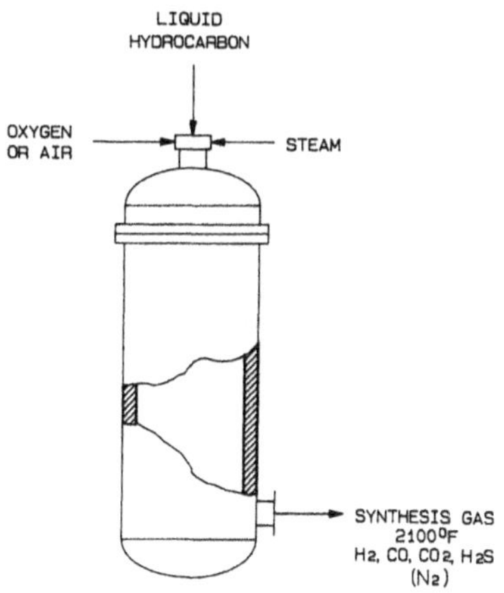

Figure 4.20. Partial oxidation reactor.

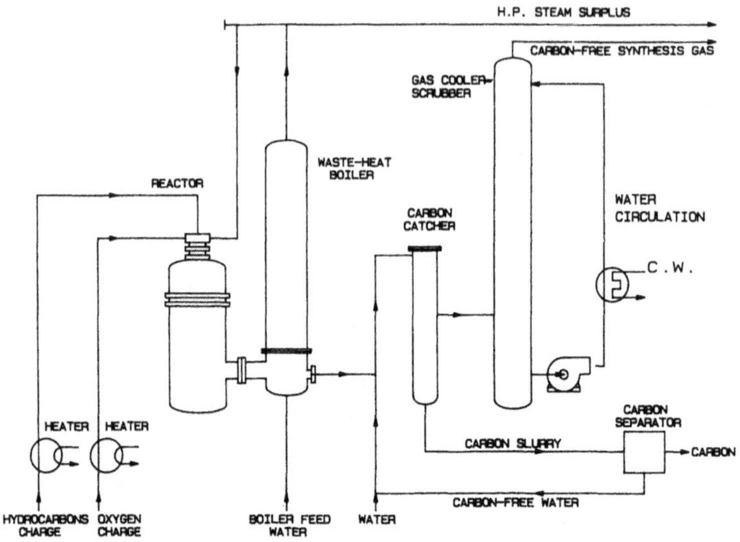

Figure 4.21. Basic Shell partial oxidation process configuration.

For the reaction to proceed from left to right it is necessary that heat be transferred at high temperatures.

Most coal gasification processes are based on the concept of direct combustion of a portion of coal feedstock with oxygen or air introduced in the same vessel where the gasification takes place: combustion and gasification occur at the same time, the first reaction providing heat for the second to occur. At any steady-state condition the energy balance of the system is thus zero (balanced reaction), and the combination of the reactions listed in Table 4.2 theoretically gives no net flow of energy.

Peculiar characteristics of coal, such as its ash content and composition, tendency to agglomerate, sulfur content, reactivity, and so on, make coal gasification a complex process with many alternative routes. Figure 4.23 summarizes the available coal gasification processes grouped per type of reactor. Moving-bed reactors have been utilized for almost a century to generate

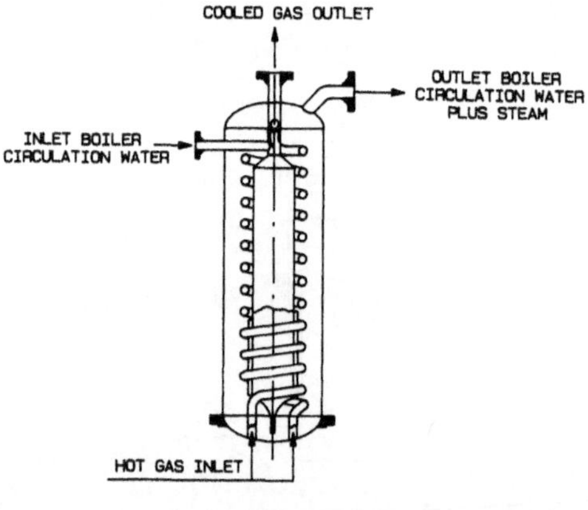

Figure 4.22. Helical tube waste heat boiler.

Table 4.2. Basic Reactions Occurring Inside a Coal Gasifier

Reaction	Energy absorbed (kcal/mol)
$C + \frac{1}{2}O_2 \Leftrightarrow CO$	−26.4
$C + O_2 \Leftrightarrow CO_2$	−94.2
$C + H_2O \Leftrightarrow CO + H_2$	+32.2
$C + 2H_2O \Leftrightarrow CO_2 + 2H_2$	+23.1
$3C + 2H_2O \Leftrightarrow 2CO + CH_4$	+44.3
$2C + 2H_2O \Leftrightarrow CO_2 + CH_4$	+2.9

hydrogen-rich gas, from coal or lignite, by injecting steam and air, from the bottom of a vessel filled with coal. Continuous rather than batch operation has been achieved by use of a rotating grid on the bottom which removes the ash. Improvements in metallurgy, process control, and flow distribution made it possible to develop oxygen-blown pressurized gasifiers based on moving-bed technology, as those utilized nowadays.

The need to utilized caking coals (which tend to agglomerate in the upper portion of moving-bed gasifiers) and the complications presented by tar and oils condensing during gas cooling, have resulted in alternative processes where gas is formed at much higher temperatures and coal is either suspended in a fluid bed or injected through special burners. Furthermore, steel production technology, which is a main user of coal, can be employed to develop even higher temperature processes, aimed at producing pure, cleaner gas from less reactive coals.

Figure 4.24 shows schematically the four classes of coal gasification routes. On the side, temperature profiles illustrate the main peculiarities of each process.

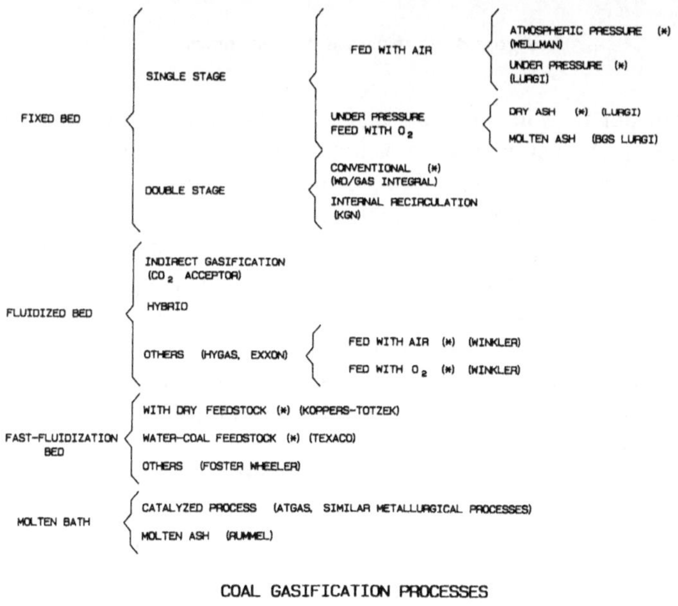

Figure 4.23. Coal gasification routes.

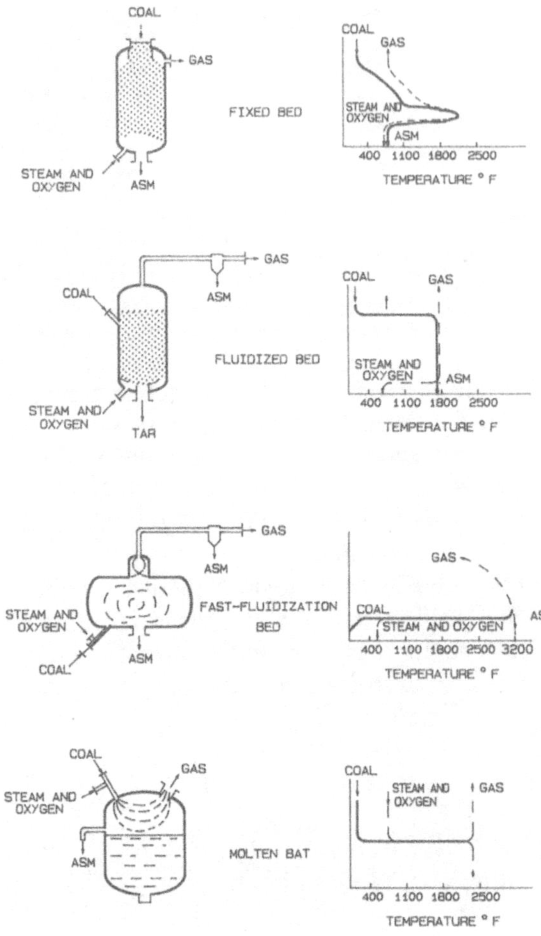

Figure 4.24. Coal gasification routes.

It is important to note that not just the temperature, but also the average residence time of gas–vapor molecules have an effect on produced gas composition. Coal is an agglomerate of carbon, heavy hydrocarbon compounds, ash, and other elements. When exposed to heat, the portion called volatile matter will evaporate and join the gas phase: if these vapors are not given enough time to react with steam, or if the temperature is not sufficiently high (say above 900°C), the vapors remain unreacted and will eventually condense when gas is cooled, forming tar and oils. For most applications tar and oils represent not just a reduction in gas yield but also a main process complication to deal with, since they tend to plug heat exchangers, pipes, and compressor blades and pollute condensed waters from coal gasification facilities.

The variety of elements present in coal results in complex side reactions of such elements between themselves and with combustion and gasification products. For instance, sulfur contained in coal reacts with oxygen, carbon monoxide, hydrogen, and water vapor to form gaseous compounds such as H_2S, COS, SO_2, and mercaptans. Similarly nitrogen and chlorinated compounds may react.

The presence of by-products such as H_2S, COS, HCl, and NH_3 in the synthesis gas poses serious problems not just for its utilization (such as in fuel cells) but also in the gasification and cooling equipment: due to high temperature, high partial pressure of hydrogen, and mechanical stress, the effect of such contaminants on every component of the coal gasification plant has to be carefully evaluated.

The specific characteristics of each coal, its ash composition, coal grindability, etc., all influence gas composition and may dictate a specific process selection. This aspect has a significant effect on potential utilization of coal gasification coupled with modular fuel cell systems. Every plant will require special design in order to accommodate for specific coal characteristics.

This may pose a limitation in the early development of small-size modularized coal-based fuel cell power plants. It is indeed very difficult to conceive such a plant in sizes below 100 MW. At such size the costs related to site-specific engineering may be compensated by economy of scale.

A coal-based fuel cell power plant is conceptually similar to a plant based on natural gas or liquid hydrocarbon feedstock. The fuel processor consists of a coal gasifier, followed by cooling and purification–CO conversion equipment (even when natural gas is utilized as the reformer) and these are coupled with desulfurization, shift conversion, and cooling equipment.

However, due to the peculiar composition of synthesis gas from a coal gasifier, which is rich in contaminants and may contain tars, oils, and ash, the type of treatment the gas needs to be subject to is much more complex. Shift conversion (for the further production of H_2 from CO and H_2O) can substantially occur, in reactors similar to those of steam-reforming plants. Ash, fines and other

Figure 4.25. 150-MW coal gasifier fuel cell power plant.

particulates, plus tar and oils must be removed, each at its own optimal temperature, in order to avoid plugging ducts, reactors, or exchangers.

Furthermore, air separation, water treatment, ash handling, and coal storage have to be included in coal gasification facilities. Such auxiliary equipment may represent over 25% of the investment cost for a coal gasification plant designed for, say, a 150-MW fuel cell stack system.

Overall, a facility such as that in Fig. 4.25, with a rated capacity of 150 MW, would require an investment in excess of $180 million, exclusive of fuel cell stack and power conditioner costs. This means that for such a coal gasification plant, including the fuel processing section, the power plant will cost over $1600 per kilowatt.

Since a large portion of such investment is for auxiliary equipment, it may be possible to reduce the cost by having coal-gasification-based fuel cell power plants built close to existing coal-handling facilities.

4.7. METHANOL STEAM REFORMING

For the manufacture of hydrogen-rich synthesis gas, there are two economically attractive routes. Natural gas is a relatively low cost source of energy, and a synthesis gas unit based on the natural gas steam-reforming route will have low variable operating costs but a relatively high investment cost.

Conversely, methanol is a relatively expensive source of energy when compared to natural gas. Thus, the methanol-reforming route for producing hydrogen-rich synthesis gas has relatively high variable operating costs but a much lower investment cost.

In the past, small hydrogen plants based on the catalytic reforming of methanol with steam have been available on the market. The main characteristics of these methanol-reforming plants is the simple reactor system for producing hydrogen, leading to a low investment cost. Another possible alternative for small-scale production of hydrogen is the catalytic dissociation of a methanol feedstock into hydrogen and carbon monoxide, commonly known as methanol cracking.

Hydrogen production by steam reforming of methanol may be a process attractive to small- and medium-scale users of hydrogen. As usual, the plant can be divided into two sections: (i) fuel processor and (ii) separation section. The first part of the unit is usually a tubular steam reformer for methanol. Due to the low temperatures involved, the construction of this equipment is simple, and the use of a thermal fluid as the heating medium is performed in one straightforward and compact reactor.

Liquid methanol and water are mixed, pumped to process pressure and vaporized in heat exchangers. In the tubular reformer the gas passes over a catalyst that promotes, at temperatures of 280–300°C, the reactions

$$CH_3OH \Leftrightarrow CO + 2H_2,$$

$$CO + H_2O \Leftrightarrow CO_2 + H_2.$$

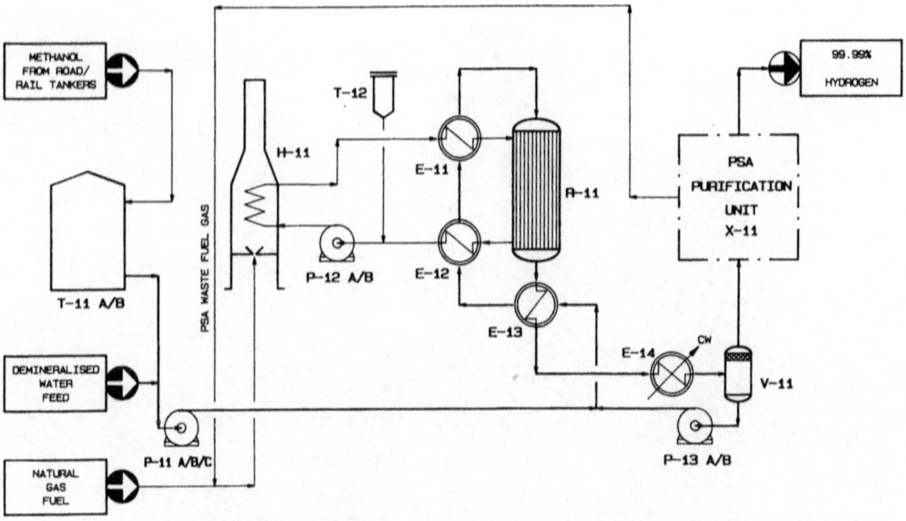

Figure 4.26. Simplified flow diagram of methanol reformer plant.

As reactor effluents, one finds H_2, CO_2, and small quantities of CO, CH_4, and CH_3OH in the gas. The condensable phase is water with some CH_3OH dissolved, which is recycled.

The second part of the unit produces the raw hydrogen, usually by applying the pressure swing adsorption technique (see Fig. 4.26).

4.8. MATERIALS FOR HYDROGEN SERVICES

The selection of materials for piping and equipment of fuel processing systems poses challenges similar to commercial hydrogen plants. The main concerns are hydrogen in syngas and carbon dioxide solubility in any condensates.

Hydrogen is a very peculiar gas, whose very small interatomic distance (0.074 nm) makes it very easy for hydrogen molecules to move within the crystal lattice of metals. Hydrogen is also highly reactive, and it can form hydrates with metals. It reacts with carbon atoms to form methane gas within the metallic structure, and it reacts with oxygen (often present in copper alloys) to produce water.

Hydrogen corrosion may take different paths, depending on operating conditions. As an example, blistering occurs preferably at low temperature and high partial pressure. It is caused by hydrogen diffusion within the metallic structure; when two atoms meet, they form a molecular hydrogen or react with carbon to form methane. The molecules may collect along the crystal boundaries. When the material is heated, pressure may build up, which will eventually crack the metal under moderate stress.

To prevent blistering, austenitic steels are employed because their homogeneous matrix structure makes it more difficult for two hydrogen atoms to migrate. In case of extreme pressures austenitic steel tends to transform into a martensitic form, which is less resistant to blistering, and special alloys then have to be employed.

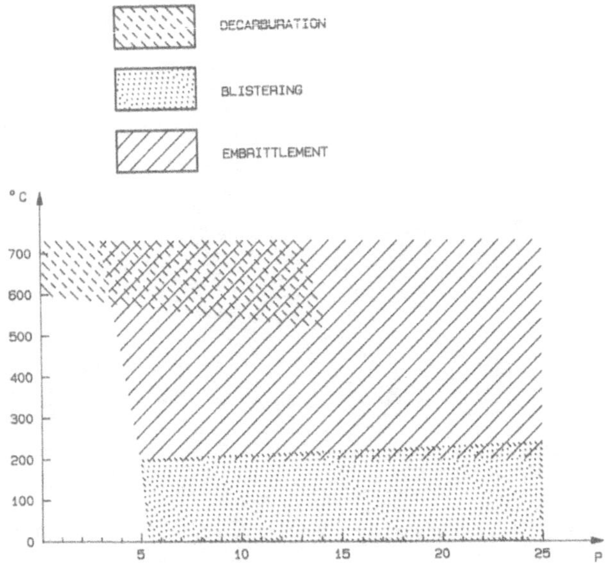

Figure 4.27. Blistering, embrittlement, and skin decarburation—typical operating conditions.

At higher temperature (>230°C) and high-pressure embrittlement takes place (e.g., carbon and titanium are very sensible to it), which is due to hydrogen reaction with metal atoms (to form hydrides) and with carbon (to form methane). Pressure builds up, and carbon subtraction results in reduced ductility.

Embrittlement is prevented by utilizing chromium and molybdenum alloys, where Cr and Mb stabilize carbon atoms, preventing methane formation. Molybdenum is specifically utilized for pressure vessels subject to slow depressurization cycles.

At high temperatures (>560°C) and low-pressure, decarburation takes place. It is due to the reaction on the metal surface of hydrogen with carbon atoms. Water vapor emphasizes this phenomenon, which can be prevented by chromium and molybdenum alloys.

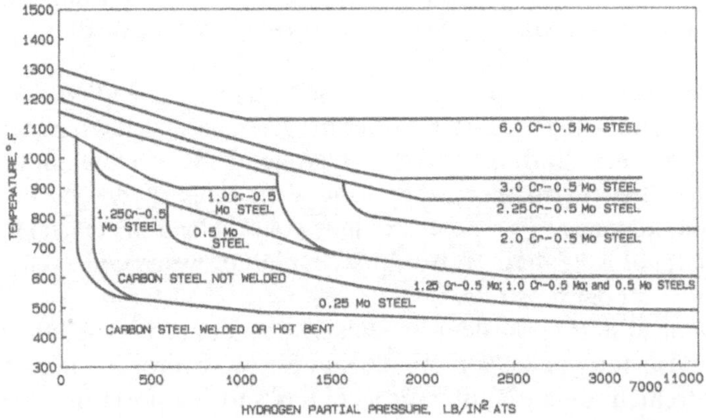

Figure 4.28. Nelson diagram for steel selection in hydrogen plant.

Figure 4.27 presents typical conditions for blistering, embrittlement, and skin decarburation. Figure 4.28 shows a Nelson diagram, which indicates proper material selection for each operating condition.

Fuel processing in fuel cell plants typically is carried out at much lower pressures than in standard hydrogen or syngas plants, and should reduce the risk of corrosion. Operating conditions of a fuel processing system can be seen on Nelson's diagram in Fig. 4.29, which can also be compared to a conventional hydrogen plant (Fig. 4.30). Comparing the two figures shows that fuel processors can still utilize common carbon steel where standard hydrogen plants cannot. It should be considered, however, that low alloys or stainless steels may still be required for high-temperature services due to better mechanical performance and because materials have to be designed for transient conditions.

Carbon dioxide solubility in water is shown in Fig. 4.31. Acid condensate is always present in a fuel processor's low-temperature condenser. To withstand acid corrosion stainless steel has to be employed.

Proper materials selection is also influenced by a third factor (besides corrosion resistance and mechanical characteristics)—cost. Figure 4.32 sum-

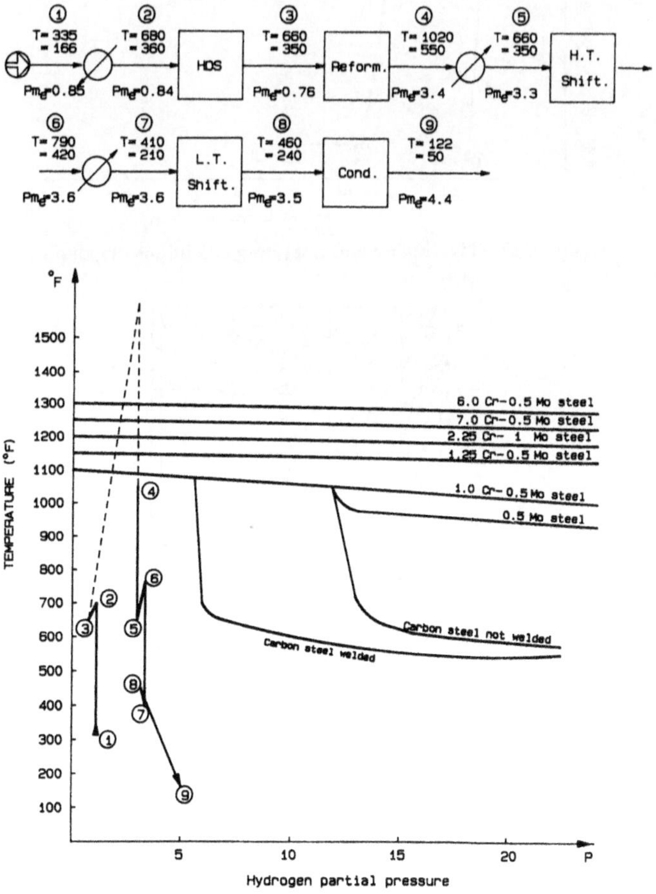

Figure 4.29. Fuel processing system operating conditions diagram.

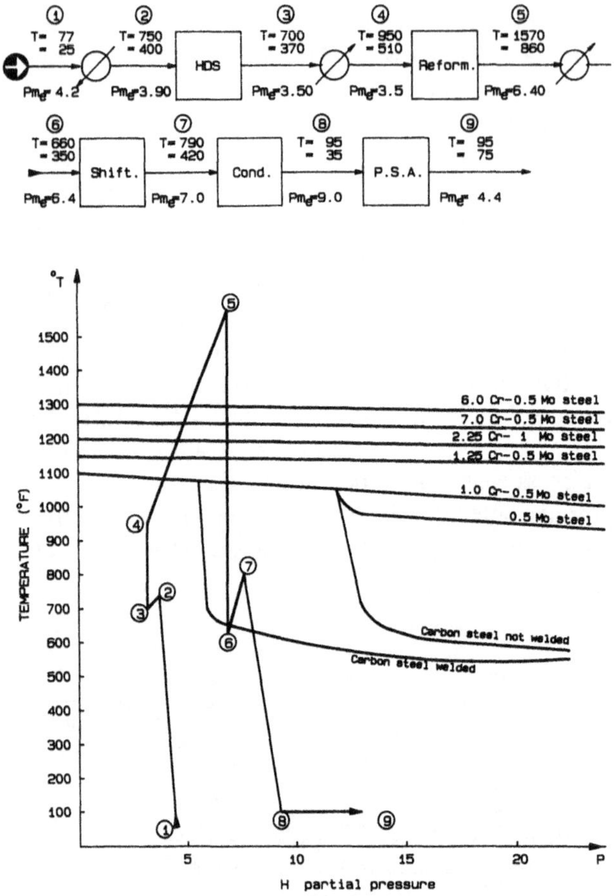

Figure 4.30. Hydrogen plant operating conditions diagram.

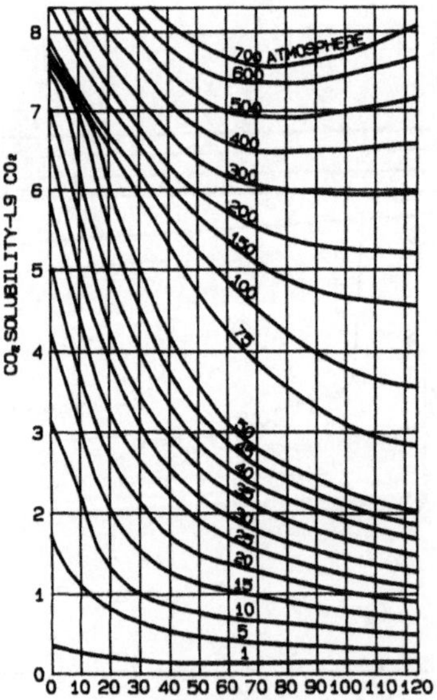

Figure 4.31. Carbon dioxide solubility in water.

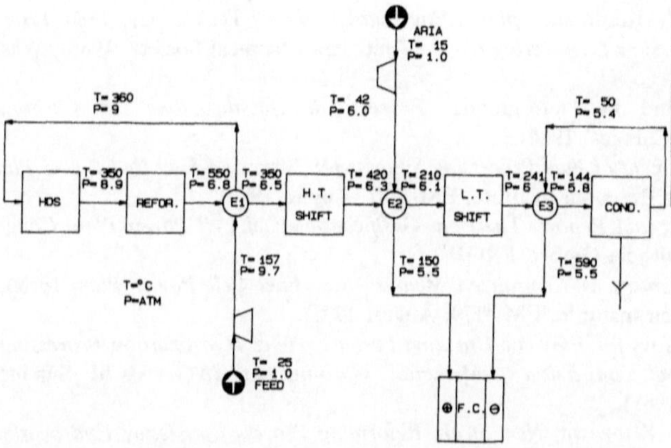

EQUIPMENT	MATERIAL	
DESULPHURIZER	A335/P11	
REFORMER TUBES	MANAURITE 36 X / A608/HK40	
H.T. SHIFT REACTOR	A335/P11	
L.T. SHIFT REACTOR	A335/P11	
DEGASIFIER	VESSEL A106/B	SEPARATOR A/TP316
CONDENSER	SHELL A335/P11	TUBES A312/TP316
HEAT EXCHANGER E1	SHELL A335/P11	TUBES A312/TP316
HEAT EXCHANGER E2	SHELL A335/P11	TUBES A312/TP316
HEAT EXCHANGER E3	SHELL A335/P11	TUBES A335/P22

Figure 4.32. FPS plan construction material.

marizes properties, chemical composition, and indicative cost ratio for different materials utilized in hydrogen plant services.

REFERENCES

1. M. Vincenzini *et al.*, *Growth and Photoproduction of* H_2 *by Rhodopseudomonas palustris under Light Dark Cycles* (Proc. 1986 Int. Congr. Renewable Energy Sources, Madrid, 1986).
2. W. Balthasar, *J. Hydrogen Energy* **9,** No. 8 (1984).
3. F. Giacobbe *et al.*, *Hydrocarbon Process.* **Oct.** (1985).
4. Logsjoen *et al.*, *Combined Cycle Power Plant Emissions* (Proc. EPA/EPRI Joint Symp. Stationary Combustion NO_x Control, Denver, CO, October 1980), EPRI WS 79 220, Vol. 1.

BIBLIOGRAPHY

W. B. Balthasar and D. T. Hambleton, in *2nd World Hydrogen Energy Conf. Proc.* (Zurich, August 1978), p. 1007–1028.

G. Baron, C. Hofke, and H. Vierratu, *The Lurgi Pressure Coal Gasification Process* (Proc. 3rd Int. Symp. Large: Chemical Plants, Antwerp, Oct. 1976), pp. 26–41.

R. E. Billings, *Hydrogen from Coal—A Cost Estimation Guidebook,* 1983.

H. G. Corneil, F. J. Heinzelmann, and W. S. Nicholson, *Production Economics for Hydrogen Ammonia and Methanol during the 1980–2000 period* (Exxon Research & Engineering Company, Linden, NJ, 1977), Report, No. BNL-50663.

D. T. Hooie and A. Kaufman, *Update: Engelhard Fuel Cell Technology, 1986* (*Proc. 21st Intersociety Energy Conversion Engineering Conf.*, American Chemical Society, 1986), Washington DC, pp. 1153–1155.

G. Iaquaniello and P. Pietrogrande, *Processi di Gassificazione del Carbone* (ICP, Rivista dell'Industria Chimica, 1984).

KTI Corp., *Assessment of Fuel Processing Systems for Dispersed Fuel Cell Power Plants* (prepared for Electric Power Research Institute, EM-1487, August 1980).

KTI Corp, *Confidential Report, The Coal Gasification–Fuel Cell Power Plant Concept* (prepared for Kohlegas Nordrhein GmbH, FRG).

KTI Corp., *Assessment of Centralized Medium Size Fuel Cell Power Plant* (prepared for Electric Power Research Institute, EM-4179, August 1985).

A. J. MacNab, *Alloys for Ethylene Cracking Furnace Tubes Hydrocarbon Processing.*

J. C. Selover, *Coal Conversion to Methanol, Gasoline and SNG* (AIChE Summer Nat. Meeting, Detroit, MI, 1981).

T. Tomita and M. Kitogawa, *New Steam Reforming Process for Heavy Hydrocarbons* (Achema 76, Frankfurt, Germany, 1976).

W. F. Van Weenem and J. Tielrooy, *How to Optimize Hydrogen Plant Designs* (AIChE Spring National Meeting, Anaheim, CA, 1982).

J. Voogd, (*Het Ingenieursblad* **42,** 579–585, 1973.

H. Yoon *et al., Energy Prog.* **5,** June 38–83 (1985).

Texaco Partial Oxidation

Dubois Eastman, *Ind. Eng. Chem.* **48,** 1118–1122 (1956).

L. W. der Haar, *Het. Ingenieursblad* **42,** 586–592, 1973.

H. Fricke, *Chem. Zeit.*, **96,** 123–124, 1972.

C. J. Kuhre and C. J. Shearer, *Hydrocarbon Process.* **Dec.,** 113–117 (1971).

G. J. v.d. Berg, E. F. Reinmuth and E. Supp, *Chem. Proc. Eng.* (**Aug.**, 53–57, 1970).

5

Characteristics of Fuel Cell Systems

A. J. Appleby

5.1. BACKGROUND

Fuel cell power systems possess certain generic characteristics, which may make them favorable for future power production compared with devices based primarily on rotating machinery using thermomechanical processes. Many of these operational characteristics of fuel cell systems are superior to those of conventional power generators. The most important are their potentially outstanding advantages compared to those of other existing or anticipated technology, namely thermodynamic efficiency, part-load characteristics, response time, emissions (including chemical emissions, noise, thermal emissions and visual or esthetic effects), modularity, and siting flexibility. Other factors which will affect their future economic viability in respect to their competition are expected to be their lifetime, on-line availability, reliability, start-up and shutdown characteristics, control, power conditioning, safety, materials, multifuel ability, and finally the waste disposal of their materials on dismantling of the plant. In this chapter, fuel cell systems will be characterized with respect to these factors. Finally, their overall economics, which will be largely dictated by their effective capital cost and lifetimes, will be considered in general terms.

5.2. EFFICIENCY

5.2.1. General Thermodynamics

The heat absorbed per mole of reactant at constant pressure for a real fuel cell can be expressed as

$$Q = \Delta H + W_{elec} \tag{5.1}$$

A. J. Appleby • Center for Electrochemical Systems and Hydrogen Research, Texas Engineering Experiment Station, The Texas A&M University System, College Station, Texas 77843-3402.

Fuel Cell Systems, edited by Leo J. M. J. Blomen and Michael N. Mugerwa. Plenum Press, New York, 1993.

where Q is the heat produced, ΔH is the reaction heat, and W_{elec} is the electrical work delivered by the cell, all in kJ/mol. Since heat is evolved in essentially all fuel cell reactions, Q will be negative (i.e., there will be a heat loss by the system). This expression can be expressed as

$$Q = T\,\Delta S - nF\left(\sum |\eta| + IR_{int}\right), \tag{5.2}$$

where $\Sigma\,|\eta|$ is the sum of the activation and concentration polarization, I is the cell output current, R_{int} is the sum of the ohmic resistances, n is the number of electrons involved in the overall process, and F is the Faraday (96,500 coulombs per equivalent). The rate of heat loss per mole in the cell also can be expressed as an equivalent power loss in watts (thermal), giving the cell cooling requirement

$$P_{th} = I\left\{\frac{T\,\Delta S}{nF} - \sum |\eta|\right\} - I^2R_{int}. \tag{5.3}$$

The theoretical reversible (or zero current) cell voltage E_T at operating temperature T is given by the free energy of the process ($-\Delta G_T$, the maximum work that can be performed by the process at the temperature of operation), expressed in electron volts (eV, volts per electron or per equivalent). Thus, the numerical value of the Faraday in joules per equivalent (96.5 kJ/equ) is equal to 23.06 kcal/equ, or 1.0 eV; i.e., $-\Delta G_T$ in kJ/mole $= nFE_T$, where n is the number of electrons involved in the overall process. This theoretical reversible potential is related to the standard state potential E_{T° by the Nernst equation:

$$E_T = E_{T^\circ} + \frac{RT}{nF}\ln\left[\frac{\Pi(r)}{\Pi(p)}\right], \tag{5.4}$$

where R is the gas constant in J/equ and $\Pi(r)$, $\Pi(p)$ are the products of the activities of the total number of equivalents of anode and cathode reactants and products involved in the overall n-electron process. Finally, the overall cell potential, V, of an operating fuel cell is given by

$$V = E_T - \sum |\eta| - IR_{int}, \tag{5.5}$$

where $\Sigma\,|\eta|$ is an increasing function of current density (see Chapter 3).

5.2.2. *Fuel Utilization*

All practical fuel cell systems intended for multikilowatt or multimegawatt power generation use hydrogen as the fuel, even if the feedstock is a carbon compound (e.g., a hydrocarbon, alcohol, or coal). The latter must be converted to hydrogen and carbon dioxide via an endothermic reaction with steam followed as necessary by water–gas shifting. In a real fuel cell, the objective is to convert as much as possible of the hydrogen in the anode feedstock to work, in order to obtain the maximum overall conversion efficiency. The same will be true for the cathode oxidant stream, unless it is atmospheric oxygen in the form of air. Even if air is used, some net oxygen conversion will take place, since the practical circulation of large amounts of air requires pumping work, which results in an energy loss. Thus, unless process air cooling is used, conversion (utilization) will be kept at the maximum consistent with good electrochemical performance. In the liquid-cooled phosphoric acid system, 50% oxygen utilization is normally chosen.

In typical systems, conversion of reactants to products occurs as the reactants move across the face of the cell, resulting in a higher current density on the inlet side and a lower current density on the outlet side. Since the cell must have high electronic current-collector conductivity, the electrodes are equipotential surfaces. The maximum theoretical potential the cell can have will be the lowest Nernst potential of the cell. In systems with coflow of reactants, this will occur at anode exit. While the situation in systems with counterflow, crossflow or more complex geometry is not so obvious, in general the cathode reactant utilization with atmospheric oxygen (i.e., with air feedstock) is usually rather small, so that the anode outlet Nernst potential is a good approximation to the maximum possible cell potential. Since the difference between the inlet and exit Nernst potentials is strongly temperature-dependent due to the presence of the RT/F term in (5.4), and since $-\Delta G_{T°}$ (i.e., $E_{T°}$) falls with temperature, the overall cell performance will be a strong function of the operating temperature of the cell. This is illustrated by Fig. 5.1, which shows the theoretical open-circuit potential with the hydrogen and oxygen reactants and gaseous water products in their standard states as a function of temperature, as well as the exit Nernst potentials, assuming hydrogen utilizations of 85% and 90% (see following) and oxygen

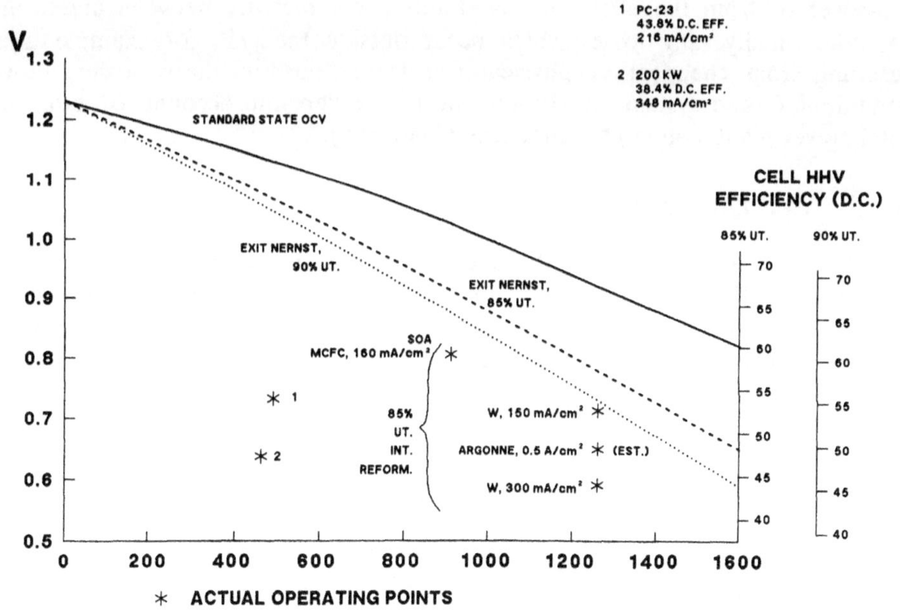

Figure 5.1. Theoretical open-circuit potentials of (a) hydrogen–oxygen fuel cells as a function of temperature (Kelvin) with reactants and products at 1 atm, and (b) exit Nernst potentials with hydrogen at 0.10 atm and 0.15 atm; water vapor at 0.90 atm and 0.85 atm (i.e., 90% and 85% hydrogen utilization), with atmospheric oxygen at 0.10 atm (i.e., 50% utilization).

utilization from air of 50%. It is clear from Fig. 5.1 that the high-temperature systems are penalized by reversible thermodynamics.

If pure hydrogen fuel is used, the system may theoretically be designed so that all the fuel can be consumed in the cell. For example, an acid electrolyte cell, in which hydrogen first ionizes at the anode to give H^+ ions, which are subsequently consumed with oxygen at the cathode to give water as product, the anode can theoretically be given a "dead-ended" hydrogen supply. In contrast, an alkaline electrolyte system, in which ionic conduction is from cathode to anode via OH^- ions, gives product water mostly at the anode. If this is condensed in a recycle feedback loop, hydrogen utilization could theoretically be 100%. However, in both systems, any inert impurities in the gas will accumulate at the anode, and must be bled off. This bleeding can be carried out by either occasional purging,[1] or by a continuous bleed.[2] Either approach will result in a small loss of hydrogen, so utilization will be less than 100% in practice.

If the hydrogen in the feedstock is diluted with inert gases which cannot be easily removed in a feedback loop, its utilization may be limited as its dilution increases either by the effect of reversible thermodynamic potential losses, or by the effect of slower kinetics, or both. Thus, the hydrogen in methane reformate after water-gas shifting, which is initially a 4:1 hydrogen: CO_2 mixture containing some steam, can only be effectively used at about 85% utilization in a low-temperature (acid) fuel cell, or perhaps 90% in a high-temperature (molten carbonate or solid oxide) fuel cell.

Since in-cell fuel utilization will never be 100% in practice, the in-cell efficiency (the voltage efficiency expressed in terms of the fuel actually consumed in the cell) must be multiplied by the utilization to give the gross cell efficiency. This will also include any parasitic losses, for example, those resulting from gas crossover or from the effect of any shunting conductivity between anode and cathode. Finally, any losses which occur outside the cell, for example those resulting from chemical or physical transformations, auxiliary power, power conditioning, and system integration, must be taken into account to obtain the total power plant system efficiency (see Chapter 6).

5.2.3. Cell Efficiency

The in-cell efficiency of the fuel cell is given by the practical voltage produced in the cell (V) under real utilization conditions, divided by the theoretical value. The latter may be expressed in various ways. As an illustration, it might be expressed on a free-energy (ΔG) basis. Since $V < E_{rev}$, the fuel cell *voltage efficiency*, or *free-energy efficiency*, ε_V, can be expressed as

$$\varepsilon_V = \frac{V}{E_{rev}}, \tag{5.6}$$

where E_{rev}, the reversible cell potential (equivalent to $-\Delta G$ for the reaction), can be defined in various ways, depending on the circumstances. For example, E_{rev} can be put equal to E_T in (5.4), i.e., the maximum theoretical open-circuit voltage of the cell at operating temperature T. This E_T-value may be expressed as the inlet Nernst potential for the unconverted reactants, so that the ideal efficiency of a reversible, zero-conversion fuel cell at operating temperature T will be 100% compared with the available free energy of reaction at that temperature. Perhaps more logically, since conversion of fuel is the object of a real cell, the value of E_T may be defined as the exit Nernst potential for converted reactants, so that ε_V indicates the *irreversible* losses in the cell. For modeling purposes (i.e., to define the variation of $\sum |\eta|$ in (5.5) with reactant concentration and temperature) the local value of E_T, corresponding to the local reactant and product concentrations in the cell, must be used to indicate the *local* irreversible losses.

While E_T falls with the operating temperature of the cell (Fig. 5.1), the irreversible losses relating V and E_T in (5.5) typically show an even stronger temperature dependence. Hence, ε_V expressed as V/E_T will normally *rise* with increasing operating temperature. However, if ε_V is to be compared with a value obtainable with the same fuel used in a different type of fuel cell operating at a lower temperature (for example if an alkaline fuel (AFC) operating on hydrogen at low temperature and 0.8 V is to be compared with a solid oxide fuel cell (SOFC) at 1273 K and 0.7 V), then it is more convenient to use the standard state value of E_T, namely, E_{298}. This obviously varies according to whether the reaction product (typically water) is in the form of water vapor or liquid water. The corresponding free-energy efficiencies are those corresponding to water vapor product (the lower free-energy value, or LFEV, efficiency) or to liquid water as product (the higher free-energy value, or HFEV, efficiency).

All practical fuel cell systems planned for electric utility application liberate water from the cell as vapor, and strictly speaking they therefore operate on the LFEV of the fuel (i.e., they do no work by condensing water vapor). In consequence, the theoretical open-circuit potential under exit reactant conditions in the phosphoric acid fuel cell (PAFC) operating at 1 atm pressure with "normal" reformate fuel gases (see Chapter 6) and utilizations is about 1.08 V at 200°C (473 K), and varies only by a negligible amount (about 20 mV) at an operating pressure of 8 atm. Thus, at an operating potential of 0.73 V, irreversible losses (almost entirely due to irreversibility of the electrochemical oxygen reduction kinetics) are 0.35 V, and the free-energy efficiency (ε_V) based on the exit Nernst potential, E_T, is 66%. In the atmospheric pressure molten carbonate fuel cell (MCFC) operating at 650°C (923 K), the exit Nernst potential has fallen to close to 0.90 V, yielding 84% free-energy efficiency at an operating potential of 0.76 V. In an SOFC operating at 1000°C, the exit Nernst potential would only be about 0.775 V, so the free-energy efficiency would be 89% in an SOFC operating at 0.69 V. The corresponding conventional free-energy efficiencies based on the maximum attainable value of E_{rev} with hydrogen fuel under standard temperature conditions (1.229 V with hydrogen and oxygen in their standard states, with liquid water as product at 298 K) would be 59.4%, 61.8%, and 56.1% respectively. Conventional free-energy efficiencies as high as 60–80% have been obtained for fuel cells,[3] and, in theory at least, there are no reasons why the maximum intrinsic free-energy conversion efficiencies (see Table 5.1) cannot be closely approached under optimal operating conditions and with steady improvements in electrocatalysis.

In practical operating systems, part of the reactants may participate in a nonproductive side reaction. While this does not occur with hydrogen and oxygen in a well-designed cell, it is seen, e.g., in the autodecomposition of hydrazine to nitrogen and hydrogen in hydrazine fuel cells, where it results in incomplete

Table 5.1. Standard Heats and Free Energies of Combustion (or reaction) for Candidate Fuels and Oxidants, and Corresponding Theoretical Standard Reversible Potentials and Theoretical Free-Energy Efficiencies ($\Delta G°/\Delta H°$)

Fuel	Reaction	n	$-\Delta H^{\circ a}$	$-\Delta G^{\circ a}$	$E^{\circ}_{rev}{}^{b}$	%
Hydrogen	$H_2 + 0.5O_2 \rightarrow H_2O(l)$	2	286.0	237.3	1.229	82.97
	$H_2 + Cl_2 \rightarrow 2HCl_{(aq)}$	2	335.5	262.5	1.359	78.33
	$H_2 + Br_2 \rightarrow 2HBr_{(aq)}$	2	242.0	205.7	1.066	85.01
Methane	$CH_4 + 2O_2 \rightarrow CO_2 + 2H_2O_{(l)}$	8	890.8	818.4	1.060	91.87
Propane	$C_3H_8 + 5O_2 \rightarrow 3CO_2 + 4H_2O_{(l)}$	20	2221.1	2109.3	1.093	94.96
Decane	$C_{10}H_{22} + 15.5O_2 \rightarrow 10CO_2 + 11H_2O_{(l)}$	66	6832.9	6590.5	1.102	96.45
Carbon monoxide	$CO + 0.5O_2 \rightarrow CO_2$	2	283.1	257.2	1.066	90.86
Carbon	$C + 0.5O_2 \rightarrow CO$	2	110.6	137.3	0.712	124.18
	$C + O_2 \rightarrow CO_2$	4	393.7	394.6	1.020	100.22
Methanol	$CH_3OH + 1.5O_2 \rightarrow CO_2 + 2H_2O(l)$	6	726.6	702.5	1.214	96.68
Formaldehyde	$CH_2O_{(g)} + O_2 \rightarrow CO_2 + H_2O_{(l)}$	4	561.3	522.0	1.350	93.00
Formic acid	$HCOOH + 0.5O_2 \rightarrow CO_2 + H_2O_{(l)}$	2	270.3	285.5	1.480	105.62
Ammonia	$NH_3 + 0.75O_2 \rightarrow 1.5H_2O_{(l)} + 0.5N_2$	3	382.8	338.2	1.170	88.36
Hydrazine	$N_2H_4 + O_2 \rightarrow N_2 + 2H_2O_{(l)}$	4	622.4	602.4	1.560	96.77
Zinc	$Zn + 0.5O_2 \rightarrow ZnO$	2	348.1	318.3	1.650	91.43
Sodium	$Na + 0.5H_2O + 0.25O_2 \rightarrow NaOH_{(aq)}$	1	326.8	300.7	3.120	92.00

[a] kJ/mol (1 kJ/mol = 4.184 kcal/mol. 23.06 kcal/electron = 1 V).
[b] Volts.

conversion of reactants to desired products. This may be accounted for by the *faradaic efficiency*, or *current efficiency*, ε_f. This will also include the in-cell parasitic losses due to some inevitable gas crossover, together with any voltage losses between the anode and cathode, (for example, electronic conductivity in the SOFC or shunt current losses in bipolar alkaline fuel cell assemblies with circulating electrolyte). For most practical utility fuel cell systems, the faradaic efficiency is very close to unity.

The overall conventional free-energy conversion efficiency, $\varepsilon_{overall}$, in the cell is given by

$$\varepsilon_{overall} = \varepsilon_{max}\varepsilon_V\varepsilon_F u, \tag{5.7}$$

where ε_V is V/E_T, ε_{max} is E_V/E_{298}, and u is the fuel utilization in the cell.

Instead of using free-energy efficiencies, it is more conventional, if thermodynamically less rigorous, to use efficiencies in terms of *heat of combustion* of the fuel, since the use of these values is standard practice when combustion machinery supplies the power. For hydrogen and fossil fuels, depending on whether water or steam is the reaction product, the heat of combustion with condensed water as the product (higher heating value, HHV) efficiency or that for gaseous water as the product (lower heating value, LHV) can be used. The HHV efficiency at normal temperature uncorrected for the pressure of the feedstock has been conventionally used in United States practice for purposes of comparison with other technologies. The $-\Delta H°$ (HHV) value is 1.48 eV for liquid water product, whereas 1.25 eV is the LHV value for gaseous water. Figure 5.1 also shows gross theoretical cell HHV efficiencies (i.e., $-nuFE_{exit}/\Delta H_{298}$, with ε_F taken as unity) as a function of temperature for hydrogen-containing fuel mixtures. It can be seen that theoretical gross cell HHV efficiencies fall from a maximum value of 69.5% at 298 K to only about 45.5% at 1273 K due to the effect of fuel utilization.

To obtain the overall system efficiency, the gross *real* cell efficiency (i.e., the value $-n(\varepsilon_F u)FV/\Delta H_{298}$), must be multiplied by the efficiency of any transformations required in the fuel pretreatment system, including parasitic power losses (for example, in steam-reforming natural gas to produce the required hydrogen), and that for any electric transformations required, e.g., dc to utility-quality ac power conversion. Any further electricity obtained by the use of waste heat in bottoming cycles must be accounted for, and so must any transformations in fuel system integration, that is, the processes involved in combining individual reactions and the corresponding heat exchanges in the so-called process flowsheet.

To obtain maximum efficiency, an ideal fuel cell should be designed to use pure reactants, with the removal of the product in a pure form, so that the maximum free energy available can be exploited. In this way, the inlet and exit Nernst potentials will be as close to each other as possible, and as much of the reactants as possible may be used, thus satisfying the optima of reversible thermodynamics. This obviously cannot be done with feedstocks containing hydrogen mixed with inert materials, such as carbon dioxide and nitrogen, nor is it possible in systems where no phase change between reactants and products occurs. However, it can be arranged in gaseous pure hydrogen–oxygen cells, which either product liquid water directly, or can separate gaseous water either via direct evaporation from a circulating electrolyte (Chapter 7 and Ref. 4) or through special membranes incorporated in each cell (certain AFC systems, Chapter 7 and Refs. 4,5). An example of the first type of cell, which produces liquid water product, is the solid polymer electrolyte–proton exchange membrane system (SPE–PEM, Chapters 2 and 11, Ref. 5). Such systems can at least in principle approach theoretical ideality from the viewpoint of reversible cell thermodynamics and fuel utilization.

An SPE–PEM cell operating on pure hydrogen produces water at the cathode, since it operates via H^+ ion conduction. Its anode can in principle be run dead-ended,[5] though it requires humidification to ensure electrolyte membrane conductivity. The product water, together with water transported through the membrane by protons, can be removed from the cathode, for example by the use of a physical process such as wicking. The AFCs are OH^- conductors, and produce product water at the anode, so the cathode can be run dead-ended with pure oxygen, as for the anode of the AFC described previously. However, for the pure hydrogen PEM anode, or the pure oxygen AFC cathode, a purge or bleed is necessary to remove any trace inert impurities which accumulate at the electrodes, as discussed earlier. The AFC anode produces gaseous water, which can be removed by condensation in a feedback loop. To obtain the maximum cell voltage, utilization per pass can thereby be kept comparatively low, and very nearly 100% of the hydrogen can be consumed.

5.2.4. Carnot Machines and Fuel Cells

Fuel cell power sources are conventionally said to be non-Carnot-limited, and therefore are more efficient than heat engines. A Carnot-limited engine uses an "ideal" working fluid whose heat storage properties depend only on temperature. This statement requires some caution, since it has arisen in the literature from an *electrochemical* consideration of the processes occurring in the fuel cell stack itself. By stack, we mean the bipolar battery of separate fuel cell elements, called by many developers the *power module.* The latter directly converts all or part of the available free energy in the fuel at the stack mean operating temperature into low-voltage dc electricity. Fuel cells are not Carnot-limited machines in the sense that they operate directly on available chemical free energy rather than on heat. However, this does not necessarily mean that they have higher overall efficiencies than heat engines. In fact, they are subject to thermodynamic limitations similar to those in heat engines, as can be shown by the following argument.

By definition, for any chemical process,

$$\Delta G_T = \Delta H_T - T\,\Delta S_T, \tag{5.8}$$

where ΔG_T, ΔH_T are the standard free energy and heat of reaction at temperature T (K) and ΔS_T is the corresponding change in entropy. Let us consider a combustion process with the reactants initially in their standard states at a given temperature, which is followed by a spontaneous reaction, leading to the production of heat, which is further followed by cooling of the product back down to the starting temperature, with transfer of the heat of reaction to a working fluid. Combustion will occur at temperature $T_{(C)}$, where $\Delta G_{T_{(C)}}$ is equal to zero, i.e., $T_{(C)} = \Delta H_{T_{(C)}}/\Delta S_{T_{(C)}}$. If this heat can be used in a Carnot-limited heat engine at a source temperature equal to $T_{(C)}$, which rejects heat to a lower temperature heat sink at $T_{(S)}$, then the maximum thermal efficiency of this heat engine can be calculated. If the system is ideal, i.e., if the specific heat of the working fluid at constant pressure (C_p), is temperature-independent, then the heat entering the engine (assuming an ideal system), will be $T_{(C)}C_p$, and that exiting will be $T_{(s)}C_p$. The difference, amounting to the useful work provided by the process, is $T_{(C)} - T_{(S)}/T_{(C)}$, which is the expression known as Carnot's theorem. However, this derivation is simplistic, since it assumes that reactants at $T_{(C)}$ undergo isothermal combustion. The reactants are normally only available at a lower temperature (for example, $T_{(S)}$), and the reaction products must similarly be used to raise the temperature of the working fluid from $T_{(S)}$ to $T_{(C)}$. Assuming reversible heat transfer from the products of combustion to the working fluid with C_p constant, the source temperature T will be reduced from $T_{(C)}$ by the ratio $\Delta H_{T_{(S)}}/\Delta G_{T_{(C)}}$. Thus the source temperature, T, will in fact become $\Delta H_{T_{(S)}}/\Delta S_{T_{(C)}}$. From (5.7), $T_{(S)} = (\Delta H_{T_{(S)}} - \Delta G_{T_{(S)}}/\Delta S_{T_{(S)}}$. Substituting these temperatures into the Carnot efficiency expression, $1 - T_{(S)}/T$, we obtain the following expression for the maximum thermal efficiency, $\varepsilon_{T_{(C)}, T_{(S)}}$, of the heat engine operating at any spontaneous heat of combustion:

$$\varepsilon_{T_{(C)},T_{(S)}} = 1 - \frac{\Delta S_{T_{(C)}}(\Delta H_{T_{(S)}} - \Delta G_{T_{(S)}})}{(\Delta S_{T_{(S)}}\Delta H_{T_{(S)}})}, \tag{5.9}$$

where $\Delta G_{T_{(S)}}$ and $\Delta H_{T_{(S)}}$ are the actual free energy and heat of the combustion reaction at the heat sink temperature, $T_{(S)}$, at the given gas composition. If $\Delta S_{T_{(S)}}$ were equal to $\Delta G_{T_{(S)}}$, this would become equal to $\Delta G_{T_{(S)}}/\Delta H_{T_{(S)}}$. Thus, the efficiency of a Carnot engine operating at the spontaneous combustion temperature of the fuel–oxidant mixture is fundamentally the same expression as that conventionally used for the ideal thermal efficiency of an electrochemical fuel cell, i.e., $\Delta G_{T_{(S)}}/\Delta H_{T_{(S)}}$, where the $\Delta G_{T_{(S)}}$ and $\Delta H_{T_{(S)}}$ values now refer to the free energy and enthalpy of the overall reaction, for the actual chemical potentials of the reactants and products, and for zero net (i.e., reversible) transformation, at $T_{(S)}$, the isothermal cell operating temperature.

In reality, the perfect Carnot engine efficiency will often be marginally different than that of the perfect fuel cell operating on the same fuel, since $\Delta S_{T_{(C)}} = \Delta S_{T_{(S)}} + \Delta C_p \ln T_{(C)}/T_{(C)}$ where ΔC_p, assumed here to be temperature-independent, is the molar specific heat change for the reaction. This is a result of the enthalpy required to raise the temperature of the fuel and oxidant from its available temperature (which was assumed above to be $T_{(S)}$ to $T_{(C)}$ for spontaneous combustion). This enthalpy may be negative or positive, depending on the value of the ΔC_p value for the reactants compared with that of those for the products, whose sensible heat is assumed to be exchanged from $T_{(S)}$ to $T_{(C)}$ and from $T_{(C)}$ to $T_{(S)}$, respectively.

In a fuel cell operating at $T_{(S)}$, no such heat exchange between reactants and products is required. However, it would be required if the fuel cell operated at the higher temperature, T, above $T_{(S)}$ at which the fuel and oxidant supply were initially available. However, in this case the heat rejected by the ideal reversible fuel cell (i.e., $T\,\Delta S$) can be used to supply this heat. Taking a simplifying case, if the ΔC_p values for the gaseous reactants and products are the same, the ideal Carnot heat engine efficiency with a heat source at the spontaneous fuel combustion temperature is *exactly* equal to the thermal efficiency of an ideal fuel cell. The reason why the two expressions are the same is that both the electrochemical converter and the heat engine are thermodynamic black boxes, which transform chemical free energy into useful work. No more work can be produced by the converter than the available free energy of the incoming fuel. The efficiency of the converter can, however, be defined in various ways. The traditional one dates from the steam engine era, and it considers the efficiency to be the "work" the fuel makes available divided by its heat of combustion. As we have seen, the latter can be based on the LHV or of the fuel.

The expression *work* is traditionally somewhat loosely defined. *Energy* is often defined as work, or the ability to perform work, which is not thermodynamically correct. It is better defined as a quantity possessed by masses in motion, such that it is always conserved by the system of masses. In contrast, *momentum* is conserved by the individual masses, and it depends directly on the mass and on its degree of motion, or velocity. Hence, energy is the integral of momentum. To more clearly define work, as distinct from energy, we introduce the concept of *exergy,* which is work, or the ability to produce or be converted to work. Thus, *potential energy* in the classical physical sense is exergy, as is the chemical free energy (with the sign changed) for the combustion of a fuel. Energy is conserved in all processes (the first law of thermodynamics), whereas exergy is only conserved in processes which are by definition reversible. Real processes are always irreversible, so that in these exergy is partly consumed to give ethalpy. This is another way of expressing the second law of thermodynamics, formulated by Gibbs in 1873.[6] The concept of exergy was first defined by Rant in 1953,[7] and more generally by Baehr in 1965.[8]

The exergy of a system is determined by its displacement from thermodynamic equilibrium. Since $dG = V\,dP - S\,dT + \sum\mu_i\,dn_i$, where V and P are volume and pressure, respectively, and μ_i is the chemical potential of the ith chemical component with a number of molecules equal to n_i, it follows that the exergy E of a system going between an initial state and a reference state (subscript 0)

$$E = S(T - T_0) - V(P - P_0) - \sum n_i(\mu_i - \mu_{i0}). \tag{5.10}$$

The symbol E for exergy should not be confused with the same symbol for internal energy in classical thermodynamics. Similarly, the available exergy in a flow of heat is equal to

$$E = (H - H_0) - T_0(S - S_0) - \sum \mu_i(n_i - n_{i0}). \tag{5.11}$$

From Ref. 9 it is apparent that a quantity of available heat equal to ST at the source temperature T, with no other system changes, leads to the Carnot efficiency given by $(T - T_0)/F$. Similarly, the exergy efficiency of a quantity of heat equal to $\int C_p\,dT$ integrated between the limits T and T_0 is given by $[T - T_0 - T_0\ln(T/T_0)]/(T - T_0)$, since $H - H_0$ is $\int C_p\,dT$, and $S - S_0$ is $\int C_p\,d\ln T$, both integrated between the same limits. The generality of the exergy method provides a powerful tool to estimate the free energy available in generalized reversible processes, always remembering that no real process will ever be reversible.

It is evident from the above discussion that so-called Carnot-limited machines are not in fact limited by free-energy considerations as such. They are rather limited by two factors: the normal irreversibilities in all real processes, and materials considerations. Available materials do not allow heat to be transferred to the working fluid and then to the moving parts in a Rankine or Stirling cycle at the spontaneous combustion temperature of the reaction, which will be more than 4000 K for pure hydrogen and oxygen. The gases must therefore be diluted and allowed to cool before heat transfer occurs. Similarly, in Brayton, Diesel, and Otto cycle machines, air is used to cool the combusted gases before they contact working parts. This represents by far the most important loss from the ideal Carnot efficiency value. For a hydrogen–oxygen system this is theoretically equal to 83.1%, based on heat rejection at 298 K with liquid water product (i.e., based on the higher heating value, HHV, heat, and free energy of combustion). This assumes that the heat source is the real flame combustion temperature for pure hydrogen and oxygen at 1 atm pressure.

In practice, a steam Rankine cycle is limited by the corrosion of the materials used at maximum turbine inlet temperatures of about 923 K. Heat rejection must be somewhat above ambient, at about 313 K, giving a theoretical Carnot efficiency of 66%. However, the Rankine cycle actually used in these engines is not ideal. In addition, heat transfer is not a reversible process, but *requires* temperature gradients to be effective. Further, other losses, such as friction, occur. In consequence, the *practical* maximum efficiency of a typical real cycle is less than two thirds of the theoretical Carnot efficiency, or just over 40% (HHV). The latest generation of gas turbines, such as the 150-MW Westinghouse 501F single-cycle machine, use air-cooled turbine blades with inlet temperatures of 1260°C (2300°F). Assuming ideal adiabatic expansion, this should result in a Carnot efficiency of 58%. Practically, the machine is capable of a heat rate of 9500 Btu/kW-h (10,000 KJ/kW-h) in LHV terms, corresponding to 60% of the theoretical value, or 35.9% overall. Since product water cannot be condensed in a simple-cycle gas turbine, it is customary to express efficiencies for such generators in terms of the LHV of the fuel. If a condensing steam cycle is added operating on the heat available in the exhaust via a heat exchanger, improved efficiencies can be obtained. In this case, a heat source temperature (the turbine exhaust) of 642 K is assumed, with a heat sink at 313 K, operating on the 64.1% of the LHV of the fuel remaining in the gas stream. Such a small, relatively low temperature steam cycle will operate at about 45% of the theoretical Carnot efficiency, so that a further 14.6% of the LHV of the fuel can be converted to work, giving an overall efficiency of 50.5% (LHV) in a combustion-turbine combined cycle (CTCC) unit developing 210 MW.

The record for efficiency in single-cycle heat engines occurs in systems of closed-circuit type, such as the hydrogen–oxygen turbine, consisting of a rocket motor as a combustion chamber together with a condensing steam turbine, with steam injection into the system for cooling purposes. The steam is raised by waste heat from the combustor. Such a system has been developed by the General Electric Co. (HOTSHOT, hydrogen–oxygen turbine superhigh operating temperature) and by Rocketdyne (NPNE, nonpolluting noiseless engine). These may be capable of an HHV efficiency apparently approaching 55% (65% LHV). It is to systems like this, and to the CTCC, that the fuel cell should be compared from the efficiency viewpoint.

Like a heat engine, the fuel cell cannot use all of the free energy available in the fuel, because of inevitable inefficiencies. These are analogous to the irreversible heat losses in a thermal engine. However, the losses due to the irreversible $T\,\Delta S$ terms in thermal engines are normally greater than those in practical fuel cells. An idealized practical engine is a combination of a high-temperature fuel cell with a combined cycle. In this, the fuel cell acts as a topping cycle at the thermal engine source temperature, so that the waste heat present in the fuel cell exhaust gases can be recovered. The high operating temperature of the fuel cell reduces irreversible fuel cell losses in the cell stack, and addition of the thermal cycle compensates for the lower free energy available from the fuel at the high fuel cell operating temperature. While the system can be treated in an exact manner using the exergy approach, for heuristic purposes it suffices to use the simplifying assumption that ΔC_p for the fuel cell reaction is zero. Then the theoretical thermal efficiency (ε_T) of the fuel cell alone is

$$\varepsilon_T = \frac{\Delta G_T}{\Delta H}, \tag{5.12}$$

where ΔG_T is the free energy (exergy) available in the fuel used at the operating temperature T of the fuel cell (i.e., at the heat source temperature of the thermal engine in the bottoming cycle) and ΔH is the heat of combustion of the fuel, generally defined at 298 K. A generalized formula for the maximum work available in any energy conversion device, whether it be a fuel cell operating alone, an ideal thermal engine operating at the maximum theoretical temperature (the combustion temperature of the fuel), or a fuel cell and thermal engine combination, is given by (5.9), resulting in

$$\varepsilon_T = \frac{\Delta G_{(S)}}{\Delta H_{(S)}}, \tag{5.13}$$

where the ΔG and ΔH values are those at the heat sink temperature of either ideal thermal engine operating alone, or of the bottoming thermal engine operating with a fuel cell. Alternative, they are those of an ideal fuel cell operating alone at the sink temperature of the above thermal engines. A corresponding expression for a Carnot-limited thermal engine operating between heat source and heat sink temperatures T and $T_{(S)}$ is

$$\varepsilon_T = \frac{\Delta G_{(S)} - \Delta G_{(T)}}{\Delta H - \Delta G_{(T)}} = \frac{T - T_{(S)}}{T}. \tag{5.14}$$

At the spontaneous fuel combustion temperature, $\Delta G_{(T)}$ is zero and (5.14) becomes identical to (5.13). If a fuel cell operating at T is used as a topping cycle for the thermal engine, its theoretical efficiency is given by (5.12), and the fraction of waste heat available for further conversion at T is $1 - \Delta G_T/\Delta H$. The overall theoretical efficiency of the combination is given by the sum of the work produced in the fuel cell [(5.13)] and that produced in the bottoming cycle [(5.14)], multiplied by the fraction of waste heat available. Rearrangement of this expression gives a result for the overall efficiency of the combination equal to (5.13). Thus, if a reversible fuel cell is used as a topping cycle in combination with an ideal thermal engine, the theoretical losses of each system cancel, and the combination behaves ideally; i.e., it has the same theoretical efficiency as that of an ideal fuel cell operating alone at low temperature, or as that of a heat engine operating at the spontaneous combustion temperature of the fuel with a heat sink at the same low temperature as that of the ideal fuel cell.

Thus, a high-temperature fuel cell combined with, for example, a steam cycle condensing close to room temperature is a "perfect" thermodynamic engine. The two components of this perfect engine also have the advantage of having practically attainable technologies. The thermodynamic losses (i.e., irreversibilities) in a high-temperature fuel cell are low, and a thermal engine can easily be designed to operate at typical heat source temperatures equal to the operating temperature of high-temperature fuel cell. Thus, the fuel cell and the thermal engine are complementary devices, and such a combination would be a practical "ideal black box"* energy system. Even a fuel cell operating at only 150°C above $T_{(S)}$, for example the PAFC, can benefit from the use of what is essentially a thermal bottoming cycle using fuel cell waste heat to operate a turbocompressor for the anode and cathode gases, which yields a higher cell voltage and system efficiency.

*Or perhaps, due to its low environmental impact, a "green box."

The efficiency of a fuel cell system is conventionally expressed in terms of its *heat rate* in most (U.S.) fuel cell literature. This may be defined as the ratio of the HHV of the fuel used in the system to the net ac electrical energy output, so that the lower the heat rate the more efficient the system. Following normal power plant usage, in the United States it is expressed in Btu/kW-h (3413 Btu = 1.00 kW-h_{th}). For the first-generation IFC 11-MW PAFC system operating on reformed natural gas or clean light distillate, the best estimate in 1983 was an overall system heat rate of about 8300 Btu/kW-h (8755 kJ/kW-h, i.e., 2.43 kW-h_{th}/kW-h or 41.1% HHV efficiency) at end of life.[9,10] With some system improvements it was considered that 7900 Btu/kW-h (8330 kJ/kW-h, 2.31 kW-h_{th}/kW-h or 43.2% HHV efficiency) could be achieved.[11]

These estimated heat rates were better than those for any fossil steam units in the United States, even those operating on clean fuels. In Europe, on the other hand, system efficiency is usually expressed as dc or ac power (kW) divided by total LHV of fuel feedstock to the plant (expressed in thermal kW). The plant efficiency is the value obtained after subtracting all parasitic power losses in the plant from the total ac output. Thus, since the European convention considers the feedstock LHV as the basis of the efficiency calculation, the result will always be slightly higher in percentage than if the U.S. convention is used. We should note that U.S. developers of utility fuel cells usually use HHV efficiencies for the net conversion of fuel to ac power, whereas developers of combustion turbines tend to quote the LHV efficiencies of their equipment.

In general, the 35–40% upper limit for the HHV efficiency of most practical fossil fuel Carnot cycle heat engine systems[12] is the *minimum* efficiency for a practical utility or on-site fuel cell power unit operating on natural gas or light distillate fuel. Future system developments are expected to reach higher values (see also Chapter 6). The highest energy and efficiency performance of any pure hydrogen–oxygen fuel cell (at the stack level) which has been published to date is that of the system developed for the U.S. Air Force by International Fuel Cells (IFC, formerly United Technologies Corporation (UTC), Power Systems Division). This alkaline system operates at 390 K and 13.6 atm(a), and it can provide 0.8 V at the previously unheard-of current density of 4.0 A/cm^2 and 1.0 V at 2.0 A/cm^2 (fuel cells intended for utility purposes presently operate at only 150–350 mA/cm^2, using air rather than pure oxygen). Since the IFC power system uses hydrogen and oxygen in the cell at close to 100% utilization, it operates at an HHV efficiency of 67.6% at 1.0 V (20 kW/m^2) and 54.0% at 0.8 V (40 kW/m^2). These values are higher than those of the pure hydrogen–oxygen HOTSHOT and NPNE heat engines referred to earlier, which have a power density (power per unit weight) of the same order of magnitude.

The figures quoted for the IFC system do illustrate an important additional point: namely that the efficiency of a fuel cell *stack*, independent of any auxiliary power requirements or other losses, will normally increase as one goes from the peak power value into the part-load regime. This is in contrast to the efficiency of a heat engine, for which mechanical and heat losses always increase at part-load, so efficiencies correspondingly decrease under part-load operation.

This is illustrated by Fig. 5.2, which compares the part-load efficiencies of various fuel cell power units with those of heat engines. It also requires some explanation, since the results for many fuel cells seem to contradict the statements made above. First, it shows practical efficiencies for many fuel cell

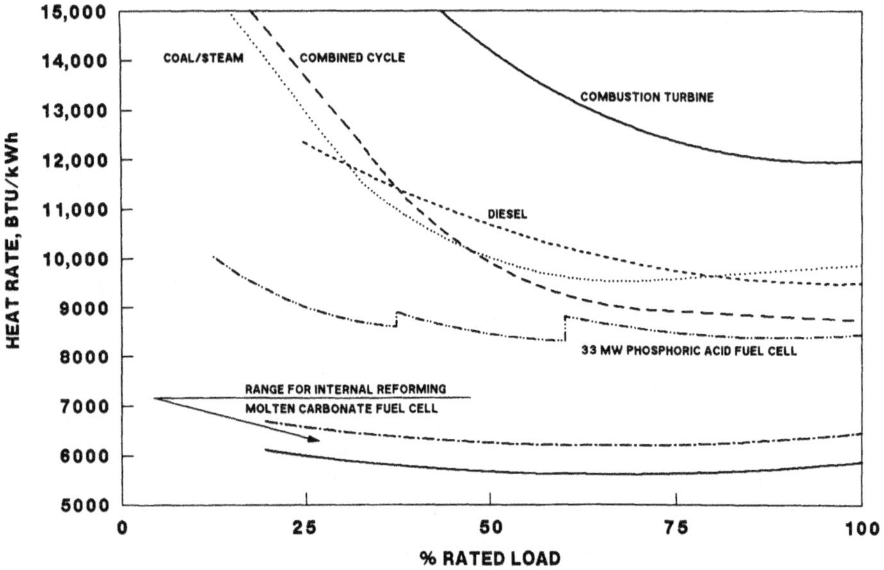

Figure 5.2. Comparison of part-load efficiencies of fuel cells and heat engines.

power units which are much lower than those already discussed; second, it shows that fuel cell power system efficiency falls with part load in many cases. The reason is that the *electrochemical efficiency* of the fuel-to-dc conversion process represents only part of the overall efficiency of conversion in a practical fuel cell system operating on readily available fuels rather than on pure hydrogen. A good example is the 4.5-MW (net ac) UTC PAFC demonstrator (see also Chapter 8), which showed a full-load heat rate (fuel-to-ac power) of 9300 Btu (HHV)/kW-h (9800 kJ/kW-h), operating on natural gas. This corresponds to 36.7% HHV efficiency, or 40.3% LHV. Its minimum heat rate was at 70% load and was equal to 9000 Btu/kW-h (9490 kJ/kW-h, i.e., 37.9% HHV efficiency, or 41.7% LHV). At lower part-load values, the heat rate increased, reaching a maximum value of about 10,500 Btu/kW-h (11,075 kJ/kW-h, i.e., 32.5% HHV efficiency, or 35.8% LHV) at 30% load, due to parasitic losses in the main turbocompressor, as shown by the plot in Fig. 5.3. In consequence, the system was designed to turn to a smaller low-flow turbocompressor at this power level, so that its heat rate would then become about 8700 Btu/kW-h (9175 kJ/kW-h, i.e., 39.2% HHV, 43.2% LHV). The resulting heat rate versus load curve is sawtoothed; however the *mechanical* inefficiencies of the system still make it show a rising trend in heat rate as load is reduced. A 33-MW IFC PAFC design, whose efficiency as a function of load is shown in Fig. 5.2, would behave in the same way. Between about 30% load and full load, its average HHV heat rate (fuel to net ac power) would be 8500 Btu/kW-h (8965 kJ/kW-h, i.e., 40.2% HHV, or 44.2% LHV). The major reason for the change in efficiency with part load for large practical fuel cell systems is the fact that the losses (heat and parasitic power) in the fossil fuel-to-hydrogen conversion part of the power plant increase as size or output is reduced, in the same way as the losses in heat engines. Depending upon system design, the increasing proportion of parasitic power may offset the gain in cell stack efficiency under part-load conditions (cf Figs. 5.2 and 5.3). We should also

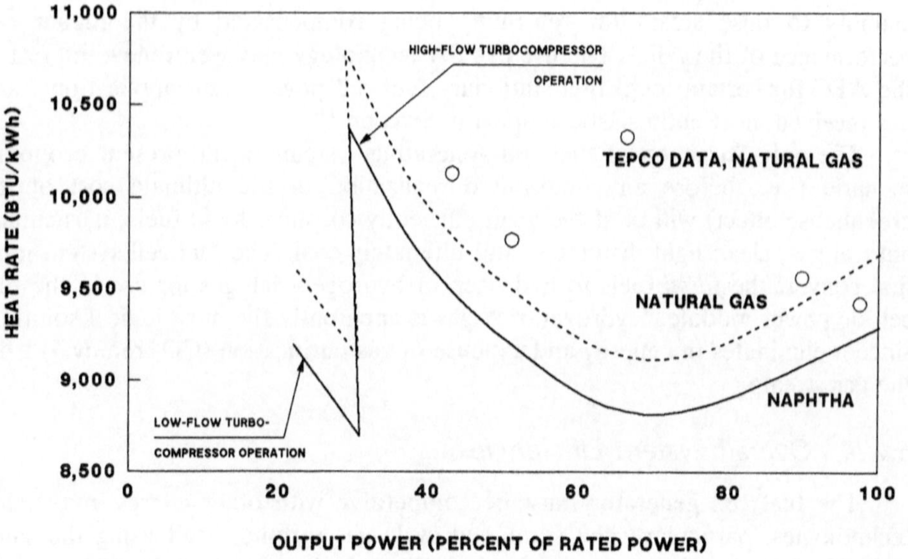

Figure 5.3. Practical higher-heating-value heat rate of the UTC 4.5-MW (ac) power plant (1980–84) as a function of load.

remember that the efficiency of the dc to ac solid-state inverter is maximum at full load.

The discussion given earlier showed that alkaline fuel cells requiring pure *hydrogen* fuel are very efficient users of the hydrogen fuel itself. However, they were eliminated from consideration for utility fuel cell power plants in the United States at an early point in the development of fuel cell systems. Production of hydrogen from natural gas steam-reforming plants, incorporating water–gas shift conversion and CO_2 separation, would have made the system more complex and less efficient than a unit producing a hydrogen–carbon dioxide mixture. State-of-the-art efficiency for pure hydrogen production from natural gas was generally considered to be quite low (typically 65%), which would have resulted in an overall natural gas fuel to electrical efficiency of no more than 35% if an AFC was used.[13] This was considered to be unacceptable compared with competing technologies. The major reason for this low overall efficiency is that low-cost alkaline fuel cells cannot operate at temperatures high enough to produce steam for reforming from fuel cell waste heat, which results in many efficiency penalties. The alkaline fuel cell is presently not being considered for use with today's fuels by most developers, which is unfortunate because little emphasis is thereby being placed on its development for use with hydrogen in the future. Moreover, hydrogen technology has developed considerably for many applications, and the use of pressure swing adsorption (PSA) for pure H_2 removal from a H_2–CO_2 gas stream has considerably increased overall H_2 production efficiencies. An 80-kW PAFC plant is presently operating in Europe, which uses H_2 purification to 99% before entering the fuel cell. The loss in efficiency by the purification of H_2 is largely offset by increased fuel utilization and higher anode performance in the PAFC cell stack, so that the resulting overall plant efficiency is the same as that of an IFC design, i.e., about 40% HHV (see Chapter 6). An AFC could be used with similar results in the same plane with a comparable overall efficiency, its inability to raise steam for reforming being compensated by the higher cell performance of the AFC. The use of PSA technology may well renew interest in the AFC for certain fossil fuel stationary fuel cell power plant applications, and has received most enthusiastic support in Sweden.[14]

The only attraction of fuel cell generating systems in the present economic scenario (i.e., before any undisputed realization of the ultimate cost of the greenhouse effect) will be if they can efficiently consume fossil fuels, particularly natural gas, clean light distillates, and ultimately coal. The fuel cell system must first convert the *fossil* fuels to hydrogen or hydrogen-rich gas for use in the fuel cell dc power module. Hydrogen-rich gas is apparently the most logical solution, since it eliminates the energy and expense of gas purification (CO_2 removal) from the gas stream.

5.2.5. Overall System Efficiencies

The fuel cell generator must be competitive with other energy conversion technologies, particularly the combined cycle gas turbine, itself using the same clean fuels (natural gas or light oil distillates). In a typical fuel cell power generator, it is therefore necessary to design a system comprising fuel gas cleanup (sulfur removal to prevent reformer catalyst poisoning), a steam reformer to produce hydrogen and carbon monoxide, and finally a shift converter to produce more hydrogen and carbon dioxide. All of the above should be connected with the fuel cell stack so that the waste heat from the stack can be used to achieve the highest possible efficiency in an integrated unit. In such a system, the fuel processing (steam-reforming) section uses steam supplied by the fuel cell stack cooling system, and therefore operates indirectly on waste heat rather than on excess fuel as would be the case in an isolated steam-reforming plant. The heat of reforming should also use as much waste energy as possible, to increase overall system efficiency. For best results in typical PAFC or high-temperature systems operating directly on reformate, the heat from burning the dilute exhaust fuel gas exiting from the anode of the fuel cell can be used, since a total fuel conversion exceeding 85% (or perhaps 90% in the high-temperature systems) cannot be used for economical energy conversion in the fuel cell. Beyond this point, power densities become too low to be economical, due to the increase in local anode polarization. Since the system must respond rapidly to load changes, and since the upstream section (the reformer) must receive energy from the downstream section (the fuel cell module), a complex software-controlled feedback mechanism to control energy flows is requried.

5.2.6. Efficiencies of Utility Fuel Cell Systems

It is instructive to examine the efficiencies of the different parts of the conversion chain of a typical fossil fuel (natural gas) to impure or pure hydrogen, then to dc electricity and ac utility power. Desulfurized fuel is mixed with, for example, 3 moles H_2O per mole of carbon (steam carbon ratio), i.e., three times the theoretical amount of steam for the reforming reaction to CO and H_2 to prevent cracking and carbon deposition, and to allow the reaction to go to equilibrium, i.e.,

$$CH_4 + 3H_2O = CO + 3H_2 + 2H_2O. \tag{5.15}$$

This endothermic reaction requires an external high-temperature heat source. For illustrative purposes, we can consider real natural gas to be equivalent in thermodynamic properties to methane, which is a good approximation. A second approximation in this illustrative calculation is to ignore the ΔC_p or ΔC_v of reaction (as is appropriate for the reaction conditions), since these changes are small compared with the chemical energy differences and the latent heat of evaporation of water (cf. the earlier argument on system efficiencies corresponding to (5.11)). Methane (1.039 eV LHV, 1.154 HHV) theoretically requires 0.265 eV of heat to upgrade it to a stoichiometric $CO + 3H_2$ mixture, which has an LHV of 1.304 eV. This heat input requires the burning of slightly more than the corresponding amount of fuel, since the irreversibilities of heat transfer imply that high-temperature heat will inevitably be lost in the process. This reaction is followed by the exothermic water–gas shift reaction at lower temperature, generally in two stages to achieve a low equilibrium level of CO to prevent poisoning of the fuel cell anode:

$$CO + 3H_2 + 2H_2O = CO_2 + 4H_2 + H_2O \tag{5.16}$$

The heat generated in this step (0.054 eV, since the LHV of H_2 is 1.25 eV) cannot be used in the reforming process, since it is generated at a lower temperature. However, it can be used to generate part of the steam for reforming. The total amount of steam, corresponding to 3 moles of water for every 4 moles of hydrogen produced, theoretically amounts to 0.173 eV for its latent heat of evaporation. After taking the heat credit for the water–gas shift reaction, this leaves a requirement of 0.119 eV, although in practice it will be larger again because heat transfer can never be 100% efficient. Finally, excess steam left over after the shift process will be wasted if it cannot be utilized for feedstock preheating or for other potential heating applications. A summation of the LHV heat inputs on this basis shows a maximum theoretical LHV efficiency of 87.8% (i.e., 1.25/[1.039 + 0.265 + 0.119]) for the entire reforming operation. If we now consider the efficiency, we must bear in mind that the heat input operations requiring fuel (not the heat credit for the shift reaction) must be multiplied by 1.11, representing the ratio between the HHV (1.154 eV) and the LHV (1.039 eV) of methane. Since the HHV of hydrogen is 1.48 eV, this yields a theoretical HHV efficiency of 93.3%. However, in practice, heat requirements are greater, heat credits should be reduced, and parasitic pumping work should be taken into account, giving an overall HHV efficiency for production of mixed H_2–CO_2 of about 75%, or a few percent lower if pure hydrogen is required.

The above illustrative calculation assumes that all heat inputs come from burning extra fuel. If "waste" heat is available from the fuel cell stack in a quantity and at a temperature sufficient to raise steam (i.e., the theoretical value of 0.119 eV, after the credit for the water-gas shift process), the theoretical maximum LHV reforming efficiency will become 1.25/[1.039 + 0.265], or 95.9%. The corresponding figure, after multiplication of the denominator by 1.11, will be 1.48/1.45, or 103%. This waste heat will be theoretically available if the cell voltage is less than 1.25 − 0.119, or 1.13 V. If the fuel cell anode is operated at less than 100% fuel utilization, energy will be available from the fuel exhaust stream. For example, at 80% utilization, 0.25 eV of spent fuel is available "free", and only 0.015 eV of methane need be theoretically burned to provide the heat of reforming. Thus, the LHV efficiency of reforming is then 1.25/[1.039 + 0.015] in this example, or 118.5%. The corresponding theoretical HHV efficiency is 126.5%. If the fuel cell operates at a sufficiently high temperature to supply the heat of reforming using its waste heat (i.e., if it operates at temperatures over 650°C or greater), at cell voltages less than 1.25 − (0.263 + 0.119), or 0.866 V, the LHV efficiency of reforming will then become 1.25/1.039, or 120%.

In practice, of course, these theoretical efficiencies must be reduced to account for heat transfer losses. However, if larger quantities of waste heat are available than one can be consumed, they can certainly be approached. Thus, reforming efficiency depends on how much "waste" heat is available from the fuel cell, which becomes a trade-off between losses in the reforming operation and losses in the fuel cell itself. In the IFC PAFC system operating at a fuel to net ac efficiency of 45.6% (LHV), the corresponding gross dc efficiency is about 48.0%. The cell stack operates at 0.73 V and at 80% hydrogen utilization. The LHV efficiency of fuel use in the stack is (0.73 × 0.8)/1.25, or 46.7%. Thus, the LHV reforming efficiency is almost 103% (109% HHV) in practice in this case, based on the heating value of the fuel actually consumed in the stack.

If the whole fuel cell generator can be regarded as a thermodynamic black box in which 80% of the methane feedstock is transformed at a cell voltage of 0.73 V, the total LHV gross dc system efficiency would be (0.73 × 0.8)/1.039, or 56.2%. The LHV reforming efficiency in this quasi-perfect system would therefore be (101 × 0.526)/0.47, or 113%. This is the case in the internal-reforming high-temperature fuel cell, in which the heat available in the anode exhaust is essentially unused, and all transformations are from waste sensible heat from the fuel cell stack.

Since methanol (LHV 1.103 eV) can be easily reformed at low temperature, the product gas is already shifted to essentially pure $CO_2 + 4H_2$, so that only 0.147 eV of heat is required to vaporize the methanol and provide the heat for the reforming process. The waste heat from a PAFC stack can provide a steam–methanol vapor mixture, and the anode fuel gas can be taken to 90% utilization and still provide the heat of reforming. Such a PAFC–methanol system also behaves as a perfect black box, with a gross LHV dc efficiency at 0.73 V of (0.73 × 0.9)/1.103, or 59.56%.

The electrochemistry of the PAFC is less efficient than that of the AFC. At atmospheric pressure, it will only operate at about 0.65 V at the same mean current density as an AFC operating at 0.8 V. In addition, its fuel utilization is limited to about 80% as the inpure H_2 becomes progressively more dilute as it goes from anode entrance to exit. However, a PAFC operating on mixed H_2–CO_2 with an integrated methane reformer whose steam is supplied by fuel cell waste heat can be much more efficient than the AFC unit operating on hydrogen from reformed natural gas. At an operating potential of 0.65 V, the PAFC HHV in-cell efficiency would be only 0.65/1.48, or 44%, and, further, only 80% of the hydrogen fuel might be consumed in the cell. However, as shown in the previous examples, the spent anode fuel is not wasted, but is used to supply a large part of the reforming heat. This and the "free" steam increase the fuel processing efficiency to more than 113% on an HHV basis, giving an overall gross efficiency before dc-to-ac conversion and parasitic electrical requirements of about 40%. The latter losses represented about 6.2% of total dc output for the UTC 4.5-MW (ac) demonstrator,[14] which was roughly equally divided between the inverter and other parasitic electrical requirements. Net natural gas HHV efficiency for the above PAFC unit was therefore about 37%. In contrast, an AFC operating at 0.8 V on pure hydrogen at 100% utilization at 70°C cannot supply waste heat to raise steam. The reforming efficiency to produce hydrogen would only be 70%; hence the overall gross dc HHV system efficiency would be $(0.8/1.48) \times 0.7$, or 37.8%. Hence, the compared electrochemically more efficient alkaline system would have an overall efficiency lower than that of the PAFC.

This simple example emphasizes the importance of the systems aspects of fuel cell power plants operating on fossil fuels (see also Chapter 6). Even so, it may be possible to design very large AFC systems in which the high-temperature waste heat is used in a thermomechanical bottoming cycle, giving efficiencies comparable to those of PAFC systems.[14] A major problem of this approach is the fact that lifetimes of inexpensive AFC cells operating on CO_2-scrubbed air are presently unknown, whereas individual PAFC cells have shown test lifetimes exceeding 100,000 h.[9]

The integrated PAFC system has sufficient waste heat to allow it to be pressurized via a turbocompressor, raising system efficiency by increasing stack voltage. For example, pressurizing a state-of-the-art system from 1 to 8.2 atm will increase cell voltage by about 80 mV at 200°C, four times the theoretical thermodynamic (Nernstian) amount of 21 mV. This reflects the kinetic, rather than thermodynamic, limitation of oxygen electrode performance in aqueous systems.[15,16] (Also see Chapters 3 and 8 of this book.) A PAFC power plant operating at this pressure and 0.73 V will therefore have a net efficiency of over 41% HHV, the estimated value for the proposed IFC 11 MW unit (often inaccurately called the PC-23, Ref. 9). It is estimated that further improvements in cell performance and system efficiency could ultimately increase this to 46–47%[17,18] (also see Chapter 6 of this book).

The natural-gas-fueled PAFC system described above represents one extreme of a fuel processor–fuel cell combination. In it, the reformer operates at a much higher temperature than the fuel cell, which introduces some energy integration problems, but it also opens opportunities for the better use of waste heat. If a system can operate with the reformer and the fuel cell at similar temperatures, much better heat integration can be achieved. A good example is the combination of the 200°C PAFC with a 250°C methanol steam reformer, as already discussed. In this case a relatively simple integrated system without any shift converter can give an effective reforming efficiency of 118% or 113% LHV, based on the hydrogen used in the cell.[19]

If the shift stages needed in a natural gas PAFC plant could be eliminated by the use of a fuel cell capable of operating on unshifted reformate (i.e., the H_2–CO mixture), the gross system efficiency would rise. The high-temperature fuel cells (MCFC, SOFC) are capable of this, since they can consume CO directly (mostly via internal shifting to hydrogen rather than by direct electrochemical reaction). This makes them intrinsically more capable than the PAFC in a system with separate reforming. However, both the MCFC and the SOFC operate at a sufficiently high temperature to be also capable of internally reforming methane using fuel cell waste heat. This in principle eliminates heat transfer requirements and the use of burners and flames in external reforming. Hence, like the PAFC system operating on reformed methanol, the high-temperature fuel cells can have a maximum fuel conversion efficiency in the internal-reforming or sensible-heat-reforming modes. While definitions differ somewhat, internal or direct reforming is normally understood to mean reforming in the anode chamber of the fuel cell, whereas sensible-heat reforming is understood to occur at the temperature of the fuel cell in a separate reactor using transferred waste heat. Since the SOFC operates at about 1000°C, 200°C above the normal steam-reforming temperature of methane, both waste heat conversion systems are simple and efficient. Indeed, internal reforming of a steam natural gas mixture in the SOFC anode chamber is rapid enough to require no special catalyst other than the electrode material itself. Since steam is formed in the hydrogen oxidation reaction at the fuel cell anode, internal reforming is "forced" and will proceed to completion, however, steam must be injected with the incoming fuel to prevent carbon deposition. Normally, a steam-to-carbon molar ratio of 2.5–3.0 is used.

Since the MCFC is limited by materials constraints to operation at about 650°C, internal or sensible-heat reforming is a relatively slow reaction. However, fuel cell space velocities are much lower than those in industrial reformers, and this permits effective forced reforming if a separate catalyst is used in the anode chamber. This must be protected from deactivation by the molten carbonate electrolyte over the lifetime of the cell stack and, of course, against carbon formation by ensuring that the steam–carbon ratio is always high enough to prevent this raction taking place.

As in the examples already discussed, the high-temperature cells operated with internal reforming can be regarded as thermodynamic black boxes which directly consume methane, so their gross system HHV efficiency is simply given by the relationship $Vu/1.154$, where V is the mean cell voltage, u is the utilization, and 1.154 eV is the HHV of methane. The corresponding LHV efficiency is $Vu/1.039$. In the direct reforming mode, the waste anode exit gas cannot be directly used in fuel processing, but it can be used for preheating, in a bottoming cycle, or for cogeneration along with other fuel cell stack waste heat, which is particularly valuable in the case of the SOFC and MCFC because of their high operating temperature. This also means less fuel pretreatment, but more posttreatment, and also less integration, with corresponding advantages and disadvantages. For example, posttreatment in the higher temperature SOFC system could convert waste heat to electricity via a steam cycle at an efficiency of 40%. In the MCFC, it could be converted at perhaps 35% efficiency uing a steam cycle, and at 25% using an air turbine in sufficiently large systems. Thus, at a mean cell voltage of 0.7 V, the high-temperature fuel cells using direct methane reforming operating without bottoming cycles could have a net HHV efficiency at 90% utilization of 51.3% after typical parasitic and power conditioning losses of 6%. Use of the high-temperature waste heat (the difference between the fuel LHV and the product of the cell voltage and utilization, i.e., 0.409 eV) could be used in steam cycles to yield further power. Total HHV efficiency would then be $(Vuc + hc')/1.154$, where c is the correction for dc-to-ac conversion and parasitic power losses, h is the waste heat, and c' is the heat to ac power conversion ratio in the bottoming cycle. Rearrangement shows that this expression is equal to $[(Vu(c - c') + 1.039c']/1.154$. For the SOFC, a total HHV efficiency of 65.5% could be obtained with the above assumptions, and 61.9% for the MCFC, the difference in efficiency being determined by the heat source temperature for the bottoming cycle.

Since the MCFC and SOFC operate at much higher temperatures than the PAFC, their electrochemical processes are much more rapid than those of the latter, particularly at the cathode. The latter is essentially reversible for both high temperature systems, so that polarizations in respect to the exit Nernst potentials are quite low. However, IR drop in these systems is higher than that of the PAFC. In the case of the MCFC, this results from oxide film contacts at the cathode and the use of relatively thick electrolyte layers for mechanical reasons. The SOFC is limited by the necessary use of doped oxides of rather low electronic conductivity as both cathode and between-cell interconnection material in most cell designs. Finally, the exit Nernst potentials are lower for the high-temperature systems, as Fig. 5.1 shows.

For the MCFC with atmospheric pressure reformate, produced by sensible-heat or internal reforming (IR-MCFC), with 85% fuel utilization at 650°C, the exit Nernst potential at 650°C and 85% fuel utilization will be 0.914 V, whereas for the SOFC at 1000°C it is 0.772 V (Fig. 5.1). These values correspond to theoretical gross HHV efficiencies of 67.3% and 56.9% respectively. At 150 mA/cm^2, state-of-the-art MCFCs will operate at total polarization and IR losses of about 150 mV, whereas for the state-of-the-art (1989) SOFC the corresponding losses are about 80 mV, compared to 120 mV in 1985.[20] These values are being constantly improved. Hence, the gross in-cell HHV efficiency of the MCFC will be about 56.0% under the above practical operating conditions (0.76 V at 150 mA/cm^2), and that of the SOFC will be about 51.0% at 0.69 V and the same current density. However, for economic reasons involving power density, the SOFC is presently being operated close to 0.6 V at 300 mA/cm^2,[21] yielding an in-cell efficiency of 44.2%. Without a bottoming cycle and assuming 6% parasitic losses, these cell voltages will result in ac system HHV efficiencies of about 52.6%, 47.9%, and 41.5% respectively. The first two are higher than that obtainable with the PAFC, which is limited by oxygen electrode catalysis to a total polarization of about 350–400 mV (at 8.2 and 1 atm(a) respectively) relative to its exit Nernst potential. With bottoming cycles producing more electric power, the high-temperature fuel cells should both have HHV efficiencies of about 64% at 150 mA/cm^2, whereas the SOFC operating at 300 mA/cm^2 would have a corresponding HHV efficiency of 60%. If fuel costs remain relatively low, the lowest cost of electricity may be reached at higher fuel cell power densities to make best use of the capital invested, hence at lower overall system efficiencies. The merits of the competing high-temperature fuel cell systems will indeed be decided largely by their relative capital costs, taken with other economic considerations such as system complexity, maintenance, tolerance to contaminants, and lifetime.

Since natural gas or gasified coal fuels would normally be available at high pressure, an alternative method of using waste heat is to compress air and thus pressurize the fuel cell, as in the electric utility PAFC system. This technique has also been used with laboratory MCFC stacks, though it is yet to be attempted with the SOFC. Increasing pressure from 1 to 10 atm with the MCFC increases stack voltage by about 100 mV. This is much greater than the thermodynamically expected value of 46 mV, and results from kinetic and diffusion limitations in the cell at low pressure. Indeed, most of the performance improvement (about 60 mV) occurs between 1 and 3 atm, which may represent an optimal economic pressure. In this case, in-cell natural gas HHV efficiency with internal reforming will be about 64% at 90% utilization. Pressurized operation is conceptually the presently preferred MCFC operating mode for large units, especially those planned to use waste heat from a coal gasifier in gas turbine and steam bottoming cycles.[22] These, using a gasifier with an efficiency of 65%, may reach a plant coal–ac power efficiency of over 50% with essentially zero emissions.

If we compare the PAFC, MCFC, and SOFC from a systems viewpoint, the main factor to consider is the complexity of high heat transfer integration in the former to achieve efficiencies of over 40%. This requires a large number of heat exchangers and is costly. In contrast, the MCFC system can be simpler and cheaper and will still achieve efficiencies of 50% or more.[23] However, to obtain significantly higher efficiencies, such systems must operate at higher pressure, which in turn leads to higher capital costs for equipment (see Chapters 6 and 12 in this book), including the complexity of pressurized fuel posttreatment in the anode-to-cathode CO_2 recycle. The SOFC can be even simpler than the MCFC, since it requires no such CO_2 transfer. However, SOFC systems require ceramic components, which are presently quite costly. In addition, the manufacturing cost of the SOFC fuel cells themselves from such parts has yet to be determined. The SOFC dc power system efficiencies will also always be somewhat lower than those of MCFCs for thermodynamic reasons, though this can be partly offset by the use of bottoming cycles. A future system operating at 800°C, which would allow operation at a potential about 90 mV higher than that of today's SOFC and give better bottoming-cycle integration, may result in the highest overall efficiency.

In summary, polarization curves for different state-of-the-art fuel cells

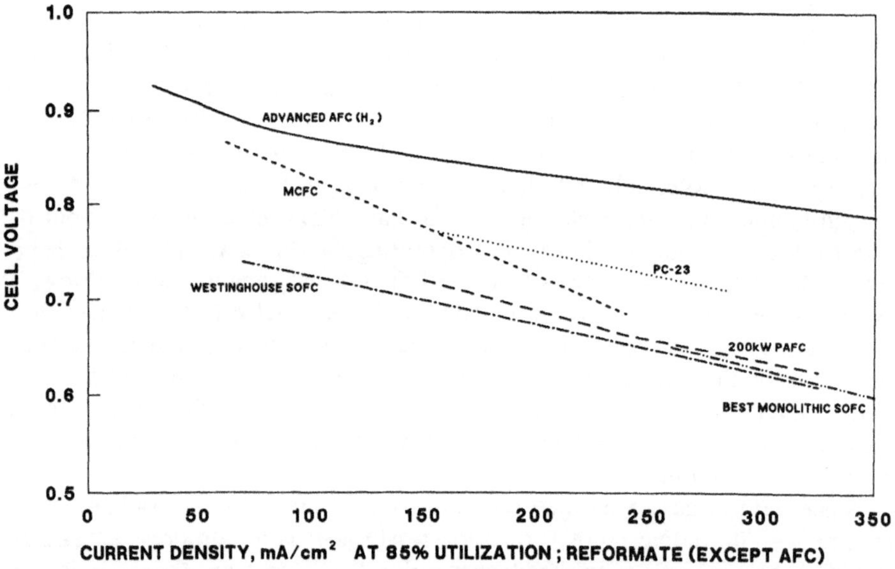

Figure 5.4. State-of-the-art performance of atmospheric pressure PAFC (IFC, 200 kW, 190°C); pressurized PAFC (IFC, 11 MW, 205°C, 8.2 atm); atmospheric pressure MCFC (650°C); SOFC (Westinghouse Tubular in 1989; Argonne Monolithic in 1989); and advanced AFC (with pyrolyzed cobalt TAA macrocyclic cathode at 80°C). All except the latter operate on reformate fuel at 80% utilization (PAFCs), or 85% utilization (MCFC, SOFC, which give identical performance on reformate and with internal steam reforming of methane). The AFC is assumed to operate on pure hydrogen at close to 100% utilization. In all cases, the oxidant is air (CO_2 scrubber in the case of the AFC) at 50% oxygen utilization. (Westinghouse had achieved a substantially lower slope in 1992.)

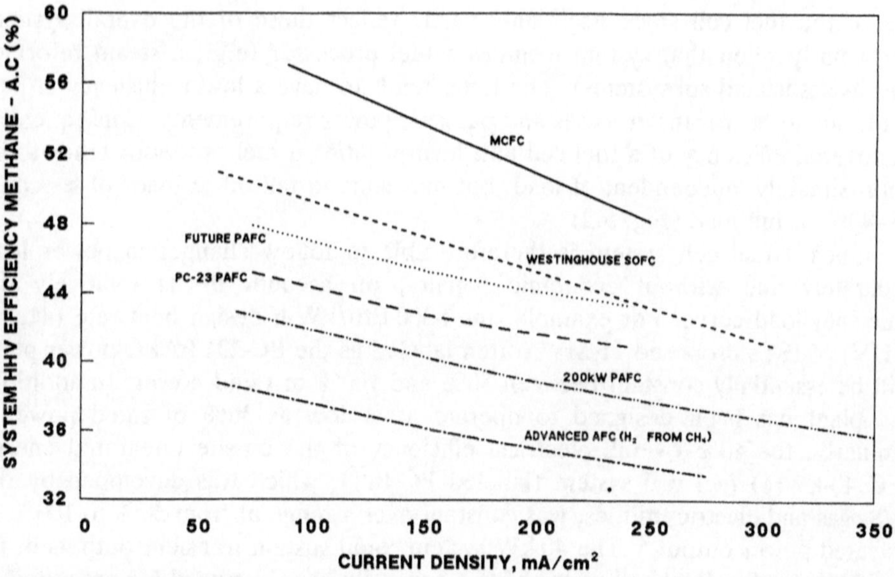

Figure 5.5. Current density as a function of system HHV efficiency for the cells shown in Fig. 5.4 with the system operating on natural gas fuel. The combined parasitic and dc–ac conversion losses are assumed to be 6.25%. The PAFCs operate using standard external reforming, whereas the MCFC and SOFC use internal reforming. The "future PAFC" is assumed to incorporate IFC system improvements. The AFC with the cell performance shown in Fig. 5.4 is assumed to operate on pure hydrogen produced at 65% efficiency from natural gas.

operating on allowable fuels derived from natural gas (i.e., pure hydrogen for the AFC, hydrogen plus CO_2 for the atmospheric and pressurized PAFC, natural gas and steam for the atmospheric pressure IR-MCFC and SOFC), are shown in Fig. 5.4. It is apparent that the atmospheric pressure AFC system is the least polarized, followed by the pressurized PAFC, the MCFC (which still has high IR drop), the on-site atmospheric pressure PAFC, and the SOFC. Figure 5.5 shows the corresponding comparison of state-of-the-art system efficiencies as a function of cell current density, using calculations along the lines of those given above for the high temperature fuel cells (85% utilization, 6% parasitic losses, no bottoming cycles) and published data for the integrated PAFC natural gas systems. The AFC is assumed to use hydrogen produced from a 65% efficient reformer at 100% utilization, with 6% parasitic losses. The advantages of the ability of the high-temperature fuel cells to use waste heat for reforming are immediately apparent, whereas the AFC, in spite of its high cell performance, suffers from the inefficiency of stand-alone transformation of natural gas to hydrogen. This situation can be expected to improve as this transformation becomes more efficient with further development.

5.3. PART-LOAD CHARACTERISTICS

As Fig. 5.2 shows, conventional heat engines operate most efficiently at full load and show a rapid falloff in efficiency at part load. In contrast, the voltage (i.e., the cell efficiency) of a fuel cell, like that of a battery, is higher at part load than at full load. However, as was stressed earlier, in most cases the characteristics of the fuel cell stack itself only partly reflect those of the overall system, particularly when that system includes a fuel processor (e.g., a steam reformer and its associated subsystems). The latter tends to have a lower efficiency at part load, due to heat transfer losses and parasitic power requirements. Consequently, the overall efficiency of a fuel cell unit incorporating a fuel processor tends to be approximately independent of load, but may start to fall off at loads of less than 25–40% of full load (Fig. 5.2).

Such a fuel cell system is therefore able to follow changes in power load accurately and without economic penalty, on account of its relatively flat efficiency-load curve. For example, the 8300 Btu/kW-h design heat rate (41.1% HHV) of IFCs proposed 11-MW (often labeled as the PC-23) PAFC power plant will be essentially constant between 50% and 100% of rated power: In addition, the plant has been designed to operate at as low as 30% of rated power.[24] Similarly, the 40% overall electrical efficiency of the on-site integrated energy IFC 40-kW(e) fuel cell system (labeled PC-18[25]), which was developed by the U.S. gas and electric utilities, was constant over a range of from 50% to 100% of its rated power output.[26] The 40 kW system could sustain transient outputs of up to 56 kW for 5 s. Fuel cell systems have generally been designed for operation at the highest efficiency consistent with the economic goal (cost of power) that best fit the mission. In general, this has been at a voltage below that of the maximum continuous power point and much below that of the maximum transient power point. Provided that the power conditioner can handle its output, any utility fuel cell system designed along the above lines should be capable of delivering high peak reserve power output for a short time.

5.4. RESPONSE TIME, SPINNING RESERVE CAPABILITY

In addition to its relatively constant efficiency over a wide range of operating conditions, a fuel cell system also has a very fast reactive power response.[11,24] For example, the 4.5-MW (ac) demonstrator units in New York and Japan were designed to respond to load transients in only 0.3 s by the use of a "feedforward" sensor and energy transfer system (steam and anode exhaust from the cell stack to the reformer, anode gas from the reformer to the cell stack). This was similar to that used in the 40-kW units. In order to simplify the control system and lower costs, the PC-23 11-MW units which were under development up to 1988 were designed to respond to each 1-MW load change in 1 s, using a "feedback" energy control and transfer system. This still represents a very fast response time for utility generating equipment. A fast power response should permit a utility to lower its need for on-line spinning reserve, which operates in a standby mode with machinery rotating. Such reserve can be provided almost instantaneously by the rapid reaction time of the fuel cell.

Because of their fast reactive power response capabilities, grid-connected fuel cell power units can provide alternatives to shunt reactors and capacitors, since they are able to react to and dampen system voltage surges and swings, such as may occur during switching operations or lightning storms. Similarly, if a short circuit occurs in part of a series-connected fuel cell stack array, the short-circuit current generated will be limited, thus reducing the necessity for high-current circuit breakers. This feature of this type of generator may enable a utility system containing a significant number of fuel cells to better survive transmission cascading, voltage disturbance, and associated frequency regulation that often have resulted in widespread blackouts in the past.

5.5. EMISSIONS

5.5.1. Chemical Emissions

The primary source of air pollutants is the consumption of fossil fuel for heat and other forms of energy. Coal and its by-products, natural gas, fuel oil, and gasoline are the chief contributors to community air pollution, which is well known to have adverse ecological effects. These are produced not only by the gases and particulates coming directly from smokestacks and automobile exhausts, but also by secondary pollutants derived from complex chemical interactions between emitted gases and atmospheric constituents over a relatively long time scale. The immediate damage caused by air pollutants, weighted according to the type of fossil fuel burned, has been determined by certain authors to be about 13% of GNP,[27] whereas fossil-fuel costs represented about 4% of the GNP in 1989. This "tax" on the GNP is paid for by society as a whole. It is even incomplete, since it does not include long-term effects that are presently difficult to quantify, such as global warming (the "greenhouse effect"[28,29]). It includes the influence of pollutants on human health, lifespan and ability to work, on structures and other capital investments, and on nature, including forest destruction by acid rain. Vegetation appears to be extremely sensitive to toxic air pollutants. Sensitivity to specific pollutants varies widely among species and sometimes among strains within a single variety, but it is often produced by long-term, low-concentration exposure to phytotoxic pollutants.

Sulfur dioxide, fluorides, nitrogen dioxide, particulates, and acid aerosols threaten certain types of vegetation where coal is burned in large quantities. Combustion of fuel oils may produce some of the same types of toxic pollutants, and natural gas under certain combustion conditions may contribute significantly to the atmospheric concentration of nitrogen oxides. Nitrogen dioxide is well known to be toxic and has been shown to be a key factor in the photochemical reaction of multiple-carbon hydrocarbons via aldehydes and, ultimately, formaldehyde, producing ozone and peroxyacyl nitrates, which are highly toxic oxidative air pollutants. Most of the above air pollutants are emitted from fossil-fuel-burning power plants and from automobile internal combustion engines. For nuclear power plants, air pollution should be minimal, as long as no major malfunctions occur. The greatest problems posed by nuclear power in the future will involve the disposal of highly radioactive wastes,[30] the danger of further nuclear proliferation, and the impossibility of fully protecting facilities from terrorists, who may themselves have nuclear explosives.

In contrast to other fossil-fuel utility or mobile power plants, the impact of fuel cell systems on the environment should be minimal. Fuel cells are catalytic devices which require clean fuel so that the reforming and anode reactions function correctly. Since they use clean fuel, they generate clean power. Moreover, since they operate at lower temperatures than thermomechanical devices relying on combustion in hot flames, they generate very small amounts of

Table 5.2. Comparison of Power Plant Emissions

Contaminant	Type of power plant: Gas-fired	Oil-fired	Coal-fired	FCG-1 fuel cell	EPA limits
SO_x	—	3.35	4.95	0.000046	1.24
NO_x	0.89	1.25	2.89	0.031	0.464
Particulates	0.45	0.42	0.41	0.0000046	0.155

From Refs. 11 and 31.
(kg/1000 kW-h) where 1 kg/1000 kW-h = 0.646 lb/million Btu.

NO_x. We should note that the very small amounts of NO_x generated in fuel cell systems result from the fuel processor, where lean gas burners are often employed for that and other reasons. These burners determine the total system emissions. Table 5.2 shows the measured exhaust emissions from the 1-MW prototype natural gas fuel cell system developed by United Technologies Corporation (UTC) with those of conventional power plants.[11,31] Its emission levels were much lower than those permitted by the U.S. Environmental Policy Act. Furthermore, a study funded by the U.S. Environmental Protection Agency (EPA) showed that not only are fuel cells themselves more efficient and environmentally benign than conventional power generators, but also that their advantages are seen from the resource extraction stage to the final end use of the electricity they generate.[32] Some of the latest available PAFC data were obtained with the UTC 4.5-MW (ac) natural gas PAFC prototype unit tested at the Tokyo Electric Power Company (TEPCO) during 1983–1985. Emissions design values were NO_x = 20 ppmv and SO_x = 0.013 g/GJ. Measured values were <10 ppmv and zero (i.e., not measurable), respectively.[33] At the Kansai Electric Power Company's 1-MW PAFC demonstrator operating on natural gas, measured values in 1988 were NO_x < 20 ppmv and SO_x < 0.1 ppmv.[34] Data obtained for high-temperature fuel cells (the 3-kW Westinghouse SOFC demonstrator units in Japan) showed NO_x = 1.3 ppmv at Tokyo Gas and under 0.5 ppmv at Osaka Gas.[35] In this connection, we should note that SO_x emissions standards for new fossil-fuel plants in many developed countries are about 200 g/GJ, more than four orders of magnitude higher than that from the fuel cell. Similarly, new source performance standards (NSPS) for combustion machinery (e.g., gas turbines) are 75 ppmv for NO_x in most parts of the United States. As stated, the very low NO_x emissions in the PAFC result from the anode tail-gas burner, which supplies heat to the fuel processing system. In the second-generation MCFC, part or all of the combusted tail gas, which is very lean, will be passed through the fuel cell cathode after mixing with air. Since the cathode is an excellent "NO_x scrubber," and provides additional residence time at 650°C for the destruction of NO_x already present, emissions of this major pollutant should be negligible.

In the shorter term, the low-emission characteristics of fuel cell power plants should allow some utilities to defer or offset the costs of installing additional scrubbers or of emission control equipment. Indeed, in many jurisdictions, reequipment (repowering) of fossil-fuel plants with advanced technology (such as fuel cells) to bring them below the legal emission limits will enable emissions from existing facilities to be decreased. In the longer term, the environmentally benign characteristics of fuel cells should help to reduce the social and material costs of pollution that are normally associated with the economic development of new regions. In the medium term, their high efficiency should result in lower fossil-fuel CO_2 emissions, which will help reduce the rate of global warming. Finally, in the long term, when hydrogen fuel is readily available from sources that do not emit CO_2 to the atmosphere, then fuel cells are the logical devices to use that hydrogen. With it as a fuel, their efficiency will be higher than of any competing system, thus lowering operating costs. In addition, their emissions (pure steam or liquid water only) will be totally benign, and their capital cost (essentially the fuel cell stack, power conversion, and control equipment only) will be low.

A decision to reduce emissions substantially will involve a heavy commitment of capital resources. Cost-benefit analyses have been made that provide a systematic framework to estimate the cost to society of committing these resources to pollution abatement and of its benefits to society. An early 1971 study estimated that the benefits of lowering air pollution by 50% in the largest North American cities would result in annual benefits six times greater than the annual investment.[36] In addition to the savings in the economic cost of illness and mortality, there also would be savings in cleaning, maintenance, and life (e.g., of structures and buildings), savings corresponding to animal and plant life, and beneficial psychological effects, including the number of sunny days, and in general the quality of life. This suggests that the involuntary tax on the GNP discussed above, perhaps equal to 13%, can be tackled by an ultimately affordable investment.

5.5.2. Acoustic Emissions

Acoustic emissions strongly interfere with our ability to operate under ordinary conditions. Since the processes that take place in a fuel cell are electrochemical, ideally there should be no moving parts to generate noise and vibration. In practice, however, pumps, blowers, transformers, and other auxiliary devices are required in any systems whose output exceeds 0.5–1.0 kW. In spite of this limitation, large fuel cell power units operate much more quietly and smoothly than any other power-generating devices.

Some types of problem noise, e.g., turbocompressor noise in prototype electric utility PAFC units, are fairly obvious. Potential noise problems of this type can be readily solved by the use of state-of-the-art soundproofing or baffles. The overall noise level (largely that of the turbocompressor) at 30 m from a fuel cell power plant of this type has been measured to be only about 55 dB.[11] Since this results from the noisiest component by at least an order of magnitude, and is in any case much lower than any competition (it is about equal to a household air-conditioning unit), fuel cell generators should be ideal for urban or suburban power generation in dispersed quiet locations.

5.5.3. Thermal Emissions

Since fuel cells are generally more efficient than Carnot-limited thermomechanical power generators, especially in small units, they produce proportionally less waste heat. Small amounts of heat would be produced even if the fuel cell system did not have a separate fuel processing subsystem to generate hydrogen from a fossil fuel. Whether this is so will depend on the future availability of hydrogen fuel or on the development of integrated fuel processing (e.g., internal reforming of methanol or natural gas in high-temperature fuel cells (see foregoing)). When the waste heat which is generated is not utilized for auxiliary processes or on-site heating, it can normally be dumped directly into the air. Hence, water requirements of fuel cell power systems are either very low or can be met by condensed product water.

5.5.4. Esthetics

"Visual emissions" are something that must be considered with any dispersed power unit which might be used in a high-property-value area covered by zoning laws. Although the flow diagram of the fuel processing subsystem of a fuel cell generating station may tend to resemble a miniature refinery, it need not have a noticeable visual impact. With noise levels lower than those of the urban background, and with no need for high smokestacks or obvious cooling towers beyond those in sizes resembling air-conditioning ducts, it should be possible to enclose the low-profile subsystems within the basements of downtown buildings or under streets, as is being proposed for urban use in Japan (see, e.g., Chapter 8).

5.5.5. Conclusion

With the possible exception of photovoltaic generators out of sight on rooftops, a fuel cell generating system is probably the only type of power plant that can be located without intrusion in downtown, residential, or scenic areas. This results from its low chemical, thermal, and acoustic emissions as well as from its small visual impact. In this connection, thanks to the experimental installation of the UTC 4.5-MW PAFC fuel cell generator in Manhattan, which received its operating license after completion of its process and control test (despite problems which have repeatedly been stressed in literature), fuel-cell-based units will be the only new utility power systems which will be allowed in New York City in the future.[37] Similar restrictions will in the future apply to all high emissions power systems in large areas of the world.

5.6. MODULARITY

Fuel cell stacks are certainly modular. In principle, like batteries, they can be built in as small a unit as desirable, and a system can be assembled in any size from these units. In practice, there are a number of limitations on module size. The first is that, unlike batteries, they have to be supplied with reactants, and reaction products have to be eliminated. In utility fuel cells, this means a multiplicity of manifolding and piping requirements, which must as far as possible be simplified and reduced in number to lower costs. Related to this is the fact that utility fuel cell stacks have a limited optimal range of operating temperature, which means that careful temperature control must be maintained over a wide range of operating conditions from standby to overload. Hence, each cell stack should have a carefully designed thermal mass and insulation and a well-controlled cooling system. Finally, the electrical requirements of the end user must be considered in regard to the power produced. For small fuel cell systems operating as substitutes for battery packs, the output requirement might be a low dc voltage (e.g., 9 V, 12 V, 24 V, or 28.5 V). This will require a system with a small number of cells, each of small area. For example, a 28.5-V, 1.5-kW (nominal net output) PAFC might operate with 50 cells, each of which has an area of $500\,cm^2$. This was the approximate specification of small methanol reformate fuel cell systems intended for military use.[38] A future 30-kW AFC for an electric vehicle may be required to produce 150-V output, so would require 200 cells. If these operate at $300\,mA/cm^2$, $700\,cm^2$ would be the required cell area. Such dc units will require an accurate match to end use because of the relatively high cost of high-power dc–dc conversion equipment. In contrast, on-site units may produce 220-V, 380-V, or 440-V ac, whereas utility units must produce a high voltage to interface with the utility grid. Both require dc–ac power conditioners producing utility-quality power; hence, cell and stack optimization can be made on the basis of least capital expenditure and lowest electricity cost. These considerations result in least-cost cell area being a function of stack size, which in turn will be a function of plant size. As an example, IFC believes that the optimal cell area for the 200-kW PAFC on-site stack will be $0.34\,m^2$ operating at $325\,mA/cm^2$ and at 0.62 V under unpressurized conditions. The reasons for this choice are that pressurization is not practical for a unit of this size, and 0.62 V (36% electrical efficiency) allows a 1:1 electricity-to-heat ratio, which is optimized for certain on-site applications. Finally, the high current density lowers unit cost. However, for a unit in the 1–3 MW class, which may be desirable for small electric utilities, a pressurized system will be required to reduce the volume (and cost) of fuel processing equipment and piping, as well as to allow the highest electrical efficiency at 0.73 V or greater for a noncogeneration unit. Lowest cost is unlikely to be achieved by simply using a pressurized $0.34\,m^2$ design, which was the cell size of the IFC 4.5-MW Tokyo Electric Power Company (TEPCO) demonstrator. The best compromise may be a 0.95-m^2 stack similar to that of the PC-23 11-MW unit. In this case, each stack would be about 650 kW rather than 200 kW. Finally, for very large units (100 MW), cell size is likely to undergo a further scale-up to about $2.8\,m^2$ to reduce the number of stacks in the unit and simplify piping requirements. Similarly, a higher operating pressure may be used for best economics. Stack size is then likely to be about 2 MW.

The lesson of the above is that fuel cell stacks are in principle modular, but to go from a plant of a certain size to one 10 times greater using the same cell stacks will not necessarily lead to the most economical plant. Operating conditions and flexibility and costs of auxiliaries may be the governing factors. A statement which may be made concerning the modularity of the dc fuel cell power section is that it may possibly be produced in the form of a number of factory-manufactured cell stacks, depending on its size. Each of these could be made on a production line and so would be, in this sense, modular. In particular, they would benefit from learning-curve cost reductions as production volumes increase. Other parts of the system may also be modular. For example, there is no reason why the steam reformer could not be manufactured in small modules, each with its own burner, of about 200 kW or even in very small units of a few kilowatts or even less. The former refers to the Haldor–Topsøe tubular "heat exchanger reformer"[39] or the KTI radiant tube reformer,[40] whereas the latter is the flat-plate Ishikawajima–Harima Heavy Industries (IHI) reformer.[41] Again, the ultimate question will be economics. A multiplicity of small mass-produced reformer or water–gas shift units may have a cost advantage over one large single custom-built shell-and-tube reformer (or it may not; this factor remains to be seen). Other parts of the system must also be considered. The water treatment plant for the PAFC cooling system is one example. It is primarily not modular, and is usually constructed on a case-by-case basis. The dc-to-ac converter may also be modular, at least above a certain size.

The above refer to the generalized modularity of a fuel cell plant, i.e., the possibility of doubling or tripling of plant size by simply adding more identical complete units. It certainly will be possible to build a fuel cell power plant this way, but whether this would be the most economical approach is not presently known. A gas turbine, for example, has a lower limit to size, and the cascading of several identical small gas turbines units to make a larger power generator does not make sense, because a single larger unit will be more efficient and will cost less per kilowatts. In contrast, a fuel cell stack should have an efficiency which is largely independent of size, and its cost per kilowatt may also be approximately size-independent. The reason for the latter is that economies of scale in large plants may be compensated by economies of mass production in smaller units. Similarly, a complete fuel cell power plant consisting of a number of different subunits (fuel processors, dc power units, power conditioners, etc.) may allow a certain redundancy of function (i.e., it may increase overall system availability). Such cascading may ultimately benefit the cost of electric power.

Individual fuel cell systems may certainly be modular, in a somewhat different understanding of the word, in which the various parts of a fuel cell power plant (fuel processor, dc power unit, power conditioner, water treatment unit, etc.) may be arranged in separate compact modules in a unit of typical size. These could be individually factory-assembled. The IFC 4.5-MW (ac) PAFC systems intended for electrical testing in the United States[37] and actually tested in Japan[33] were designed so that many major components of the various subsystems could be preassembled in modular form at the manufacturing plant. UTC (IFC) has attempted to use this same philosophy for future commmercial electric utility systems now in the planning and development stage.[10] However, it has become apparent that the modularity concept may not necessarily fit in with the needs of some individual customers or with ease of maintenance of the power plant. Having trailer-sized factory-built modules which will fit under highway bridges implies a tight internal structure. While this may be appropriate for certain parts of the plant, for example the power conditioner, it has not yet proved satisfactory for the fuel processor and power sections, which experience has shown will still require more accessibility for maintenance than was anticipated in the 4.5-MW TEPCO unit.

Further, different customers require different power footprints (see next section), so that the unit must not only be accessible but should also be very compact in area in urban locations with high land costs. This may require a multistory fuel cell unit. The concept of prefabricated factory-built modules for large electric utility fuel cell power plants (as distinct from the smaller on-site units) therefore needs to be rethought by the developers. In any case, there is a general trend in chemical plants to build modular units, and the problems associated with this approach are of a general nature and in no way applicable solely to fuel cell power plant technology. The financial advantage of spreading out investments as a function of time is generally considered to be the major asset of modularized plants, and therefore the more generalized use of modular systems may be expected to increase in the near future.

The concept of modularity is subject to limitations, but it is nevertheless true that many of the future advantages of fuel cell power systems may be attributable to their potential modularity, which derives from the inherent scalability of individual fuel cells and fuel cell stacks. Unlike heat engine technologies, the results obtained from small bench-scale fuel cells having electrode areas of 10–100 cm^2 may be directly applied to estimate the performance of prototype-size units having electrode areas of 1–3 m^2. It is therefore possible to design small peaking plants of only a few megawatts which will be as efficient as large base-load plants in the several-hundred-megawatt class. This means that after a certain unit size is reached, a fuel cell power plant may achieve high efficiency and reliability, independent of plant scale, because it will be a multiple of a certain number of identical units, which will be limited in scale by physical considerations (for example, there is an upper limit to cell area). This characteristic should lower technical risk and also lower the capital cost burden.

Because large parts of the system, including the cell stacks (the only item which goes beyond the known art and involves appreciable new technological risk) should arrive at the site preassembled and pretested, construction times may be very short. Installations will ideally be repetitive, so that site preparation can be standardized, allowing low installation costs per megawatt compared with those for conventional power plants. In essence, all that will be required is the laying down of electrical and other services and concrete pads for the system units, which can then be hooked up.

5.7. SITING FLEXIBILITY

Owing to the potentially modular construction of a fuel cell power plant, the amount of land required for the installation of a single unit may be relatively small. For example, the IFC 11-MW utility design was intended to occupy about 0.5 ha (1.2 acres).[10,24] The unit was also intended to be independent of a water supply. These characteristics, together with very low environmental impact and low maintenance requirements, should permit fuel cell power plants to be located in remote, relatively inaccessible sites which otherwise might prove to be unsuitable for conventional generation units. In addition, its "neighborliness" and high level of safety will allow it to be sited close to the point of use (e.g., in urban areas), where its waste heat may be used in cogeneration applications.

Recently, it has become clear that some of the developers' assumptions concerning urban siting of PAFC units require review. First, the accessibility requirement for maintenance of early multimegawatt-class PAFCs requires an even more open structure than that of the PC-23 design, which in turn was much more open than that of the 4.5-MW demonstrators. This has required 25% more land area than does the PC-23. However, for many urban installations this is impractical. The proposed Palo Alto, CA, PC-23 unit would have had a site cost of \$3.5 million (i.e., \$320 per kilowatt, about equal to the potential credits for dispersed urban generation; see below). This land cost was low for the city, since it was to be built on a landfill site. In Manhattan, land costs would have been so high for the PC-23 that Consolidated Edison estimated that a practical fuel cell unit would require a power footprint of 0.09 m^2/kW for a greenfield site and 0.04–0.07 m^2/kW for a repowering site, compared with 0.34 m^2/kW for the PC-23 design. In consequence, the urban approach to fuel cells will be to use the parking garage rather than the parking lot, and Consolidated Edison has proposed a three-story fuel cell design.[42] The above argues that dispersed urban fuel cells should not use purpose-designated real estate, but should be incorporated in, for example, the basements of structures or under highways.

This dispersed siting capability of the fuel cell should not only permit a reduction in transmission and distribution losses but should also provide for power control close to the load center. It should therefore enable a utility to defer transmission and distribution investments. The lack of siting restrictions and the fact that on-site generation will reduce the need for the right-of-way for transmission corridors. With careful planning, this may result in highly acceptable land-use characteristics for utility fuel cell power systems. These should result in economic credits for a fuel cell power system compared with its immediate competition, particularly the gas turbine combined cycle unit operating on natural gas. The latter has higher emissions and will generally not be permitted in urban environments. In addition, the combined cycle can only achieve comparable HHV efficiencies to the PAFC (over 45% for future designs) in large units over 100 MW in size. However, the attraction of the advanced combined cycle is the fact that its capital cost will be only about half of that of a mature PAFC system, which will certainly ensure it of a significant part of the market for intermediate-load utility power units as long as its emissions are considered acceptable.[43] Whether the paper credits of the PAFC, which may lower the effective capital cost of the latter to the same level as that of the advanced combined cycle, will serve to open up a broad market and large market share remains to be seen. However, the second-generation MCFC promises to be simpler, cheaper, and more efficient that the PAFC, and thus should prove to be a much better competitor to the advanced combined cycle.

5.8. LIFETIME

Whereas the balance of the plant should have a 20–30 year lifetime (depending upon the load factor), the life of fuel cell stacks will be limited by materials considerations. For the PAFC, this is largely because component corrosion increases with rising electrochemical potential. The PAFC cathode has been pushed to the maximum operating potential consistent with economic lifetime at a guaranteed final system heat rate. Conventionally, this has been assumed to be a lifetime of 40,000 h.[9] However, cathode component corrosion is sufficiently slow that lifetimes over 100,000 h have been achieved in the laboratory,[9] and it seems possible that commercial cells may achieve a useful life exceeding this value provided that a further 1–2 percentage points degradation in HHV efficiency can be tolerated before repowering is needed. This will depend on economic circumstances and the provision of either a large enough reservoir of the slightly volatile electrolyte in each cell or a provision for supply of electrolyte as a function of time.

The present MCFC cell will almost certainly have a design life limitation in the 25,000–40,000 h range due to the slow dissolution of the nickel oxide cathode and its reprecipitation as nickel nodules in the electrolyte close to the anode. The rate of dissolution–precipitation depends on the cathode CO_2 partial pressure, and hence on the total cell pressure.[23] Present economic assessments of the MCFC take this into account. While complete loss of the cathode is not a factor to be considered, the nickel nodules may eventually result in cell shorting. A great deal of research is presently aimed at either reducing nickel oxide dissolution or finding a substitute with improved properties. It is generally felt that at least one of these approaches will prove successful and that the MCFC cell lifetime can be extended, and should be, to at least 40,000 h, even under pressurized conditions, with an acceptable decay rate. Less is known about this than the corresponding decay profile of the PAFC.

Components for the SOFC have operated for 35,000 h,[44] but little is known at the present time about the real decay rate of cells produced in a pilot plant.

5.9. WASTE DISPOSAL OF MATERIALS ON DISMANTLING OF PLANT

No special hazards or difficulties were encountered in dismantling and clearing of the New York and Tokyo Electric 4.5-MW PAFC demonstrators. Component recovery, rather than waste disposal, is likely to be the issue in dismantling a fuel cell plant. In the PAFC, nickel from the reformer catalyst and platinum from the anode and cathode will require recovery for their "scrap value". For the MCFC, stainless steel can be recovered from bipolar plates, and nickel from the electrodes and reforming catalysts. In the SOFC, nickel and all zirconium- and lanthanum-containing ceramic components are likely to be recovered.

5.10. ON-LINE AVAILABILITY

The power generating processes which occur in a utility or on-site or transportation fuel cell power unit are mechanically static, though they require auxiliary circulation devices. These operate under relatively benign conditions, so they can be designed such that maintenance is required only at infrequent intervals. Above all, they can be used for unmanned operation with automatic or remote dispatch of electricity generated, which is vital for the economics of dispersed generators, for which labor costs would otherwise predominate. The same factors do not necessarily apply to very large units incorporating bottoming cycles or units using coal gasifiers. Unmanned operation was the philosophy used for the IFC 40-kW cogeneration PAFC units, which have shown trouble-free automatic runs of over 2000 h.[45]

Depending on system requirements, the overall hardware can be complex or can be as simple as is needed. For example, pressurized units with computerized feedback or feedforward subsystems between the heat- and energy-integrated fuel cell stack and fuel processor will be quite complex. Such systems are necessary to obtain the highest possible efficiency and rise time for electric utility use. On the other hand, atmospheric pressure units intended for continuous use can have more modest characteristics, particularly in developing countries. They may indeed be designed so that only the simplest tools and skills will be needed for maintenance activities, provided that plug-in replacement modules (for example, for the fuel cell stack) are available at short notice. The latter will be an important feature because the limited availability of technically trained personnel for the installation, operation, and maintenance of power supply equipment is a characteristic common to both frontier and rural areas, where the use of transported skilled labor will add significantly to operating cost. Indeed, the lack of skilled personnel is often cited as the reson for the failure of rural electrification programs involving the use of small, conventional on-site generators. Thus, fuel cell power systems which require few technically trained personnel or which permit less expensive, more rapid training of indigenous labor should gain wide acceptance in developing and rural areas.

5.11. RELIABILITY

The U.S. aerospace program has proved the reliability of fuel cells for steady, uninterrupted power generation. Design reliabilities of greater than 95% have been achieved consistently under the stringent conditions of space flight. No fuel cell battery in NASA's program has yet failed (the often-quoted failure on board *Apollo 13* in April 1970 was that of a cryogenic oxygen tank, not a fuel cell). The design reliabilities (mean time to failure) required for most terrestrial uses are lower than those needed for man–qualified aerospace applications. However, terrestrial units require reliability during much longer total operating periods. It is expected that the first generation of commercial fuel cell power generating systems will have an on-stream availability of about 90%.[11] Ultimately, fuel cell plants can have on-stream availabilities of 98–99%, values which are currently achieved in conventional hydrogen plants. It is interesting that in the 4.5-MW U.S.-built PAFC demonstrators and their 1-MW Japanese counterparts, the fuel cell stacks have proven to be more reliable than the conventional balance of plant (see Chapter 6).

The high reliability of a fuel cell system will largely result from the modularity of the stacks and stack components, but should also be attributable to their lack of highly stressed moving parts operating under extreme conditions and to their ease of maintenance. Because the modules are repetitive, any problems which arise should be similar instead of one-of-a-kind. This characteristic permits a rapid buildup of competent well-trained personnel. Similarly, testing can be done on a repetitive basis, which enables rapid feedback of any potential problems to the manufacturer or designer in time to be useful. The use of modular units permits a site layout that can be designed to permit the replacement of complete modules, which not only allows for a more economical use of spare parts but also minimizes lost output. Also, by replacing spare modules, a plant could be operated at full power during periods of routine maintenance, should this be necessary. Even without spare parts, plants could be designed so that only partial shutdown will be necessary in the event of failure.

5.12. START-UP AND SHUTDOWN

Start-up and shutdown procedures are specialized, but they can be automated and performed by remote dispatch. For the PAFC, the major precaution to be taken on shutdown is to prevent the cathode reaching potentials more positive than about 0.8–0.85 V versus the hydrogen anode while the cell is hot, since this can cause damage due to corrosion. To prevent this, the cathode is flooded with nitrogen. The cell can with certain limitations be shutdown on standby (at operating temperature) or stored in a cold shutdown mode. Precautions taken at IFC then include maintaining cell stacks above 45°C to avoid any danger of freezing the concentrated acid electrolyte, which can damage Teflon-bonded electrode structures. Electric heaters are used to maintain temperature and for start-up. During start-up, hydrogen is used at the cathode until temperatures are reached where carbon monoxide poisoning is no longer a problem. Start-up and shutdown procedures for the PAFC are available from the developers.

For the MCFC, thermal cycling is the crucial problem on start-up and shutdown, which means that cooling and heating rates must be slow. However, recent stacks at IFC and at other developers have been given systematic thermal cycles to room temperature without cracking, gas crossover, or failure. In the SOFC, cells seem to be sufficiently robust that no special precautions need be taken on start-up or shutdown beyond slow cooling and heating rates to prevent thermal cycle failure. Indeed, this may more likely occur in ceramic components other than the cells themselves. In all systems, provision must be made for automatic controlled shutdown if any equipment failure occurs.

5.13. CONTROL

Procedures for internal energy control, involving feedforward and feedback sensors to allow rapid response of the equipment to load changes, vary according to the plant design and particularly on the ramping rate required for a given application. Output beyond the maximum rated level is prevented by the power conditioner.

5.14. POWER CONDITIONING

Utility and on-site systems generally will require the conversion of gross dc fuel cell output to utility-quality ac power, unless, for example, the output of the fuel cell is required for electrolysis applications in which dc power can be used directly. A solid-state power conditioner is therefore required. This may be line-commutated (for equipment connected to the utility grid) or electronically self-commutated. Developments in electronic components and circuitry over the last 10 years have been such that self-commutated systems are now preferred. These use gate-turnoff (GTO) thyristors, which can now individually handle several thousand amps and volts. In 1978, power-conditioner costs were predicted to fall from several hundred dollars per kilowatt to $70 per kilowatt in 1978 and $120 per kilowatt in 1988. Suitable power conditioners, which are still not in full production, cost about $150 per kilowatt in 1988. Their development has resulted from the advance of the state-of-the-art of GTO thyristors and from their wider recent use in other applications, such as variable-speed drives for electric motors. According to Japanese sources, the price is expected to fall to $70 per kilowatt by 1995.[46] Like all solid-state devices, power conditioners are likely to possess high reliability.

5.15. SAFETY

Fuel cells certainly are as safe as coal-, oil-, and natural-gas-burning central power stations, and they should be safer than nuclear power plants, because of the possibility of disaster in the latter resulting from systems or materials failure, neglect, and, more particularly, sabotage or acts of war. In this connection, distribution systems using dispersed power generation will always be safer and inherently more reliable than those using centralized generation. Compared to other power generation systems operating on natural gas, fuel cell systems are at least as safe. However, fuel cell systems convert the natural gas to hydrogen as an intermediate fuel, and many people appear to have a conceptual fear of the flammability of hydrogen, especially in confined spaces, resulting from the *Hindenburg* disaster. The odorant in natural gas does give some protection, whereas the hydrogen-rich gas after cleanup and Carnot processing has no odor. Thus, any leakage of hydrogen-rich gas from a fuel cell power plant will require sensors for detection. Without going into detail on the safety aspects of hydrogen, note that protection against any hydrogen leakage is relatively easy, since its high diffusivity and lower molecular weight causes it to disperse rapidly. Additionally, a hydrogen leak from the same hole or crack is less dangerous than involving natural gas at the same pressure: while the number of molecules lost in a given time is greater, the overall energy lost will be lower. Some aspects of hydrogen safety are given in Ref. 47. Since hydrogen and CO were important constituents of many gaseous fuels in the earlier decades of this century (producer gas, water gas, town gas, coke oven gas, coal gas, etc.), the infrastructure for handling those compounds represents proven (and partly even nineteenth-century) technology.

5.16. MATERIALS

Like other emerging technologies, fuel cells are basically materials-limited. However, their problems in this respect are certainly much less critical than those associated with magnetohydrodynamics, nuclear fission, and nuclear fusion (in which much larger amounts of R&D funding have been spent). Certain materials that are presently necessary (platinum in the PAFC, lanthanum and zirconium in the SOFC) may impose constraints on widespread use. This question has been reviewed for platinum in acid electrolyte fuel cells.[48] The MCFC and, potentially, the advanced AFC have virtually no materials constraints.

5.17. MULTIFUEL CAPABILITY

5.17.1. General

Although the first generation of electricity utility fuel cells can consume only hydrogen-containing gases as fuel, the hydrogen does not need to be pure and

Table 5.3. Effect of Switching Fuels in a PAFC Plan Design to Operate on Natural Gas–Naphtha on Cost of Plant and on Plant Efficiency

	Naphtha or methane	Medium BTU coal gas[a]	Methanol	hydrogen
Cost Impact, % above or (below) baseline	0	(1) ~ 5	1	(11)
Efficiency, %	41	41	41.5	42

[a]Depends on specific gas composition.
From Ref. 49.

may be from a variety of potential sources, including steam-reformed natural gas or light petroleum distillates such as JP-4 or naphtha, or from steam-reformed alcohols such as methanol. These fuel cells also can use the hydrogen contained in synthetic fuels such as remotely manufactured high-, medium-, or low-heating-value coal gas. This versatility of the fuel cell will enable it to be readily adapted to future changes in energy sources, and should ensure that it will not become redundant or obsolete upon the unavailability of certain fuels. In addition, large additional capital investments are not required to switch from one steam-reformable fuel to another. This aspect of fuel cell operation has been studied by UTC under contract to the Tennessee Valley Authority (TVA). Tables 5.3 and 5.4 show some of the results.[49] Table 5.3 shows the effects on plant cost and efficiency of using alternative fuels in a phosphoric acid fuel cell power plant designed for use with natural gas or naphtha. The baseline configuration includes fuel cleanup, a steam reformer, a shift reactor to reduce CO to less than 1.5% to prevent anode catalyst poisoning, and the heat exchanger equipment required to integrate the waste heat produced from the fuel cell power sections to the fuel treatment subsystem. Some attention is also given to the effect of using different fuels in fuel cells in Chapter 6.

It is noteworthy that the electrical energy conversion efficiency of the plant (input fuel to ac power) is almost independent of the fuel used, since the heat requirements for the fuel processor derive from the waste heat from the power section (steam and anode tail-gas) and not from the burning of extra fuel (see also Chapter 6). In this connection, we should remember that upgrading the HHV of propane or methane (about 111 kJ/equivalent) or of methanol (about 121 kJ/equivalent) to that of hydrogen (143 kJ/equivalent) requires a net energy

Table 5.4. Effect of Retrofitting a Natural Gas–Naphtha PAFC to Operate on Alternative Fuels

Fuel	Heat rate (kJ/kW-h)	Heat rate (BTU/kW-h)	Capital cost (% change)
Methane (baseline)	8808	8350	0
Medium BTU gas			
with methane	8818	8360	+0.4
without methane	8808	8350	+4.3
Methanol	8660	8210	+1.0
Hydrogen	8544	8100	+1.1

From Ref. 50.

input. If hydrogen were available on-site as a fuel (e.g., as a byproduct from the chloralkaline industry), a large capital savings in the system could be realized. This is further discussed in the following section.

There are special cases where a more efficient and cheaper plant would result if it were designed for only one fuel. Two of these are a dedicated methanol-fueled system (which has the advantage of simplified, low-temperature reforming), and a plant designed only for use on pure hydrogen, should this be available. For the methanol plant, an overall HHV efficiency of about 49% should be possible, with a capital savings per kilowatt of about 15%, resulting from the higher efficiency. If a complete plant were designed for use with hydrogen only, then much more significant savings could be realized, because the entire fuel processing sections and water recovery subsystems could be eliminated. In this latter case, a plant cost reduction of at least 25% and an absolute HHV efficiency of almost 50% may be expected. Another case is a dedicated coal-gas plant. In the latter, the integrated steam-reforming unit is uneccessary and the fuel cell stack can be directly supplied with remotely manufactured gas or eventually integrated on-site with a coal gasifier. For use in a PAFC, the gas must be water-gas-shifted to be low in CO, whereas this is not necessary for the MCFC and SOFC.

Table 5.4 shows the effects of retrofitting a plant that has been designed for use with natural gas to enable it to run on alternative fuels. As indicated in this table, and concluded by the TVA–UTC study, the equipment modifications required to retrofit a phosphoric acid fuel cell power plant to an alternative fuel are not extensive, within limitations such as discussed above. Furthermore, the use of coal-derived fuels will have almost no impact on power rating and only a very slight impact on heat rate or capital cost.

5.17.2. *Fuel Cells and Pure Hydrogen Fuel*

All present generation fuel cell units first convert carbonaceous fuels to hydrogen-rich gases for direct use in the fuel cell stack. The major difference between a fuel cell power plant operating on a fuel such as natural gas and one operating on hydrogen is that no fuel processing is required in the latter case. A fuel cell system operating on hydrogen will consist only of the stack itself, a feedback loop for cooling and water removal, associated subsystems such as pumps, valves, and heat exchangers, and power conditioning equipment. The feedback loop could be arranged in such a way that all the hydrogen in the system is consumed. Hence, with hydrogen, the fuel cell power unit is likely to be much less expensive than one using natural gas. In addition, hydrogen fuel cell power units will be much lighter and more compact than those operating on fuels which must be reformed, such as natural gas or methanol.

It is therefore not inconceivable that high-power-density hydrogen fuel cell systems will eventually be used for future transportation needs. However, when (and if) pure hydrogen fuel does become available in quantity, widespread use of fuel cell power units for vehicles will require systems without expensive, low-availability platinum catalysts.[48] As an example, if 25% of the platinum mined today (100 metric tons/year) is to be used in phosphoric acid fuel cells requiring 5 g Pt/kW, this will represent only 5000 MW or 120,000 automobiles per year, even assuming a modest 20 kW per vehicle. Thus the most attractive approach which can be foreseen at present for vehicle use is a small mass-produced AFC with 2500–5000 h lifetime, using the approach of the Alsthom alkaline system of the 1970s,[4] but with non-noble metal cathodes and with the elimination of the circulating electrolyte. These relatively short stack lifetimes (for example, 5000 operating hours on an intermittent basis) would be satisfactory for private automobile use. Unlike the proposed monolithic SOFC, such a system has the advantage of starting easily from cold. As envisaged today, an AFC along these lines would use lightweight components (porous graphite for the electrolyte reservoir plates, graphite–plastic composite or nickel-plated magnesium bipolar plates, plastic frames, and other cell parts) with pyrolized macrocyclic cathodes and could produce probably 50 kW in a 90-kg stack package. One approach which would allow the use of a noncirculating electrolyte might use one which is effectively replaced by capillary action through the porous electrolyte reservoir plates to avoid carbonate buildup. Cell stack costs on the order of $20 per kilowatt can be ultimately targeted.

For utility application, heavier, more reliable, and longer-lived equipment will be necessary. The cost for a H_2 driven system might in the end be 10 times that of automobile units, or about $300 per kilowatt, including the dc–ac inverter. Hydrogen-powered electric utility units of this type should be capable of producing electricity from hydrogen more cheaply than fuel cell systems with a fuel processor using natural gas. For example, if future natural gas costs are $6.00/per gigajoule as is predicted by the Electric Power Research Institute for the year 2000, with hydrogen costs (from coal) equal to $10.00 per million Btu, then with natural gas fuel cells costing $800 per kilowatt and hydrogen fuel cells at $200 per kilowatt, both in mature production, the expected cost of electricity will be 6.84 (dollar) cents per kilowatt hour for the natural gas unit and 6.78 cents per kilowatt hour for the hydrogen-powered unit. The reason for this is the lower capital cost of the hydrogen unit, which favors the final electricity cost, despite the higher fuel cost.

It is generally stated that a fuel cell power unit intended for on-site or utility use, with certain rare exceptions, cannot use hydrogen today because of its cost. The exceptions may be waste hydrogen from chlorine-caustic plants in certain locations, and perhaps inexpensive hydrogen made by electrolysis from hydropower, provided that the hydrogen can be economically transported to the site of use. However, the latter is not possible today because of the general absence of any infrastructure to transport hydrogen except in relatively small amounts as a liquid (by barge, tank car, or truck) or as a compressed gas in cylinders, all of which are relatively expensive. The costs of transporting hydrogen by these means is illustrated by the fact that the cheapest hydrogen available today, made by steam reforming of natural gas costing $1.85/per gigajoule ($2.00/MM Btu) has an on-site cost of about $5.00 per gigajoule. Liquid hydrogen made from this natural gas now costs $16.00 per gigajoule at the plant and about $20 per gigajoule delivered, whereas compressed hydrogen (delivered) can cost up to $100 per gigajoule, due to the weight of cylinders transported along with it.

In consequence, any hydrogen available must normally be used on site. The only immediate possibility where excess hydrogen (beyond any requirements of on-site chemical use) might be available for electricity production may be in some chlorine-caustic plants, where the dc power can be directly used in the electrolysis cells. Even if cheap hydrogen may be theoretically available at a hydroelectric plant, it clearly makes no economic sense to use it on-site to make electricity at the hydroelectric plant. Pipelines are by far the cheapest way to transport hydrogen,[47] and they will be needed if future hydrogen use is to become widespread. Hydrogen pipelines do presently exist in parts of Germany, Belgium, and France. Even in the United States, there is a small and underused pipeline in South Texas joining industrial plants. However, hydrogen distribution for general use by this method would require a capital-intensive pipeline network similar to (and largely the same as) that used for natural gas at present.

5.18. FUEL CELL POWER PLANT ECONOMICS

Whether fuel cell power plants will be widely used in the future will depend on their economics compared with competing systems, which particularly include the CTCC. This latter will have a capital cost advantage over the utility fuel cell power plant systems envisaged today. Since the latter must convert carbonaceous fuels into hydrogen-rich gas, they are complex and capital-intensive. The fuel processing system in future fuel cell power systems which use external reforming or coal processing is likely to comprise at least 75% of the total cost and bulk of the unit as a whole. It is now estimated that PAFC systems in mature production will cost something on the order of $800 per kilowatt, of which the stack will represent perhaps 20%. In the PAFC, the stack cost will include platinum catalyst worth about $80 per kilowatt, so that the total stack cost minus the catalyst will also be approximately $80 per kilowatt. High-temperature systems, especially those using internal reforming, are likely to be less expensive and more efficient than the PAFC, and therefore more competitive with advanced CTCC systems, once those price levels will have been approximated.

However, it is apparent from the earlier discussion that fuel cell power units have unique features which make them more attractive than CTCCs or conventional large power plants in many cases. These include their dispersibility, high efficiency in small units, and negligible emissions, all of which will allow them to be placed in locations close to the end use of the power produced, where it will be impossible to site other power plants. They are also uniquely adapted to cogeneration. The USA Fuel Cell Users Group (FCUG) was incorporated in the early 1980s to determine in more detail the benefits to be obtained from the use of fuel cells. After performing this task it disbanded in the spring of 1988, as was planned in the original articles of incorporation in the spring of 1988. It had members representing about 50 utility organizations, who worked closely with the Electric Power Research Institute (EPRI). The results of these efforts, obtained after close consultation with 25 privately owned U.S. utilities, representing about 15% of total U.S. electrical generating capacity, are summarized in Table 5.5. This table indicates in a quantitative manner the savings in 1982 dollars which may be made by installing fuel cells instead of conventional power plants. Another report has estimated that the use of fuel cell power plants in dispersed siting applications could result in savings of up to $154 per kilowatt in deferred transmission and distribution costs, up to $26 per kilowatt in reduced line losses, $10–$20 per kilowatt in reactive power control, and up to $200 per kilowatt in environmental credits.[50] The most complete and up-to-date assessment of fuel cell benefits to utilities was published in April 1986.[51] Interest in both gas and utility

Table 5.5. Perceived Valuation of Fuel Cell Advantages by Utility Planners

Benefit	Value ($/kW capacity)[a]		
	Low	High	Average
Air emission offset	0	217	11
Spinning reserve	0	33	8
Load following	0	52	7
VAR control	11	22	12
Transmission and distribution capacity	0	247	59
Transmission and distribution energy losses	8	35	15
Cogeneration	0	442	56

[a] Average of values selected by 25 utility planners.

fuel cells in the United States and elsewhere is high. For example, in late 1988 the American Public Power Association (APPA), representing small municipal utilities, issued a request for proposals (Notice of Market Opportunity for Fuel Cells, 1988). This has led to new demonstration activity for the MCFC system.[52]

5.19. CONCLUSIONS

The rule in the power industry has been that both power generation systems and transmission networks should be as large as possible to achieve economies of scale. Because of this, buildup of capacity has tended to occur stepwise in anticipation of future demand. Hence, large amounts of system capacity have often remained idle until a demand materializes. This problem is exacerbated by the fact that in developing rural and frontier areas demand growth is unpredictable, especially in the early stages of power availability. Because of the overriding importance of the efficient use of capital, there is a need for an approach permitting capacity addition that is in phase with demand growth, thereby allowing the capital input and power capacity in a system to be utilized at all stages of its development.

The use of fuel cells can provide such an improved use of construction capital because system capacity can be added in small increments that are consistent with the growth in actual demand. On account of their modular nature, the lead times for fuel cell system construction will be short, typically less than three years from the initial planning stages to final installation, which would even become less than one year for smaller-capacity units. This is in sharp contrast with the 10- to 14-year lead times required for conventional power plants. Hydrocarbon-fueled simple-cycle combustion turbines in the 10-MW class can be installed within two to three years, but they are still inefficient compared with fuel cell power plants. They have typically had efficiencies in the 23–28% LHV range, although 33% LHV efficiencies are available in the latest large units (150 MW) with high turbine inlet temperatures. Since they are combustion machines producing relatively high nitrogen oxide levels, their environmental acceptability is much less than that of fuel cells, as mentioned earlier.

The short lead times for fuel cell power plant installation, combined with their high efficiency even at submegawatt-class ratings, should permit them to be added annually to the generating mix, whereas conventional large central station systems are normally not fully utilized for 5–10 years after building commences. This may result in capital charges which are 30–50% lower for fuel cells[31] than those for conventional central station plants.

The ability to add capacity in small increments, as required by increased demand, minimizes the penalties and business risks that result from inaccurate load forecasting and provides a significant degree of financial flexibility for planners. For example, by gradually adding a few smaller units per year, a relatively constant annual cash flow can be achieved. During short periods of relatively high interest rates it is possible for a utility to limit its short-term debt by delaying the installation of one or two small incremental fuel cell units. This type of planning flexibility is possible with fuel cell power plants, whereas making even modest changes in the installation schedule in a system consisting of a large base-load conventional power plant has proven to be very capital intensive.

Because of high construction costs, budget constraints, and uncertainties in load growth, few of the large U.S. utilities are currently building new central power plants, and most of these now appear to feel that they will not need fuel cell power plants for some time to come. In spite of this, many projections indicate that utilities will require additional generation capacity, especially in urban and suburban areas, by the mid-1990s, especially in the northeastern United States. If these predictions materialize, utilities will have no time to install conventional large-scale power plants. It appears likely that early commercial fuel cell plants will acquire a share in filling that gap.

REFERENCES

1. B. J. Crowe, *Fuel Cells, A Survey* (National Aeronautics and Space Administration, Washington, D.C., 1973), SP-5115.
2. D. Gidaspow, R. W. Lyczkowski, and B. S. Baker, U.S. patent No. 3,823,038 (1974).
3. R. Roberts, in *The Primary Battery* (G. W. Heise and N. C. Calhoun, eds.) (Wiley, New York, 1971), Ch. 9.
4. J. O'M. Bockris and A. J. Appleby, *Energy* **11,** 95 (1986).
5. A. J. Appleby and E. B. Yeager, *Energy* **11,** 137 (1986).
6. J. W. Gibbs, *Trans. Connect. Acad.* **2,** 382 (1873).
7. Z. Rant, *Forschung Ing.-Wesens* **22,** 36 (1956).
8. H. D. Baehr, *Energie und Exergie* (VDI Dusseldorf, Germany 1965).
9. A. J. Appleby, *Energy* **11,** 13 (1986).
10. L. M. Handley, *Description of a Generic 11 MW Fuel Cell Power Plant for Utility Applications* (Electric Power Research Institute, Palo Alto, CA, 1983), EM-3161.
11. B. E. Curry, *National Fuel Cell Seminar Abstracts* (Electric Power Research Institute, Palo Alto, CA, 1981), p. 13.
12. J. H. Horton, *Proc. 25th Power Sources Symp.* p. 92, 1972.
13. A. J. Appleby, in *Fuel Cells, Trends in Research and Application* (A. J. Appleby, ed.) (Hemisphere, New York, 1987), p. 19.
14. O. Lindstrom, in Ref. 13, p. 191.
15. A. J. Appleby, in *Modern Aspects of Electrochemistry*, Vol. 9 (B. E. Conway and J. O'M. Bockris, eds.) (Plenum, New York, 1974), p. 369.
16. A. J. Appleby, in *Comprehensive Treatise of Electrochemistry*, Vol. 7 (B. E. Conway, J. O'M. Bockris, S. U. M. Khan, and R. E. White, eds.) (Plenum, New York, 1983), p. 173.

17. E. W. Hall, L. M. Handley, and G. W. May, *Capital Cost Assessment of Phosphoric Acid Fuel Cell Power Plants for Electric Utility Applications* (Electric Power Research Institute, Palo Alto, CA, 1988), AP-5608.
18. A. J. Appleby and F. R. Foulkes, *Fuel Cell Handbook* (Van Nostrand Reinhold, New York, 1989), pp. 613–616.
19. T. G. Benjamin, *Handbook of Fuel Cell Performance* (Institute of Gas Technology, Chicago, IL, 1980).
20. J. T. Brown, *Energy* **11,** 209 (1986).
21. W. G. Parker, *Fuel Cell Seminar Abstracts,* 1988, p. 248.
22. M. Krumpelt, V. Minkov, J. P. Ackerman, and R. D. Pierce, *Fuel Cell Power Plant Designs, A Review* (U.S. Department of Energy, Washington D.C., 1985), DOE/CC/4994 1-1833.
23. J. R. Selman, *Energy* **11,** 153 (1986).
24. L. M. Handley and R. Cohen, *Specification for Dispersed Fuel Cell Generator* (Electric Power Research Institute, Palo Alto, CA, 1981), EM-2123.
25. Anon. (UTC Power Systems Div.), *On-Site Fuel Cell Power Plant Technology Development Program* (Gas Research Institute, Chicago, IL, 1983), 1982–1983 Annual Rep., GRI 82099; *On-Site* 40 *kw Fuel Cell Development Program,* (U.S. Department of Energy, Washington, D.C., 1985), DOE/NASA-0255-1 NASA CR-174988.
26. W. C. Racine and T. C. Londos, *40-kW On-Site Fuel Cell Field Test Summary Utilities Activities Report* (Gas Research Institute, Chicago, IL, 1987), GRI 87/0205.
27. T. N. Veziroglu, *Int. J. Hydrogen Energy* **12,** 99 (1987).
28. B. Kileman, *Global Warming, Chem. Eng. News Special Report* March 13, 1989.
29. S. H. Schneider, *Global Warming, Are We Entering The Greenhouse Century* (Sierra Club Books, San Francisco, CA, 1989).
30. D. R. Inglis, *Nuclear Energy: Its Physics And Its Social Challenge* (Addison-Wesley, Reading, MA, 1973).
31. J. M. King, *Proc. 8th Intersoc. Energy Eng. Conf.*, 1973, p. 111.
32. G. L. Johnson, *Proc. Nat. Fuel Cell Seminar Abstracts,* 1979, p. 38.
33. L. M. Handley, M. Kobayashi, and D. M. Rastler, *Operational Experience with Tokyo Electric Power Company's 4.5 MW Fuel Cell Demonstration Plant* (Electric Power Research Institute, Palo Alto, CA, 1986).
34. N. Itoh, T. Kimura, M. Ogawa, H. Kaneko, and H. Kawamura, *Fuel Cell Seminar Abstracts,* 1988, p. 226.
35. H. Harada and Y. Mori, *Fuel Cell Seminar Abstracts,* 1988, p. 18; Y. Yamamoto, S. Kaneko, and H. Takahashi, *Fuel Cell, Seminar Abstracts,* 1988, p. 25.
36. L. B. Lave, *Proc. 6th. Intersoc. Energy Eng. Conf.*, p. 337, 1971.
37. Anon. (UTC Power Systems Div.), *4.5 MW Fuel Cell Development Program* (Electric Power Research Institute, Palo Alto, CA, 1984), EM3856-LD.
38. In Ref. 18, pp. 151–160.
39. Anon., Haldor-Topsøe Inc., Commercial Brochure (Electric Power Research Institute, Palo Alto, CA, 1986).
40. KTI Radiant Tube Reformer, U.S. Patent 4,692,306, September 8, 1987, KTI Corp.
41. S. Sato, IHI, Private Communication; Y. Miyazaki, H. Okuyama, T. Kodama, A. Fukutome, and Y. Kurihara, in *Molten Carbonate Fuel Cell Technology* (J. R. Selman, H. C. Maru, D. A. Shores, and I. Uchida, eds.) (The Electrochemical Society, Pennington, NJ, 1990), p. 50.
42. In Ref. 18, p. 615.
43. D. W. Boyd, O. E. Buckley, C. E. Clark, Jr., R. B. Fancher, and J. R. Selman, *EPRI Roles in Fuel Cell Commercialization* (Electric Power Research Institute, Palo Alto, CA, 1987), AP-5137.
44. F. J. Rohr, *Proc. Workshop High Temperature Solid Oxide Fuel Cells* (Brookhaven National Lab., 1977), p. 122.
45. In Ref. 18, p. 620.
46. R. Anahara, Fuji Electric Co., private communication, June 1986.
47. D. P. Gregory, *A Hydrogen Energy System* (American Gas Association, Washington D.C., 1973), L 21173.
48. A. J. Appleby, in *Precious Metals 1986* (U. V. Rao, ed.) (International Precious Metals Institute, Allentown, PA, 1986), p. 1; *Proc. 10th. IPMI Precious Metals Conf.*, p. 379, 1986.
49. L. J. Henson and S. B. Jackson, Contract TV 53900A, Final Report FCR-2948, Tennessee Valley Authority (April 1981); See A. J. Appleby, in *Proc. Renewable Fuels and Adv. Power Sources for*

6

System Design and Optimization

Michael N. Mugerwa and Leo J. M. J. Blomen

6.1. INTRODUCTION

New power generating technologies need to be clean, highly efficient at both full and partial load, highly reliable, and competitively priced. Moreover, they need to be constructed as modules to enable stepwise capacity addition at reasonable cost and to increase planning flexibility. They must also have low maintenance requirements and low noise levels. It is generally expected that fuel cell power systems of the future will be able to meet all these criteria.

The objectives of this chapter include an explanation of the fuel cell *system* concept, its various components, and how they interact. Further, to show how an integrated system design may be systematically derived and optimized for both electrical efficiency and installed cost, including an assessment of all critical design parameters and options. In addition, this chapter will present brief discussions on, *inter alia*, expected trends in fuel cell power plant design, the effect of changing feedstocks, and system reliability issues.

In principle, a fuel cell system comprises fuel cell and fuel processor subsystems, and, depending upon battery-limit conditions and user requirements, a power conditioner and other ancillary subsystems, such as a water treatment plant. In the design of most fuel cell systems one endeavors to develop a flowsheet whose overall electrical efficiency exceeds that of the fuel cell subsystem itself (see Chapter 5). Such synergy is obtained if the "waste" products of one subsystem can be utilized in another subsystem, resulting in a reduction of the input requirements to the entire system. However, this approach does demand that a balance be struck between the quantity and quality of the available "waste products" and the needs of the complementary subsystem. This is easily illustrated by considering that the "waste heat" from a molten carbonate fuel cell stack at 650°C can be easily employed in a typical fuel processor, which operates between 400 and 800°C, whereas the low-level waste heat of c. 80°C from an

Michael N. Mugerwa • Kinetics Technology International S.p.A., Via Monte Carmelo 5, 00166 Rome, Italy. Leo J. M. J. Blomen • Mannesmann Plant Construction, 2700 AB Zoetermeer, The Netherlands. *Present address*: Blomenco B. V., Achtermonde 31, 4156 AD, Rumpt, The Netherlands.

Fuel Cell Systems, edited by Leo J. M. J. Blomen and Michael N. Mugerwa. Plenum Press, New York, 1993.

alkaline fuel cell (AFC) stack cannot. One should also consider that process conditions, process equipment, and working fluids must be selected such that the quality of the waste products is not excessively degraded.

The primary purpose of a fuel processor is the conversion of commercially available fuels, such as natural gas or liquified petroleum gas (LPG), to a hydrogen-rich gas that can readily be utilized in a fuel cell stack for the production of electricity (see Chapter 4). The performance of the fuel processor, in terms of amount of hydrogen-rich gas produced for a given amount of primary fuel feedstock, can be optimized by the adjustment of parameters such as reforming pressure and temperature, steam-to-carbon ratio, combustion pressure, reformer geometry, shift reactor configuration, and heat exchanger network. The optimum values for all these and other parameters vary considerably depending upon the fuel cell type and its method of cooling, the available primary fuel, and the amount and quality of cogeneration heat required. It is also important to note that the optimal parameter settings are in general quite different to those considered as optimal for a hydrogen- or syngas-producing plant.

Fuel cell operating conditions may also be subject to optimizations, and include operating pressure and temperature, current density, fuel and oxidant compositions, and degree of fuel and oxidant utilizations.

As is evident from the discussion above, a host of parameters may be adjusted in order to obtain an optimized fuel cell system for a specific application. However, it must be emphasized that optimum operating conditions within the fuel cell stack alone may not necessarily result in an optimized system design. Integration between the fuel cell and fuel processing subsystems takes place through materials and heat and power recovery subsystems (see Fig. 6.1).

So, for example, *materials recovery* might be the use of anode off-gas as fuel in the fuel processor combustor, the use of depleted cathode air in the reformer burners, and the use of fuel cell product steam in the fuel processor. *Heat recovery* might be the use of fuel cell reaction heat and reformer flue gas heat for process-steam-raising purposes. Finally, *power recovery* could be the use of pressurized effluent streams to satisfy the plant's shaft power demands. The greater the degree of interaction possible between the fuel cell and fuel processing subsystems the better the combined performance of the system design will be.

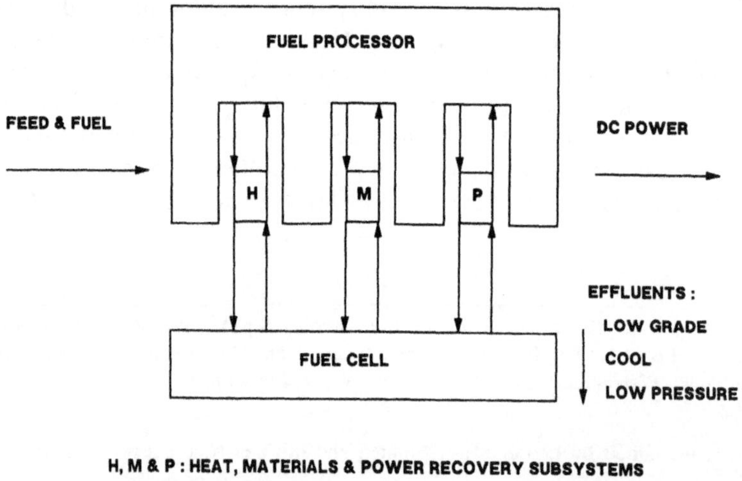

Figure 6.1. The fuel cell system integration concept.

In addition to optimizing operating conditions in a fuel cell system it is also worth considering several process options, which could include oxygen enrichment, carbon dioxide removal methods (especially for molten carbonate fuel cell (MCFC) systems), and differing cell cooling mechanisms. The application of such options can prove to be economical for specific battery-limit conditions or for multi-MW_e system capacities. Technical optimizations as outlined above are always constrained by cost considerations and therefore plant cost and performance trade-offs must be adequately quantified. One method of achieving this balance is by the application of pinch technology.[1,2]

In the following sections, the considerations outlined above necessary for the design and development of economically competitive fuel cell systems are discussed and illustrated by way of examples. Critical aspects of plant design such as control and instrumentation requirements, partial load capability, dynamic behavior, and plant emissions will also be given coverage. Emphasis is placed upon phosphoric acid fuel cell (PAFC) and MCFC system technologies, though some attention is also given to other fuel cell system technologies. Most discussion centers on natural-gas-based fuel cell systems, as this is generally expected to be the fuel upon which fuel cell technology will launch itself into commercialization, and for which most operating experience exists. Nevertheless the requirements and performance of fuel cell systems on alternative fuels such as biogas, LPG, naphtha, and coal are also reviewed.

6.2. THE IMPORTANCE OF FUEL CELL SYSTEM INTEGRATION

Fuel cells are electrochemical devices capable of continuously transforming the chemical energy of a fuel, typically hydrogen, and an oxidant, typically oxygen from air, directly to electrical energy. This process is in fact the reverse of electrolysis. It is an isothermal process that involves the conversion of the Gibbs free energy of reaction to electricity. This implies that straightforward Carnot cycle limitations do not exist, which gives rise to the possibility of combining fuel cells and fuel processors into systems with high fuel-to-electricity conversion efficiency (see Chapter 5). Based upon the lower heating value (LHV) of a fuel fed to a fuel cell, efficiencies of up to 60% have been reported.[3] The inefficiency of the process is in the form of heat that must be removed from the fuel cell in order to maintain an isothermal environment. The recovery and utilization of this waste heat in other parts of a fuel cell system help to build up the overall system efficiency.

The basic fuel cell structure consists of an electrolyte within a support matrix sandwiched between two electrodes (the anode and the cathode). Figure 6.2 is a schematic showing the principle of fuel cell operation typical for an acid electrolyte cell.

Each electrode contains a catalyst to enable an appropriate electrochemical reaction to take place:

At the anode,

$$H_2(g) \rightarrow 2H^+ + 2e^-, \tag{6.1}$$

and at the cathode,

$$2H^+ + \tfrac{1}{2}O_2(g) + 2e^- \rightarrow H_2O(g). \tag{6.2}$$

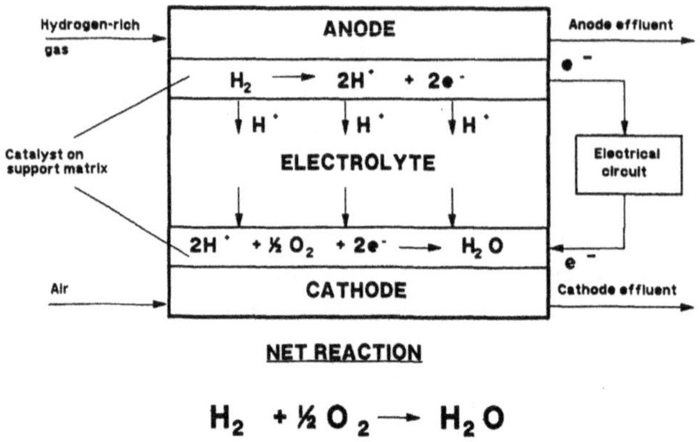

Figure 6.2. Schematic showing the principles of fuel cell operation.

The ions are transferred from the anode to the cathode via the electrolyte, and the electrons via an external load circuit. The overall reaction is therefore

$$H_2(g) + \tfrac{1}{2}O_2(g) \rightarrow H_2O(g) + \text{dc electricity} + \text{heat}. \qquad (6.3)$$

Although the same overall reaction occurs in a molten carbonate cell, there are significant differences in the mechanism of ion transfer compared to that in an acid cell (see Chapter 3). MCFC operation involves the transfer of carbonate ions (CO_3^{2-}) from the cathode to the anode, whose effect on the system design is such that carbon dioxide produced at the anode must be transferred back to the cathode inlet in order to prevent carbonate ion depletion of the electrolyte.

A fuel cell under typical operating conditions will generate between 0.6 and 0.8 V, which is not by any means a practical working voltage.[4,5] Therefore, in practice, fuel cells are connected in series to form a stack to which a hydrogen-rich fuel gas and oxidant are fed by means of a manifold system.[3]

The provision of a hydrogen-rich gas to the fuel cell stack requires the conversion of commercially available fuels, such as natural gas and naphtha, in a fuel processor. In addition, the end-user requirements are often such that the dc electric power produced by the fuel cell stack has to be converted to ac electricity and transformed to a higher voltage in a power conditioner. Therefore, a typical fuel cell power plant comprises three major subsystems: a fuel processing section, a fuel cell power section, and a dc-to-ac inverter or power conditioning section, as shown in Fig. 6.3.[6]

Furthermore, an energy recovery system must be provided to recover heat and power for process heating and gas compression requirements, respectively, thereby enabling the achievement of maximum overall system efficiencies. In some process schemes a bottoming cycle, which typically comprises a steam turbine unit, may be employed to produce electric power from the energy recovery system.[4,7]

Typical net electrical efficiency figures for well-designed fuel cell power plants on a LHV basis are in excess of 45% for PAFC system and can attain about 60% for MCFC systems. By contrast only the most modern combined gas and steam turbine (combined cycle) power plants can achieve maximum efficiencies of around 50%.[8]

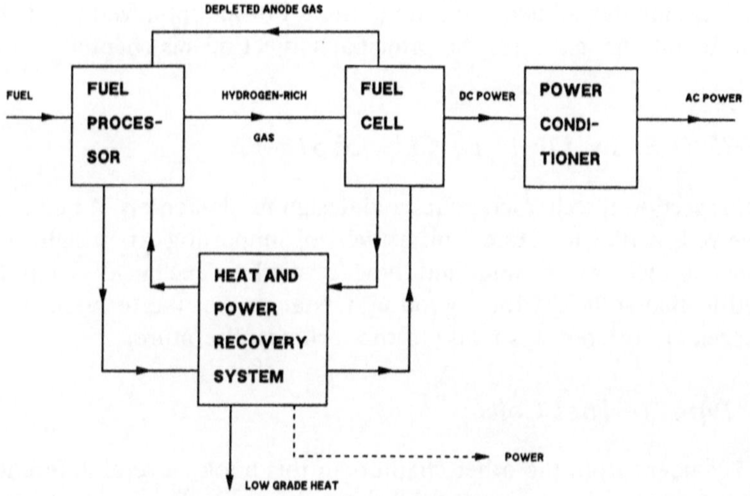

Figure 6.3. Fuel cell power plant.

It must be emphasized that overall fuel cell system efficiency is far more sensitive to the performance of the fuel processor and its proper integration with the fuel cell stack than to the performances of either the electrochemical or electrical components that constitute the balance of the system. This important principle is best illustrated quantitatively.

For PAFC systems, a typical fuel cell stack will have an efficiency range of 40–45% (based on LHV anode feed to dc electricity), dependent upon manufacturer and operating conditions.[4,5] The efficiency of dc-to-ac power conditioning will typically be 90–99%, mainly subject to the required capacity and the inverter technology employed by the manufacturer.[9] However, the efficiency of a fuel processor integrated with a fuel cell stack can vary from 70% to 130% subject to the degree of system optimization, type of fuel, and the fuel processor design.*[6]

In principle the overall system performance is computed as the product of the fuel processor, fuel cell, and power conditioner efficiencies, yielding a gross system efficiency of gross ac power produced for a given LHV feed and fuel (see Chapters 3 and 5). However, plant parasitic power requirements must also be accounted for; thus, the net system efficiency for the total PAFC power plant can vary between 25% and 42% for unpressurized systems, and rise to 47% for pressurized systems. In pressurized fuel cell operation, further enhancements to system performance are possible with pressurized combustion in the fuel processor because this enables tighter integration between the fuel cell and fuel processor operating conditions. Applying a similar analysis to MCFC systems can generate overall system efficiencies ranging from 32% to 60%.

The significance of these analyses is that while R&D efforts on fuel cell stack and inverter technologies are only likely to yield improvements of a few percent in overall system efficiency, the fuel processor and system designs will most

*Fuel processor efficiencies of greater than 100% are possible as they are defined as the ratio of LHV of the anode feed to the LHV of the feed and fuel to the fuel cell system. A consideration of the LHVs of the various components in the feed and product mixtures can yield a fuel processor efficiency of up to ca. 130% (see Chapter 5).

certainly have greater influence on the efficiency of an optimized plant. It is these latter considerations that form the principal subject of this chapter.

6.3. THE DESIGN OF FUEL CELL SYSTEMS

In this section the characteristics and design fundamentals of fuel cell systems are reviewed, with the stated objectives of imparting an insight into what constitutes a fuel cell system and how it works. Technical detail has been restricted to that sufficient for a good understanding of the technology, although where necessary reference is made to the technical literature.

6.3.1. Types of Fuel Cells

As is evident from the other chapters in this book, several different types of fuel cells are under development, and these are generally classified according to the type of electrolyte or ion conduction media and temperature of operation. These include the phosphoric acid fuel cell (170–210°C), the molten carbonate fuel cell (≈650°C), the alkaline fuel cell (≈80°C), the solid polymer fuel cell (15–100°C), and the solid oxide fuel cell (≈1000°C).[4,5]

Although the issue of system design and optimization is of importance for all types of fuel cells, the emphasis of this chapter will center on the development of phosphoric acid and molten carbonate fuel cell systems. This is because it is widely believed that it is these fuel cell types that will enjoy the most widespread use in stationary power plant applications in the near future (i.e., within 10 years).

The reader will also be aware from Chapters 5 and 9 that, in principle, two types of MCFC are possible: one based upon the internal reforming principle in which the methane fuel is reformed and thereby utilized directly within the fuel cell itself, and the other based upon the external reforming principle. It is the latter type of MCFC that forms the basis for discussion on MCFC system design and optimization in this chapter.

6.3.2. Phosphoric Acid Fuel Cell System Design

Phosphoric acid fuel cells are the most mature fuel cells in terms of technological advancement and readiness for commercialization. Intensive development has been ongoing for about 25 years, and total investment to date is estimated at more than \$500 million (U.S.).[3] Significant improvements in the performance, durability, and cost of PAFCs have been realized during their development (see Chapter 8). State-of-the-art PAFC power plants range from 25 kW_e to 4.6 MW_e, and the nominal economic target of a 40,000-h lifetime has been achieved for the fuel cell stacks. Demonstration systems of 2 MW_e, 5 MW_e, and 11 MW_e have been offered to the electric utility market. Improvements have been realized from the basic electrochemistry to system optimization,[3] with the upper limit to overall system efficiency currently at about 45% (LHV basis), with about 50% (LHV basis) expected for the future.

The PAFC is well down the technology learning curve and is at a critical stage in establishing its commercial viability. The major issues confronting PAFC

Table 6.1. Maximum Levels of Impurities Permissible in the PAFC[4]

CO_2	Diluent
CH_4	Diluent
N_2	Diluent
CO	<1% at 175°C <1.5% at 190°C <2% at 200°C
H_2S, COS	<100 ppm
C_2^+	<100 ppm
Cl^-	<1 ppm
NH_3	<1 ppm
Metal ions (Fe, Cu, etc.)	Nil

commercialization today are system reliability and installed-cost reduction.[10] Demonstration and early production units at a multi-MW_e scale are essential to effectively overcoming these market entry barriers.[10,11] These problems are probably also crucial to the sustained development of the next generation of fuel cells. In other words, PAFC demonstration power plant experiences that involve activities such as fuel processor development, modular design, control system design, advancements in power conditioning, and installed cost reduction efforts are likely to advance the commercialization of other fuel cell types, especially in utility applications.

Acid electrolytes have been considered for use in fuel cells because acids are tolerant to carbon dioxide which enables the use of impure hydrogen and air. For a detailed description of the principles of operation of the PAFC and a review of the developments of the technology, see Chapter 8. In designing PAFC systems critical consideration must be given to the impurities that adversely affect the performance of the PAFC. Table 6.1 outlines the maximum levels of impurities allowed in a PAFC.[4]

Typical PAFC voltage characteristics, for a given fuel composition, as a function of pressure are shown in Fig. 6.4.

The actual voltage used in the evaluation of the intrinsic fuel cell efficiencies varies according to the compositions of the fuel and oxidant feeds to the anode and cathode, respectively, at the given fuel cell operating conditions. The fuel cell efficiency describes the conversion of chemical energy in the anode feed gas to dc power and is defined as

$$\varepsilon_{FC} = \varepsilon_E \times \varepsilon_H \tag{6.4}$$

or

$$\varepsilon_{FC} = \varepsilon_{Th} \times \varepsilon_V \times \varepsilon_I \times \varepsilon_H. \tag{6.5}$$

A summary of the efficiency equations and the terms involved is given in Table 6.2.

Note that only the fuel cell power section of a fuel cell system is described by ε_{FC}. The gross electrical efficiency for a fuel cell system without taking into account parasitic load requirements is defined as

$$\varepsilon_{GROSS} = \varepsilon_{FP} \times \varepsilon_{FC} \times \varepsilon_{PC}, \tag{6.6}$$

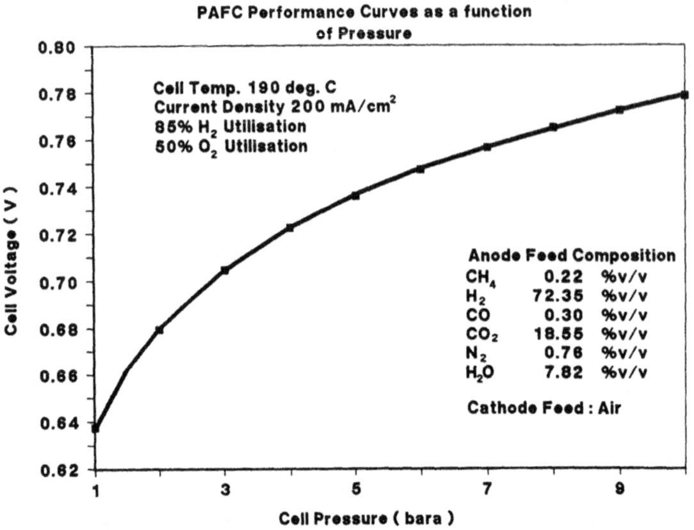

Figure 6.4. PAFC performance curves as a function of pressure.

where

ε_{FP} = LHV of anode feed gass–LHV of feed and fuel to fuel processor
ε_{FC} = dc power produced/LHV of anode feed gas
ε_{PC} = ac power/dc power*
ε_{GROSS} = ac power/LHV feed and fuel to fuel processor

Table 6.2. Summary of Parameters and Efficiency Equations[4]

V	Cell load, volts	
E	Open-circuit voltage	
ΔG_r	kJ/kmol anode feed gas	$-nFE\|_{T,P}\Gamma_{H_2}$
n	Number of electrons transferred in a reaction	
F	Faraday constant	
Γ_{H_2}	mol fraction of available hydrogen	
ΔH_r	kJ/kmol anode feed gas	$LHV_{H_2}\|_T\Gamma_{H_2}$
ΔH_c	kJ/kmol anode feed gas	$\sum LHV_i\|_T\Gamma_i$
LHV_i	Lower Heating value of component i, kJ/kmol$_i$	
ε_V	Voltage efficiency	V/E
ε_I	Current efficiency	Fuel utilization
ε_{Th}	Thermodynamic efficiency	$\Delta G_r/\Delta H_r$
ε_E	Electrochemical efficiency	$\varepsilon_V\varepsilon_I\varepsilon_{Th}$
ε_H	Heating value efficiency	$\Delta H_r/\Delta H_c$
ε_{FC}	Fuel cell efficiency	$\varepsilon_E\varepsilon_H$

*In this chapter all reported optimizations use the following power conditioner efficiencies:

System capacity (kW$_e$)	ε_{PC} (%)
25	93
250	96
3,250	98
100,000	99

The net electrical efficiency including parasitic losses is calculated as

$$\varepsilon_{NET} = \text{(ac power} - \text{parasitics)/LHV feed and fuel to fuel processor.} \quad (6.7)$$

The design of a PAFC system must observe the limitations on impurities to the fuel cell given in Table 6.1. It is the specification for carbon monoxide in the anode feed gas that has the most significant effect upon the design of the fuel processing system. Figure 6.5 depicts a typical flow scheme of a PAFC system using natural gas as feed and fuel.

Desulfurization of the natural gas feed is required since the sulfur compounds present in the natural gas would poison the reformer, shift, and fuel cell catalysts (see Chapter 4). The natural gas is compressed and preheated before entering the desulfurizer. The desulfurizer would typically contain a zinc oxide adsorbent which removes hydrogen sulfide and some reactive organic sulfur compounds from the feed gas stream. During the process the zinc oxide is converted to zinc sulfide, and when the bed is saturated with sulfur (ca. 20% wt) it is replaced with fresh adsorbent. In cases where organic sulfur compounds are present in the feed gas in small quantities (<10 ppm), the zinc oxide is sufficient for desulfurization requirements. However, if the natural gas contains nonreactive (e.g., thiophene) or larger amounts of reactive organic sulfur compounds, a hydrodesulfurization catalyst is required prior to the zinc oxide bed to convert these higher sulfur compounds to hydrogen sulfide. Finally, an alternative method of desulfurization especially suited for small (i.e., multi-kW_e) fuel cell systems, uses activated carbon to adsorb all sulfur compounds from the natural gas feed.

The desulfurized natural gas is mixed with superheated process steam and enters a regenerative reformer. Fuel gas for the reformer is anode off-gas and makeup natural gas. Preheated combustion air is supplied by an air compressor that also supplies the PAFC cathode with air. In the reformer, the process gas passes through tubes containing a nickel-based reforming catalyst. The methane and higher hydrocarbons are reformed in a strongly endothermic reaction to a hydrogen and carbon monoxide mixture, commonly called synthesis gas or syngas. By virtue of the so-called shift reaction the syngas is converted to still more hydrogen and carbon dioxide before exiting the catalyst bed. Exit temperatures may range from 700 to 900°C. The reformed gas then passes along a regenerative tube annulus simultaneously returning heat to the process gas passing over the reforming catalyst. This is a highly efficient arrangement which reduces the fired heat duty of the reformer by up to 25% when compared to a conventional reformer.[12–15] The process gas leaves the reformer at about 500°C.

The reformer effluent contains an equilibrium mixture of hydrogen, carbon monoxide, carbon dioxide, steam and methane. Typically, all higher hydrocarbons and more than 98% of methane are converted in the reformer. This process gas is subsequently sent to high- and low-temperature shift reactors to reduce the carbon monoxide level to that acceptable in the PAFC. The hydrogen-rich gas stream is fed to the fuel cell anode where up to 85% of the hydrogen is consumed by an electrochemical reaction to produce dc electricity. The anode off-gas is routed to the reformer burners.

Process optimization has shown that the operating pressure levels of a fuel cell system have a marked effect not only on fuel cell performance but also on the net system efficiency, ε_{NET}. The higher partial pressures of hydrogen and oxidant

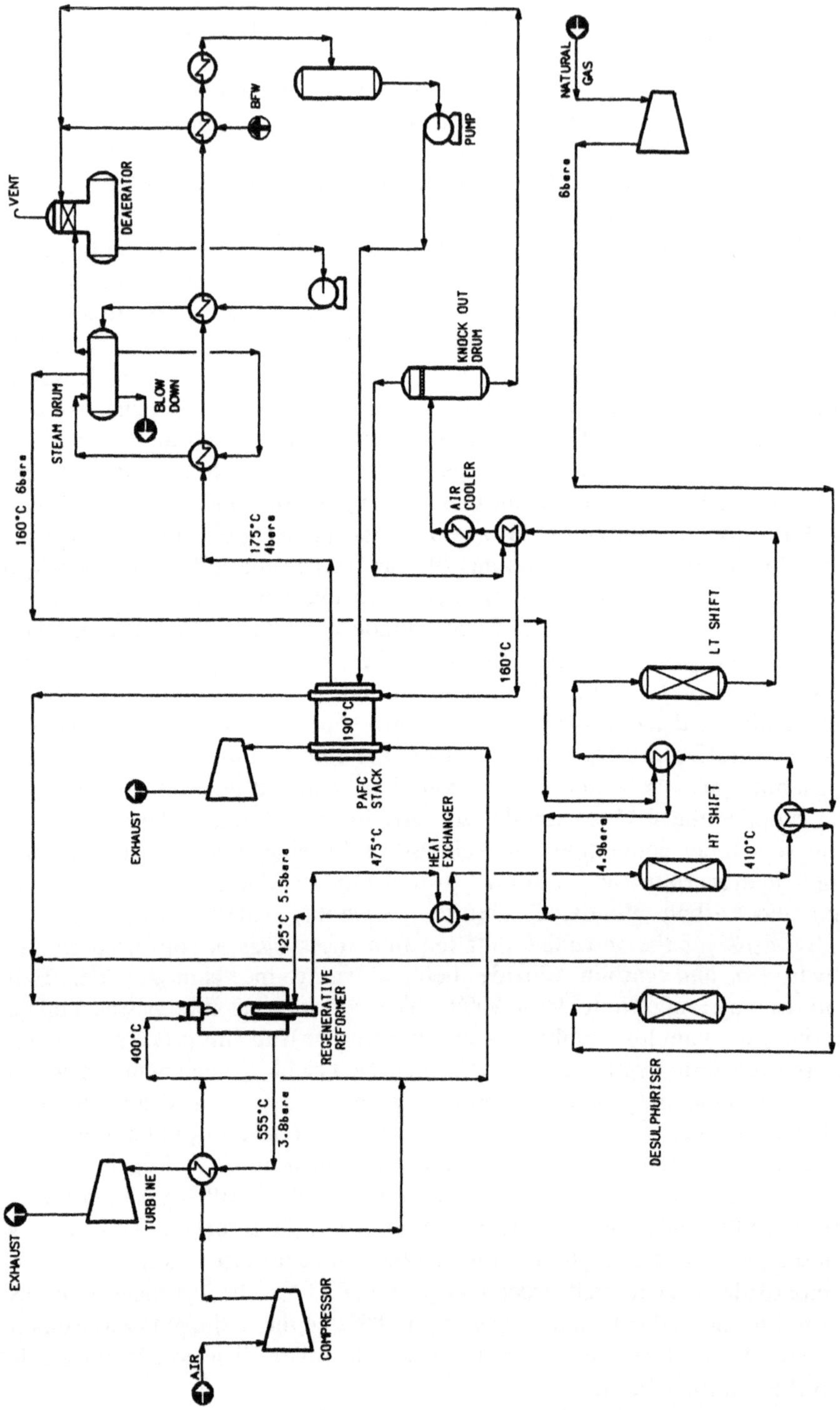

Figure 6.5. Flowsheet of a 250-kW PAFC power plant.

in the fuel cell translate directly to improvements in fuel cell efficiency, ε_{FC}. It is possible to enhance integration between the fuel cell and the fuel processor by employing pressurized combustion in the reformer firebox.[6] Pressurized combustion improves the heat transfer characteristics from the flue gas to the process due to increased radiative and convective heat transfer, which has the added benefit of making for a compact reformer design.[14,16] In addition, pressurized combustion makes possible the use of a turboexpander to recover shaft power to drive the air compressor. This power recovery principle makes the arrangement similar to that employed in gas turbine cycles. Shaft power may also be recovered from the cathode off-gas from fuel cell stacks operating at superatmospheric pressures.

Finally, a determining factor in the design of a highly efficient fuel processing arrangement is the systems for heat recovery. The fuel cell stack waste heat is removed using an indirect or direct cooling circuit and used to raise the process steam for the reforming reaction. The heat contained in the reformer flue gas is used for steam superheating, boiler feedwater preheating, feed gas preheating, and air preheating.

6.3.3. Molten Carbonate Fuel Cell System Design

The MCFC has a typical operating temperature of about 650°C and can use impure hydrogen as fuel and air as oxidant. The MCFC uses non-noble metal electrodes and has an electrolyte of molten alkali metal carbonates in a porous ceramic tile. Technically, the MCFC is at least five years behind the PAFC at current rates of development. The largest operating fuel cell stacks are in the 25-kW_e class.[17] However, when compared to the PAFC, the MCFC has an intrinsically higher fuel cell efficiency, ε_{FC}, and is generally expected to have a lower cost per kilowatt at commercialization. In addition, the significantly higher temperature of operation implies a higher level waste heat which enhances the possibility of tighter integration between the fuel cell and the fuel processor. In general, net system efficiencies (ε_{NET}) in the range of 50–60% are expected to be readily achievable with MCFC systems. For a detailed description of the principles of operation of the MCFC and a review of the developments of the technology, see Chapter 9. In designing MCFC systems critical consideration must be given to the impurities that adversely affect the performance of the MCFC.

Table 6.3 details the maximum levels of impurities expected to be allowable in the MCFC.[4,18,19] It is worth elaborating on two features in the table. Not only hydrogen but also carbon monoxide and methane are consumed as fuel in the MCFC. And in an MCFC it is the carbonate ions that transfer their charge across the electrolyte. The reactions at the cathode and anode are, respectively,

$$CO_2 + \tfrac{1}{2}O_2 + 2e^- \rightarrow CO_3^{2-} \qquad (6.8)$$

and

$$H_2 + CO_3^{2-} \rightarrow H_2O + CO_2 + 2e^-. \qquad (6.9)$$

The overall reaction in the MCFC is

$$H_2 + \tfrac{1}{2}O_2 \rightarrow H_2O + \text{electricity} + \text{heat}. \qquad (6.10)$$

Table 6.3. Maximum Levels of Impurities Permissible in the MCFC[4,18,19]

CO_2	Diluent at anode, produced at anode, and consumed at cathode
CH_4	Consumed at anode
N_2	Diluent
CO	Consumed at anode
H_2S, COS, CO_2	<1 ppm
C_2^+	<100 ppm
HCl	<0.1 ppm
NH_3	<10,000 ppm
Cd	<30 ppm
Hg	<35 ppm
Zn	<15 ppm
Pb	Maximum limit unknown, though tolerance expected to be low

These electrochemical reactions imply that carbon dioxide is consumed at the cathode and produced at the anode. In order to ensure that carbonate ions are not depleted in the electrolyte the fuel cell system has to be designed such that the carbon dioxide produced at the anode is returned to the cathode.

Typical MCFC voltage characteristics, for a given fuel composition, are shown as a function of pressure in Fig. 6.6. The actual voltage used in the evaluation of the intrinsic fuel cell efficiency, ε_{FC}, varies depending upon the compositions of the fuel and oxidant feeds to the anode and cathode, respectively, at the given fuel cell operating conditions. Therefore, in addition to the partial pressure of components in the fuel and the oxygen in the oxidant, the partial pressure of carbon dioxide will also affect the efficiency, ε_{FC}, of the MCFC. However, the definitions accorded to the intrinsic fuel cell efficiencies remain the same as those given in Section 6.3.2.

The most important limitation to the configuration of a MCFC system concerns the need to transfer carbon dioxide from the anode off-gas to the cathode inlet stream. This necessitates the use of carbon dioxide removal technology that can vary from the very simple to an entire carbon dioxide removal plant consisting of absorbers, regenerators, and the like. In addition, extra equipment is required to recycle the recovered carbon dioxide stream to the cathode, which adds to the system's shaft power requirements and can lead to reduction in the net system efficiency, ε_{NET}.

Figure 6.7 depicts a basic flow scheme developed for the simulation of an

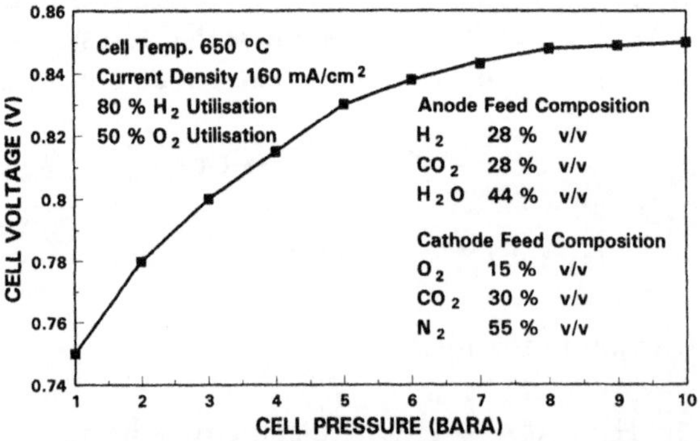

Figure 6.6. MCFC performance curves as a function of pressure.

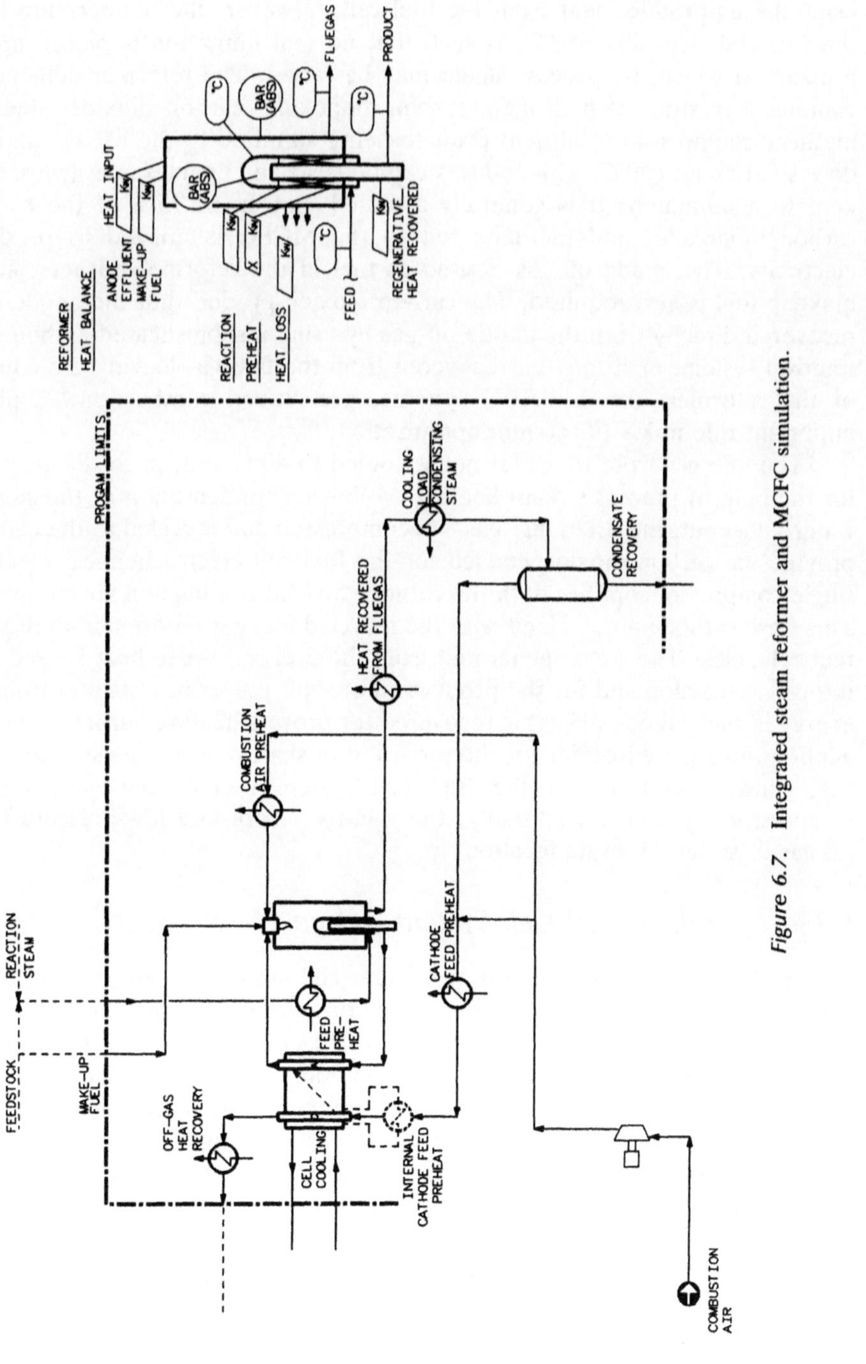

Figure 6.7. Integrated steam reformer and MCFC simulation.

MCFC system using natural gas as feed and fuel.[6] The desulfurization and steam-reforming requirements are much the same as those for PAFC systems as detailed in Section 6.3.2. As before, steam for the reforming reaction is raised by using the by-product heat from the fuel cell; however, the temperature level of the fuel cell, typically 650°C, is such that no real limitation is placed upon the pressure at which the process steam may be raised. The reformer effluent which contains a mixture of hydrogen, carbon monoxide, carbon dioxide, steam, and methane requires no treatment prior to being admitted to the MCFC anode and does so at about 650°C. This reduces capital costs and enables energy losses to be kept to a minimum. It is generally assumed that up to 75% of the hydrogen, carbon monoxide, and methane fed to the MCFC is utilized to produce dc electricity. The anode off-gas is used as fuel in the reformer burners, although makeup fuel is also required. The carbon dioxide produced at the anode may be recovered directly from the anode off-gas by using a sophisticated carbon dioxide removal system, or it may be recovered from the flue gas leaving the combustor of the reformer. As in PAFC systems, pressurized combustion[14,16] plays an important role in MCFC system optimization.

The flue gas from the reformer is cooled to 40°C and, in so doing, provides for the bulk of process stream heating. Following condensation of the steam and water knockout, the dried flue gas is recompressed and recycled to the cathode to provide the carbon dioxide needed for the fuel cell electrochemical reaction. A single compressor supplies both the cathode air and combustion air requirements. This fresh cathode air is mixed with the recycled flue gas before admittance to the fuel cell stack. The steam generated from the fuel cell waste heat is used for the reforming reaction and for the production of shaft power in a steam turbine. The energy in the cathode off-gas is recovered for process heating purposes and is also admitted to a gas expander for the provision of shaft power. The sum total of the shaft power produced in the fuel cell system is balanced with the plant compressor requirements. Finally, the relatively cool and low-pressure cathode off-gas is vented to a safe location.

6.3.4. *Trends in Fuel Cell System Design*

In this chapter, most attention to fuel cell system design and optimization centers on what could be termed "generic system designs" as described in Sections 6.3.2 and 6.3.3. It is, however, worth outlining some of the trends that may be expected to take place in fuel cell system design. For instance, in a bid to realize the onset of commercialization of fuel cell technology, Japanese fuel cell manufacturers are placing significant emphasis on bringing relatively low cost, nonpressurized fuel cell systems to the market as quickly as possible.[20] This first generation of PAFC power plants is characterized by a relatively low net system efficiency, ε_{NET}, of about 40%,[21] but possesses that most endearing of fuel cell characteristics of ultralow NO_x emissions (<10 ppm).[22]

In general, however, most international efforts to commercialize fuel cell technology within the next five years are targeted at what might be termed the intermediate power generation market in the capacity range of several hundred kW_e to several MW_e. A combination of economy of scale and cost reduction through the construction of several similar power plants in a single order are expected to enable the penetration of this potentially very lucrative market.[10,23]

Aside from cost reduction the other major thrust of fuel cell commercialization efforts are aimed at demonstrating the reliability of fuel cell power plants by utilizing as many standard off-the-shelf proven components as possible (see Section 6.10). Despite this, there is also an increasing amount of effort being devoted to the development of composite gas turbine–fuel cell systems.[24]

In general, these composite gas turbine–fuel cell systems are based upon commercially proven gas turbine technology combined with PAFC, MCFC, and SOFC technologies, with expected net system efficiencies (ε_{NET}) around 55–65%.[25] Despite the apparent increase in plant complexity by the extensive use of rotating equipment in these systems, the extremely high efficiencies, low NO_x emissions, and the projected low installed capital costs per kW_e are expected to make these composite systems highly attractive in the intermediate and larger capacity power generation markets. The relatively low capital costs of below \$1000/kW_e and relatively high reliability expected from these composite systems is due to use of commercially proven gas turbine technology.[6,26b] However, due to the recent nature of these developments, and in view of the promising commercial potential of these processes, little or no literature references are available.

Another trend of potential interest for fuel cell system designs of the future concerns the use of the latest generation of hydrogen plant technology. These process developments incorporate the latest regenerative reformer technology, and most importantly allow for the number of plant process units to be significantly reduced.[27] The simpler system design implicitly reduces the plants instrumentation requirements and process control complexity, with a commensurate reduction in the installed capital costs. Finally, further cost reductions are made possible through the extensive use of the latest techniques in modular construction.[28] (See Chapter 12.)

Finally, in the longer term it seems reasonable to expect that fuel cells will be better able to accommodate the wide range of feedstocks available, initially by enabling greater flexibility in the process conditions and compositions of streams entering or leaving the fuel cell stacks. This increased flexibility will enable fuel cell system developers to design even-better-performing power plants at lower cost. Further down the line, the development of internal-reforming MCFC and SOFC technologies which enable the fuel cell to accept directly certain hydrocarbon feedstocks, such as natural gas, biogas, or coal gas, will also permit the design of high-efficiency and low-cost power plants for specific applications.

6.3.5. Other Fuel Cell Systems

Comparatively little work has been done in the area of fuel cell system development for the other types of fuel cell in stationary power plant applications. AFC has a well-proven track record in space applications, with a world AFC stack production capacity of several megawatts per year.[29] AFCs are highly sensitive to poisoning by both carbon monoxide and carbon dioxide, even the carbon dioxide content of air at 353 ppm (in 1990!)[30] is unacceptably high for the AFC. As a consequence the AFC can only accept pure hydrogen as anode feedstock. In addition the relatively low temperature of operation of the AFC means that essentially no fuel cell waste heat can be used in a fuel processor being used to convert a commercially available fuel to pure hydrogen.

Nevertheless, the performance of the AFC on pure hydrogen is high, with a fuel cell efficiency, ε_{FC}, approaching 60% (see Chapters 3 and 7). Such relatively high fuel cell efficiencies have encouraged some efforts to investigate the performance and costs of AFC systems for stationary power plant applications.[31] It has been concluded that such a fuel cell system must incorporate pressure swing adsorption (PSA) technology to produce the required hydrogen purity. On natural gas fuel, the gross system efficiencies (ε_{GROSS}) for these power plants are high at 48–60%;[32] however, the net system efficiencies (ε_{NET}) tend to be lower due to relatively high parasitic power requirements. If the capital cost projections of \$1000–2000/kW$_e$ can be realized in a commercialization program of AFC systems directed at stationary power plant applications, then this technology may be able to find a competitive niche in the market.[33]

Similar results to the above have been realized for the other type of low-temperature fuel cell, the solid polymer fuel cell (SPFC). However, this type of fuel cell does not have hydrogen- or air-purity requirements as strict as those for the AFC.[34]

6.4. THE OPTIMIZATION OF FUEL CELL SYSTEMS

In the development of a fuel cell system the designer must be able to determine the set of design parameters that give rise to optimum plant performance, for a given type of fuel cell, capacity, and plant battery-limit conditions. Such design parameters include process variables critical to the performance of the fuel processor and the fuel cell alike, fuel cell cooling requirements, the reformer geometry, the heat exchanger network design, and the shaft power balance of the plant rotating equipment. Evidently, a host of design options are open to the process developer, and the purpose of this section is to explain their significance. In so doing the crucial role of system development to the successful commercialization of fuel cell technology will be outlined.

6.4.1. Process Parameters

The most critical process variables affecting the performance of a steam reformer are the steam-to-carbon ratio at the reformer inlet, the outlet pressure, and the outlet temperature. The steam reforming of methane is a strongly endothermic reaction whose conversion is favored by a high steam-to-carbon ratio, high outlet temperature, and low pressure. The integration of a steam reformer with a fuel cell tends to define limits over which these and other process parameters may be varied. Further, the fuel cell itself has several critical parameters that may be modified to optimize its performance. Clearly, the overall fuel cell system performance is dependent upon the relationships between all such parameters, and detailed studies are necessary to enable the design of efficient and cost-effective fuel cell power plants. This section will give an overview of the various process considerations and their effect on both PAFC and MCFC system performance.

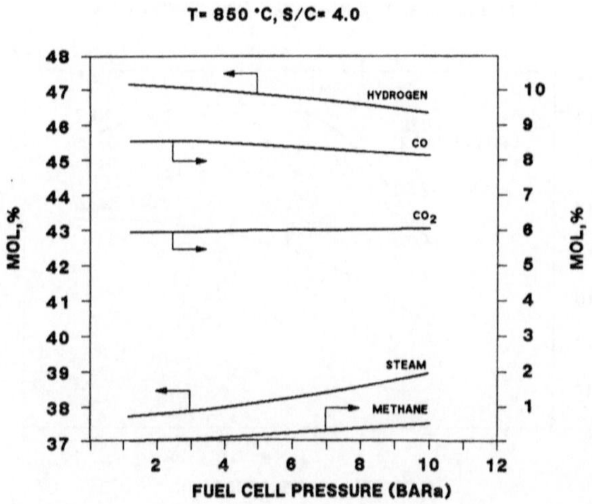

Figure 6.8. 3.25-MW MCFC: effect of pressure on anode inlet composition.

6.4.1.1. Pressure

The reforming process converts a mixture of hydrocarbons and steam into hydrogen, methane, and carbon oxides, according to the following overall reaction schemes in the presence of a catalyst:[35–39]

$$C_nH_m + nH_2O \rightarrow nCO + (n + m/2)H_2, \quad (6.11)$$

$$CH_4 + H_2O \leftrightarrow CO + 3H_2, \quad (6.12)$$

$$CH_4 + 2H_2O \leftrightarrow CO_2 + 4H_2, \quad (6.13)$$

$$CO + H_2O \leftrightarrow CO_2 + H_2. \quad (6.14)$$

The conversion of methane in a steam reformer is favored by the lowering of the reformer outlet pressure. Figure 6.8 clearly shows the effect on changing pressure on the reformer effluent gas composition.

These process simulations were carried out on a natural-gas-fueled 3.25-MW_e MCFC system in which the reformer effluent is fed directly to the anode compartments of the MCFC. The results show that, for a given steam-to-carbon ratio and reformer outlet temperature, an increase in the reformer outlet pressure results in a lower methane conversion (i.e., an increase of the methane slip*) and a consequent reduction in yield of hydrogen and carbon monoxide. However, due to Nernst effects,[4,5] an increase in hydrogen partial pressure at the fuel cell anode with a commensurate rise in oxygen partial pressure at the cathode leads to an increase of the fuel cell efficiency (ε_{FC}) as shown in Fig. 6.9.

The effect of increasing pressure on the performance of the fuel processor is initially to increase the fuel processor efficiency,† ε_{FP}, but thereafter to lower it. This is in part due to the beneficial effect of pressurized combustion[14,40] by which

*Note that the methane slip from the reformer increases more markedly at pressures above 5 bar abs.
†See Section 6.2 for an explanation as to how ε_{FP} may exceed 100%.

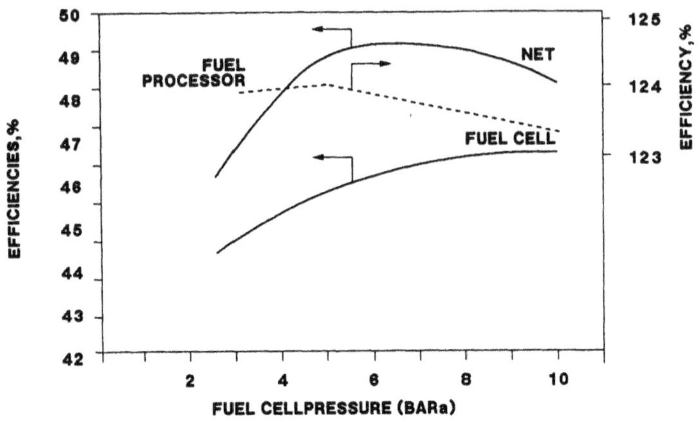

Figure 6.9. 250-kW MCFC: effect of pressure on efficiencies.

improved radiative and convective heat transfer reduce the reformer-fired duty requirement for a given reforming duty. Eventually, however, a pressure level is reached at which no makeup fuel is needed, and the pressurized anode off-gas supplies the total fired duty. Further increases in pressure result in a decrease in the amount of steam reforming, i.e., an increased methane slip, and an excess of anode off-gas fuel that cannot be utilized. It is evident, therefore, that a trade-off occurs between the beneficial effect of pressurized combustion and an increase in methane slip which eventually results in a drop of ε_{FP}.

It should be noted that the point of maximum fuel processor efficiency does not necessarily coincide with the point of maximum net efficiency (ε_{NET}) since this depends on the combination of the fuel processor and fuel cell efficiencies and on the parasitic power demands of the power plant. This technical optimum also varies according to the values of other process parameters, feedstock, system capacity, rotating equipment efficiencies, etc., and detailed investigations have shown that for MCFC systems the optimum pressure level can lie between 6 and 8 bar abs as measured in the fuel cell.*[6]

In PAFC systems a similar result is obtained when, for a given set of process parameters, the system pressure is progressively increased from an approximately atmospheric pressure level in the fuel cell and the reformer combustor. However, when compared to an MCFC operating at 650°C, the relatively low level of waste heat available from the PAFC limits the maximum level at which process steam may be effectively raised to around 160°C (ca. 6 bar abs). A consideration of the pressure drops through equipment in the fuel processing system means that the effective fuel cell working pressure is limited to about 4 bar abs. It is certainly possible to raise higher level process steam by placing steam-raising coils in the convection section of the reformer. However, not only does this approach increase the capital investment requirement, it also demands extra fuel firing in the reformer to meet the higher convection bank heat load and negates the fuel cell waste heat contribution to the fuel processing section. The net result is that

*Note that there exist "nonconventional" MCFC system configurations for which the technically optimum pressure level may be as low as 3 bar abs (see Section 6.5).

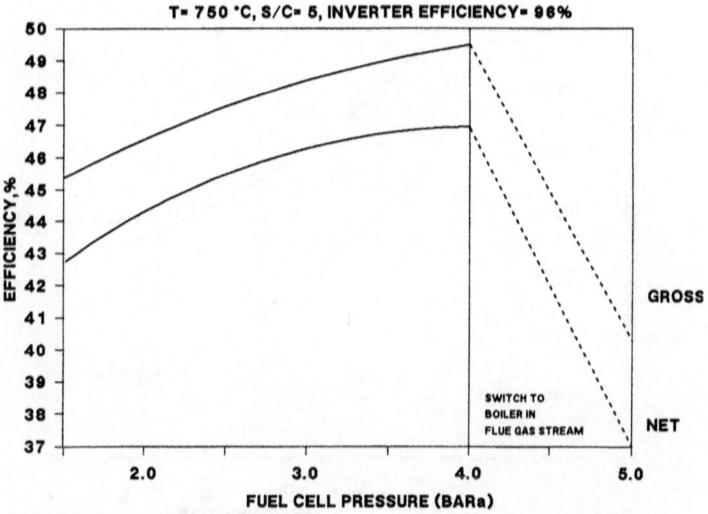

Figure 6.10. 250-kW PAFC: effect of pressure on system efficiencies.

above a fuel cell operating pressure of 4 bar abs the net system efficiency falls off sharply, as shown in Fig. 6.10.

As before the technically optimum value for the fuel cell operating pressure is dependent upon a number of factors, and detailed investigations have shown that maximum net system efficiencies of PAFC systems can be obtained by operating the fuel cell at a pressure between 2.5 and 4 bar abs.[6]

6.4.1.2. Steam-to-Carbon Ratio

The conversion of methane in the reformer is favored by an increase in the steam-to-carbon ratio. This parameter is commonly expressed as the molar ratio of steam to the atomic carbon in the hydrocarbon constituents of the feed. Although an increase in the steam-to-carbon ratio yields more hydrogen, it also raises the fired duty of the reformer due to increased endothermic heat of reaction and sensible heat requirements. In fuel cell systems the effect of an increase in the steam-to-carbon ratio manifests itself in a lowering of the fuel processor efficiency, as shown in Table 6.4.

However, Table 6.4 also shows that for PAFC systems increasing the steam-to-carbon ratio raises the efficiency of the fuel cell itself. This is due to the increasing partial pressure of hydrogen at the anode inlet. The gain in fuel cell efficiency more than offsets the falloff in fuel processor efficiency, such that in PAFC systems a high steam-to-carbon ratio of between 5 and 6 is generally

Table 6.4. Influence of Steam-to-Carbon Ratio on System Efficiencies[6]

S:C Ratio	2.5	3.0	3.5	4.0	6.0	7.0
Efficiency (%)						
ε_{FC}	39.86	41.52	42.47	43.04	43.92	44.07
ε_{FP}	115.43	112.10	110.32	109.29	107.69	107.35
ε_{GROSS}	43.70	43.30	43.60	43.70	44.00	43.90
ε_{NET}	38.40	38.80	39.10	39.10	39.50	39.60

P = 1.4 bar abs; reforming temperature = 750°C; inverter efficiency = 93%.

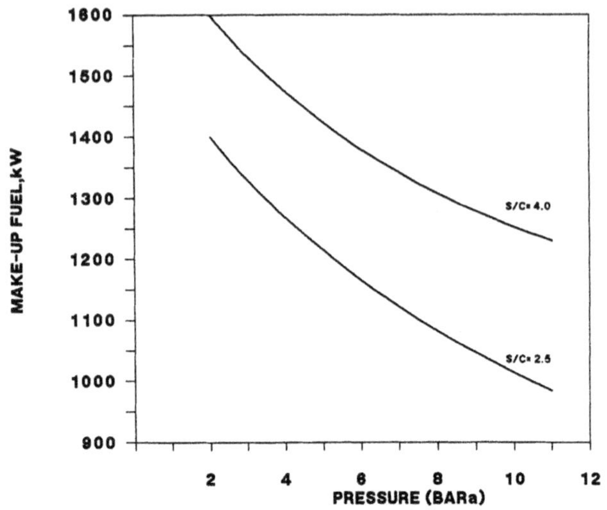

Figure 6.11. 3.25-MW MCFC: effect of pressure on makeup fuel load. (T = 850 °C).

preferred to maximize system efficiencies. However, the available fuel cell waste heat for process-steam-raising purposes may limit the maximum possible steam-to-carbon ratio. This problem is especially acute for small multi-kilowatt PAFC systems where heat losses from the fuel cell and process equipment can limit the maximum steam-to-carbon ratio to 3.5.[41]

Changes in the steam-to-carbon ratio of MCFC systems have a different effect on system performance. Figure 6.11 clearly shows that an increase in steam-to-carbon ratio hikes the reformer makeup fuel necessary to maintain a fixed electric power output; in other words, the MCFC system efficiency falls.

Closer inspection of component efficiencies has shown that, as with PAFC systems, the fuel processor efficiency decreases as the steam-to-carbon ratio is raised. However improvements in the performance of the MCFC itself are only marginal. This occurs because the operating conditions at the MCFC anode enable rapid equilibration of the syngas mixture via the steam reforming and shift reactions (see reactions (6.12) to (6.14)), such that the carbon monoxide and methane components sustain an indirect source of hydrogen for utilization by the electrochemical reaction. The net effect is that an increase in the steam-to-carbon ratio does not significantly increase the availability of both "direct" and "indirect" hydrogen at an MCFC anode, but still demands a substantial increase in reformer-fired duty.

There is, however, a practical lower limit to the steam to carbon ratio which is necessary to prevent carbon deposition in the reformer and pore plugging in the anode compartments of MCFCs.[5] The formation of carbon may take place via the following reactions:[18,35,41]

$$2CO \leftrightarrow C + CO_2, \tag{6.15}$$

$$CH_4 \leftrightarrow C + 2H_2, \tag{6.16}$$

$$CO + H_2 \leftrightarrow C + H_2O, \tag{6.17}$$

$$CO_2 + 2H_2 \leftrightarrow C + 2H_2O, \tag{6.18}$$

$$CH_4 + 2CO \leftrightarrow 3C + 2H_2O, \quad (6.19)$$

$$CH_4 + CO_2 \leftrightarrow 2C + 2H_2O, \quad (6.20)$$

In addition, the likelihood of carbon formation resulting from the above equilibrium reactions becomes progressively more critical at lower temperatures. This suggests that the selection of operating conditions is crucial to minimizing the risk of carbon formation in either the reformer or the fuel cell.[42]

PAFC systems that use a high-temperature-shift (HTS) reactor have an additional minimum steam-to-carbon-ratio requirement in order to prevent the Fischer–Tropsch reaction[43,44] taking place on the iron-based catalyst, in which hydrocarbons are synthesized from carbon oxides and hydrogen.

6.4.1.3. Reforming Temperature

The conversion of methane in a steam reformer is favored by an increase in the reformer outlet temperature. On the other hand, the exothermic shift reaction (6.14) is disfavored by an increase in reformer outlet temperature. As the reforming temperature is raised, the increased degree of steam reforming combined with the lower availability of exothermic heat from the shift reaction increases the fuel firing requirements of the reformer. Furthermore, higher reformer outlet temperatures imply higher reformer tube-skin temperatures such that, for a given fuel firing, the quantity of heat transferred by radiation* is reduced. This means that the proportion of heat exiting the radiant section by way of the flue gas increases, and hence the radiant box efficiency† falls. To some extent, the greater degree of convective heat transfer implicit in a pressurized combustion furnace means that the drop in radiant box efficiency can be somewhat ameliorated; however, this trend cannot be reversed.

Although the fuel cell efficiencies of both phosphoric acid and molten carbonate fuel cells rise with increasing reformer outlet temperature, these increases are not sufficient to compensate for the considerable drop in the corresponding fuel processor efficiencies. Consequently, the net fuel cell system efficiencies tend to decline as the reformer outlet temperature is raised (see Table 6.5 and Fig. 6.12).

Table 6.5. Influence of Reformer Temperature on System Efficiencies[6]

Reforming temp. (°C)	650	700	750	800	850	900
Efficiency (%)						
ε_{FC}	42.39	43.66	44.01	44.10	44.13	44.13
ε_{FP}	112.02	108.58	107.51	107.06	106.70	106.57
ε_{GROSS}	44.20	44.10	44.00	43.90	43.80	43.70
ε_{NET}	39.70	39.65	39.60	39.50	39.40	39.30

P = 1.4 bar abs; S:C = 6.5; inverter efficiency = 93%.

*Radiant heat transfer is governed by the Stefan–Boltzmann law[45] in which the heat transferred, Q, is proportional to $T^4_{\text{heat source}} - T^4_{\text{heat sink}}$. This means that a small decrease in the heat sink temperature leads to a disproportionately large increase (to the fourth power) in the radiant heat transferred, and vice versa.

†Typically, the radiant box efficiency is defined as the ratio of the sensible and reaction heat load of the process gas in the reformer tubes to the fuel fired in the radiant box.

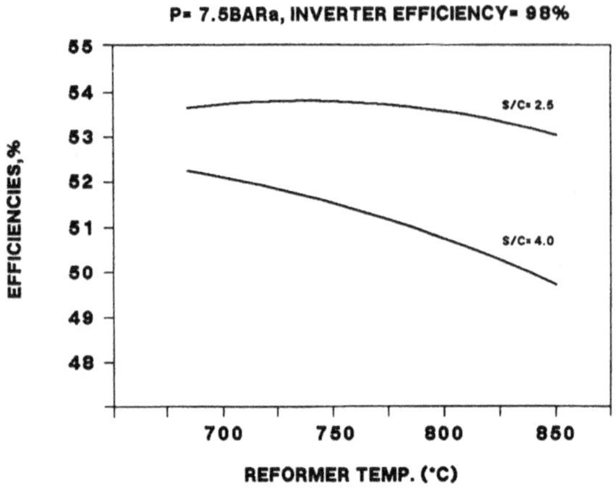

Figure 6.12. 3.25-MW MCFC: effect of reformer temperature + steam-to-carbon ratio on gross efficiencies.

It is worth noting that the high-severity operating conditions often employed in conventional steam reformers (in which cash credits are often realized for the additional export steam raised in the convection section) are constrained by maximum allowable reformer outlet temperatures due to mechanical considerations in the design and configuration of reformer tubes.[43] Conversely, there also exists a practical minimum reformer outlet temperature of 700°C, below which the danger of carbon formation via the Boudouard reaction (6.15) becomes a real possibility (see Section 6.4.1.2).

6.4.1.4. Shift Reactors

The water-gas shift reaction has been recognized as a means of producing hydrogen from carbon monoxide and steam since the end of the nineteenth century.[46] This conversion is equilibrium-limited and, being an exothermic reaction, is favored by lower temperatures. However, at lower temperatures equilibrium is not attained for kinetic reasons, and the content of carbon monoxide remains considerably higher than the calculated value. The addition of a large excess of steam to the shift reactor feed gas enables high-equilibrium conversions of carbon monoxide and raises the heat capacity of the reaction mixture, which restricts the adiabatic temperature rise through the reactor. In general, one may distinguish two types of shift reactors, namely, those containing an iron–chromium-based HTS catalyst (340–420°C), and those with a more expensive copper–zinc-based low-temperature shift (LTS) catalyst (180–260°C).*[44,46]

The operation of the PAFC at ≈190°C is constrained by a maximum carbon monoxide concentration at the anode inlet of approximately 0.5–2% vol. In a typical PAFC system, several shift reactor schemes are possible that will allow this level of conversion. This conversion level usually cannot be achieved with a single HTS reactor, but it is possible to do so in a scheme consisting of two HTS reactor beds with intermediate steam injection. The disadvantage of this method

*Medium-temperature shift (MTS) catalysts (200–320°C) based on a copper-zinc formulation are also produced on an industrial scale.[47,48]

is that very often insufficient process steam is available from the fuel cell stack waste heat.

A single HTS bed reactor in combination with a PSA purification system will ensure that the impurity levels at the fuel cell anode are met, though the economics of such a fuel processing scheme is usually only justified for application in alkaline or solid polymer fuel cell systems, or perhaps in large plants for the other fuel cell types.

A single LTS bed reactor in combination with a PSA purification system is also possible. This arrangement is marginally less efficient than the HTS–PSA combination and is best applied in cases where system wide heat losses preclude the use of an HTS. An example of this scheme has been the 80-kW$_e$ PAFC power plant designed and built by Kinetics Technology International for Solar-Wasserstof-Bayern's (SWB) solar power demonstration project in 1991.[49] The selected design is flexible and permits SWB to produce at will electric power from hydrogen generated by solar cells and electrolysis units or from hydrogen produced by a natural-gas–steam-reforming plant. The latter source of hydrogen can also be stored for use elsewhere in the facility, if the solar-generated hydrogen supplies are insufficient. Moreover, the plant is expected to be more reliable, since the level of integration with the fuel cell is less critical.

The preferred shift reactor scheme involves passage of the feed gas through an HTS catalyst bed, followed by interstage cooling and passage through an LTS catalyst bed. The advantage of this scheme is that most of the carbon monoxide conversion is performed by the cheaper and less sensitive iron–chromium oxide catalyst, with the more expensive and sensitive, but more active, copper–zinc oxide catalyst being used to convert the residual carbon monoxide to specification.

The use of a single LTS bed reactor is also possible, although only in multi-kW$_e$ (typically <100 kW$_e$) PAFC systems,[41] where the significantly higher priced LTS catalyst can be economically more preferable to using a conventional double-reactor HTS–LTS scheme, which requires more process equipment, with commensurately higher heat losses and increased control complexity.

6.4.1.5. Condensate Recovery

In a typical PAFC system, steam is consumed by reaction in the reformer and in HTS and LTS reactors, but can still constitute up to 40 vol% of the LTS effluent. When this stream is cooled to 40°C, more than 95% of the steam may be recovered as condensate. The resulting synthesis gas volumetric flow rate to the fuel cell is reduced, but the increased hydrogen partial pressure at the anode leads to higher fuel cell and gross system efficiencies. However, since the anode off-gas is used to fuel the reformer, a reduction in the flue gas flow implies that less power can be recovered by the flue gas–cathode off-gas expander (see Fig. 6.5). As a consequence, the net system efficiency tends to fall; for example, in an optimized 250-kW$_e$ PAFC system, ε_{NET} has been shown to decrease from 47.8% to 47.1%.[6]

It may nonetheless be preferable to implement condensate recovery in a PAFC flowsheet, since this reduces the demineralized water makeup requirement and its corresponding cost. It is even considered necessary for packaged-type plants for remote locations with limited or no water supply. In addition, with the nitrogen that is present in many natural gas feedstocks, there is a tendency to form up to 100 ppm of ammonia in the reformer and HTS reactors. As outlined in Section 6.3 the performance of the PAFC is sensitive to as little as 1 ppm ammonia impurity at the fuel cell electrodes. Fortunately, the cooling of the process gas to a dew point of 40°C enables the extraction of the ammonia with relative ease since it is condensed with and dissolved in the process condensate. To be absolutely certain of satisfying the ammonia impurity specification, a phosphoric acid guard bed at 60°C may also have to be installed.

6.4.1.6. Fuel Cell Operating Temperature

In general, an increase in the temperature of operation of a fuel cell improves cell performance because of increases in reaction rate and mass transfer and a decrease in cell ohmic resistance due to higher ionic conductivity of the electrolyte[4] (see Chapter 3). These factors combine to yield higher cell operating voltages. In addition, the tolerance of PAFCs to carbon monoxide poisoning improves at higher cell operating temperatures. Such improvements in cell performance translate directly to higher net system efficiencies. An example for a biogas-fueled 250-kW_e PAFC system is provided in Table 6.6.

For a PAFC designed to sustain operation at 205°C the optimum operating pressure level of the fuel cell system may, in principle, be raised to above the range of 2.5 to 4 bar abs indicated in Section 6.4.1.1. This is because the pressure level of process steam may be increased as the fuel cell waste heat becomes available at higher temperature levels.

Improvements in system performance may also be realized for MCFC systems. These gains are most pronounced when the cell operating temperature is raised from 600 to 650°C, whereas only modest gains in performance are obtained at temperatures above 650°C.

There are, however, disadvantages associated with operating fuel cells at higher temperatures, including increased corrosion rates, catalyst sintering and recrystallization, and higher electrolyte loss through evaporation.[4,5]

6.4.1.7. Fuel Utilization

Fuel utilization refers to the fraction of the total anode feed that reacts electrochemically. In the PAFC this fuel utilization refers only to the hydrogen in the anode feed. However, in the MCFC the utilization of carbon monoxide and

Table 6.6. Biogas-Fuel 250-kW_e PAFC System at 4 bar abs Cell Pressure[6]

Operating temperature (°C)	190	205
Open-circuit emf (V)	1.172	1.166
operating voltage (V) (i = 200 mA/cm^2)	0.704	0.720
ε_V (%)	60.1	61.8
ε_{FC} (%)	37.6	38.5
ε_{FP} (%)	116.4	116.5
ε_{GROSS} (%)	43.8	44.9
ε_{NET} (%)	40.4	41.5

methane must also be considered because cell operating conditions are such that these fuels can be utilized indirectly via the steam-reforming and shift equilibrium reactions (see reactions (6.12) to (6.14)).

However, increasing fuel utilization causes cell potential loss, and a compromize is required, since although low fuel utilization improves cell performance, it decreases the current efficiency and increases the fuel cell cost per kW_e produced. System optimization studies have determined that both acid- and carbonate-type fuel cells running on synthesis gas at between 75 and 85% fuel utilization do so at essentially optimum conditions for maximum system efficiency and minimum cost.[6]

6.4.2. Fuel Cell Cooling

In fuel cell stacks provision must be made for the removal of heat generated during operation of the cell. This fuel cell waste heat can be removed in a number of ways that to a certain extent depend on the type of fuel cell. In PAFC stacks, heat is removed by using boiling water, dielectric liquids, or gas (especially air) coolants. No MCFC stacks to date exceed 28 kW_e[17] capacity, and these stack designs are generally cooled by the reactant gases themselves. For MCFC stack designs of the future to be economically viable, it will probably be necessary to use either specially developed organic heat transfer oils or eutectic liquid salt mixtures to cope with the high temperatures of fuel cell operation at about 650°C.

In order to keep mechanical stresses within fuel cell stacks as low as possible, it is generally desirable to minimize temperature differences in all directions within the fuel cell stack. This consideration forms a critical constraint to both the design and operation of any cell stack cooling system and, consequently, the design and optimization of the entire fuel cell system.

In PAFC stacks, the coolants are passed through cooling channels located at regular intervals within the cell stack. Liquid cooling requires complex manifolds and connections, but destroys less exergy* during heat removal than does air cooling.[51–53] If boiling water is used as coolant, very high purity levels must be maintained to minimize corrosion, to prevent blockage of cooling channels, and to minimize leakage of electric current.[54] This means that a water treatment plant consisting of filtration and demineralization units must be included in the system design.[41] Conversely, gas cooling has the advantage of simplicity and reliability, but the associated gas compression requirements tend to inflate the parasitic power demand of the fuel cell power plant, and the stacks can become larger.

System optimization studies have been carried out on multi-MW_e MCFC systems to determine the benefits of connecting the fuel cells in series, and use a system of heat exchangers to remove the fuel cell reaction heat from the effluent gases.[6] In this work, an allowable temperature difference between inlet and outlet streams of 100°C was assumed for both anode and cathode streams, i.e., T_{in} = 550°C and T_{out} = 650°C. The outlet streams from the nth fuel cell stack

*_Energy_ analysis is based on the principle of conservation of energy, i.e., the first law of thermodynamics. However, _exergy_ analysis is based on the principle of nonconservation of entropy, i.e., the second law of thermodynamics, in addition to the principle of conservation of energy. Hence, exergy is consumed (or entropy produced) due to irreversibilities in a process, and for an ideal process exergy consumption is zero. Hence, exergy analysis enables losses within a fuel cell (or any other) system to be pinpointed.[50]

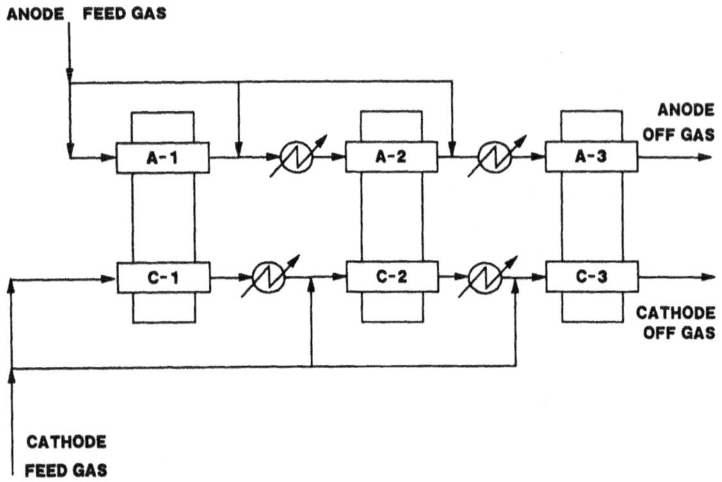

Figure 6.13. 100-MW MCFC: external cooling (cocurrent).

were cooled by a fresh feed mix and a heat exchanger before being fed to the $(n + 1)$th fuel cell stack (see Fig. 6.13).

The main advantage of this system of external cell cooling is that problems related to the design and use of integral cooling plates and external coolant may be avoided. However, process evaluations showed that the system possesses considerable disadvantages, such as the fact that the inlet anode and cathode streams become increasingly diluted along the sequence of the fuel cell stacks. This leads to reductions in fuel cell efficiency and increases the amount of waste heat to be removed. In addition, the increasing volumetric flows being fed to each stack in the series implies that differing MCFC stack designs would have to be supplied for each power plant, resulting in an escalation in capital cost requirements.

Optimization studies on a 100-MW_e MCFC power plant showed that when 44% of the fuel cell heat is removed using external cooling, the fuel cell efficiency along the sequence of MCFC stacks falls from 49.8% to 42.9%. Increasing the number of cell stacks in the series enables up to 75% of the waste heat to be removed, but the fuel cell efficiency falls even further to only 37%.

By contrast, an optimal design employing a parallel gaseous feed system to the fuel cell electrodes and an integral cooling arrangement permits an average fuel cell efficiency of 47.1%, with only 15% of the fuel cell waste heat being removed by the effluent gases.[55] In this work, the temperature difference between the feed and effluent streams had been assumed to be 50°C, i.e., T_{in} = 625°C and T_{out} = 675°C, with a molten salt eutectic mixture of KNO_3 (53% w/w), $NaNO_2$ (40% w/w), and $NaNO_3$ (7% w/w) assumed for the removal of the majority of the MCFC stack waste heat.

6.4.3. Reformer Geometry

Parametric studies, as outlined in Section 6.4.1, are used to enable the optimization of process variables such as the steam-to-carbon ratio, reforming pressure, and reforming temperature. However, an important ingredient of the optimization of these process variables is the optimization of the reformer geometry itself. This is because the reformer design has a considerable impact upon the radiant box efficiency and on the heat and shaft power requirements of the surrounding fuel cell system.

Simulation studies[6] have been carried out using a regenerative type of reformer, as developed by United Technologies Corporation,[12,14] whose geometry is depicted in Fig. 6.14. In a regenerative-type reformer, the process gas exiting the catalyst bed passes through a regenerative tube annulus and is typically cooled from 800 to 650°C. The sensible heat provided by this process gas cooling is used to provide a part of the endothermic heat of the steam-reforming reaction duty, thereby reducing fired duty needs. The combustion side of the reformer is pressurized with the two main mechanisms of heat transfer being by radiation and convection. Pressurized combustion enables a substantially increased convective heat transfer contribution amounting to up to 50% of the heat

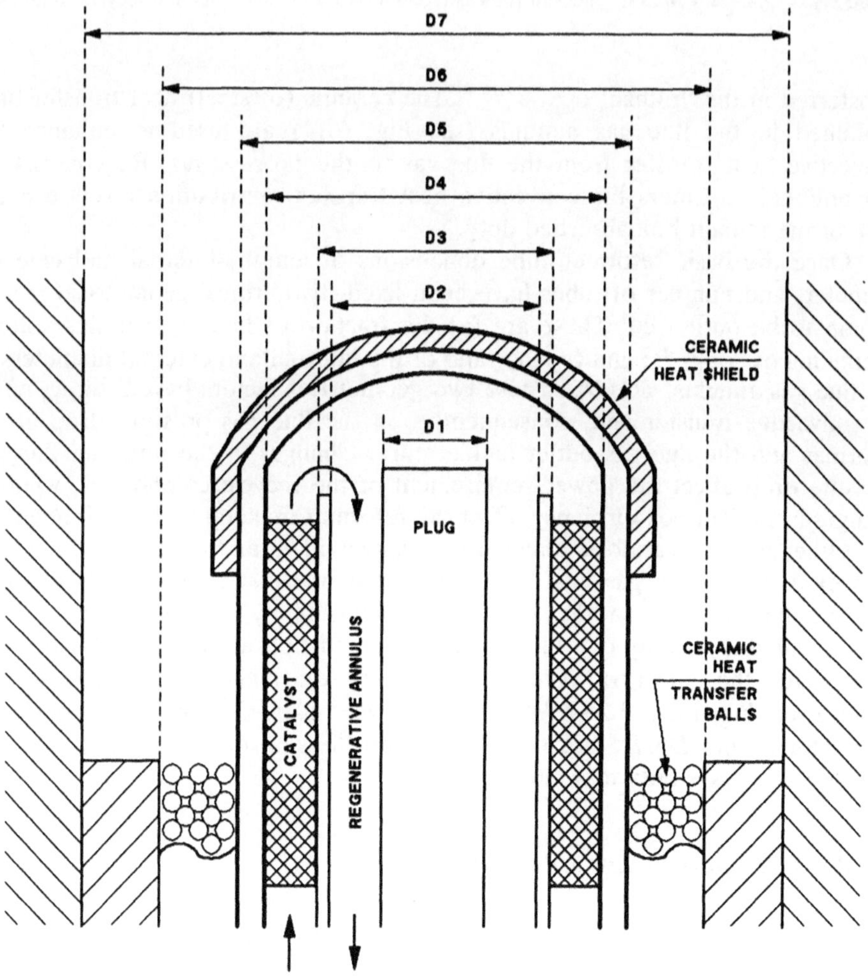

Figure 6.14. Single regenerative tube reformer configuration (with reactor dimensions used in computer modeling).

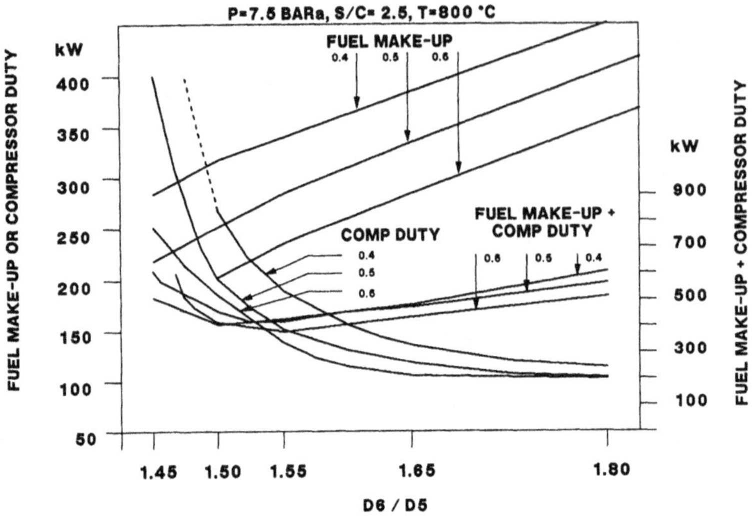

Figure 6.15. 3.25-MW MCFC: effect of reformer geometry on makeup fuel and recycle compressor duty.

transferred in the "radiant box."[14,16,40] The ceramic (or steel) heat transfer balls contained in the flue gas annulus (see Fig. 6.14) are used to enhance the convective heat transfer from the flue gas to the process gas. By contrast, in conventional reformers the convective heat transfer contribution rarely exceeds 10% of the radiant box absorbed duty.[56]

Once the basic reformer tube dimensions of length, internal and external diameter, and number of tubes have been fixed, two critical geometrical factors remain to be optimized. These are (a) the fraction of flue gas annulus volume containing packing, H; and (b) the ratio of the internal and external diameters of the flue gas annulus, $D6/D5$. These two geometrical factors have a direct effect on convective transfer and, consequently, on the flue gas pressure drop in the reformer and the flue gas outlet temperature. Changes in the reformer flue gas pressure drop affect the power requirement of the recycle compressor, whereas changes in radiant box efficiency affect the reformer makeup fuel requirement.

Other process variables affected by changes in H and $D6/D5$ include heat recovered in the flue gas heat exchanger train, preheat requirements of dried recycled flue gas up to the fuel cell cathode inlet, and differences in air compressor duty mainly due to changes in combustion airflow rate. Figure 6.15 summarizes the effect of reformer geometry on makeup fuel and recycle compressor duty for a 3.25-MW_e MCFC power plant. Obviously, the optimum values for H and $D6/D5$ are those that keep the sum of the makeup fuel and recycle compressor duty to a minimum.

6.4.4. Heat and Power Integration

As is evident from the discussion so far, the reformer and the fuel cell generate excess heat which can, in principle, be utilized elsewhere in the fuel cell system. The aim of any good system design must be to ensure a high degree of heat and power integration, thereby enabling attainment of maximum fuel cell system efficiencies. It is, however, important to keep the degradation of the quality of the surplus heat (exergy)* to a minimum by utilizing this energy at the maximum temperatures possible.[50,51,57] To take an example, steam is a favored heat transfer medium for transferring energy from the fuel to the reformer and, in some cases, to a steam turbine for the generation of shaft power. In this process exergy loss occurs as heat is transferred from the fuel cell electrochemical reaction to the stack coolant fluid,† and during the heat transfer for process steam raising. Since exergetic losses are irreversible, each occurrence reduces the maximum attainable system efficiency.

Pinch technology is a methodology originally developed as a tool for defining energy-saving opportunities, particularly through improved heat exchanger network design.[1,58,59] This section will illustrate how the methodology can also be used to understand the heat and power integration issues of fuel cell system design. In general, pinch technology enables the identification of targets for heat exchanger network design. These targets are calculated from the given flow rate, enthalpy, and temperature data for the streams defining the heat exchanger network problem.[2] These targets can include the minimum energy (hot and cold utility) requirements, the minimum number of heat exchanger shells, and the minimum heat exchanger network surface area. Such targets enable the objective assessment of actual designs with a view to identifying the optimum trade-off between maximum energy efficiencies and minimum capital cost targets.

Extensive system optimization work has been carried out on a 3.25 MW_e MCFC power plant in which pinch technology was used to help determine the optimum steam system configuration and heat exchanger network design.[6] Two design parameters were discovered to be of prime importance in this optimization, namely, the degree of cathode feed preheat and the steam pressure level. It has been established that these two parameters are strongly interrelated.

The optimization assumes that for a fuel cell operating temperature of 650°C, the preheated cathode air may be fed to the fuel cell at temperatures in the range of 580 to 620°C without any detriment to the fuel cell performance. This means that the cathode stream also acts as a means of removing heat from the fuel cell stack. However, as more cathode feed preheating takes place within the fuel cell stack,‡ less waste heat is carried by the stack coolant loop for steam generation. The level to which cathode air is preheated also strongly affects the steam pressure level and degree of steam superheat possible.§ This is because of the limited availability of highest quality heat from the reformer flue gas, which is, to a greater or lesser extent, also used for cathode air preheating.

These effects are most clearly seen by examining the heating and cooling curves of two MCFC system designs, one at 40 bar abs steam pressure and a cathode inlet temperature of 580°C, and the other at 10 bar abs steam pressure and a cathode inlet temperature of 620°C (see Figs. 6.16 and 6.17, respectively).

*See Section 6.4.2. and Chapter 5 for an explanation of the principle of exergy.

†Note that the exergy loss for boiling-water cooling (at a typical vaporization per pass of only a few percent) is approximately 2% higher than the exergy loss encountered when using a single-phase coolant, such as triethylene glycol.[52]

‡At a cathode feed inlet temperature of 580°C, about 20% of the fuel cell stack cooling requirement can be met by the cathode gas stream.

§The performance of steam turbines improves appreciably at high inlet steam pressures and at corresponding maximum possible levels of steam superheat.[6]

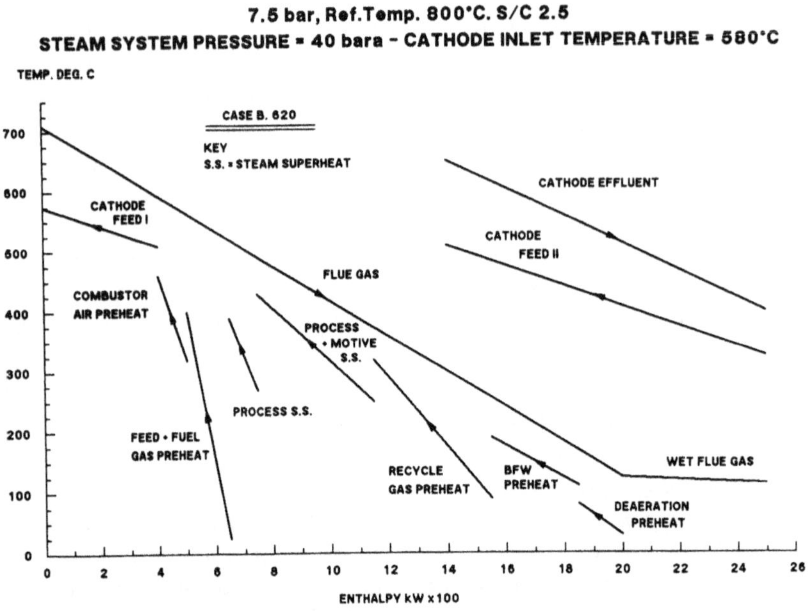

Figure 6.16. MCFC: heat recovery profile 3.25 MW with high pressure steam generation.

These heating–cooling curves give a clear and concise picture of the amount and quality of heat available in a fuel cell system, and on the number and positioning of heat exchangers. A minimum temperature difference, ΔT_{min}, for heat exchange between hot and cold streams* of 50°C has been used. It is now quite apparent that decreasing the level of cathode air preheat from 620 to 580°C

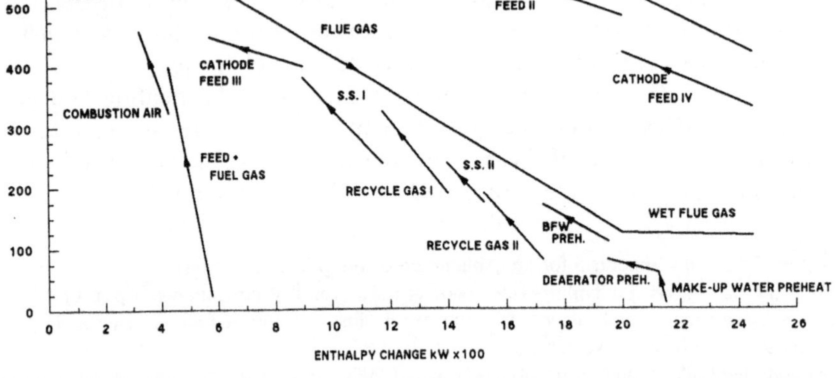

Figure 6.17. MCFC: heat recovery profile 3.25 MW with medium pressure steam generation.

*A cold stream is defined as one that requires heating, whereas a hot stream requires cooling.

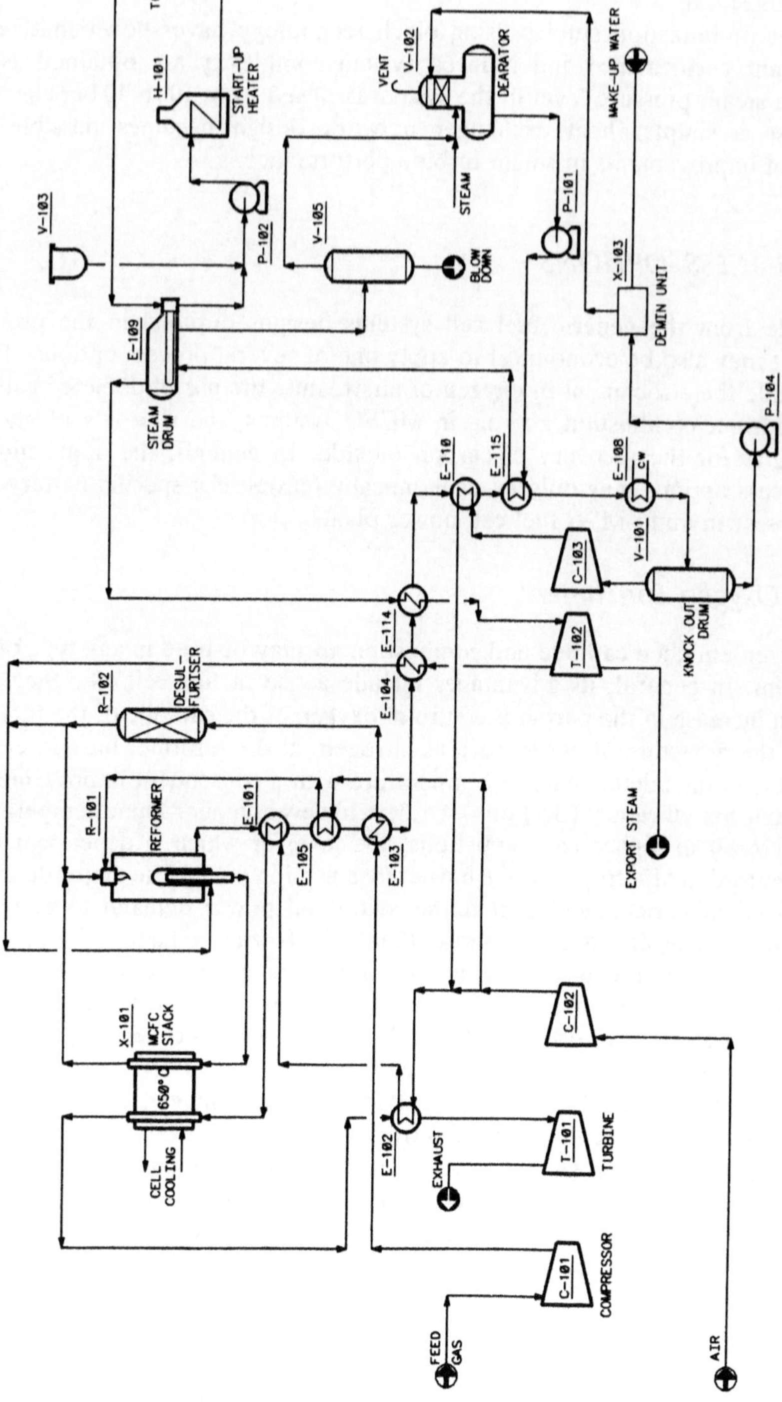

Figure 6.18. 3.25-MW MCFC power plant (580°C cathode inlet temp.).

makes more high-quality heat available to all other cold streams, and thus enables a simpler and cheaper heat exchanger network design. In fact, a reduction of the number of heat exchangers from 12 to 9 is achieved, resulting in a flowsheet depicted in Fig. 6.18.

Other optimization studies using pinch technology have shown that better power plant performance and reduced system complexity are obtained as the maximum steam pressure level in the system is raised from 10 to 40 bar abs. This is because a simpler heat exchanger network design becomes possible and because of improvements in steam turbine performance.

6.5. PROCESS OPTIONS

Aside from the generic fuel cell systems designs outlined in the previous section, it may also be economical to apply one of several process options. These may include the enrichment in oxygen of airstreams, the use of depleted cathode air as reformer combustion air, or, in MCFC systems, the use of sophisticated technologies for the recovery of carbon dioxide. In general, the application of such process options may only be economically feasible for specific battery-limit conditions or in multi-MW_e fuel cell power plants.

6.5.1. Oxygen Enrichment

Oxygen-enriched cathode and combustion air may be used in any type of fuel cell system. In general, its advantages include a rise in fuel cell efficiency, ε_{FC}, due to an increase in the partial pressure of oxygen at the cathode of the fuel cell. Further, the reduction of inerts, such as nitrogen, at the reformer burner leads to an increase in the adiabatic flame temperature with a consequent improvement in the radiant box efficiency (Section 6.4.1.3). However, higher flame temperatures can also result in higher NO_x emissions, the level of which is dependent upon the concentration of nitrogen at the burner (see Section 6.11). The major disadvantage of oxygen enrichement is that the additional power demand to effect the separation of oxygen from air more than offsets the reduction in plant air compression requirements on account of diminished gas flows. In addition, the reduction of circulating inerts in the fuel cell system results in lower volumetric flows of hot flue gases to drive the fuel cell system power recovery expanders.

Oxygen-enriched airstreams may be sourced from PSA, membrane, or cryogenic units dedicated to oxygen production. Alternatively, oxygen-enriched waste gases can also originate from a nitrogen PSA plant.

System optimization studies have been performed using a PSA system that produces a 90% purity oxygen stream from air for an electric power consumption of $0.6\ kW \cdot h/Nm^3$ of oxygen,[60] and a membrane-based system producing a 40% purity oxygen stream at $0.175\ kW \cdot h/Nm^3$ of oxygen (i.e., state-of-the-art membrane technology).[61,62] The oxygen-enriched stream may be mixed with fresh air to produce oxidant gases with oxygen concentrations ranging from 20.9% to 90%. Figure 6.19 shows the effect of changing oxygen content in the cathode airstream on the system efficiencies of a 250-kW_e PAFC power plant.

It is concluded that with state-of-the-art power consumption figures, oxygen enrichment is economically unattractive for stand-alone PAFC systems. Natu-

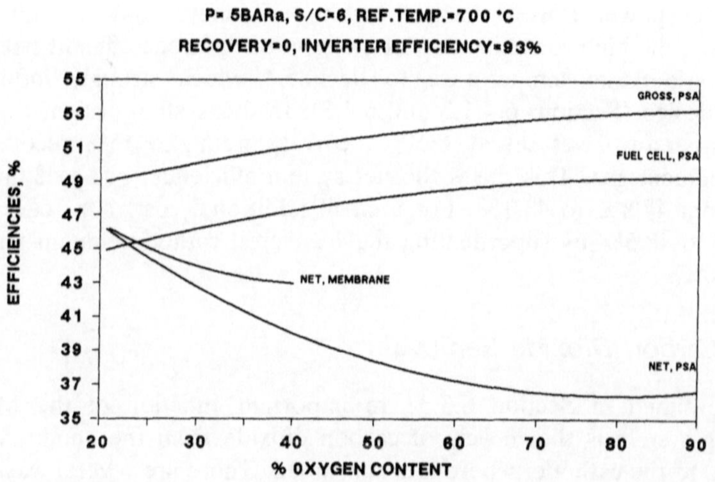

Figure 6.19. 25-kW PAFC: effect of oxygen enrichment of cathode air on efficiencies.

rally, if a relatively inexpensive oxygen-enriched airstream is available at plant battery limits, for example, as a waste gas from a nitrogen PSA plant, the fuel cell system economics will improve.

6.5.2. Cathode Air as Combustion Air

The use of depleted cathode air to supplement the combustion air requirements of the reformer can be an attractive alternative process option for the improvement of fuel cell system performance. Although the cathode off-gas is depleted in oxygen, it is available as a preheated and pressurized process stream. Figure 6.20 shows quite clearly that the use of depleted cathode air increases the gross system efficiency, ε_{GROSS}.[6]

When 30% of the cathode effluent is fed to the reformer combustor, fresh combustion air is no longer required. At 40% cathode effluent recovery, the loss

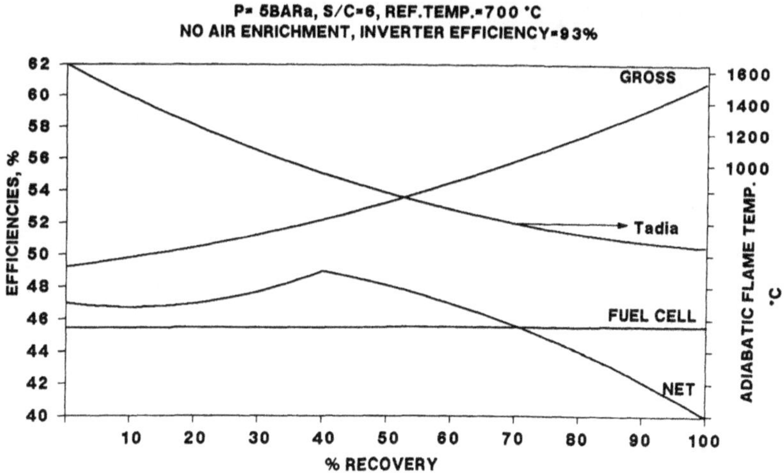

Figure 6.20. 25-kW PAFC: effect of cathode effluent recovery on system efficiencies.

of expander power causes the net system efficiency, ε_{NET}, to fall off. Most significantly, the high concentration of inerts in the cathode effluent stream causes the adiabatic flame temperature to decline,* which strongly influences the reformer design (Sections 6.4.1.3 and 6.4.3). In this system design, the adiabatic flame temperature was set at 1350°C, corresponding to 15% recovery of the cathode effluent gas. This raises the net system efficiency, ε_{NET}, of the 250-kW$_e$ system from 47.8% to 48.1%. For a small additional cost, ε_{NET} can be further increased to 48.5% by superheating the recovered cathode effluent stream from 190 to 400°C.

6.5.3. Carbon Dioxide Removal

As outlined in Section 6.3.3, an important function of the MCFC fuel processing system is the recycle of carbon dioxide from the anode, where it is produced, to the cathode, where it is consumed. There are several ways by which this may be accomplished, and the complexity of the carbon dioxide recovery system can vary from simply recycling the reformer flue gas following removal of the process condensate to the use of sophisticated carbon dioxide removal technologies. These technologies include those based on the principles of PSA, physical absorption, chemical absorption, or synthetic permeable membrane purification.[53,55]

In general, the choice of carbon dioxide removal technology is determined by the selected process arrangement and the capacity of the fuel cell system under consideration. For example, studies have shown that an activated MDEA solution process[63] is economically preferable for MCFC power plants in the range of 6.25 to 25 MW$_e$, whereas PSA technology is attractive for larger capacity plants from 25 to 100 MW$_e$.[55] It is important to note that the absorption-based processes consume large quantities of low-grade heat to effect the separation of carbon dioxide from the anode off-gas. This means that little or no cogeneration

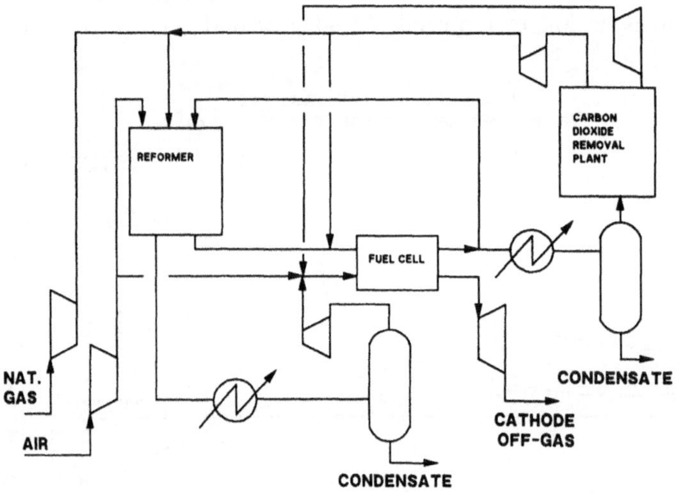

Figure 6.21. 100-MW MCFC: carbon dioxide removal plant.

*Note that a lower flame temperature results in lower NO_x emissions (see Section. 6.11).

heat is available from the power plant (see Section 6.9). By contrast, when PSA technology is employed, significant quantities of export steam are produced by the MCFC power plant.

A typical processing arrangement using carbon dioxide removal technology involves the recovery of carbon dioxide from the anode off-gas. The carbon dioxide is mixed with the fresh cathode air feed, and the carbon dioxide removal plant off-gas is sent to the reformer combustors to provide a part of the reforming heat of reaction (see Fig. 6.21).

For the processing scheme outlined above, a net system efficiency, ε_{NET}, of up to 56.6% is achievable for a 100-MW$_e$ MCFC power plant using an absorption-based carbon dioxide removal technology.[6]

6.6. FEEDSTOCKS

An important characteristic of fuel cell systems is their ability to utilize a variety of hydrocarbon feedstocks. Naturally the fuel processing section of the fuel cell system must be designed and subsequently optimized to handle the specific feedstock available at plant battery limits. Aside from natural gas, alternative feedstocks include landfill gas (also known as biogas), LPG, which is generally a mixture of propane and butanes, and naphtha.* In general, these feedstocks may be characterized in terms of their increasing fuel-carbon-to-hydrogen ratio, and it is of interest to uncover what effect such a parameter would have on fuel cell system performance.

The effect of changing the type of fuel fed to a 250-kW$_e$ PAFC power plant is illustrated in Fig. 6.22.

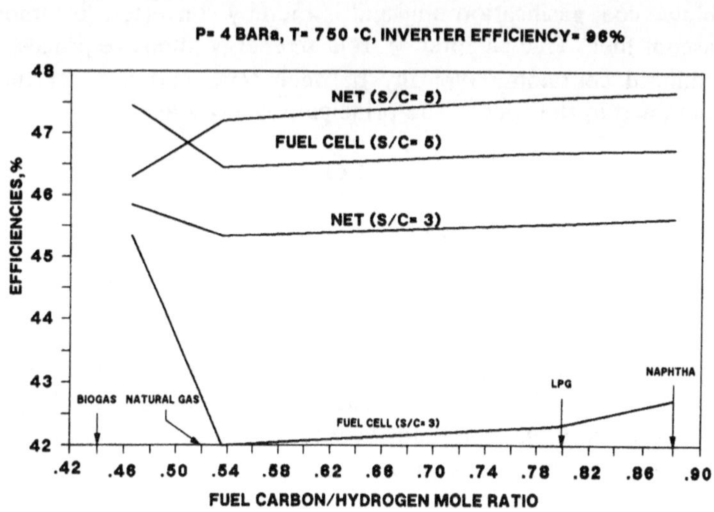

Figure 6.22. 250-kW PAFC: effect of varying fuels on efficiencies.

*Note that fuel processors for fuel cell systems have also been developed to handle methanol[64] and high-sulfur-content distillate fuels.[65,66]

This optimization has been performed at two steam-to-carbon ratios. The fuel cell efficiency, ε_{FC}, is, especially at low steam-to-carbon ratios, clearly dependent upon the feedstock being utilized. The particular sensitivity of ε_{FC} to biogas fuel is due to the high concentrations of carbon dioxide and hydrogen, 33.3% and 5%, respectively, in the study feedstock.*[6] The presence of carbon dioxide at the reformer inlet accentuates the effect of the following reactions over the reformer catalyst:[35]

$$CH_4 + CO_2 \leftrightarrow 2CO + 2H_2, \tag{6.21}$$

$$CH_4 + 3CO_2 \leftrightarrow 4CO + 2H_2O. \tag{6.22}$$

These reactions occur in addition to reactions (6.11)–(6.14) (Section 6.4.1.1). As a consequence, the biogas-reforming process yields more carbon monoxide and has a lower methane slip than that produced by, for example, natural gas reforming.[42,67] Hence, especially at relatively low steam-to-carbon ratios, hydrogen constitutes a greater proportion of the anode feed heating value, leading to an increase in the fuel cell efficiency, ε_{FC} (Section 6.3.1 and Chapter 5).

Nevertheless, Fig. 6.22 also shows that once a system design has been optimized for a given fuel, the net system efficiency, ε_{NET}, is significantly less sensitive to the feedstock type.† These conclusions have been confirmed in practice by the performance of an International Fuel Cells PC-18 40-kW$_e$ PAFC power plant designed for natural gas fuel. The unit, owned by Southern California Edison, was modified by Kinetics Technology International to run on biogas for 80 h of continuous operation. The work showed the power plant performance to be unaffected by the change in feedstock, despite operating at essentially design process conditions.[42,68]

Coal, or coal gas piped in from remote locations, is generally expected to be another important feedstock to fuel cell systems.[7] The coal may be fed to an air- or oxygen-blown coal gasification unit and is thereby converted to a more easily utilizable gaseous fuel[69] (see Chapter 4). It is an energy-intensive process with the coal gas produced containing typically between 60% and 80% of the energy originally contained in the coal.[70] The principal reactions are

$$C + O_2 \rightarrow CO_2, \tag{6.23}$$

$$C + H_2O \rightarrow CO + H_2, \tag{6.24}$$

$$C + CO_2 \rightarrow 2CO. \tag{6.25}$$

Reaction (6.23) is the combustion reaction providing the endothermic heat of reaction for reactions (6.24) and (6.25). The resulting coal gas, which is primarily a mixture of carbon monoxide and hydrogen but also contains a host of contaminants, requires processing prior to being fed to fuel cell stacks. Typical

*By contrast the feedstocks of natural gas, LPG, and naphtha contain predominantly paraffinic hydrocarbons.

†Note that the calculations for net system efficiency, ε_{NET}, for the liquid fuels, such as LPG and naphtha, incorporate reduced plant parasitic power demand since pumping instead of compression work is required for the primary fuel.

contaminants from coal gasifiers include fly ash particulates, sulfur and sulfur-containing compounds, halogens, trace metals such as arsenic and zinc, ammonia, aromatics, and polynuclear compounds.[18,19]

In the fuel processing system, such contaminants must be brought down to levels acceptable for fuel cell operation. In addition, the coal gas must be mixed with superheated steam and admitted to a shift reactor system so that the carbon monoxide and steam may be shifted to hydrogen and carbon dioxide by reaction (6.14). In PAFC systems the carbon monoxide level required is approximately 1%, and in MCFC systems a maximum level of between 6% and 10% is usually required prior to admittance to the fuel cell anode. Fuel cell waste heat is used to raise process steam. The unutilized fuel gases leaving the fuel cell stacks are fed to an adiabatic pressurized combustor, with the resulting flue gas being used for process heating needs prior to being mixed with the cathode off-gas and let down through a power recovery turbine.[6]

Fuel cell systems may either accept pipeline coal gas as feedstock at battery limits or be integrated with a coal gasification facility. Studies have shown that system efficiencies for multi-MW_e integrated coal gasifier–MCFC power plants are about 50% (higher heating value (HHV) basis, or approximately 54% LHV basis).[4] Multi-MW_e MCFC and PAFC systems operating on piped coal gas are expected to realize net system efficiencies, ε_{NET}, of about 50.6% and 41.7% (LHV basis), respectively.[6,71]

As is evident from this discussion, one may conclude that fuel cell system economics will be more dependent upon the installed plant cost and fuel pricing than on variations in the system performance due to the use of different fuels. At current pricing, natural gas, biogas, and coal gas will have the edge over the more expensive fuels of LPG and naphtha (Chapter 12).

6.7. EFFECT OF SCALE

In general, increasing the capacity of a fuel cell system allows for an increase in the complexity of design in the knowledge that system investment costs are not directly proportional to scale. In fact, the only system components with a scale-up factor of 1 are the fuel cell stacks themselves.[54] All other components of the fuel cell system have a scale-up factor of less than 1, so, for example, typical reformer scale-up costs are subject to an exponential cost-factor in the range 0.6–0.8.[56] Hence, the proportion of the cost of electricity (COE) attributable to cost of capital decreases with an increase in the system capacity.

The implications of the effect of scale are that it is economically attractive to squeeze extra fuel efficiency out of a fuel cell system at the expense of increased investment as the system capacity is raised. Table 6.7 summarizes the results of extensive optimization studies, which included an evaluation of the effect of scale on the allowable system complexity and, hence, net system efficiency, ε_{NET}.[6,72]

Extensive economic evaluations[6] (see Chapter 12) fully justify the thinking that the complexity of optimized fuel cell systems may increase with scale for the benefit of gains in system efficiencies.

Table 6.7. Effect of Scale on ε_{NET} for PAFC and MCFC Systems[a]

	25 kW$_e$	250 kW$_e$	3.25 MW$_e$	100 MW$_e$
PAFC (%)	41.2	47.1 36.9[b]	41.7[c]	—
MCFC (%)	51.5[d]	49.1 50.1[c]	53.4 50.6[e]	56.6 40[f]

[a]Natural gas fuel unless otherwise stated; each system is fully optimized.
[b]Alternative design with steam raising in convection section.
[c]Coal-gas fuel.
[d]Oxygen enrichment process option.
[e]Naphtha fuel.
[f]Conventional reformer (no pressurized combustion and no regenerative reformer).

6.8. LOAD RESPONSE

The control and operation of a fuel cell power plant is carried out by segregating it into four major subsystems, namely, the hydrogen production system, the steam system, the fuel system, and the power generation system. The latter includes the fuel cell stacks and the dc–ac power conditioner. In general, a fuel cell system may be designed to operate safely down to 25–30% of its nominal capacity.[41] Unlike other power generation technologies the net system efficiency, ε_{NET}, and fuel processor efficiency ε_{FP}, of the plant are largely constant over the turndown range[56] (see Fig. 6.23).

However, control of fuel cell system performance during dynamic load conditions and during plant start-up and shutdown is by no means straightforward. This is because the fuel cell, being an electrochemical device, reacts to load variations more or less instantaneously, whereas the fuel processing system has relatively slow response times.

In a typical control scheme, the power plant receives a signal for an increase in capacity from the electrical load, for example, the electricity grid. This load is not drawn from the fuel cell instantaneously, since this would cause the fuel cell stack to become starved of hydrogen and oxidant. In practice, therefore, the capacity signal instructs the natural gas feedstock controller to admit more fuel to the reformer while engaging the steam system and the steam-to-carbon controller.

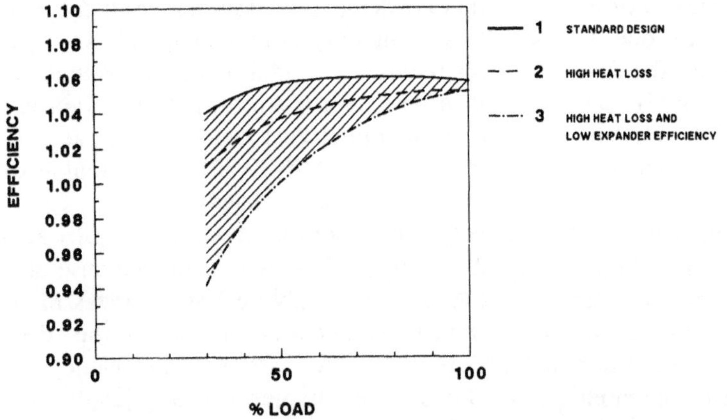

Figure 6.23. Fuel processor efficiencies vs. load 250-kW unit.

The latter ensures that sufficient steam is made available for mixing with the reformer feed and that a minimum steam-to-carbon ratio is maintained. Although the combustion system acts right away to maintain the reformer outlet temperature, there is usually sufficient thermal inertia in the reformer (especially in multi-kW_e capacity systems) to prevent excessive fluctuations in synthesis gas composition during load changes. Once the fuel processing system can ensure sufficient supplies of hydrogen and air at the fuel cell electrodes, the control system permits the electrical grid to draw power from the fuel cell stacks. Load changes of up to 64% per minute have been recorded for a 50-kW_e pressurized PAFC power plant.[73]

An important contribution to the optimum and safe operation of a fuel cell power plant can be made by the use of process simulation software packages. Such dynamic simulation models may be used to increase one's understanding of the interactive nature of various power plant components, and thereby enable development of basic and advanced control strategies for fuel cell power plants.[41] For example, the system load change response or the effect of malfunctions on performance may be studied. In fact, such dynamic simulation packages may also be interfaced with the control computer of an actual plant, enabling data acquisition, model parameter tuning facilities, and full operator training facilities.[74]

6.9. COGENERATION

The major emphasis of this chapter so far has been directed at the design and optimization of fuel cell systems for the highest possible system electrical efficiencies. However, in some applications heat in the form of hot water or steam is also required. In these cases any waste heat available from a power plant can have significant value, since the alternative is to generate the heat separately at cost. The relative demand for heat compared to electricity varies widely from a low of 14% to a high of 2760% of the electricity demand, depending upon the application[75,76] (see Chapter 12). However, it should be stressed there are also many locations and large market segments where heat is not required or desirable. The amount and quality of heat available from a power plant may not always match the demand, so some "usable" cogeneration heat may have to be lost, or, in the other extreme, a supplementary boiler may have to be installed.

Once optimized for maximum electrical efficiencies, both PAFC and externally reformed MCFC systems only have waste heat available at relatively low temperatures of below 140°C and 150°C, respectively. Generally speaking, this heat can only be usefully used in cogeneration applications down to 70–100°C, depending upon the application. If cogeneration heat is required at higher temperature levels, this is possible either by "deoptimizing" the design or by installing a boiler for the purpose. For example, a fuel cell system operating at essentially atmospheric pressures could enable utilization of reformer exhaust gas for high-temperature cogeneration (see Section 6.4.1.1.). Since the value ascribed to cogeneration heat varies considerably from application to application, the generalized evaluation of cogeneration economics is notoriously difficult. Nonetheless, in cogeneration applications, fuel cell technology can be competitive with the more traditional cogeneration technologies such as gas turbines and gas engines (see Chapter 12).

6.10. RELIABILITY

The reliability of the current generation of fuel cell power plants is one of the most critical issues facing the commercialization of this technology today. As far as most potential users are concerned, reduced reliability means increased maintenance costs and an increased dependence upon more costly backup supplies of electrical energy or a need for redundant power supply capacity (extra investment). In fact, in most instances, diminished reliability makes the application of such technology decidedly unattractive, this despite all the advantages attributed to fuel cell systems. Most of the world's fuel cell demonstration programs have managed to highlight those well-known attributes of fuel cell technology, namely, high-efficiency, high-part-load efficiency, environmental friendliness, and modular characteristics. However, many of these programs have been unable to initiate the round of expected commercialization of this technology by singularly failing to demonstrate that fuel cell systems can indeed be highly reliable.

Analyses of the reasons for the reduced reliability of many fuel cell power plants in the United States and Japan have shown that by far the vast majority of forced shutdowns (>95%) have been caused by the failure of the balance of plant components. These stoppages have involved electrical and electronics components (including sensors and cabling), mechanical controls, leakages, reformer malfunctions, and rotating equipment shortcomings.[77,78] In short, those elements that represent established and well-known technologies, which are, of course, in stark contrast to relatively high availability of the "new" fuel cell stack technology itself.

Modern hydrogen plants, which are in widespread use in the electronics, food, metallurgical, and petrochemicals industries, as well as in refineries, require constant availability, 24 h/day, over 8400 h/year, without any unexpected shutdowns. Industry uses hydrogen like individuals use tap water or electricity from the grid. Reliability of plant operation is a matter of design philosophy, engineering procedures, and quality control. In fuel cell power plants, standard components should be used that have proven operation in existing hydrogen plants. This fact favors larger capacity fuel cell plants, where many more proven components can be used than in small-capacity fuel cell plants (<100 kW). In the latter case, dedicated development of many components complicates the matter, though once proven, mass production can more quickly reduce the costs. In recent years, many of the world's fuel cell manufacturers have begun to involve and utilize the experience of specialized engineering contractors. The objectives of such joint developments are to improve the reliability and reduce the cost of the current generation of fuel cell power plant technology.

6.11. EMISSIONS

By combusting essentially hydrogen with oxygen to water, the fuel cell stacks themselves are clean sources of energy, and produce only high-purity water as well as electricity and waste heat. The fuel processor itself requires heat input, which is normally supplied by furnace burners, although if methanol is to be converted into a hydrogen-rich gas then lower grade heat sources may be used.

Table 6.8. NO_x Emissions of Fuel Cell Systems g/GJ_e (LHV)

	Equilibrium[a] emissions	Standard conv. burners	Modern burners	Best low NO_x burners
PAFC				
25 kW	207	7.1	3.4	0.6
250 kW	218	7.4	3.7	0.6
MCFC				
25 kW	17 (with oxygen enrichment)	0.6	0.3	0.05
250 kW	32	0.9	0.9	0.3
3.25 MW	55	1.8	0.9	0.09
100 MW	202 (with carbon dioxide removal)	6.6	2.8	0.53

[a]Equilibrium evaluations assume that the NO_x formation reactions proceed until they approach their equilibrium states. The reality is, however, quite different, as kinetic and mixing effects must be accounted for.

The burners are a major source of nitrogen oxides (NO_x), carbon monoxide, unburnt hydrocarbons, and particulates. Nevertheless, modern burner design can guarantee acceptable levels of carbon monoxide, unburnt hydrocarbon, and particulate emissions. In the case of fuels with low nitrogen content, NO_x is mainly formed from nitrogen contained in the combustion air, and such emissions from a fuel cell system may be extremely low due to the following:

1. Fuel cell system burners mainly use the anode off-gas, which is a very lean gas containing the unutilized hydrogen, carbon monoxide, and some traces of unreacted hydrocarbons. This lean gas burns with low flame temperatures, and hence there is a low tendency to form NO_x.[6]
2. Unconventional, but low-cost, burners may be used, which, by limiting the actual flame temperature, produce very low NO_x levels. These burners can be based on catalytic combustion, ceramic fiber, or metallic fiber matrix technologies.[79,80]
3. Many fuel cell systems of the future are expected to employ pressurized combustion in the reformer, which is known to further suppress NO_x formation.[14]

Table 6.8 details the results of recent studies which, based on computer simulation of equilibrium NO_x formation in different type of burners, have predicted very low NO_x emissions from PAFC systems (between 1 and 10 ppm at 3% oxygen), and as much 5 to 10 times lower from MCFC system.[6] Actually measured emissions from PAFC systems have confirmed these predictions.[77,81]

Fuel cell systems are also expected to produce environmentally acceptable noise levels, an issue that has already been confirmed by demonstration systems.[41,77] Noise is mainly generated by rotating equipment, and by good system design and the application of relevant noise-limiting technology, noise may be kept to an absolute minimum.

6.12. CONCLUSIONS

This chapter has shown the importance of understanding how the various components of a fuel cell system interact. It may be evident that system optimization is no simple matter and requires a systematic study of many parameters.

Three process variables, namely, the pressure in the reformer and fuel cell, reforming temperature, and steam-to-carbon ratio are most critical in determining fuel cell system performance. In general, the reforming conditions most favored by MCFC systems tend to be less severe (that is, lower reforming temperatures and steam-to-carbon ratios) than those required for optimum performance of PAFC systems. This is primarily due to the fact that the MCFC stack can utilize carbon monoxide and methane in addition to hydrogen.

It has also been demonstrated that the fuel processor itself must be properly optimized for each case under study. The reformer geometry impacts directly on the reformer makeup fuel requirement and the power plant parasitic power demand, and an optimum design must be identified.

Various process design options may be implemented in fuel cell system flowsheets, such as oxygen enrichment and the use of depleted cathode air as reformer combustion air. In addition, MCFC system complexity may be further increased by reformer carbon dioxide removal technologies. The economic benefits of applying such process options have been comprehensively assessed and the results presented in Chapter 12.

REFERENCES

1. B. Linnhoff *et al.*, *User Guide on Process Integration for the Efficient Use of Energy* (IChemE, 1985).
2. A. P. Rossiter *et al.*, *Hydrocarbon Process.* **Jan.** 63–66 (1991).
3. S. S. Penner, ed., *Energy* **2**(1/2) (1986).
4. T. G. Benjamin, E. H. Camara, and L. G. Marianowski, *Handbook of Fuel Cell Performance* (Institute of Gas Technology, 1980).
5. K. Kinoshita, F. R. McLarnon, and E. J. Cairns, *Fuel Cells: A Handbook* (Lawrence Berkely Laboratory, 1988), DOE/METC-88/6096.
6. M. N. Mugerwa *et al.*, *Fuel Cell Systems, Design Optimization and Environmental Aspects of Fuel Cell Systems,* KTI BV report (Dutch Management Office for Energy Research (PEO), Utrecht, 1988).
7. Fluor Engineers and Constructors, *An Economic Comparison of Molten Carbonate Fuel Cells and GasTurbines in Coal Gasification-Based Power Plants* EPRI AP-1543, 1980.
8. R. H. Williams and E. D. Larson, in *Electricity, Efficient End-Use and New Generation Technologies and their Planning Implications* (T. B. Johansson *et al.*), 1989.
9. Holec Nederland B. V., private communication, 1988.
10. L. J. M. J. Blomen and M. N. Mugerwa, *J. Power Sources* **29** 71–75 (1990).
11. American Public Power Association (APPA), *Notice of Market Opportunity (NOMO) for Fuel Cells,* 1988.
12. United Technologies Corp., *Fuel Processor Developments for 11-MW Fuel Cell Power Plants* EPRI EM-4123, 1985.
13. G. G. Elia, *Fuel Processing Analyses for Fuel Cell Power Systems* (AIChE Symp. Ser., Heat Transfer-Denver, 1985), pp. 13–18.
14. L. J. M. J. Blomen *et al.*, *Pressurized Combustion,* KTI BV report (Dutch Management Office for Energy Research (PEO), Utrecht, 1987).
15. Haldor Topsoe Inc. Westinghouse Electric Corp., *Demonstration of a High-Efficiency Steam Reformer for Fuel Cell Power Plant Applications* EPRI AP-5319, 1987.

16. S. Bakhsh *et al.*, in 7th Int. Symp. Large Chemical Plants, (Brugge, Belgium, 1988), pp. 193–201.
17. T. Kahara *et al.*, in Fuel Cell Sem. (Phoenix, AZ, 1990), pp. 95–98.
18. Pigeaud *et al.*, in (Int. Fuel Cell Sem., The Netherlands, 1987), pp. 309–317.
19. M. C. Williams and D. A. Berry, in (Fuel Cell Sem., (Phoenix, AZ, 1990), pp. 306–309.
20. *Fuel Cell News*, **8**(4) (1990).
21. R. Anahara, in Fuel Cell Sem. (Phoenix, AZ, 1990), pp. 486–490.
22. R. Anahara, *J. Power Sources* **29**, 109–117 (1990).
23. M. K. Bergman, R. W. Claussen, D. M. Rastler, in Fuel Cell Sem. (Phoenix, AZ, 1990), pp. 499–502.
24. M. R. Fry, *Fuel Cells—R&D Review and Assessment of Composition Generation Options*, confidential report (W. S. Atkins Energy, Surrey, UK, 1990).
25. Confidential sources.
26. KTI confidential report, *Turbomachinery: General Investigation into Present High Efficiency Equipment and Prediction of Future Trends*, 1990.
27. KTI Group, European patent appl. No. 89250073.7, 1989.
28. R. Clement, *Chem. Eng.* **Aug.**, 169–170 (1989).
29. H. van den Broeck, ELENCO, private communication.
30. H. U. Dütsch, in *Proc. 2nd Ann. CMDC Conf.* (Zürich, 1990).
31. Exxon Enterprises Inc., *Application of the Alstrom/Exxon Alkaline Fuel Cell System to Utility Power Generation*, EPRI EM-384, 1977.
32. O. Lindström, in Fuel Cell Sem. (Long Beach, CA, 1988), pp. 273–176.
33. S. Thyberg *et al.*, in Fuel Cell Sem. (Phoenix, AZ, 1990), PP. 411–414, 467–510.
34. Poster presentation, Int. Fuel Cell Sem. (The Netherlands, 1987).
35. E. S. Wagner, *Kinetic Study of Carbon Formation and Gasification under Steam Reforming Conditions*, Ph.D. thesis, (Rijks Universiteit Gent, 1990).
36. K. S. Raghuraman and T. Johansen, in *Reforming Technology*, Symp. Science of Catalysis and Its Application in Industry (FPDIL, Sinari, 42, 1979), pp. 418–427.
37. K. C. Khurana, *Chem. Age India*, **20**, 777–783 (1969).
38. P. M. Plehiers and G. F. Froment, in 1st European Conf. Industrial Furnaces and Boilers (Lisbon, 1988).
39. M. A. Hossain, *Hydrocarbon Process.* **May**, 76-A–76-C (1988).
40. P. F. van den Oosterkamp, J. E. Hille and L. J. M. J. Blomen, *Werktuigbouwkunde*, **no. 7/8**, 37–41 (1987).
41. KTI BV, *Design and Operating Manual 25 kW PAFC Power Plant TUD*, 1988.
42. P. Pietrogrande, in Int. Fuel Cell Sem. (The Netherlands, 1987), pp. 76–95.
43. K. S. Raghuraman, J. F. Nomden, and S. Ratan, *Scouting Study on Hydrogen Manufacturing Unit*, KTI Confidential Report for Shell, 1986.
44. M. V. Twigg, ed., *ICI Catalyst Handbook*, (Wolfe Publishing Co., 2nd ed. 1989).
45. A. J. Chapman, *Heat Transfer*, 3rd ed. (MacMillan Publishing Co., New York, 1974.
46. T. van Herwijnen, *On the Kinetics and Mechanism of the CO-Shift Conversion on a Copper/Zinc Oxide Catalyst*, Ph.D. thesis (TH Delft, 1973).
47. ICI, private communication, 1990.
48. Süd-Chemie, private communication, 1990.
49. Das Solar-Wasserstoff-Projekt, *Gewinnung von Wasserstoff aus Solarenergie, Speicherung, Energetische Anwendung*, Publicity Brochure, 1988.
50. Institute for Hydrogen Systems Report, *Energy-Exergy analyses of Current and Future Electricity Generating Facilities*, 1986.
51. M. A. Rosen, *The Development and Application of a Process Analysis Methodology and Code Based on Exergy, Cost, Energy and Mass*, Ph.D. thesis (University of Toronto, 1986).
52. A. A. Goorse, *Energie-, en Exergieanalyse van een PAFC Brandstofcelsysteem*, Chem. Engineer's thesis (TU Delft, 1990).
53. KTI BV confidential report, *The Economics of Carbon Dioxide Recovery and Recycle in a 6 MW_c to 8 MW_c MCFC Power Plant*, 1990.
54. Fuji, private communications 1988.
55. KTI BV confidential report, *The Influence of MCFC Power Plant Capacity on the Choice of Technology for Carbon Dioxide Recovery and Recycle*, 1990.
56. KTI, private communications, 1991.
57. M. A. Rosen and D. S. Scott, in 7th World Hydrogen Energy Conf. (Moscow, 1988), pp. 761–772.

58. K. M. Lee *et al.*, *Hydrocarbon Process.* **April,** 49–53 (1989).
59. S. M. Ranade *et al.*, *Hydrocarbon Process.* **July,** pp. 39–43 (1989).
60. Air Products, *Chem. Eng.* **Oct.** (1987).
61. V. P. Belyakov and L. N. Chekalov, *Membrane Gas Separation in the Industry* (V/o Chimmachexport, USSR, 1987).
62. S. G. Kimura and W. R. Browall, *J. Membrane Sci.* **29,** 69–77 (1986).
63. BASF A. G., FRG, U.S. patent 4537753 (Aug. 27, 1985).
64. T. Nogi, T. Toma, and H. Yamazaki, in Fuel Cell Sem. (Phoenix, AZ, 1990), pp. 322–325.
65. R. G. Minet, P. G. Cronin, and D. Warren, in Int. Gas Research Conference, 1981.
66. G. Steinfeld, *Diesel Fuel Processing for the PAFC Process Demonstration,* Interim technical report (Belvoir Research, Development and Engineering Center, 1986), DAAK 70-85-C-0090.
67. S. Mullick, P. Pietrogrande, and D. Warren, *Peformance Evaluation for a 11 MW_e* Fuel Cell Power Plant which Utilizes a Generic Landfill Gas Feedstock, KTI Corp. report (EPRI, 1986).
68. D. Warren *et al.*, *Performance Evaluation for the 40 kW_e Fuel Cell Power Plant Utilizing a Generic Landfill Gas Feedstock,* KTI Corp. report (Southern California Edison, 1986).
69. J. G. Speight, ed., *Fuel Science and Technology Handbook* (Marcel Dekker, New York, 1990), pp. 661–734.
70. KTI Corp. report, *Assessment of Small Scale Power Generation Systems Based on Coal Fuel,* 1984.
71. B. R. Krasicki and B. L. Pierce, *Int. J. Hydrogen Energy* **87,** 499–508 (1983).
72. L. J. M. J. Blomen *et al.*, *MCFC Power Generation,* KTI B. V. report (Haskoning, 1985).
73. T. Saito, K. Kubota *et al.*, in Fuel Cell Sem (Phoenix, AZ, 1990), pp. 256–259.
74. Pyrotec, private communication, 1988.
75. Tokyo Gas Co. Ltd., *On-site Fuel Cells,* Lecture, May 20, 1987.
76. Institute of Gas Technology, *Fuel Cell Support Studies, Onsite Molten Carbonate Systems,* Final Report, Project No. 30514, 1979.
77. Gas Research Institute, *40 kW_e On-site Fuel Cell Field Test Program (Jan. 1982 –May 86),* Final report, 1986.
78. Tokyo Electric Power Company, private communications, 1987.
79. B. E. Enga, *Catalyst Systems for the Control of Nitrogen Oxide Emissions* (IChemE Symp. Ser. no. 57, Salford, 1979), pp. S1–S4.
80. P. T. Gilmore, M. N. Mugerwa, and Q. Valenti, *Erdöl + Kohle—Erdgas – Petrochemie/Hydrocarbon Technology* **43,** 147–149 (1990).
81. M. Kaizaki *et al.*, in Fuel Cell Sem. (Phoenix, AZ, 1990), pp. 330–333.

7

Research, Development, and Demonstration of Alkaline Fuel Cell Systems

Hugo Van den Broeck

7.1. INTRODUCTION

Alkaline fuel cells are among the first fuel cells to have been studied and taken into development. They are the first cells to have achieved successful routine application, although in a very specific environment, namely the space shuttle missions in the United States. They quite probably also represent the type of fuel cells on which the largest number of development programs have begun in the world, particularly in Europe, resulting in almost as many discontinued programs.

In the past, work on alkaline fuel cells has covered various fuels as well as low and medium electrolyte temperature ranges. At the present time, these areas have been narrowed considerably.

The direct use of a liquid fuel in a fuel cell has always attracted potential users because of the advantages in terms of logistics as compared to gaseous fuels. However, in conjunction with alkaline cells, which usually use a KOH solution as their electrolyte, the chemical reaction of the KOH with some fuels make that approach impossible, and a fuel processing step before the fuel cell becomes a necessity. The most well-known example of such a liquid fuel is methanol.

A liquid fuel which has, however, appeared to be compatible with an alkaline electrolyte is hydrazine (N_2H_4). It can be considered as liquid storage of hydrogen, since it decomposes easily into hydrogen and nitrogen at the anodes. Work on hydrazine fuel cells has been carried out intensively in the 1950s and 1960s. Most efforts were in the scope of defense projects, in Europe (UK, France, Germany) as well as in the United States. Depending on the projects, air, pure oxygen, or H_2O_2 were used as the oxidant. Major disadvantages, such as the highly poisonous nature and the high cost of the hydrazine, as well as material

Hugo Van den Broeck • Elenco N.V., Gravenstraat 73 bis, B-2480 Dessel, Belgium.

Fuel Cell Systems, edited by Leo J. M. J. Blomen and Michael N. Mugerwa. Plenum Press, New York, 1993.

problems, brought work on hydrazine fuel cells to an end by the beginning of the 1970s, and it does not seem likely that these developments will be started up again in the near future. Therefore hydrazine fuel cells are not discussed further. The focus instead is on the use of hydrogen itself. For more information on hydrazine fuel cells several interesting reviews exist.[1,2,3] As for the operating temperature ranges to be considered for alkaline fuel cells, the emphasis will be on low temperatures (under 100°C), although the possibility of higher temperatures will be discussed briefly. This chapter also discusses the utilization of both air and oxygen as oxidants.

7.2. MANUFACTURING TECHNIQUES AND MATERIALS

7.2.1. Electrodes and Catalysts

Even limiting the discussion to alkaline cells, electrodes and catalysts could fill a book on their own, given the impressive amount of work carried out until now not only by industrial companies but also at universities and research centers. Within the present scope, only a short review will be presented, and in the selection of examples the focus has been on those electrodes currently used in alkaline fuel cell programs.

The type and manufacturing of anodes as well as cathodes are linked to the catalysts chosen. Unlike phosphoric acid fuel cells, alkaline fuel cells offer the possibility to use not only precious metal catalysts, but also nonprecious ones. The latter are most commonly based on Raney nickel powders for the anodes and silver-based catalyst powders for the cathodes, whereas platinum or platinum-alloy catalysts are either based on the use of precious metal particles deposited on carbon supports or are part of metallic electrodes, generally based on nickel substrates.

It is clear that there are some general requirements for all types of electrodes:

1. Good electronic conductivity, in order to reduce ohmic losses
2. Adequate mechanical stability and suitable porosity
3. Chemical stability in the alkaline electrolyte environment
4. Electrochemical stability with time, which means stability of the catalysts as such and stability of the way they are incorporated in the electrodes

Another important characteristic is the hydrophobic or hydrophilic nature of electrodes. The latter usually are metallic electrodes. The former are only partly wetted, generally because of the presence of hydrophobic PTFE (polytetrafluoroethylene) in carbon-based electrodes. It is important for the long life of hydrophobic electrodes that their hydrophobicity be adequately preserved by an appropriate structure of the PTFE-containing catalytic layer.

Usually, electrodes consist of several layers with different porosities, so as to organize within and through the electrodes the respective flows of liquid electrolyte and gaseous fuel (hydrogen) or oxidant (air or oxygen). Widely different techniques can be used for manufacturing such electrodes or some of their layers. Usually powders are mixed and then pressed or calendered into layers. Sedimentation as well as spraying techniques can be used, and in many cases high-temperature sintering operations are used to ensure good mechanical stability.

As for other electrochemical power sources, such as batteries, homogeneity within electrodes and reproducibility of the electrodes are key elements for good operation and long life. This also means that adequate quality control measures must be implemented for each electrode production, either in the laboratory or when setting up serial production.

As a good example of two quite different approaches, those followed by the two European companies most active at present on alkaline fuel cells can be presented.

Elenco electrodes Anodes and cathodes, developed for operation with H_2 and air, consist of three layers: a current-collecting nickel mesh, a catalytic layer, and a hydrophobic gas-porous PTFE layer at the gas side of the electrodes. The catalytic layer is calendered, using a mixture of platinized carbon particles with PTFE, the only difference between anodes and cathodes being the type of carbon substrate used for the catalyst, in order to cope with the different electrocatalytic environments (reduction or oxidation). The catalyst loading is relatively low: some 0.6 mg Pt per square centimeter of cell, anode and cathode taken together. The manufacturing process only uses ambient temperature steps and no wet steps, with a view to cost effectiveness of mass production. The described electrodes operate at atmospheric pressure. Their thickness is only 0.4 mm.

Siemens electrodes Siemens electrodes have been developed for operation with H_2 and O_2. They are gas-diffusion electrodes 1 mm thick. The catalyst, a binding agent, and some asbestos fibers are put on a supporting nickel mesh by a sedimentation technique. For the anodes, the catalyst is titanium-doped Raney nickel with a loading of $110\,\text{mg} \cdot \text{cm}^{-2}$; for the cathodes the catalyst is doped silver with a loading of $60\,\text{mg} \cdot \text{cm}^{-2}$. At the electrolyte side of the electrodes, a thin asbestos layer is pressed onto the electrodes, its function being to stop the gases from going into the electrolyte compartments. The electrodes are used in a slightly pressurized environment: 2.3 bar at the H_2 side and 2.1 bar at the O_2 side.[4,5]

Another typical electrode development has been that of the VARTA company: they are called bipolar "Janus" electrodes and are based on the use of Raney nickel in the anodes and Raney silver in the cathodes.[6]

For more detailed and general information on alkaline fuel cell electrodes, their composition, manufacturing, see the literature.[1,2,3,7]

7.2.2. Electrolytes

By far the most common alkaline electrolyte is a KOH water solution, with normalities of 6 to 8. The KOH used must be sufficiently pure so that no impurities can cause catalyst poisoning.

In most European developments (Elenco, Siemens), the electrolyte is mobile, which means that it is pumped through the cells. Another approach is to contain the KOH in a matrix layer, as is done with the phosphoric acid in PAFC systems. This so-called immobile electrolyte arrangement has not yet been successfully developed for alkaline cells, except for IFC's alkaline fuel cell used in the American space shuttles, where the KOH is contained in the asbestos matrix and a so-called electrolyte reservoir plate acts as a KOH buffer.

As an alternative for KOH, NaOH has been envisaged. On balance, its performance characteristics are somewhat less interesting than those of KOH, and the cost advantage is not really that important because the KOH electrolyte can be used for a long time in the fuel cell so that its contribution to the overall cost is almost negligible.

7.2.3. Alkaline Fuel Cell Stacks

In order to achieve practical voltages, many electrodes have to be built into stacks. The requirements on stack materials and manufacturing technologies are demanding. The material must show excellent behavior in terms of chemical stability (not only versus KOH, but also versus KOH + O_2) and thermal stability (at nominal operating temperatures, with allowance for local temperature peaks). Electrical properties must be adequate, and compatibility must exist with suitable manufacturing techniques to achieve a cost-effective and reproducible product within the narrow geometrical specifications defined in view of space and weight minimization requirements. Thus, the number of qualifying materials is quite limited. Some epoxy resins may be used, but they do not lend themselves to cheap mass production. Polysulfone is an excellent material (e.g., used for the Siemens cell and the U.S. shuttle fuel cell), as well as ABS (acrylonytril–butadiene–styrene, used by Elenco), provided the temperature peaks can be kept under 100–110°C. In recent years, studies in the United States (NASA) and Europe (Elenco) have pointed to PPS as an extremely interesting material with, for different reasons, higher potential than polysulfone and ABS.

Building up a stack consists of two essential steps: framing the electrodes

Figure 7.1. Frames and components for a Siemens alkaline fuel cell stack.

Figure 7.2. Elenco alkaline fuel cell stack unit.

and, if necessary, other components such as separators, and then building the frames together. The framing can be done by injection molding (Elenco). The frames can be pressed together by a filter press technique (Siemens) or welded if the plastic material used allows it (Elenco).

It is obvious that both the design and the construction techniques of the stacks have to be performed to ensure complete tightness between the electrolyte, the hydrogen, and the oxidant compartments in the stack.

To illustrate stack technology, Fig. 7.1 shows an arrangement of frames and components of a Siemens fuel cell before being incorporated in a filter press stack, whereas Fig. 7.2 shows an Elenco stack unit which actually is one solid block after welding operations. For some more information and further illustrations in relation to fuel cell stacks we refer to Refs. 6, 7.

7.2.4. Alkaline Fuel Cell Systems

The stacks are just the core of a system. This system further consists of equipment which must ensure adequate circulation of all reactants (fuel, oxidant, electrolyte, or some other coolant) and reaction products (electricity, water, and waste heat). When selecting and/or designing the components for this equipment, many requirements have to be fulfilled: compatibility with reactants and/or reaction products, compatibility with the operating parameters (such as temperatures, pressures, pressure differences), and, of course, reliability. Some components may be found as currently available commercial products, whereas others must be designed and manufactured.

Unfortunately, all these peripherals add substantially to the cost of fuel cell systems; in particular, for low-power units (a few kilowatts) the ratio between the cost for the peripherals and that for the stacks can be surprisingly high, especially when stacks are being mass produced.

For a more detailed review of the fuel cell systems structure see Ref. 4 for the Siemens fuel cell system and to Ref. 8 for the Elenco fuel cell system.

7.3. ELECTROCHEMICAL PERFORMANCE

7.3.1. General Considerations

The electrochemical performance of a fuel cell system can be judged in terms of specific weight ($kW \cdot kg^{-1}$) or specific volume ($kW \cdot L^{-1}$). Both values consist of a contribution from the stacks and of one from the peripheral equipment. The stack contribution will directly relate to the current density ($mA \cdot cm^{-2}$) and the cell voltage (V) at which the electrodes are operated. The stack being modular, its $kW \cdot kg^{-1}$ values are the same for all fuel cell power sizes, whereas those for the peripheral equipment will depend on the power size because of scale effects.

On the other hand, the stack performance does depend on several parameters, such as electrode compositions (and, e.g., catalyst amount and cost), nature of the oxidant (air or oxygen), operating pressure, operating temperature, and, to some extent, KOH concentration of the electrolyte.

All this makes it necessary to know all relevant data when comparing different fuel cell systems and/or stacks in terms of their performance.

7.3.2. Influence of Oxidant

Some alkaline fuel cell stacks have specifically been developed for oxygen (by IFC in the United States and Siemens in Germany), whereas others have basically been developed for air (e.g., Elenco in Belgium). The latter stack is also able to use pure oxygen or any form of oxygen-enriched air, in which case performance will be increased. As an example, it can be stated that an Elenco stack shows a 50% increase of its current at constant nominal voltage when going from air to pure oxygen.

On the other hand, it often appears quite uninteresting to operate some stacks, developed and optimized for oxygen, on air, the reason being that the electrodes cannot cope with the large amount of inert gas (N_2) to be transported, and they may be too sensitive to some impurities present in ambient air, even after most of these, such as CO_2, have been largely removed.

It is clear that the use of either air or oxygen will depend on the application envisaged. Oxygen is, of course, the only choice for space and some defense (submarines) applications, whereas for many commercial applications, such as electric traction, the use of air is certainly to be recommended for several reasons (e.g., safety and cost).

7.3.3. Influence of Pressure

It results from the theory described in Chapter 3 that an increase of pressure has a positive effect on fuel cell performance, and it is well known that many fuel cell systems developed have operating pressures of a few bars. This is also the case for some of the alkaline fuel cell systems, such as the fuel cell developed by IFC for the orbiter or shuttle utilization. The increased pressure (3–4 bar) permits a quite compact stack to be made, which obviously is a major requirement for space utilization. Increasing the pressure much higher would only yield minor improvements from the electrochemical side, which then would be compensated by the weight increases associated with the higher mechanical strength needed because of the higher pressure. A major concern with reactant pressure increase is the fact that the Δp values between gases on the one hand and the electrolyte on the other must at all times be kept within given limits, dictated by the nature of the electrodes and their composition. If this is not respected, gas would burst into the electrolyte compartments, disturb the good operation, and, if the phenomenon happened simultaneously at both gas sides, dangerous hydrogen–oxygen mixtures in the electrolyte loop could be generated.

7.3.4. Influence of Temperature

Because the conductivity of a KOH solution is relatively good at low temperatures (in comparison with that of phosphoric acid), an alkaline fuel cell with a nominal operating temperature around 70°C will at room temperature show a power level about half that at its nominal temperature. The power level increases almost linearly between room temperature and 50–60°C, and the power increase with further increasing electrolyte temperature depends on other parameters, such as the KOH concentration. Pressure may also play a role. The information given is related to Elenco's experience with air and at atmospheric pressure conditions. For six to seven normal KOH concentrations. 70–80°C appears to be a kind of optimum, whereas for eight or nine normal KOH concentrations somewhat higher temperatures (90°C) give better performance results.

One of the oldest alkaline fuel cells, namely the one for the Apollo space flights, itself based on the historical Bacon cell (see Chapter 1) operated at 200°C, but to prevent electrolyte boiling the pressure had to be at least 4 bars.[1]

7.3.5. Some Examples of Performance Data

Siemens has developed 6–7 kW pressurized H_2–O_2 fuel cell systems, fully incorporating well-engineered peripheral equipment. Such a unit has a weight of 85 kg, which means a specific weight of $12\ kg \cdot kW^{-1}$ (being defined at the nominal operating point at $420\ mA \cdot cm^{-2}$).[4] The contribution of the stack to that figure is not mentioned in the reference, but a reasonable estimation points to a value of $7 \pm 1\ kg \cdot kW^{-1}$ at the same current density value of $420\ mA \cdot cm^{-2}$ and 0.78 V cell voltage.

For Elenco's alkaline H_2–air fuel cell, working under atmospheric conditions and using low-precious-metal-content electrodes, a value of $12\ kg \cdot kW^{-1}$ at a rated current density of $110\ mA \cdot cm^{-2}$ has been achieved. However, within the framework of space application, further work has been done to check a combination of increased catalyst content, utilization of oxygen, increased pressure, and further improved stack compactness. This has shown the feasibility of a figure of $6.5\ kg \cdot kW^{-1}$ at $200\ mA \cdot cm^{-2}$ and a cell voltage of 0.88, which corresponds to 3.5 kg · kW^{-1} at 400 mA · cm^{-2} and 0.78 V. This low value is achieved because of the light and thin electrodes.

Recently for special purposes (American space defense projects, requiring only extremely short operating times of fractions of minutes) possible performance figures for very advanced alkaline cells of 0.15 to 0.25 kg · kW^{-1} have been quoted. Such values, however, seem far beyond what is realistic for other applications when cost and endurance have to be taken into account.

7.3.6. Endurance

Achieving a technical endurance in balance with the required economic life is necessary. For some applications 5000 and even 1000 h may be sufficient, whereas for others 25,000 or more may be required. For eventual application in electric power plants, lifetimes of at least 40,000 h with acceptable performance decay should be considered.

The technical stack stability depends mainly on the stability of the catalyst and its performance. In that area, a lot can still be done to improve the technology.

Quite generally, it can be said that at present both H_2–O_2 and H_2–air alkaline cells have stacks sufficiently stable over at least 5000 h, with degradation rates of 20 $\mu V \cdot h^{-1}$ or less. Experience with complete systems is still more limited. For its 6-kW unit, Siemens has recently achieved a total of over 8000 operating hours with some 20 units.[9] The American space shuttle fuel cells have, on average, accumulated some 2000 operating hours each.

7.4. FUEL CELL OPERATION

7.4.1. Introduction

A fuel cell system has to operate so that all conditions allow for faultless access of all reactants to, and elimination of all reaction products from, the stacks. This has to be achieved by adequate peripheral equipment selection, taking into account the proper characteristics of the stacks, which essentially means the electrodes.

We briefly discuss the gas feeding system, current collector, water removal, cooling, and system controller.

7.4.2. Gas Feeding

Because of the risk of carbonate formation within the KOH electrolyte, both H_2 and the oxidant should not contain high amounts of CO_2. Actually, the amount allowed seems to depend on the nature of the electrodes used. Earlier publications have practically put the limit at 0 ppm of CO_2, but, e.g., Elenco has shown that with 50 ppm of CO_2 in air its stacks can be run for more than 6000 operating hours without showing any extra degradation as compared to running such a stack with completely CO_2-free air.[8]

Tests at 75 and 100 ppm, however, showed adverse effects in the course of time, which means that the 300–350 ppm of CO_2 in ambient air certainly is too high, so that scrubbing of the air before letting it into the fuel cell stacks has to be done, say, with soda lime or another technique.

Generally speaking, other impurities have to be avoided for alkaline cells as much as possible, although they may cause quite different effects. The presence of Hg in hydrogen originating from some chlorine plants may lead to accumulation effects in the anodes, with irreversible and continuously aggravating degradation consequences. On the other hand, some amount of CO in the hydrogen will somewhat decrease the power level, but the latter will remain stable with time and the decrease is reversible.

When H_2 and O_2 are used, the gases will normally be circulated in closed loops. In that case a regular purge must be performed in order to avoid accumulation of inert gases in the stacks; if this is not done, gas concentrations may be generated with detrimental results for a good, long operation.

7.4.3. Current Collection

As in other types of fuel cells, current collection can either be guaranteed by edge collection, using unipolar electrodes or bipolar arrangements. Both arrangements have their advantages and disadvantages, which again are not typical for alkaline fuel cells.

In the ELENCO stacks, edge collection is used, as can be seen in Fig. 7.2. The Siemens technology is based on a bipolar arrangement, whereas the American shuttle fuel cell uses a combination of bipolar and edge collection.

7.4.4. Water Removal

The product water formed in the fuel cell process has to be removed constantly from the stack, so that the KOH concentration remains within the desired range. Depending on the type of electrodes used, water removal can be organized in different ways. When using, e.g., hydrophobic carbon electrodes of the ELENCO type, a given gas circulation rate allows the water to be brought outside the stack by entraining water vapor through the electrodes and into the gas outlet stream. For its H_2–air fuel cell, ELENCO uses the airflow (stoichiometric ratio 2.5) to remove the water, whereas for the H_2–O_2 fuel cell H_2O removal at the H_2 side can be achieved with a H_2 circulation rate of 4 to 6. In both cases the water vapor is condensed outside the stack and separated from the gas carrier.

In the H_2–O_2 fuel cell of Siemens the product water remains in the electrolyte. As the latter is pumped out of the stack, it passes through a so-called electrolyte regenerator or evaporator, in which a hydrophobic diffusion membrane setup eliminates the product water and returns the electrolyte at its normal concentration to the stacks. This evaporator practically forms a kind of second stack, arranged in one line with the electrochemical one.[4]

7.4.5. Waste Heat Cooling

In the case of an alkaline H_2–O_2 fuel cell, the waste heat of the fuel cell is to a very large extent removed with the electrolyte if this is circulating, as for both the Elenco and Siemens technologies. For its H_2–air cell, however, Elenco has determined that there are two major ways of removing the heat: by the (open) air circuit and by the electrolyte. In both cases the electrolyte has to pass through a heat exchanger, which is part of the peripheral equipment of the fuel cell system. When the electrolyte is immobilized in a matrix and an electroyte reservoir plate, such as in the American shuttle fuel cell, then some special cooling compartments have to be built in the stack at regular intervals between a few groups of cells so that the waste heat can be eliminated by a cooling loop.

7.4.6. Fuel Cell System Controller

The alkaline fuel cell systems considered here do not include a reformer nor a dc–ac inverter. They are systems using pure feeding gases and providing their electricity essentially to dc users. As such, those systems must, of course, be steered and controlled. This is done by a controller that is completely programmed to perform all start-up, operating, and stop procedures in normal operation, on the one hand, and that, on the other hand, is able to intervene with a number of actions to protect the fuel cell system and its environment in case of abnormal behavior or failures.

7.5. ACTIVITIES AND DEVELOPMENT STATUS IN THE WORLD

7.5.1. Introductory Remarks

Reviewing the past activities and the present development status of alkaline fuel cells in the world cannot be complete. Even excluding the earliest work (see Chapter 1), a time span of some 40 years has to be covered. It is not possible to include all the, often very valuable, fundamental work carried out at universities. Starting as early as the 1930s Professor Bacon in Cambridge, United Kingdom, and after World War II, Professor Justi in Braunschweig, Germany, and Professor Kordesch in Vienna, Austria, have performed pioneering work on alkaline fuel cells and their components. Nowadays, many more scientists are involved at universities. This review mainly focuses on the largest industrial projects which have been and still are carried out in the area.

In view of the numerous efforts and their distribution, as well as to the mostly individual character of those programs (being no part of any well-organized national or international programmatic approach), it is impossible to estimate the accumulated worldwide research, development, and demonstration costs of these programs. Occasionally a figure can be found, but often it is just an estimate. As an example, the development cost for the space shuttle fuel cell has been estimated at $100 to $150 million (U.S.).[1]

In the following sections, alkaline fuel cell work in Europe, the United States, and Japan will successively be discussed.

7.5.2. Alkaline Fuel Cells in Europe

As in the United States, many important programs were started in the 1950s and 1960s, but many were discontinued, mostly in the early 1970s. Retrospectively, it can be said that in most cases the technical problems to be solved have been underestimated very seriously, and, on the other hand, neither the cheap energy prices at that time nor the lack of concern about environmental issues have been of any help to stimulate continuation of such efforts.

In Germany, the greatest efforts on alkaline hydrogen fuel cells have been carried out by Siemens and Varta, companies that also have been exploring several other types of fuel cells. Both have been working on hydrazine fuel cells, as well as methanol fuel cells, but that work was stopped quite some time ago. In the early 1970s Siemens selected the alkaline H_2–O_2 fuel cell to concentrate further efforts upon. Their electrodes, based on the use of nonprecious catalysts (Raney nickel in the anodes, silver in the cathodes) and a sedimentation manufacturing process, have been described in Section 7.2.1. The gas-diffusion electrodes have a gas-stop layer at the side of the electrolyte, which is circulating and used as a coolant. The electrodes are compressed isostatically over relatively flexible metallic contact sheets to improve the electric contact on the electrolyte and current-collecting sides. Electrodes are arranged in a bipolar way, so that the cells of a stack are electrically connected in series. Operation pressure in the cells is 2.3 bar at the hydrogen side and 2.1 bar at the oxygen side. The compression pressure on the collectors is 2.4–3-bar at the nominal operating temperature of 80°C. Product water is separated from the electrolyte in a electrolyte regenerator (see also Section 7.4.4). Fuel cell system development at Siemens has been concentrated since the second half of the 1970s on a 6–7 kW unit, composed of the fuel cell stack itself, the electrolyte regenerator, the electromechanical controlling unit, the supply system, and the electronic controller. This unit has been engineered in quite a compact way, as can be seen in Fig. 7.3. Its weight is 85 kg. The nominal power of 6 kW at 48 V corresponds to a current density of

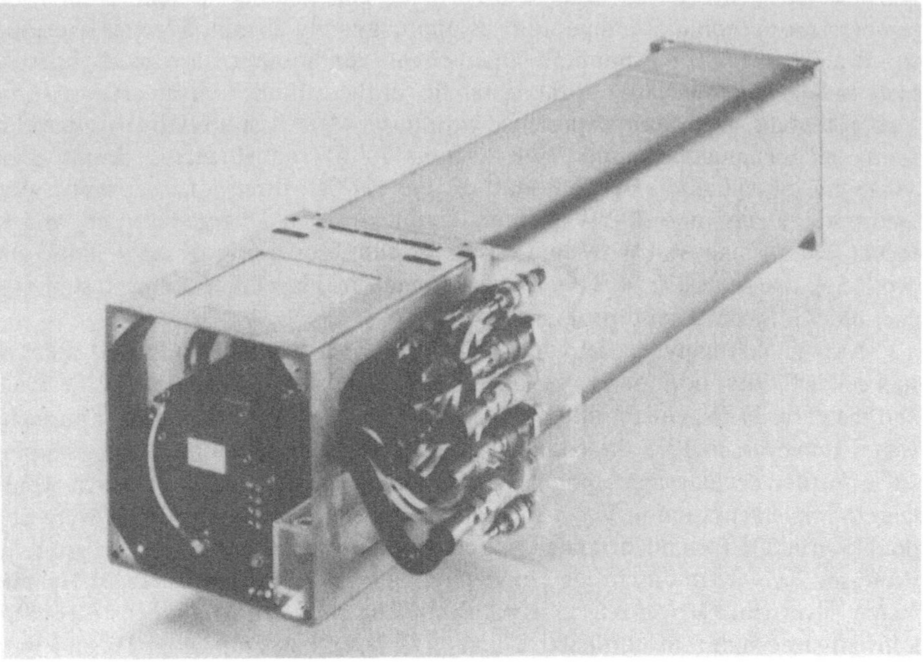

Figure 7.3. 6-kW alkaline H_2–O_2 fuel cell system developed by Siemens.

Figure 7.4. 48-kW alkaline H_2–O_2 fuel cell system (Siemens).

400 mA · cm^{-2} and a cell voltage of 0.78 V. The heating-up time from room temperature to nominal temperature is approximately 20 min. Electric overloading is allowed for some minutes. Open-circuit conditions or low loads (<20 A) have to be avoided because of the instability of the cathode's silver catalyst at too high potentials. Operating experience with this system is still relatively limited in terms of accumulated hours (see Section 7.3.6), but Siemens claims good reliability potential.[4,5] By the end of the 1970s, three of the units were incorporated into one 20-kW system, designed as an emergency unit, and in recent years a first 48-kW system has been built, consisting of eight units, and producing 250 A at 192 V. This unit, presented in Fig. 7.4, is being tested as a part of a German defense program.

Also in Germany, battery manufacturer Varta has had important alkaline fuel cell activities, both on H_2–air and H_2–O_2, and during some time in the 1960s and the early 1970s, collaboration efforts with Siemens were undertaken for some years. However, in 1973 the company decided to stop its fuel cell work, keeping some further developments going on air electrodes, the results of which would lead to implementation in Varta's metal–air batteries. Varta electrodes were of a double structure (so-called Janus) type, resulting from early work by the group of Professor Justi at Braunschweig University, Germany. Raney nickel (for H_2) and Raney silver (for O_2) were used as catalysts. The work led to the construction of kilowatt-sized units, in particular a unit of 3.5 kW,[10] operating on H_2 and pure oxygen. Such a unit was also tested in a forklift truck, the H_2 coming either from a compressed-gas bottle or from an ammonia reformer.

In France the three most important program on alkaline fuel cells have been those of Alsthom, CGE, and IFP. Early work up to the mid-1960s is described in Ref. 3. At CGE, electrodes for low-temperature alkaline cells were developed, using nickel and silver catalysts for the anodes and cathodes, respectively; at CGE the development led to the construction of 2-kW fuel cells. A similar choice of catalysts was made by IFP, which continued its fuel cell development until the beginning of the 1980s, unlike CGE where fuel cell activities were more reoriented toward advanced primary and secondary metal–air batteries. The IFP approach was technically quite different from the one at Siemens, this being related clearly to the different application area envisaged—electric traction. IFP worked on atmospheric H_2–air cells, with relatively low catalyst loadings (20 mg of Raney nickel/cm^2 and 0.9 mg Ag on C/cm^2) so as to achieve an acceptable kW cost. The targets in terms of lifetime were relatively modest, given the traction application—approximately 3000–4000 operating hours. It appeared, however, that the company was not able to come near the targets, especially in terms of stability, so the program was discontinued.

Another relatively big effort in France was conducted by Alsthom and focused on the development of low-cost electrodes and components. Ultimately, this work led to an association with Exxon, followed by one with Oxy (Occidental Petroleum). The cells used air, low loadings of precious metal catalysts in the electrodes, and atmospheric pressure conditions. After having considered in the earlier Alsthom days a wider variety of applications, the focus during the later period of the Oxy collaboration was on the installation of megawatt-sized fuel cell systems near chlorine plants, using their by-product hydrogen. The later technical results that became available on the life of the cells were, however, quite insufficient for that application. At the time of writing, it is not known if there is still some further work going on at Alsthom. If not, the only, relatively moderate, effort in France is at a small private electrochemical research company, SORAPEC, which has been conducting some high precious-metal-loading electrode development work for alkaline cells, and which is also involved in other fuel-cell-type developments.

Both the United Kingdom and Sweden had alkaline fuel cell development projects in the 1950s and the 1960s, but they came to an end relatively soon. The Swedish effort was at the ASEA company. Electrodes were based on nickel catalysts, and the technical scope was very ambitious: develop relatively large units (a few hundreds kilowatts) with a view to submarine utilization. Because of technical problems and accidents, however, the program was stopped.

In Belgium, alkaline fuel cell work was started in the early 1960s by a consortium of Belgian companies; this work was taken over in 1969 by SCK/CEN, the Belgian Nuclear Research Centre, which concentrated on the development of a cost-effective carbon electrode concept. On the basis of that successful fundamental work, a joint program with industry was organized, leading to the formation in 1976 of the Elenco company, a Belgian–Dutch collaboration. Elenco's development of electrodes and stacks was summarized in Sections 7.2.1 and 7.2.3. Also see Ref. 3. The basic stack unit of Elenco, as presented in Fig. 7.2, produces 0.45 kW at 110 $mA \cdot cm^{-2}$ and 0.67-V cell voltage, using air and H_2 under atmospheric conditions. Larger power levels are reached by building these units in rows. Figure 7.5 shows a 4.5-kW stack consisting of 10 units. Manufacturing techniques have been chosen and developed so as to reach a

Figure 7.5. 4.5-kW alkaline h_2–air Elenco fuel cell stack.

product which can be mass-produced in a cost-effective way. As part of the approach to commercialization, an automated pilot plant has been developed and built for the production of the electrodes (annual capacity 2000 kW). Since the completion of the basic development work in 1983, Elenco has been working on the development of some fuel cell systems. The largest one so far is a 40-kW system constructed on a trailer for field utilization in connection with geological activities. This system is in its final starting up phase at this writing and is pictured in Fig. 7.6. On the other hand, 1–1.5 kW fuel cell systems have been designed, and six prototypes have been built in 1987 and 1988 for evaluation by several users. In connection with the company's envisaged traction applications, an 11-kW fuel cell (later extended to 15 kW) has been successfully tested in an electric van.[8] All these fuel cell systems are characterized by circulating electrolyte, water removal by the outgoing air stream, and a nominal electrolyte temperature of approximately 70°C. These fuel cells can be overloaded to some extent, since they do not suffer from open-circuit or very low load voltages because a stable precious metal catalyst is used in the cathodes.

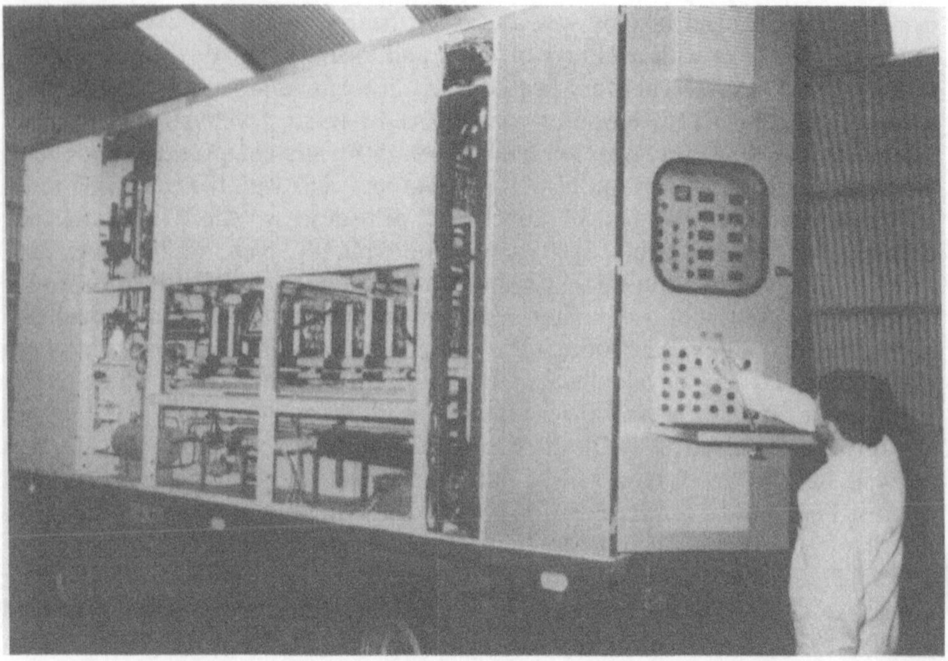

Figure 7.6. 40-kW alkaline H_2–air fuel cell system built by Elenco.

7.5.3. Alkaline Fuel Cells in the United States

While awaiting for forthcoming commercial breakthroughs of alkaline fuel cells, the 7-kW H_2–O_2 alkaline system developed by IFC (at the time Pratt and Whitney) for the American space shuttle or orbiter stands as a landmark for the successful and reliable utilization of fuel cell technology. This fuel cell was developed as a kind of second generation of the already discussed Apollo fuel cell, based on the very early work by Professor Bacon in the United Kingdom. According to Ref. 12 the technical progress between the Apollo and shuttle fuel cells can be characterized by a power output eight times larger and a weight saving of some 22 kg. The space shuttle carries three units with a nominal power output of 7 kW each and a peak power of 12 kW. At this power 436 A are provided at 27.5 V. The total weight of one 7-kW unit is 112 kg. The main technical features of this fuel cell are the following: metallic electrodes consisting of precious metal alloys (Au–Pt), KOH electrolyte immobilized in a matrix, and an electrolyte reservoir plate and a liquid-fluorinated hydrocarbon coolant. Operation is at 3–4 bar. Most power plants have already accumulated a few thousand operating hours. They are scheduled to achieve an accumulated 10,000 h of on-line service. These units are using technology frozen in 1973. In the meantime technical improvements have been made, which may be implemented in a new generation. Apart from the space fuel cell development, the largest program on alkaline cells in the United States have been those of Union Carbide and Allis-Chalmers.[1] The latter company, which worked on alkaline fuel cells until 1969, used electrodes consisting of a porous nickel plate catalyzed with noble metals facing a porous asbestos sheet retaining the electrolyte. An early demonstration project in 1959 was a tractor, incorporating a 15-kW H_2–O_2 fuel cell weighing 917 kg with a voltage of 750 V and a current of 20 A, corresponding to only 20 mA · cm^{-2}! This very low performance was improved in the following decade, and served as a basis for some space-oriented developments. None of these activities, however, were successful enough for the company to continue.

After some early work on H_2–O_2 alkaline fuel cells with baked carbon tubes in bundle arrangements or in concentric cell designs, the Union Carbide development focused, from 1960 onward, on parallel-plate designs. At first, surface-activated 3–6-mm-thick carbon plates catalyzed with silver or noble metals were used, but were later replaced by thinner (1–1.5 mm) composite porous metal–carbon electrodes. These electrodes formed the basis for both H_2–air and H_2–O_2 developments. Operating temperature was 65°C, KOH normality 9, and, with air, current density was 100 mA · cm^{-2}. Water removal was through the electrodes. On the basis of this technology, a 32-kW H_2–O_2 fuel cell was built in an electric van as part of a cooperation project with General Motors. Peak power for acceleration could be provided by the fuel cell itself, up to some 150 kW for short periods.

Some time later, a 6-kW H_2–air fuel cell was successfully, tested in an experimental electric car, this time, however, in a hybrid setup with a lead-acid battery. In the last years of its fuel cell activities, in the early 1970s, Union Carbide still introduced further improvements in its electrodes, adopting a double-porosity nickel structure on the electrolyte side. Lifetimes up to 5000 h were shown at increased current densities for atmospheric air operation, the limiting current density became 250 mA · cm^{-2} instead of 150 mA · cm^{-2}.

7.5.4. Alkaline Fuel Cells in Japan

Alkaline fuel cell work in Japan has been extremely limited in comparison with the huge efforts in that country on other fuel cell types. In fact, the alkaline fuel cell work is essentially done at one company, Fuji, which focused most of its fuel cell work in the period 1965–1975 on the alkaline cell. Although from then onward, the emphasis was gradually shifting toward phosphoric acid fuel cell work, that on alkaline cells was pursued. This was done with support from the Ministry of International Trade and Industry (MITI) until 1984. Since then Fuji is continuing on its own some laboratory developments, which are modest in comparison with the company's other fuel cell work. After having tried nonprecious and precious metal catalysts for their electrodes, it seems that the latter have been preferred for the work in the 1980s.[13] Fuji's alkaline cells use a circulating 6 N KOH solution at 65°C with almost atmospheric pressures for the gases. Water removal is through gas circulation. Precious metal loadings are 0.8 mg · cm^{-2} for anode and cathode combined. Electrodes are built in the stack in a bipolar arrangement, using a filter press system. Operation is possible with air and oxygen. A 1-kW stack (on H_2–air) and a 7.5-kW unit (on H_2–O_2) have been built and tested. The H_2–air stacks operate at a current density of 100 mA · cm^{-2}, at which 10,000 operating hours have been shown in small cells.

7.6. APPLICATION AREAS FOR ALKALINE FUEL CELLS

7.6.1. Introduction

Application areas will become accessible to alkaline fuel cells because of some genuinely favorable operating characteristics, and provided the fuel cell cost is acceptable for the economics of the application envisaged. In most areas competition with alternative solutions must be faced. At least for some applications, reasonable cost penalties will be acceptable provided there will be sufficient positive trade-offs with advantages in other areas. For alkaline fuel cells, these advantages are not only those of all fuel cell types (high conversion efficiency, quiet operation, no harmful combustion products) but also the low temperature of operation and the fast start-up potential.

In the following sections the most important possible market segments for alkaline fuel cells will be shortly addressed: space, defense, electric vehicles, and stationary applications. Introduction in those markets will include the task of making available to the fuel cell user the hydrogen fuel to be used. Depending on the type and mode of application, this hydrogen will have to be brought to the site or it should be generated on the site. In fact, this is so for all types of presently available fuel cell types, but the requirements on the H_2 purity are relatively high for alkaline cells. Therefore, this aspect should not be neglected when assessing alkaline fuel cell applications.

7.6.2. Space Applications of Alkaline Fuel Cells

Space is a well-known application field, which first saw an alkaline cell in the Apollo program and an SPE fuel cell in the Gemini program. For the space shuttles, NASA preferred the alkaline cell to the SPE cell, leading to the units described already. The use of fuel cells for space shuttle flights is undisputed. According to IFC,[12] the only alternative—batteries—would in their best available form weigh about 10 times as much as the space fuel cell for the same amount of energy. During its preparative assessments for the power system in the European space plan HERMES, the agencies ESA and CNES came to similar conclusions and preferred a fuel cell system to batteries for reasons of weight.

For these space applications, the main requirements for fuel cells are obvious: reliability, light weight, and low volume. To achieve this, high catalyst loadings, compact stack construction, pressurization, and use of oxygen all are evident measures adopted by IFC as well as for the preparative development activities for the European HERMES fuel cell. A major advantage of such advanced alkaline fuel cells is the high cell voltage and, hence, high electric efficiency (well over 70%). The fast start-up time of the alkaline cell is also an advantage for this application. The reactants, pure H_2 and O_2, are taken with the shuttle in cryogenic containers because this is the most efficient way in terms of volume and weight. The product water, which in principle is clean, can be used by the astronauts for several purposes.

Another, although less advanced, possible application of the alkaline fuel cell in space is as part of a regenerative fuel cell system, to be used in satellites rotating around the earth for many years. Here, of course, the only primary energy is solar, to be converted by photovoltaics into electricity. During the satellite's dark period, however, electricity is needed too. This can be generated by a fuel cell, using to that effect hydrogen and oxygen generated in a water electrolyzer during the light period of the satellite's rotating cycle. Such fuel cell–electrolyzer systems, which perhaps could be integrated to some extent (e.g., a common electrolyte loop, transportation of the water), are evidently quite complex. Their potential advantage compared with the alternative way to store solar energy in batteries, is their lifetime; the needed 15,000 fuel cell operating hours, spread over five years, is a quite realistic target, whereas for the same total life batteries should be able to reach some 25,000 charge–discharge cycles, which is much higher than existing technology. For a further discussion of these regenerative fuel cell systems (RFCS), see publications by NASA[14] and ESA.[15]

7.6.3. Defense Applications of Alkaline Fuel Cells

Low temperature, silent operation, and no poisonous exhausts are the big selling characteristics of alkaline fuel cells for defense applications. The combination of those characteristics makes such a fuel cell generator very difficult to detect by common acoustic and infrared detection techniques.

There are quite a few examples of potential uses in defense areas, where the fuel cell offers the unique combination of silent operation (compared to combustion engine–generator sets) and long running time (compared to batteries). Some of these have become generally known, whereas some others are treated with more secrecy. This section is not exhaustive in terms of possible defense applications. Some examples will be presented.

Diesel-powered nonnuclear submarines, for strategic reasons, may need to remain underwater for some time completely undetectable, which means with engines stopped—a period of which is too long for the batteries to provide all the power needed. Insertion of fuel cell systems of a few hundred kilowatts is a valid contribution in such a situation, multiplying the number of hours of silent underwater operation significantly, depending of course on the reserves of hydrogen and oxygen available. One such program in Germany uses alkaline fuel cells from Siemens and cryogenic hydrogen storage and gaseous hydrogen storage in metal hydrides.

Another possibility for fuel cell generator utilization lies in much smaller sets, in the kilowatt range, which can be used to recharge batteries silently at isolated places where no grid is available and where the noise of combustion engine generators could be harmful. In that case, the alkaline fuel cell will preferable use ambient air so that only the hydrogen fuel, stored, e.g., in lightweight bottles, has to be transported to the place of utilization.

As for fuel cell cost, the defense market seems somewhat intermediate between the space market, where cost is a very minor consideration in comparison to other criteria and requirements, and the normal commercial applications market. If the fuel cell really is the answer to a strategic defense need for which no comparable technical solutions exist, then a relatively high price for the fuel cell equipment is acceptable. Seen in this light, defense applications can be considered as a help to get out of the "chicken-and-egg-situation" which fuel cell technology often faces and to get into normal commercial markets. This explains why several fuel cell developers and manufacturers welcome such opportunities.

7.6.4. Electric Vehicle Applications of Alkaline Fuel Cells

Under the pressure of the first oil crisis (1973) and under that of the increasing environmental problems associated with combustion engine vehicles, much effort has been devoted during the last 10–15 years to the development of electric vehicles of all types: cars, vans, trucks, buses. The basic aim is to achieve electric alternatives that can offer the same, or at least similar, performance as their combustion engine counterparts. These efforts have included work on batteries, power electronics, electric motors, and electronic controllers, and exciting technical progress has been made in most areas. It is clear though that the weak point is still the traction battery. Many different electrochemical couples have been investigated, but inevitably the results have been disappointing when comparing them to the expectations when the projects were started. The basic lead–acid battery could be improved in terms of energy density by some 50%, allowing for equally increased ranges for the vehicle, say 90–120 km instead of 60–80 km on one battery charge. Quite often, announced "superbatteries," five or more times better than lead, saw their potential gradually decrease with technical development advancing from the small cells in the laboratory to the real-life engineered battery, so that in most cases an improvement with a factor of 2 compared to lead–acid is now already considered quite successful. Furthermore, however, all these batteries (take, e.g., Ni–Fe, Ni–Zn, Zn–Br_2, Na–S) are still stuck with other problems, such as high cost (even when mass-produced) or unreliability. So the general conclusion can be that with batteries electric vehicles will always remain handicapped by intrinsically limited ranges and long recharging times.

This situation offers a big opportunity to the fuel cell, and the alkaline fuel cell in particular. The latter's low temperature, high electric efficiency, and fast start-up capability are indeed advantages for vehicular application. This was recognized many years ago and has been one motivation for the start of several alkaline fuel cell programs (e.g., IFP in France, Union Carbide in the United States, and Elenco in Belgium). Because the fuel cell is a conversion device and, unlike the battery, not an energy storage device, the range of a fuel cell electric vehicle is determined only by the amount of fuel available, whereas the fuel cell size is defined by the required power level.

The basic technical feasibility of the use of an alkaline fuel cell in a vehicle was first proven by Professor Kordesch at Union Carbide and then, in the early 1980s, by Elenco in Belgium (see Sections 7.5.3 and 7.5.2 respectively).

Of great importance is the basic economical feasibility. In that respect a significant difference has existed in the respective approaches. In France, IFP followed the same reasoning as Shell did in the 1960s when launching its direct methanol acidic fuel cell development: in order to have a huge market, attention should be focused on everybody's car. The consequence is obvious: the cost requirement for the investment in the fuel cell is so dramatically low (40$ $\cdot$ kW^{-1} was mentioned in 1980[16] by IFP, that reaching the target did not seem possible within the foreseeable future. Unfortunately, this conclusion was only reached after many years of sustained technical development work, leading to a frustrating decision to stop development.

Other companies, such as Elenco, followed a quite different approach from the start, trying to be more pragmatic. Private cars were the last vehicles to be considered, not only because of the economical and technical problems to be solved, but also because of the as yet nonexisting hydrogen distribution "gas" stations. Although their number is much smaller, attention was focused on vehicles used in fleets (which means with "centralized" fueling facilities) and in urban environments (which means a greater environmental impact). On that basis the city bus was selected as the number one potential application, to be followed by refuse-collecting trucks, some urban vans, and perhaps eventually taxicabs.[17] In relation to the city bus application the following important points should be stressed:

1. The motivation from the user's viewpoint (i.e., the public transportation authorities) is primarily because of environmental concerns; for the longer term, the use of a fuel which can be obtained from sources other than oil is also an advantage.
2. A hybrid setup of not too large a fuel cell (some 70 kW) and a buffer-type second power source (a battery set or a flywheel) should be used to reduce cost, weight, and volume.
3. The impact of the fuel cell depreciation cost on the overall kilometer cost of a city bus is relatively small and completely different from the situation one can calculate for a private car with the same fuel cell cost; this implies that market introduction of the fuel cell is much easier.
4. In terms of fuel, making hydrogen available for a bus system in a city can technically be envisaged in different ways: by *in situ* production from energy resources available on the spot, such as natural gas or electricity, or transportation from central plants, either in liquid form by trucks or in gaseous form through pipelines.

If market introduction for such specific vehicles is successful, and if the fuel cell can be developed further, then a more generalized application, also in cars, is possible. This will certainly be the case if environmental legislation becomes more stringent. Forklift trucks may also constitute an interesting application area for alkaline fuel cells, using ambient air and hydrogen. The latter would then preferably be stored in iron–titanium metal hydrides, which offer excellent compactness. Their high weight might even be a welcome feature in the truck.

7.6.5. Stationary Applications of Alkaline Fuel Cells

Electricity generation of hundreds of kilowatts or multimegawatt scale in utilities has been the main objective of most fuel cell developments in the United States and Japan. In the mid-1970s, the respective merits of the different fuel cell types (at that time mainly the phosphoric acid and alkaline cells) have been assessed for utility power generation,[18] and the phosphoric acid fuel cell is preferred. A crucial element in that decision has been the claim that the purity requirements for the alkaline fuel cell gases, especially hydrogen, were such that the cost of pure hydrogen, obtained in reformers out of naphtha or natural gas, would be much higher than in phosphoric acid fuel cells. This decision was taken on the basis of gas purification technology of the time, which is now 15 years old. According to some specialists, however, these purification technologies have changed so much since that time that the conclusion may well be different now.[19] In fact, a plant producing high-purity H_2 for fuel cells has been constructed by KTI in Bavaria (Germany) for SWB's hydrogen technology demonstration project, where this hydrogen will be used in alkaline and phosphoric acid fuel cells with similar overall efficiencies.

Another example of an interesting stationary application for the alkaline fuel cell is in conjunction with chlorine plants, where very pure hydrogen is obtained as a by-product. This hydrogen can be converted in the fuel cell, and the dc power produced can be used to feed the chlorine-producing electrolytic cells, thus reducing the power taken from the grid by some 15–20%. This application has been studied carefully by several chlorine manufacturers, e.g., for the Alsthom/Oxy and the ELENCO cells. It is clear that sometimes the environmental advantages of the fuel cell are of little significance and that only pure economics decide. The conclusion of the studies mentioned was that, assuming a low but reasonable kilowatt-cost for the fuel cell, an operating life of 20,000 h would still be required to reach a break-even point.

In terms of stationary applications, the alkaline fuel cell has further potential in some specific market segments. This is the case in some specialized geological and oceanographic areas where the characteristics of the fuel cells are highly valued, where the "hydrogen logistics" are not considered a problem. Other possibilities are in the camping and caravaning areas where again the silent operation and absence of polluting combustion gases would be important selling arguments. On the other hand, however, this would require achieving suitable and adequate hydrogen logistics, with lightweight bottles or metal hydride containers being made available to the public at large. Also, fuel cell sets should become very easy to handle and "maintenance-free" for the user.

Another area often mentioned, and considered by Siemens and Fuji, is emergency current supply, the fuel cell being continuously in standby. For this application, the fast start-up capacity of the alkaline fuel cell certainly is an advantage, although for uninterrupted power supply (UPS) units a short bridging period by batteries should be foreseen in the system.

7.7. DEMONSTRATION AND COMMERCIALIZATION

7.7.1. Introduction

It is not easy to discuss commercialization of the alkaline fuel cell in a specific, and yet general, way. The approach to commercialization is often linked to the policy of a given industrial company, to the financial means available to prepare for commercialization, and to the degree of collaboration received from potential users and customers.

Given the relatively new product character of the fuel cell in general, and the alkaline fuel cell in particular, commercialization has to be preceded by some precommercial demonstration projects. It appears generally possible to involve later users with such demonstration projects, if not financially, then certainly technically and for the purpose of testing.

The detailed process of demonstration and commercialization may also depend on the application areas envisaged as they have been described in the previous sections. They are indeed quite different for each application, and it is clear that the business and technological mentalities may be substantially different and hence require specific approaches.

Therefore the remainder of this chapter should not be seen as general but rather personal considerations by the author.

7.7.2. Specific Markets

Aerospace is the most specific market. If the technology is selected at the start, and provided no major technical problems arise, it can be assumed that the market, which is financially interesting already in the development phase but remains relatively limited, can be kept for a long time. The situation is different for defense applications. In general, defense authorities will financially contribute, partly or fully (depending on the project and also on the country), to the demonstration projects and hence facilitate the start of the following commercialization phase, provided the demonstrations are successful. A market thus penetrated can be kept and expanded, but in general competition will play a role, if not in the beginning then in a later phase, which will necessitate regularly improving the fuel cell. For alkaline fuel cells, in Europe, the two companies having the largest activities, Elenco and Siemens, are both involved with defense projects.

With regard to these specific markets it can be said that they will be affected very little by the way the hydrogen fuel is obtained. Quantities will indeed be limited, and there will be no problem for the specialized gas-selling companies to follow the demand, using their existing industrial hydrogen sources, either hydrogen from chloride plants or made from natural gas in central production plants.

7.7.3. Vehicle Markets

As explained earlier, it can be expected that market penetration will start in a few selected areas. Serious demonstration projects in all of them will be required to validate the concept and convince potential consumers. At the present time, the increase of the environmental concern in society in general, and with political authorities in particular, is a helpful stimulus to set up demonstration projects and will certainly speed up commercialization.

As an example, Elenco's efforts in the direction of the city bus application can be mentioned. After successive technical–economical assessment, in cooperation with urban transportation authorities in several Belgian and Dutch cities and with bus manufacturers, Elenco has achieved a realistic appraisal of this alkaline fuel cell application. On that basis, a specific project has been prepared, and started up in 1991. In the frame of EUREKA and in collaboration with Dutch, Italian, and French industrial partners, a large size articulated city bus has been developed and is being built for demonstration. It will be tested in the cities of Amsterdam and Brussels in the year 1994. This approach should allow conclusions to be reached on the overall feasibility of a hybrid fuel cell bus, and, in case of success, should bring closer the following phase in the market penetration process, which will normally consist of the realization of a small fleet of maybe 10 or 12 vehicles to serve one line in a city.

As for hydrogen fuel, the market itself will point to the cheapest way of providing a city with fuel cell buses with all the hydrogen required. In terms of global energy resources, this will have little practical impact, and it would certainly be premature to speak of a hydrogen economy. It, however, may be a stepping-stone toward such an economy on a small scale.

For the fuel cell manufacturer, the application in city buses would soon constitute quite a sizable market. Even adopting modest fractions of the number of public buses in any European country and calculating the corresponding market (including replacement market, because the lifetime of a bus is a multiple of the 5000 or even 10,000 operating hours of the fuel cell stacks, which hence will have to be replaced), quite rewarding turnover figures can be obtained.

7.7.4. Stationary Markets

If H_2–air alkaline fuel cells can achieve sufficiently long lifetimes, they may start to appeal to users in the chlorine business or in the utility industry. The size of these markets is relatively big, but is has to be expected that only a fraction will be available. For chlorine factories, local considerations and alternative utilization modes for hydrogen influence the decision to install a fuel cell plant.

For utility application, the situation will evidently be more complicated. Studies should be carried out, similar to a number of others carried out in the past 10 years or so in the United States, Japan, and Europe for the utilization of other fuel cell types for utility purposes, in order to identify more exactly the most interesting market segments. There is no doubt that the general energy resources background of the country or area considered will play an important role. Any scenario that will favour nonfossil energies, be it nuclear, hydro, solar, or other renewables, will encourage the possible role of pure hydrogen generated via electrolysis. On the other hand, improvements of gas separation and purification techniques can lead to cheaper production of pure hydrogen from fossil fuels, especially from natural gas and, perhaps, coal.

7.7.5. Competition with Other Fuel Cell Types

Possible competition between the alkaline fuel cell and other fuel cells has to be looked at separately for almost every type of application. For megawatt-size applications in the chlorine and utility sectors, the alkaline cell has a higher electric efficiency than the phosphoric acid fuel cell, and if the primary objective is indeed to generate electricity the alkaline cell should have an advantage. If, on the contrary, a heat source at a temperature well over 100°C is an additional benefit to the user, then the phosphoric acid fuel cell may turn out to be the most interesting one (see Chapter 8). It can indeed be said that the alkaline fuel cell is not attractive for cogeneration (exception made for heating up the interior of a city bus with the waste heat!).

For vehicles, and supposing that H_2 is taken as the fuel, the alkaline cell is certainly more attractive than the phosphoric acid one, due to lower temperature, faster start-up, higher electric efficiency. Possible competition in that respect could maybe come from SPE cells, which are operating at the same temperature and require relatively pure hydrogen. At present they are less developed than alkaline cells, and it remains to be seen if a number of their typical cost and operational problems will be solved to the same level reached by the alkaline cells (see Chapter 11).

7.7.6. Further Improvements Needed

Modern alkaline fuel cells have reached a maturity which makes them suitable for demonstration projects. Even those for the American space applications, but certainly all others, can still be improved. The technology is only at the beginning of its learning curve, and many improvements remain possible. They can be widely varying, as illustrated by the following listing (in this respect space applications should be excluded, the criteria being too different from those for other applications):

1. Further reduce the $kg \cdot kW^{-1}$ value, mainly by increasing the specific catalytic activity, as expressed in $mW \cdot cm^{-2}$.
2. Increase performance stability with time, mainly by introducing more stable catalysts.
3. Try to limit and simplify the components in the peripherals of the fuel cell system.
4. Design and manufacture all components of the system so that assembly, particularly for serial production, becomes easy and fast.
5. Apply instrument quality control on all components so as to achieve maximum reliability for the system.

ACKNOWLEDGMENT

The author is grateful to Siemens for permission to use their photographs.

REFERENCES

1. K. V. Kordesch, *J. Electrochem. Soc.* **125,** (3), 77C–91C (1978).
2. G. Sanstede (ed.), *From Electrocatalysis to Fuel Cells* (The University of Washington Press, Seattle and London, 1972).
3. Institut Français du Pétrole, ed., *Les piles à combustible* (Editions Technip Paris, 1965).
4. K. Strasser, *Electrochem. Soc.* **127,** 2173–2177 (1980).
5. K. Strasser, in *BAKWVT-Symp. "Elektrochemische Energiequellen"* (Mannheim, Germany, 1988).
6. A. Winsel, in *Ullmanns Encyklopädie der technischen Chemie,* Band 12 (Verlag Chemie GmbH, Weinheim/Bergstr., 1976), pp. 113–136.
7. K. V. Kordesch, *Brennstoffbatterien* (Springer-Verlag, New York, 1984).
8. H. van den Broeck, G. van Bogaert, G. Vennekens, L. Vermeeren, F. Vlasselaer, J. Lichtenberg, W. Schlösser, A. Blanchart, in *Proc. 22nd IECEC Meeting* (Philadelphia, 1987), 1005.
9. K. Strasser, L. Blume, and W. Stühler, *Program and Abstracts 1988 Fuel Cell Seminar,* Long Beach, 1988.
10. D. Sprengel, Entwicklung eines Brennstoffzellen-Aggregats für 3.5 kW Leistung, *Chemie-Ing.-Technik* (MS 156/74, 1974).

11. H. van den Broeck, L. Adriaensen, M. Alfenaar, A. Beekman, A. Blanchart, G. van Bogaert, and G. Vanneste, *Inst. J. Hydrogen Energy* **11,** 471 (1986).
12. International Fuel Cells, Orbiter Fuel Cells, Fact Sheet.
13. K. Koseki, T. Kobayashi, and Y. Tsuji, *Abstracts 1985 Fuel Cell Seminar,* Tucson, 1985, p. 186.
14. M. A. Hoberechts and D. W. Sheibley, *Abstracts 1985 Fuel Cell Seminar,* Tucson, 1985, p. 145.
15. F. Baron, *Program and Abstracts 1988 Fuel Cell Seminar,* Long Beach, 1988.
16. Y. Bréelle, B. Choffe, and A. Grehier, *Incidence Economique des Piles à Hydrogène,* EEC report EUR 6960 FR, 1980.
17. H. van den Broeck, in *Proc. Workshop on Fuel Cells: Trends in Research and Applications,* (Ravello, 1985).
18. Exxon, *Application of the Alsthom/Exxon Alkaline Fuel Cell System to Utility Power Generation,* EPRI report EM-384, 1977.
19. Private communications by L. Blomen (KTI) and Prof. O. Lindström (Stockholm).

8

Research, Development, and Demonstration of Phosphoric Acid Fuel Cell Systems

Rioji Anahara

8.1. INTRODUCTION

The phosphoric acid fuel cell (PAFC), as its name implies, uses phosphoric acid as the electrolyte. Electric power is produced by feeding hydrogen-rich and CO_2-containing reformate gas to the anode and air to the cathode. Tolerance for CO_2 is one of the characteristic features of PAFCs. Basic technical features of PAFC have been reviewed in Chapter 2.

PAFC is ideally suited for installation in urban areas (near user concentrations) as an on-site cogeneration power source. PAFC's high efficiency, compactness, and low pollution characteristics are key features. The advantages and disadvantages of PAFC compared to the other types of fuel cells are shown in Table 8.1.

PAFC is the most advanced type of fuel cell available to date and is closest to commercialization. Numerous demonstration plants have been constructed and are or have been operating in the United States, Japan, and western Europe in the range of several kilowatts to megawatts.

The applications of PAFC range from electric utilities, on-site cogeneration, to vehicular applications where unique characteristics of PAFC exceed those of conventional power plants.

Key components of PAFC systems are (1) fuel processing systems (converting fossil fuel to hydrogen-rich gas), (2) fuel cell stacks (converting hydrogen-rich gas and air to electricity), (3) inverters (converting electricity from dc to ac), and (4) control systems (controlling all components to match the electrical and/or thermal load demand).

Rioji Anahara • Technology Development Management Center, Fuji Electric Company, 1-12-1 Yurakucho, Chiyoda-ku, Tokyo 100, Japan.

Fuel Cell Systems, edited by Leo J. M. J. Blomen and Michael N. Mugerwa. Plenum Press, New York, 1993.

Table 8.1. Comparison of Various Types of Fuel Cells

Characteristics	First generation	Second generation	Third generation	
Fuel cell	Phosphoric acid fuel cell (PAFC)	Molten carbonate fuel cell (MCFC)	Solid oxide fuel cell (SOFC)	Alkali fuel cell (AFC)
Electrolyte	Phosphoric acid (H_3PO_4)	Molten carbonate (Li_2CO_3–K_2CO_3) (Na_2CO_3)	Solid oxide (stabilized) (ZrO_2)	Alkali (KOH or NaOH)
Working temperature	180–210°C	600–700°C	900–1000°C	Amb–100°C
Fuel	H_2 (tolerant to CO_2)	H_2 and CO	H_2 and CO	Pure H_2 (no CO_2 is allowed)
Raw material of fuel	Natural gas, methanol, naphtha, LPG	Natural gas, coal, methanol, naptha, LPG		Water electrolysis or by-product of chloralkali industry or others
Advantages	(1) Tolerant to CO_2 (2) Most advanced stage (3) Applicable to small-capacity plants and vehicular uses	(1) No noble metals required (2) CO is usable fuel (3) Internal reforming in cells is feasible (4) High-grade heat is available	(1) No noble metals required (2) CO is usable fuel (3) Internal reforming in cells is feasible (4) CO_2 recycling not required (5) No electrolyte management problem	(1) Lower cost electrocatalysts than PAFC (3) Low working temperature (3) Better O_2 electrode kinetics than PAFC
Disadvantages	(1) Catalysts require noble metals (2) CO is an anode poison (3) Low conductivity electrolyte	(1) Material problems related to life and mechanical stability (2) CO_2 source required for cathode (3) Phase changes of electrolyte between working and ambient temperatures	(1) High temperature presents severe constraints on cell materials (2) Relatively high electrolyte resistivity	(1) Cannot accept CO_2 (2) Low-grade heat only available

The working range of a PAFC stack is 180–210°C, 1 to several atm, and around 40–45% electrical conversion efficiency of a PAFC system.

The quality of waste heat generated by a PAFC is relatively high and clean. Total system efficiency will achieve up to 80% of combined electrical and thermal efficiency, dependent on useful lowest temperature of the heat.

Though the technological developments of PAFC over the years have been significant, improvements in the reliability and durability of PAFCs still need to be realized. Achieving cost reduction of the total plant and demonstrating long-term stable operation (~40,000 h) through the accumulation of real operating experience are essential for successful commercialization, which is targeted for the mid-1990s.

8.2. HISTORICAL REVIEW OF PAFC DEVELOPMENTS

8.2.1. Review of PAFC Technology Developments

The development history of fuel cells is very long, and the concept of the fuel cell appeared at the beginning of the nineteenth century (see Chapter 1). Since then, however, until the end of World War II, key interests were only limited to basic laboratory research.[2]

As described in Chapter 1, the first practical application of fuel cells was developed as part of the NASA space program in which General Electric and Pratt and Whitney Aircraft Company (currently International Fuel Cells Corporation, IFC) have supplied the fuel cells for space application. The types of fuel cells used were based on solid polymer and alkaline electrolytes. Stimulated by NASA's commitment, other developers began research activities on several types of fuel cells, including alkaline, solid polymer, and molten carbonate fuel cells.

Through the preliminary development phase, Allis Chalmers constructed an experimental farm tractor powered by a 20-kW alkaline electrolyte fuel cell. Union Carbide in the United States delivered a 30-kW alkaline fuel cell to General Motors to power an experimental delivery van. Monsanto Research Corporation provided a 40-kW fuel cell to the U.S. Army to power an experimental truck (see Chapter 7).

The first research work on PAFC started in the early 1960s. A larger scale demonstration program of PAFC systems was undertaken in the late 1960s and known as TARGET (team to advance research for gas energy transformation). This program was initiated by a group consisting of gas and electric utilities, the principal objective of which was the development of small combined heat and electricity power plants for residential use. PAFC was selected due to the need of a CO_2-tolerant fuel cell technology (reformate from fossil fuel gas contains about 20% of CO_2). In a cooperative venture with Pratt and Whitney Aircraft Company, 65 experimental natural-gas-fueled PAFC power plants of 12.5-kW rating were operated by utilities across the United States, Canada, and Japan between 1971 and 1973. TARGET was no doubt the first attempt of practical terrestrial application of PAFC.

Fuel cells had been selected for on-board space capsules because of their superior energy density compared to batteries and thermal cycle generators. However, for TARGET and other PAFC projects aiming for commercial terrestrial markets, factors such as cost, high efficiency, low pollution, long life, easy maintenance, etc., were important considerations.

Stimulated by TARGET, many PAFC development programs were initiated under the sponsorship of ERDA (now DOE, Department of Energy), EPRI (Electric Power Research Institute), and GRI (Gas Research Institute) in the United States. These new research, development, and demonstration programs had the common objective of satisfying *commercial* requirements. This can be viewed as the first development phase of PAFC. Specifically, (1) the construction and operation of 48 sets of 40-kW PAFC cogeneration plants was initiated, starting from 1976, under the support of GRI and DOE (GRI–DOE Project), (2) 1-MW and 4.5-MW plants for electric utilities (water-cooled) were promoted by UTC (United Technologies Corporation, now IFC) and utility companies, and (3)

R&D activities for air-cooled and oil-cooled PAFC plants for on-site uses were promoted by Westinghouse and Engelhard Corporation respectively.

The so-called energy crisis in 1973 stimulated and accelerated the development of PAFC power plants, not only in the United States but also in Japan. In Japan, the Moonlight Program, aiming at developing energy saving technologies with Japanese government support, adopted the development of fuel cell technology in 1981. This national fuel cell development program included the construction and operation of two sets of 1-MW PAFC plants as well as basic R&D on various types of fuel cells. Many private PAFC projects have since been promoted. These are explained more in detail in the following sections.

8.2.2. PAFC Demonstration Programs in the United States

8.2.2.1. TARGET Program

As described in the previous section the TARGET PAFC Program was initiated in 1967. Twenty-eight gas and electric companies in the United States, Japan, and Canada started the development of small combined heat and electricity PAFC power plants using natural gas for on-site applications. TARGET contracted with Pratt and Whitney Aircraft Company for the manufacturing of sixty-four 12.5-kW plants. These plants were field-tested at 35 different sites, in various factories, apartment buildings, restaurants, and other locations in the United States, Japan and Canada. Figure 8.1 shows a schematic flow diagram of such a 12.5-kW system.

In this system the reformate gas (from natural gas) is supplied to the PAFC stack and dc electricity is generated. Fundamentally similar flow diagrams as the one developed under TARGET have been incorporated into most of the present commercial PAFC plants.

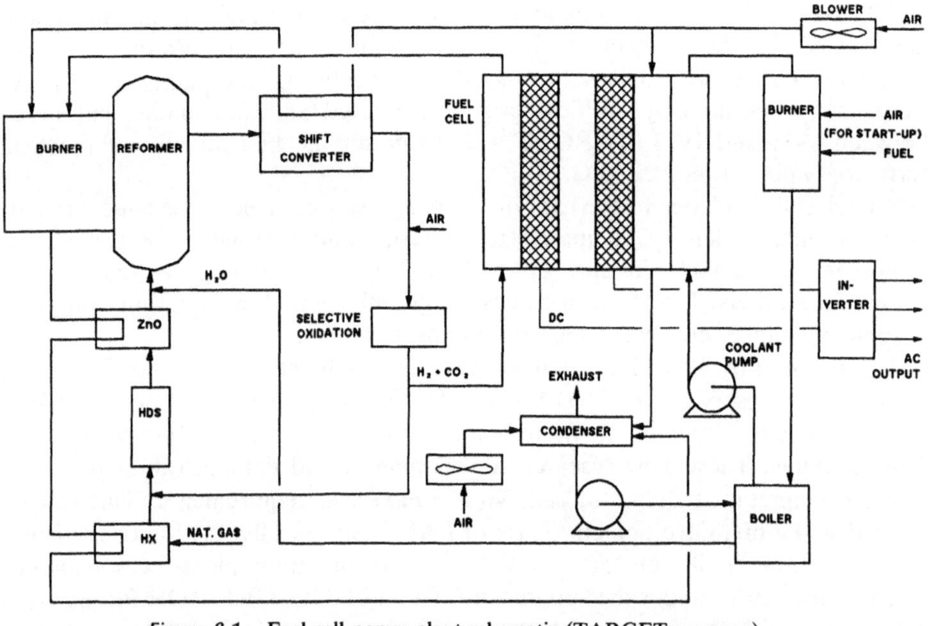

Figure 8.1. Fuel cell power plant schematic (TARGET program).

The TARGET program, promoted by private funding of $58 million from 1967 to 1976, was followed by the GRI–DOE project.

The technical developments in the TARGET program were significant. For example, graphite was selected as cell construction material.[6] Other proven technology developments included a new structure of electrolyte-containing matrix and the two-phase (boiling) water cooling system.[7] Superior plant characteristics such as low emissions and high-efficiency performance (especially under partial load conditions) were also proven during the periods of plant operation. These results further stimulated many utility companies to support PAFC developments.

Though TARGET provided positive directions for further development, these plants could not reach the goal of cost reduction and very reliable operation. This was mainly due to its high platinum content and short stack lifetime.[38]

8.2.2.2. GRI–DOE Project

The GRI–DOE project had two principal goals: to demonstrate the stable operation of PAFC plants and to demonstrate cost competitiveness of PAFCs. The project was promoted from 1976 to 1986, and the results obtained from the operation of the project are summarized as follows:[3]

1. Forty-six plants were manufactured and located on 42 sites in the United States and two in Japan. These plants operated both in grid-dependent and independent modes. Heat generated was used for domestic hot water, space heating, pool heating, and low-temperature process loads.
2. The average plant operated for over 6500 h. Fourteen power plants exceeded 8000 h of operation, with the longest running unit operating for over 15,000 h.
3. The successful operation demonstrated highly feasible performance of small-capacity PAFC systems.

The simplified system diagram of this unit is shown in Fig. 8.2.[1,8] Also, the key specification is shown in Table 8.2.[8]

Field tests were completed by 1986, showing satisfactory technical performance. From the commercial production point of view, however, the 40-kW capacity unit still suffered from high capital costs. IFC since then decided to develop commercial units with a capacity of 200 kW, and are currently seeking potential customers to purchase this capacity with a semicommercial price of $500,000 (U.S.) for the initial plants. This price will drop to $300,000 with manufacturing experience.[64]

8.2.2.3. Electric Utility Application

The so-called FCG-1 project started in 1971 under a contractual arrangement between nine electric utility companies in the United States and UTC (now IFC). FCG-1 aimed at developing PAFC power plants for electric utility companies. EPRI later joined the FCG-1 project in 1973.

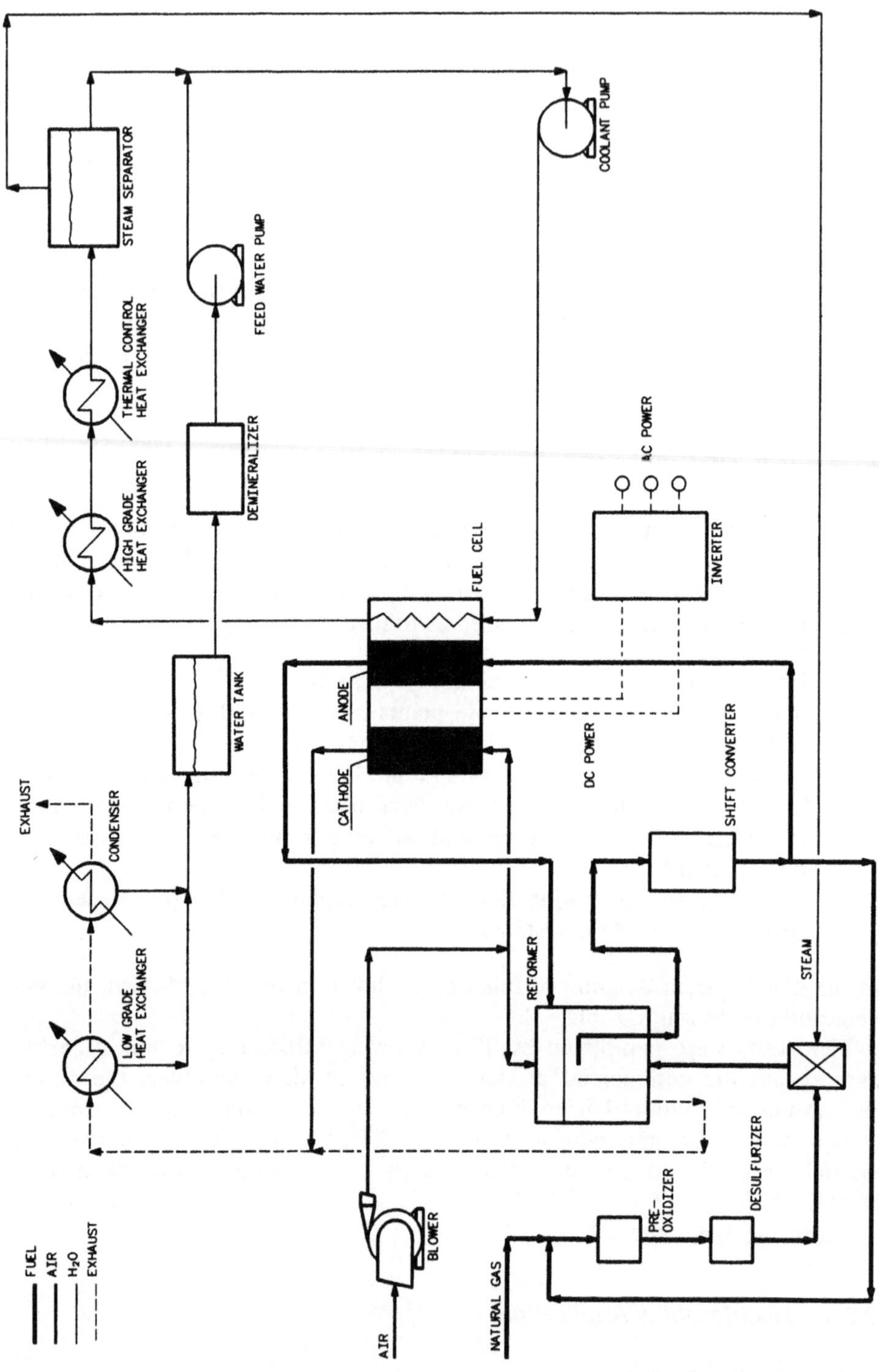

Figure 8.2. System diagram of 40-kW PAFC system (IFC).[8]

Table 8.2. Specification of 40-kW PAFC Plant of GRI–DOE Project[8]

Rated power	40 kW
Electrical efficiency	40% (LHV)
Total efficiency	80% (LHV)
Frequency	60 Hz
Voltage recovery	Instantaneous from 0 to full load (2 cycles or less)
Start-up time	4 h
Fuel	Natural gas (8.6 m^3/h)
Fuel gas pressure	100–350 mm H_2O
Operation	Automatic
Installation	Outdoor–indoor type
Weight	3600 kg (dry weight)
Dimensions	2.80 × 1.60 × 2.0 (h) m
Noise level	60 dB(A) at 4.6 m distance
Fuel cell size	45 × 45 cm (ribbed substrate)
Number of cells	270
Cell working conditions	190°C, 1 atm
Cooling method	Water cooling

The FCG-1 project plan called for the construction of a 4500-kW demonstration plant in New york City by UTC. Consolidated Edison Company conducted the process and control (PAC) test, but, unfortunately, due to many unexpected problems (not related to the fuel cells themselves), the plant was dismantled before generating electric power.

In parallel with the FCG-1 program, Tokyo Electric Power Company (TEPCO), Japan, purchased a similar plant from UTC with their own funds, and began the construction on the premises of their thermal power station located along Tokyo Bay.

This plant successfully operated from 1983 to 1985 and demonstrated the feasible performance of a PAFC plant as the new electric power station for electric utility companies. The results obtained from the construction and operation of this 4.5-MW plant are summarized as follows:[4]

(a) Total accumulated generation experience turned out to be 2400 h, including 500 h of uninterrupted operation.
(b) The emission and noise level were very low and are acceptable within the environmental regulations.
(c) The plant overall efficiency, which is one of the most important performance criteria, comprises various subsystem efficiencies as shown in Fig. 8.3. The overall efficiency reached was 36.7% (on the basis of ac power (kW/HHV (higher heating value) of fuel)) and is the resultant of 82% of reactant supply subsystem, 48% of cell stack assembly, 96% of inverter, and 97% of mechanical power under full-load conditions.

The performance demonstrated was by far the best experienced in large-scale PAFC plants so far. However, operating performance under partial load was not satisfactory. Based on operation and construction experience, IFC and Toshiba (Japan) jointly developed an improved, semiconventional 11-MW PAFC plant for electric utility application during 1985–1986. TEPCO requested Toshiba to supply

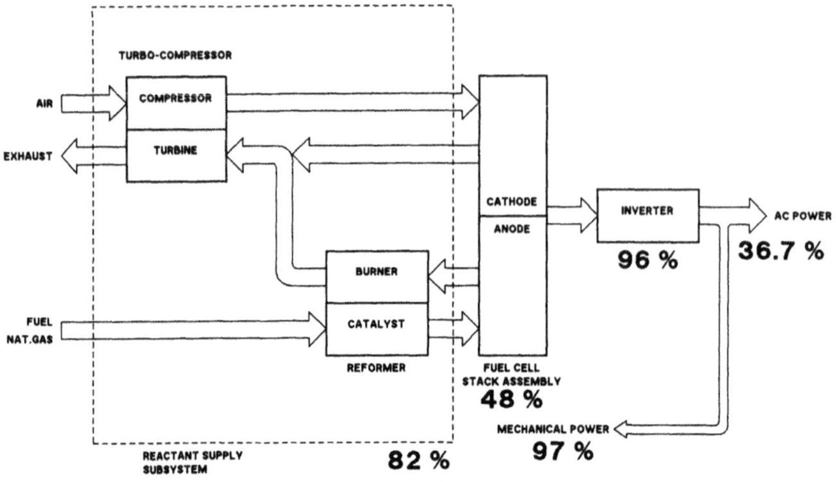

Figure 8.3. 4.5-MW fuel cell efficiencies (HHV base).[4]

the 11-MW plant, the world's largest. It began operation in April 1991. A summary of the performance is shown in Table 8.3.[9]

8.2.2.4. Development at Engelhard Corporation

Engelhard Corporation was engaged in the development of PAFC technology for more than 20 years.[50–52] Their interests covered not only the development of the Pt catalyst, but included small-capacity PAFC plants suitable for on-site cogeneration and forklift truck application. In other words, Engelhard showed no direct interest for large-scale electric utility power plants, but concentrated their efforts on the development of natural-gas-reformed on-site plants of 25–50 kW and methanol reformed PAFC power systems for forklift trucks.

The fuel-cell-powered forklift truck was developed to replace the conventional battery-powered forklift. The main advantage of the fuel cell forklift lies in multishift operation: battery recharging is not required.

From technical point of view, Engelhard's technology is characterized by dielectric liquid cooling (oil cooling) systems and special electrode structures. The dielectric liquid cooling system, using synthetic oil (as organic heat transfer medium), shows a performance intermediate between water and air cooling: it

Table 8.3. Comparison of Performances of 4.5-MW Plant and 11-MW Plant[9]

Item	4.5 MW	11 MW
Rated power (MW)	4.5	11
Electrical efficiency (HHV) at sending end (%)	36.7	41.1
Power range (%)	25–100	30–100
Plant operating pressure (kg/cm^2g)	2.5	7.4
Fuel cell active area (cm^2)	3440	9300
Plant footprint (m^2)	3240	3300
Start-up time (h)	4	6
NO_x emission (ppm)	<10	≤10
Noise level at site boundary (dB)	≤55	≤55

has a better heat transfer characteristic than air, but lower than water. Handling is easier and simpler than water-cooled, but not air-cooled, systems. Since the forklift truck application requires compact design and easy maintenance, oil cooling is preferred.

The second unique feature of Engelhard technology is its electrode structure, intended for simple and cheap fuel cell stack assembly. This structure consists of two substrate plates bonded together by dense carbonized resin (the ABA structure), the resin acting as the separator plate (hence, ABA is a kind of bipolar plate structure.)

Engelhard has designed, manufactured, and tested a PAFC experimental forklift fueled by methanol for several years. More recently Engelhard's technology was transferred to Fuji Electric Company in Japan. Fuji intends to extend the results of Engelhard to the commercial level.

For the on-site cogeneration plant, Engelhard earlier introduced a natural-gas-reforming system from KTI and subsequently built and operated a KTI-designed total system on natural gas, utilizing their own PAFC stack technology. In 1988 Engelhard left the fuel cell development field, due to a change in corporate strategy.

8.2.2.5. Developments at ERC

ERC (Energy Research Corporation) has been developing PAFC stack technology since 1969. Their technology is characterized by the use of air cooling, compared to the water-cooling system of IFC or liquid cooling of Engelhard.

ERC's focus was to develop small PAFC units (0.5–30 kW) for use primarily by the U.S. Army. Based on their air-cooled PAFC technology,[54] ERC applied their plant technology to vehicular application. In 1988, ERC joined the DOE–DOD-sponsored fuel-cell-powered bus program.[55]

The basic design configuration of ERC's air-cooling cell structure has been adopted by Westinghouse. Westinghouse intended to apply air-cooled technology for large-scale pressurized PAFC plants for electric utilities. ERC's air-cooled stack technology was also introduced to Sanyo Electric Company in Japan. Sanyo is expanding ERC technology to plants in the several hundred kilowatt range (refer to Table 8.6).

8.2.2.6. Developments at Westinghouse Electric Corporation

Westinghouse Electric Corporation (WE) started development activities of PAFC in 1978.[56,57] The primary focus of Westinghouse was to supply a simple PAFC power plant for electric utility application. For this purpose, unlike UTC and Engelhard, WE decided to apply the air-cooling system stack technology developed by ERC. WE further extended the ERC technology to higher pressure and larger electrodes. A 7.5-MW PAFC unit was planned as a first demonstration plant.

Unfortunately WE development encountered many difficulties, and the completion schedule of the plant as well as the commercialization forecast were strongly delayed. WE development activities have been supported by DOE and GRI primarily because air-cooled stacks offer the opportunity for a simplified system. (The comparison of water-, liquid-, and air-cooling systems will be further dealt with in Section 8.2.3.)

8.2.3. PAFC Demonstration Programs in Japan

The PAFC development program in Japan is divided into government-sponsored programs (*vis-à-vis* DOE, DOD programs in the United States) and private programs.

8.2.3.1. Overview of Government-Sponsored Development Project

Stimulated by the active development activities in the United States and faced with ever-escalating energy costs triggered by the oil shock in 1973, the Japanese government decided to start a comprehensive development program of various types of fuel cells in 1981 under the Moonlight Project. This program is sponsored by the Japanese government and its goal is to develop energy conservation techniques.

The fuel cell R&D program had originally been structured as a 10-year project with an overall PAFC budget of about Y11,500 million (about $77 million, June 1990). In March 1987 the basic plan was extended to a 15-year program (until fiscal year 1995) owing to new development and research activities on the molten carbonate fuel cell. The total R&D budget also increased to about Y57,000 million.[10] The overall R&D timetable of fuel cell is shown in Fig. 8.4.[10] Under this Moonlight Project two sets of 1-MW and 200-kW PAFC plants were constructed respectively. The key specification of the two 1-MW plants is shown in Table 8.4.

Principal goal for the 1-MW plants was to realize stable operation and performance through real-time operation. This was almost achieved in 1988.

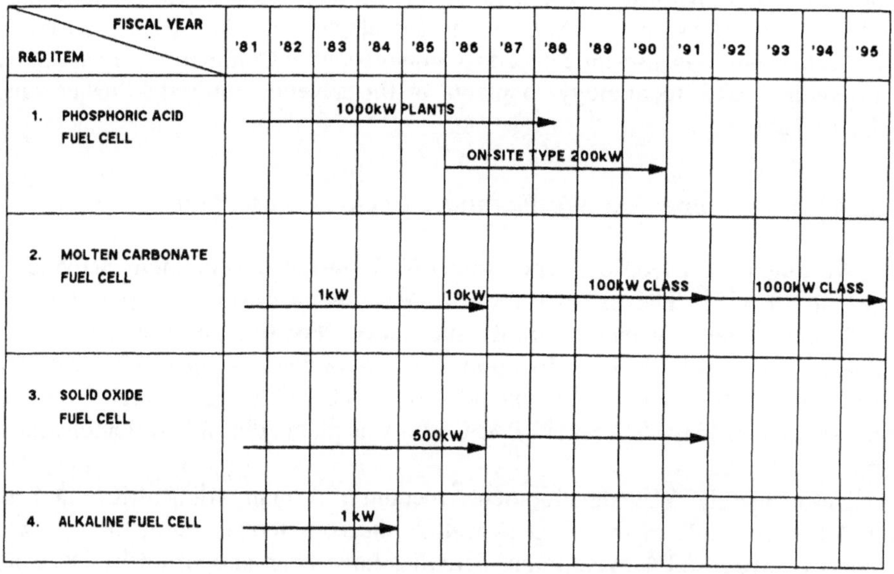

Figure 8.4. Fuel cell R&D program (Moonlight Project).[10]

Table 8.4. Key Specification of 100-kW PAFC Plants (Moonlight Project)

Type	Dispersed type	Central station type
Output power (generating end)	1000-kW ac	1000-kW ac
Efficiency (generating end)	40% (HHV)	42% (HHV)
Operating condition of fuel cell	5 ata, 190°C	7 ata, 205°C
Pt loading	<6.5 g/kW	<6.5 g/kW
Cooling method of stack	Boiling-water cooling	
Fuel	Natural gas	
Start-up time (cold start)	4 h	

In addition to the 1000-kW plants, the Moonlight Project expanded to include two 200-kW PAFC plants for on-site cogeneration and remote island application in 1986. These two plants entered into on-site operation at the beginning of 1990, and after the very satisfactory operation for two years, these were dismantled.

The key specifications of these 200-kW PAFC plants for a remote island and for on-site cogeneration use are shown in Table 8.5.

Following the above projects the new grid interconnection demonstration program was planned by the government in 1987. This program intended to demonstrate feasible performance of various types of new generation systems such as wind power, photovoltaic solar cell, and fuel cell systems.

The total capacity of fuel cell systems in this project is 900 kW, composed of 14 sets of 50-kW and 1 set of 200-kW PAFC plants which were supplied from Fuji Electric and Mitsubishi Electric respectively, from 1990 to 1992 (also refer to Section 8.5.3). In addition, the Japanese MITI and electric and gas utility companies started construction of a 5-MW dispersed and a 1-MW on-site PAFC plant in 1991 under contract with Fuji Electric and Toshiba, respectively.

8.2.3.2. Developments in Private Industries

As shown in Table 8.6[10] Japanese electric utility companies, gas utility companies, and several petroleum companies (Petroleum Energy Center) interested in PAFC systems have contracted manufacturing companies with private funding.

Table 8.5. Key Specification of 200-kW PAFC Plants (Moonlight Project)

Items	Key specifications	
Output power (sending end)	200-kW ac	200-kW ac
Efficiency (sending end)	36% (HHV)	37.6% (HHV)
Operating conditions of fuel cells	1 ata, 190°C	1 ata, 190°C
Pt loading	4.5 g/kW	4.5 g/kW
Cooling method of stack	Water cooling	Water cooling
Fuel	LNG	Methanol
Start-up time (cold start)	3 h	3 h
Purpose	On-site cogeneration plant	Power source in remote island

Table 8.6. Outline of Private Development Program of PAFC Plants in Japan

Company	Plant description	Manufacturer	Operation dates
Tokyo Electric Power Co.	220 kW, NG, 1 ata, air	Sanyo Elect.	9/87–10/89
	2 × 200 kW, NG, 1 ata, water	IFC	10/88–11/90 3/89–
	11 MW, NG, 8.4 ata, water	Toshiba	4/91–
	4.5 MW, NG, 3.5 ata, water	IFC	4/83–12/85
Kansai Electric Power Co.	30 kW, NG, 1 ata, air	Fuji Elect.	12/82–12/83
	5000 kW, NG, 7 ata; water	Fuji. Elect.	4/94–(plan)
Tohoku Electric Power Co.	50 kW, NG, LPG, 2 ata, water	Fuji Elect.	5/87–10/89
Hokkaido Electric Power Co.	100 kW, methanol, 1 ata, water	Mitsubishi Elect.	10/87–5/89
Tokyo Gas Company	2 × 12.5 kW, NG, 1 ata, water	IFC	3/72–8/72 for both
	2 × 40 kW, NG, 1 ata, water	IFC	3/82–5/87 12/84–2/87 2/90–5/92
	3 × 50 kW, NG, 1 ata, water	Fuji Elect.	2/91– & 5/92–
	100 kW, NG, 1 ata, water	Hitachi	8/90–5/92
	100 kW, NG, 1 ata, water	Fuji. Elect.	7/92–
	1000 kW, NG, 1 ata, water	Toshiba	4/95 (plan)
	10 × 200 kW, NG, 1 ata, water	IFC	92–
Osaka Gas Company	2 × 12.5 kW, NG, 1 ata, water	IFC	2/73–8/73 for both
	2 × 40 kW, NG, 1 ata, water	IFC	3/82–2/85 11/84–10/87
	200 kW, NG, 1 ata, water	IFC	9/89–9/90
	6 × 50 kW, NG, 1 ata, water	Fuji Elect.	from 10/90 succ.
	100 kW, NG, 1 ata, water	Fuji Elect.	4/92–
	10 × 200 kW, NG, 1 ata, water	IFC	92–
Toho Gas Company	50 kW, NG, 1 ata, water	Fuji Elect.	11/91–
	200 kW, NG, 1 ata, water	IFC	92–
Tokyo Osaka, & Toho Gas Companies (jointly)	9 × 50 kW, NG, 1 ata, water	Fuji Electric	92–
	7 × 100 kW, NG, 1 ata, water	Fuji Electric	92–
Petroleum Companies & Petroleum Energy Center	200 kW, naphtha, 1 ata, water	IFC	2/90–6/91
	50 kW, naptha, 1 ata, water	Fuji Elect.	7/90
	80 kW, LPG, naptha, kerosene, 1 ata, water	Fuji Elect.	3/90
	100 kW, naptha, 1 ata, air	Sanyo Elect.	10/89–2/91
	2 × 4 kW, methanol, 1 ata, air	Fuji Elect.	5/88–12/90 11/88–3/91

Except for the 11-MW and 4.5-MW units mentioned before, four sets of 40-kW and several 200-kW plants, which Tokyo Electric Power Company, Tokyo Gas Company, and Osaka Gas Company have been introducing from IFC (or UTC) in the United States respectively, many PAFC plants have been developed and demonstrated in a cooperation between domestic electric and gas utility companies, petroleum companies, and manufacturing companies in Japan.

Utility companies are interested in supporting these demonstration plants to accumulate operational experience and to evaluate the applicability and suitability of PAFC plants for their own network.

As also shown in Table 8.6 natural gas, LPG, methanol, and naphtha are used as primary fuels. Natural gas is most widely used for on-site cogeneration plants and large-scale power plants, but methanol is used for vehicular, mobile, and special applications, such as remote islands. Methanol is liquid at ambient temperature and is easier to reform to hydrogen-rich gas than natural gas and LPG (see also Chapters 4 and 6).

Generally speaking, the PAFC development in Japan is quite diversified, with wide ranges of capacities, application fields, and cooling methods.

The primary development objective was to develop non-oil-based substitution energy technology, but, as shown in Table 8.6, the Petroleum Energy Center and some petroleum companies have exhibited continued interest to expand the effective uses of petroleum products. In the future, technical programs will encompass the development of petroleum-reforming systems is conjunction with fuel cell manufacturing companies.

Not only the above-described stationary plant application but also the vehicular application is under development at Fuji Electric. Based on the dielectric cooling technology and fuel cell–battery hybrid system know-how transferred from Engelhard (see Section 8.2.2.4), Fuji Electric is intending to apply the fuel-cell-powered forklift truck technology to the bus system which is now promoted as a DOE and DOD project in the United States. The estimated PAFC capacity will be more than 50 kW (see Section 8.2.2.5).

8.2.4. PAFC Demonstration Program in Europe

Construction of PAFC demonstration plants in Europe is currently being promoted with support from the Commission of European Communities (CEC) and governmental and private organizations. Under this program the construction of several PAFC plants was committed as shown in Table 8.7.[11]

The basic development philosophy of some European firms for PAFC systems is to integrate fuel cell stacks manufactured in the United States or Japan with their own systems. Overall responsibility of plant performance is that of the European firms. Due to the fact that the development of PAFC stacks is close to the commercial stage, it would be counterproductive for Europeans to enter into R&D on basic PAFC development.

A 25-kW on-site cogeneration plant was constructed in the Netherlands in 1988. KTI Group has taken overall responsibility of this first PAFC plant in Europe. The PAFC stack was supplied by Fuji Electric Company in Japan.

A 1000-kW plant being constructed by Ansaldo, Italy, for the Milano Municipal Electric Power Company is due to be completed by 1992. The reformer and stacks, two key components of the plant, were supplied by Haldor Topsoe (Denmark) and IFC (United States) respectively.

As a further effort, KTI has recently built an 80-kW PAFC plant (with hydrogen being purified) which can operate on natural gas and on pure hydrogen. This plant is part of Solar–Wasserstoff-Bayern's demonstration project (SWB Project) which demonstrates total hydrogen technology development. (The 80-kW stack was supplied by Fuji Electric, Japan.)

Table 8.7. European Fuel Cell Demonstration Program[11] (PAFC only)

Capacity	1 & 5 kW	25 kW	1000 kW
Number of sets	1 each	1	1
Coordinator	ENEA	Netherlands	ENEA
Constructor	Ansaldo	KTI	Ansaldo
Supplier			
fuel reformer	OTB Participazioni	KTI[a]	Haldor Topsoe (Denmark)
fuel cell	Ansaldo	Fuji Electric (Japan)	IFC (U.S.)
Operator	Ministry of Defence	[b]	AEM (Milan Municipal Electric Utility)
Fuel	Methanol	Natural gas	Natural gas
Schedule of operation		1989, 1990	1992

[a]Kinetics Technology International (in the Netherlands and Italy).
[b]Operator varies.

Electric and gas utility companies in European territories are also expressing interest to introduce the PAFC cogeneration plants ranging from 50-kW to 200-kW capacity from Japan and the United States. If in the future the plants can demonstrate satisfactory operation and competitive prices, the market of PAFC plants in Europe may expand very rapidly. For example, the potential market in Italy in the period of 1995 to 2000 is estimated to be about 6000 MW.[11]

8.3. PAFC CELL PERFORMANCE AND STACK COMPONENTS

8.3.1. Key Features of PAFC Stacks

PAFC stacks use the phosphoric acid as its electrolyte and is tolerant to CO_2 contained in the reformate gas and in air. For these reasons PAFC has been recognized as a very flexible fuel cell for many environments, in comparison to low-temperature cells such as alkaline fuel cells (CO_2 or CO is not allowed) or proton exchange membrane fuel cells (CO_2 is allowed, but CO is prohibited).

As described in the previous section, a great deal of development activities have been completed in the United States and in Japan, and it is generally recognized that PAFC presently has the best chance for commercialization. PAFC is envisioned to be the first commercial generation type of fuel cell.

This section briefly describes PAFC cell technology in order to provide a deeper understanding of PAFC systems.

8.3.1.1. Working Conditions

The following working conditions of PAFC are shown as follows:

(a) Working temperature range: 180–210°C (average). Stacks show higher efficiency at higher temperature than at lower temperature.

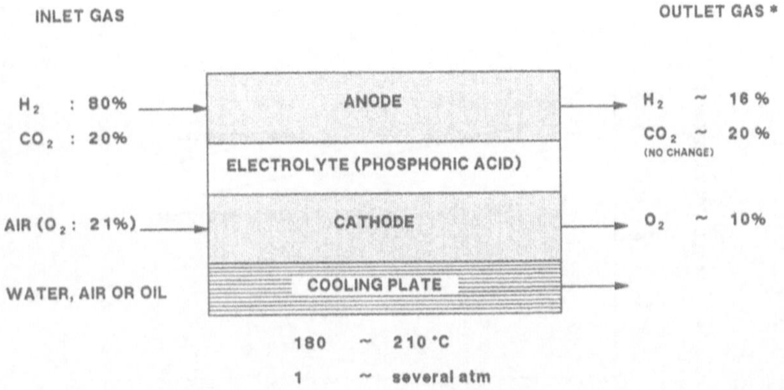

Figure 8.5. Reaction in fuel cell (assumed utilization factor: 80% for anode, 50% for cathode).

(b) Working pressure range: 1 to several atmospheres. One atm is generally adopted for smaller capacity range and several atm for larger capacity range. Stack efficiency is improved at higher pressure compared to lower pressure.
(c) Cooling method: air, water (or water and steam mixture), and dielectric liquid (oil) (refer to Section 8.3.3 for details).
(d) Fuel utilization factor: 70–80% (typical case). The fuel utilization factor is a ratio of how much hydrogen is consumed in fuel cells. A utilization factor of 70–80% means that 70–80% of the hydrogen contained in the reformate gas is consumed in the stacks to generate electricity.
(e) Oxidant utilization factor: 50–60% (typical case). Oxygen content in air is about 21%, and a 50–60% utilization factor means that 50–60% of the oxygen contained in air is consumed in the fuel cells.
(f) Reactant gas composition: Typically, reformed gas contains about 80% H_2, 20% CO_2, and small amounts of other impurities such as CH_4, CO, sulfur compounds, etc., on dry gas basis. Concentration of such gases may have impact on performance, as will be described later.

Figure 8.5 shows some data in the cell reaction as explained above.

8.3.1.2. Typical Performance of PAFC

A typical voltage–ampere performance curve to characterize a PAFC is shown in Fig. 8.6.[5] As seen from this performance curve, the output voltage is very much dependent on the current. The V–I curve indicates that PAFC has a fairly high internal resistance. Working current density in PAFC generally ranges from 150 to 350 mA/cm^2.

8.3.1.3. Efficiency of the Fuel Cell

In this chapter, the efficiency of the fuel cell, η_{FC}, is defined as follows:[44]

$$\eta_{FC} = \frac{\text{electric energy generated in the cell}}{\text{hydrogen energy consumed in cell}},$$

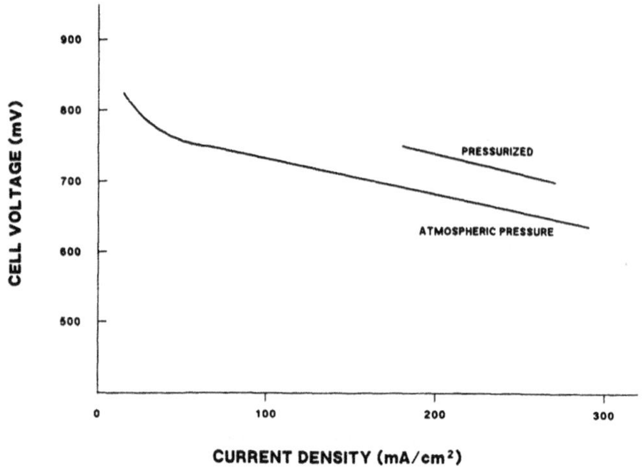

Figure 8.6. Voltage–ampere characteristic of PAFC unit cell (an example).[5]

so η_{FC} is based on the utilized hydrogen only (see Chapters 5 and 6). The hydrogen energy consumed in the cell is $G_2 \times H_2 \times \eta_H$, where

η_{FC} = efficiency of the fuel cell

G_2 = hydrogen quantity flowing into the anode (mol/h)

H_2 = heating value of hydrogen (kcal/mol) (LHV or HHV base)

η_H = utilization factor of hydrogen in the anode (in %)

$$\therefore \eta_{FC} = \frac{0.86 \times V \times I}{G_2 \times H_2 \times \eta_H},$$

where

0.86 = conversion factor from W·h to kcal

V = output voltage of fuel cell (dc volts or V_{dc})

I = output current of fuel cell (dc amps or A_{dc}).

The consumed hydrogen is converted to electric current following Faraday's law. In the reaction

$$H_2 + \tfrac{1}{2}O_2 \rightarrow H_2O,$$

one mole of hydrogen (2 g atom) is consumed and releases 2 × 96,522 A·s or 2 × 26.81 A·h.

The output current I is therefore calculated as (assuming that all consumed H_2 is converted into electricity)

$$I = G_2 \times \eta_H \times 2 \times 26.81 \text{ A·h},$$

$$\therefore \eta_{FC} = 46.11V/H_2.$$

The heating value of hydrogen, H_2 = 68.32 kcal/mole at 25°C, HHV based. Therefore,

$$\eta_{FC} = 0.675V \quad \text{(HHV base)}.$$

Alternatively, on LHV base, H_2 heating value (at 25°C) = 57.71 kcal/mole and

$$\eta_{FC} = 0.799V \quad \text{(LHV base)}.$$

From these equations it is apparent that the efficiency of the fuel cell is determined only by the output voltage of the cell. Reducing the current density is an effective way to increase the efficiency as output voltage increases for lower current density. However, reducing the current density results in an increase in fuel cell capital cost. In practice, the PAFC current density is usually 150–350 mA/cm^2. Under atmospheric pressure, the output voltage of one unit cell, at atmospheric pressure, is approximately 700–600 mV, and the cell efficiency on this base is approximately 40–47% (HHV) or 48–56% (LHV).*

From the above mentioned reasons, it is evident that improvement of cell performance is critical for improving plant efficiency. To reduce plant cost, keen efforts have been undertaken worldwide to increase the current density without sacrificing the output voltage. However, in the future the cost of mass production stacks will be about one third or one fourth of the total system cost, so the current density should not be maximized without limitation, but an optimum value will result.

Since the output voltage of a cell can be increased by increasing the working temperature and pressure, proper selection of working conditions is essential to increase stack efficiency. For example, under pressurized conditions (e.g., 5 ata) the cell voltage increases by approximately 100 mV. Cell voltages then are 800–700 mV, and the cell efficiency on HHV basis is 47–54% (increasing the working pressure has some reverse effects; see Section 8.3.2.1).

Fuel cells also perform with a higher efficiency under partial load conditions, since the cell voltage increases at lower current density. This is a special feature of fuel cells which cannot be realized with any other conventional technology.

8.3.2. Factors Affecting Stack Performance

The *V–I* characteristics shown in Fig. 8.6 depend on various working conditions such as temperature, pressure, utilization factor, and others. This dependence will be briefly reviewed below (as to utilization factors see Section 8.4.4).

*Note that cell efficiencies are also often expressed on the basis of total heating value of all feed to the anode (including the unutilized part). In that case, numerical values of the fuel cell efficiencies are lower. In overall system design, this results in lower stack efficiency numbers, but higher reformer efficiencies, giving the same overall system values (see Chapters 5 and 6).

8.3.2.1. *Effect of Temperature* [13]

Increasing temperature enhances mass transfer and thus leads to higher reaction rates, lower cell resistance, reduced polarization, and, hence, improved cell performance. The relationship between voltage changes and temperature changes depends on various factors: for example, cell designs, electrode type, current density, number of operating hours, and others.

The following examples are described in *Handbook of Fuel Cell Performance* (*1980*)[13] and the Japanese MITI-AIST report (1984):[33]

Example 1. Experimental formula in *Handbook.*[13] The experimental formula showing the relationship between voltage and temperature is given by

$$\Delta V_T \quad (\mathrm{mV}) = K(T_2 - T_1)(^\circ\mathrm{C})$$

between 190°C and 218°C, where $K = 1.15$ (for H_2–air supplied at 3.5 ata and 322 mA/cm^2 conditions). Here

ΔV_T (mV) = voltage difference under various working temperatures

T_1, T_2 = two different working temperatures.

For example, the voltage at 300 mA/cm^2 and 1 ata will increase from 600 to 617 mV by raising the temperature from 190 to 205°C, i.e.,

$$\Delta V_T\,(\mathrm{mV}) = 1.15(205 - 190) = 17.25\ \mathrm{mV}.$$

The coefficient K, is dependent on several operating conditions such as current density, pressure and the operating time of stacks. 1.15 is the experimental value for relatively high pressure and high current density ranges, such at 3–4 ata, around 300 mA/cm^2, and at the beginning stage of the operation (new stack). If current density decreases, K will decrease (e.g., with a current density of 100 mA/cm^2, the coefficient will be about 1.05). At ambient temperature the coefficient decreases to 0.8.

Example 2. Experimental formula in MITI-AIST report.[33] This report indicates that correlation formulas for the changes of temperature and pressure are given by

$$V = V_0 + \Delta V_P + \Delta V_T,$$

where

V = cell voltage under different temperature and pressure (mV)

V_0 = cell voltage under standard working temperature and pressure

ΔV_P = voltage change under different pressure (mV)

ΔV_T = voltage change under different temperature (mV).

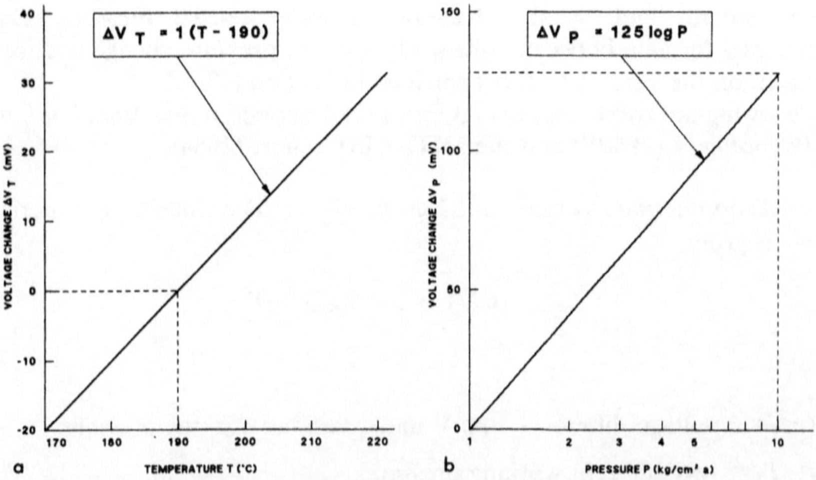

Figure 8.7. Effect of pressure and temperature on cell voltage: (a) temperature dependence; (b) pressure dependance.

Assuming that the standard working temperature is 190°C, ΔV_T is given experimentally as follows (refer to Fig. 8.7a):

$$\Delta V_T = 1.0(T - 190) \quad (\text{mV}),$$

where

$$T = \text{cell working temperature (°C)}.$$

Taking the V–I characteristics shown in Fig. 8.6 as the standard performance at 190°C and 1 ata, V_0 is 680 mV at 200 mA/cm^2. For the preceding case (T = 205°C)

$$\Delta V_T = 1.0(205 - 190) = 15 \quad (\text{mV}).$$

(This means the coefficient K is equal to 1.0.)

However, increased temperature results in increased corrosion, catalyst sintering and recrystallization, and electrolyte loss from evaporation, all of which adversely affect stack life. Conditions must be carefully chosen for optimum design and performance.

From the point of view of material selection, the highest working temperature based on current technology is about 220°C for peak and 210°C for continuous operation. It is not recommended to operate the PAFC continuously when the temperature exceeds 210°C.

Therefore, the uniform temperature distribution in each cell in horizontal direction and in vertical direction is an important factor to increase the cell performance.

8.3.2.2. *Effect of Pressure*

Generally speaking, increased pressure gives higher cell performance.[12,33] This is mainly because of the increased reaction at the cathode by increased oxygen contents and by the decrease in water partial pressure. But the experimental formula between voltage change and pressure change is once again dependent on the various factors explained in Section 8.3.2.1.

Following are two experimental formulas described in the *Handbook of Fuel Cell Performance* (*1980*)[12] and the MITI-AIST report (1984).

1. *Experimental formula in Handbook.*[12] The following experimental formula is given:

$$\Delta V_P \quad (\text{mV}) = 142 \log_{10}(P_2/P_1),$$

where

$\Delta V_P\,(\text{mV})$ = voltage difference in mV under various working pressures

P_1, P_2 = two different working pressures.

For example, if $P_1 = 1$ ata and $P_2 = 5$ ata, then

$$\Delta V_P \quad (\text{mV}) = 142 \log_{10}(5/1) = 99.3\ \text{mV}.$$

That is, the effect of pressure increase on the voltage change is quite large compared to the temperature effect (cf. Section 8.3.2.3).

2. *Experimental formula in MITI-AIST report.*[33] Using the same experimental formula described in Section 8.3.2.1 (Example 2), then

$$V = V_0 + \Delta V_P + \Delta V_T.$$

ΔV_P is given experimentally (see Fig. 8.7b) as

$$\Delta V_P = 125 \log P,$$

where

$$P = \text{cell working pressure (ata)}.$$

Using the same example as above where $P = 5$ ata,

$$\Delta V_P = 125 \log 5 = 87.4\ \text{mV}.$$

That is, V at 200 mA/cm^2, 5 ata, and 205°C will be

$$V = 680 + 15 + 87.4 = 782.4\ \text{mV}.$$

These experimental factors are the resulting effects of pressure and temperature on the activity of cell electrodes. As described before, these factors depend on cell structure, kind of catalysts, current density, utilization factors of reactant gases, and other various factors.

Therefore, the practical experimental factor should be determined on the developed cells and stacks by each manufacturer.

8.3.2.3. *Effect of Reactant Concentration*

Since pure hydrogen and pure oxygen are not usually available at a plant site, reformed gas from fossil fuels (such as natural gas, methane, LPG, etc.) and air are used as fuel and oxidant gases, respectively. In some cases, such as the chloroalkaline industries, pure hydrogen is available as a by-product.[14]

The phosphoric acid fuel cell performance under pure hydrogen and oxygen is greatly improved compared to the case of reformed gas and air. The change of cell performance with respect to the partial pressure of hydrogen

$$\Delta V_{H_2} = 77 \log \frac{\{(P_{H_2})_2}{(P_{H_2})_1\}} \quad (\text{mV})$$

where ΔV_{H_2} is the voltage difference between different pressures 1 and 2 in the anode and P_{H_2} is the average hydrogen partial pressure in the anode. As an example, a numerical comparison follows between reformed gas and pure hydrogen. Assumptions are as follows:

Composition of reformed gas H_2 80%, CO_2 20%*; fuel utilization factor 80%; composition of H_2 in the cell is (mole fractions).

	Inlet	Exit	Average
H_2 mol	0.8	0.16	0.48
CO_2 mol	0.2	0.2	0.2
Total mol	1.0	0.36	0.68
$(P_{H_2})_1$	0.8	0.44	0.70

Therefore

$$(P_{H_2})_1 = 0.70\ \text{ata}, \qquad (P_{H_2})_2 = 1.0\ \text{ata (pure } H_2),$$

$$\Delta V_{H_2}\,(\text{mV}) = 77 \log(1.0/0.7) = 11.96\ \text{mV}.$$

So cell voltage will increase 11.96 mV utilizing pure hydrogen versus reformed gas of said composition.

For oxygen, the following formula has been derived experimentally:

$$\Delta V_{O_2} = 103 \log \left\{ \frac{\{(P_{O_2})_2}{(P_{O_2})_1} \right\} \quad (\text{mV}),$$

where $(P_{O_2})_1$ and $(P_{O_2})_2$ are average oxygen partial pressures in the cathode at the two pressures being compared.

Similar example calculations can be made. Assuming the O_2 content in air is 21% (N_2: 79%), air utilization factor is 50%, and the composition of air in the

*Usually some steam (water) is contained in the anode inlet gas, but such small contents of water are neglected for the sake of simplification.

cell is

	Inlet	Exit	Average
O_2 mol	0.21	0.105	0.16
N_2 mol	0.79	0.79	0.79
H_2O mol	0	0.21	0.105
Total mol	1.00	1.105	1.055

Then for air and pure hydrogen (1 ata)

$$\frac{(P_{O_2})_2}{(P_{O_2})_1} = \frac{1.00}{0.16/1.055} = 6.60.$$

$$\therefore \Delta V_{O_2} \quad (\text{mV}) = 103 \log 6.60 = 84 \text{ mV}.$$

The difference in cell voltage between pure oxygen and air is quite large, much greater than in the case of changing anode feed gas from synthesis gas to pure hydrogen.

8.3.2.4. *Effect of Impurities*

As explained earlier, reformed gas is composed of about 80% hydrogen and 20% carbon dioxide and other impurities.[15] The following is a typical example of a reformed gas composition acceptable for PAFC operation (refer to Section 8.3.1 and Table 8.16): H_2 78%, CO_2 20%, CO < 1%; H_2S, Cl, NH_3 < 1 ppm.

For PAFC, CO_2 has no impact on cell performance other than the effect of partial pressure. Other impurities, however, affect cell performance in various ways.

a. Effect of CO. The existence of CO (carbon monoxide) decreases the activity of the Pt catalyst and cell performance.[15] The effect is larger for lower temperature and smaller for higher working temperature. It is recognized that the performance loss is almost proportional to the CO content. But the loss is reversible and can be offset by increasing the temperature.

The permissible level of CO is dependent on the working cell temperature. At a working temperature of around or over 190°C (as is the case in present PAFCs), 1% of CO is acceptable without noticeable adverse effects on the performance (for comparison, CO content must be limited to less than several ppm for alkaline and/or proton exchange membrane fuel cells, because their working temperatures are about 80°C or less).

b. Effect of Sulfur Compounds. Since raw fuel is supplied to the reformer after passing through a desulfurizing process, sulfur compounds generally have little effect on cell performance.[15] However, if sulfur-containing compounds in the fuel are converted to H_2S and not removed upstream, cell performance can be greatly affected.

The performance loss is again reversible and can be recovered by increasing the temperature. Diagnostic tests indicated that no permanent performance damage occurred at the anode, operating in the presence of H_2S.

c. Effect of Nitrogen Compounds. Nitrogen compounds, especially ammonia (NH_3), impact cell performance negatively.[16]

Nitrogen is usually contained in hydrocarbon feedstock, and some nitrogen compounds (NH_3, HCN, NO_x) are by-products of the reforming process, and hence are supplied to the fuel cells.

The effect of ammonia, however, is generally reversible by removing the ammonia. For good operation, the ammonia content should be limited in practice to less than 1 ppm. No data are available on the effect of nitrogen compounds other than NH_3 at this time.

d. Effect of Internal Resistance. The output cell voltage is reduced by reduced ionic flow in the electrolyte and electronic conduction in the electrodes, current collectors, and interfaces.[18] The magnitudes of such losses are approximately 15–20 mV at 100 mA/cm^2. The internal resistance can be estimated as

$$\Delta V_{iR} \quad (\text{mV}) = -0.20i$$

where i is the current density of the electrode (mA/cm^2).

8.3.3. Cooling Method

There are three different cooling methods for PAFC stacks: air cooling, dielectric liquid cooling (synthetic oil cooling), and water cooling.

Generally speaking, the water-cooling method is superior to the others from the cooling performance point of view and is appropriate for larger plants. On the contrary, the air-cooling method is very simple compared to the water-cooling system and is recognized to be appropriate for relatively small plants. However, air-cooling systems generally require relatively large amounts of auxiliary power for circulating cooling air and inevitably result in a loss in net efficiency.

From the cooling performance and system complexity point of view, the dielectric liquid (oil) cooling system has intermediate performance between that of air and water. However, it can be compact and corrosion is not likely. Figure 8.8 shows a general comparison of the three cooling methods for PAFC systems,[62] which will now be described.

8.3.3.1. Water-Cooling System

Water cooling is the most popular cooling method, especially for large-capacity plants. Water cooling can be achieved by boiling-water cooling and pressurized-water cooling. In boiling-water cooling the heat generated in the cell is removed via the latent heat of vaporization. Since the average working temperature of the fuel cell is around 180–210°C, the temperature of the cooling water is about 160–180°C. Almost all heat generated in the fuel cell is removed by this cooling process. The temperature difference between inlet and outlet of cooling water is generally low. The volume of water for cooling can be kept low. All these factors cause lower power consumption. Uniform cell temperature distribution can be easily achieved and, in effect, leads to increased fuel cell efficiency.

	WATER COOLING	OIL COOLING	AIR COOLING
CONFIGURATION	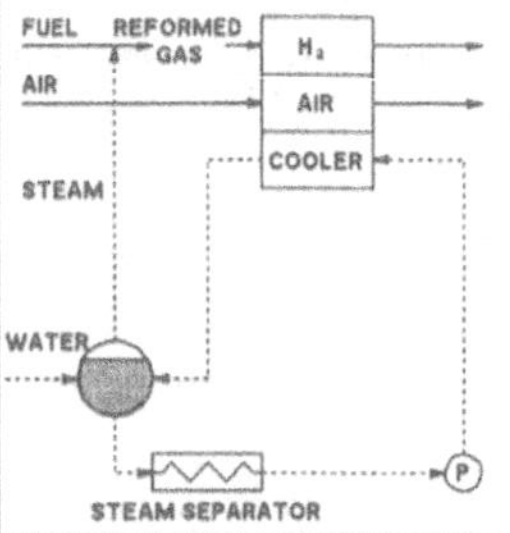	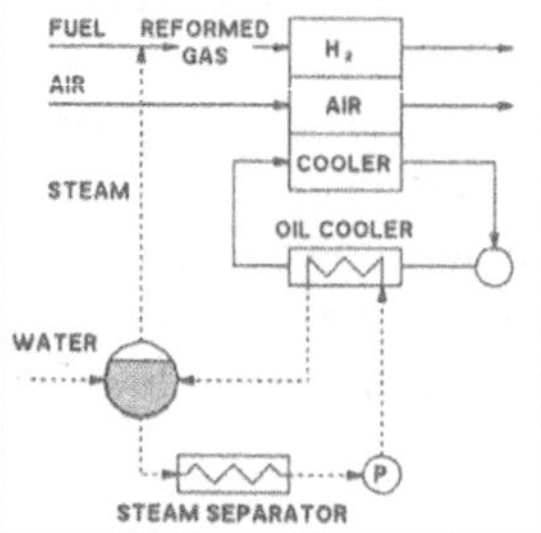	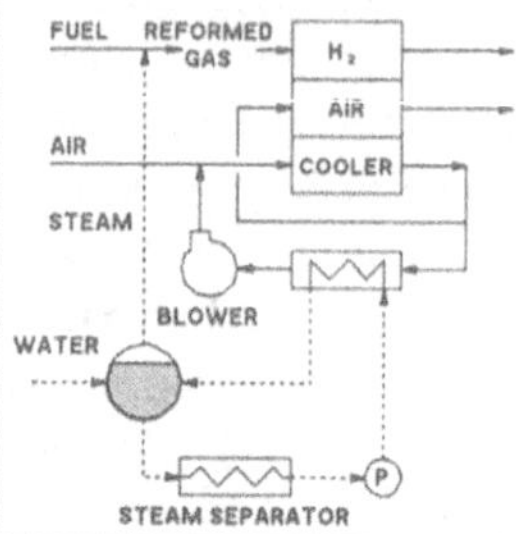
COOLING EFFICIENCY	EFFICIENT	MEDIUM	LESS EFFICIENT OR MEDIUM
SYSTEM STRUCTURE	COMPLEX	MEDIUM	SIMPLE
COOLANT TREATMENT	COMPLEX	MEDIUM	SIMPLE
WORKING PRESSURE	HIGH (SEVERAL atm)	LOW	FLEXIBLE
COGENERATION	EFFECTIVE	MEDIUM	LESS EFFECTIVE OR MEDIUM
AUXILIARY POWER	SMALL	MEDIUM OR SMALL	LARGE OR MEDIUM
COOLING PLATE	EVERY SEVERAL CELLS	EVERY SEVERAL CELLS	EACH CELL OR EVERY SEVERAL CELLS
MANUFACTURER	IFC, FUJI, HITACHI, TOSHIBA, MITSUBISHI	FUJI	ERC, WH, SANYO

Figure 8.8. Comparison of cooling systems.

In a pressurized-liquid-water-cooling system, heat is removed by the "heat capacity" of the cooling water. The temperature difference between inlet and outlet is higher than that of boiling-water cooling but still lower than the dielectric-liquid and air-cooling systems due to higher specific heat.

The insulated cooling pipes embedded in the graphite plates are usually inserted between cells, as shown in Figures 8.9.[19] Figures 8.10 and Fig. 8.11.[19] show

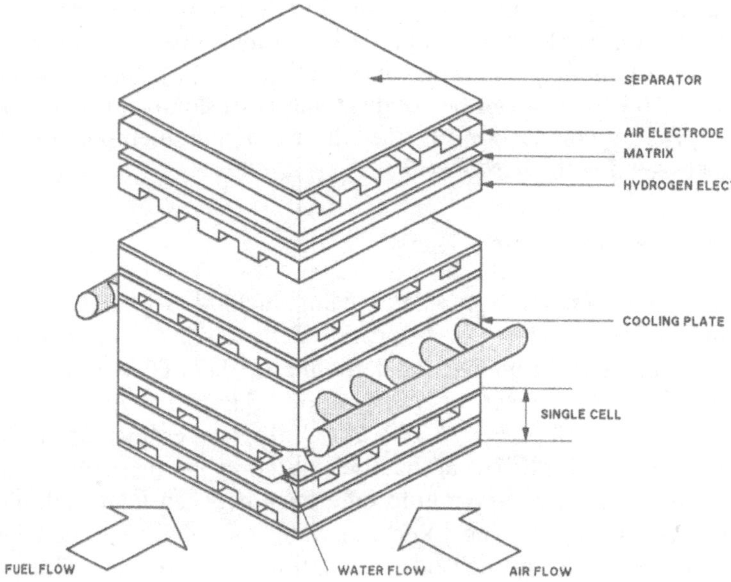

Figure 8.9. Model of PAFC stack configuration (water cooling system).[19]

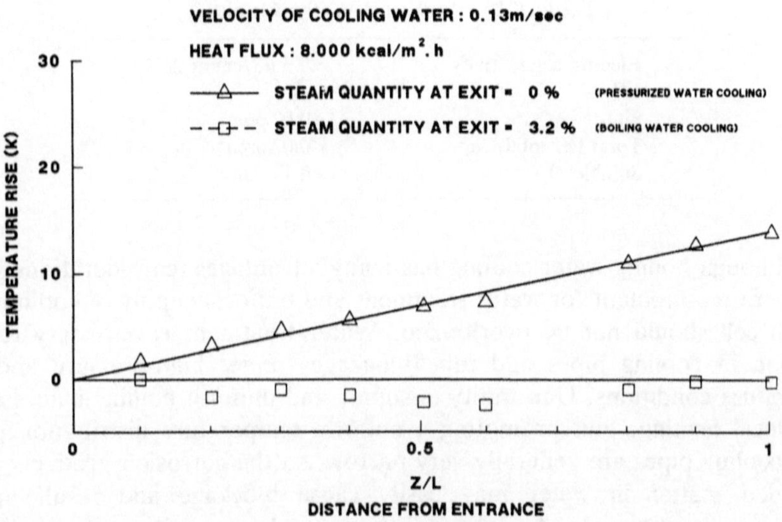

Figure 8.10. Temperature distribution along cooling pipes (horizontal direction) (an example).[19]

examples of the horizontal and vertical stack temperature distribution of boiling-water cooling and pressurized-water cooling. As shown in Fig. 8.10, the temperature distribution is proportionally increased to the heat transferred for pressurized-water cooling, but is almost constant for boiling-water cooling. Figure 8.11 shows an example of the cell temperature distribution along the stacking direction (vertical direction).

Cooling plates have been inserted in each group of five cells in this design. The middle cell layer generally exhibits the highest temperature, and the end cell layer the lowest. The difference is about 15°C in this example. Of course, the temperature difference can be reduced by reducing the number of cells between each cooling plate.

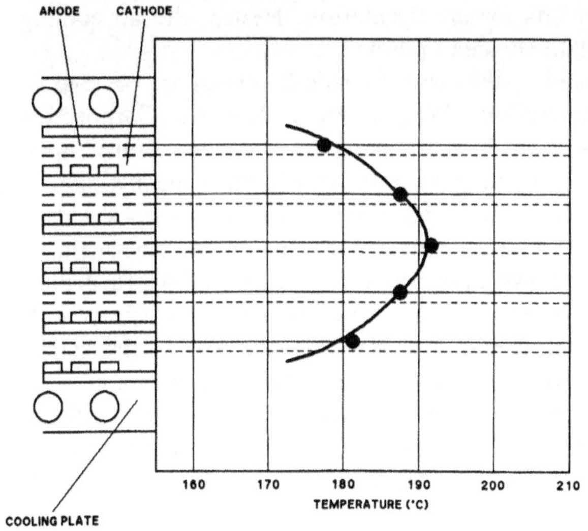

Figure 8.11. Temperature distribution along vertical direction (an example).[19]

Table 8.8. Required Water Quality[42]

Electric conductivity	<0.5 μS/cm at 25°C
PH	6–8
SiO_2	<0.02 ppm
Total Fe, total Cu	<0.02 ppm
Soluble O_2	<0.1 ppm

Although boiling-water cooling has many advantages, considerations such as the severe requirement for water treatment and uniform supply of cooling water to each cell should not be overlooked. Water treatment is necessary to avoid corrosion of cooling pipes and tube blockages under high-pressure and high-temperature conditions. Uniformity in supply and uniform boiling is necessary to avoid local heating, thus promoting a uniform temperature distribution profile. Metal cooling pipes are generally very narrow, so the corrosion products and all suspended matter in water may easily cause blockage and result in local overheating. An example of water quality required for cooling water is shown in Table 8.8, and is similar to boiler feedwater for modern thermal power stations.

Special design considerations are also required to provide a uniform temperature distribution along the vertical direction, as large-capacity stacks may span several meters in height.

The required installation of a water treatment system increases the capital cost and operating cost, including the maintenance. This is the reason why water-cooling systems can only be justified in relatively large installations, and for cogeneration plants.

8.3.3.2. Air-Cooling System

Air cooling removes all heat generated in the fuel cell by forced air. Generally speaking, the features of an air-cooling system are relatively simple, but the system is less efficient due to lower heat removal rate and large auxiliary power requirements for air circulation. Hence, the air-cooling system is usually prefered for relatively small plants.

An air-cooled system with its simple structure, however, can provide stable and reliable operation. Westinghouse Electric Corporation has adopted a pressurized-air-cooling PAFC plant for electric utility application, a unique application of air cooling to relatively large capacity plants. Unlike water or dielectric liquid cooling, a large-capacity air-cooled stack is designed by increasing the number of stacked small area cells, not by increasing the cell area.

Figure 8.12 shows an example of ERC's DIGAS (distributed gas) air-cooling cell structure.[26] The cell stack structure has a crossflow configuration, and the airstream is diverted into two types of channels in the stack: channels in each individual cell, with relatively small cross-sectional area; and channels in the cooling plates (approximately one for every five cells) with larger cross section. The channel area and distributions are calculated based on required process airflows, oxygen utilization, current density, and heat transfer requirements.[26] One advangage of this system is that the cooling portion of the air supply does not come into contact with the electrolyte and therefore does not contribute to

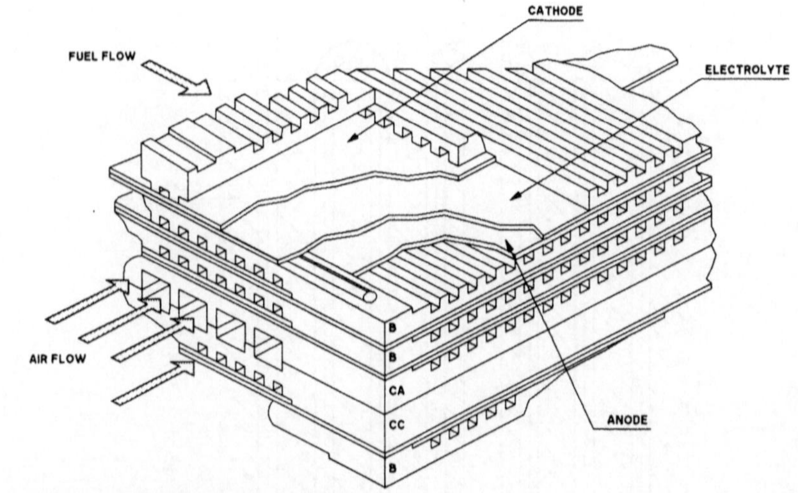

Figure 8.12. ERC–WE DIGAS structure.[26]

evaporation. Westinghouse Electric has developed the SGC (separated gas cooling) system for their pressurized-air-cooling stacks for electric utility applications.

8.3.3.3. Dielectric-Liquid-Cooling System

The dielectric-liquid-cooling system uses a single-phase dielectric liquid (synthetic oil) to remove heat from the stack and offers some unique advantages.

The basic development work was conducted at Engelhard Corporation, funded by DOE, primarily for small on-site units and forklift truck application. The advantages and disadvantages of a dielectric cooling system compared with water and air cooling are summarize in Table 8.9. The major advantages of dielectric liquid coolant compared with water cooling are its lower coolant

Table 8.9. Advantages and Disadvantages of Liquid-Cooling System

	Compared with water-cooling system	Compared with air–cooling system
Advantages	(1) Low coolant pressure Water cooling: 6–7 kg/cm^2g Liquid cooling: up to 1 kg/cm^2a (2) Less complex coolant treatment	(1) Better heat transfer performance (2) Low temperature difference between inlet and outlet (3) Higher average cell temperature (4) Low auxiliary power
Disadvantages	(1) Lower heat transfer performance (2) Bigger temperature difference between inlet and outlet (no boiling cooling is applied) (3) Disadvantage for cogeneration (4) Coolant leakage can cause problems	(1) Complicated system (2) Coolant leakage can cause problems

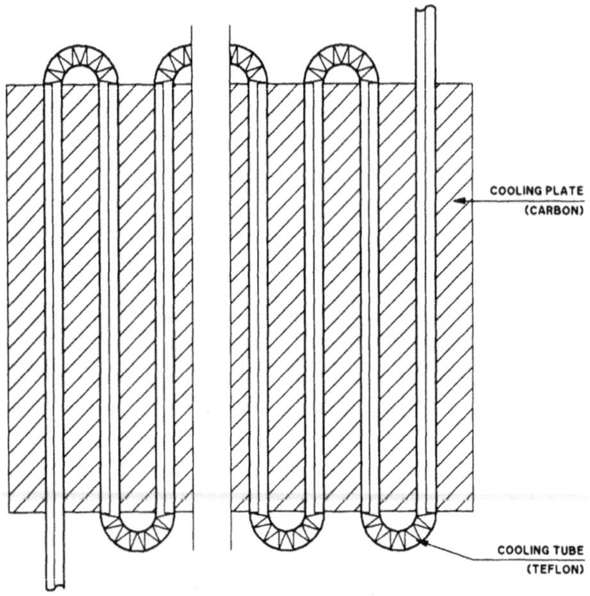

Figure 8.13. Model of dielectric liquid cooling tube.[27]

pressure and simple coolant treatment process, both of which are effective in reducing plant capital and operating costs. Cooling tubes, heat exchangers, and piping do not need to be made of special materials, since the operating pressure is low.

Figure 8.13 shows an example of a cooling plate in which corrugated PTFE cooling tubes are embedded.[27] The coolant circulates in intercell cooling plates, usually inserted at four- to six-cell intervals throughout the stack. Heat taken up by the coolant loop is rejected externally. Therminol 44 (a kind of synthetic oil) or high-temperature mineral oil is used as the dielectric coolant. Dielectric-liquid-cooling systems are suitable for relatively small compact units requiring simple design and easy operation.

In summary, a water-cooling system is most suitable for large units and cogeneration plants from an efficiency point of view, whereas air-cooling systems are superior for small units from the point of view of simplicity. Dielectric-liquid-cooling systems are also suitable for relatively small units such as vehicles, and on-site and other special applications where compactness is necessary.

8.3.4. Structure of Cell Stacks

8.3.4.1. General Description of Stack Structure

PAFC stacks are composed of the following key components:

1. Electrode (anode and cathode)
2. Matrix containing phosphoric acid
3. Separator
4. Cooling plate
5. Manifolds
6. Other small components

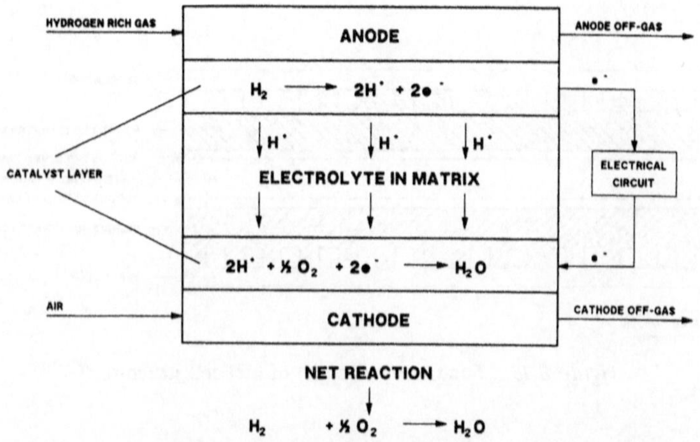

Figure 8.14. Schematic showing the principle of fuel cell operation.[66]

The basic fuel cell structure consists of an electrolyte (phosphoric acid) contained in the matrix sandwiched between two electrodes, the anode and cathode, as shown in Fig. 8.14.

The purpose of the matrix is not only to hold phosphoric acid as an integral part of the cell structure, but also serves to prevent crossover of reactant gases into the opposite electrode compartments.

Each electrode contains a catalyst layer in which the following electrochemical reactions take place:

At the anode:

$$H_2(g) \rightarrow 2H^+ + 2e^-.$$

At the cathode:

$$2H^+ + \tfrac{1}{2}O_2(g) + 2e^- \rightarrow H_2O(g).$$

Hydrogen ions are transferred from the anode to the cathode through the electrolyte, and the electrons from anode to cathode bia an external load circuit.

The net reaction is therefore

$$H_2(g) + \tfrac{1}{2}O_2(g) \rightarrow H_2O(g).$$

The matrix, containing the electrolyte, is ionically conductive but electrically nonconductive.

Typically, a single cell will generate 0.6–0.7 V at 200 mA/cm^2 current density (the output voltage depends very much on the working conditions as shown in Section 8.3.2). Fuel cells are connected in series to form a stack. With proper manifolding, hydrogen-rich gas and air are supplied to the stack. Figure 8.15 shows a fundamental model of a fuel cell structure. The electrochemical reactions take place in the porous catalyst layer attached to the porous substrate.

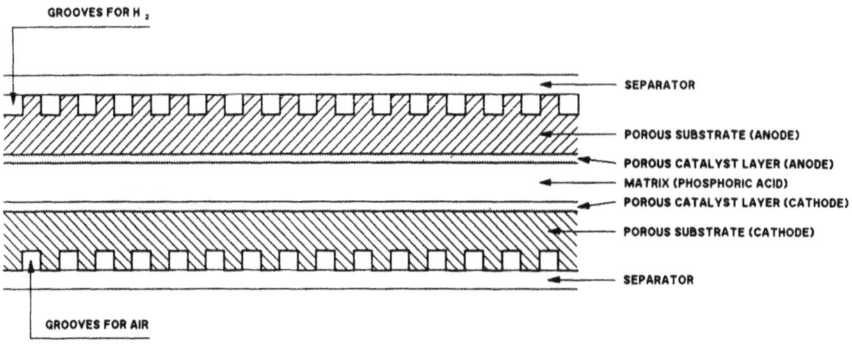

Figure 8.15. Fundamental model of fuel cell structure.[39]

Hydrogen-rich gas supplied to the grooves of the anode substrate is diffused through the porous substrate adjacent to the porous catalyst layer. The hydrogen ion H^+, ionized in the catalyst layer, is transferred to the cathode porous catalyst layer adjacent to the ribbed cathode substrate through the electrolyte. Oxygen gas diffuses to the cathode catalyst layer through the porous cathode substrate.

Reactions at the anode and cathode occur at a so-called three-phase zone in the catalyst layer (see Fig. 8.16). Three-phase means the liquid phase (phosphoric acid), solid phase (Pt catalyst), and gas phase (hydrogen or oxygen). In the porous catalyst layer numerous minute pores exist, and in these pores three-phase contacts are made among the electrolyte film, gaseous reactant, and catalyst. The reactant gases, hydrogen and oxygen, diffuse into the porous electrodes and react on the surface of the three-phase zone. At the same time, the electrolyte penetrates from the matrix into the catalyst layer and contact zone.

For this reaction to take place efficiently, the porous catalyst layer should be

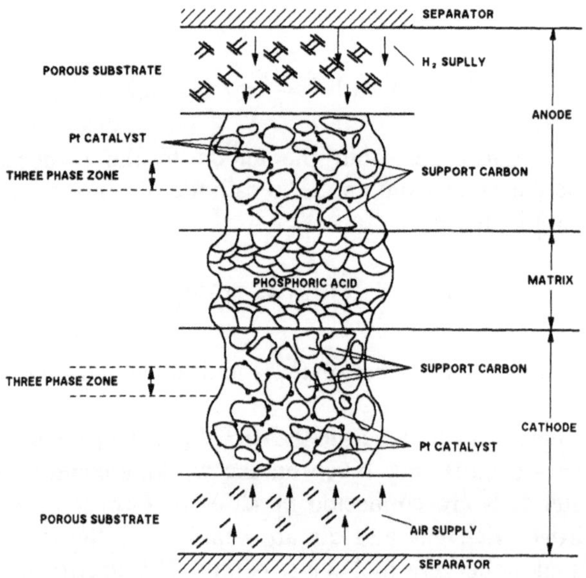

Figure 8.16. Model of three-phase zone in cell structure.[40]

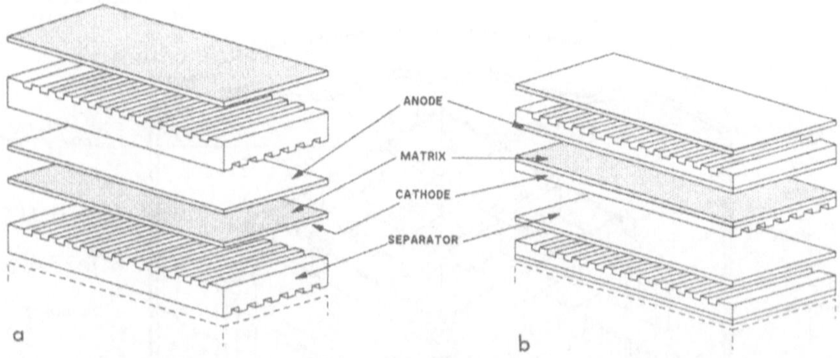

Figure 8.17. Typical cell structure:[40] (a) ribbed separate type; (b) ribbed substrate type.

fabricated to provide sufficient pores through which the gas diffuses freely and the acid properly penetrates, providing sufficient contact on the surface of the Pt catalyst. Penetration of gas and acid should be controlled in the three-phase zone within the catalyst layer, and the water produced in the three-phase zone of the cathode catalyst layer should be taken away by the process airflow through the porous substrate. Substantial research activities have been concentrating on how to make a cell structure in which the above two reactions and the water removal can be promoted effectively.

Figure 8.17 shows two typical cell structures adopting either a ribbed separator or a ribbed substrate. The ribbed separator structure is the structure whose separator has ribs grooved on its surface. The ribbed substrate type is the structure whose substrate has ribs into which the reactant gas is supplied.

The ribbed substrate type is generally used in larger stacks and has the following advantages compared to the ribbed separator type:

a. Flat surface contacts between catalyst layer and substrate promote better and uniform gas diffusion to the electrode. (In the ribbed separator type, the ribbed separator surface contacts the catalyst layer and the diffusion of reactant gases to the catalyst is not as uniform as the flat contacts.)
b. Phosphoric acid can be stored in the porous substrate.

Therefore many manufacturing companies in the United States and Japan are now adopting the ribbed substrate type, although other structures have been proposed by several organizations.[28]

The cell stack substructure is equipped with "manifolds" which are usually attached to the outside of the stacks. Typical manifolds for reactant gas, hydrogen-rich gas, and air, respectively, are shown in Fig. 8.18. (For more details see Section 8.3.4.2f).

The primary function of each inlet manifold is to supply the reactant gas uniformly to each cell. This is especially important for the supply of the hydrogen and CO_2 mixture, due to the difference in densities. The upper portion of the stack has the tendency to become hydrogen-rich. Therefore it is common practice to divide the whole stack into several blocks in order to supply the gas uniformly through each manifold attached to each block.

From a manufacturing point of view, sealing between the manifold and the

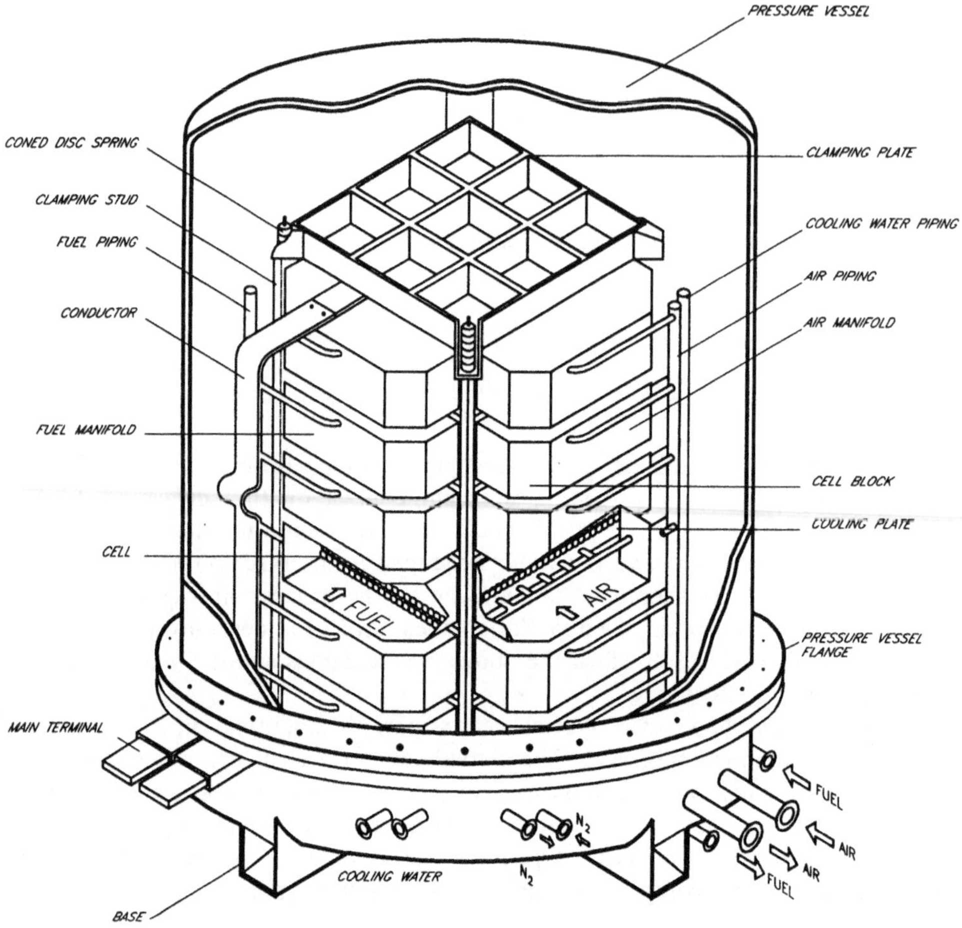

Figure 8.18. Cutaway view of fuel cell stack (an example).

side surface of the stack is no trivial task, since the side surface of the stack is composed of many stacked cell edges.

After stacking the necessary numbers of cell, clamping plates are placed on the top and bottom of the stack. Clamping studs are used to clamp the whole cell structure together.

If the fuel cell is to be operated under high-pressure conditions, the fuel cell assembly is placed in a pressure vessel filled with pressurized nitrogen gas (see Fig. 8.18). Since the internal pressure of the vessels is usually selected to be slightly higher than the working pressure of fuel cells, pressurized PAFC stacks do not require special pressurized manifold designs.

*8.3.4.2. Detailed Description of Each Key Component**

a. *Electrode and Catalyst.* The electrode is composed of the catalyst layer where the electrochemical reaction takes place and the substrate on which the thin catalyst layer is mechanically supported. The key components of the catalyst

*In this section the explanation is based on the ribbed-substrate-type structures.

layer are carbon-supported, highly dispersed platinum catalyst and hydrophobic agents such as PTFE (for example, Teflon). The thickness of the catalyst layer is about 0.1 mm.

Platinum in high-surface-area form is currently the preferred catalyst material, and carbon is preferred as supporting material (a noble metal catalyst is not necessary for higher temperature fuel cells such as MCFC, SOFC, etc., but for lower temperature fuel cells like PAFC and alkaline type such a catalyst is essential to promote electrochemical reaction).

As previously described, electrochemical reactions take place at the surface of the so-called three-phase zone in the electrode (see Fig. 8.16). To increase the current density the number of contact sites must be maximized, the partial pressure of reactant gases be kept high, and the diffusion pathway be minimized. The electric conductivity of the catalyst layer should also be high to reduce ohmic losses in the electrode. Furthermore the electrode hydrophobicity must be optimized to maximize gas diffusion and control electrolyte wettability. Flooding of acid disturbs the effective diffusion of reactant gas. Overdiffusion of reactant gases prevents the free access of acid to the contact area. Both effects are undesirable.

To prevent the flooding of acid, the PTFE-bonded carbon support is incorporated into the catalyst layer, utilizing the waterproofing characteristics of PTFE. To prevent overpenetration of reactant gases, the adjacent layer to the matrix in the porous electrode should be designed to have high capillary action. Electrolyte held tightly in the pores can protect the overpenetration of reactant gases.

Another important function of proper electrode structure is the ability to remove water produced at the cathode. At high working temperature, water can be removed by natural vaporization in the form of steam through the porous electrode and finally taken away with the process airflow. Although the vapor pressure of water at the working temperature is higher than that of the acid, some acid is generally entrained with the produced water vapor.

The activity of Pt catalyst depends on the kind of catalyst, crystallite size, and specific surface area. Smaller crystallite size and larger specific surface area lead to higher catalyst activity. Crystallite sizes of down to the order of 20 Å and specific surfaces areas of up to 100 m^2/g or more of platinum have been achieved.

The carbon support structure is another key component, the choice of which can affect the performance and life of electrodes. The main functions of the carbon support are to

a. Disperse the Pt catalyst
b. Provide numerous micropores in the electrode
c. Increase the electrical conductivity of the catalyst layer

A typical micrograph of carbon-supported Pt catalyst is shown in Fig. 8.19.

Currently two types of carbon black are used as support carbon: acetylene black and furnace black. Acetylene black has generally lower electrical conductivity, higher corrosion resistivity, and lower specific surface area than furnace black. This performance affects the initial and long-term performance of electrodes, so some additional treatments are required to realize the desirable characteristics for both support materials. For example, with acetylene black, a

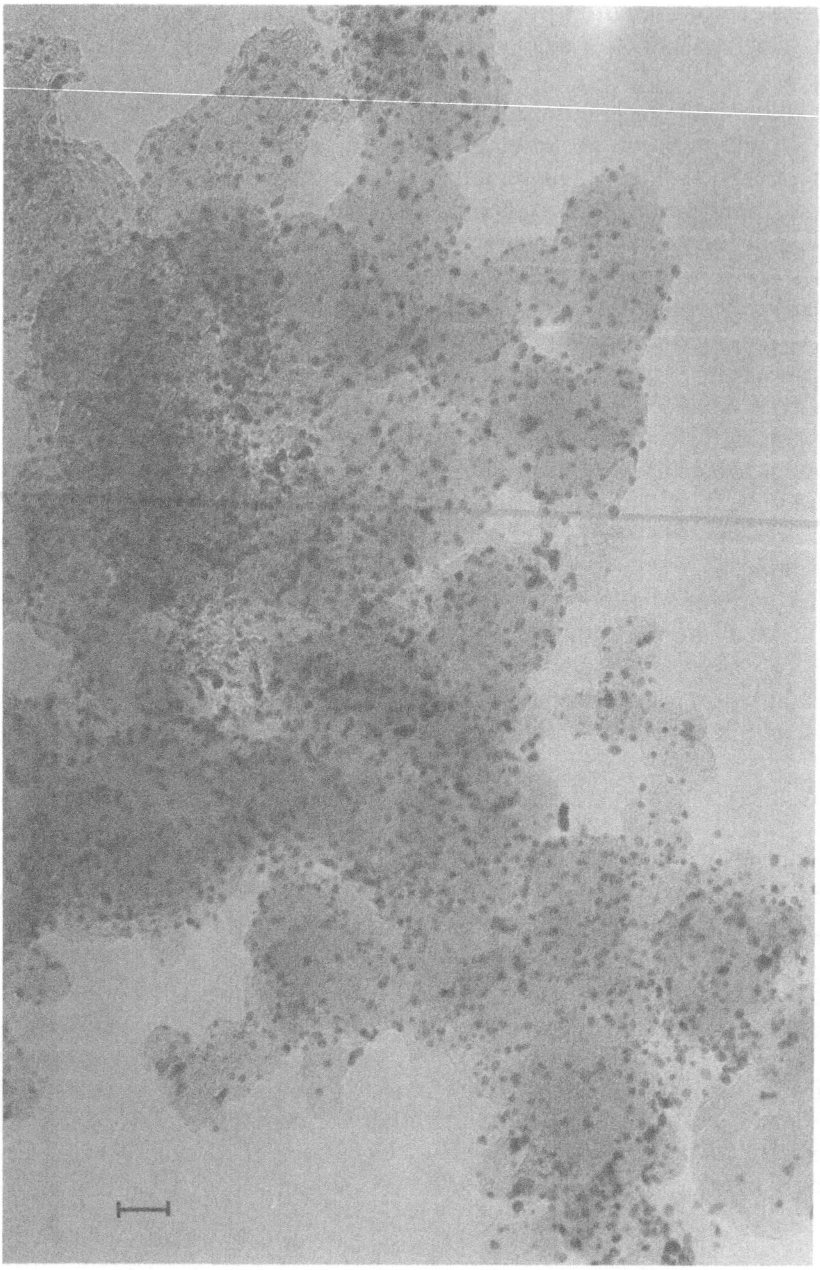

Figure 8.19. Microscopic photograph of carbon-supported Pt catalyst. (Scale bar = 2×10^{-6} cm.)

steam-activated process is implemented to increase the specific surface area; with furnace black, some heat treatment is required to increase its corrosion resistivity.

The following conditions are important to make high-performance electrodes.

a. To provide numerous and fine electrolyte networks in the electrode to minimize diffusion paths, allowing reactant gas easy access to the Pt catalyst

b. To produce sufficient networks of diffusion paths of reactant gases adjacent to the electrolyte network
c. To minimize the catalyst particle sizes and maximize mutual contact between support carbon particles

Electrode performance generally decays with operation. Performance decay is attributed mainly to the sintering of Pt catalyst and the obstruction of gas diffusion due to flooding of the catalyst layer. A more detailed explanation of this decay will be discussed later (see Section 8.3.5).

Pt alloy catalysts have been developed/or are under development to increase the catalyst activity and to minimize the sintering effect.

b. Substrate. The electrode substrate as a reinforced material adjacent to the catalyst layer permits passage of electrons as well as the reactant gases. From a performance point of view, the substrate should be.

a. Stable under working conditions and in phosphoric acid environment
b. Electronically and thermally conductive
c. Porous, through which the reactant gas can diffuse effectively
d. Mechanically strong under pressurized working conditions

At the working temperature, 100% phosphoric acid is very corrosive. Graphitized carbon material is usually recognized as the material of choice.

Substrates are made by molding from graphite fibers and inexpensive phenol resins, followed by baking with binder at very high temperature. Porosity and pore size are generally around 60–65% (porosity) and around 20–40 μm diameter.[58] In the case of a ribbed substrate structure, the surfaces of the molded ribs are finished by sanding or machining.

The porous substrate also has the ability to reserve acid. A ribbed substrate structure is preferred for this reason. The thickness of substrates is about 1 to 1.8 mm, which contains around 0.6–1.0 mm rib thickness for each electrode. The ribs are in contact with the flat separator. The catalyst layer is positioned on the other side of the substrate. The directions of ribs on anode and cathode substrates are set orthogonally for ease in manifolding.

In the ribbed separator configuration, on the contrary, the anode and cathode substrates are made from very thin carbon fiber paper. The ribs in this case are located on the separator and made of impermeable graphite materials.

A so-called integrated electrode substrate is one of the advanced configurations of cell structure, and basically consists of two substrates and a sandwiched separator (see Fig. 8.23). These three components are fabricated into a single structure by the components' supplier. Electrical and thermal resistance between layers (compared to the individually stacked structure) can be reduced, resulting in improved performance. Assembly of the integral structure is also greatly simplified.

c. Phosphoric Acid. Phosphoric acid is selected as the electrolyte for the following reasons:

a. Workability under high temperature
b. Tolerance to CO_2

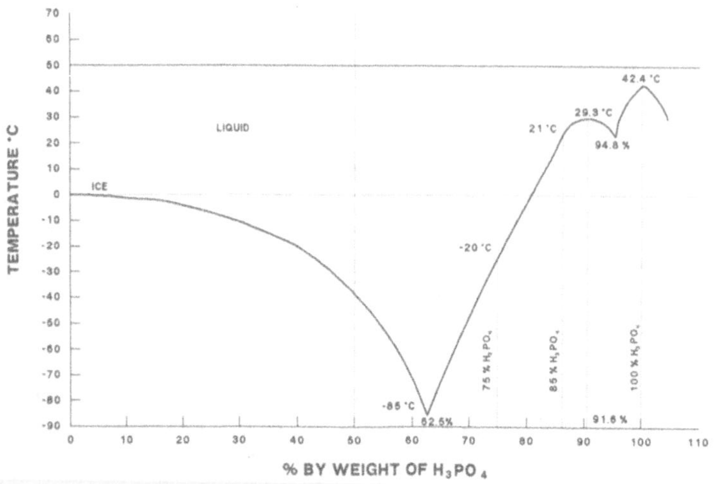

Figure 8.20. Solidifying temperatures of phosphoric acid.[35]

c. Low vapor pressure
d. High O_2 solubility
e. Good ionic conductivity at high temperature (no electronic conductivity)
f. Low corrosion rate at high-temperature conditions
g. Large contact angle (>90°)

Phosphoric acid is a colorless, viscous, and hygroscopic liquid. The acid is not used in a free liquid form in the PAFC cell structure but is contained in a porous matrix structure made of SiC. The matrix concept is adopted widely in PAFC stack design and is described in the next section. Generally, 100% phosphoric acid (H_3PO_4),* which has a high solidification point (so-called freezing point) of 42°C, is used in PAFC stacks. If the stack is left at ambient temperature, the phosphoric acid contained in the matrix will solidify, with consequent volume increase. A volume change of acid also occurs between on-load conditions and off-load conditions. Frequent volume changes can result in damage of electrodes and matrix, leading to decreasing cell performance. Therefore it is necessary to keep the stack at a temperature of at least 45°C, even under off-load conditions. This is a shortcoming of PAFC, and imposes market limitations on several applications.

The solidification point of phosphoric acid is very much dependent on the density of H_3PO_4, as shown in Fig. 8.20. Solidification points for concentrated phosphoric acid of around 100% weight are higher. For lower density H_3PO_4, solidification points decrease rapidly. For example, the solidification point of 62.5% (weight) H_3PO_4 is −85°C (75% H_3PO_4 has a −20°C solidification point, but for 85% H_3PO_4 it is 21°C). Lower density H_3PO_4 is generally used to avoid electrolyte solidification during transportation from factory to the site. Of course, it is necessary to replace the lower density acid with strong phosphoric acid before entering into operation. Once the stack is in operation, the stack must be kept

*One hundred percent phosphoric acid (H_3PO_4) contains 72.43% phosphoric anhydrid (P_2O_5); 100% H_3PO_4 has a density of 1.863 g/mL at 20°C, and the density increases as P_2O_5 content decreases.

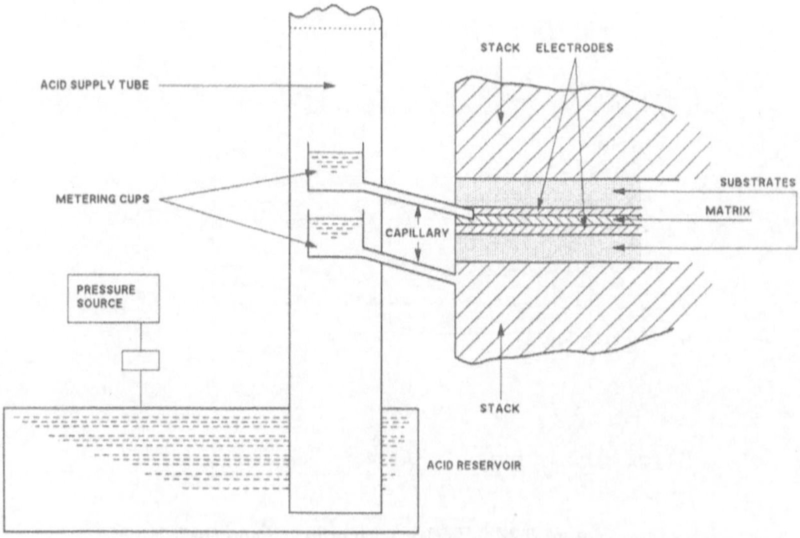

Figure 8.21. An example of acid replenishment equipment.[29]

warm continuously, including off-load conditions. For this purpose the stack must be equipped with an appropriate heating device.

Some synthetic acids have been developed for their promise to show performance superior to phosphoric acid, but they are still in the developmental stage.

Since some phosphoric acid is lost from the matrix during operation, appropriate acid replenishment systems or equipment is necessary.* The acid replenishment system feeds acid to each matrix uniformly at certain intervals under off-load conditions. Figure 8.21 shows an example of such replenishment equipment (U.S. patent 4463066).[29]

d. Matrix. The phosphoric acid is held in a so-called matrix. The function of the matrix is to hold the acid by capillary action. The matrix currently used is composed of fine silicon carbide (SiC) powder bound with a small amount of PTFE.

The matrix thickness should be kept as low as possible to minimize internal cell resistance, approximately 0.1–0.2 mm.

The basic requirements for the matrix containing electrolyte are

a. High capillary action to retain acid
b. To act as an electronic insulator
c. To prevent the crossover of reactant gases within the cell structure
d. High thermal conductivity
e. Chemical stability at high-temperature working conditions
f. Sufficient mechanical strength

*If the cell can retain sufficient acid which can cover the total amount of losses throughout the operation period, replenishment equipment is not necessary. Currently several companies are adopting such a design philosophy.

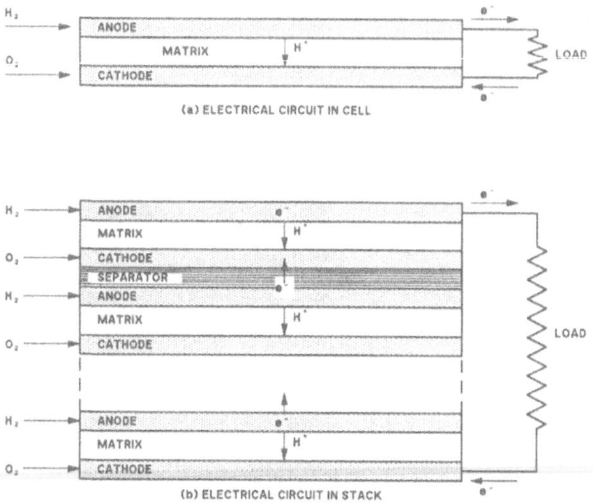

Figure 8.22. Electric current in cell and stack.

Current matrix structures can satisfy all these requirements except mechanical strength. The electric circuit is completed by the electron path (outer circuit) and the ion path in the matrix (see Fig. 8.22a, b). (Electrons which are necessary to the cathode within the stack are supplied from the adjacent anode through the conductive separator.)

The vapor pressure of phosphoric acid itself is very low, but the acid loss due to evaporation cannot be neglected during long operation periods at high-temperature working conditions. The amount of electrolyte dissipation depends on the reactant gas velocity in the cell and also on the current density; that is, acid loss rate increases at higher gas velocity and higher current density. (Higher current density results in the generation of a larger amount of water and higher steam pressures. The acid also scatters in the form of mist via entrainment with steam.)

Since too much loss of acid in the matrix causes the crossover of reactant gases through the matrix and results in decreased performance, acid feed to the matrix is necessary. One countermeasure is to reserve a sufficient amount of acid in the cell structure at the beginning, and the other is to replenish acid to the stack externally. Usually a combination of these methods is necessary. Another possible way is to use the pores in the substrate as an electrolyte reservoir.

The reserve of excess electrolyte in the porous substrate is also useful to expand space for acid–volume change during load variation. For this purpose the pore size of the silicon carbide matrix should be designed to be small compared to that of the substrate. This ensures the matrix will always be correctly wetted by the electrolyte, thus reducing the possibility of crossover of reactant gases under conditions of differential pressure.

According to an EPRI report,[30] ribbed substrate cells with 40% of acid could accommodate electrolyte evaporation over 40,000 h of operation with throughput at 8.2 bar pressure and 205°C temperature. In an atmospheric pressure on-site system, the lower operating temperature (190°C) compensates, at least partially, for the effects of the higher gas–volume throughput on the evaporation rate.

Finally some explanation on the prevention of mixing of the two reactant gases may be necessary. As shown in Fig. 8.15 the matrix is sandwiched between the anode and cathode. Therefore, the matrix must have the role of preventing the crossover of gases supplied to the anode and cathode under any transient operation conditions. Under some transient conditions, however, the pressure difference between the two electrodes can swing to several hundreds millimetres aqueous. The matrix should be able to endure such pressure difference. Sufficient bubble pressure of the matrix is required to reduce normal and transient operating conditions. The SiC-made matrix presently used is not sufficiently strong from this point of view, but will endure at least about 1000 mm Aq. Each PAFC plant should be designed to suppress the pressure differences between the two electrode components to less than 1000 mm aqueous even under any transient condition.

e. *Separators.* The purpose of separators is to prevent mixing of hydrogen-rich gas and air in the cell structure and to connect the two electrodes electrically (Fig. 8.15). Leakage would of course lead to performance losses and hazardous situations.

Key requirements for the separators are summarized as follows:

a. Sufficient nonpermeability to prevent mixing of the two reactant gases
b. Chemical stability at high temperature, high pressure, and in phosphoric acid environment
c. High electric and thermal conductivity
d. Sufficient mechanical strength

Thin glassy carbon plate (or vitreous carbon, polymer carbon) is generally adopted. The thickness should be as thin as possible to reduce electric and thermal resistance, typically less than 1 mm. The surface should be flat and smooth to ensure uniform contact with other cell components.

Glassy carbon, manufactured through carbonization of a thermosetting resin such as phenol resin and/or epoxy resin at 1000–2000°C, has high strength and low permeability. Glassy carbon had been commercialized more than 10 years ago but could not find an appropriate market, except for application as material

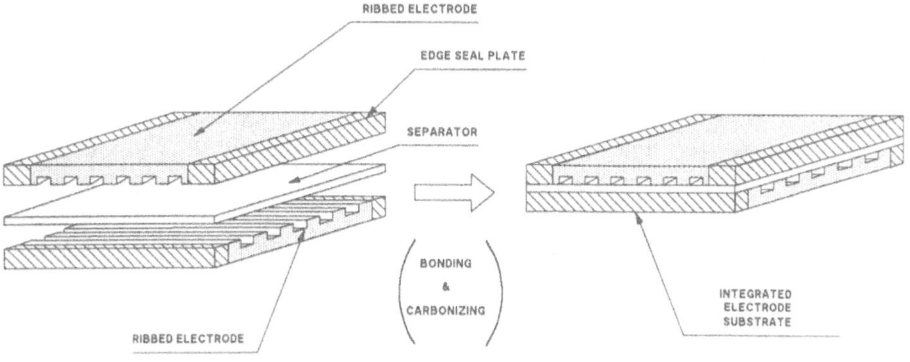

Figure 8.23. Integrated electrode substrate.[31]

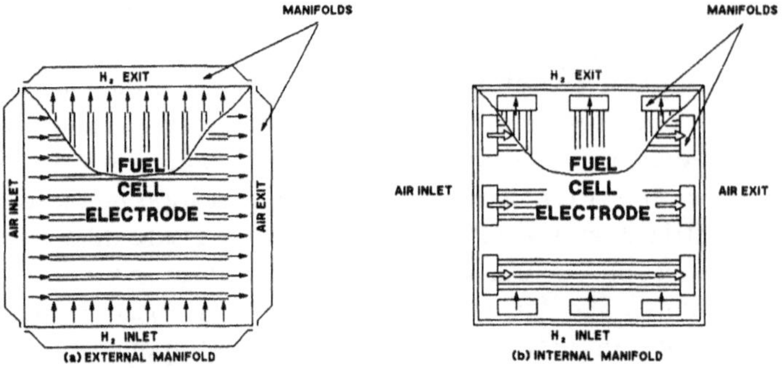

Figure 8.24. Conceptual drawings of external and internal manifolds.

for melting pots and guard tubes of thermocouples. Therefore, PAFC separators could be the new market for glassy carbon.

f. Gas-Supplying Structure (Manifold). The appropriate gas-supplying structures must be equipped to supply two reactant gases to fuel cell stacks.

There are two types of gas-supplying equipment for fuel cells: internal and external manifold structures. In the internal manifold system, the gas-supplying manifold is incorporated in the cell component in the form of holes through the stack itself, perpendicular to the cell plane. In external manifolds, the more popular, the manifold box is usually attached to the sides of the stack.

Figure 8.24 shows conceptual drawings of external and internal manifolds. (Also Fig. 8.18 shows an example of an external manifold.) There is no special material requirement for the external manifold structure other than acid resistance.

The design criteria of manifolds are as follows:

a. The pressure drop through the manifold should be kept as low as possible.
b. The material should be electrically insulating and be chemically and mechanically stable.
c. Sealing through manifolds should be kept tight under any working conditions including transients, and therefore the manifold should also have a low, matching thermal expansion coefficient.

8.3.5. *Cell Life*

8.3.5.1. *End of Life*

Cell life is not yet understood completely, but the lifetime of a fuel cell has been defined as "the time of operation after which the output voltage at rated current has decreased by 10% of the initial voltage."[62] The initial voltage is recognized to be the output voltage after about 100 h of trial operation. (Generally it takes at least 100 h of operation to reach a steady-state initial condition.)

Cell life is estimated to be about 40,000 h, which is equivalent to about five years of continuous operation. So cell performance is projected to decay from, for example, 0.7 V, 200 mA/cm^2 at the initial stage to about 0.63 V, 200 mA/cm^2

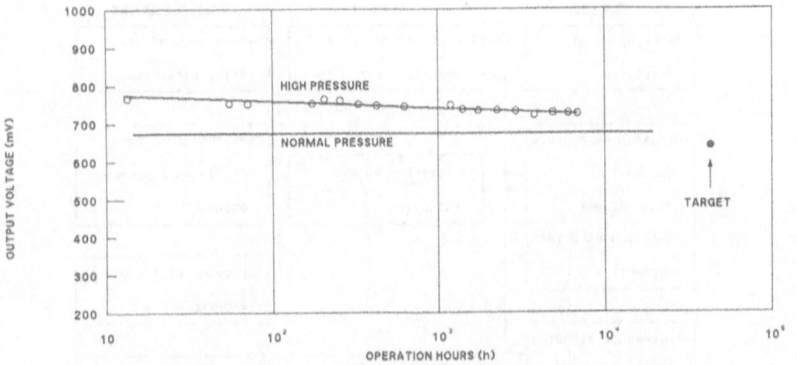

	TEMPERATURE	PRESSURE	CURRENT DENSITY
HIGH PRESSURE	190 °C	4kg/cm²g	200mA/cm²
NORMAL PRESSURE	190 °C	0kg/cm²g	160mA/cm²

Figure 8.25. An example of output voltage for various operation hours.[23]

after 40,000 h operation. Figure 8.25[23] shows an example of output voltage for various operating periods. Cell life depends very much on operating conditions, such as working temperatures, voltages, pressures, and modes of operation, such as start-up and shutdown conditions of the plant. These will be described in more detail later.

It is generally accepted that performance decay of a fuel cell is the result of sintering of Pt catalyst particles, corrosion of support carbon, and progressing of electrolyte flooding. The relation between these factors and steps to minimize the decay are shown in Fig. 8.27.[45]

8.3.5.2. Agglomeration of Platinum and Corrosion of Support Carbon

The sintering of Pt particles results in a decrease in active surface area of catalyst and leads to decreased cell performance. Models for sintering and corrosion are summarized in Fig. 8.26.

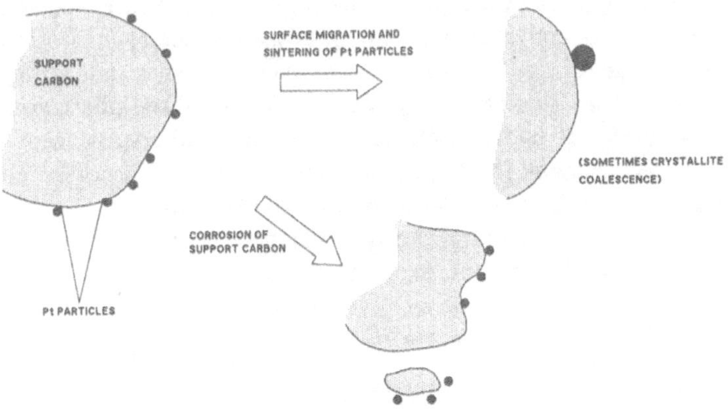

Figure 8.26. Models of Pt sintering and support carbon corrosion.

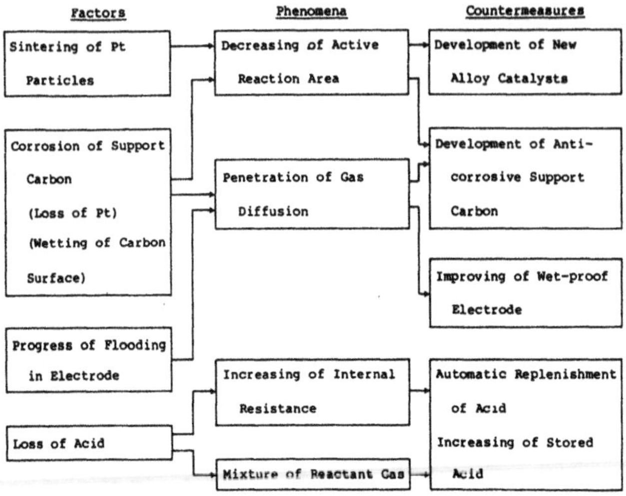

Figure 8.27. Performance decay of cell.

During operation, Pt particles have the tendency to migrate on the surface of the support carbon and to coalescence (agglomerate) into larger size particles, thus decreasing the active surface area. The rate of sintering is proportional to the logarithm of time and depends mainly on the working temperature. At temperatures of about 150°C surface migration of Pt particles on support carbons begins to occur, and at higher temperature in the working range it appears that crystallite coalescence is predominant.[32]

The corrosion of support carbon will also cause a loss of Pt particles and accelerate wetting on the carbon surface. These phenomena not only decrease the active area of catalyst, but also prevent gas diffusion to the catalyst layer. It is believed that the corrosion rate of the carbon support depends on voltage and working temperature. The reaction rates accelerate at higher voltage and higher temperature conditions. The corrosion rate also is dependent on the type of carbon used (see Section 8.3.4.2a).

8.3.5.3. Electrode Flooding

As mentioned, the hydrophobic electrode is manufactured by adding appropriate amounts of waterproof material, e.g., PTFE powder, to the carbon-supported Pt catalyst powder. Well-designed electrodes possess sufficient hydrophobic characteristics. However, these characteristics deteriorate gradually over time. The progressive electrolyte flooding of the three-phase zone is mainly caused by the decay of hydrophobicity and the gradual corrosion of the carbon support. Countermeasures for these phenomena are to incorporate noncorrosive support and improved flooding characteristics of electrodes.

The various factors affecting the performance decay, their phenomena, and their countermeasures are summarized in Fig. 8.27.[33]

8.3.5.4. Requirements for Operation

To maximize cell life, the following considerations are important:

1. The cell voltage should never exceed 0.8 V. This implies that the cell should not be left in an open-circuit condition at 180–190°C or above.

Experimental data show that the deterioration rate measured at 180°C with an open-circuit voltage of up to 1.0 V is eight times that tested at <180°C and 100 mA/cm^2 load condition.[23] In case of plant shutdown, it is recommended to decrease the stack voltage by appropriate methods. Reactant gases are generally purged with nitrogen at shutdown as a precaution. Short-circuiting is one possible way to reduce the terminal voltage. During start-up, a similar procedure should be adopted to extend the cell life. Decay rates are generally recognized to be higher for cathodes than anodes.

2. Severe transient conditions should be avoided, including rapid start–stop operations. Rapid shutdown of plants causes disturbance of reactant gas flow and may result in excessive pressure difference between electrodes. This effects acid distribution at the three-phase zones. Such disturbances may lead to electrode flooding or electrode starvation (of acid). A proper control system is necessary to suppress this undesirable effect.
3. Proper supply of reactant gases to cells must be ensured. Insufficient supply of gases will produce deviations in the output voltage, leading to deterioration of cell life. Sufficient amounts of gases should be supplied to cell stacks even under transient conditions. As the reformed gas takes at least several seconds (to several minutes even for larger units) from the reformer to the stack, it is generally difficult to supply sufficient gas instantaneously from the reformer at instant power demand. Appropriate design is necessary.

Extension of cell lifetime remains an area of intense R&D, since it directly affects plant economics.

8.3.6. *Cell Stack Design Configuration*

Key items necessary for proper design of PAFC stacks are

Capacity (kW)
Rated voltage and current
Type of cell configuration (ribbed substrate or ribbed separator)
Cell size (effective)
Number of cells
Working conditions (temperature and pressure)
Cooling method
Utilization factor of reactant gases

8.3.6.1. *Cell Size*

The cell voltage is determined by the requirement of fuel cell efficiency (see Section 8.3.1.3). The current density of a cell is determined corresponding to the cell voltage from the V–I characteristics curve, and the cell size is fixed by the rated current of the stack, which is selected from the plant capacity. Hence, the rated voltage and number of cells required in the stack are known.

In 1000-kW PAFC dispersed-type plants developed within the framework of the national project in Japan in 1985, the rated performance of 0.7 V and

Table 8.10. Key Specifications of Stacks for 1000-kW PAFC Plant[49] (Moonlight Project)

	Dispersed type	Central generation type[a]
Contractor	Fuji, Mitsubishi	Hitachi, Toshiba
Dc output	260 kW	260 kW
Cell configuration	Ribbed substrate (Fuji) Ribbed separator (Mitsubishi)	Ribbed substrate
Operating pressure	4 kg/cm^{2}g	6 kg/cm^{2}g
Operating temperature	190°C	205°C
V–I characteristics	200 mA/cm^2, 0.70 V	220 mA/cm^2, 0.72 V
Pt loading	6.5 g/kW	6.5 g/kW
Cooling method	Water cooling	Water cooling
Cell size	3600-cm^2 class	3600-cm^2 class

200 mA/cm^2 at 190°C, 4 kg/cm^{2}g, had been achieved through basic research activities of manufacturing companies in Japan. Four sets of 260-kW stacks were selected, and a 3600-cm^2 class cell size was offered by each manufacturing company, as shown in Table 8.10.

In the other examples of the 4.5-MW and 11-MW plants, 3440-cm^2 and 9300-cm^2 cell sizes were offered by IFC as shown in Table 8.3.

Fuji Electric (Japan) designs 2000-cm^2 cell size for 50-kW and 100-kW stacks for on-site plants and for the DOE's bus applications, which have 25- and 50-kW capacity (see Table 8.11.)

Each manufacturer currently develops its own standard series of cell sizes and applies a corresponding cell size to the desirable stack capacity. It is obviously not cost-effective to prepare a new cell size to meet every capacity.

As shown from these examples, however, cell size can be 1000–10,000 cm^2, depending on stack capacity.

Table 8.11. Examples of Cell Sizes

Manufacturer	Plant capacity (kW)	Stack capacity (kW)	Number of stacks	Cell size (effective) (cm^2)
IFC	11000	670	18	9300
Fuji	5000	860	6	8000
IFC	4500	240	20	
Fuji, Mitsubishi	1000	260	4	3440
Fuji Electric	50, 100	50, 100	1	3600-cm^2 class
Fuji Electric	25[a], 50[a]	25[a], 50[a]	1	2000
IFC	40	40	1	2000
				2025[b]

[a]Bus application.[65]
[b]Apparent size.

8.3.6.2. Working Pressure

As mentioned, higher working pressure yields better cell performance and higher efficiency. This advantage is partially offset by the increase in power requirement (compressor). Furthermore the pressurized system is much more complicated than a nonpressurized system, especially from the operation and control points of view.

As a guideline, for large plants of more than 1-MW, a pressurized system has higher overall efficiency than the nonpressurized system. For those systems, an increase in efficiency at higher pressure cell generally more than offsets the increase in energy consumption for the compressor.

8.3.6.3. Working Temperature

Elevated working temperature increases the cell efficiency and, unless properly designed, may cause severe corrosion problems of components, which then shorten cell life. The maximum temperature currently adopted is 220°C.

The number of cells between cooling plates should be determined from the maximum allowable temperature of the cell and the cooling-water temperature.

Increasing the cell numbers between cooling plates decreases the total stack heights, which is preferable for reducing stack cost. Four to eight cells between cooling plates are generally adopted for water-cooled stacks.

8.3.6.4. Stack Configuration

The process of stack assembly usually consists of stacking the unit cells, inserting cooling plates, assembling current-collecting plates, insulating clamping plates at the top and bottom of the assembly, and clamping the total assembly with clamping bolts (see Fig. 8.18).

The clamping pressure must be appropriate to minimize the electrical and thermal resistances between layers within the limit of material strength. Figure 8.28 shows an example of how internal resistance decreases by increasing surface pressure tested with a subscale stacked model.[41] As shown in this example, decreasing the internal resistance has its limit.

There are two types of cell assemblies. One is to assemble all unit cells, cooling plates, manifolds and other components in one unit. The other is the "block assembly," in which the total number of cells is divided into several groups; and each group is equipped with its own manifold and endplates. One cell block is equivalent to an independent ministack or substack (Fig. 8.18), but the full stack assembly is taller (and heavier) than a one-unit stack structure. Block assembly, however, has the following advantages compared to single-stack assembly:

a. Each block has its own manifold, and hence the reformed gas is uniformly distributed to all cells easily.
b. If one cell is found defective, the block can easily be replaced by a new one.

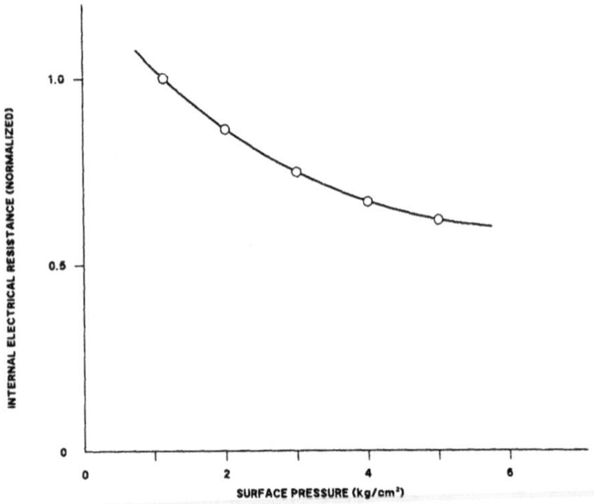

Figure 8.28. Relation between surface pressure and electrical resistance (an example).[41]

c. Confirmation and testing of cell performance can be done block by block; this is easier and simpler. In addition, the capacity of a testing stand is sufficiently small compared to a full-stack testing stand, and the same stand can be used for blocks for each stack capacity, which is evidently cheaper.

8.4. KEY COMPONENTS OF PAFC SYSTEMS

There are four key components in PAFC systems: the fuel processing system (reformer), the fuel cell stack(s), the inverter, and the control system. These key components are only briefly reviewed with respect to the PAFC system. More general treatments can be found in Chapter 4 on reforming and Chapter 6 on system design.*

8.4.1. Fuel Processing System (Reforming System)

Steam reforming is an established technology for chemical industry applications. Fuel cell reformers differ from conventional ones since

a. The load characteristics are different; that is, the load for a chemical plant is very stable, and the reformer is usually operated at constant load. On the other hand, a fuel cell plant reformer operates under variable load. Dynamic responses are critical considerations for fuel cell reformers.
b. Compactness, which is important for fuel cell plants, is less of a consideration for many chemical plants, especially those of large capacity. In particular, for package-type PAFC plants, the size of the reformer is critical.

*Besides the four key components, the air supply system should be described briefly. In the pressurized system the air is supplied by the compressor, driven by the turbine or the motor; in the atmospheric pressure system the air is supplied by the blower.

Table 8.12. Operating Condition of Each Process in Fuel Processing System

	Desulfurizing process	Steam-reforming process	CO-shift process
Role	Desulfurization	Reforming of natural gas to H_2 and CO	Converting CO to hydrogen-rich gas and CO_2
Reaction formula	$R{-}SH + H_2 \rightarrow R{-}H + H_2S$ R: Hydrocarbon $H_2S + ZnO \rightarrow ZnS + H_2O$	$CH_4 + H_2O \rightarrow CO + 3H_2$	$CO + H_2O \rightarrow CO + H_2$
Operating condition (PAFC system)	Temperature: 200–400°C Pressure: 0–10 kg/cm^2	Temperature: 750–850°C Pressure: 0–10 kg/cm^2 Steam–carbon ratio: 2–4	Temperature: High temperature: 320–480°C Low temperature: 180–280°C Pressure: 0–10 kg/cm^2
Catalysts	Co–Mo catalysts, or Ni–Mo catalysts, ZnO	Ni catalysts	Fe–Cr catalysts, Cu–Zn catalysts

The fuel processing system (or the reforming system) of natural-gas-fueled PAFC plants usually comprises three processes:

a. Desulfurization
b. Steam reforming
c. CO shift

The role, reaction formula, operating condition, and catalysts of these processes are shown in Table 8.12.

Figure 8.29 and Fig. 8.30 show the composition of a fuel processing system and the cross section of a 1-MW steam reformer.[22] Figure 8.31 shows the flow diagram of a 1-MW fuel processing system, which has been constructed and

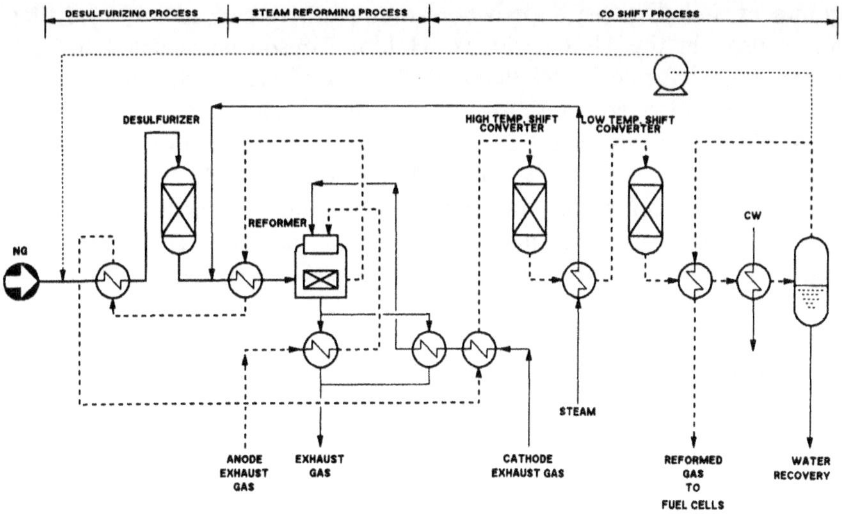

Figure 8.29. Composition of fuel processing system (Japanese 1000-kW PAFC plant).[22]

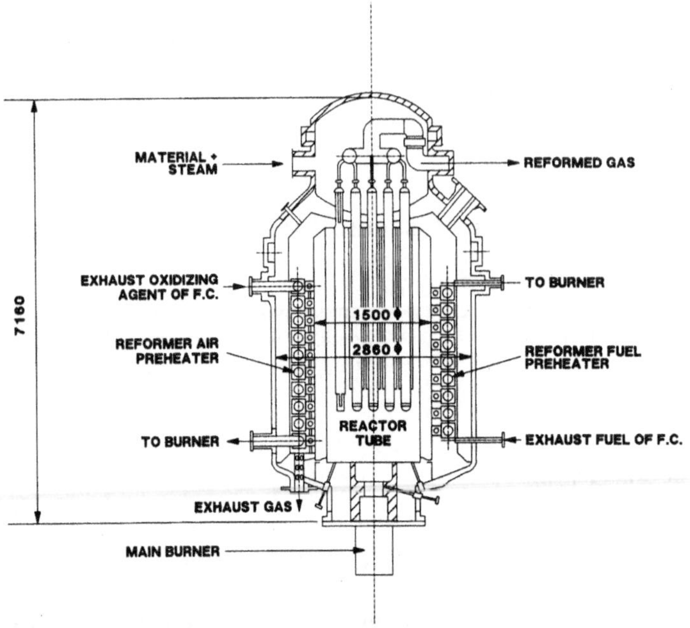

Figure 8.30. Cross section of reformer (Japanese 1000-kW PAFC plant).[42]

operated under the Moonlight Project in Japan.[42] The key specification of this 1-MW reformer is shown in Table 8.13.[22]

Demonstration operation of this 1-MW plant has been carried out almost satisfactorily, but some dynamic performance characteristics, such as short start-up time and quick load response, turned out to require further improvement. This implies that the reformer should be improved to realize a short start-up (4 h) and quick load response (25–100%/min).[59] The latter have been demonstrated in subsequent simulated operation of the 1-MW plant.

The most difficult design criterion, as envisaged from the foregoing experience, is the improvement in dynamic response of the reformer itself. Development and demonstration were among others carried out by EPRI and Haldor Topsoe in the United States. Haldor Topsoe constructed a 1.25-MW "heat exchange reformer" and demonstrated satisfactorily the dynamic characteristics inclusive of rapid transients and start-up shutdown.[20,21]

In Table 8.12 the resultant overall reaction of the fuel processing system is $CH_4 + 2H_2O \rightarrow 4H_2 + CO_2$. Theoretically, about 80% H_2 and 20% CO_2 are supplied to the fuel cell stack, but, in practice, residual methane and CO, a small amount of excess steam, and small impurities are included. The allowable limits of impurities specified under current practice are shown in Table 8.14.

8.4.2. *Inverter*

The role of the inverter is to convert dc power to ac power. Since the inverter itself represents a well-established electrical technology, a detailed description will not be included here. Only its working principle and its special features for fuel cell plants will be briefly discussed.

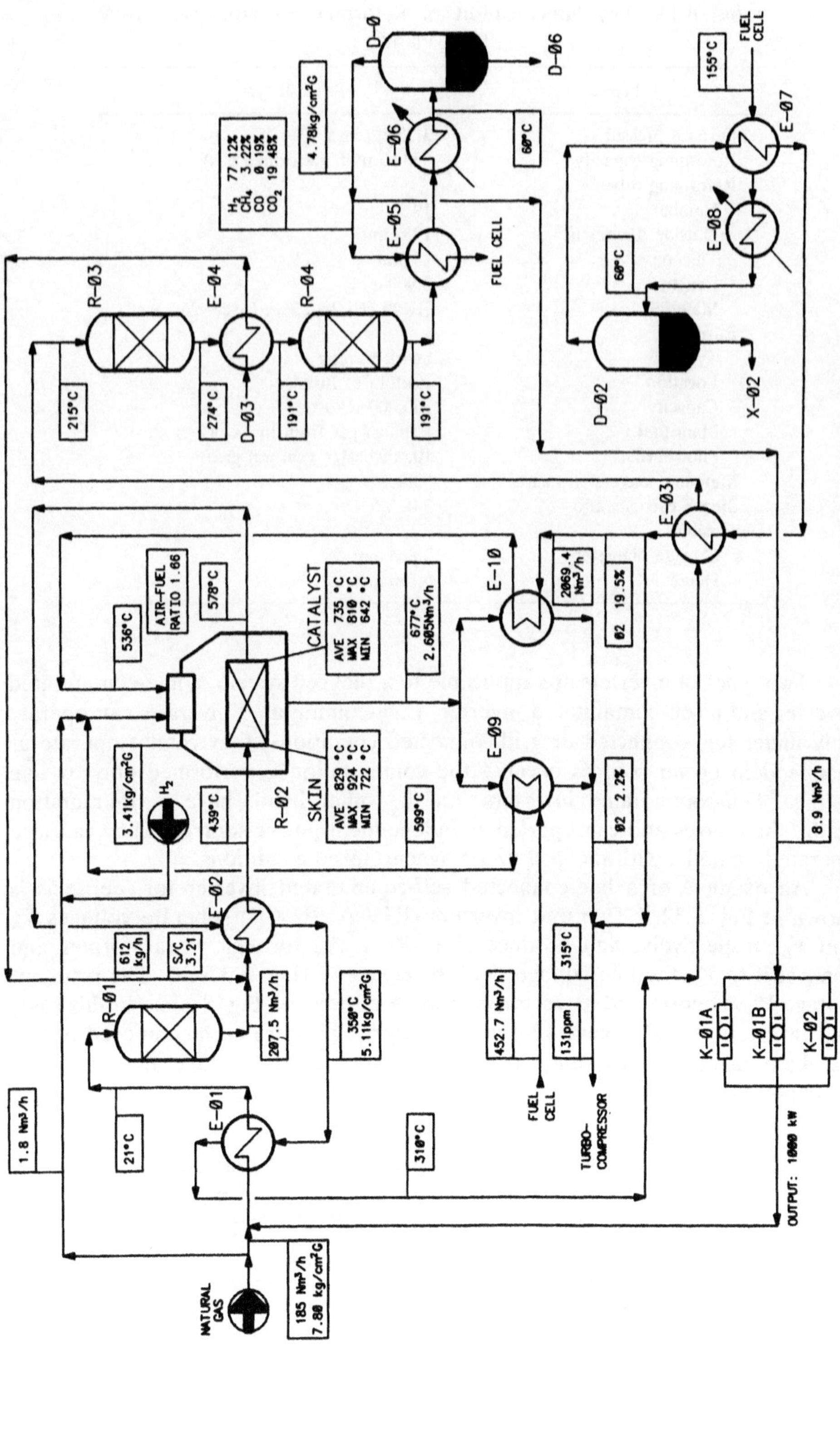

Figure 8.31. Flow diagram of fuel processing system for 1000-kW PAFC plant (test results).[42]

Table 8.13. Key Specification of Reformer for Japanese 1-MW PAFC Plant[22]

Type	Vertical type
Furnace pressure	3.35 kg/cm^{2}a
Reforming capacity	650 N·m^3/h (reformed gas)
Reforming tube	
Number	19
Outside diameters	133 mm
Thickness	14 mm
Length	3,900 mm
Material	HK40 (Ni 20–23%, Cr 23–26%)
Burner	
Type	Swirl burning
Location	Bottom of furnace
Capacity	700,000 kcal/h
Main fuel	Exhaust gas from fuel cell
Pilot burner	10,000 kcal/h (natural gas)
Methane conversion factor	92.4%
Steam–carbon ratio	3.0–3.5
Dimension	
Outside diameter	2,884 mm
Height	6,860 mm

Two types of inverters are applicable to a fuel cell system: a line-commutated inverter and a self-commutated inverter. Line-commutated inverters can operate only under line-connected or grid-connected conditions; they cannot operate as independent power sources because the commutation is performed only by line voltage. Self-commutated inverters, on the other hand, have a commutation ability themselves and can operate as independent power sources. They can also operate in parallel with the grid under synchronized conditions.

An example of a line-connected self-commutated inverter for fuel cells is shown in Fig. 8.32.[43] Two unit inverters (INV-A, INV-B) generate voltages V_A and V_B, respectively, and produce V_i $(=V_A + V_B)$ through a transformer and connect it to V_S, the line voltage, with reactance X (Fig. 8.32a,b). The resultant voltage V_i waveform and its vector diagram are shown in Fig. 8.32c. (In this case, unit inverters A and B have 12-phase inverters which are each composed of two six-phase inverter units, and thus have better output voltage than a single six-phase unit inverter.)

Table 8.14. Allowable Limits of Impurities in Reformed Gas

Impurities	Limiting amounts
H_2O	10–20%
CO	<1.0%
Fe, Cu, metal irons	Nil
H_2S, COS	<1 ppm
C_2^+	<100 ppm
Cl^-	<1 ppm
NH_3	<1 ppm

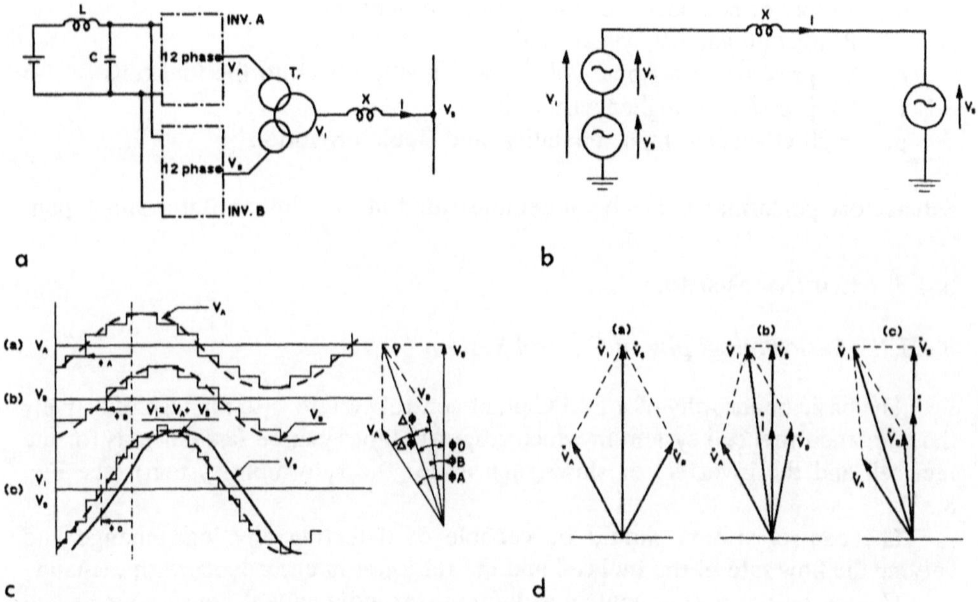

Figure 8.32. Self-commutated inverter: (a) function of inverter; (b) equivalent circuit model; (c) waveforms of output voltage; (d) vector diagram (grid connected (power factor = 1).

Since INV-A and INV-B are self-commutated types, phases and amplitudes of the 12-phase inverter output voltages can be controlled by each inverter phase angle control and therefore V_i, the resultant voltage of V_A and V_B, can also be adjusted so as to connect it with the line voltage V_S. A special operational condition for a power factor of 1 is shown in Fig. 8.32d for three cases.

An inverter as a component of a PAFC plant has special features.

1. The output voltage of the fuel cell stack decreases sharply as the current increases. In the worst case, the output voltage at rated current is almost half of that at no load. It is uneconomical to design an inverter to meet a very wide range of input voltages and constant output. This would result in a large equivalent capacity and poor performance. To avoid this, a chopper is connected to convert the broad range of voltages to constant voltage, which is to be fed to the inverter. The role of the chopper is to convert the input dc voltage to another dc voltage, and is classified into two types: a step-up chopper and a step-down chopper. However, the total system cost including chopper and inverter, might be higher and the system efficiency might be lower compared to that of a single inverter unit; hence, the system design should be determined from the comparison of merits and demerits of these systems.

2. In many PAFC plants for electric utility application the inverter is the interface equipment between the fuel cells and the electrical network. (For comparison, many inverters now currently used are usually installed at the end terminal of networks and act as the voltage and frequency adjusters to the final load, e.g., motors, fans, and so on.)

Interface conditions require the following characteristics for PAFC inverters:

a. Synchronization to the network
b. Conformation with the network voltage regulation

c. Supply of necessary reactive power to network
d. Protection against system faults
e. Suppression of a ripple voltage which is fed back to the fuel cells
f. Suppression of higher harmonics
g. High efficiency, high reliability, and stable operation

Satisfactory performance has been demonstrated in the United States and Japan.

8.4.3. Control System

8.4.3.1. Basic Philosophy of Control System

The basic philosophy of a PAFC plant control system is to manage effectively the combined fuel cell system in which response times can be fast (such as for the fuel cell and the inverter) or slow (such as for the reforming system) (see Fig. 8.33).

The control system should be capable of detecting any load change and varying the flow rate of the fuel cell and the reformer in accordance with demand.

The method of output control is different for independent power sources and grid-connected systems. For independent power sources, generally applied to small-scale on-site (cogeneration) plants, (1) the inverter is controlled by detecting the change in output voltage due to load change, and (2) the gas flow to the fuel cell is controlled by the corresponding value of the output dc current, which is equal to the input of the inverter (see Fig. 8.33a).

For example, when the output current of a fuel cell increases, the inlet valve of the reformer is correspondingly opened to increase the supply of natural gas (raw fuel) to the reformer. If the fuel cell output current increases rapidly, only the hydrogen stored in the piping between reformer and fuel cell must be fed

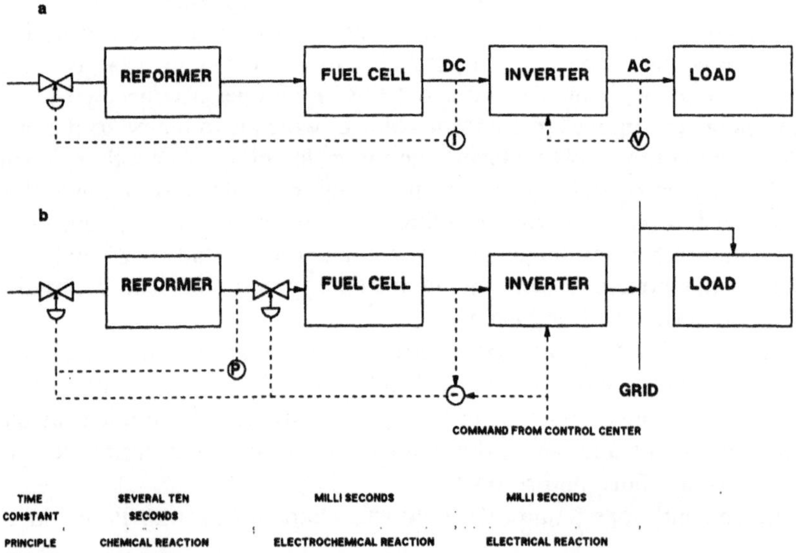

Figure 8.33. Principle of PAFC plant control: (a) independent power plant; (b) grid-connected power plant.

instantaneously to the fuel cell, because the reformed hydrogen cannot reach the fuel cell quickly enough to meet the required response to the load. Therefore, in some cases, an appropriate storage tank of hydrogen should be equipped in front of the fuel cell. This must be determined in each design.

In case of electric-utility applications, a large-capacity plant is connected to the grid and the inverter output is usually controlled by the load demand from the central control center. In most cases the control of gas inflow to the reformer and fuel cells is performed simultaneously. An additional valve is necessary to control the flow balance between the inlet and outlet of the reformer. The control system controls opening of the valve by detecting the outlet pressure of the reformer (see Fig. 8.33b).

8.4.3.2. Control of Pressurized Air Supply System

In pressurized large-scale PAFC plants, the control systems are complex compared to atmospheric pressure systems. The compressor is the key component for pressurizing the air system, and there are currently two different power sources to drive the compressor: the first is a turbocompressor system, and the second is a motor-driven compressor system (see Fig. 8.34). The turbocompressor is driven by the exhaust gas from the reformer and has a common shaft for the turbine and compressor. The motor-driven compressor, on the other hand, is driven by a motor powered by electricity generated by a steam or gas turbine, driven in turn by the reformer exhaust gas.

Table 8.15 shows an approximate comparison between these two systems from control point of view.

8.4.3.3. Pressure Difference Control

The pressure difference between the reactant gases in the fuel cell should be limited to a relatively small value. It is currently recommended in most PAFCs to

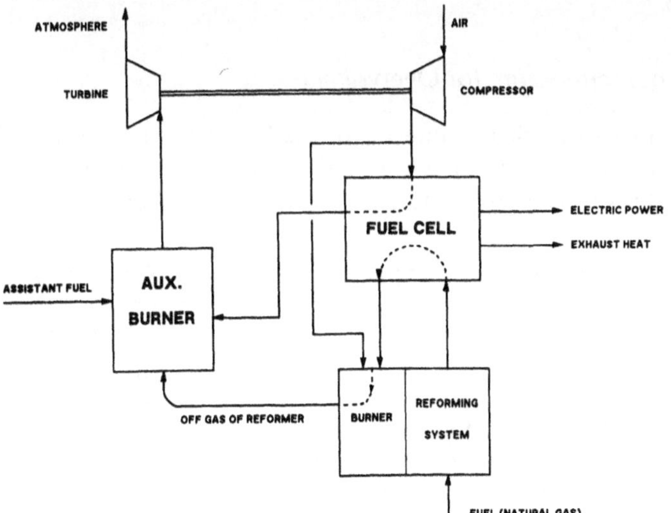

a

Figure 8.34. Simplified fuel cell system: (a) turbocompressor system; (b) motor-driven compressor system.

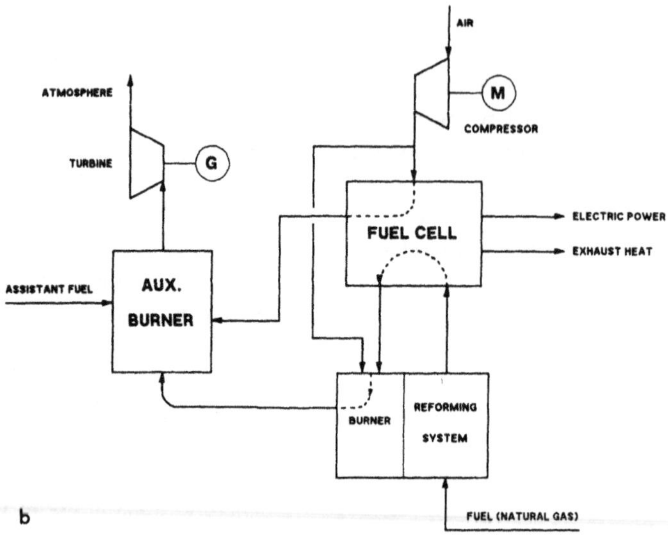

Figure 8.34. (*continued*)

keep this pressure difference to a value less than about 500-mm Aq under transient conditions. From safety reason the anode side (hydrogen) pressure should be set higher than the cathode side. For this reason, the system must be equipped with pressure difference controls.

8.4.3.4. *Warming Control*

As already described in the previous section, the phosphoric acid solidifies at approximately 42°C. Therefore it is necessary to maintain the stack temperature at a higher level than the solidification temperature. Operating the stack at low-load condition and heating the stack body by self-powered electricity are both possible.

8.4.3.5. *Countermeasure for Overvoltage*

As previously described, sintering of the Pt catalyst and corrosion of support carbon are accelerated very much at high temperatures (above 180°C) and high voltages (>0.8 V). Therefore the stack should be connected to an appropriate

Table 8.15. Approximate Comparison between Two Compressor Systems

	Turbocompressor	Motor-driven
Total efficiency	Superior at full load but not at partial, load	Lower at full load, but in many cases better at partial load
Controll-ability	Reformer, FC, and compressor systems are closely connected and might be complex to control, especially under transient conditions (turbine and compressor have common shaft)	Controls of motor-driven compressor system are independent of the gas flow condition of reformer and FC Controllability is improved

load (or load bank) to keep the voltage below 0.8 V at all times. This is an important control issue to extend cell life. (Lowering the cell temperature or decreasing of the partial pressures of reactant gases are also effective ways to realize the same purpose.)

8.4.4. Overall Plant Efficiency and Utilization Factor

The overall plant efficiency of fuel cell systems (η_{total}) is defined as the product of the following four efficiencies (see Fig. 8.3 and Chapters 5 and 6):

1. Fuel processing efficiency η_p*
2. Fuel cell efficiency η_{FC}
3. Inverter efficiency η_{inv}
4. Auxiliary power factor η_m

$$\therefore \eta_{\text{total}} = \eta_P \eta_{FC} \eta_{\text{inv}} \eta_m.$$

Each efficiency is defined as follows:

$$\eta_P = \frac{\text{heating value of consumed } H_2 \text{ in fuel cell† } (W_{dc})}{\text{heating value of natural gas to reformer } (W_T)},$$

$$\eta_{FC} = \frac{\text{electric energy generated in fuel cell}}{\text{heating value of consumed } H_2 \text{ in fuel cell†}},$$

$$\eta_{\text{inv}} = \frac{\text{ac output power from inverter}}{\text{dc input power to inverter}},$$

$$\eta_m = \frac{\text{ac output power to load}}{\text{ac output power from inverter}},$$

$$\eta_{\text{total}} = \frac{\text{ac output power}}{\text{heating value of natural gas to reformer}}.$$

In Fig. 8.3 the following efficiencies are shown as an example showing the values for the Tokyo 4.5-MW PAFC plant (HHV base):

$$\begin{aligned}
\eta_P &= 82\%,\\
\eta_{FC} &= 48\%,\\
\eta_{\text{inv}} &= 96\%,\\
\eta_m &= 97\%,\\
\eta_{\text{total}} &= \eta_P \eta_{FC} \eta_{\text{inv}} \eta_m = 36.7\%.
\end{aligned}$$

*If the turbocompressor is used to send the pressurized air to the fuel cell, the exhaust gas of the reformer burner is sometimes used to drive the turbocompressor. Also the high-temperature off-gas from the cathode is supplied to the burner. These effects are all included in the fuel processor efficiency η_P (see Fig. 8.3).

†In some cases the total heating value of all feed gas to the anode (including the unconsumed part) is used (see the footnote in Section 8.3.1.3).

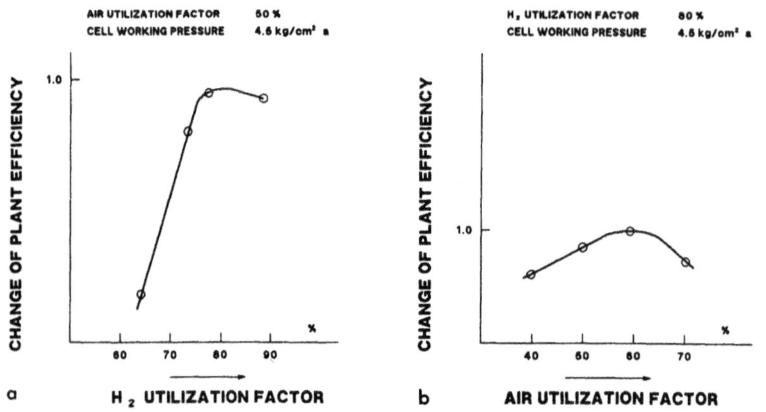

Figure 8.35. Relation between plant efficiency and utilization factor (an example)[44]: (a) effect of H_2 utilization factor; (b) effect of air utilization factor.

As mentioned previously, the reformer burner is usually fed by the exhaust gas from the anode, and no additional fuel of natural gas is necessary under normal operation conditions. Therefore all the natural gas supplied to the reformer is usually used for reforming gas only.

Values of η_{FC} and η_{total} are dependent on utilization factors. Generally, increasing the fuel utilization factor results in lower cell efficiency; lower utilization of hydrogen in the anode enhances cell efficiency. However, the latter results in lower efficiency for the reforming system.

Too low a utilization factor for hydrogen will result in an increase in the hydrogen content in the off-gas from the anode. This increases the cell efficiency, but decreases the reformer efficiency because too much hydrogen is supplied to the reformer (the excess hydrogen is exhausted to the atmosphere).

On the contrary, too high a utilization factor for hydrogen results in a decrease in the hydrogen content in the off-gas, and this decreases the stack efficiency but increases the reformer efficiency.

As a result of these tendencies the overall plant efficiency generally exhibits a maximum value at a certain utilization point (see Fig. 8.35a).[44]

The air utilization factor has similar effects on plant efficiency (see Fig. 8.35b).[44] At lower utilization of air (oxygen) the air feed rate should be increased, and this increases the compressor (or blower) power. Higher fuel cell efficiency at lower air utilization is offset by increased compression (or blowing) power consumption.

8.5. APPLICATION AND OPERATION OF PAFC PLANTS

8.5.1. Available Fuels (Comparison of Natural Gas, LPG, and Methanol)

Natural gas, LPG, methanol, and naphtha presently constitute the principal fuels for PAFC plants. Table 8.16 shows a comparison of these fuels from a performance point of view.

Table 8.16. Comparison of Various Fuels for PAFC[37,47]

	Natural gas	Methanol	Naphtha	LPG[a]
Chemical formula	CH_4	CH_3OH	C_6H_{14} (average)	C_4H_{10} (Butane)
Total heating value (kcal/kg)	13,080	5,420	11,700	11,830
Boiling point (°C)	−161.5	64.7	37–200	−0.5
Chemical reaction				
Steam reformer	$CH_4 + H_2O = CO + 3H_2$	$CH_3OH + H_2O = CO_2 + 3H_2$ (25%) (75%)	$C_6H_{14} + 6H_2O = 6CO + 13H_2$	$C_4H_{10} + 4H_2O = 4CO + 9H_2$
Shift converter	$CO + H_2O = CO_2 + H_2$		$6CO + 6H_2O = 6CO_2 + 6H_2$	$4CO + 4H_2O = 4CO_2 + 4H_2$
Resultant	$CH_4 + 2H_2O = CO_2 + 4H_2$ (20%) (80%)		$C_6H_{14} + 12H_2O = 6CO_2 + 19H_2$ (24%) (76%)	$C_4H_{10} + 8H_2O = 4CO_2 + 13H_2$ (24%) (76%)
Reforming temperature (°C)				
Steam reformer	750–850	250–300	870–980	750–850
Shift converter	150–400	(not necessary)	150–400	150–400
Reforming catalyst				
Steam reformer	Ni metal (noble metal)	Cu–Zn	Ni metal, noble metal	Ni metal noble metal
Shift converter	Fe–Cr, Cu–Zn		Fe–Cr, Cu–Zn	Fe–Cr, Cu–Zn

[a]LPG is usually the mixture of propane (C_3H_8) and butane (C_4H_{10}). Butane is more popular than propane for industrial use.

The results can be summarized as follows:

1. LPG reformer and natural gas reformer have similar performance.
2. Methanol, on the other hand, has only half the heating value of natural gas and LPG, and the reforming temperature of methanol is quite low. Shift conversion (for converting CO to H_2 and CO_2) is not necessary for methanol because of the low reforming temperature. This makes the methanol-reforming system simpler and the fuel processor and system efficiencies higher.
3. Methanol is a liquid at room temperature and can be easily handled and stored in a vessel. Therefore methanol is recognized as an appropriate fuel for trransportation or for use in remote areas. Natural gas is the preferred fuel for large-scale PAFC power stations.

Generally speaking, the system efficiency of methanol-reforming PAFC plants is 1–2% higher than of natural gas plants. If methanol can be supplied and stored economically on a large scale, methanol-reforming PAFC plants do constitute an attractive alternative.

8.5.2. Grid-Connected and Grid-Independent Configuration

In on-site cogeneration PAFC plants there are two configurations in relation to the grid: grid-connected and grid-independent plants (see Fig. 8.36).

In a grid-independent plant, which is the most popular configuration for small-capacity systems, the plant capacity must be selected to correspond to the maximum system load, and hence the load factor of such a PAFC plant is inevitably decreased. Furthermore, if the plant is used for cogeneration purposes, both heat and electricity generation should match the load demand. Suitable equipment for storing "overenergy" must be installed. In practice, an electrical storage system is not practical (except for a battery, which however is inefficient and expensive). Thermal storage is usually preferred.

For a grid-connected plant, in the case of not being able to supply electricity to the grid, the system only receives electric power from the grid when the PAFC plant's output is insufficient.

Even if it is not allowed to supply electricity to the grid, a shortage of electricity can be supplied from the grid, and hence the plant capacity can be chosen to be reasonably small and the load factor can be raised. A higher load factor increases efficiency and lowers the total cost of electricity (COE). If the plant is used for cogeneration, the PAFC plant capacity can be designed to match electrical and thermal energy requirements. A shortage of electricity can be met by the grid if necessary. This also allows higher operability and higher efficiency, as the plant can be operated at a higher load factor.

If the plant is allowed to supply electricity to the grid, much more effective and economical plant operation is possible. For this reason market size is larger for grid-connected PAFC units than for grid-independent units. Figure 8.37 shows a market estimate by GRI of the United States.[24]

At plant unit cost of \$1500/kW, the expected market size is 10^4 MW for the grid-independent application but 4×10^4 MW for the grid-connected case. This figure also shows that there is a fairly large market even at higher installed cost of \$2500–3000/kW (for the grid-connected case). Market consideration will be further dealt with in Chapter 13.

When PAFC plants are connected to the grid, the plants must meet stringent safety regulations. The following conditions are required for grid-connected plants:

a. To maintain high quality of electricity (higher harmonics, voltage regulation, frequency change, and so on)
b. To provide suitable protection in the event of system faults
c. In the case of "no supply of electricity to the grid" the plant should be equipped with a "reverse power protection system," which may be a dummy load or other appropriate equipment.

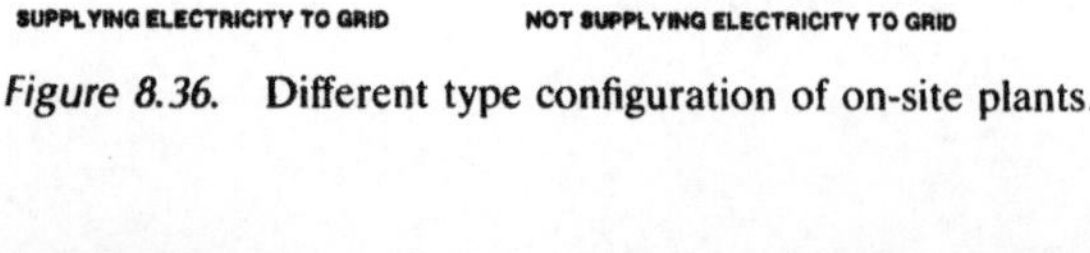

Figure 8.36. Different type configuration of on-site plants.

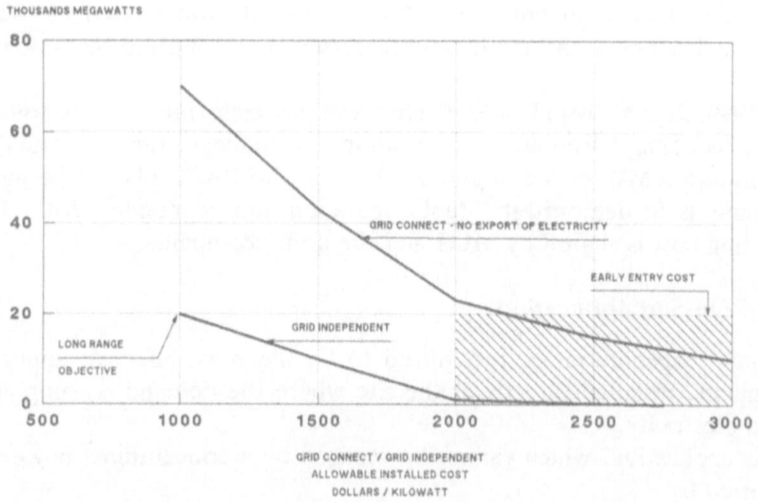

Figure 8.37. On-site fuel cell economic market size vs. cost.[24]

8.5.3. Application of PAFC Plants

PAFC plants are widely applicable, from kilowatt to megawatt range, for stationary uses (electric utility plants, cogeneration plants) and for vehicular uses.

8.5.3.1. Electric Utility application

The higher efficiency and the environmentally benign nature of a PAFC plant compared to a conventional power station are key factors for considering PAFC for electric utility application. This application of PAFC plants can be divided into two categories: dispersed type and central station type. The dispersed-type plant has a typical capacity of 10–20 MW and is to be installed in distribution substations. The central station type plant may constitute an alternative to medium-range thermal power stations and has a typical capacity over about 100 MW.

The advantages of applying PAFC plants for electric utilities are summarized in the following:

a. Higher efficiency for partial load
b. Can be installed in a suburban area because of excellent environmental performance
c. Simple construction activities and short construction period on site because of its modularized nature (transportation of skid to site after prefabrication in factory)
d. Easy expansion of plant capacity because of its modularity

As mentioned, IFC in the United States has constructed and operated a 4.5-MW plant and Toshiba–IFC have jointly constructed an 11-MW plant for Tokyo Electric Power Company. Two 1-MW plants have been constructed and operated in Japan under the Moonlight Project. Key specifications for these plants have been shown in Tables 8.3 and 8.4, respectively.

The raw fuels supplied to dispersed and central power stations are natural gas, naphtha, and methanol. Natural gas is usually transported to the site by pipeline, and naphtha and methanol are currently transported by tanker or tank truck.

In 1990, Japan's MITI, and electric and gas companies jointly founded the PAFC Technology Research Association to manage the construction and operation of a 5-MW dispersal and a 1-MW on-site PAFC plant. The purpose of these plants is to demonstrate stable operation and economical cost. The total construction cost is shared by MITI and the utility companies.

8.5.3.2. On-Site Application

On-site cogeneration is recognized to be the most effective application of PAFC plants, installed directly at the site where the demand is, supplying both heat and electricity.

This application, which cannot be realized by a conventional power station, is supported by

1. High-quality heat, which can be used effectively because the plant can be installed at the demand site (heat is the most difficult medium to transport; therefore, unless the plant is installed at the demand center, effective usage of heat cannot be realized).
2. A PAFC plant, even a small-capacity plant of 50 or 100 kW, has similar efficiency to that of modern large thermal plants, and there is no efficiency penalty for partial-load operation, even under 50% load. This is a key feature of PAFC on-site cogeneration plants.
3. Since there are practically no transportation losses of natural gas and no transmission losses of electricity, PAFC on-site plants can supply the energy efficiently.

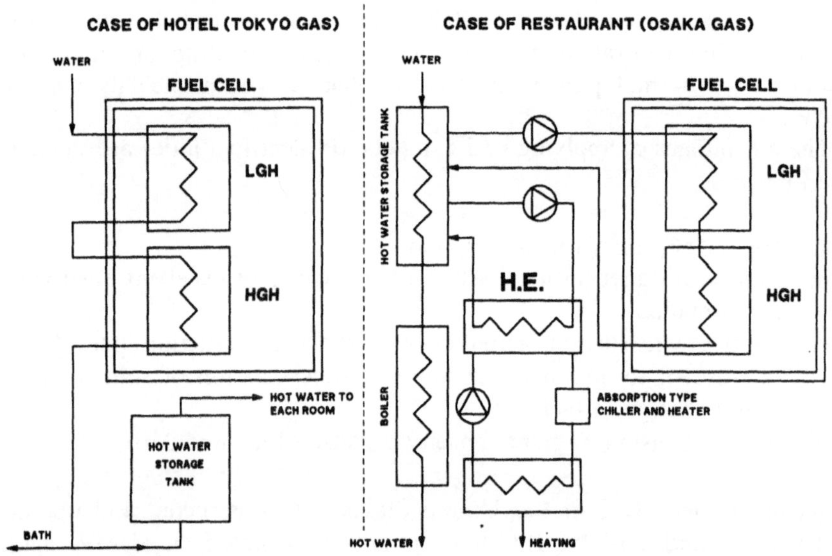

Figure 8.38. Application systems of cogenerated heat from PAFC on-site plant (experienced in Japan) (40 kW plant, IFC).

Because electricity and heat should be consumed at the same time, special attention should be paid to the plant design. Site selection is also important. Hotels, hospitals, restaurants, and laundries are probably the most appropriate application fields of PAFC cogeneration plants.

Figure 8.38 shows some examples of such application, with systems using cogenerated heat from a PAFC on-site plant. (In these cases, Tokyo Gas and Osaka Gas, in Japan, applied two UTC PC-18 (40-kW) plants to a hotel and a restaurant respectively).[60]

8.5.3.3. Vehicular Application

A hybrid system of a methanol-fueled PAFC and a battery is now under development for vehicular application in the United States and Japan, such as for a forklift truck and a bus.

The advantages of PAFC-powered forklift trucks are summarized as follows:

1. The system can offer long-term operation without any loss time for charging the battery, which is inevitable in a battery-powered forklift. As long as the methanol is supplied to the system, the truck can operate continuously without any off-line recharging of the batteries.
2. The role of a methanol-fueled PAFC is to supply base-load power, and the battery can be charged under light-load conditions, if necessary. The hybrid battery serves to supply the power for peak demand and during start-up period (about 15 min) prior to fuel cell operation. Figure 8.39 shows a hybrid system for a forklift truck.[48]

The capacities being developed for forklift trucks and buses are about 5 kW and 50–100 kW, respectively. The main purpose of the fuel-cell-powered bus is evidently to develop an environmentally clean public transportation system.

8.5.3.4. Small-Capacity Transportable System Application

The application of small-capacity PAFC systems for transportable uses is unique for PAFC. Methanol-fueled small-scale PAFC plants equipped with fuel processor, fuel cell, inverter, and control system have the advantages of low noise, low pollution, low heat dissipation, and high efficiency.

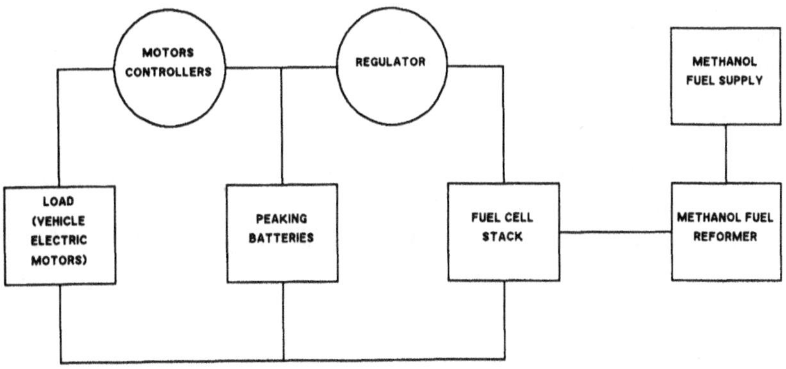

Figure 8.39. Block diagram of fuel cell battery hybrid system.[48]

Application examples are as telecommunication power sources for military use, as movable power sources for emergency power supply and recreational vehicles, etc. Low noise levels and low heat dissipation characteristics are attractive for military use compared to conventional power sources such as diesel generators, because of the low infrared signature. The capacity range for such military applications is about 1–5 kW, but the requirements are very severe compared to ordinary PAFC plants. Very rapid start-up from cold atmosphere, easy maintenance (such as the absence of any heating requirements under off-load conditions, no N_2 purging, etc.), shockproof and vibration-proof performance are some of the requirements.*

The necessity for the development of such small-scale PAFC transportable power sources is very strong. Once such power sources are supplied at a competitive price, market potential is attractive. Cost reduction is critical for the developers to achieve success.

8.5.3.5. Other Special Applications

There are a number of industrial plants (such as chloroalkaline) which produce high-quality hydrogen as a by-product. In such cases the PAFC plants installed on site can generate electricity by consuming such off-gas hydrogen. No reforming system and no inverter are necessary. This makes the plant simple and economical, and certainly a potential application for PAFC. Preliminary study has indicated that if the off-gas hydrogen is consumed in a PAFC plant, one third of the total electricity used in the factory can be saved.[36]

Another special application is the connection of a PAFC plant to an ordinary steam turbine generating plant.

8.5.4. Operation of PAFC Plants

Taking an example of a natural-gas-reforming, pressurized, and water-cooled PAFC system for electrical utility application, the operation procedures (including steady state and start-up) shall be briefly described, however, this procedure is only an example: there are many variations of operational procedures, depending on the designs.

8.5.4.1. Steady Operation

Figure 8.40 shows an example of the flow diagram of a PAFC plant, in which the following items are referred to:

A. Reforming subsystem

1. Natural gas (NG) is mixed with hydrogen in the desulfurizer to remove small amounts of sulfur.
2. After the desulfurizer, steam is added and supplied to the reformer, where NG is converted to H_2 and CO over a Ni-containing catalyst: $CH_4 + H_2O \rightarrow 3H_2 + CO$.

*It is reported that the U.S. Army has decided not to use methanol as the raw fuel for its PAFC plants because it prefers to use the same fuel currently used for general purposes.

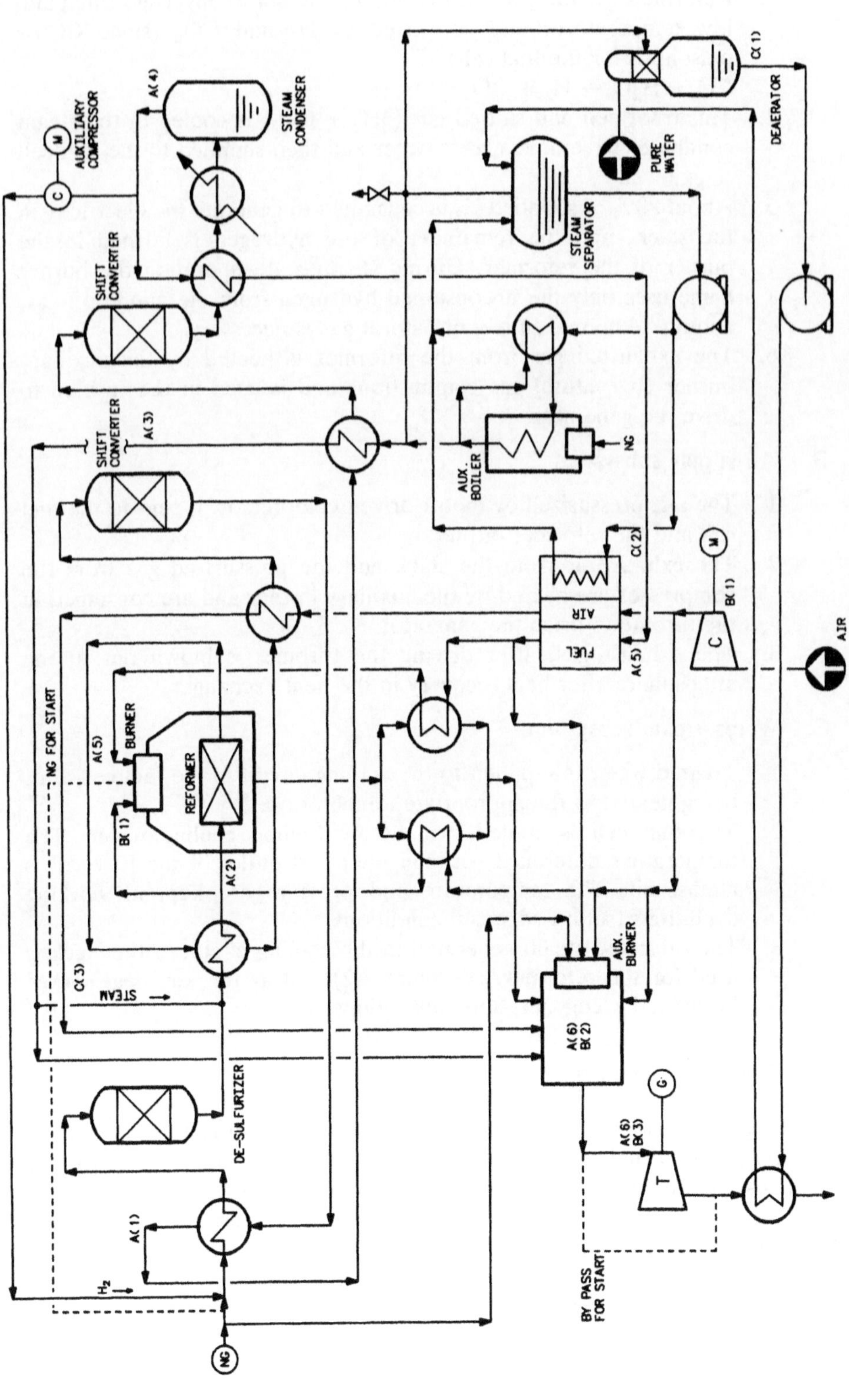

Figure 8.40. Flow diagram of PAFC plant (steady state) (an example).

3. Reformed gas ($3H_2 + CO$) is supplied to shift converters (high and low temperature) and converted to H_2 and CO_2 (since CO is poisonous for the fuel cell):
$CO + H_2O \rightarrow H_2 + CO_2$.
4. This reformed and shifted gas ($4H_2 + CO_2$) is cooled in the steam condenser to remove excess water and then supplied to the fuel cell stack.
5. About 80% of supplied H_2 is consumed to generate the electricity in the stack, and the remainder of the hydrogen is burned in the burner of the reformer. During steady-state operation the burner flame uses only the unconsumed hydrogen from the anode off-gas, and no additional supply of natural gas is necessary.
6. The exhausted gas from the reformer is heated by the auxiliary burner (by natural gas combustion) and is used in the turbine to drive the generator.

B. Air supply subsystem

1. The air, pressurized by motor-driven compressor, is sent to the fuel cell and the reformer burner.
2. The exhaust air from the stack and the pressurized gas from the compressor are heated by the auxiliary burner and are consumed in the turbine to drive the generator.
3. The exhaust gas, after driving the turbine, is blown out to the atmosphere after heat recovery in the heat exchanger.

C. Water–steam subsystem

1. Treated water is supplied to the cooling circuit in the fuel cell after being heated to the appropriate temperature.
2. The fuel cell is cooled, usually by boiling cooling water. The temperature difference between inlet and outlet of the fuel cell is almost nil. The temperature and pressure are kept at working conditions (saturated steam conditions).
3. The saturated steam generated in the cooling circuit of the stack is used for the reformer (see point A2) and as the key heat source available for cogeneration applications.

8.5.4.2. Start-up Procedure (see Fig. 8.41)

Under off-load conditions the system is filled with nitrogen (inert gas) at atmospheric pressure and kept at room temperature. The fuel cell stack only, however, is kept at about 40–80°C (by electrical heating and/or by the circulation of warm cooling water of the stack to protect the phosphoric acid from solidification).

An example of a start-up procedure for the plant in Fig. 8.41 (each item number is shown in the figure) is as follows:

1. The reforming system containing N_2 is pressurized, usually by supplying N_2 from high-pressure N_2 sources (not shown in Fig. 8.41).
2. Pressurized N_2 is circulated by the auxiliary motor-driven compressor.

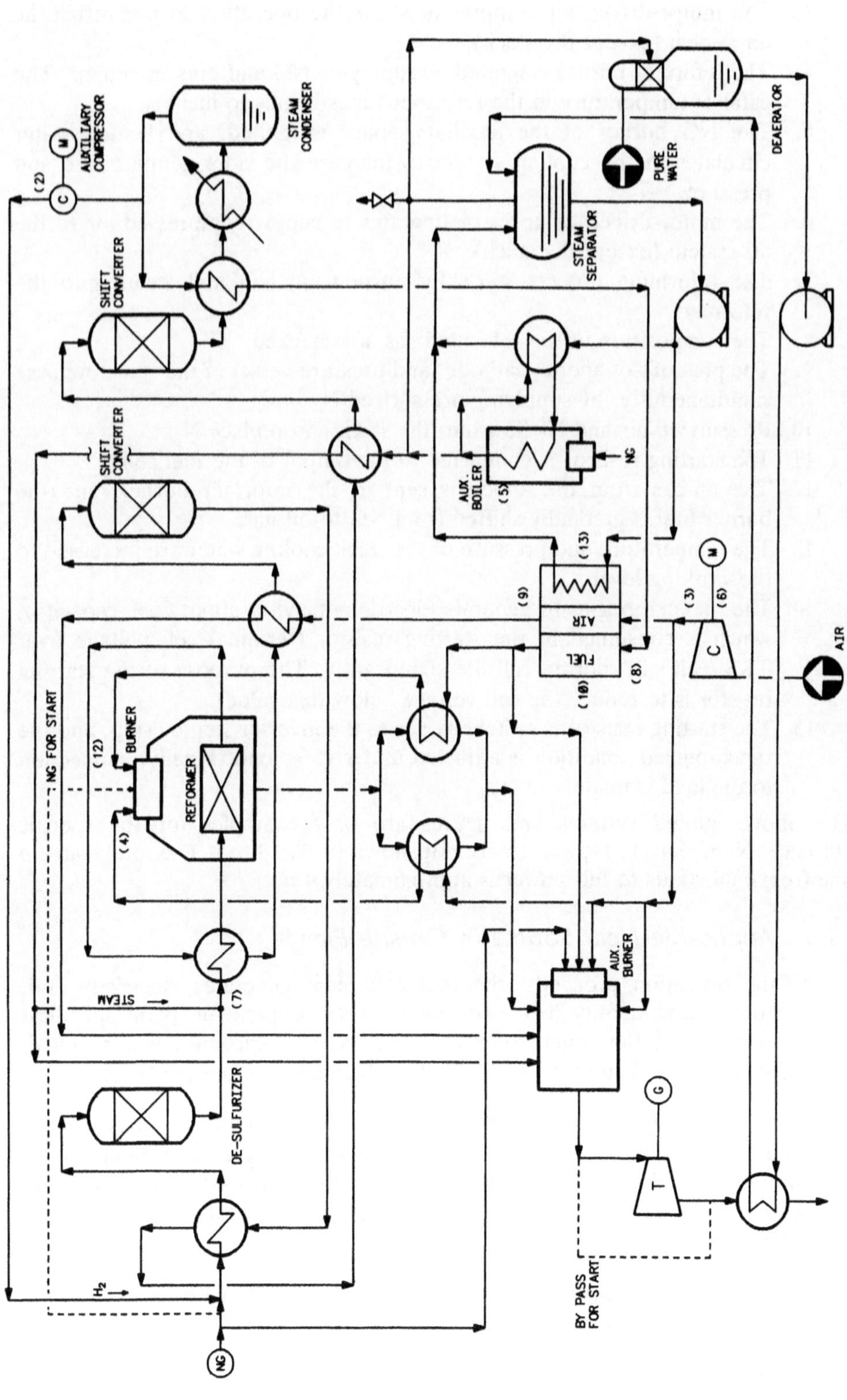

Figure 8.41. Flow diagram of PAFC plant (start-up) (an example).

3. The motor-driven air compressor starts the operation to pressurize the air system (except the stack).
4. The reformer burner is ignited by supplying NG and pressurized air. The catalyst temperature in the reformer tubes begins to increase.
5. The NG burner of the auxiliary boiler is ignited, and heated water circulates in the cooling system to increase the stack temperature and pressure.
6. The motor-driven compressor operates to supply compressed air to the air system (except the stack).
7. The reforming process starts by introducing NG and steam into the reformer.
8. The composition of the reformed gas are checked.
9. The pressure of anode, cathode, and pressure vessel of the stack increase simultaneously, by supplying pressurized N_2.
10. Pressurized air and H_2 flow into the stack and replace N_2.
11. The starting resistor is connected to the output of the fuel cell.
12. The off-gas from the anode is sent to the reformer burner, and the burner fuel is gradually shifted from NG to off-gas.
13. The temperature and pressure of the stack cooling water are increased to the working level.
14. The fuel cell begins to generate electricity (say, at about 25% capacity), which is consumed in the starting resistor (terminal cell voltage over 0.8 V/cell will shorten cell life significantly. The purpose of the starting resistor is to reduce the cell voltage below this value).
15. The starting resistor is switched over to the inverter step by step, and the synchronized condition is adjusted under 25% load. Finally, connection to the load is made.

(The above quoted systems and figures are only examples for the specific reference. Numbers 11, 14, and 15 are not shown in Fig. 8.41). The total start-up time from cold status to full power is approximately 4 h.

8.5.5. Applicable Heat Source in On-site Plants

As the operation proceeds, the fuel cell plant generates electricity and, simultaneously, high-quality heat. The heat sources depend on plant specifications, capacity, and the other conditions. For a small-capacity, water-cooled, on-site PAFC power plan (see Fig. 8.42) the following example can be given:

Example 1. High-temperature heat source. The outlet of stack cooling water is usually the highest temperature heat source. Since the outlet temperature is about 170°C (in the form of saturated steam), the heat usable from the heat exchanger is at 130°C or higher, and can be used for air-conditioning purposes as well as hot water.

Example 2. Medium-temperature heat sources. Usually the following medium-temperature heat sources are available in the PAFC plant:

Inlet gas of the anode
Outlet gas of the cathode
Exhaust gas of the reformer

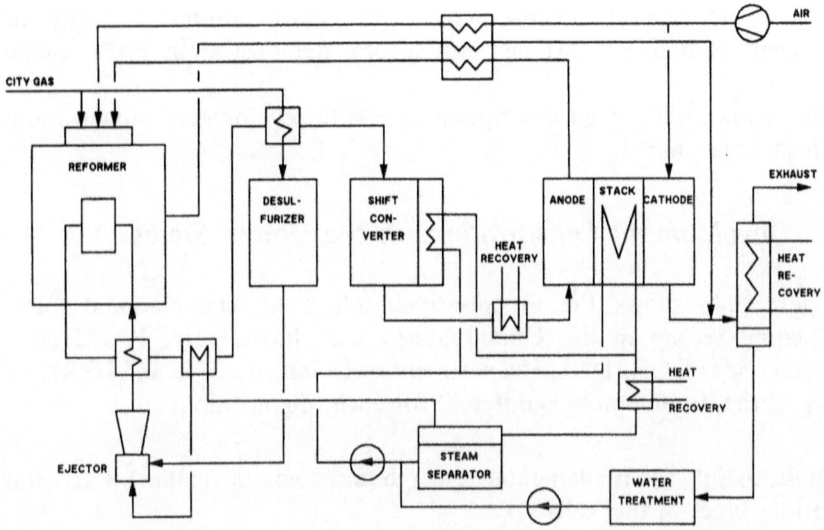

Figure 8.42. Process flow diagram of small-capacity on-site PAFC plant (simplified) (an example).[63]

The temperature level of these heat sources is about 60°C (dependent upon design), and this heat is usually used for hot-water supply.

The amount of high- and medium-temperature heat is approximately 15–30% and 25–45% of input energy respectively. Total available energy of electricity and heat can attain over 80% of total input heating value of natural gas.

As mentioned, hotels, restaurant, and hospitals are preferable application areas for cogeneration plants because those consumers require hot water simultaneously with the electric power.

Figure 8.38 shows two examples of 40-kW cogeneration plants installed in a restaurant and in a hotel in Japan.

8.6. *PAFC SYSTEM DEVELOPMENT IN THE WORLD*

8.6.1. *General*

The development of PAFC systems has been undertaken in the United States, Japan, and Europe. The first development started at Pratt and Whitney, and two big demonstration programs (the TARGET Program and FCG-1 Project) were started in 1967 and 1971, respectively. These projects were afterwards pursued with the support of the government (now DOE). EPRI and GRI were also cosponsors. Presently, Toshiba and IFC are intending to supply 11-MW plants for electric-utility companies and 200 kW for on-site applications. The first 11-MW plant is now operating. The contract was made between Tokyo Electric Power Company and the Toshiba–IFC Group.

After the oil crisis in 1973, Japan has expanded research and development activities of fuel cell technologies very actively under governmental support, and now many private industries are competing to improve plant performance and reduce the cost.

In Europe several organizations have shown interest recently in the development of fuel cell systems, and several demonstration PAFC plants are now operating.

The following is a brief description of PAFC development organizations and budgets in each country.

8.6.2. Development Organizations in the United States

Figure 8.43 shows the organizational chart of The National Fuel Cell Coordinating Group in the United States. As shown here, PAFC has been developed under DOE, DOD (Department of Defense), GRI, and EPRI, but the purpose of the development is different for each organization:

DOE: mainly for fundamental research and some development activities for various types of fuel cells
DOD: mainly for the small-scale, high-energy-density power source
NASA: mainly for space application (alkaline, PAFC)
GRI: focusing on on-site cogeneration application (PAFC, MCFC, and SOFC)
EPRI: focusing on electric utility application (PAFC, MCFC, and SOFC)

With the support of these organizations IFC, ERC, Westinghouse Engineering Corp., and others are now promoting their own development programs, which have been described in Section 8.2.2.

The budgets of DOE, EPRI, and GRI are shown in Table 8.17. The total budget for the five-year period is about $223 million,* and about 55% of this is spent for PAFC. The budget trend for PAFC is decreasing, since the development phase of PAFC has arrived at the semicommercialization stage. No large support can be expected from government and utility organizations in the future.

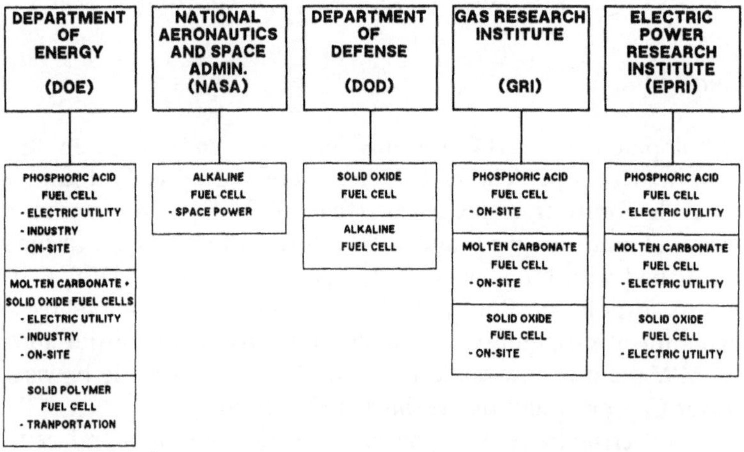

Figure 8.43. National Fuel Cell Coordinator Group (U.S.A.).

*The $223 million includes some amount of budget for nonspecified use.

Table 8.17. Budget for Fuel Cell Development in the United States ($ million)

Organization	DOE					EPRI					GRI				
Year	'85	'86	'87	'88	'89	'85	'86	'87	'88	'89	'85	'86	'87	'88	'89
PAFC	28.6	20.4	15.5	13.2	9.5	4.5	2.7	2.2	3.2	0.6	5.0	4.0	6.0	2.1	—
MCFC	8.9	9.9	7.6	11.1	7.1	2.0	1.6	1.9	2.2	2.9	0	0	0	0.3	0.75
SOFC	3.0	3.8	5.0	8.4	8.4	0	0.1	0.3	0.2	0.4	0	1.0	2.3	2.1	2.0
Total	40.5	34.1	28.1	32.7	25.0	6.5	4.4	4.4	5.6	3.9	5.0	5.0	8.3	4.5	2.8

The total budget for TARGET (1967–1976) was $58 million (see Sections 8.2.2.1 and 8.2.2.2) and $35 million for the 46 sets in the 40-kW GRI–DOE project (DOE cost-shared $12 million and GRI $23 million). The expenses for field testing were shared by all organizations.

All figures shown in the table do not include any expenses by the private organizations, IFC, ERC, Westinghouse, and others, because they are difficult to estimate. But during the R&D stage in the United States, the government and utilities usually bear the largest share of the total budget.

8.6.3. Development Organizations in Japan

Stimulated by the oil crisis in 1973, the Japanese government set up a comprehensive fuel cell technology development program (Moonlight Project) in 1981. Four PAFC plants have been constructed and operated under the Moonlight Project, as shown in Table 8.18. This government program further stimulated many private organizations.

The developmental organization of governmental projects and outline of private development status are shown in Fig. 8.44 and Table 8.6, respectively.

The total national budget for the development of PAFC technologies from 1981 to 1990 is about $115 million, including the construction of four demonstration plants (see Table 8.19). The government subsidies for R&D in PAFC under the Moonlight Project were terminated by 1990, and after that MCFC and SOFC are the main development tasks subsidized by the government. This implies that,

Table 8.18. Demonstration Plants Constructed under NEDO Project (Japan)

Capacity (kW)	Object	Constructor	Operator	Completed	Location
1,000	Electric utility (dispersed)	Fuji Electric Mitsubishi Electric	Kansai Electric Power Company	1987	Near Osaka
1,000	Electric utility (central station)	Hitachi Toshiba	Chuba Electric Power Company	1987	Near Nagoya
200	On-site cogeneration	Mitsubishi Electric	Osaka Gas Co. Kansai EPC	1990	Osaka
200	Remote Island	Fuji Electric	Okinawa Electric Power Company	1990	Okinawa Island

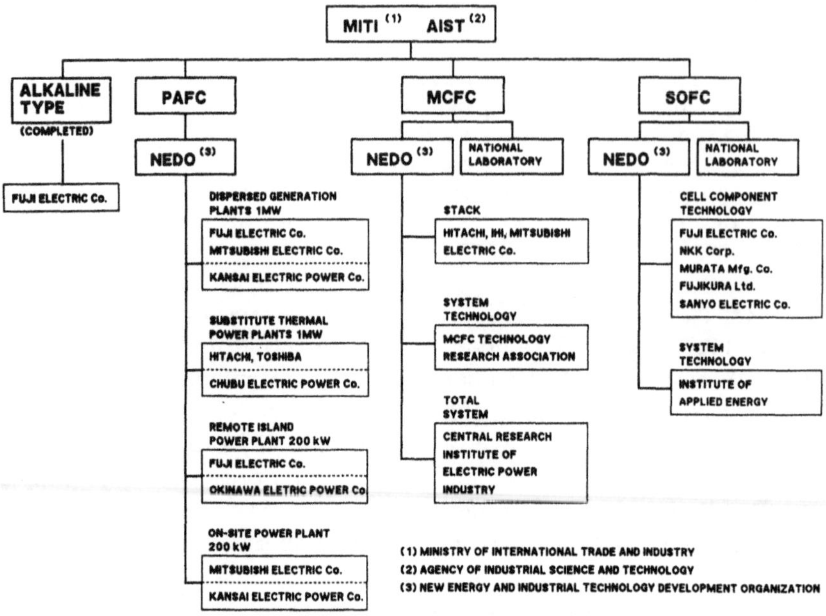

Figure 8.44. Development organization of fuel cell technologies in Japan (government projects).

as in the United States the development phase of PAFC plants enters into industrial improvement level and further progress for commercialization should be promoted by the joint efforts of the government, utility groups, and manufacturers.

Apart from the Moonlight Project, which aims at the development of PAFC technologies, another MITI project has started from 1987 to demonstrate the feasible grid interconnection technologies of PAFC plants. For this purpose Kansai Electric Power Company (the second largest electric utility company in

Table 8.19. Budget of FC Development in Japan (Moonlight Project) ($ million)

	'81	'82	'83	'84	'85	'86	'87	'88	'89	'90	Total
PAFC											
R&D on Components	0.56	2.61	11.0	4.34							
											115
Plant costruction & operation			0.76	17.93	27.36	15.90	18.13	8.46	5.06	2.72 (finish)	
MCFC	0.40	0.88	1.69	3.22	6.07	6.39	5.30	15.43	19.39	17.86	59
SOFC	0.39	0.55	0.71	0.49	0.66	0.44	0.49	0.56	1.11	1.47	−7
Alkaline	0.28	0.31	0.31	0.21	—	—	—	—	—	—	−1
System assessment	0.06	0.06	0.04	—	—	—	—	—	—	—	−0.2
Total[a]	1.71	4.41	14.51	26.21	34.11	22.78	24.16	24.61	26.39	23.01	203

$1 = ¥140
[a]Total figures include small amount of additional fee.

Japan) has developed a fairly large scale testing station near Osaka. Fourteen PAFC plants of 50 kW and one PAFC plant of 200 kW (total 900 kW) will be connected to an artificial grid, dispersely, and this project entered into a test program from 1992. Dispersed photovoltaic generators for 400 kW and 2 × 15 kW wind power generators also are connected in parallel with PAFC plants (see Section 8.2.3.1). As described in Section 8.5.3.1, the PAFC Technology Research Association was established under NEDO in 1990 to promote the construction of a 5-MW dispersed and a 1-MW on-site PAFC plant. These plants should begin operation in 1994 and 1995, respectively.

Moreover, to accelerate the commercialization plan of the environmentally benign PAFC plants, a new government subsidy plan is now being promoted in Japan. This plan apparently aims at an early establishment of cost-competitive, fully reliable PAFC plant technologies. These are necessary to smoothly introduce a new technology such as PAFC on the market and at an early stage.

8.6.4. Development Organizations in Europe

As already mentioned in Section 8.2.4, the basic development philosophy for PAFC plants of European firms is to integrate fuel cell stacks manufactured in the United States or Japan with their own systems. Therefore, there is no special PAFC development organization in Europe, but the Netherlands and Italy are promoting their own demonstration plants sponsored by governments, CEC (Commission of the European Communities), and private funds.

In the case of Italy 70% of total funds from 1986 to 1989 were shared by ENEA (Renewable Sources and Energy Saving Department) and the remainder by the Ministry of Defense and EEC. PAFC development programs are also managed by ENEA.

A 1000-kW demonstration plant is now under construction by Ansaldo (Italy) with ENEA funding. One 25-kW PAFC pilot plant was constructed by KTI with contributions from Dutch and CEC, as well as one 80-kW plant.

In addition to these governmentally supported programs, some industrial firms intended to participate privately, as described in Section 8.2.4. The intention of these companies is apparently to first enter a demonstration phase, followed by market development. For example, some electric and gas utility companies in Sweden, Spain, Italy, Germany and Switzerland intend to introduce 50-kW and 200-kW PAFC pilot plants from Japan and the United States with their private funds.

8.7. FUTURE IMPROVEMENTS AND COMMERCIALIZATION FORECAST

The major attraction of a fuel cell power plant is its low environmental impact. Encouraged by the recent worldwide clean air movements, fuel cell development is of great importance. Developmental activities for on-site cogeneration plants and/or the vehicular application, targeted toward commercialization, are stimulated by this new trend. The biggest barrier for commercialization, however, is the cost issue. The target cost, which is considered competitive with the conventional power plant, is around \$1800/kW for small-scale cogeneration plants[61] and around \$1000/kW for large-scale electric utility application. The present prices for these applications are several times higher than the target prices. Developers expect to reach the target cost by improving the stack performance and by mass production of standard design products. Cost reduction is volume-dependent. Developers and users should share the initial burden.

The manufacturers must demonstrate operability and reliability of PAFC plants. Note that the expectation for early commercialization of clean generating plants is now becoming a worldwide requirement.

REFERENCES

1. UTC, *On-Site 40kW Fuel Cell Power Plant Model Specification* (US-DOE and GRI, 1979).
2. E. A. Gillis, *Fuel Cell Research, Development and Demonstration Progress in the United States* (CEC-Italian Fuel Cell Workshop Proc., 1987).
3. J. M. King, R. Falcinelli, and R. R. Woods, *Abstracts Fuel Cell Seminar,* 1986, p. 23.
4. T. Iino, in *Int. Symp. Fuel Cell and Advanced Battery* (Tokyo, 1986).
5. S. S. Penner, *Assessment of Research Needs for Advanced Fuel Cells* (Pergamon, New York, 1986), Fig. 2.5.1, p. 21.
6. S. S. Penner, *Assessment of Research Needs for Advanced Fuel Cells* (Pergamon, New York, 1986), p. 27.
7. K. Fueki and M. Takahashi, Eds., *Fuel Cell Design Technologies* (Japanese) (Science Forum Japan, 1987), p. 20.
8. Osaka Gas, *40kW On-site Fuel Cell PC-18* (catalogue).
9. T. Asada and Y. Usami, *J. Power Sources* **29**, 97–108 (1990).
10. N. Itoh, *J. Power Sources* **29**, 29–36 (1990).
11. R. Vellone, *National R/D and Demonstration Program on Fuel Cells* (CEC-Italian Fuel Cell Workshop Proc., 1987).
12. T. Benjamin, E. H. Camara, and L. G. Marianowski, *Handbook of Fuel Cell Performance* (DOE, 1980), p. 35.
13. T. Benjamin, E. H. Camara, and L. G. Marianowski, *Handbook of Fuel Cell Performance* (DOE, 1980), p. 33.
14. T. Benjamin, E. H. Camara, and L. G. Marianowski, *Handbook of Fuel Cell Performance* (DOE, 1980), pp. 37, 47.
15. T. Benjamin, E. H. Camara, and L. G. Marianowski, *Handbook of Fuel Cell Performance* (DOE, 1980), pp. 39–40.
16. T. Benjamin, E. H. Camara, and L. G. Marianowski, *Handbook of Fuel Cell Performance* (DOE, 1980), p. 41.
17. T. Benjamin, E. H. Camara, and L. G. Marianowski, *Handbook of Fuel Cell Performance* (DOE, 1980), p. 43.
18. T. Benjamin, E. H. Camara, and L. G. Marianowski, *Handbook of Fuel Cell Performance* (DOE, 1980), p. 45.
19. T. Hirota, Y. Yamazaki, and Y. Yamakawa, *Fuji Electric J.* **61**(2), 133–187 (1981).
20. N. R. Udengaard, L. J. Christiansen, and D. M. Rastler, *1988 Fuel Cell Seminar Abstracts,* 1988, p. 47.
21. H. Stahl, N. R. Nielsen, and N. R. Udengaard, *1985 Fuel Cell Seminar Abstracts,* 1985, p. 83.
22. H. Kaneko, K. Yamaguchi, and T. Katagiri, *Fuji Electric J.* **61**(2), 138–142 (1988).
23. M. Sakurai and K. Kondo, *Fuji Electric J.* **61**(2), 159–163 (1988).
24. Onsite Fuel Cell Users Group, Market/Business Assessment Task Force, The Gas Powercel National Market Report, Dec. 1985, p. 5.5.
25. MITI, *Guide Line for Grid Connected Technology of Cogeneration Plant* (Japanese), 1988.
26. S. S. Penner, *Assessment of Research Needs for Advanced Fuel Cells* (Pergamon, New York, 1986), p. 32.
27. Canadian patent #P1229377.
28. S. S. Penner, *Assessment of Research Needs for Advanced Fuel Cells* (Pergamon, New York, 1986), p. 64.
29. US patent, 4463066, *Fuel Cell and System for Supplying Electrolyte Thereto.*

30. L. M. Handley, *Integrated Cell Scale-up and Performance Verification,* EPRI EM-1134, R. D. Breault, *Improved FCG-1 Technology,* EPRI EM-1566, 1980.
31. H. Fukuda and M. Shigata, in 1986 Fuel Cell Seminar, 1986, p. 295.
32. M. Sakurai, K. Ito, and H. Enomoto, in Abstract No. 163, Fall Meeting Honolulu of The Electrochemical Society, 1987.
33. MITI-AIST, *Future Prospect of PAFC Generating Technologies* (*Japanese*) MITI-AIST-M/L FC-1, 1984, pp. 14–15.
34. TEPCO, Fuji Electric, and JGC, *Future Prospects of PAFC Generating Technologies* (*Japanese*), MITI-AIST Working Group, p. 2.1.
35. *Meller's Comprehensive Treatise on Inorganic and Theoretical Chemistry,* Vol. VIII, Suppl. III, p. 669.
36. I. Hori (Chairman), *Electrochemical Industry and Electric Storage System,* survey report (Japanese) (Electrochemical Institute, 1980), p. 118.
37. *Effect of Alternate Fuels on the Performance and Economies of Dispersed Fuel Cells* EPRI EM-1936, 1981.
38. A. J. Appleby and F. R. Foulkes, *Fuel Cell Handbook* (Van Nostrand Reinhold, New York, 1988), p. 63.
39. T. Benjamin, E. H. Camara, and L. G. Marianowski, *Handbook of Fuel Cell Performance* (DOE, 1980), p. 30.
40. J. Hiramoto and R. Anahara, *Fuji Electric J.* **9,** 555 (1982).
41. NEDO Annual Technical Report no. 3, 1983, p. 425.
42. K. Shinobe, K. Suzuki, and H. Keneko, *Fuji Electric Rev.* **34**(2) (1988).
43. N. Eguchi, T. Imura, and N. Kuwayama, *Fuji Electric J.* **34**(2), 143–146 (1988).
44. H. Kaneko and Y. Date, *Fuji Electr J.* **55**(9), 588–594 (1982).
45. K. Okano, *Future Prospects for Commercialization and Present Development Status of Fuel Cells* (Fuji Electric Seminar for Electric Utility Companies, 1989).
46. D. M. Rastler, *J. Power Sources* **29,** 47–57 (1990).
47. MITI-AIST-M/L-FC-03 Working Group, *Future Prospects of PAFC Generating Technologies* (Japanese), 1986.
48. Booz Allen and Hamilton Inc. Project Team, *Fuel Cell/Battery Powered Bus System—Fuel Cell Bus System,* 1985.
49. R. Anahara and I. Mieno, *Fuji Electric Rev.* **34**(2), 80–84 (1988).
50. A. Kaufman, *Fuel Cell Seminar Abstracts,* 1982, p. 42.
51. M. J. Brand, J. J. Early, A. Kaufman, A. Stawsky, and J. Werth, *Fuel Cell Seminar Abstracts,* 1983, p. 115.
52. A. Kaufman, *Fuel Cell Seminar Abstracts,* 1985, p. 33.
53. S. Abens, M. Farooque, and T. Schneider, *Fuel Cell Seminar Abstracts,* 1983, p. 111.
54. S. Abens and M. Farooque, *Fuel Cell Seminar Abstracts,* 1985, p. 139.
55. C. Chi, D. Glenn, and S. Abens, *Fuel Cell Seminar Abstracts,* 1988, p. 365.
56. M. K. Wright and L. E. Van Bibber, *Fuel Cell Seminar Abstracts,* 1982, p. 69.
57. W. A. Summers, J. M. Feret, and M. K. Wright, *Fuel Cell Seminar Abstracts,* 1988, p. 210.
58. Y. Suzuki, *Electrode Material, Latest Technology and Future Problems* (FCDIC Workshop, 18 (Japanese)).
59. R. Anahara, *J. Power Sources* **29,** 109–117 (1990).
60. *Demonstration Test of PC-18 Type PAFC Plant,* FCDIC Seminar, 1987, presented by Tokyo Gas and Osaka Gas (Japanese).
61. N. Hashimoto, *J. Power Sources* **29,** 87–96, 1990.
62. MITI-AIST, *Future Prospect of PAFC Generating Technologies* (Japanese), MITI-AIST-M/L FC-1, p. 21.
63. Tokyo Gas Information, *Development of On-site Fuel Cell at Tokyo Gas,* 1990.
64. *Fuel Cell News,* issued by Fuel Cell Association, June 1990, p. 1.
65. *Research, Development and Demonstration of a Fuel Cell/Battery-Powered Bus System Phase I,* Booz, Allen and Hamilton, 1990 (DOE contract).
66. KTI BV, *Design Optimisation and Environmental Aspects of Fuel Cell Systems,* Report for PEO, 1988, p. 1.

9

Research, Development, and Demonstration of Molten Carbonate Fuel Cell Systems

J. R. Selman

9.1. INTRODUCTION

The aim of this chapter is to give an overview of the technology of the molten carbonate fuel cell (MCFC) and to assess the status of MCFC performance. The MCFC is generally considered a "second-generation" fuel cell, whose entry into the power generation market will follow that of the phosphoric acid fuel cell (PAFC), discussed in Chapter 8. In spite of the greater technical difficulties in its development, the MCFC has undeniable advantages over the PAFC, because of its higher electrical efficiency, the possibility of using natural gas without external reforming, and the high-grade waste heat generated. These characteristics allow a spectrum of applications varying from central power generation to industrial or commercial cogeneration. They are directly connected with the higher operating temperature (typically 650°C).

Similar advantages also apply in the case of the solid oxide fuel cell (SOFC), discussed in Chapter 10, which in its present form operates at 1000°C and consists exclusively of solid components. The lower temperature of the MCFC and its relatively simple method of construction and sealing currently appear to give the MCFC an edge over the SOFC in the prospects for commercialization. However, alternative SOFC concepts capable of lower operating temperature are being studied intensively, and both SOFC and MCFC fabrication methods will undergo further evolution. Eventually there may be room for both systems, each with particular areas of application.

The concept of the MCFC dates back to the late 1940s, and the first MCFC cells were demonstrated by Broers and Ketelaar in the 1950s.[1] The first

J. R. Selman • Illinois Institute of Technology, Department of Chemical Engineering, Illinois Institute of Technology, Chicago, Illinois 60616-3793.

Fuel Cell Systems, edited by Leo J. M. J. Blomen and Michael N. Mugerwa. Plenum Press, New York, 1993.

pressurized MCFC stacks were operated in the early 1980s.[2] Realization of the MCFC, therefore, has been relatively slow. A strong incentive for its development as a utility fuel cell did not exist until the early 1970s, and its effectiveness and reliability were not demonstrated until later in that decade. The high operating temperature, the corrosivity of the molten salt, and the rather complicated electrode processes of the MCFC put many obstacles in the way of a rapid development. Even now some aspects of its chemistry are not fully understood.

In this respect, MCFC technology is not different from other technologies in which empirical discovery of a new concept predated understanding and improvement by research. In the case of the MCFC, several decades of research have certainly contributed much to improved performance. The effectiveness of the MCFC is now beyond doubt, and its development is successfully approaching the 100-kW pilot plant stage.

Section 9.2 presents a brief overview of the chemistry and electrochemistry of the MCFC. More detailed accounts of its early history, and reviews of pioneering MCFC research are available elsewhere.[3–5] In this chapter we emphasize the design and performance of cells and cell stacks.

The factors that determine cell performance are discussed in Section 9.3. This section deals with the design and development of single cells. Much of our knowledge about cell performance was established during the period 1975–1985, in extensive R&D sponsored by the U.S. Department of Energy (DOE) and the Electric Power Research Institute (EPRI). It is available in the form of reports, as well as journal publications. Several conference and symposium proceedings,[6,7] containing detailed references to these reports, provide an overview of the technology in the early and mid-1980s. Particularly valuable is a 1985 review of the technology prepared for the U.S. DOE by the Penner Committee.[8]

Section 9.4 deals, in as much detail as space allows, with the design, operation, and performance of MCFC stacks.

Section 9.5 summarizes briefly the most prominent current R&D topics. It also lists major research and development programs and demonstration projects currently underway. Because of the constant change in detailed objectives and direction which is typical for a growing enterprise, programs, and objectives are only very succinctly sketched.

9.2. CONCEPT AND COMPONENTS

9.2.1. Cell Chemistry

The chemistry of molten carbonate makes it a uniquely suitable electrolyte for fuel cells using carbonaceous fuels, at moderately high temperatures (600–800°C). The carbonate ion, CO_3^{2-}, plays the role of ionic current carried from cathode (positive electrode) to anode (negative electrode), as shown schematically in Fig. 9.1.

The main electrode reaction at the anode is oxidation of hydrogen, which is the principal reactant in the fuel gas:

$$H_2 + CO_3^{2-} \rightarrow H_2O + CO_2 + 2e^-. \tag{9.1}$$

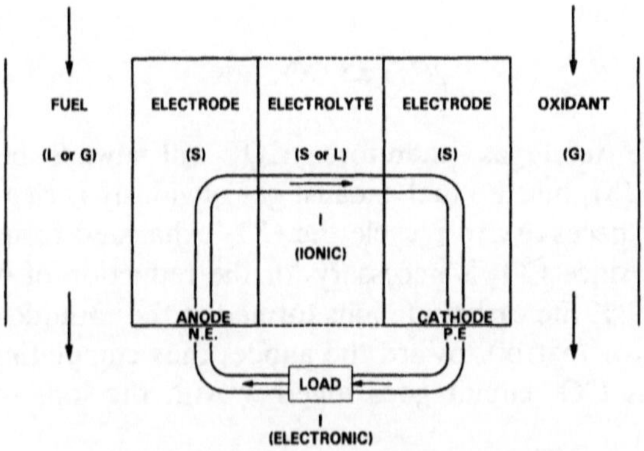

Figure 9.1. Schematic of MCFC operation.

The electrode reaction at the cathode is, overall, the reduction of oxygen and CO_2, which are the reactants in the oxidant gas:

$$\tfrac{1}{2}O_2 + CO_2 + 2e^- \rightarrow CO_3^{2-}. \tag{9.2}$$

The net electrochemical reaction, therefore, is the oxidation of hydrogen, with transfer of two electrons from anode to cathode:

$$H_2 + \tfrac{1}{2}O_2 \rightarrow H_2O. \tag{9.3}$$

As shown in Fig. 9.1, two electrons transferred in the net reaction must be recycled through the external circuit. Likewise, one molecule of CO_2 must be recycled from the fuel gas exit to the cathode gas (oxidant) entrance, or in some other way supplied to the cathode.

At the anode, other fuel gases such as CO, CH_4, and higher hydrocarbons, etc., are oxidized by conversion to H_2. Although direct electrochemical oxidation of CO is possible, it occurs very slowly compared to that of hydrogen. Carbon monoxide oxidation, therefore, occurs mainly via the water-gas shift reaction

$$CO + H_2O \rightleftarrows CO_2 + H_2, \tag{9.4}$$

which, in the temperature range of the MCFC, equilibrates very rapidly at catalysts such as nickel.

Direct electrochemical reaction of CH_4 appears to be negligible. Therefore, CH_4 and other hydrocarbons must be steam-reformed (methanation equilibrium):

$$CH_4 + H_2O \rightleftarrows CO + 3H_2. \tag{9.5}$$

This can be done either in a separate reformer (external reforming) or in the MCFC itself (internal reforming), as discussed in Section 9.2.5 (see also Chapter 4).

Water and CO_2 are important components of the fuel gas. Water, produced by the main anode reaction (9.1), helps to shift the equilibrium reactions (9.1) and (9.5) so as to produce more hydrogen. However, H_2O must be present in the feed gas, especially in low-Btu (high-CO) mixtures, to avoid carbon deposition in the fuel gas channels supplying the cell, or even inside the cell (Boudouard equilibrium):

$$2CO \rightleftarrows CO_2 + C. \tag{9.6}$$

In the anode feed gas, therefore, CO_2 will always be present due to equilibria (4) and (5), but the fuel exhaust gas obviously is rich in CO_2 as well as H_2O. It is usually necessary to recycle the CO_2 exhausted from the anode to the cathode feed gas, since CO_2 is necessary for the reduction of oxygen, Eq. (9.2). As shown in Fig. 9.1, the carbonate ions formed at the cathode move through the electrolyte "tile" (or matrix) toward the anode, thus completing the CO_2 circuit. In the MCFC this CO_2 circuit goes together with the ionic–electronic current circuit.

9.2.2. *Electrodes and Electrolyte*

The electrodes, shown schematically in Fig. 9.2, are porous gas-diffusion electrodes, i.e., electrodes whose porous structure allows extensive contact between reactant or product gas, liquid electrolyte, and the conducting electrode material itself.[9a]

Molten carbonate is, in general, an extremely corrosive medium, but the stability of specific metals and alloys depends strongly on the gas atmosphere. Only a few noble metals are stable under an oxidizing atmosphere such as air or oxygen–CO_2 mixtures. Therefore, from the viewpoint of cost, semiconducting oxides are the only practical cathode materials. The currently used material is lithiated nickel oxide, formed from porous nickel by oxidation and lithiation. Lithiation, as well as oxidation, occur spontaneously when nickel is in contact with a melt containing lithium carbonate under an oxidizing atmosphere, during initial cell operation.

However, nickel oxide is slightly soluble, which severely limits lifetime in various ways (discussed in Section 9.3.9). Alternative cathode materials are being explored.

The MCFC anode operates under a reducing atmosphere, at a potential typically 700 to 1000 mV more negative than that of the cathode. Many metals are stable in molten carbonate under these conditions. Several transition metals are suitable as electrocatalyst for hydrogen oxidation under these conditions. (CO oxidation occurs mainly indirectly, via the shift reaction, Eq. (9.4).) Nickel, cobalt, and copper are currently used as anode materials, often in the form of powdered alloys and composites with oxides. The anode, too, is a porous electrode; as a porous metal structure, it is subject to sintering and creeping

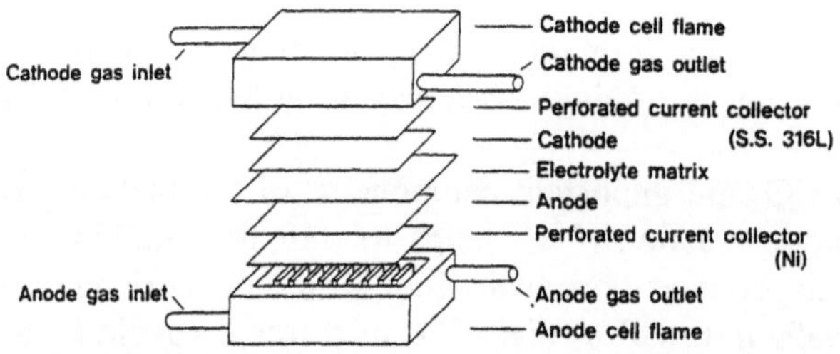

Figure 9.2. Schematic of cell structure. (From Selman and Marianowski.[5])

under the compressive force necessary for stack operation. Additives such as chromium or aluminum form dispersed oxides and thereby increase the long-term stability of the anode with respect to sintering and creeping.

The electrolyte currently used by most developers is a mixture of lithium carbonate (Li_2CO_3) and potassium carbonate (K_2CO_3), and possibly smaller amounts of sodium carbonate (Na_2CO_3) and carbonates of earth–alkaline metals, with a melting point of approximately 500°C. Much effort has been made, and is still being made, to optimize the composition of the electrolyte and the cell operating temperature. Electrolyte composition and cell temperature have a strong effect on the ohmic resistance of the cell and on those factors which determine cell polarization, such as the solubility of various gases and the kinetics of oxygen reduction. In addition, nickel oxide solubility, which limits cell lifetime, is affected by electrolyte composition. This is further discussed in Section 9.3.9.

9.2.3. Cell Structure

The structure of an individual cell is shown schematically in Fig. 9.3. Table 9.1 shows the characteristics of state-of-the-art components used in such cells.

One of the unique features of the MCFC is the use of an *electrolyte tile or matrix* consisting of a mixture of ceramic powder (usually lithium aluminate, $LiAlO_2$) and carbonate electrolyte. In this semisolid mixture, the electrolyte is effectively immobilized. In a fuel cell in general, the electrolyte serves both to separate the reactant gases (fuel gas oxidant) and to conduct the carrier ions (in this case, CO_3^{2-}). In an operating MCFC, the pastelike mixture of ceramic powder ("support") and electrolyte held by capillary force in the interstices between ceramic particles forms an impermeable and stiff, though deformable, layer separating the porous electrodes. This electrolyte tile or matrix concept was adopted after early experiments with frits consisting of sintered ceramic particles showed that they are unstable in molten carbonate and naturally tend to disintegrate.

The plasticity of the tile or matrix is made use of in the MCFC to provide a gastight seal, which is a major challenge in high-temperature fuel cells. Because it

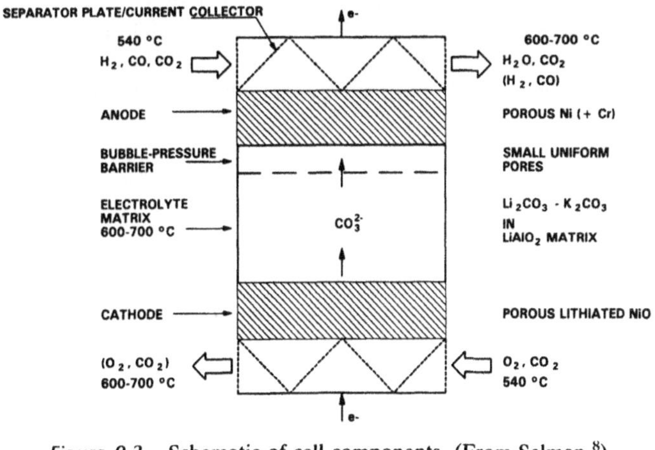

Figure 9.3. Schematic of cell components. (From Selman.[8])

Table 9.1. Characteristics of State-of-the-Art MCFC Cell Components

Property	Anode	Cathode
Material	Ni with 2–20% Cr	Nio with 1–2 wt % Li
Thickness	0.5–1.5 mm	0.4–0.75 mm
Porosity	0.50–0.70	0.70–0.80
Pore size	3–6 μm	7–15 μm
BET surface area	0.1–1 m^2/g	0.15 m^2/g (Ni pretest) 0.5 m^2/g (posttest)

Property	Electrolyte Matrix		
Thickness	1.8 mm (hot-pressed), 0.5 mm (tape-cast)		
Support	γ-$LiAlO_2$		
Support surface area	0.1–12 m^2/g		
Composition	$LiAlO_2$	45.0 wt%	38 vol%
	Li_2OC_3	26.2 wt%	
	K_2CO_3	28.8 wt%	

Property	Anode	Cathode
Current collector	Ni or Ni-plated steel (perforated) 1 mm thick	Type 316 (perforated) 1 mm thick

Adapted from Pigeaud *et al.*[10] and Selman.[11]

is plastic as well as impermeable to gas, the electrolyte matrix can provide an edge seal which prevents escape of gas to the ambient atmosphere. In this "wet seal," shown schematically in Fig. 9.4, the electrolyte tile functions as a gasket between plane metal surface. Figure 9.5a illustrates schematically the wet seal of a single cell, where the electrolyte is held between the plane edges of the metal shells forming the housing. The cell hardward is shown in Fig. 9.5b. The wet seal in a MCFC stack follows the same principle, with variations discussed in Section 9.4.2.

Wet sealing of the cell is practically the only way to achieve gastightness when the cell housing is made of metal. Theoretically, of course, it is possible to use ceramic materials as structural components, in which case a conventional seal could be applied. However, very few ceramics are sufficiently stable in lithium-containing carbonates to function as structural materials. In practice, only high-density alumina is suitable; moreover, a cell made from alumina (or other

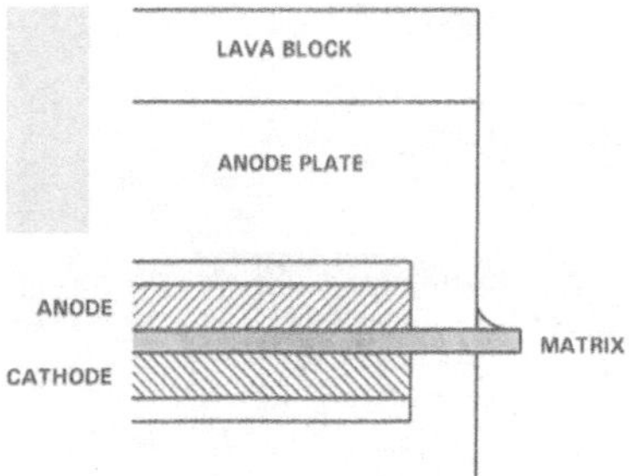

Figure 9.4. Schematic of the wet seal of a molten carbonate fuel cell. (Modified after Ref. 8.)

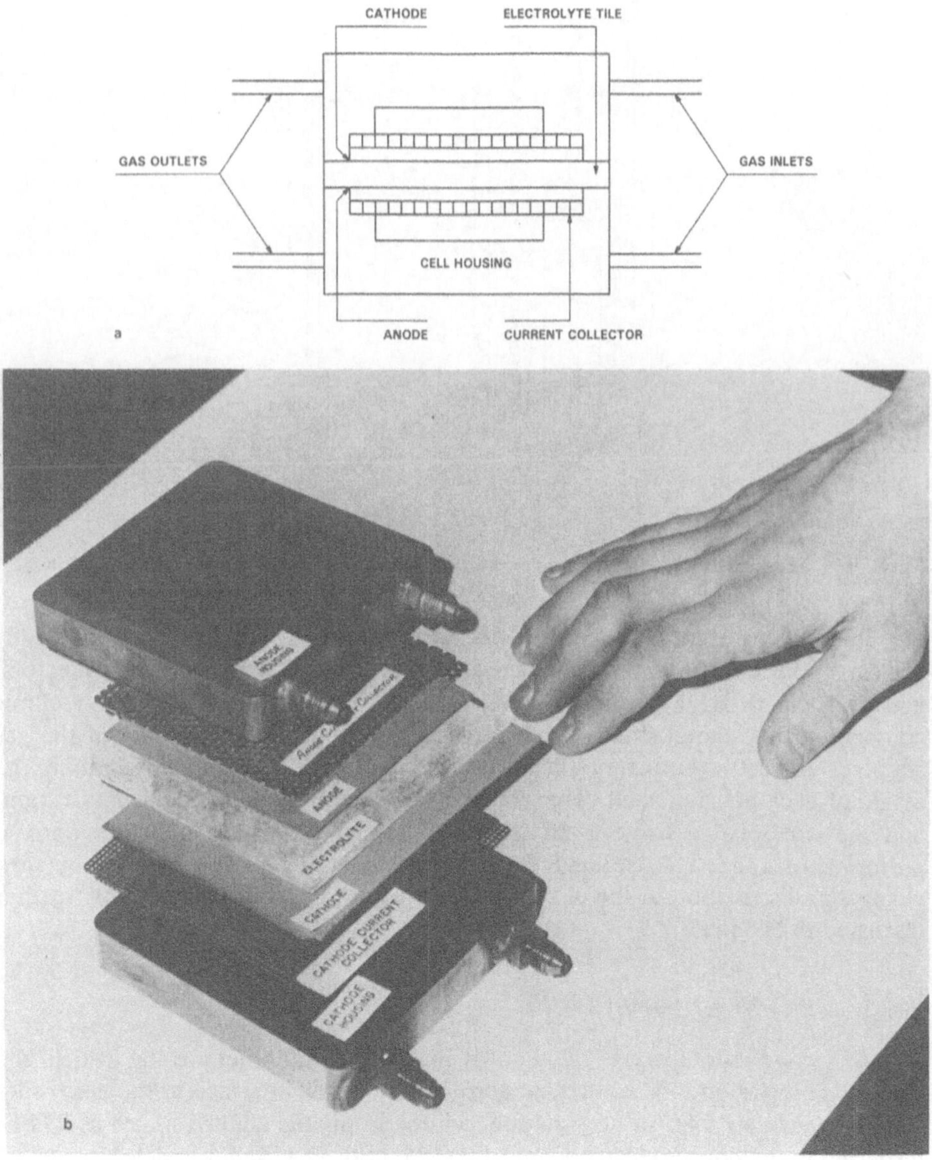

Figure 9.5. Bench-scale molten carbonate fuel cell: (a) schematic (From Ref. 8); (b) single-cell hardward. (Courtesy Institute of Gas Technology, Chicago, IL.)

dense ceramics) would be very sensitive to thermal shock. Nevertheless, small cells with completely ceramic housing are sometimes used for testing electrode materials and structures; a schematic of such a lab cell is illustrated in Fig. 9.6. The all-ceramic housing has certain advantages, in particular, that cell operation is not affected by hardware corrosion, which in small cells tends to skew electrode and cell performance. The lab cell of Fig. 9.6 also incorporates a liquid junction and reference electrode. This makes it possible to determine individual electrode polarizations.

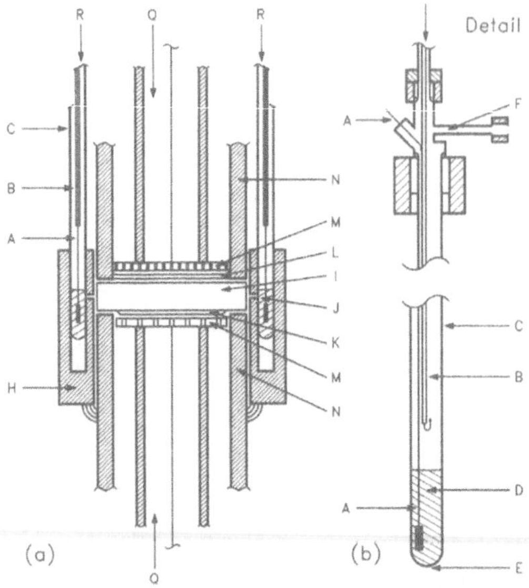

Figure 9.6. Schematic of a laboratory molten carbonate fuel cell. (From Selman and Marianowski.[5])

In a "bench-type" single cell, as shown in Fig. 9.5 (a cell of typically 100 cm^2), the electrodes are contacted on the back side by current collectors, usually stainless steel or nickel metal screens. These ensure a uniform ohmic contact with the metal shells which serve as cell housing. Uniformity of the gas supply to the cell is ensured by means of the manifolds and channels forming the inside of each housing shell. The entire cell assembly is put under compression, and the compressive force is adjusted from time to time to maintain optimal performance under electric load. Such adjustments are made necessary by slow changes in distribution of the electrolyte between the porous components (further discussed in Section 9.3.9).

9.2.4. Electrolyte Distribution

A second key feature of the MCFC cell is the method by which the electrolyte distribution is controlled and the liquid–gas interface in the electrodes established. In a PAFC or alkaline fuel cell, hydrophobic materials such as PTFE, dispersed in the porous electrode matrix, serve both as a binder and as a means of establishing a stable gas–liquid interface. In the MCFC such a method cannot be used since no dewetting material is known that has sufficient stability in molten carbonate under oxidizing conditions. Therefore, a different approach is chosen: *capillary equilibrium as a means of controlling electrolyte distribution.*

The concept of this "fixed-volume capillary equilibrium" method is illustrated in Fig. 9.7. It relies on a balance in capillary pressures to establish the relative amounts of electrolyte in each component (Fig. 9.7a). At thermodynamic equilibrium, the diameters of the largest flooded pores in the porous components are related by the equation

$$\gamma_c \cos \theta_c / d_c = \gamma_e \cos \theta_e / d_e = \gamma_a \cos \theta_a / d_a, \tag{9.7}$$

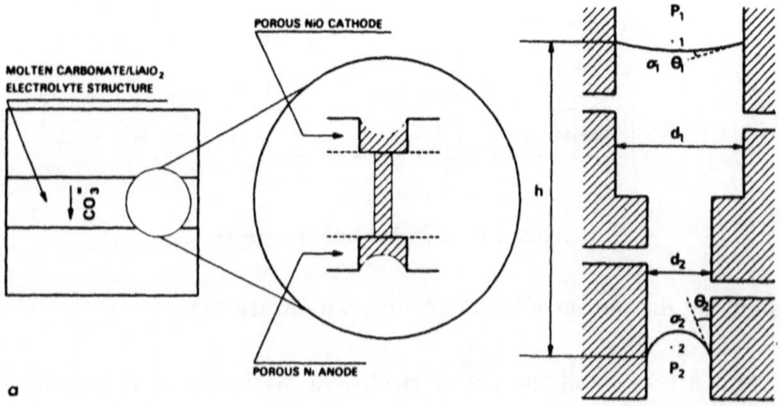

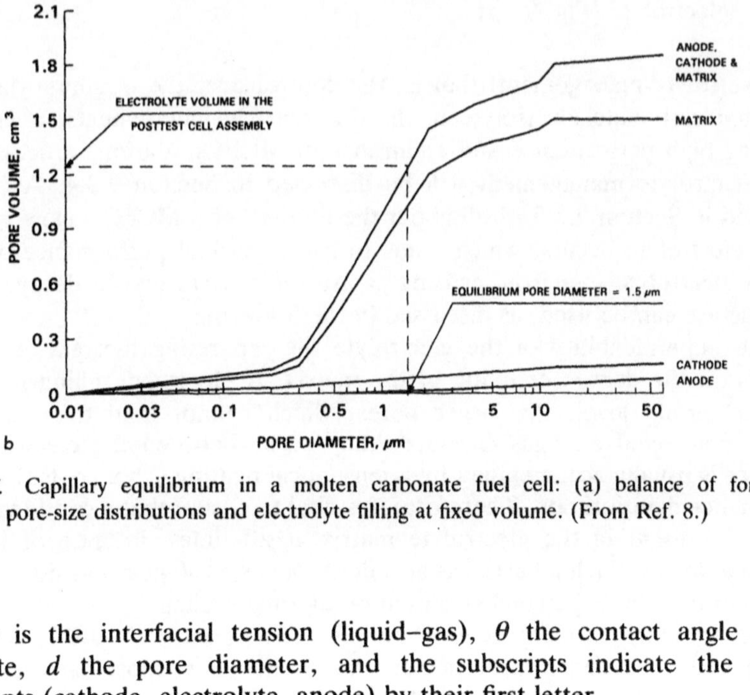

Figure 9.7. Capillary equilibrium in a molten carbonate fuel cell: (a) balance of forces; (b) component pore-size distributions and electrolyte filling at fixed volume. (From Ref. 8.)

where γ is the interfacial tension (liquid–gas), θ the contact angle of the electrolyte, d the pore diameter, and the subscripts indicate the porous components (cathode, electrolyte, anode) by their first letter.

When a completely filled tile is first contacted with empty porous electrodes, a part of the electrolyte wicks out of the tile into the electrodes until the above dynamic equilibrium of capillary forces is established between the three components. If the interfacial tensions of the solid components (with respect to the electrolyte) would be equal, and the wetting angles also equal, then Eq. (9.7) would imply that all components would be filled completely up to the same pore size, as indicated in the schematic of Fig. 9.7b. Therefore, the volume of electrolyte in each component can be determined easily from cumulative pore-volume curves, as in the figure. Actually, the components have different interfacial energies and wetting angles. These must be accounted for, which is usually done by semiempirical correction factors. Thus, by a proper choice of the pore-size distributions of the electrodes as well as the electrolyte matrix, and by

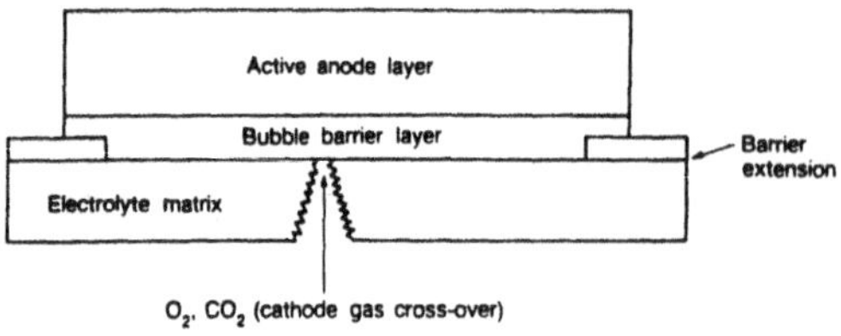

Figure 9.8. Bubble barrier (concept).

partly prefilling the porous electrodes, one can ensure that

1. only a very small amount of electrolyte wicks out of the electrolyte tile, and
2. the amount of electrolyte in each electrode is sufficient to yield the desired liquid–gas interface position, i.e, the largest pore size flooded by electrolyte (Fig. 9.7b).

Electrolyte management, that is, the control over the optimum distribution of molten carbonate electrolyte in the different cell components, is critical for achieving high performance and endurance of MCFCs. Various processes which affect electrolyte management will be discussed in Section 9.3.4. As discussed there and in Section 9.3.9, throughout the lifetime of a MCFC a slow but steady loss of electrolyte occurs, which tends to cause gradual performance decay. To prevent electrolyte loss from causing catastrophic performance decay, a special cell structure can be used, as discussed in the following.

The impermeability of the electrolyte tile separating fuel gas and oxidant depends on the degree of filling of the matrix. If electrolyte filling locally falls below a certain level, the larger pores, which cannot hold their electrolyte, become connected and gas crossover may occur between the electrodes. Gas crossover is usually accompanied by intense local heating ("hot spots") and rapid performance deterioration. Therefore, it is vital to ensure that a continuous liquid barrier is present in the electrolyte matrix at all times, in spite of long-term electrolyte losses. Such a barrier is also desirable as a safeguard against structural damage to the tile (e.g., cracks) caused by thermal cycling.

Consequently, in many cells a "bubble pressure barrier" (BPB) is inserted. This consists of a dense layer of very fine metal or support particles, usually located between the anode and the electrolyte tile (Fig. 9.8). The operation of the bubble barrier, and its characteristics, are further discussed in Section 9.3.4.

9.2.5. Internal Reforming

In a conventional fuel cell system, a carbonaceous fuel is fed to a fuel processor where it is steam-reformed to produce a mixture of H_2, H_2O, CO, and CO_2, which is fed to the fuel cell, usually after further manipulation, and electrochemically oxidized. It was realized early in the development of the MCFC that a suitable fuel cell design might be able to eliminate the need for a separate fuel processor by providing for the reforming of carbonaceous fuels in the fuel cell itself, near the electrochemically active sites. This "internal reforming"

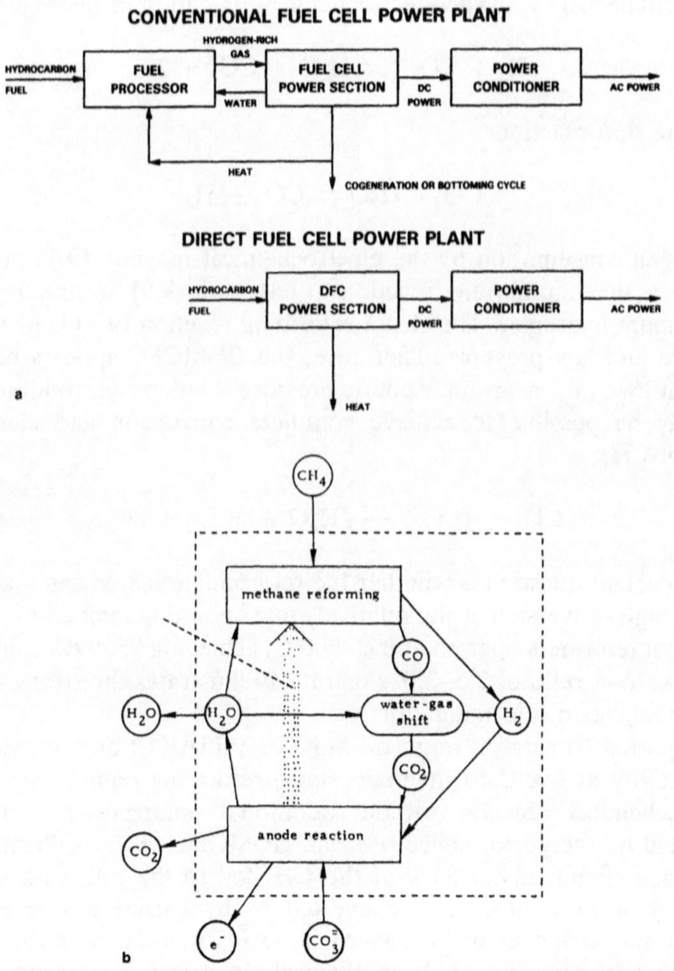

Figure 9.9. Schematic of internal reforming concept: (a) system simplification; (b) processes taking place in direct internal reforming (DIR). (From Ref. 8.)

concept appears feasible in high-temperature fuel cells because the steam-reforming reaction can be sustained by the continuous production of H_2O in the anodic oxidation of hydrogen, with the help of a separate catalyst, as in the MCFC, or without it, as in the SOFC where the nickel–cermet anode doubles as catalyst.

By integrating the reforming reaction with the electrochemical oxidation reaction within the fuel cell, one arrives at the concept of the internal-reforming MCFC (IR-MCFC), illustrated in Fig. 9.9. The IR-MCFC eliminates the need for the external fuel processor with its ancillary equipment, and it is believed to provide a highly efficient, simple, reliable, and cost-effective alternative to the conventional MCFC system.[12]

Methane is the major constitutent of natural gas commonly used to fuel the IR-MCFC. The methane steam-reforming reaction

$$CH_4 + H_2O \rightleftharpoons CO + 3H_2 \tag{9.5}$$

occurs simultaneously with the electrochemical oxidation of hydrogen

$$H_2 + CO_3^{2-} \rightleftharpoons H_2O + CO_2 + 2e^- \tag{9.1}$$

and with the shift reaction

$$CO + H_2O \rightleftharpoons CO_2 + H_2. \tag{9.4}$$

Hydrogen consumption by the electrochemical reaction (9.1) produces H_2O which drives the equilibrium reactions (9.4) and (9.5) to the right, thereby producing more hydrogen. The steam-reforming reaction (9.5) is favored by high temperature and low pressure. Therefore, the IR-MCFC appears best suited to operation at low, i.e., near-atmospheric pressure. Under such conditions it should theoretically be possible to achieve complete conversion and electrochemical oxidation of CH_4:

$$CH_4 + 4CO_3^{2-} \rightarrow 2H_2O + 5CO_2 + 8e^-. \tag{9.8}$$

An important question is whether the reforming reaction can take place with sufficiently high conversion at the relatively low operating temperature of the cell. Conventional reformers operate at 800–900°C, achieving 95–99% conversion with a steam-to-carbon ratio of 2.5–3.0. Figure 9.10 illustrates the strong temperature dependence of steam reforming kinetics.

A supported Ni catalyst (e.g., on MgO or γ-$LiAlO_2$) does provide sufficient catalytic activity at 650°C to maintain steam-reforming equilibrium and sustain the electrochemical reaction without additional polarization. This has been demonstrated by the performance of small IR-MCFC cells, as illustrated in Fig. 9.11. At open circuit, about 83% of the CH_4 fed to the cell, with steam in the steam-to-carbon ratio of 2.0, is converted to hydrogen; this is close to the equilibrium conversion at 650°C, which is 85% for a steam-to-carbon ratio of 2.5.[15] When current is drawn from the cell, hydrogen is consumed and H_2O

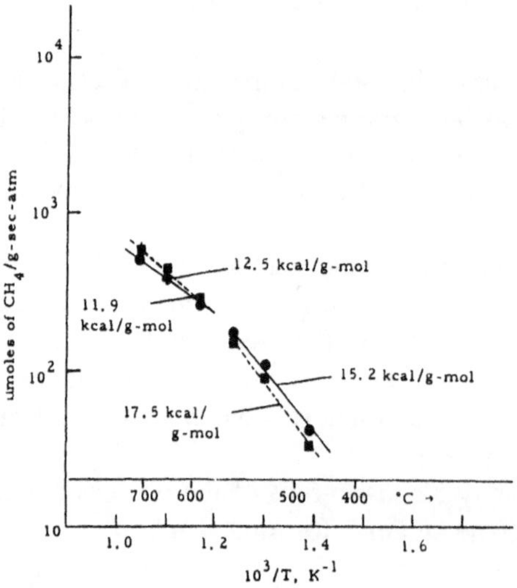

Figure 9.10. Methane-reforming activity of nickel catalyst. (From Maru[13].)

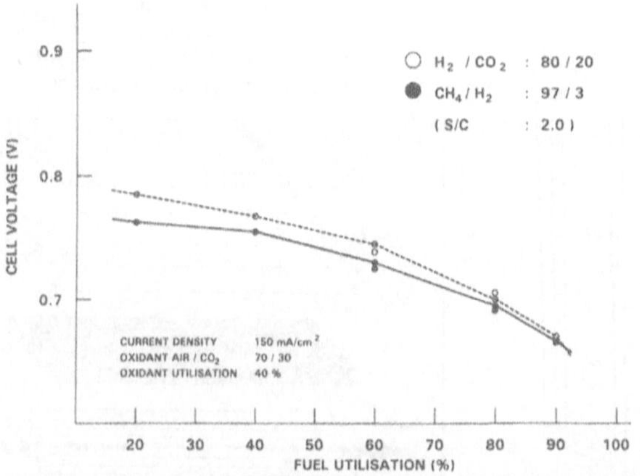

Figure 9.11. Performance of a lab-scale IRMCFC. (From Murahashi.[14])

produced. The conversion of CH_4 then increases and approaches 100% at fuel utilization greater than approximately 50%.

Thus, the concept of internal reforming has been verified convincingly using small cells. If a similar CH_4 conversion and utilization can be achieved in MCFC stacks, IR-MCFC systems of 1.8-MW size should be capable of a heat rate of 6450 Btu/kW · h, or 52.9% chemical-to-ac efficiency (electric).[16]

However, the thermal management of scaled-up IR-MCFC cells presents special challenges to stack design. This will be discussed in Section 9.4.4.

9.2.6. *Stack Structure*

The structure of the single cell (Fig. 9.3) is simply repeated in the MCFC stack (Fig. 9.12). The separator plate (also called bipolar plate) now takes the place of the single-cell housing and provides cell-to-cell electronic contact. Usually one of the current collectors is omitted, or both are, if there is sufficient electronic contact between electrode and separator plate. The separator plate is usually corrugated or profiled to channel gas flow on either side, as illustrated in Fig. 9.13 for the case of a crossflow stack with external manifolding. The corrugation or profile provides electronic contact between electrodes and separator plate, but a constant compression must be applied to the stack to minimize contact resistance between components as the stack ages.

An important advantage of the MCFC is that the cell area can be scaled up to large size without generating excessive mechanical stress. This is due to the plastic nature of the electrolyte tile and the elasticity of the metal separator plate

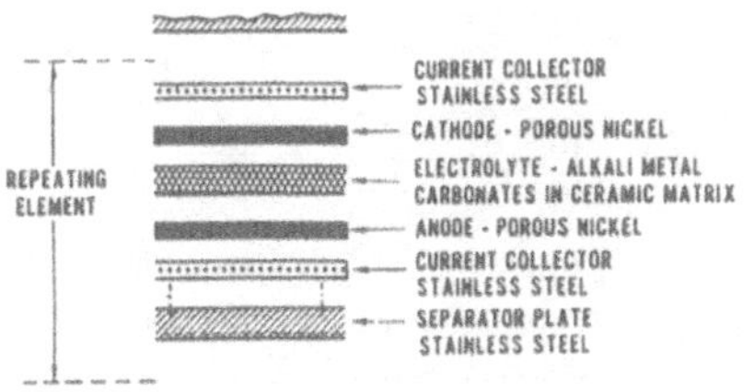

Figure 9.12. Repeating elements in MCFC stack.

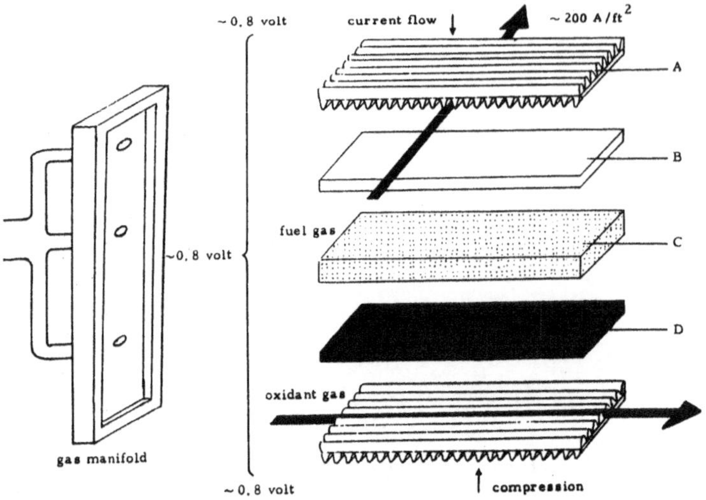

Figure 9.13. View of a crossflow stack. (From Ref. 8.)

which is the main structural support element in the MCFC stack. The stack structure thus absorbs to a considerable extent the strains developed due to nonuniform heating of an MCFC in stationary operation (see also Section 9.2.3). The all-solid SOFC, on the other hand, has a ceramic support structure and is vulnerable to thermoelastic stresses, especially in its planar or monolithic embodiments which require exact matching of thermal expansion between the electrode and electrolyte layers in the ultrathin cell package. These types of SOFC, therefore, appear to be limited to stack lengths of perhaps 20–25 cm.

The fully developed MCFC stack, in contrast, consists typically of cells which, though less than 0.5 cm thick, can each have an area more than 1 m^2 in large-capacity stacks. Therefore, the stack retains, to a high degree, the two-dimensional characteristics of the individual MCFC cell.

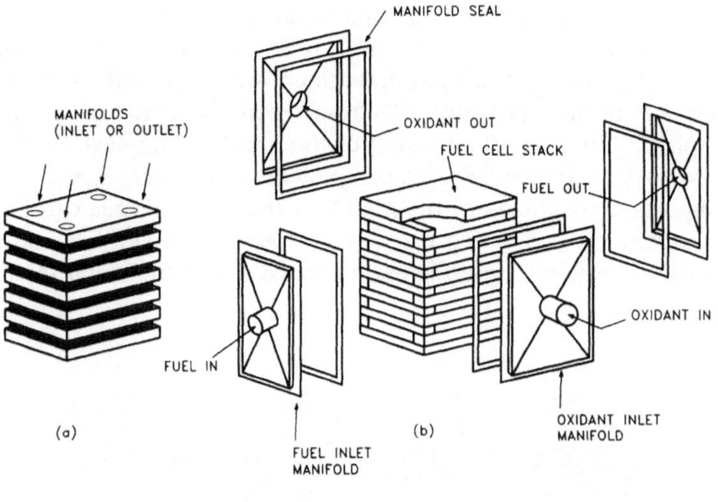

Figure 9.14. Manifolding concepts. (From George and Mayfield,[17] Fig. 5.) (a) internal manifolding, (b) external manifolding. Note that active area of stack (a) is smaller than that of (b) for stacks of apparently equal dimensions.

As suggested by Fig. 9.13, the overall structure of a stack is determined by the method of supplying gas to individual cells. The structure illustrated is that of a crossflow stack, in which the flow directions of fuel and oxidant gas are perpendicular to each other. This configuration is convenient for the external type of gas distribution (manifolding) shown in Fig. 9.13. The two basic forms of manifolding (internal and external) are shown schematically in Fig. 9.14. They will be discussed further in Section 9.4.1.

In a large-area stack operating at high fuel conversion the gradients of current density and temperature can be quite large, and the optimal design of such a stack from the viewpoint of thermal management is a technical challenge. Another important issue in stack design is the choice of material for the separator plate, its structure, and fabrication. These and other aspects of stack design are discussed in more detail in Section 9.4.

9.3. CELL DESIGN AND PERFORMANCE

9.3.1. Gas Utilization

The performance of a single fuel cell, or fuel cell stack, is characterized by voltage-versus-current data, for specified conditions. These conditions include the temperature and pressure of operation and the inlet compositions of the fuel and oxidant gases. In addition, the utilization of the fuel gas and that of the oxidant gas are key parameters.

Utilization of the fuel gas, or fuel utilization (U_f), refers to the fraction of the fuel gas (or of the main reactant in the fuel gas) that reacts electrochemically in the fuel cell. Fuel utilization is easy to define if hydrogen is the only reactant present in the fuel gas. In that case,

$$u_f = \frac{N_{H_2,in} - N_{H_2,out}}{N_{H_2,in}}, \tag{9.9}$$

where $N_{H_2,in}$ and $N_{H_2,out}$ are the molar fluxes (molar flow rates per unit cross-sectional area) of H_2 at the inlet and outlet of the fuel cell, respectively. If hydrogen is consumed by other reactions (e.g., corrosion of cell components) or lost by leakage out of the cell, this will increase the apparent utilization of hydrogen without contributing to the electrical energy derived from the fuel cell.

Oxidant utilization is defined similarly. The MCFC cathode has two reactant gases, oxygen and CO_2, and therefore two ways to define oxidant utilization. Utilization is usually based on the limiting reactant. This may be oxygen, in which case,

$$u_{ox,O_2} = \frac{N_{O2,in} - N_{O2,out}}{N_{O_2,in}}. \tag{9.10}$$

In practice, however, CO_2 is likely to be the limiting reactant, in order to minimize the slow NiO cathode dissolution (discussed in Sections 9.3.9 and

9.3.12). In that case, utilization is based on the consumption of CO_2:

$$u_{ox,CO_2} = \frac{N_{CO_2,in} - N_{CO_2,out}}{N_{CO_2,in}}. \tag{9.11}$$

Since MCFCs are capable of utilizing CO as fuel via the shift reaction (9.4), fuel utilization can exceed the value for consumption of hydrogen based on Eq. (9.9). For example, a typical anode gas composition corresponding to low-Btu gas (see Table 9.2) would be 23.1% hydrogen, 28.8% H_2O, 9.3% CO, 22.8% CO_2, and 16% nitrogen. A fuel utilization of 80% can be achieved with this gas composition, although this would require 12.2% more H_2 than is available in the inlet gas. The shift reaction provides the necessary additional hydrogen for oxidation at the anode:

$$u_f = \frac{N_{H_2,in} + N_{CO,in} - N_{H_2,out} - N_{CO,out}}{N_{H_2,in} + N_{CO,in}}. \tag{9.12}$$

Because the water-gas shift reaction (9.4) is very rapid in the presence of metals such as nickel, the equilibrium relationship for this reaction

$$K_{shift} = \frac{(p_{H_2})(p_{CO_2})}{(p_{H_2O})(p_{CO})} \quad (\approx 2 \text{ at } 650°C) \tag{9.13}$$

is always satisfied. Therefore, it is not difficult to establish the actual gas composition at any point in a fuel cell, if the utilization at that point is specified.

Table 9.2. Fuel Gas Compositions and Their OCV at 650°C

Type	T_{hum} (°C)	Composition H_2	H_2O	CO	CO_2	CH_4	N_2	OCV_{exp}[a] (mV)
Dry gas								
High-Btu	53	0.80			0.20			1116
Med.-Btu	71	0.74			0.26			1071
Low-Btu-1	71	0.213		0.193	0.104	0.011	0.479	1062
Low-Btu-2	60	0.402			0.399		0.199	1030
V.-low-Btu	60	0.202			0.196		0.602	1040
Shift equilibrium								OCV_{theor}
High-Btu	53	0.591	0.237	0.096	0.076			1122
Med.-Btu	71	0.439	0.385	0.065	0.112			1075
Low-Btu-1	71	0.215	0.250	0.062	0.141	0.008	0.326	1054
Low-Btu-2	60	0.231	0.288	0.093	0.228		w.160	1032
V.-low-Btu	60	0.128	0.230	0.035	0.123		0.484	1042
Shift and steam-reforming								OCV_{theor}
High-Btu	53	0.555	0.267	0.082	0.077	0.020		1113
Med.-Btu	71	0.428	0.394	0.062	0.112	0.005		1073
Low-Btu-1	71	0.230	0.241	0.067	0.138	0.001	0.322	1059
Low-Btu-2	60	0.227	0.290	0.092	0.229	0.001	0.161	1031
V.-low-Btu	60	0.127	0.230	0.035	0.123	0.000	0.485	1042

[a]With respect to 33% oxygen–67% CO_2 oxidant.
From Lu and Selman.[18]

Table 9.3. Outlet Gas Composition as a Function of Utilization (650°C, 1 atm)

Gas	Utilization[a] 0	25	50	75	90
Anode					
H_2	0.645	0.410	0.216	0.089	0.033
CO_2	0.064	0.139	0.262	0.375	0.436
CO	0.130	0.078	0.063	0.033	0.013
H_2O	0.161	0.378	0.458	0.502	0.519
Cathode					
O_2	0.300	0.290	0.273	0.231	0.158
CO_2	0.600	0.581	0.545	0.461	0.316

[a]Same utilization for fuel and oxidant. Fuel 80% H_2, 20% CO_2, T_{hum} 25°C (composition based on water-gas shift equilibrium); oxidant 30% O_2, 60% CO_2, 10% N_2.
From Kinoshita *et al.*[19]

Table 9.3 illustrates the gas composition as a function of utilization for reformed natural gas.

In a cell with internal reforming of natural gas (i.e., mainly CH_4), the gas composition does not usually satisfy the equilibrium relationship of the reforming reaction (9.5)

$$K_{\text{ref}} = \frac{(p_{\text{CO}})(p_{\text{H}_2})^3}{(p_{\text{CH}_4})(p_{\text{HO}_2})} \qquad (\approx 0.02 \text{ at } 650°\text{C}) \tag{9.14}$$

because the reforming reaction occurs at a finite rate. This is illustrated in Fig. 9.15.

9.3.2. Cell Voltage

The zero-current or open-circuit voltage (OCV) of a fuel cell is identical to the reversible cell potential, provided the cell is in thermodynamic equilibrium. This is usually a good assumption in high-temperature cells except, for example,

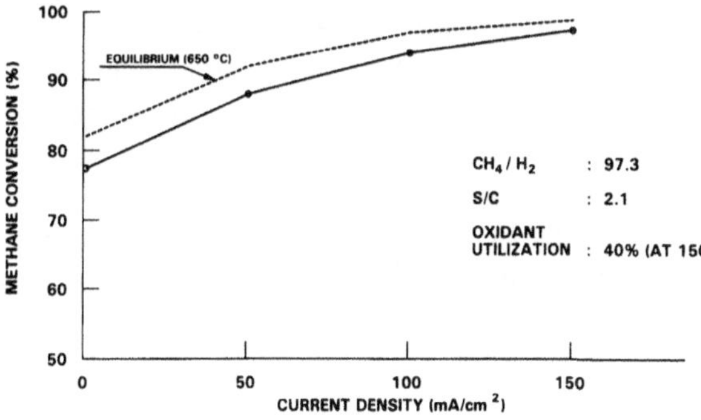

Figure 9.15. Methane conversion as a function of current density. (From Murahashi.[14])

Table 9.4. Cell Parameters and Operating Conditions for Fig. 9.29

Fuel gas (mol%)	H_2 30.4, CO 21.0, H_2O 12.9, CO_2 35.7
Oxidant (mol%)	O_2 15, CO_2 30, N_2 55
Inlet temperature gas	800 K
Pressure	1 atm
Utilization	Fuel 80%, Oxidant 25%
Cell size	40 × 40 cm
Cell voltage	0.75 V
In-plane thermal conductance	0.117 W/K
Gas channel heat transfer parameter	30 cm^{-1}

From Wilemski and Wolf.[45]

Table 9.5. Cell Parameters and Operating Conditions for Fig. 9.31

Fuel gas (mol%)	H_2 2.4, CO 2.7, H_2O 23.2, CO_2 56.1, $CH_4$15.6
Oxidant (mol%)	O_2 15, CO_2 30, N_2 55
Inlet temperature gas	900 K
Pressure	1 atm
Utilization	Fuel 80%, Oxidant 25%
Cell size	40 × 40 cm
Cell voltage	0.75 V
In-plane thermal conductance	0.117 W/K
Gas channel heat transfer parameter	30 cm^{-1}

From Wilemski and Wolf.[45]

when corrosion reactions occur at a very high rate, in which case the mixed potential established between main electrode reaction and corrosion reaction may deviate considerably from the equilibrium value for the main reaction.

The reversible cell potential for a MCFC depends on the gas composition at the fuel electrode (partial pressures of H_2, H_2O, and CO_2) and at the cathode (partial pressures of O_2 and CO_2):

$$E_{eq} = E^0 + \frac{RT}{2F}\ln\left\{\left[\frac{pH_2 \cdot pO_2^{1/2}}{pH_2O}\right]\left[\frac{pCO_{2,c}}{pCO_{2,a}}\right]\right\}, \tag{9.15}$$

where the subscripts a and c refer to the anode and cathode gas, respectively. When the CO_2 partial pressures in cathode and anode gas are identical, the reversible potential depends only on the partial pressures of the net reactants and product (cf. Eq. (9.3)). The values of E^0 (the standard cell potential) for hydrogen oxidation and for several other oxidation and reforming reactions, at 650°, are shown in Table 9.6.

The effect of gas composition on the reversible potential is illustrated in Table 9.2. Between the most hydrogen-rich and hydrogen-poor compositions the difference in OCV is only 70–80 mV. Using Eq. (9.15) one can calculate the cell voltage which thermodynamically corresponds to the outlet gas composition when a cell is operated under load, at a given utilization (cf. Table 9.3). The difference between the OCV of the cell and this "outlet equilibrium voltage" is an important quantity in the analysis of large-cell performance. It is sometimes termed the

Table 9.6. Thermodynamic Characteristics and Voltages of Fuel Cell Reactions at 650°C

Reaction	$\Delta G°$ (kJ · mol^{-1})	E° (V)	KJ · mol^{-1})	$E°_{tn}$ (V)
$H_2 + \frac{1}{2}O_2 \rightarrow H_2O$ (9.3)	−196.92	1.020	−247.45	1.282
$CO + \frac{1}{2}O_2 \rightarrow CO_2$	−202.51	1.049	−283.01	1.467
$CH_4 + 2O_2 \rightarrow CO_2 + 2H_2O$ (9.44)	−800.89	1.038	−800.64	1.037
$CH_4 + H_2O \rightarrow CO + 3H_2$ (9.5)	−7.62	0.010	+224.72	−0.291
$CH_4 + CO_2 \rightarrow 2CO + 2H_2$	−2.04	0.003	+260.62	−0.338
$CO + H_2O \rightarrow CO_2 + H_2$ (9.4)	−5.58	0.029	−35.56	+0.184
$2CO \rightarrow C + CO_2$ (9.6)	−14.62	0.076	−171.36	+0.888

From *JANAF Thermochemical Tables*, 3rd ed. (M. W. Chase, *et al.*, eds.), NSBS, 1985.

Nernst loss. As illustrated in Fig. 9.16, the Nernst loss for reformed natural gas is approximately 200 mV, for a typical utilization of 80%. Note that in Fig. 9.16 oxidant utilization equals fuel utilization. In most large fuel cells, oxidant utilization is approximately half the fuel utilization, and the Nernst loss is smaller than in Fig. 9.16.

At a given current density, utilization depends on the gas flow rates. When the current density in a large-area cell or cell stack is kept constant while the flow rates at one or both electrodes are varied, the cell voltage can be determined as a function of utilization. For commonly used fuel gas compositions and standard oxidant it is found that voltage depends on utilization almost linearly.[20]

To characterize cell or stack performance, current–voltage curves are determined for one or more fuel and oxidant inlet compositions. Each data point on such current–voltage curves, or set of curves, corresponds to a certain fuel utilization at given flow rate and to a certain oxidant utilization at fixed flow rate. The performance of large fuel cells and fuel cell stacks is usually characterized by giving current–voltage data in one or the other of two modes of operation: constant utilization (with variable flow rate) or constant flow rate (with variable utilization).

Figure 9.17 illustrates *constant-utilization performance* curves. Such curves

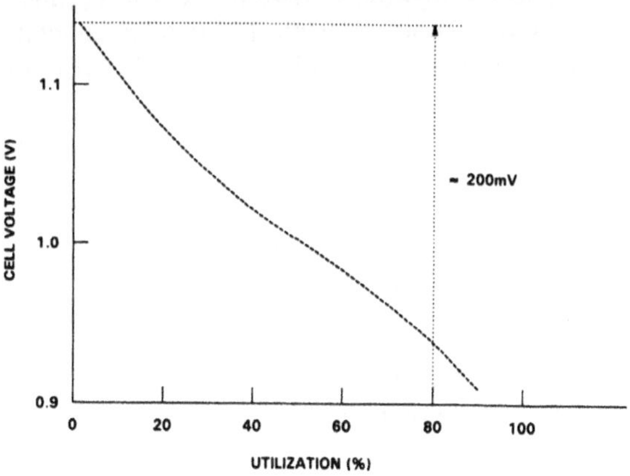

Figure 9.16. Reversible cell voltage as a function of utilization. (From Kinoshita *et al.*[19])

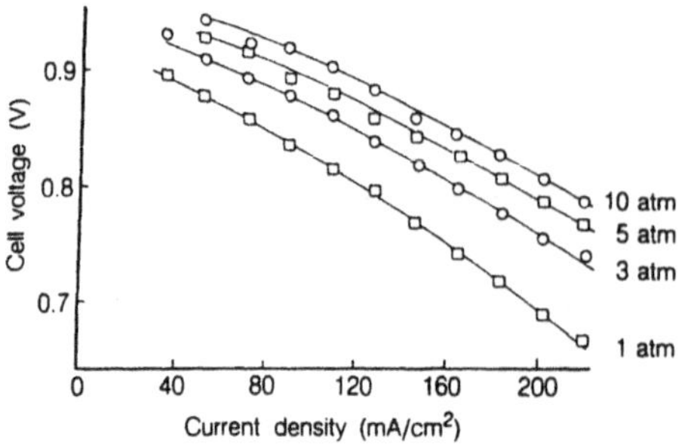

Figure 9.17. Constant-utilization performance curves. (From Baker.[21])

characterize the performance of a fuel cell for more or less similar average concentrations of reactant and product gases. For commonly used fuel gas compositions and standard oxidant it is found that, at constant utilization, voltage depends on current density almost linearly in the current density range of interest for fuel cells.[20] This is not surprising in view of the fact that ohmic resistance of the electrolyte is a major contribution to voltage loss, as discussed below; moreover, the other source of voltage loss, polarization of the electrodes (see Section 9.3.6), is for fixed gas composition practically linear with current density over a wide range of gas compositions.[21]

Constant–flow rate performance curves, though less common, are more practical when the range of current densities applied is very wide. It is practically impossible to extend measurements at constant utilization to very low current density: the flow rates would have to be decreased to such low values that cell operation might become unstable or poorly defined. Conversely, constant-utilization measurements at very high current densities are impractical due to excessively high flow rates, with the risk of temperature gradients or thermal instability.

Not infrequently the performance of different fuel cells is compared to assess the effectiveness of various electrode designs or the catalytic activity of various electrode materials. In that case it is necessary to account for the Nernst loss due to different utilizations. Therefore, polarization curves must be compared at the same utilization (as in Fig. 9.17). Alternatively, small cells may be used which are easily operated at very low utilization, so that the known inlet gas compositions may be assumed to control their operation. Such small cells have been used extensively to explore the polarization behavior as a function of gas composition, pressure, and temperature.

From Eq. (9.15) it follows that pressure increase, e.g., from P_1 to P_2, causes an increase in equilibrium cell voltage

$$\Delta V_p = \frac{RT}{4F} \ln\left(\frac{P_2}{P_1}\right). \tag{9.16}$$

Thus, a fivefold increase in pressure should yield, at 650°C, a gain in OCV of 32 mV, while a 10-fold pressure increase should yield 46 mV. As will be discussed in Section 9.3.5, the actual increase in cell voltage under load is more than twice as large. This suggests the major role that polarization plays in cell performance.

Similarly, the effect of temperature on the performance of an actual cell

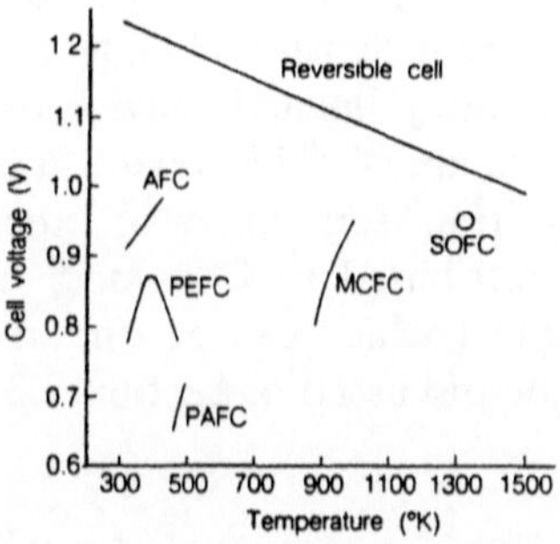

Figure 9.18. Dependence of OCV and initial operating voltage on temperature. (From Simons *et al.*[22])

under load is not satisfactorily predicted by thermodynamics. This is illustrated in Fig. 9.18, which shows that actual cell voltage of the MCFC, although always less than the thermodynamic equilibrium voltage, increases steeply with temperature, while the equilibrium voltage decreases somewhat. Figure 9.17 also demonstrates that MCFC and SOFC are competitive in electrochemical performance, in spite of the difference in operating temperature and equilibrium voltage.

The trends in this figure clarify why for the SOFC 1000°C is a temperature ceiling beyond which performance would improve only marginally. Figure 9.18 also suggests that for both SOFC and MCFC an operating temperature of 800–900°C might be optimal for performance. This temperature range would also be optimal for MCFC system efficiency, since it would provide an optimal match with coal gasifiers. However, MCFC materials problems are markedly magnified at these temperatures; in fact, to ensure better long-term performance there is some interest in exploring lower operating temperatures than 650°C.

The effect of gas composition on performance is very important for the optimal design and choice of operating conditions of fuel cell stacks. Of particular interest is the effect of CO_2 partial pressure on cathode performance, since CO_2 partial pressure must be minimized to decrease nickel oxide solubility and increase cell lifetime (see Sections 9.3.9 and 9.4.8). Figure 9.19 shows the effect

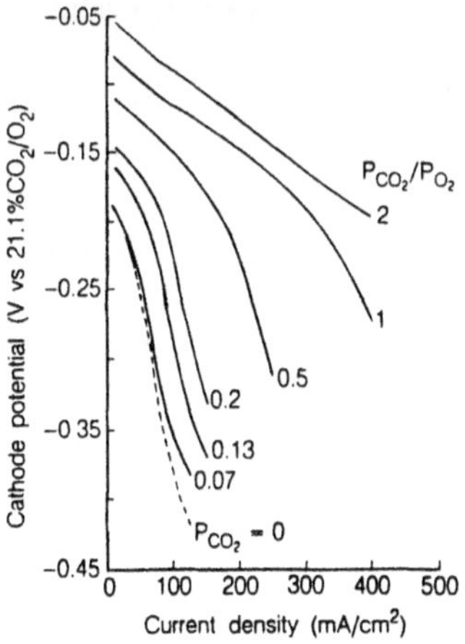

Figure 9.19. Effect of CO_2/O_2 ratio at cathode on performance. (From Ref. 23.)

of CO_2 partial pressure at constant O_2 partial pressure, for total pressure 1 atm. Polarization is linear with current density at high-to-moderate CO_2/O_2 ratio, in agreement with the approximately linear behavior of current–voltage curves at constant utilization, referred to above.[20] However, at CO_2 partial pressures below 0.03 atm a marked polarization sets in when the current density exceeds 100 mA/cm^2. This may be attributed to CO_2 mass transfer resistance. It is an example of a polarization effect which can be understood from the structure of the porous cathode, as will be discussed in Section 9.3.4.

9.3.3. *Efficiency*

To quantify the performance of a fuel cell various definitions of efficiency are used. Some of these compare the theoretical output of electrical energy, i.e., thermodynamic equilibrium work, with the thermal energy release of a completely irreversible reaction. Thus the *thermal efficiency* or *maximum efficiency* of a fuel cell (ε_{th}) is equal to the ratio of free-energy change to enthalpy change:

$$\varepsilon_{th} = \frac{\Delta G_r}{\Delta H_r} = \frac{\Delta H_r - T\,\Delta S_r}{\Delta H_r}. \tag{9.17}$$

Here $T\,\Delta S_r$ is the reversible heat produced, and ΔH_r is the heat of combustion or heating value of the fuel, i.e., the maximum heat energy that can be produced in a reaction at constant temperature and pressure. If the fuel gas contains several reactants, such as H_2 and CO, the quantities ΔG_r, ΔH_r, and ΔS_r are calculated on the basis of 1 mole of fuel gas. The Gibbs free energy of reaction is related to the equilibrium voltage or emf (Eq. (9.15)) by the thermodynamic equation

$$E_{eq} = \frac{-\Delta G_r}{nF}. \tag{9.18}$$

If the cell reaction takes place under standard conditions (i.e., all gaseous reactants and products pure at 1 atm pressure, and all liquid or solid reactants and products likewise pure, under 1 atm pressure), then E_{eq} equals E^0, as can be seen in Eq. 9.17. Table 9.6 illustrates E^0 values for a number of reactions which may occur in the MCFC.

The actual output of electrical energy by a fuel cell is the product of cell voltage and current applied to an external load. This corresponds to the power produced, which is often plotted as a function of current (I) or as a function of the external load (R_e). When the internal resistance is constant, as is often approximately the case for isothermal cells (linear voltage–current behavior, corresponding to internal resistance R_i), maximum power production would be expected at a current corresponding to $E_{eq}/2R_i$, i.e., half the current at zero load (shorting). Large MCFCs and MCFC stacks show indeed an approximately parabolic dependence of power produced on the current (Fig. 9.20), but the relationships are complicated by utilization and temperature distribution effects.

The *energy efficiency* of the cell is defined with respect the product of maximum cell voltage and maximum cell current. This makes the energy efficiency the product of two further efficiencies. The ratio of actual cell voltage,

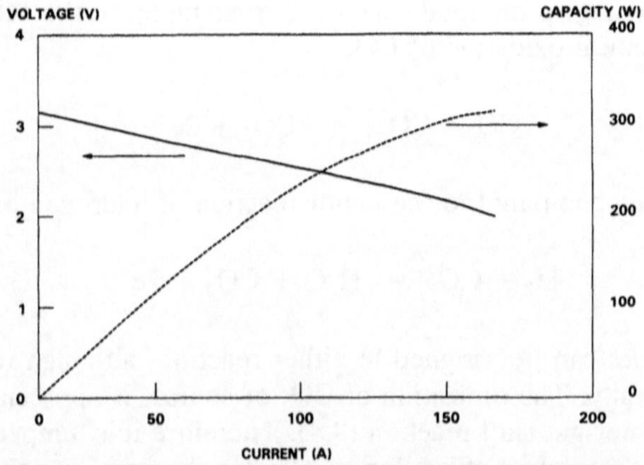

Figure 9.20. Dependence of power produced on cell current in a three-cell stack (area of each cell $1000\,cm^2$). The abscissa represents current, the left ordinate stack voltage, and the right ordinate power. (From Boersma.[24])

E, to maximum cell voltage, E_{eq}, is termed the *voltage efficiency*:

$$\varepsilon_v = \frac{E}{E_{eq}}. \tag{9.19}$$

The other efficiency is that of reactant conversion. As noted earlier, the fuel gas input is usually not completely converted to electric current; i.e., the *utilization* is not 100%. Therefore

$$\varepsilon_E = \varepsilon_v u_f. \tag{9.20}$$

Utilization is actually linked to the *current efficiency* or *Faradaic efficiency* of the fuel cell. In electrochemical cells one or more minor electrode reactions may occur in addition to the main reaction. Corrosion reactions are an example of such minor or parasitic electrode reactions. The current efficiency of the main reaction is then defined as

$$\varepsilon_I = \frac{I_{mr}}{I}, \tag{9.21}$$

where the subscript mr denotes the main reaction, and I_{mr} is the current corresponding to the main reaction. In a fuel cell the major reaction is maintained by a supply of gaseous reactants. Therefore, denoting by N_r the reactant flux toward the electrode, one may write

$$\varepsilon_I = \frac{u_{mr} n F N_r}{I}. \tag{9.22}$$

Here u_{mr} stands for the utilization of the reactant(s) toward the major fuel cell reaction. This implies that these same reactants may also undergo side reactions contributing to the total current. For example, in the MCFC anode the direct electrochemical oxidation of CO,

$$CO + CO_3^{2-} \rightarrow 2CO_2 + 2e^-, \qquad (9.23)$$

is a minor reaction compared to the major reaction of hydrogen oxidation

$$H_2 + CO_3^{2-} \rightarrow H_2O + CO_2 + 2e^-. \qquad (9.1)$$

Current efficiencies can be assigned to either reaction, although this may not be important practically. The utilization of CO, of course, is appreciable but occurs mainly via the water-gas shift reaction (9.4). Therefore it is lumped together with that of hydrogen in the fuel utilization, u_f. Under abnormal conditions, however, corrosion reactions or other parasitic electrode reactions may generate a significant part of the cell current.

9.3.4. Electrode–Electrolyte Structure and Fabrication

Since the mid-1970s, the materials used for electrolyte and electrode structures have been essentially unchanged (Table 9.1), but important developments have taken place in the technology of fabrication.[9] Some of these developments were reviewed earlier by Maru and co-workers,[10,25] Petri and Benjamin,[26] and Selman.[11]

The strength and stiffness of the electrolye structure at cell operating temperature depend on the relative amounts of carbonate and lithium aluminate. At low carbonate content the structure is rigid (essentially particle-to-particle contact), at high carbonate contact it is fluid. In a narrow intermediate region the structure is plastic.[27] The extent of the plastic range depends on the $LiAlO_2$ particle size distribution. The smaller the particles, the higher the carbonate content at which plastic behavior is found to occur.

The typical electrolyte structure used nowadays contains 40 wt% $LiAlO_2$ and 60 wt% carbonate. The electrolyte support is usually finely divided submicron $LiAlO_2$ particles. The size, shape, and distribution of the particles control carbonate retention, mechanical properties, and effective conductivity of the electrolyte tile. The relative stability of the phases in which $LiAlO_2$ can exist determines long-term retention characteristics. Of the three allotropic forms of $LiAlO_2$ (α, β, γ) the γ form is the most stable in Li_2CO_3–K_2CO_3 electrolyte.[28] The most desirable shape for the particles appears to be an elongated rod or fiber shape of <1 μm diameter. Hence "papermaking" technology has been explored, especially by Japanese developers, and found effective.[29]

The conventional process used to fabricate the electrolyte tile (matrix) structure prior to the 1980s involved hot-pressing a mixture of $LiAlO_2$ and alkali carbonates (typically 60–65 wt%, >50 vol% of the latter) at approximately 5000 psi (3.4 MPa) and at temperatures just below the melting point. (For the commonly used eutectic composition 62 mol% Li_2CO_3–38 mol% K_2CO_3 the mp is 488°C.) These electrolyte structures were relatively thick (1–2 mm) and difficult to produce in sizes exceeding 1 m^2 with adequate dimensional and structural uniformity.

The electrolyte structures produced by hot-pressing suffered from excessive porosity (>5%) and high ohmic resistance, as well as poor uniformity of microstructure and, therefore, high frequency of crack formation under stress. To solve this problem, MCFC developers in the early 1980s explored alternative processes for producing thin, uniform structures, such as tape casting[25] and electrophoretic deposition.[30]

Tape casting, which is a commonly used processing technique in ceramic and microelectronic technology, has proved to be the most successful and has been generally adopted. Electrode structures can also be fabricated by tape casting. This greatly helps to streamline the fuel cell production process, since the cell package is assembled from similarly produced tapes, i.e., electrolyte, carbonate, and electrode tapes.

The tape-casting process involves dispersing the ceramic ($LiAlO_2$) or metal (Ni) powder in a solvent (usually organic) containing organic binders, plasticizers, and additives. The formulation must yield a suitable consistency and viscosity of the resulting mix ("slip"). The slip is cast over a smooth substrate, and the desired thickness of the spreading film is controlled by a knife-edge device ("doctor blade"). In large-scale production the slip is cast continuously onto a plastic sheet forming a belt conveyor and is sheared to uniform thickness by flowing under the stationary doctor-blade.

After drying, the slip becomes a semistiff "green" structure, which is assembled into the fuel cell structure. The organic binder is decomposed by heating to 250–300°C ("burnout"). Electrode tapes (e.g., nickel anodes) must be subsequently reduced by heating in a reducing atmosphere at approximately 900°C.

After assembly of the various tapes, the cell package is slowly heated to the fuel cell operating temperature, during which the alkali carbonate melts at approximately 490°C and is absorbed into the ceramic structure. Obviously, the heating process must take place at a carefully programmed rate to avoid excessive degassing and ensure uniform impregnation by electrolyte.

The tape-casting and electrophoretic deposition processes are amenable to scale-up and capable of producing very thin electrolyte structures (0.25–0.5 mm). From the viewpoint of short-term performance only, the electrolyte matrix should be as thin as possible to reduce ohmic resistance. At a current density of 160 mA/cm^2, the ohmic voltage drop of a 1.8-mm-thick electrolyte structure, with a specific conductivity of 0.3 ohm$^{-1}\cdot$cm^{-1} at 650°C (approximately one third of that of the carbonate mixture, for a tile containing 55 wt% carbonate, 45 wt% $LiAlO_2$[31]), was found to obey the relationship

$$\Delta V^{\mathrm{ohm}} \quad [\mathrm{V}] = 0.533t \quad [\mathrm{cm}], \tag{9.24}$$

where t is the thickness (in cm).[30] According to Eq. (9.24) the ohmic drop of the fuel cell could theoretically be reduced from 96 mV to a mere 13.5 mV by employing an 0.15-mm-thick electrolyte structure. However, such a thin electrolyte structure would increase the risk of enhanced gas crossover by liquid-phase diffusion as well as via opening of pores due to long-term electrolyte loss. In addition, the thinner the cell the greater the likelihood of shorting due to cathode dissolution and nickel reprecipitation (see Sections 9.3.9 and 9.3.11).

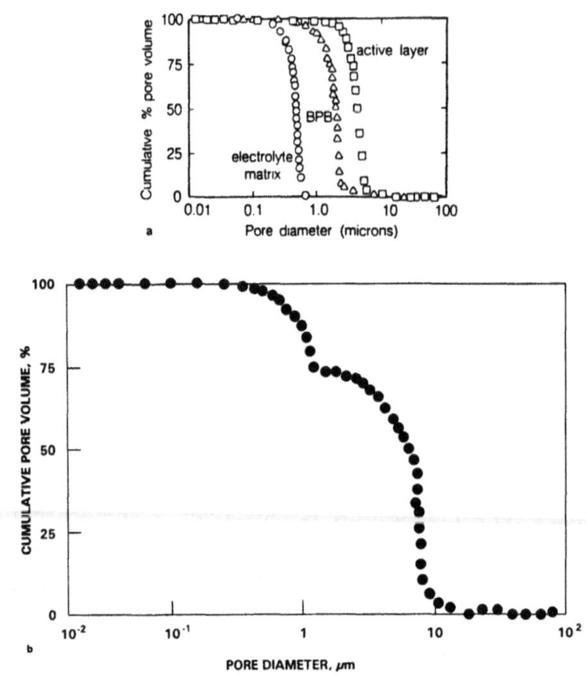

Figure 9.21. (a) Schematic of bubble-pressure barrier (BPB) action in a MCFC; (b) typical pore-size distributions of BPB and other components. (From Iacovangelo and Karas.[32])

Therefore, electrolyte structures in recent cells are usually not thinner than approximately 0.4 mm.

An important feature added to MCFC technology in the 1980s is the bubble pressure barrier. The concept of the BPB was discussed in Section 9.2.2. It consists of a dense, fine-pore-size metal (nickel) or ceramic ($LiAlO_2$) structure usually located between the anode and electrolyte matrix.[32] Due to its very fine pore size, it remains always filled with electrolyte. Its purpose is to prevent gas crossover from one electrode to the other and to reinforce the electrolyte matrix.

The BPB can conveniently be fabricated as an integral part of the tape-cast anode structure; in that case it is made of very fine nickel power. A schematic representation of such a structure and the pore-size distributions of the components are shown in Fig. 9.21. To design an effective BPB requires careful matching of its pore-size distribution with those of the anode and electrolyte matrix. If the median pore size of the anode is 4–8 μm and that of the electrolyte matrix 0.5–0.8 μm, then an appropriate size for the BPB pores would be 1.0–1.5 μm. Such a combination prevents drainage of electrolyte from the BPB part of the matrix (tile) and therefore prevents gas crossover due to long-term electrolyte loss or in the event a crack develops in the matrix.

9.3.5. Electrolyte Management

"Electrolyte management" is the term used to denote control over the quantity and distribution of molten carbonate electrolyte among the different cell components, throughout the life of a cell or stack, so as to optimize performance and extend lifetime. In the design of a cell the quantity and initial distribution of electrolyte among cell components is determined by properly coordinating the pore diameter distribution of the electrodes with that of the electrolyte tile (or matrix). The principle on which this is based, the "fixed-volume capillary equilibrium method," was discussed in Section 9.2.3 and is illustrated in Fig. 9.7a,b.

During the lifetime of a MCFC various processes occur which cause loss of a small part of the electrolyte and redistribution of the electrolyte remaining in the cell. This has consequences for the long-term performance of the cell.

Electrolyte is lost, slowly but continually, due to corrosion processes and volatilization. Electrolyte may have a tendency to flow very slowly out of or into a cell via the wet seal. This liquid displacement process (discussed in more detail in Section 9.3.9) is driven by capillary forces ("creepage"), galvanic corrosion cells, and migration, i.e., ionic motion due to an electric field. In cell stacks, such electrolyte flows may be powerfully reinforced, and electrolyte loss from some cells in a stack may become quite significant (Section 9.4.6).

Whenever electrolyte loss occurs, the remaining electrolyte will be redistributed to satisfy the capillary equilibrium among cell components. However, even without electrolyte loss, redistribution would be likely to occur. This is due to changes in the pore-size distributions of the electrodes, which are quite pronounced in the early stages of cell operation but continue to occur throughout the life of the cell. (Changes in the particle-size distribution of the electrolyte

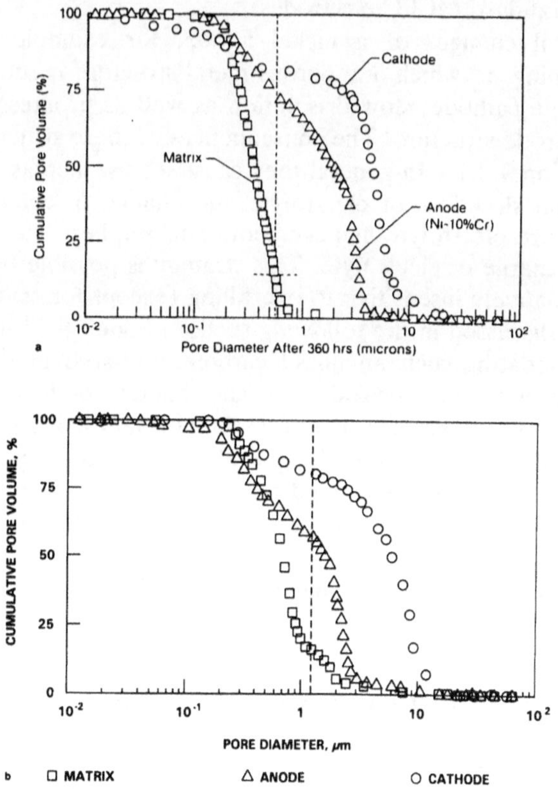

Figure 9.22. Evolution of pore-size distribution of standard MCFC components: (a) after 360 h; (b) after 1766 h. (From Ref. 33 and Ref. 8.)

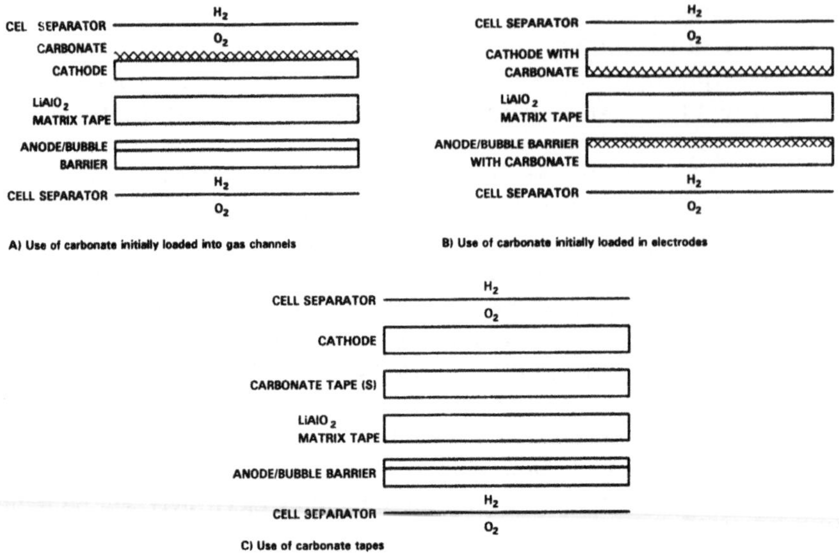

Figure 9.23. Methods of loading carbonate in the cell package. (From Marianowski *et al.*[34])

support, i.e., $LiAlO_2$, are possible but seem to be negligible with presently used formulations for tile fabrication.) Figure 9.22 illustrates the evolution of pore-size distributions of standard MCFC electrodes.

Morphological changes of a nickel anode, for example, are caused by sintering or creeping, to which any porous metal structure is subject. In the case of the nickel oxide cathode, slow dissolution as well as progressive compression changes the electrode structure. The consequences of these structural changes are discussed in Section 9.3.10. In general they cause a slow increase in polarization.

The continual slow loss of electrolyte and change in component pore-size distributions require electrolyte management. This implies, first of all, providing an excess initial charge of electrolyte. This strategy is possible because electrode polarization is relatively insensitive to overfilling (except for severe overfilling of the cathode), as discussed in the following section. Figure 9.23 illustrates various ways of accommodating such an initial carbonate charge in the tape-cast cell composite: adjacent to the electrode in a gas channel, or loaded in one of the electrodes, or as one, or more, separate carbonate tapes. Intermittent additions of electrolyte, if the cell (and stack) designs allow it, are also a possible method of electrolyte management. In the second place, the compression of the cell stack must be adjusted continually so as to compensate for the effect of structural changes.

These aspects of electrolyte management are discussed in detail by Maru *et al.*[35] and Kunz.[36]

9.3.6. Electrode Microstructure and Polarization

As discussed in Section 9.2, the choice of materials for the MCFC electrodes is rather limited. Moreover, the absence of a wet-proofing material resistant to molten carbonate makes it necessary to maintain careful control over the pore-size distributions of the electrodes as well as the electrolyte tile. As

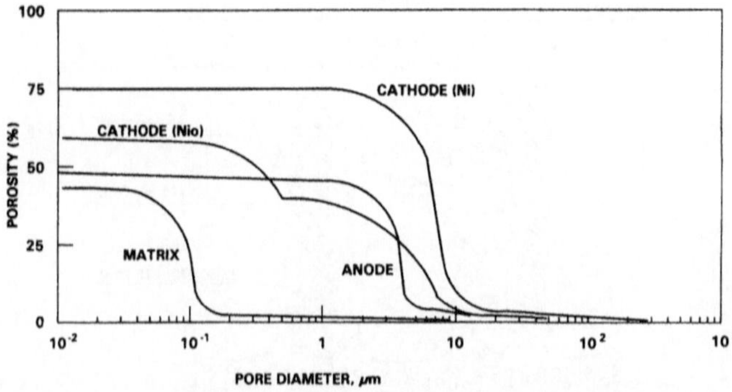

Figure 9.24. Pore-size distribution of NiO cathodes. (After Itoh *et al.*[37])

illustrated in Fig. 9.7 the pore-size distributions determine the distribution of electrolyte among these three components and, therefore, the degree of gas–liquid contact inside the porous electrodes. The performance of the electrodes, i.e, their polarization depends on this.

MCFC cathodes consisting of lithiated NiO start as nickel plaques or sinters which typically have a preoxidation porosity of 70–80%. This is reduced to 55–65% by *in situ* oxidation. They have initially a mean pore size of 10 μm but develop, as Fig. 9.24 shows, a bimodal pore-size distribution with median 5–7 μm diameter. The small pores are flooded with electrolyte and provide thereby an extended reaction surface as well as a cross-sectional area for the ionic conduction path. The large pores remain open and provide a cross-sectional area for the diffusion of gas into the interior of the electrode.

Lithiation of the *in situ* formed oxide takes place to approximately 2 at%. This yields an electronic conductivity of approximately 5 $ohm^{-1} \cdot cm^{-1}$, which is adequate for the cathode operation. Alternatively it is possible to fabricate NiO *ex situ* from nickel oxide powers.[38]

The thickness of the NiO cathode used in MCFCs ranges from 0.4 to 0.8 mm. Cathode polarization is affected by cathode thickness, since ohmic losses in both the liquid conductor (electrolyte) and solid conductor (nickel oxide) increase as the thickness increases. Diffusional losses in the gas phase also increase with thickness; however, losses due to liquid diffusion resistance and kinetic activation decrease as the small pore area increases. Therefore, minimum polarization occurs at a certain cathode thickness; this optimum thickness depends on gas composition and current density.[39]

Another factor which controls polarization is the extent to which the pores in the cathode are filled by electrolyte. On closer inspection, the nickel oxide cathode consists of a network ("agglomerate") of mostly very small NiO particles, with the electrolyte filling the crevices (micropores) between individual particles. The agglomerates are separated by gas-filled channels (macropores), which correspond to the pores of the original nickel plaque or sinter (Fig. 9.25). Mathematical models based on this concept yield performance predictions in good agreement with experimental data for small cells.[40,41]

Lithiated NiO is wetted completely by molten carbonate, so the agglomerates are entirely covered with a thin film of electrolyte. This causes some diffusion resistance even under optimal operating conditions, i.e., at 15–30% filling of the

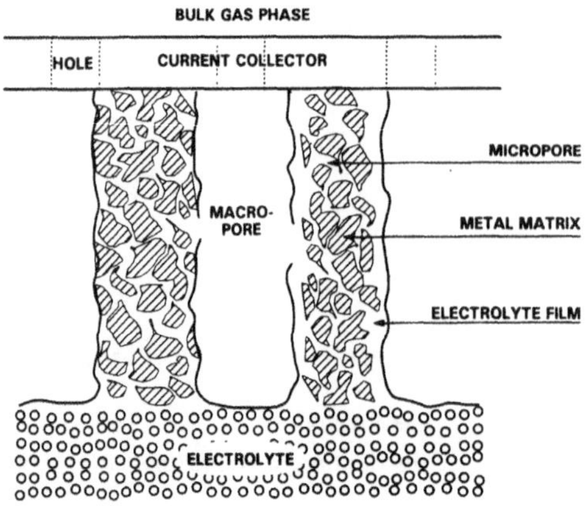

Figure 9.25. Schematic of agglomerate structure of MCFC anode (left half: nonwetted agglomerate) and cathode (right half: film-covered agglomerate). (From Selman.[8])

total pore volume by electrolyte. However, when the macropores become filled with electrolyte ("flooding"), diffusion resistance increases steeply. The nickel oxide cathode, by the nature of its strucuture, is especially sensitive to the degree of filling. This is illustrated in Fig. 9.26.

MCFC anodes are at present made from porous sintered nickel, containing a few percent chromium or aluminum which is oxidized *in situ* and forms submicron particles of $LiCrO_2$ or $LiAlO_2$ on the surface of the nickel particles forming the sinter. The function of the oxides is to prevent sintering of the nickel particles and to stabilize the sinter against creeping, which tends to occur in a stack under compression. Sintering results in increased particle size and therefore reduced capacity of the anode to hold electrolyte. Creep, i.e., microdimensional deformation under mechanical load, results in decreased porosity, increased contact resistance, and increased risk of anode-to-cathode gas leaks. Dispersed oxide strengthening, such as by chromium or aluminum addition, has been shown to be effective against both sintering and creep.

Anodes have a porosity of 55–70%, with a mean pore size of approximately

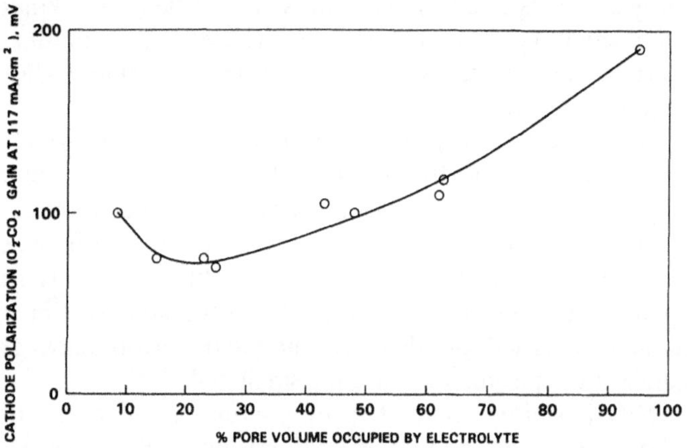

Figure 9.26. Effect of electrolyte filling on cathode polarization. (From Selman and Marianowski.[5])

5 μm. The anode *in situ* has a small fraction of micropores which are filled with electrolyte, and a majority of medium-to-large pores which function as gas channels or are partly filled with electrolyte. It is likely that the walls of the larger pores are not covered with an electrolyte film, since nickel is not wetted completely by molten carbonate in a reducing atmosphere, but this depends largely on the distribution of wettable oxide particles over the pore walls.

MCFC anodes typically have a thickness of 0.8–1.0 mm. The anode needs less internal surface area than the cathode, because the electrode kinetics of hydrogen oxidation as well as mass transfer in the micropores of the nickel anode are more rapid than the corresponding processes in the cathode. Therefore, the thickness of the anode could be less than that of the cathode; the minimum might be 0.4–0.5 mm for atmospheric operation. Nevertheless, the anode is usually made thicker and filled to 50–60% of its pore volume with electrolyte.

The anode is much less sensitive to overfilling than the cathode. As in the cathode, polarization is minimal for a certain degree of filling, but the minimum covers a fairly wide range (Fig. 9.27). Likewise, polarization does not increase markedly with thickness. The reason is that, as discussed above, electrode kinetics and mass transport are relatively rapid at the anode. This concentrates the electrode reaction close to the meniscus, wherever that may be. Moreover, the level of polarization is not high because of the rapid kinetics. Therefore the

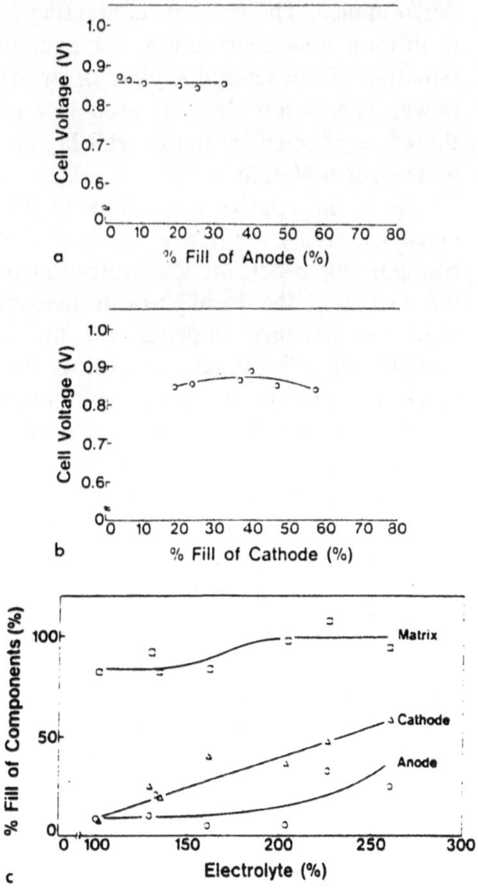

Figure 9.27. Comparison of (c) electrolyte filling effect on cell components and of effect on performance of (a) anode and (b) cathode. (From Itoh *et al.*[37])

anode can serve as an electrolyte reservoir. Excess electrolyte will be wicked from the anode to the electrolyte tile and cathode if the latter become depleted, to restore the capillary equilibrium (see Section 9.2.4).

In the last few years, an intensive effort has been made to identify and develop alternative materials for both the anode and the cathode. Some promising candidate materials have emerged and these are discussed in Section 9.3.12, in the context of long-term MCFC performance.

9.3.7. Current and Temperature Distribution

As discussed in Section 9.2.3 a cell consists basically of a thin cathode–electrolyte–anode package contacted on the anode side by fuel gas and on the cathode side by oxidant. In a cell stack it is separated from each adjacent cell by a thin impervious but conducting separator; in a single cell the cell package is held by a cell housing, which is usually made of metal and consists of two rather massive "holders." (Only small single cells may have a single ceramic housing; see Fig. 9.6.) Inside the holders, current-collector plates enhance the electric contact with the electrodes, which is further ensured by compression.

Thus far, the structure and characteristics of individual cell components have been discussed. These determine ohmic resistance and polarization and are indeed very important for cell and stack performance. Also, clearly, intimate contact of fuel gas and oxidant with the electrodes is critical for good performance. The gas flow contacting each side of the cell is channeled to ensure a uniform flow distribution, for example by means of a grooved or corrugated structure of the separator plate or of the housing. However, in fuel cell stacks for power generation the cell area is typically 0.1–1 m^2, and cell performance is therefore affected by the overall layout of the gas flows (flow configuration) and by the gas flow rate.

Even in relatively small cells ("bench-scale" cells, typically 100–300 cm^2 electrode area), the performance characteristics depend strongly on the gas flow configuration when the gas utilization is significant. This can be understood from the fact that the local current density across the thin electrode–electrolyte–electrode package depends on the thermodynamic driving force, i.e., the equilibrium cell voltage, as well as the local "resistance" (ohmic and polarization). The driving force, i.e., the equilibrium voltage, is different from point to point in the cell due to the increasing utilization of fuel gas and oxidant as they each flow from inlet to exit. The actual cell voltage, i.e., the equilibrium voltage minus the losses due to ohmic potential drop and polarization, must, of course, be everywhere the same since the cell housing (or separator plate) is, by design, a good ohmic conductor and therefore at a uniform potential.

A quantitative development of these considerations has been presented elsewhere, as part of an overall approach to performance modeling applied to fuel cells.[42] Mathematical models are useful to understand an predict the distribution of the current density and the temperature in single cells and in cell stacks. The potential balance concept described above is the basis for one of the key equations in cell performance models.

The following potential balance must be satisfied at any point in the cell plane:

$$V = E - iR_t - iZ_{an} - iZ_{cath}, \tag{9.25}$$

where V is the actual (uniform) cell voltage, E the local equilibrium cell voltage, i the local current density, R_t the ohmic resistance, and Z the "polarization resistance" at either electrode. The polarization resistance is, of course, not a true constant, but can be treated by approximation as an effective constant.

The local equilibrium cell voltage, E, is given by the Nernst Eq. (9.15), which contains the partial pressures of reactant and product gases. The local values of these partial pressures are obtained by applying a mass balance to each component. The local current density is related to the change of reactant flow rate, e.g., for hydrogen,

$$i = -nF\left(\frac{dN_{H_2}}{dx}\right)h_a, \tag{9.26}$$

where n is the number of electrons transferred, F Faraday's constant, N_{H_2} the hydrogen molar gas flux, and h_a the height of the fuel gas channel. Using the definitions of utilization presented in Section 9.3.1., and integrating Eq. (9.26), the equilibrium cell voltage E can easily be related to the local fuel utilization and oxidant utilization. As utilization increases, E and, therefore, i decrease at a rate which depends also on the importance of ohmic resistance [R_t in Eq. (9.25)] and polarization (Z_{an}, Z_{cath}).

To quantify the polarization, one can use either porous electrode models or empirical correlations based on measurements using small laboratory cells at negligible utilization. As an example of the latter, Yuh and Selman[43] report

$$Z_{an} = 2.27 \times 10^{-5}(p_{H_2})^{-0.42}(p_{CO_2})^{-0.17}(p_{H_2O})^{-1.0}\exp\left(\frac{53{,}500}{RT}\right), \tag{9.27}$$

$$Z_{cath} = e^{-11.8}(p_{O_2})^{-0.43}(p_{CO_2})^{-0.09}\exp\left(\frac{77{,}300}{RT}\right), \tag{9.28}$$

where Z is in $\text{ohm}\cdot\text{cm}^{-2}$, the partial pressures are in atm, and R is $\text{J}\cdot\text{g mol}^{-1}\cdot\text{K}^{-1}$.

From these expressions it is clear that temperature plays an important role in polarization resistance. This is mainly due to the large activation energy of electrode kinetic reactions, but also to the temperature dependence of transport processes. Both work to decrease polarization resistance. Ohmic resistance, mainly due to the electrolyte, also decreases with temperature. Local heat generation, on the other hand, increases more than linearly with the current density, since irreversible (ohmic) heat generation is proportional to the square of current density. Therefore, if heat is not conducted or convected away from high current density areas, the current density will be amplified, leading to "hot spots." Figure 9.28 illustrates schematically several ways in which heat is generated and transferred, by conduction and convection, in one cell of a fuel cell stack bounded on both sides by a ribbed separator plate.

In most single cells, especially in small cells such as bench-scale cells, a uniform cell temperature is established due to the large thermal capacity and thermal conductivity of the metal housing. Therefore, the operation of such cells can be considered isothermal. Their performance may be characterized by a

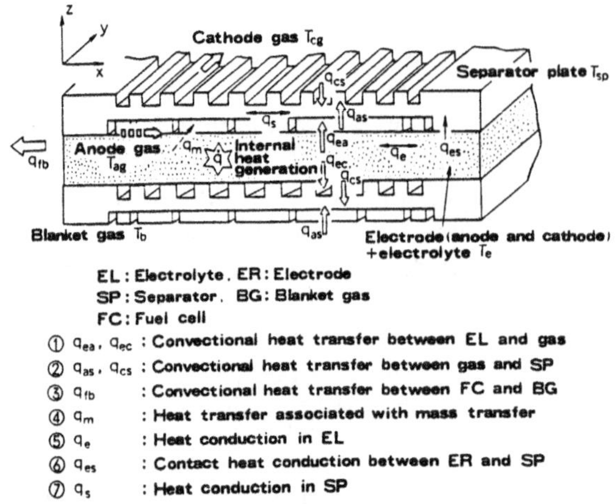

Figure 9.28. Schematic of heat generation and heat transfer processes in one cell of a MCFC stack. (After Takashima *et al.*[44])

single temperature and, of course, by the fuel and oxidant utilization. Performance data for such cells are given in the next section.

However, in fuel cell stacks, where cells typically have a surface area of 0.1 to 1 m^2 and the separator is an 0.5-mm-thick plate, temperature uniformity is by no means ensured. Therefore, the current and temperature distributions in such stacks interact in a rather complicated way and it is necessary to apply both energy and mass balances to analyze the distributions in detail. A key role in the temperature distribution is played by the in-plane conductance of the solid components, i.e., the electrode–electrolyte–electrode package and the separator. Figure 9.29 illustrates that the current distribution in a nonisothermal cell, with limited in-plane conductance, is quite different from that in an isothermal cell.

The nonisothermal cell performance model is a useful design tool in stack dimensioning and optimization of the operating conditions. For a complete analysis it is in principle necessary to apply four energy balances (for the electrode–electrolyte–electrode package, the fuel gas, the oxidant gas, and the separator) and at least two mass balances (one for the fuel gas and one for the oxidant gas), with additional equations to account for water-gas shift (Eq. (9.4)) and methanation (Eq. (9.5)) if applicable. As Fig. 9.30 shows for a representative externally reformed fuel gas, the fuel gas and oxidant temperatures are usually close to the solid temperature; i.e., local thermal equilibrium prevails.

Only in the case of in-cell reforming of CH_4 ("direct reforming") is there a strong thermal effect of the gas-phase reaction on solid temperature. The reforming reaction (Eq. (9.5)) is strongly endothermic and its occurrence within the fuel cell provides a major sink for waste heat. The reaction takes place on catalyst loaded directly in the anode gas channel, and a steep temperature drop near the cell entrance occurs, as shown in Fig. 9.31, even if the fuel gas inlet temperature is 900 K rather than 800 K as in Fig. 9.30. The calculation of Fig. 9.31 assumes complete reforming equilibrium; however, results for a finite kinetic rate and a graded catalyst loading show significantly improved uniformity of temperature and current near the inlet.

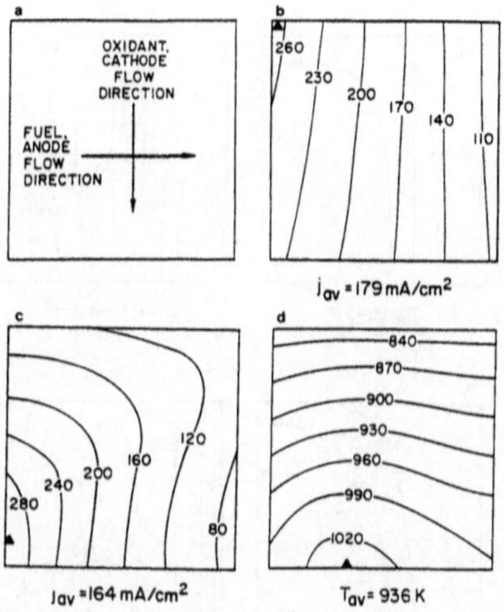

Figure 9.29. Calculated current and temperature distributions for the same 1600-cm^2 cell under isothermal and nonisothermal conditions. Cell and operating parameters: see Table 9.4. (a) Flow directions; (b) isothermal current distribution at 936 K, assuming cell resistance 0.262 ohm · cm^{-2} (peak current density located at Δ is 272 mA/cm^2); (c) nonisothermal current distribution assuming cell resistance 0.260 ohm · cm^{-2} (peak current density at Δ is 324 mA/cm^2); (d) temperature distribution of solid under conditions of (c) (peak temperature located at Δ is 1030 K). (After Wilemski and Wolf.[45])

9.3.8. State-of-the-Art Performance

Single-cell MCFC technology has steadily progressed during a period of 20 years and the performance of cells has improved correspondingly. Figure 9.32 gives an impression of the generic progress under conditions of very small utilization. The performance of state-of-the-art laboratory cells under such conditions can be predicted quite reliably on the basis of correlations such as Eqs. (9.27) and (9.28), in combination with the estimated ohmic resistance of the electrolyte.

The performance of bench-scale cells using state-of-the-art components, at the baseline current density of 150 or 160 mA/cm^2 and under atmospheric pressure at 650°C, is now well established as a benchmark. This performance depends of course strongly on the method of preparation and the quality of the electrolyte tile. Figure 9.33 shows the significant improvement achieved by tape casting and other techniques superseding the traditional hot-pressing method.

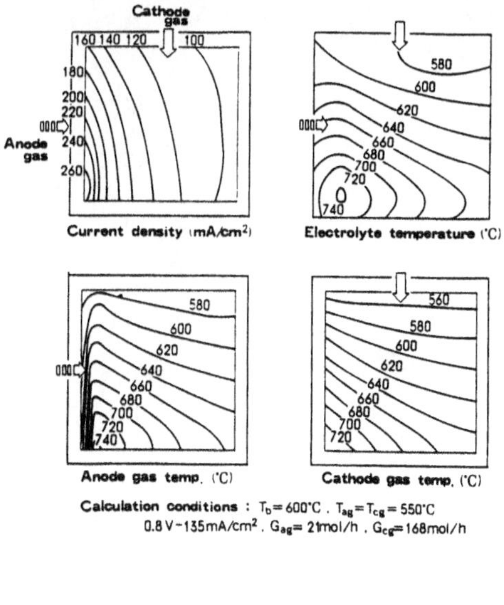

Figure 9.30. Calculated current and temperature distributions for a 3600-cm^2 cell, showing temperature of the electrolyte as well as that of the fuel gas and of the oxidant. Gas composition anode side $H_2/CO_2/H_2O$ = 77/18/10, cathode side $O_2/CO_2/N_2$ = 15/30/55; utilization fuel 60%, oxidant 60%, atmospheric pressure. Other operating conditions as stated in the figure. (After Takashima *et al.*[44])

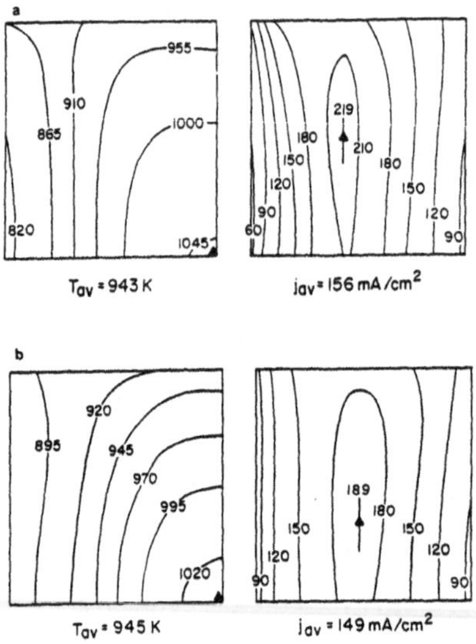

Figure 9.31. Calculated current and temperature distributions for a 1600-cm^2 cell under reforming conditions. Cell and operating parameters: see Table 9.5. Flow directions as in Fig. 9.29. (a) Full local equilibrium assumed. (b) Kinetic rate limitation assumed. (After Wilemski and Wolf.[45])

The effect of pressurization on bench-scale cells, briefly referred to in Section 9.3.2, is illustrated in Fig. 9.34. The Nernst equation predicts an increase in equilibrium voltage, at 650°C, of

$$\Delta V_p \quad [\text{mV}] = 45.8 \log\left(\frac{P_2}{P_1}\right), \tag{9.29}$$

i.e., 46 mV for a 10-fold increase in pressure. However, actual cell voltage increases by more than this amount, since polarization decreases, which is ascribed to increase of gas solubility and acceleration of the electrode kinetics. The gain in voltage of pressurized cells operating at 160 mA/cm^2, at 650°C, has been expressed[47] as

$$\Delta V_p \quad [\text{mV}] = 104 \log\left(\frac{P_2}{P_1}\right). \tag{9.30}$$

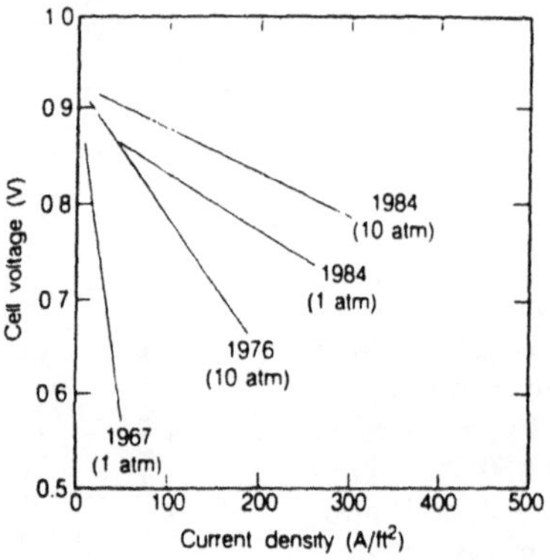

Figure 9.32. Progress in the generic performance of MCFCs on reformate gas and air.[46]

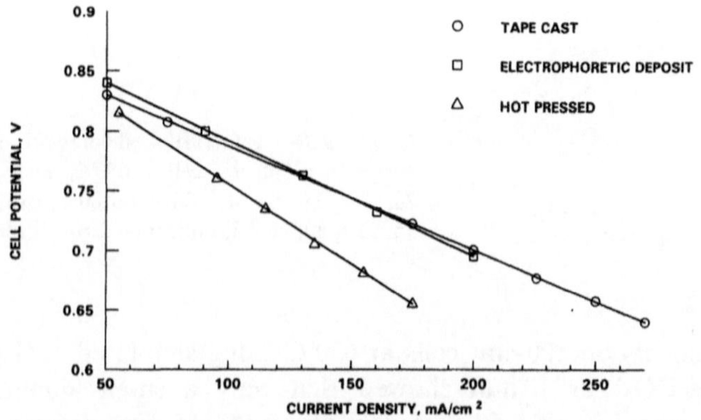

Figure 9.33. Cell performance curves for cells using tiles fabricated by various procedures, at 650°C. Composition in mol%: fuel gas, H_2 62, CO 11, CO_2 7, H_2O 20, utilization 75%; oxidant gas: air 75%, CO_2 25%, utilization 50%. (After Maru.[47])

This appears to be based on the data of Fig. 9.35, for 300-cm² cells, and the performance enhancement of Fig. 9.34 is also reasonably well represented by it. However, it is rather too simple an expression since it is unlikely that pressure effects on polarization follow a logarithmic dependence. On the other hand, Fig. 9.34 suggests that the performance enhancement by pressurization is not very sensitive to gas composition.

Recently, Mugikura *et al.*[50] carried out a systematic analysis of the performance of bench-scale cells using both tape-cast electrolyte tiles (cells fabricated by Mitsubishi Electric Co.) and electrolyte plates (cells fabricated by Fuji Electric Co. using the "papermaking" method). This analysis took current distribution effects into account. Selected data are represented in Fig. 9.36, together with predictive correlations based on the complete sets of data. The authors conclude that the decrease of cell polarization by pressure is due almost exclusively to the cathode.

That the effect on anode polarization is small is surprising since the anode is generally considered to be mass transfer controlled. Therefore, increased gas solubility should have a notably positive effect on polarization. This positive effect may be cancelled by consumption of reactants via gas-phase reactions such as methanation (Eq. (9.5) in reverse) or carbon deposition (Eq. (9.6)), both of which are favored by increased pressure. (Carbon deposition boundaries for the C–H–O system have been calculated a.o. by Cairns and Tevebaugh[51a] and determined experimentally by Pigeaud and Klinger.[51b])

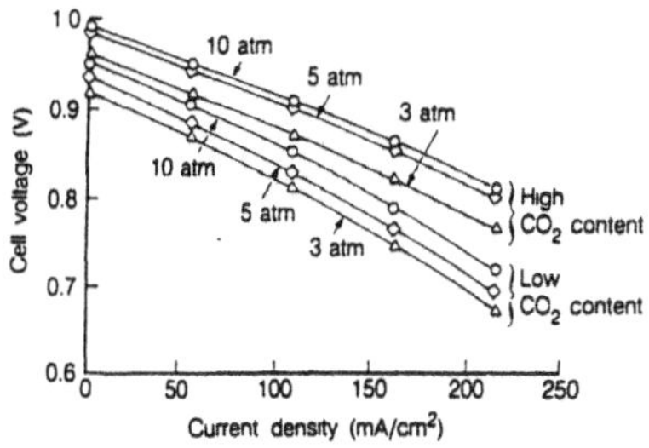

Figure 9.34. Effect of cell pressure on performance curves of a 70.5-cm² MCFC at 650°C. Anode gas not specified, cathode gas: (high CO_2; mol%) O_2 9.2, CO_2 18.2, N_2 65.3, H_2O 7.3; (low CO_2; mol%), O_2 23.2, CO_2 3.2, N_2 66.3, H_2O 7.3. Both gases 50% CO_2 utilization at 215 cm². (From Kunz and Murphy.[48])

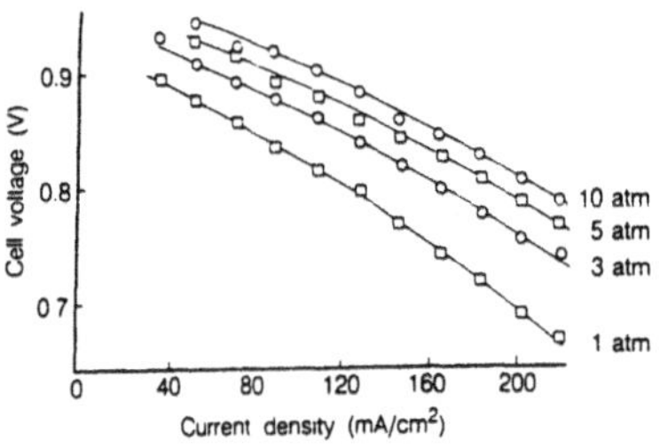

Figure 9.35. Effect of cell pressure on performance curves of a 300-cm^2 cell at 650°C. Fuel gas (mol%): H_2 28, CO_2 28, N_2 44, utilization 80%; oxidant (mol%): O_2 15, CO_2 30, N_2 55, utilization 50%. (From Baker.[49])

Measurements on 100-cm^2 cells at 650°C using simulated coal gas (38% H_2, 56% CO, 6% CO_2) at 10 atm showed that only a small amount of CH_4 is formed.[52] At open circuit 1.4% CH_4 (on dry-gas basis) was detected, and at fuel utilizations up to 50–85%, 1.2–0.5% CH_4 was measured. Carbon deposition is entirely absent, even at 10 atm pressure in a high-CO gas, if the gas is humidified sufficiently (e.g., saturated at 163°C under 10 atm).

The effect of temperature on the voltage of MCFCs may, like the pressure effect, be separated in a reversible and an irreversible part. The reversible (equilibrium) voltage calculations are not trivial since the shift and reforming reactions taking place at the anode must be considered. However, several publications report such thermodynamic calculations, which are also necessary to determine the carbon deposition boundaries of the C–H–O system.[53] Table 9.7 illustrates the effect of temperature on a typical high-Btu gas mixture saturated with water vapor at 25°C. The reversible voltage decreases with temperature.

The actual cell voltage depends strongly on polarization, which decreases with temperature. Therefore, the net temperature coefficient is positive. This has discouraged attempts to operate MCFCs at temperatures lower than 650°C, which would be attractive from the viewpoint of corrosion. In fact, the temperature coefficient decreases with increasing temperature. For small cells operating at 200 mA/cm^2 on constant flow rates (i.e., variable but low utilization) of steam-reformed natural gas and standard oxidant, the temperature coefficient of

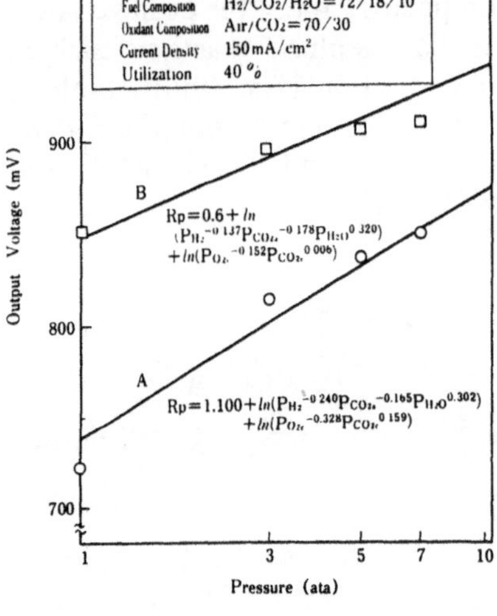

Figure 9.36. Comparison of experimental and predicted pressure effect on cell voltage. (From Mugikura *et al.*[50])

Table 9.7. Temperature Effect on Equilibrium Composition and Cell Voltage

Component	Partial pressure (atm)			
	298 K[a]	800 K	900 K	1,000 K
H_2	0.775	0.669	0.649	0.643
CO_2	0.194	0.088	0.068	0.053
CO		0.106	0.126	0.141
H_2O	0.031	0.137	0.157	0.172
Cell voltage (V)[b]		1.155	1.143	0.133
Equilibrium constant		0.2472	0.4538	0.7273

[a] H_2O saturated at 1 atm
[b] vs. oxidant: O_2 30%, CO_2 60%, N_2 10%.

cell voltage is +1.4 mV/K between 600 and 650°C, and +2.2 mV/K between 575 and 600°C.[54]

A recent study of the performance of bench-scale cells in atmospheric and pressurized operation between 600 and 650°C led to a correlation of effective resistance as shown in Fig. 9.37. This suggests a temperature coefficient of +2.8 mV/K, which is even higher than those given above.

9.3.9. *Corrosion and Cathode Dissolution*

Corrosion in the MCFC takes two principal forms:

1. Corrosion of hardware (separator plate or cell housing, electrodes), which prevails in the initial stages of cell or stack life
2. Dissolution of electrode materials, in particular, NiO from the cathode.

Both types of corrosion are major issues in the optimization of the electrolyte composition (Section 9.3.10), in which contaminant effects play only a minor role.

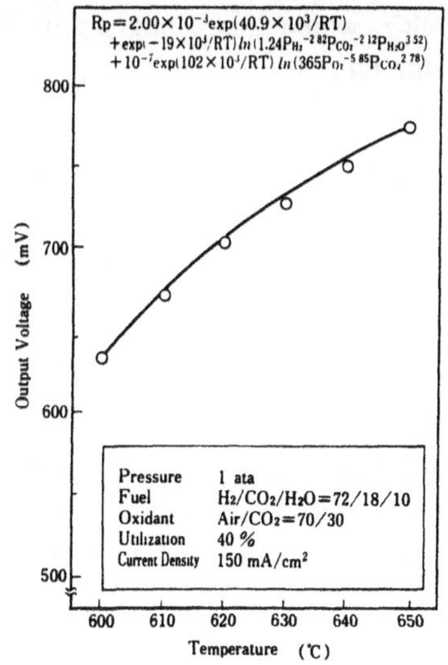

Figure 9.37. Comparison of experimental and predicted temperature effect on cell voltage. (From Mugikura *et al.*[50])

Electrolyte composition is not believed to have a strong influence on the mechanism by which contaminants, such as H_2S, SO_2, HCl, and trace contaminants affect performance. On the other hand, the presence of contaminants such as H_2S and HCl in the fuel gas may cause a strong acceleration of corrosion; this has certainly proved to be the case for H_2S. However, this synergistic effect has not been extensively studied as yet.

In this section we discuss, first, the general corrosion tendencies of the electrodes, second, the dissolution of NiO from the cathode, and third, the mechanism and extent of hardware corrosion.

9.3.9.1. *Corrosion of Electrodes*

Corrosion of the anode material, i.e., mainly nickel or Ni–Cu alloy, is not likely since the anode is the negative electrode and its environment reducing (fuel gas). At the anode potential only less noble metals such as chromium or aluminum are unstable, and these metals, used as sinter- or creep-preventing additives (see Section 9.3.12), are present as oxides which form *in situ* or are predispersed in the anode material. Current passage though the anode always entails oxidation; therefore, oxidation of nickel and/or copper may also occur, but only under severe oxidizing conditions, e.g., if the anode is polarized excessively or if the fuel gas flow is interrupted.

Structural materials adjacent to the anode (such as the anode-side wet-seal area) are also significantly at risk of corrosion, as discussed below. The closer the component is to the main ionic current path from anode to cathode, the greater its tendency to corrosion. Conversely, metal parts adjacent to the cathode or in the ionic current path toward the cathode are not significantly threatened by corrosion.

The cathode itself, however, is at a high positive potential, therefore, the cathode material is usually an oxide. The adjacent materials also are oxidized or, as in the case of certain stainless steels, they passivate, i.e., form a dense adherent oxide layer. At the cathode potential only gold, palladium, and, in lithium-poor metals, silver are stable in carbonate. Nickel is oxidized *in situ* to NiO, which has a slight solubility in carbonate (like silver and CuO but much smaller).

9.3.9.2. *Dissolution of NiO from the Cathode*

The solubility of NiO is dependent on CO_2 partial pressure, according to the equilibrium

$$NiO + CO_2 \rightarrow Ni^{2+} + CO_3^{2-}. \tag{9.31}$$

As discussed in the next section, the dissolution of NiO may be correlated with the acid–base properties of molten carbonate, since the first step in the above "acid dissolution" mechanism is

$$NiO \rightarrow Ni^{2+} + O^{2-}. \tag{9.32}$$

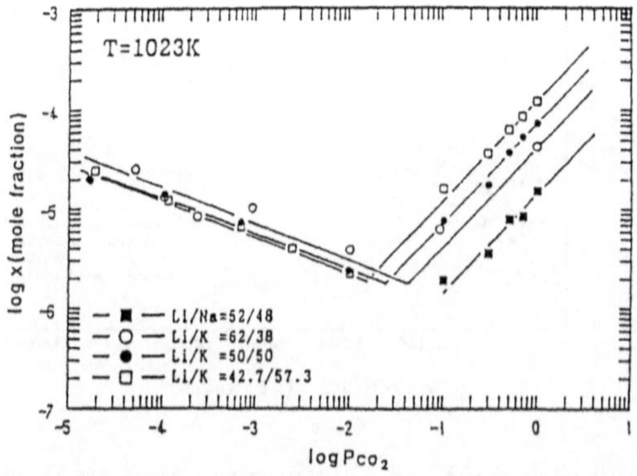

Figure 9.38. Solubility of NiO in various carbonate melts as a function of p_{CO_2} at 1023 K. (From Ota *et al.*[55])

This applies in low-O^{2-} ("acid") melts, while in "basic" (high-O^{2-}) melts the initial step is

$$NiO + O^{2-} \rightarrow NiO_2^{2-}, \tag{9.33}$$

or perhaps

$$2NiO + O^{2-} + \tfrac{1}{2}O_2 \rightarrow 2NiO_2^{-}. \tag{9.34}$$

This leads to the overall "basic" dissolution reactions

$$NiO + CO_3^{2-} \rightarrow NiO_2^{2-} + CO_2 \tag{9.35}$$

or

$$2NiO + CO_3^{2-} + \tfrac{1}{2}O_2 \rightarrow 2NiO_2^{-} + CO_2. \tag{9.36}$$

The equilibrium solubility of nickel oxide is minimal under conditions where these acid and basic branches of dissolution behavior intersect. This is illustrated in Fig. 9.38. According to recent data of Ota and co-workers[55] the minimum solubility of NiO in Li–K carbonate eutectic at 650°C is 1 molar ppm, at a partial pressure of 0.01 atm CO_2; both the minimum solubility and the associated CO_2 partial pressure increase with temperature. Under normal MCFC operating conditions, NiO dissolution is usually acidic; i.e., it increases linearly with CO_2 partial pressure.

NiO dissolution has serious implications for long-term cell performance. As discussed in Section 9.3.12, this is not primarily due to the loss of active material from the cathode, but to the process whereby dissolved nickel is reduced to metal particles and precipitated in the electrolyte tile, eventually leading to anode–cathode shorting.

9.3.9.3. *Corrosion of Cell Hardware*

Stainless steel and various nickel alloys are the principal structural materials of the MCFC. In a not too strongly oxidizing environment the corrosion of these types of steel is controlled and limited by the formation of protective layers, similarly as in passivation. The layers, however, are much thicker than the passive layers formed on transition metals in aqueous solutions.

Corrosion of cell hardware is illustrated in Figs. 9.39 and 9.40. A high corrosion rate is observed initially, after which corrosion slows down appreciably. The rapid initial corrosion is mostly due to the formation of $LiCrO_2$ on steel parts

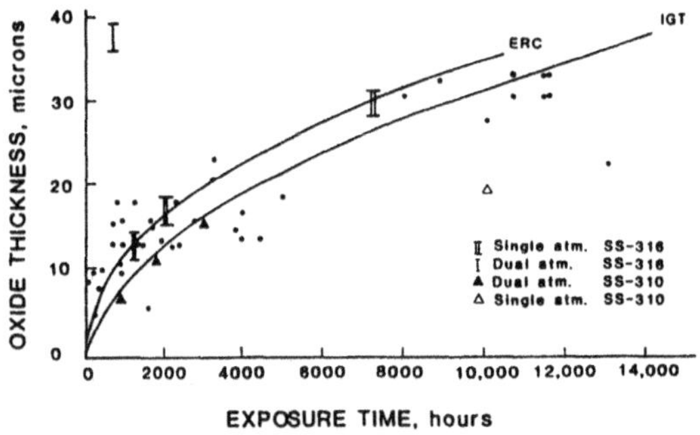

Figure 9.39. Corrosion of ss 310 and 316 exposed to standard oxidant and Li–K carbonate eutectic at 650°C. Points represent IGT corrosion tests on ss 316 cathode current collectors. Other symbols represent ERC corrosion tests on ss 310 and 316. (From Appleby and Foulkes.[56])

and of $LiAlO_2$ on aluminum-coated steel parts (usually in the wet seal area). This initial corrosion is associated with a high initial rate of electrolyte loss, mostly of Li_2CO_3, as discussed in Section 9.3.12.

Figure 9.39 shows the oxide thickness on current collectors or bipolar test plates of stainless steel 310 and 316 exposed to oxidant.[56] Most of the data points in this figure represent one-sided corrosion ("single-atmosphere test"), but a number of points represent corrosion of test plates whose backside was exposed to hydrogen gas ("dual-atmosphere test"). This is meant to simulate more closely the corrosion environment of bipolar plates in a stack. (The results are not very different, except in the case of ss 316.)

The initial corrosion rate is typically 8 μm/kh in the first 2000 h but slows to an average 2 μm/kh for the next 10,000 h. On ss 316 current collectors an 0.1-mm-thick layer of $LiFeO_2$ is formed after 40,000 h of operation, provided contaminants are absent. The thickness of the oxide layer is given approximately by

$$y = 0.134t^{0.5} \tag{9.37}$$

(shown as a solid line in Fig. 9.39, "IGT").

The anode-side corrosion of stainless steel 316 bipolar plates shows a similar time-dependent behavior, but the corrosion rate is much higher. The corrosion layer thickness also increases approximately as the square root of time

$$y = ct^{0.5} \tag{9.38}$$

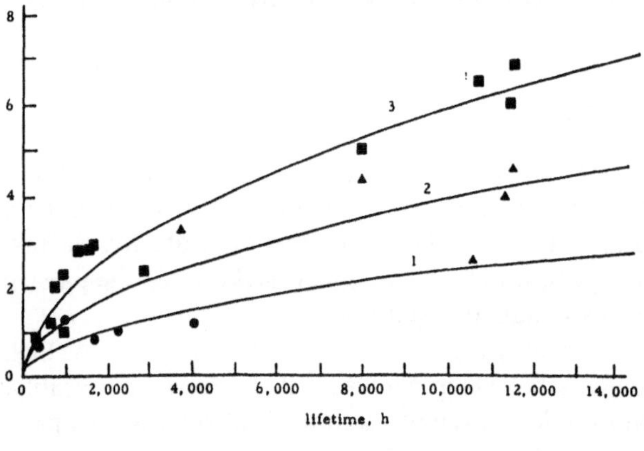

Figure 9.40. Corrosion of ss 316 anode current collectors exposed to fuel gas atmosphere, as a function of H_2O content of the gas: (1) 16% H_2O (c = 0.023); (2) 28% H_2O (c = 0.039); (3) 43% H_2O (c = 0.058). The constant c is the fitting parameter for the solid lines $y = ct^{0.5}$. (Data from IGT.[57])

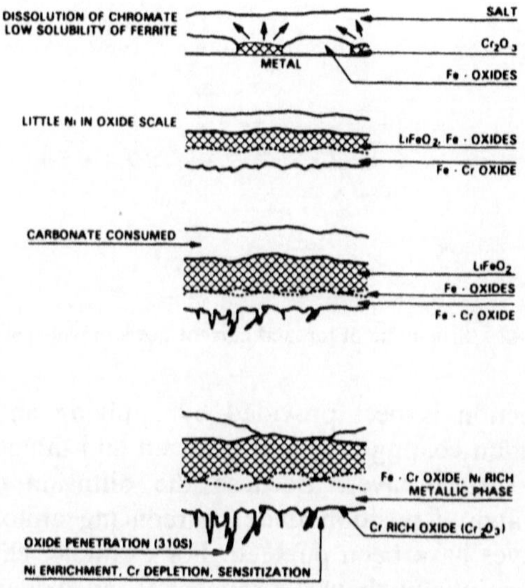

Figure 9.41. Dual-layer formation in the corrosion of austenitic steel in the MCFC cathode environment. (After Yuh.[59])

but c depends on the partial pressure of H_2O, as shown in Fig. 9.40. Since H_2O content of the fuel gas increases strongly in the downstream direction, the anode-side conditions range from reducing to highly oxidizing. In general the corrosion rate is two to five times higher than on the cathode side.

That the corrosion layer grows approximately with the square root of time suggests a mass-transfer-controlled corrosion mechanism. However, the corrosion rate cannot easily be related to particular type of oxide or diffusing species. Although the main constituents of the layers are $LiCrO_2$ and $LiFeO_2$,[58] their detailed structure is quite complex.[59] In the oxidant atmosphere a dual-layered structure is formed (Fig. 9.41). A Fe-rich outer layer (or Ni-rich outer layer in the case of Ni-based alloys) protects the Cr-rich inner layer against attack by carbonate. The inner layer limits cation diffusion and therefore the overall corrosion rate. However, the Cr-rich oxide layer is in general not compact enough to protect the substrate against further low-level attack.

The alloy composition can be selected such that a compact layer with maximum protection is formed. Unfortunately, it is difficult to create a dense layer which is also sufficiently conductive; for example, Al-containing steels are well protected but suffer from excessive contact resistance due to the aluminate layer on the surface. Thus, the selection of an optimal alloy composition and various methods to pretreat the substrate to enhance protection[59] are subjects of continuing R&D.

The wet-seal area, especially on the anode side, is especially vulnerable to corrosive attack. The mechanism of this corrosion is fairly well understood.[60] As discussed above, stainless steel and nickel alloys are not stable if adjacent to the anode. If the metal in the wet-seal area adjacent to the anode is not protected by an insulating layer, "corrosion cells" form. These dissolve metal from the wet-seal face near the anode and also cause a slow but accelerating electrolyte flow ("creepage") out of the cell, through the wet seal (Fig. 9.42). The ensuing corrosion and electrolyte loss invariably lead to rapid performance decay.

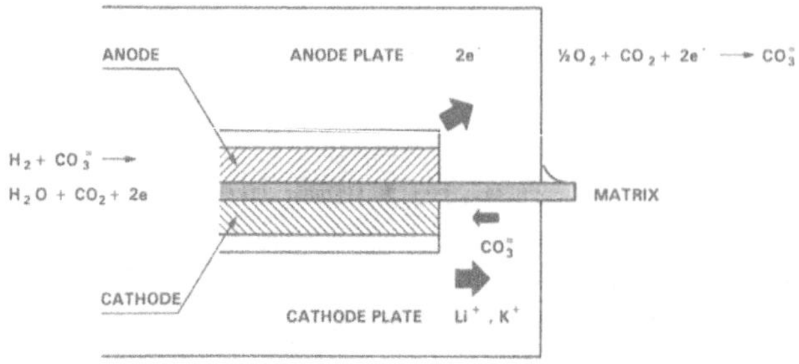

Figure 9.42. Schematic of ion and current flows in wet-seal corrosion.

Wet-seal protection is best provided by applying an aluminum diffusion coating. The aluminum coating is *in situ* oxidized and lithiated to form a dense, insulating $LiAlO_2$ surface layer. Because the diffusion coating process (by high-temperature vapor deposition under a reducing atmosphere) is slow and expensive, alternatives have been pursued. For example, chromium-based steels containing aluminum intially showed promise. In an oxidant atmosphere these steels form a compact layer of Al_2O_3 over an inner scale of Cr_2O_3, which affords good protection; however, in inert or oxygen-lean atmospheres, a porous $LiCrO_2$ layer is formed and severe corrosion occurs.[61] The experience thus far is that only aluminum diffusion coating and perhaps high-aluminum stainless steel yield satisfactory wet-seal protection for long-term performance.

9.3.10. *Contaminant Effects*

Corrosion of metal components in the MCFC is strongly affected by contaminants, as is electrode performance. The various contaminants found in coal-derived fuel gases and their potential effects are listed in Table 9.8.[62] The main contaminants are sulfur, as H_2S (in fuel gas) and SO_2 (in recycled anode exhaust), and chlorine, as HCl in fuel gas.

Table 9.8. Contaminants from Coal-Derived Fuel Gas and Their Possible Effects on MCFs

Class	Contaminant	Possible effect
Sulfur-containing compounds	H_2S, COS, CS_2, C_4H_4S	Voltage loss Sulfation electrolyte Increased corrosion
Halogens	HCl, HF, HBr, $SnCl_2$	Cathode corrosion and performance decay Increased corrosion Reaction with electrolyte
Nitrogeneous	NH_2, HCN, N_2	NO_x formation and compounds reaction with electrolyte
Trace species	Zn	ZnO precipitation
	As	Poisoning Ni Catalyst
	Pb, Hg	Surface alloy formation
	Sn (as $SnCl_2$)	Sn precipitation
	Se	Poisoning Ni

Based on Anderson and Garrigan[63] and Pigeaud *et al.*[64]

In single-cell experiments[65–68] it has been shown that, to avoid significant degradation of performance, the concentration of these contaminants must be at or below the ppm level. In the following the conditions and mechanism of performance degradation and corrosion are briefly discussed.

Sulfur compounds in more than ppm concentrations are detrimental to MCFC performance. H_2S in particular causes serious performance degradation, due to a number of factors:

Chemisorption on nickel surfaces and blocking of active electrochemical sites
Poisoning of catalytic reaction sites for the water-gas shift reaction
Oxidation to SO_2 by anode-exhaust combustion and recycling to the cathode, where reaction with the electrolyte forms sulfate

The level of H_2S tolerance depends to some extent on temperature, pressure, gas composition, cell components, and system operation (i.e., recycle, venting, gas cleanup). At atmospheric pressure and with gas utilization as high as 75%, up to 10 ppm H_2S is acceptable, provided initial hydrogen levels are high. In the oxidant gas 1 ppm of SO_2 may be tolerated. These concentration limits increase with temperature, but decrease with increasing pressure.[68]

The adverse effect of H_2S on cell performance is illustrated in Fig. 9.43. The voltage of a bench-scale cell under load decreases when 5 ppm H_2S is added to the low-Btu fuel gas. At low concentrations of H_2S the open-circuit potential is not affected, but performance degradation under load increases with increasing current density. Below a certain threshold concentration of H_2S, the performance degradation can be reversed by introducing clean fuel gas.

According to Remick *et al.*,[69] this reversibility may be explained by a mechanism in which H_2S in contact with nickel is first oxidized chemically,

$$H_2S + CO_3^{2-} \rightarrow H_2O + CO_2 + S^{2-}, \tag{9.39}$$

and then electrochemically,

$$Ni + xS^{2-} \rightarrow NiS_x + 2xe^-. \tag{9.40}$$

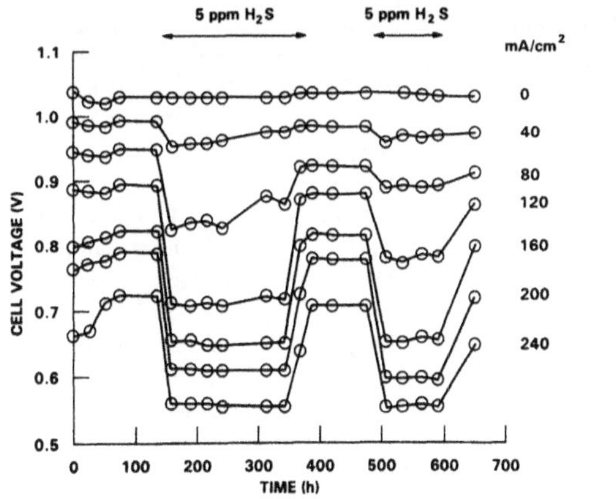

Figure 9.43. Effect of H_2S on cell potential at various current densities: 5 ppm H_2S is added to fuel gas (10% H_2, 5% CO_2, 10% H_2O, 75% He) at 25% utilization, in a 10 cm × 10 cm cell at 650°C. (From Remick.[68])

Reaction (9.39) is an equilibrium, and reaction (9.40) is also reversible at low degrees of conversion, as long as the sulfur is largely chemisorbed on nickel. A poisoned anode can then be reduced simply by reverting to open circuit or by switching to a fuel gas without H_2S. The degree of recovery is determined by the chemical equilibrium

$$NiS_x + xH_2 \rightarrow Ni + xH_2S. \tag{9.41}$$

However, performance degradation becomes irreversible at higher degrees of sulfidation, which at temperatures above 625°C causes NiS–Ni eutectic formation and virtual destruction of the anode.

The susceptibility to sulfur poisoning of the electrodes, and in particular the nickel anode, has led to a search for alternative cell designs and alternative anode materials capable of tolerating moderately small levels of sulfur. One cell design[70] is based on the concept of separating hydrogen from the contaminated fuel gas *in situ* by means of a micrometers-thick foil of hydrogen-permeable metal (such as Ni or Ni–Ti alloys) adjacent to the electrolyte matrix and acting as anode. In spite of the basic soundness of the concept, difficulties were encountered due to hindered permeation when H_2S was present in the fuel gas. In another alternative approach,[71] a number of materials in porous form were tested for sulfur tolerance when operated as anode. Lithium ferrate ($LiFeO_2$) was found to have the best sulfur tolerance, with performance unaffected even at 30 ppm H_2S. However, its kinetics in operation on hydrogen are significantly slower than those of nickel or copper.[72]

Current work on the effect of contaminants addresses not only H_2S, but also HCl and trace contaminants in coal gas. The effect of contaminants is to a

Table 9.9. Typical Fuel Gas Compositions and Contaminants from Air-Blown Coal Gasifier after Hot-Gas Cleanup, and MCFC Tolerance Limits to Contaminant

Fuel gas[a] (mol%)	Contaminants	Content	Remarks	Tolerance limits
19.2 CO	Particulates	<0.5 mg/L	Incl. ZnO from H_2S cleanup stage	<0.1 g/L (>0.3 μm)
13.3 H_2	NH_3	2,600 ppm		<5,000 ppm
2.6 CH_4	As	<5 ppm		<1 ppm
6.1 CO_2	H_2S	<10 ppm	After first-stage cleanup	~1 ppm
12.9 H_2O	HCl	500 ppm	Incl. other halides	<10 ppm
45.8 N_2	Trace metals	<2 ppm	Pb, Hg, Sn, etc.	<1 ppm
	Zn	<50 ppm	From H_2S hot cleanup	<20 ppm
	Tar	4,000 ppm	Formed during desulfurization cleanup stage	<2,000 ppm[b]

[a]Humidified fuel gas entering at 650°C.
[b]Benzene.
After Pigeaud[73,74] and Kinoshita *et al.*[19]

considerable extent dependent on stack design; it is particularly severe for internal-reforming MCFCs (see Section 9.4.4). Table 9.9 summarizes MCFC tolerance limits according to recent studies.

9.3.11. Electrolyte Optimization

Although early R&D which established the MCFC concept was carried out with various electrolyte compositions, a.o., the ternary eutectic of Li_2CO_3, Na_2CO_3, and K_2CO_3, most development after 1975 made use of the 62 mol% Li_2CO_3–40 mol% K_2CO_3 eutectic (first adopted by UTC). It was not certain, however, that this composition would yield the best short-term or long-term performance. In the DOE-sponsored programs since then, and more recently also in the Japanese MCFC development program, considerable emphasis has been given to further optimization of the electrolyte composition.

The electrolyte composition affects the short-term (e.g., 2000 h) performance of the MCFC ("performance," for short) as well as the long-term performance ("endurance"). In general, as discussed in Sections 9.3.2 and 9.3.8, performance, i.e., the cell voltage at a given current density, depends on the ohmic resistance of the cell and on the polarization of the electrodes. These, in turn, depend on many factors most of which are directly tied to the chemistry and physical properties of the electrolyte. Thus, Li-rich electrolytes have higher ionic conductivities, and hence lower ohmic polarization, due to the relatively high ionic conductivity of Li_2CO_3 compared to Na_2CO_3 and K_2CO_3. On the other hand, the solubility and diffusivity of reactant gases, such as hydrogen, oxygen, H_2O, and CO_2, appear to be lower in Li_2CO_3-rich melts. Thus, the electrolyte composition can be optimized roughly on the basis of physical properties such as these. Thus, the region of the Li–Na–K ternary phase diagram having 60–70 mol% Li_2CO_3 and O–20% Na_2CO_3 was considered promising in the early DOE optimization efforts, and indeed remarkably improved performances were obtained with a few ternary compositions in short-term small-cell tests, as shown in Fig. 9.44.[75]

However, the overall process taking place in the electrodes and electrolyte is very complicated, and a more comprehensive approach to electrolyte optimization is desirable. It is necessary to take kinetics as well as transport processes into account, and to project the effect of long-term changes in structure and composition of electrodes and electrolyte. Fortunately, the DOE program since 1975 has greatly expanded our knowledge base of electrode kinetics in carbonate and of the physical properties of binary and ternary Li–Na–K carbonate mixtures.[77,78]

Carbonate electrolyte undergoes dissociation,

$$CO_3^{2-} \rightarrow CO_2 + O^{2-}, \tag{9.42}$$

and, in the presence of H_2O, hydrolysis,

$$CO_3^{2-} + H_2O \rightarrow CO_2 + 2OH^-. \tag{9.43}$$

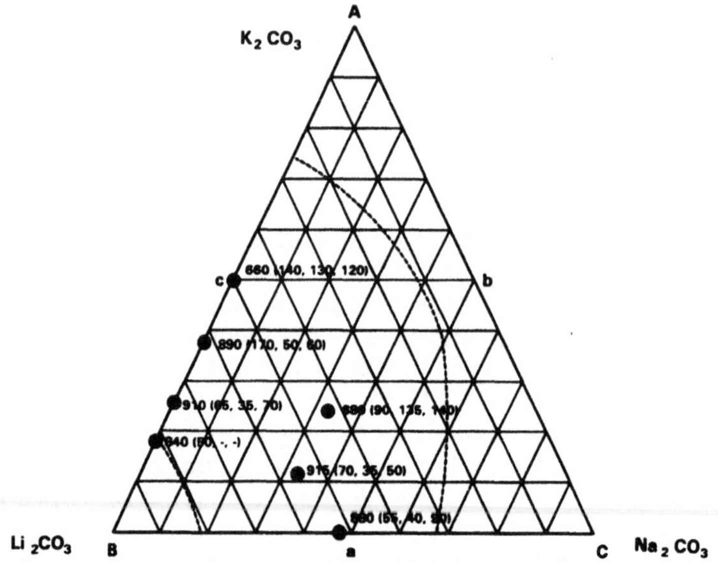

Figure 9.44. Effect of electrolyte composition on the performance of 3-cm^2 laboratory cells, operating at 650°C, and utilization less than 10%. The numbers indicate cell potential (*IR* drop, node polarization, cathode polarization), all in mV at 160 mA/cm^2. (Adapted from IGT data[75] by Appleby and Foulkes.[76])

These processes are strongly influenced by the cationic composition of the electrolyte. For example, in Li_2CO_3-rich melt the oxide content (basicity) is much higher than in Li_2CO_3-poor melts, and the hydrolysis equilibrium also lies farther to the right. Although these processes take place to a very limited extent, they play an important part in the kinetics of the electrodes and in corrosion processes.

For example, oxygen, which is reduced at the cathode according to Eq. (9.2), is not present as physically dissolved molecules of O_2 but is chemically dissolved as peroxide (O_2^{2-}) and superoxide (O_2^-) ions. This is the result of interaction with

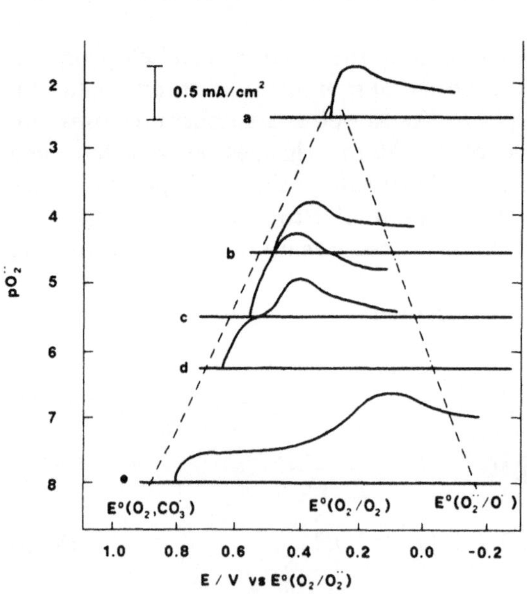

Figure 9.45. Comparison of polarization curves for oxygen reduction in bulk alkali carbonate melts of decreasing basicity (a: Li_2CO_3, b: Li_2CO_3–Na_2CO_3, c: Li_2CO_3–K_2CO_3, d: Li_2CO_3–Na_2CO_3–K_2CO_3, e: Na_2CO_3–K_2CO_3). The potential is plotted on an absolute scale, i.e., with respect to O_2/O^{2-}-potential as zero. (Adapted from data of Uchida.[77,78])

CO_3^{2-} and O^{2-} ions whose concentration, controlled by Eq. (9.42), depends strongly on melt composition. Therefore, the polarization of the cathode, which contributes significantly to the overall polarization of the MCFC, depends strongly on the electrolyte composition (Fig. 9.45). Based on the extensive Japanese work of the last decade, Uchida[77] has recently presented an excellent review of the kinetics and reaction mechanism of both the anode and the cathode of the MCFC. Most of the mechanistic aspects of the cathode reaction are now well understood; only the role of CO_2 in the electron transfer steps is an unresolved question of considerable practical importance.

The effect of kinetics and reaction mechanism on MCFC cathode polarization is also an important question, and cannot be answered without computation. Yuh and Selman[79,80] and Lee[81] have presented results of porous electrode models which take detailed kinetics into account.

Long-term performance issues, discussed in the next section, now dominate MCFC research. These issues also drive electrolyte optimization. Thus, a recent electrolyte optimization study[82] considered four major endurance-related parameters (cathode dissolution, hardware corrosion, $LiAlO_2$ matrix stability, and volatility) in addition to several primary performance parameters (electrolyte conductivity, cathode polarization, melting point, electrolyte wetting tendency). The electrolyte compositions studied were

- Standard 62/38 Li–K eutectic
- Lithium rich (72/28) Li–K binary
- Li–Na eutectic
- 62/38 Li–K eutectic with 0.5 mol% MgO (sparingly soluble)
- 62/38 Li–K eutectic with 5 mol% BaO (highly soluble)

This work concluded that Li–Na in eutectic or off-eutectic composition is a viable alternative to Li–K eutectic, yielding superior performance with respect to NiO cathode dissolution, electrolyte creepage, and volatilization, as well as conductivity and cathode polarization. Figure 9.46 illustrates this with respect to NiO dissolution: NiO solubility, as measured by Ota and co-workers,[83] is lower in Li–Na than in Li–K carbonate melt.

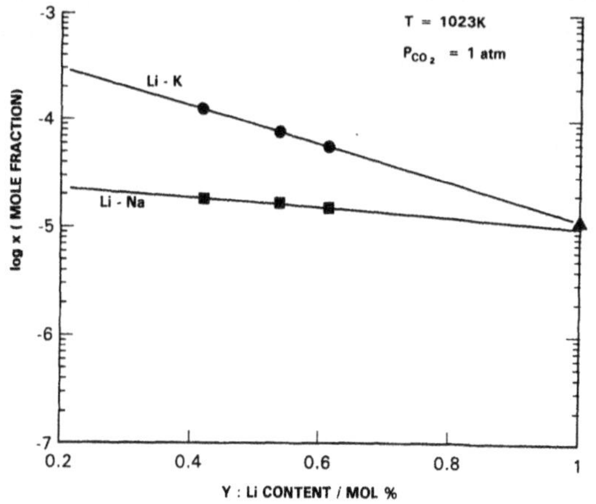

Figure 9.46. Effect of Li content on the solubility of NiO in Li–K and Li–Na eutectic. (From Ota *et al.*[83])

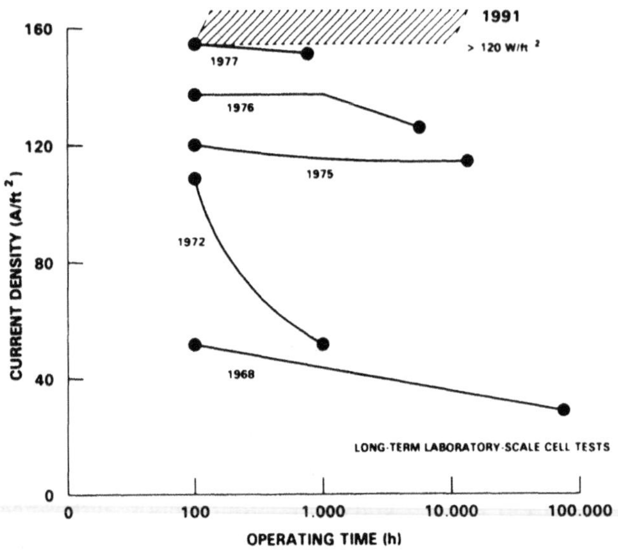

Figure 9.47. Long-term cell performance: 25-year progress. (Based on IGT data.[86])

As the preceding list shows, there has recently been renewed interest in alternative electrolyte compositions, in particular the Li–Na carbonate eutectic. Systematic effort to optimize alkaline–earth addition to Li-K and Li-Na electrolytes have been undertaken by the Japanese MCFC program and are reported by Tanimoto *et al.*[84,85]

9.3.12. Long-Term Performance

Although the focus of attention in MCFC technology has shifted to stack testing and commercialization, single-cell performance is still important. It is considered an indicator of the level and sophistication of component technology. It is now relatively easy to construct, based on published single-cell designs, a cell which has good initial performance. However, long-term performance depends on matching of overall cell design to component specifications; the optimal way to do this is learned from many more or less successful trials. In other words, successful cell design is still a trial-and-error process, but so is design in general, in spite of tools like CAD–CAM.

Early MCFC cells suffered a dramatic collapse in performance after a few hundred hours of operation. Figure 9.47 illustrates the significant advances made compared to cells of the early 1970s. As this shows, state-of-the art performance decay is now as low as approximately 5 mV/1000 h for small cells.

The principal causes of long-term performance decay are

- Electrolyte loss, leading to increased ohmic resistance and polarization
- Increase in contact resistance, due to formation of poorly conducting surface layers on cell hardware
- Electrode (especially anode) creep and sintering, i.e., a coarsening and compression of electrode particles, leading to increased ohmic resistance and electrode polarization

- Shorting of the electrolyte tile by nickel particles which precipitate as dissolved nickel from the cathode diffuses into the tile and is reduced by hydrogen

Among these factors, electrolyte loss is the most important and continuously active factor. Electrolyte is lost from a cell due to

- Consumption of electrolyte by corrosion processes
- Potential-driven migration of ions at different rates (due to differences in mobility), causing net movement of the electrolyte out of the cell, via the wet seal
- Volatilization of electrolyte (i.e., usually Li_2CO_3 and/or K_2CO_3) by evaporation either directly (as carbonate) or indirectly (as hydroxide)

These processes occur at different rates during the various stages of the life of a cell. They may also occur combined.

For example, during the initial 500–2000 h of cell operation, corrosion of hardware predominates as an electrolyte loss factor. This leads to a steep initial loss rate, since corrosion causes appreciable electrolyte loss, mostly of Li_2CO_3, as discussed in Section 9.3.12. As shown in Fig. 9.48 the inital electrolyte loss may be treated as a one-time loss followed by a constant-rate loss.

This constant-rate loss had been identified, early on, with volatilization of carbonate via the fuel and/or oxidant gas. The fuel gas in particular is effective in volatilizing carbonate, in part as hydroxide. However, from accurate out-of-cell measurements of evaporation losses in various gas atmospheres,[88] in part illustrated in Fig. 9.49, it became clear that the actual loss rate could not be caused only by volatilization. In fact, it seems that the gas streams in the MCFC are far from saturated when they leave the cell.

The amount of electrolyte lost to corrosion depends of course on the surface area of those hardware parts which are directly exposed to molten carbonate as well as the fuel and oxidant gas channels, since an electrolyte layer forms on the inner surfaces of the channels. In most single bench-type cells the ratio of hardware area to electrolyte mass is larger than in a fuel cell stack. Therefore, in a stack the initial electrolyte loss per cell will be correspondingly less.

In a properly operating single cell, as represented by Fig. 9.48, migration losses do not play an appreciable role. This is because the wet-seal passage is relatively narrow and electrolyte creepage, i.e., movement out of the cell onto the exterior of hardware enclosing the cell, is limited. This movement is driven by

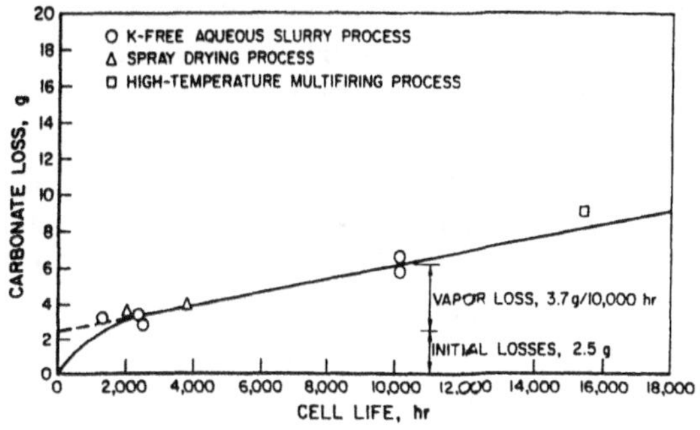

Figure 9.48. Carbonate loss from 94-cm^2 cells at atmospheric pressure, for various tile-fabrication procedures. (From Marianowski and O'Sullivan.[87])

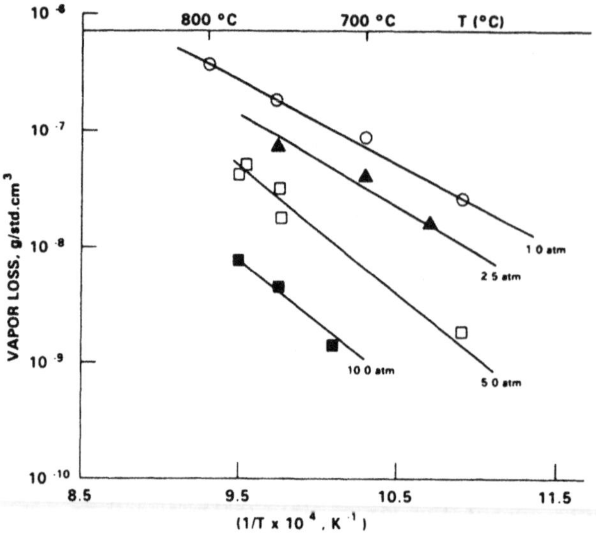

Figure 9.49. Vapor loss of 62–38 Li_2CO_3–K_2CO_3 electrolyte in fuel gas environment at different pressures, as indicated. Gas composition at 1 atm (in %): 22 H_2, 13 H_2O, 5 CO_2, 4 CO, 56 N_2; at higher pressures: 19 H_2, 12 H_2O, 4 CO_2, 4 CO, 61 N_2. (From Ong and Claar.[88])

the electric field in the wet seal, which makes potassium ions (in spite of their greater size) move faster than lithium ions. A net motion of the electrolyte results. Its rate depends on oxygen and CO_2 activity differences inside and outside the cell, which can be controlled to a great extent. Creepage therefore is ordinarily very slow. It becomes catastrophic only if corrosion protection of the wet seal, for example by an aluminum coating, breaks down and severe wet-seal corrosion creates large potential driving forces.

Electrolyte displacement by migration, however, can be important in MCFC stacks with external manifolding. (See Section 9.4.6.)

In pre-1975 MCFC cells *sintering of the nickel anode* was a serious problem, which limited cell life. This was overcome by the use of alloys such as Ni–10% Cr; the chromium component segregates in a lithium chromite ($LiCrO_2$) phase which prevents the solid-state diffusion leading to coarsening. The introduction of this alloy as anode material has contributed to the significant improvement of cell life shown in Fig. 9.44.

More recently, other dispersed-oxide–metal composites have been explored, such as Ni with $LiAlO_2$ particles and corresponding copper or copper–nickel alloy composites. Such dispersed-oxide composites are not only immune to sintering but are also creep-resistant. Electrode creep, i.e, deformation of the particulate structure of the electrode under compressive load, has been identified as an important factor in the decay of stack performance (see Section 9.4.7). According to recent reports,[89] Ni–Al alloy is very effective as a creep-resisting anode material and under compression performs better than Ni–$LiAlO_2$. NiO cathodes have quite satisfactory sinter and creep resistance; however, they are often made by *in situ* oxidation of pure nickel sinters, so the heat-up process is critical.

Creep resistance of electrode materials plays an important role in maintaining low contact resistance of the cells stack. This will be discussed in particular in Sections 9.4.3 and 9.4.6. Another factor in minimizing contact resistance is the

Figure 9.50. Comparison of soft- and hard-seal structures with respect to long-term cell performance (expressed as internal resistance). The 256-cm^2 cell is operated at 650°C under atmospheric pressure. Composition fuel gas (in %) 72 H_2, 18 CO_2, 10 H_2 (40–80% utilization); oxidant: 70 air, 30 CO_2. Anode: Ni–al alloy; cathode: NiO with CaO additive; electrolyte: 62–38 Li_2CO_3–K_2CO_3, tape-cast, $LiAlO_2$ matrix. (From Urushibata and Murahashi.[90])

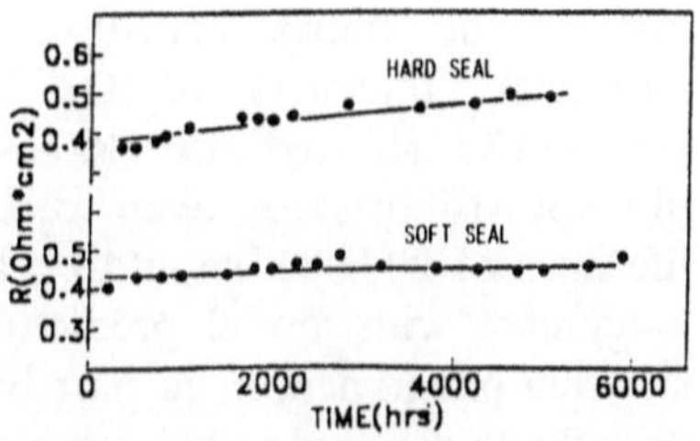

corrosion resistance of the alloy used for the separator plate. As discussed in Section 9.3.9, a suitably tailored alloy forms a dense surface layer; however, such a compact layer may have insufficient conductivity. The oxidant side of the separator plate is prone to build up poorly conducting oxide layers, particularly if the alloy contains too much aluminum. A final factor in minimizing contact resistance is the flexibility of the cell structure. Flexible structures more successfully accommodate slow changes in the electrodes and corrosion layers. Thus, it was found[90] that the increase in internal resistance of soft-seal cells is only one eight of that of hard-seal cells (Fig. 9.50). Under the most favorable conditions, the internal resistance of a soft-seal cell increases by 3 Ω-cm^2/1000 h.

As discussed in Section 9.3.9, NiO dissolution from the cathode occurs continuously during the life of a cell. This is an important life-limiting factor. However, this is not because of the effect on the cathode itself, but because of electrical shorts caused by nickel particles precipitated in the tile.[91] It has been estimated that one fourth to one third of the mass of a NiO cathode could be dissolved before the ensuing structural changes would cause serious increase in cathode polarization (the effect of compression was assumed to keep the cathode porosity constant, in these early projections).

However, the dissolved nickel diffuses into the tile, where it is reduced by dissolved hydrogen and precipitates as metal particles. These particles, due to a nucleation–dissolution–reprecipitation mechanism that is still not well understood, occur in bands whose location slowly shifts toward the anode during the life of the cell. Moreover the individual particles often exhibit an alignment which enhances the chances of shorting between cathode and anode.

It has been shown[89] that the time at which shorting starts depends not only, via the solubility of NiO, on the CO_2 partial pressure and the temperature but

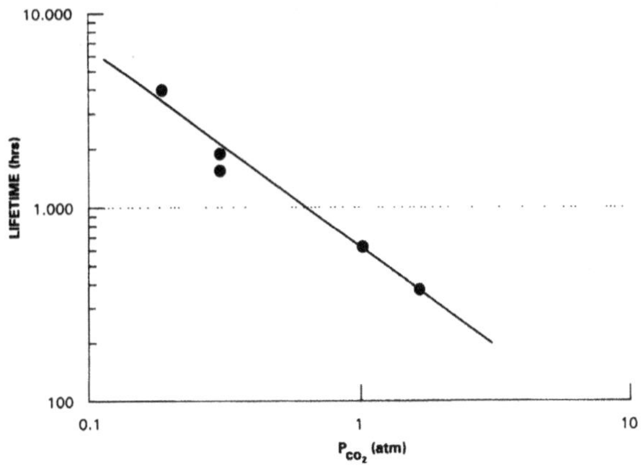

Figure 9.51. Correlation of lifetime to shorting based on single-cell results. The cell is similar to that of Fig. 9.50. Matrix thickness is 0.5 mm. (From Urushibata and Murahashi.[90])

also on the matrix structure, i.e., on the porosity, pore diameter, and, in particular, thickness of the matrix. Thus, a 0.5-mm-thick matrix under 0.3 atm CO_2 shorted after 1000–2000 h (see Fig. 9.51); however, a 1-mm matrix did not visibly short even after 10,000 h. It is estimated that at 0.1 atm CO_2 lifetimes of 20,000 h should be achievable with the present structure. This is in agreement with model predictions based on a more extensive analysis of the shorting phenomenon, in part based on empirical correlations.[92] Several aspects of cathode dissolution nevertheless remain poorly understood.

NiO dissolution can be suppressed to a considerable degree by additives which increase the basicity of the electrolyte *within* the cathode. One of the most effective additives is CaO, which itself has a very limited solubility in the Li–K electrolyte. An additive with even lower solubility is MgO. Such a limited solubility is desirable to prevent rapid diffusion of the additive out of the cathode into the tile. Alternatively, the tile can be formulated with sufficient $CaCO_3$ or $MgCO_3$ to slow down the solubilization and diffusion of CaO or MgO from the cathode.

9.4. STACK DESIGN AND DEVELOPMENT

9.4.1. Stack Configuration and Sealing

The overall structure of a fuel cell stack was introduced briefly in Section 9.2.6. It is in the literal sense a stack of thin cells and the processes taking place in the stack therefore retain strongly the two-dimensional characteristics they have in single cells. As discussed in Section 9.3.7, the distribution of current density and temperature may be quite nonuniform, even if the gas flow distribution is uniform. This in-plane distribution is the primary concern in the design of stacks, especially when scaling up the cell area. Moreover, the larger the cell area, the more difficult it is to ensure that the heat generated is convected away so as to prevent excessive temperature rise in the cell hardware.

In a fuel cell stack this problem is compounded by heat exchange between cells. The distribution of temperature, current density, etc., *among* the cells in the stack is also important. It depends largely on the distribution of the reactant gas flows among cells. Stack design aims to control this distribution to achieve stable performance without, for example, hot spots and excessive wetness of the fuel gas, which cause rapid performance decay.

The main technical concerns in the design and operation of large-scale stacks, therefore, are gas flow distribution and thermal management. The latter will be discussed further in Section 9.4.4. The key to control of the gas flow distribution is the manifolding design.

Reactant gases are fed in parallel to all cells in the stack via common manifolds. Some stack designs use external manifolds; others rely on internal manifolding (both illustrated schematically in Fig. 9.14). Hybrid designs (with fuel gas externally manifolded, oxidant internally) have been developed but abandoned.

In *external manifolding* the reactant gas mixtures are supplied each from a central source to large fuel and oxidant inlet and the exhaust gas removed via

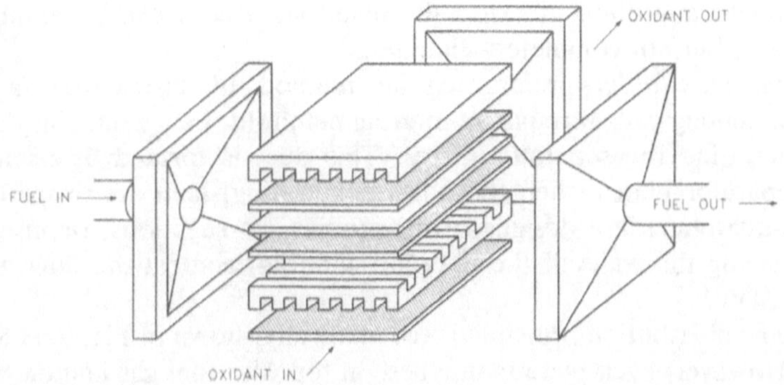

Figure 9.52. Schematic of gas flow configuration and stack sealing in external manifolding. (Modified from Ref. 93.)

large outlet manifolds, each on the opposite side of the corresponding inlet. The flow configuration is necessarily crossflow (Fig. 9.52). Details of external manifolding and crossflow stack design are discussed in reports by UTC, which pioneered this stack type.[94]

One advantage of external manifolding is simplicity: a simple and symmetrical cell design with low projected manufacturing costs. Furthermore, external manifolds allows large manifold cross sections with a minimum amount of material. This keeps the pressure drop in the manifold low and makes it easier to maintain uniform gas distribution, which is usually desirable.

Because of these obvious advantages, external manifolding was first adopted for scaled-up MCFC stacks and continues to be used by many developers (IFC, ERC, Mitsubishi Electric, Sanyo). However, external manifolding also has some disadvantages, which have emerged in practice. The primary problems are gas leakage and electrolyte displacement by migration ("ion pumping"); both are associated with the manifold gasket.

Each external manifold, covering one side of the stack, must have a gasket to form a seal with the edges of the stack. The gasket material must be insulating to prevent shorting of cells via the metal manifold. It must be elastic but stiff. In practice, zirconia felt has been used, filled with ceramics and a small amount of electrolyte. This is a compromise between the need for rigidity and isolation.

Porous cell components shrink appreciably during initial operation of the stack and continue to shrink at a slow rate under the compressive force exerted on the stack. However, the manifold and relatively stiff gasket prevent the cells from accommodating this shrinkage, so that gas leakages start to occur. Ohmic resistance may also increase, due to loss of contact between cell components.

The manifold gasket could be made more flexible, to minimize gas leakages, by using a more fluid ceramic paste. (The zirconia felt gasket contains less than 1 vol% of electrolyte). However, this would increase electrolyte displacement via the gasket. The electrolyte in the gasket provides a thin but continuous ionic path between all cells and is exposed to the full stack voltage. As mentioned in Sections 9.3.9 and 9.3.12, ion movement in the electric field (migration) causes a net bulk motion of the electrolyte. The net result in an operating stack is a slow but continuous displacement of electrolyte from cells near the positive end toward cells near the negative end.

Electrolyte migration in externally manifolded cells can be minimized (see Section 9.4.4) but not completely eliminated.

Internal manifolding refers to the method of distributing a gas flow distribution among cells through an internal manifold, i.e., a duct inside the stack itself, penetrating the separator plates. This duct is formed by profiles in the stacked separator plates when the holes penetrating them are properly aligned. The separator plates are designed with profiles, corrugations, or inserts which, besides forming the walls of the internal manifold, control the flow distribution over each plate.

The flow distribution principle is schematically shown in Fig. 9.53 for a single gas flow. However, each plate is supplied on top with fuel gas and on the bottom with oxidant gas. Therefore, each plate must be shaped to have four plenums (source–sink areas) of which only two are active on each side. In the case of counterflow, the plenums on one side of the plate function as, e.g., fuel inlet and oxidant outlet, while the complementary set of plenums on the other side function as oxidant inlet and fuel outlet. In the case of coflow, one side serves as fuel and oxidant inlet, the other as fuel and oxidant outlet (Fig. 9.54). The separator plate, consequently, is larger than in external manifolding, since each plate must contain these plenum areas in addition to the active cell area.

Internal manifolding can utilize either a penetrated-electrolyte or non-penetrated-electrolyte approach. In the penetrated-electrolyte approach, the electrolyte matrix forms the gas seal in the manifold areas, as it surrounds the gas ducts in this area, forming a wet seal as discussed in Section 9.2.3. One particular design is illustrated in Fig. 9.55. In the penetrated-electrolyte approach, the entire periphery of the cell may be wet-sealed. In the non-penetrated electrolyte approach, the electrolyte matrix extends only over the "active" cell area (the electrodes). Separate gaskets, e.g., dry ring-type gaskets, are used to form the gas seals in the manifold area.

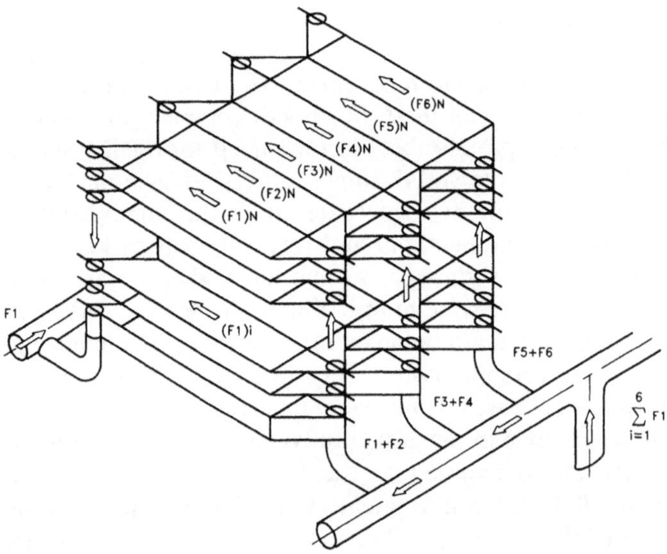

Figure 9.53. Principle of flow distribution by internal manifolding. Only a single reactant gas is shown. (From Sato.[95])

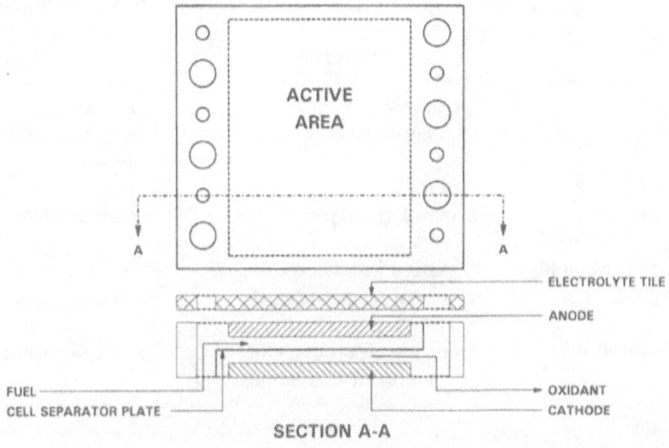

Figure 9.54. Gas flows in internal manifolding (cross section perpendicular to flow direction, in coflow). (From Marianowski *et al.*[96])

For a detailed description of internal manifolding, which applies specifically to the IHI design, refer to Sato.[95]

It is easily seen that, as far as flow configuration is concerned, internal manifolding lends itself naturally to parallel flow (coflow or counterflow) in the active cell area. However, crossflow is also possible. Internal manifolding was first seriously developed by IGT as a way to escape the external manifolding problems previously cited; this has been fairly successful, as discussed later. However, the flexibility of flow configuration has also proven to be a key attraction. This is discussed below, under "Flow configuration."

Enlargement of stack capacity by stacking is easier with internal manifolding than with external manifolds, where the manifolds become heavier with size. (Full-size IFC stacks are 86 cm wide across the flow direction; ERC plans to scale up to 85-cm width eventually. The height of the manifold depends on stack height. See also Fig. 9.80.) In operation, uniaxial compression is applied, whereas externally manifolded stacks require balanced compression along all three axes. The mechanics of stack construction and operation therefore become major concern. Several developers (e.g., MCPower, IHI, Hitachi, ECN) have now committed themselves to internal manifolding.

Internal manifolding does have disadvantages, especially in manufacturing and operation, due to its complicated design. These will be discussed further in

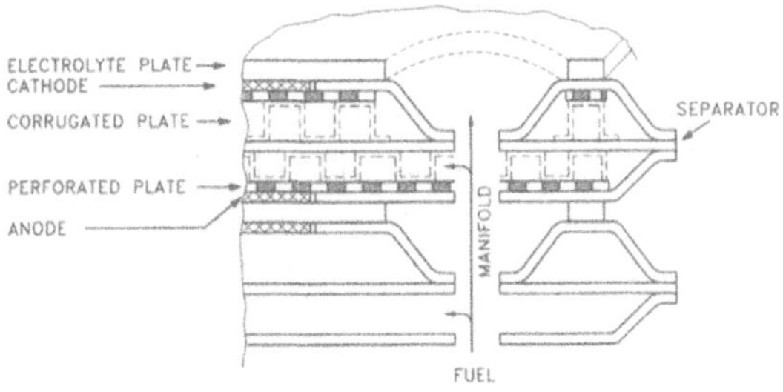

Figure 9.55. Cross section of the manifold area illustrating the penetrated-electrolyte approach to internal manifolding. (From Sato.[95])

Table 9.10. Comparison of External and Internal Manifolding

	External	Internal
Flow configuration	Crossflow	Any flow
Flow distribution	Straightforward	Requires separator plate design
Stack compression	Triaxial	Uniaxial
Critical component in fabrication	Electrolyte matrix	Separator plate
Critical step in stack assembly	Manifold installation	—
Electrolyte migration	Depends on porosity of gasket	Practically zero
Gas leakage out-of-cell	Depends on accuracy of component dimensions	No major problem

Adapted from Sato.[95]

the next section. Table 9.10 summarizes advantages of internal and external manifolding.

Various types of *flow configuration* for internal manifolding are illustrated in Fig. 9.56. Type A has a large manifold for each gas inlet and outlet, and an octagonal flow field. Crossflow takes place in the triangular fields, parallel flow in the central, active zone. Type B has multiple manifolds and a crossflow configuration, while Type C has multiple manifolds but a parallel-flow mode. Type A, which borrows much from plate-type heat exchanger technology, has been adopted by IGT (IMHEX) and M-C Power Corp. Hitachi uses a combination of four Type B configurations in its multiple stack, as discussed below. IHI and ECN have adopted Type C for their large-area stack designs. It is noteworthy that the manifolds for the oxidant flow can be larger than for the fuel (as shown in Fig. 9.56 Type C), because a larger than stoichiometric ratio of oxidant is used for cooling purposes.

Crossflow tends to give rise to complicated temperature distributions (as discussed in Section 9.3.7), which make thermal control difficult. When parallel flow is chosen, one can expect relatively small current and temperature gradients in the direction perpendicular to flow. This makes it attractive to maximize the length transverse to the flow, as in the "tatami-size" separator of IHI (Fig. 9.57).

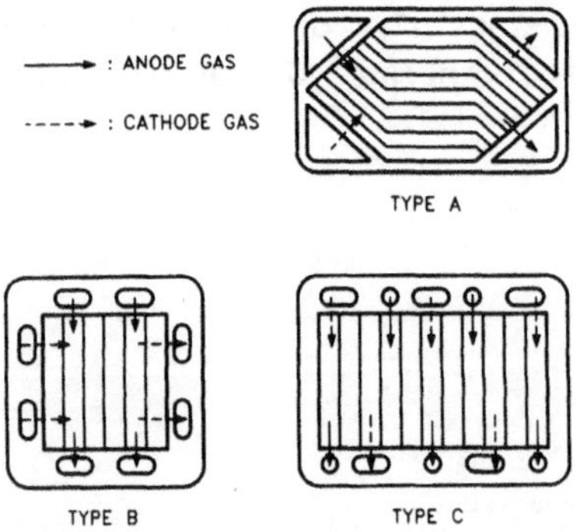

Figure 9.56. Types of flow configuration in internal manifolding. (From Sato.[95])

Figure 9.57. Large-scale separator plate for internally manifolded IHI stack. Dimensions are 1.8 m by 0.8 m. (Courtesy IHI.)

Alternatively, modules such as MCPower's IMHEX type (Fig. 9.58) can be arranged side by side.[96]

Coflow has the most uniform temperature distribution, and the maximum temperature occurs at the outlet, which facilitates thermal control. The thermal stress patterns closely follow the temperature distributions caused by flow configuration and operating conditions, and it is important to keep these as low as possible since the separator plate and its inserts are fabricated from sheet metal.[95,97]

Scale-up of stacks by increasing cell area is limited by increasing nonuniformity of current and temperature distribution in the flow direction. Although scale-up by increasing the height of an internally manifolded stack is easy, by simply stacking more cells, uniformity of flow distribution among the cells is essential. As larger and larger stacks are designed and tested, attention is increasingly given to measuring and predicting the effects of flow maldistribution on stack performance.[98–100]

It is of course possible to minimize current and temperature distribution

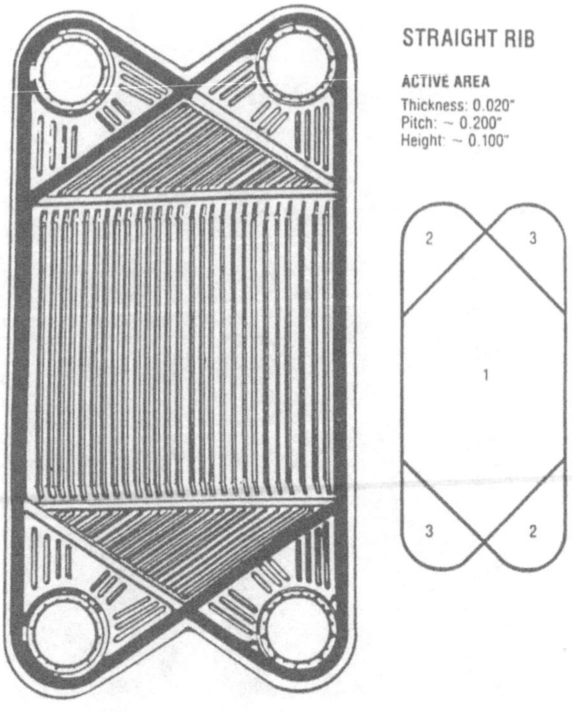

Figure 9.58. Schematic of IMHEX separator plate.[96]

problems by keeping individual stack (cell) area relatively small and combining these stacks in stack blocks or modules. In principle, the gas flows through individual stacks could be connected in series to form a desirable overall flow configuration, but this is not easy to accomplish. Hitachi uses a block of four stacks (of 3600 cm^2 active area each), which allows various combinations of parallel flow, crossflow, and U-flow (multiple large-capacity-type stack, Fig. 9.59).[101]

9.4.2. Separator Plate

In this section the separator plate, introduced in Sections 9.2.6 and 9.4.1, is discussed with respect to structure and choice of material. The selection of

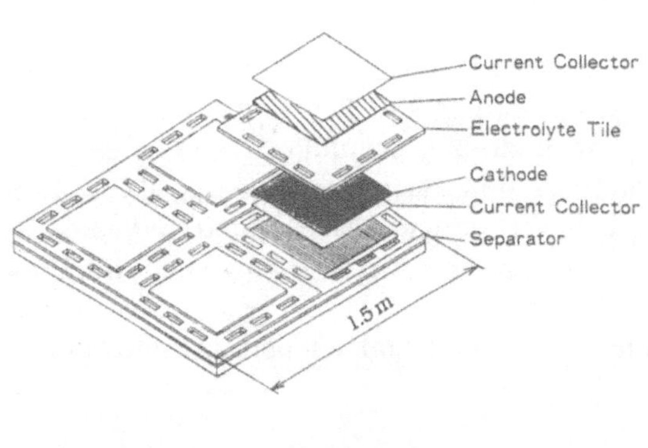

Figure 9.59. Layout of multiple stack of Hitachi MLCT type. The connections between gas manifolds are not shown (From Takashima *et al.*[101])

separator material, or materials, is a problem in common to both externally manifolded and internally manifolded stacks. However, the structure of the externally manifolded stack being somewhat simpler, the separator plate also is of a simpler design. This tends to limit the corrosion problems likely to be encountered. But this is only one aspect. The consequences for fabrication are another aspect and are of capital importance because the cost of separator fabrication controls to a large extent the total cost of the stack.

In external manifolding, where the gas flow inside the cells is straight throughout, a simple channel structure is sufficient. This can be accomplished by using a corrugated separator plate or a flat plate with either ribbed electrodes or a separate ribbed insert (current collector). It is essential that the corrugation or the ribbed insert give good access to the electrode; therefore, the pitch of the corrugation cannot be very wide (typically a few millimeters). Alternatively, a permeable insert with large void structure (e.g., Kyntex) is used (Fig. 9.60).

The internally manifolded separator has a complicated structure compared with external manifolding. The type A separator shown in Fig. 9.56 resembles a flat-plate heat exchanger in its technology. This type of separator, applied in the IMHEX concept, allows the use of a one-piece separator plate which functions simultaneously as current collector and flow guide. Such a one-piece separator is obviously most advantageous for mass production since it involves the fewest parts and processes. But, from the limitation of pressing force and the flatness of the separator, shaping by press would be very difficult for large separators, say 1 m^2 area or larger.

To avoid the 40,000-ton-force press operations necessary for one-shot pressing of a single-piece 1-m^2 separator, one can use smaller separator plates joined together or stacks connected in series gas flow. However, M-C Power

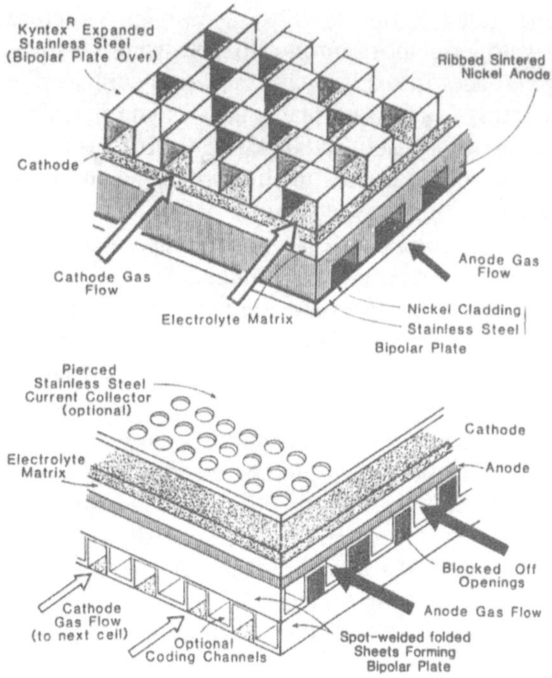

Figure 9.60. Schematic of UTC cross-flow stack design using ribbed anode, flat bipolar plate, and a current collector made from Kyntex three-dimensional expanded metal. (From Appleby and Selman.[102])

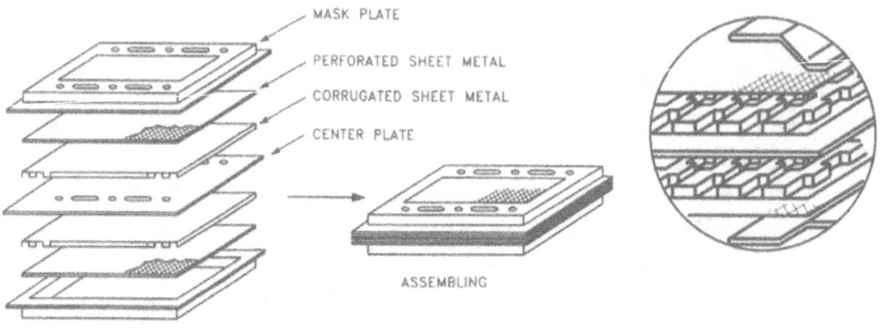

Figure 9.61. Schematic of the structure of an internally manifold separator, according to IHI. (From Sato.[95])

Corp. has fabricated and successfully tested a prototype single-piece 1-area separator. (See Fig. 9.109.)

As an alternative, some developers use a large-scale (e.g., 1-m^2) plate but insert corrugated or specially shaped metal sheets to generate the flow field. For example, the separator used by IHI has seven parts (Fig. 9.61), as discussed by Sato[95] and others.[97,103,104] These parts are Al-coated mask plates containing the manifold holes; perforated metal sheets made from 310 ss and Inconel 600 (or Ni); and a center plate of Ni–310 ss bimetal. ECN has discussed its FLEXSEP concept,[105] which is designed to make the interplate distance in the manifold area variable by the flexing of specially shaped inserts. This "soft-rail" design allows the manifold area to respond to the shrinkage of the active components of the cell, which is not possible in "hard-rail" separator designs.

The choice between a corrugated plate or a pressed-type flow-guiding structure has important implications for corrosion of the separator. Whereas corrugated plates lead to a simpler structure, they are, according to some developers, more prone to localized corrosive attack, perhaps due to local differences in oxidant concentration (differential aeration) or perhaps to residual stresses. Corrugated plates allow some flexibility in design (for example, inclined rib angles as used by Hitachi), but the profile of corrugated separators must also be carefully designed for uniform accessibility to minimize localized corrosion. In general, pressed separator parts allow greater flexibility in this regard;[106] however, pressed separator plates, like corrugated plates, require adequate mechanical strength and compliance.

The separator plate contacts the fuel gas channel on one side and the oxidant gas channel on the other side. However, the carbonate electrolyte, by creepage and by evaporation and condensation, forms a film which covers the walls of the flow channels. Therefore the separator plate is subject to corrosion under conditions very much resembling those at the current collectors in a single cell.

In Section 9.3.9 the corrosion of hardware, in particular stainless steel, in a MCFC cell was reviewed. Stainless steels 316 and 310 were found to be adequately protected against corrosion except in the wet-seal area or in direct contact with the anode, where very aggressive corrosion couples can form.

The separator plate, however, must be stable in a reducing environment as well as in an oxidizing environment. The challenge is to find a material or combination of materials which satisfies this requirement economically. It should not be expensive and should lend itself to a well-proven method of fabrication.

As mentioned in Section 9.3.9, dual-atmosphere tests (oxidant on one side and fuel gas on the opposite side of the test sample) have been used to screen commercially available Fe- and Ni-base alloys for applicability as separator material. This is necessary since the separator plate is thin and the processes on either side may influence each other in the long term. The dual-atmosphere test is aimed at specific high-risk exposure conditions, for example, high hydrogen fuel and lean oxidant. Although conditions on the anode side are more severe than on the oxidant side, especially near the fuel gas exit, the high hydrogen partial pressure near the fuel gas entrance can cause significant dissolution and diffusion of hydrogen in the plate. This may lead to carburization of the steel; it is also possible that hydrogen diffusion through the bipolar plate may prevent passivation on the oxidant side. The likelihood that this occurs depends on the gas flow configuration of the MCFC stack as well as on the separator plate material. High-chromium materials such as Type 310 stainless steel have been shown to be less susceptible to carburization than other low-carbon steels.[59]

Some of the results of a DOE-sponsored screening study[107] are shown in Table 9.11, together with corrosion tests on a series of specially designed steels by Japanese researchers.[108] ECN also carried out a systematic screening of materials for its separator plate, and concluded[105] that ss 310S, Avesta 600 (Fe–28Cr–4Ni–

Table 9.11. Candidate Alloys for Separator Material

							Corrosion rate	
Alloy (type)	Fe	Cr	Ni	Si	Al	Other	Fuel	Ox.
Fe base								
316L (FeNiCr)	bal	17	14				450/825	150/200[a]
310 (FeNiCr + Si)	bal	25	20	1.5			125/175	150/100[a]
Crutemp25 (FeNiCr)	bal	25	25	0.6			175/150	175/100[a]
Ferrilium255	bal	25	6	3	2	Cu, 0.2N	175/150	200/150[a]
FeNiCr (duplex)					1.7	3 Mo		
18SR	bal	17.5		0.6			200/150	75/175[a]
(FeNiCr + Al + Si)								
Avesta 600	bal	28	4			2Mo	[b]	
(special) . .	bal	18	8		5		100[c]	
. .	bal	18	8		3		1,000[c]	
. .	bal	18	8				1,200[c]	
. .	bal	10			5		4,600[c]	
. .	bal	5			10		4,300[c]	
. .	bal	5			5		4,200[c]	
. .	bal						11,000[c]	
Ni base								
In-601 (NiCr + Al)	14	23	bal			1.4	125/100	100/100[a]
825 (NiCr + Mo)	30	22	bal	3		2Cr, 1Ti	250/225	75/100[a]
Ra-333 (NiCr + Si)	18	25	bal	3	1.3	3W, 3Co	200/175	50/50[a]

The numbers indicate μm/year in "single/dual" atmosphere test, at 650°C, in Li_2CO_3–K_2CO_3 eutectic, with standard oxidant gas.
[a]Ref. 107.
[b]Ref. 93, 105.
[c]Ref. 108.

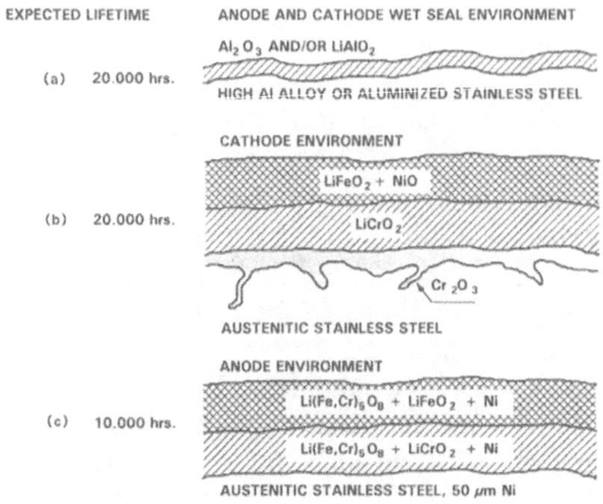

Figure 9.62. Formation of protective layers on austenitic stainless steel in the MCFC: (a) wet-seal area; (b) cathode environment; (c) anode environment. (From van der Molen.[109])

2Mo) and Inconel 601 have the best corrosion resistance in both atmospheres. High-Cr (>21%) austenitic steels such as ss 310 should be able to withstand attack in the cathode environment (non-wet-seal) for at least 20,000–30,000 h. The layer formation on such steels in the cathode environment is illustrated in Fig. 9.62.

In the anode environment, however, ss 316L is severely attacked. Though more stable, ss 310 is also significantly corroded. Corrosion tends to concentrate near the anode gas outlets. A nickel cladding or other protective coating is therefore necessary. Figure 9.62c illustrates the formation of protective layers in that case.

It is perhaps possible to develop an alloy with the requisite corrosion resistance, so the cladding operation could be avoided. However, even in that case the different requirements on either side of the plate make a graded composition, or special surface treatment, desirable. Therefore, various surface treatment techniques are being explored (see Section 9.4.3), and further research on alloy corrosion (see Section 9.5.1) is an important subject of current R&D.

The most critical area of corrosion is the wet-seal area, as discussed in Section 9.3.9. Wet-seal corrosion is especially critical in internally manifolded stacks with penetrated electrolyte, since their total wet-seal surface area is extensive due to the multiple manifolds. A good wet seal is necessary to minimize corrosion, and thereby carbonate loss and gas leakage. The most severe corrosion occurs in the wet-seal areas on the anode side (especially around the fuel gas outlets).

Wet-seal corrosion is most effectively prevented by a coating of aluminum oxide, which *in situ* is transformed to $LiAlO_2$ (Fig. 9.62a). However, the vapor deposition process in a reducing atmosphere, which is costly but feasible for small single cells, cannot be applied to large-scale separators. Therefore, other ways of aluminum coating are being studied, and much attention is also given to Al-rich alloys and alloys which form insulating surface layers (Section 9.5.1).

As an example, chromium-rich alloys as substrate have the advantage that chromium effectively reduces the corrosion rate in the anode atmosphere. FeCrAl alloy shows good protection in most environments because a protective coating of

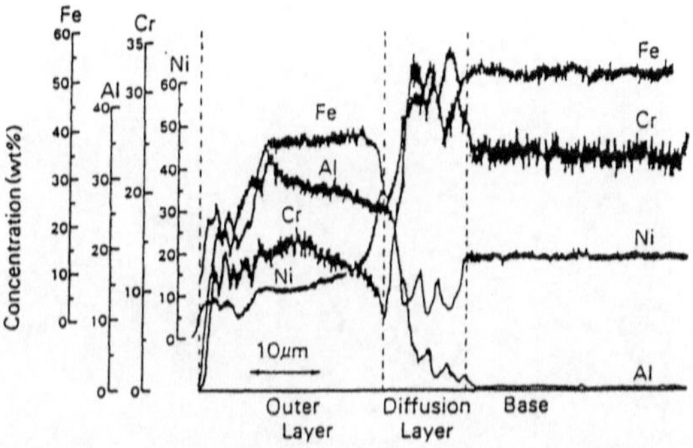

Figure 9.63. Formation of outer and inner protective layers (Al_2O_3 and Cr_2O_3, respectively) on Al-coated separator substrate. (From Sato.[95])

Al_2O_3 appears to be formed immediately on the surface (Fig. 9.63). However, this same alloy is ineffective under inert gas or very-low-oxygen-activity atmospheres because a stable inner layer of Cr_2O_3 layer does not develop; the $LiCrO_2$ layer which forms is porous and does not protect the substrate adequately.[95,97]

9.4.3. Manufacturing Issues

MCFC stack design, whether incorporating internal or external manifolding, employs flat cell components in the cell package (anode, electrolyte matrix, cathode, current collector, separator plate). Currently, all designs employ thin, tape-cast, porous electrodes and matrices. Fuel and oxidant gas are supplied to the respective electrode either through metal current collectors or through corrugated separator plates.

The material in the cell package may be as much as 50% to 70% sheet metal by weight. The electrochemical components weigh less but have the greatest potential for cost reduction. Cell package designs must reduce the amount of sheet metal required as much as possible, and the technology of manufacturing the separator plate is also a key issue in cost projections for the MCFC.[110]

Manufacturing issues *common to internally and externally manifolded stacks* concern the production of cell components, by tape casting and sintering, and the operations necessary to mass-fabricate the separator.

9.4.3.1. Tape Casting and Sintering Technology

Tape casting (see Section 9.3.4) is now accepted as the technology of choice to produce the anode (annex bubble barrier) electrode–electrolyte–cathode cell package. Since each component is cast in a single operation, the size of the tape-casting equipment (Fig. 9.64) limits the stack area. Similarly, the burnout of organics and reduction of the anode must take place in flow-through furnaces whose size may limit the stack area.

The largest externally manifolded stacks have a cell area of 8 ft^2 (7432 cm^2).[112] The largest internally manifolded stacks are 1.4 m^2 (12.60 ft^2) in area.[113] The fabrication of components for such large-area stacks, in quantity, and the assembly and operation of such stacks are remarkable achievements. The need for more and more stringent quality control increases with stack size. Satisfactory performance and endurance require rigorous uniformity of tape thickness and composition, and, finally, of electrolyte distribution in the tape.

Figure 9.64. Continuous tape-casting machine, capable of producing 0.7-m-wide matrices. (From Maru *et al.*,[111] (Courtesy ERC.)

The stability and uniformity of the casting operation must be constantly monitored and improved. Tape casting is a batch operation (using typically 0.2-m^3 batches), and the limited shelf life of the casting slurry can cause quality control problems. If the shelf life is exceeded, slurries may agglomerate and tapes crack. Therefore, it is necessary to develop a solvent-base binder system that is compatible with continuous tape casting.[112]

It is also necessary to develop a continuous electrode tape debinding (burnout) process. If the debinding occurs unevenly, the electrode tape has excessive variations in thickness as well as porosity, which inevitably cause performance deterioration.[112]

Another key technology is that of sintering large (and, for example, rectangular electrodes) at high temperature to produce uniform characteristics. In particular, accurately controlled sintering techniques are required to produce the fine-particulate bubble barrier (see Section 9.3.4), which has a thickness of 0.1 mm or less. Sintering models help to analyze the effect of the various forces (friction and elastic forces as well as sintering) which control the shrinkage process during passage through the furnace.[95]

The development of a satisfactory creep-resistant anode which can be made by tape casting is another important manufacturing issue. The Ni–10Cr alloy used in single cells (see Section 9.3.4) is not creep resistant enough in stack compression. It has been supplanted by Ni–8Cr (which has creep deformation of maximum 5% after 10,000 h[114]) and by Ni–Al.[115] Ni–5Al has been reported to have very good sintering resistance after 3200 h in-cell tests, with performance equal to that of Ni–10Cr; Ni–6Al has shown good creep resistance in a 2000-h test.[116] These alloys can be tape-cast as alloy powers, but it would be more attractive to tape-cast mixtures of metal powders. However, the extremely fine powders necessary are difficult to realize. Further development of Ni–Al–Cr and other ternary composites can be expected to solve this problem.[114]

The creep strength and mechanical stability of the microstructure of the electrolyte matrix are also issues with important manufacturing aspects. Long-term electrolyte loss and performance decay should be correlated with changes in the particulate structure of the $LiAlO_2$ matrix. This is indeed found to be the case,[117] but a simple cause-and-effect relationship cannot always be established.[118] For the mechanical stability of the electrolyte matrix, e.g., in thermal cycling, the use of inert fibers or particles as "crack arrestors" remains a possibility,[119] although most developers appear to find such devices unnecessary for thin tapes, if quality control over tape uniformity is sufficient.

Finally, the method of loading carbonate in the electrode–electrolyte package is a manufacturing issue. Carbonate electrolyte is needed in excess of that needed to fill the electrolyte matrix; this excess is needed to fill the porous electrodes partially with electrolyte. In single cells, the matrix and electrolyte are usually supplied as a powdered mixture (either hot-pressed or tape-cast) with a slight excess of carbonate so as to yield the desired electrolyte content of the matrix after capillary equilibrium has been reached (Section 9.2.4). However, this procedure is not very reliable and tends to produce large variations in electrolyte content of the matrix; hence improvement was called for.

There are basically three ways of adding the carbonate electrolyte needed in the matrix and electrodes:

(a) in the gas passages
(b) in the electrode tapes
(c) as one or more carbonate tapes

These methods are illustrated in Fig. 9.65. The method of using carbonate tapes has been adopted by most developers using internal manifolding. The

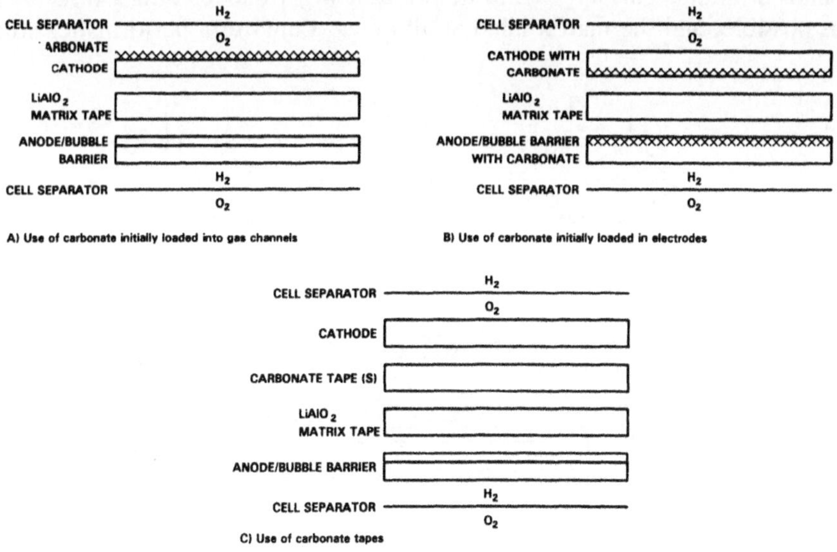

Figure 9.65. Alternative methods of loading carbonate in the electrolyte matrix and electrodes *in situ*. (From Marianowski *et al.*[120])

carbonate tape and the $LiAlO_2$ matrix extend to the edges of the cell, and although the intercell separation decreases in proportion to the thickness of the carbonate tape, sealing and conformity are maintained at all times during the heat-up of the stack. Before the carbonate melts, gas seal is maintained because the carbonate as well as the $LiAlO_2$ tapes extend to their respective sealing surfaces and contain a rubbery binder, which does not decompose until shortly before the melting point of the carbonate is reached. Once the binder is burned off and the cell temperature has reached the melting point of the electrolyte, the molten carbonate is absorbed by the pores in matrix and electrodes. The intercell spacing thereby decreases approximately by the thickness of the cabonate tape(s). This initial decrease is accounted for in the dimensioning of the stack.

Externally manifolded systems cannot use this method because the changes in intercell spacing during carbonate melting cannot be accommodated by the rigid external manifolds. It will be clear from the above, though, that the start-up of an internally manifolded stack, no less than an externally manifolded stack, must follow a carefully controlled heating protocol.[113]

9.4.3.2. Separator Fabrication Technology

Several important aspects of this have already been discussed in Sections 9.4.1 and 9.4.2.

The basic choice in separator fabrication is between the *hard-rail* concept, in which intercell separation is fixed by the height of stiff profiles or inserts, and the *soft-rail* concept, in which sheet metal inserts or masks, at the edge of the separator plate, are folded back upon themselves (for example as ECN's FLEXSEP design, Fig. 9.66, which also features spring-loading). The prevailing tend has been toward the soft-rail approach. In this approach, the separator is designed to accommodate changes in the thickness of the electrolyte–electrode package, during start-up as well as later, when slow structural changes occur over thousands of hours. This appears to be the best way to keep contact resistance as low as possible, and the matrix filled at all times. Long-term performance studies

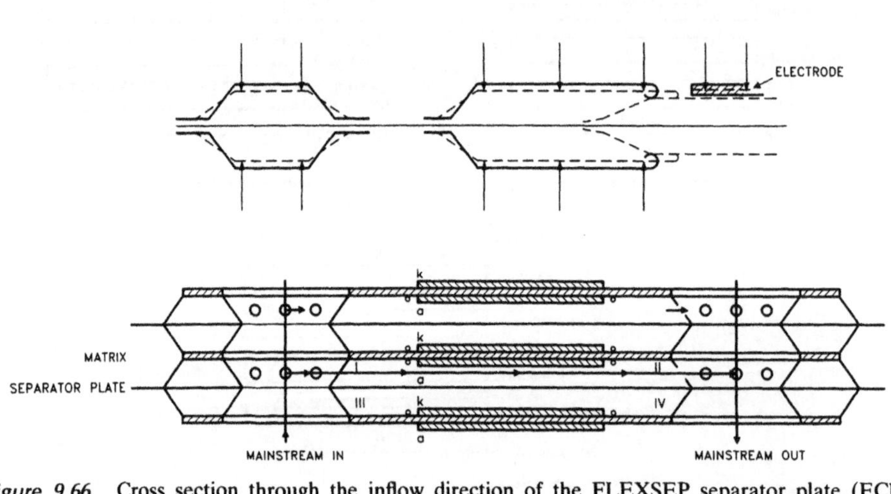

Figure 9.66. Cross section through the inflow direction of the FLEXSEP separator plate (ECN) showing how gas chamber dimensions follow the shrinkage of cell components. (From Boersma.[121])

(Section 9.4.8) show that the gradual rise in effective internal resistance (which includes contact resistance) is much less in cells with "soft package" cell design than in cells with fixed interelectrode dimensions.

The externally manifolded stack, however, requires a certain level of rigidity in cell-to-cell distance because the external manifolds have fixed dimensions. If too much sagging occurs in the stack, gas seals may become defective and leaks will occur, both to or from the ambient atmosphere and from one gas channel to another (fuel to oxidant and vice versa). Fixed dimensions, however, as in a hard-rail separator design require tight control over (1) the amount of electrolyte in each cell (to prevent long-term rise in ohmic resistance) and (2) the thermal expansion of cells (to prevent discrepancies in lateral expansion, which cannot be accommodated by stiff manifolds and gaskets, thus causing intercell gas leakage). The design and fabrication of externally manifolded stacks, therefore, must rely on accurate thermal and mechanical stack models.

These different separator designs have of course manufacturing implications. The hard-rail design requires greater stiffness and consequently more material. The soft-rail design may appear more economical in material but finally requires more pressing operations. Welding and annealing, which are inevitable in hard-rail separator manufacture, play also a key role in manufacturing internally manifolded separators, such as the seven-part separator of IHI (Fig. 9.61). In assembling such a separator three sheet metal parts (two mask plates and a center plate) have to be welded together in a single operation, using electrical resistance seam welding.

No single inexpensive material is stable under all anode and cathode conditions. Stainless steel types 310 and 316 corrode at an unacceptably high rate near the fuel gas exit. Therefore, the anode side of the separator plate may have to be clad with nickel or copper, which are thermodynamically stable. Such a cladding is also an effective protection against carburization. However, cladding tends to be expensive, and alternative surface treatments or modifications of conventional stainless steels, in particular Type 310, are being pursued. (See Section 9.5.1.)

Finally, the separator plates of both internally and externally manifolded stacks require a special *wet-seal area treatment* to protect these areas against corrosion, as discussed in Section 9.3.9. The most effective and, thus far, generally used method of protecting the wet-seal areas is by coating with aluminum (aluminizing) or Al_2O_3. Alternatively, high-Al stainless steel might be used in the sealing areas, but this would create too high an ohmic resistance if used for the remainder of the plate. Therefore, a composite or graded-composition plate would be necessary, which poses special manufacturing difficulties.

The aluminum coating techniques used thus far include[95]

1. Aluminizing by pack cementation. This is the technique generally used for single cells, where it works very well. However, the high temperature required for a significant length of time (1200°C, for 3–4 h) tends to deform the sheet metal separator.
2. Hot-dip coating in molten aluminum. This also requires a high temperature and tends to deform the separator.

3. Spraying of aluminum or aluminum alloy, using plasma spray. This is a rapid, low-temperature (<300°C) process, suitable for extended surface areas. However, Ni–Cr–Al–Y coatings were not completely dense, and corrosion eventually occurred.[95]
4. Physical vapor deposition of aluminum. This is a room-temperature process. It produces good, but weakly adherent, coatings whose thickness cannot easily be increased above 10 μm.
5. Cermet coating (by proprietary procedures). Such coatings have good corrosion resistance and are being explored; some have shown to be excessively brittle in handling.
6. Electroplated coating (from room-temperature or low-melting molten salts, <200°C). This coating technique[122] is promising and produces dense, adherent coatings, but requires further development.
7. Spray painting of aluminum slurry. This requires careful attention to substrate pretreatment to ensure adhesion.

For both internally and externally manifolded stacks, the design and production of the separator plate assembly (separator plate, current collector, flow guides) has important manufacturing implications. In-stack flow distribution and gas recycling is accomplished by low-pressure-drop flow fields incorporated in the current collectors for both reactant gases. This requires special manufacturing steps.

For *externally manifolded stacks,* the most important specific manufacturing issue[112] is the design and production of the gas manifolds and manifold gaskets. As mentioned earlier and discussed in more detail in Section 9.4.6, externally manifolded cells require a manifold design which minimizes the carbonate flow by migration.

Finally, the issue of *reformer placement* is important for manufacturing. In MCFCs with external reforming, reforming takes place outside the cell stack but adjacent to it so the heat produced by the stack can be transferred efficienctly to the reformer. The reformer would be external to the stack but in very close proximity. IFC has proposed such as integrated pressurized system (Fig. 9.67). It has the advantage of keeping stack design and manufacturing simple, and isolating reformer operation somewhat from the stack. This may be advantageous from the viewpoint of system upsets (for example, fuel contamination) and maintenance.

Some developers, notably IHI,[123,124] have adopted plate-type reformers (Fig. 9.68) of dimensions similar to the stack, which can conveniently be configured with the stack, especially if an internal manifolding design is adopted. The concept is very similar to that in indirect internal-reforming stacks, as discussed in the next section. A plate-type reformer is, of course, most appropriate in operation of MCFC stacks on natural gas, which does not require pressurization, although operation at moderate pressurization is planned by some developers. However, reformed natural gas and low-Btu coal gas under pressure may contain residual methane, which an integrated plate reformer could control to a certain extent.

A plate-type reformer is also attractive from a manufacturing viewpoint. Although current fabrication of plate reformers requires several machining operations, future mass fabrication may be possible by means of pressing and forming operations identical to those for the separator.

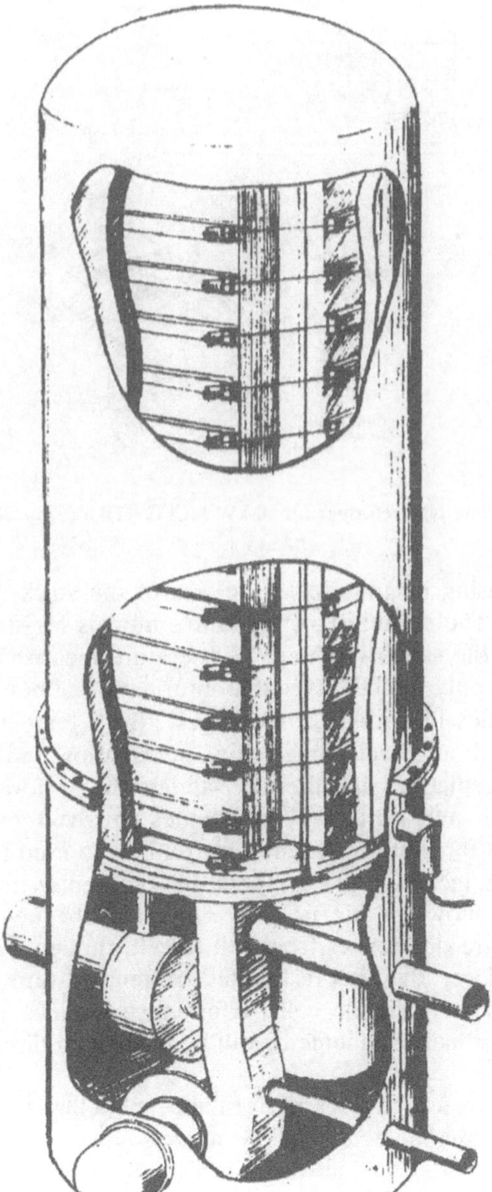

Figure 9.67. Product-integrated stack unit according to IFC. (From Ref. 112, Courtesy IFC.)

9.4.4. Thermal Management and Internal Reforming in MCFC Stacks

9.4.4.1. Temperature Distribution Modeling

An extensive discussion of temperature and current distribution in single cells was presented in Section 9.3.7, and it was shown that cell performance models play an important role in scale-up. In the design of MCFC stacks they play an even more important role since the dissipation of heat in the stack

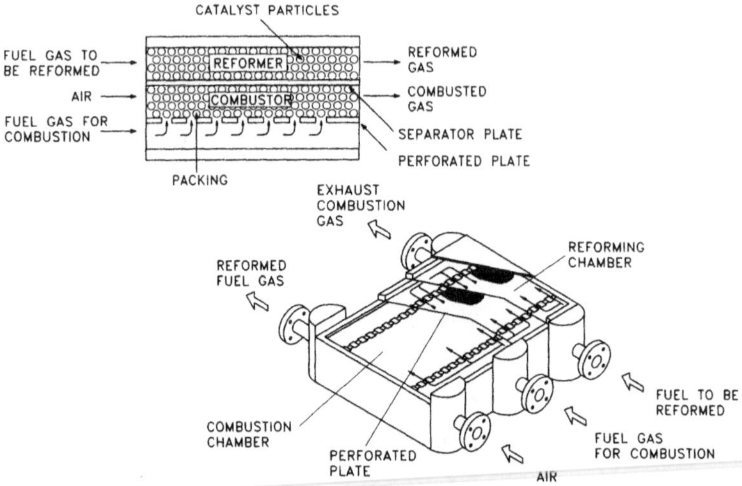

Figure 9.68. Plate-type reformer for 10-kW MCFC (IHI). (From Refs. 123 and 124.)

becomes of increasing magnitude as the size of the stack (i.e., its area and its height) increases. The reliability of predictive models for stack performance is a great concern for the stack designer. Temperature measurements within a stack are practically the only directly available information about the internal working of the stack (besides individual cell voltages). Such temperature measurements therefore serve both as direct "diagnostic" information and as calibration points to check against predictions and thereby validate the performance model used.

The principles and numerical techniques involved in stack performance modeling are, thus far, not much different from those used in single cells. This is mainly due to the fact that the temperature differences that develop between hardware and gas flows *in the flow direction* are the key design parameters, whereas temperature differences from cell to cell, though not negligible, are less important. Therefore, the first estimates of temperature distribution can be obtained from relatively simple cell performance models, which are essentially one- or two-dimensional. Refinements can be made by a three-dimensional model where necessary.

An important review of stack performance modeling applied to the MCFC is presented by Takashima *et al.*[125] Few models are available in the literature because successful models are powerful tools and contain significant information about design and operational parameters. Kobayashi *et al.*[126] developed a complete cell performance model and extensively compared the calculated results with actual data measured in several Hitachi Corp. stacks.

In Kobayashi's model the thin-film cylindrical pore model was used as a porous electrode model to determine the local overpotentials of the electrodes. The values of the film length and its thickness were set to obtain a required mean current density. The heat transfer processes were included as discussed in Section 9.3.7. Periodic boundary conditions were used at the cell-to-cell contact planes; i.e., the same gas temperature distribution was assumed in each cell, in the stacking direction.

As discussed in Section 9.3.7, the results (see, e.g., Fig. 9.30) show that the temperature distribution is dominated by the cathode gas stream, which acts as a

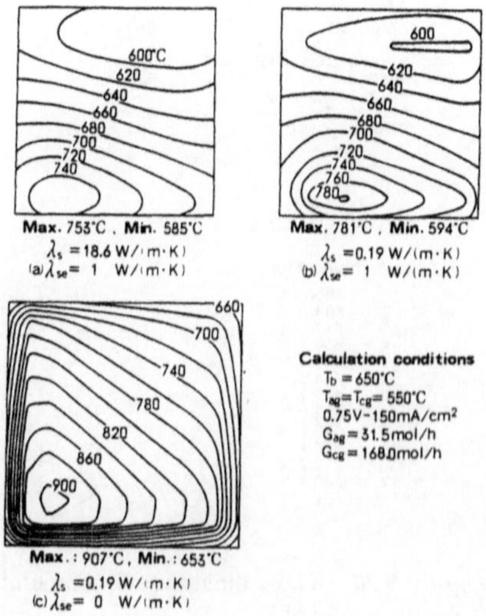

Figure 9.69. Effect of separator conductivity (as effective conductivity λ_s) and cross-plate thermal conductivity (as effective conductivity λ_{se}) on electrolyte temperature distribution in a 3600–cm^2 cell. (From Kobayashi *et al.*[126])

cooling medium, and that the current distribution is determined by the concentration of hydrogen in the anode gas stream. A steep temperature gradient is found to occur in the cathode gas streamwise direction, and a steep current density gradient in the anode gas streamwise direction.

In the case of a stack, significant deviations can occur from these single-cell patterns. This is illustrated in Fig. 9.69. The temperature distribution is strongly influenced by the thermal conductivity of the separator plate (λ_s in Fig. 9.69) and the thermal conduction between separator plate and electrode–electrolyte package, via a current collector or other structural elements. The latter conducting path can be thought of as an "effective cross-plane conductivity" (λ_{se} in Fig. 9.69). If it is very conductive, the separator will have a strong smoothing action on the temperature gradient.

Although the local temperatures measured in stacks indicate distributions in qualitative agreement with predictions, sometimes significant deviations were found.[126] These are caused by

1. The massive end plates which caused heat flows in the stacking direction
2. Contact resistances

For a better prediction of the three-dimensional distribution, the end effects can be taken into account rather easily. Figure 9.70 shows a three-dimensional temperature distribution calculated for a crossflow 12-cell stack.[125] As expected the central cell has the maximum temperature (near cathode outlet, fuel cell inlet), but the range of temperatures is appreciable.

Much more activity can be expected in the area of three-dimensional stack modeling, though the results may not be published initially. Therefore, the issue of validating models, especially those used to design stacks for pressurized operation, can be expected to increase in importance.

9.4.4.2. *Internal-Reforming Stack Design*

Energy balance considerations, as presented earlier (Sections 9.2.5, 9.3.3, and 9.3.7) are especially important for the IR-MCFC stack. This is an MCFC

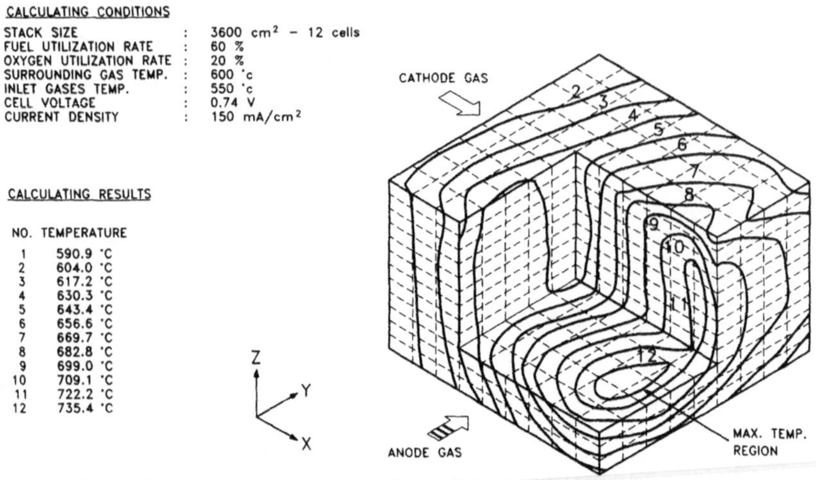

Figure 9.70. Three-dimensional temperature distribution in 12-cell stack with crossflow. (From Takashima *et al.*[125])

stack in which natural gas can be simultaneously reformed and electrochemically reacted. Natural gas, with methane, CH_4, as main constituent, is the principal fuel for which the IR-MCFC is being developed, but other light hydrocarbons (propane, naphtha) and alcohols (methanol) may be also used if carbon deposition can be avoided.

In conventional MCFC stacks operating on externally reformed natural gas, removal of heat is necessary to maintain the stack at 650°C. The cooling load is significant, since the overall fuel cell reaction (9.3) is strongly exothermic ($\Delta H_{650°C} = -247.5\,\text{kJ/mol}$). The steam-reforming reaction (9.5), however, is endothermic ($\Delta H_{650°C} = +225\,\text{kJ/mol}$), which makes the net reaction of CH_4 oxidation,

$$CH_4 + 2O_2 \rightarrow CO_2 + 2H_2O, \qquad (9.44)$$

much less exothermic. This is especially clear if the heat effect is expressed as a cell voltage, i.e., the so-called thermoneutral voltage

$$E_{tn} = -\Delta H_r/nF. \qquad (9.45)$$

Table 9.6 shows thermoneutral voltages for various cell reactions under standard conditions. A cell operated at the thermoneutral voltage has a net heat effect of zero; i.e., it operates adiabatically. For the net reaction (9.44) the thermoneutral voltage and the emf are practically equal, whereas E_{tn} for hydrogen oxidation is substantially higher. This means that the exothermic heat effect of a fuel cell operating on CH_4 is much less than that of a H_2-fed fuel cell. The difference is equivalent to 244 mV, while the electric output of the two fuel cells differs by only 18 mV.

Thus, the heat produced by the fuel cell reaction in an IR-MCFC suffices, in principle, for steam reforming. In the conventional MCFC system, with external reforming, the heat required for the reforming reaction is usually supplied by burnup (burnout) of the anode exhaust gas followed by heat exchange. In the

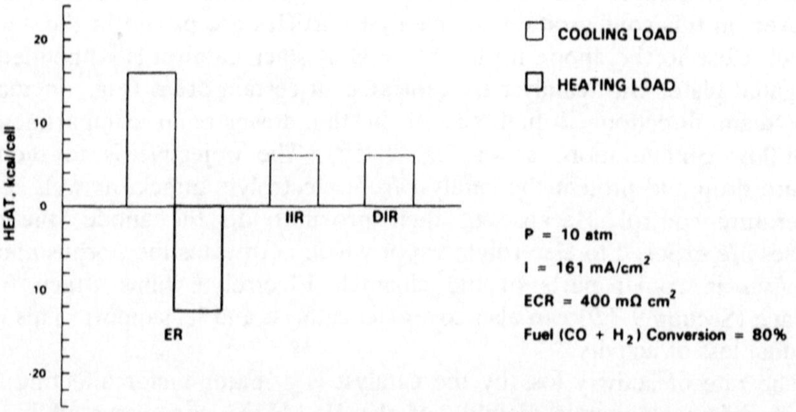

Figure 9.71. Heat load in external and internal reforming. (From Murahashi.[15])

case of the IR-MCFC, however, heat is provided by the fuel cell reaction. If the heat from the cell reaction is directly transferred within the cell and made available to the reforming reaction, the cooling load may be reduced by a significant fraction, perhaps as high as 50% (Fig. 9.71). This results in a reduction in the number and area of the heat exchangers, which in turn leads to plant simplicity and reduced capital cost.

To accomplish this internal reforming–heat exchange, two stack design concepts have been adopted as illustrated in Fig. 9.72. In the direct internal-reforming (DIR) stack, the catalyst is included in the anode chamber of the fuel cells. In the indirect internal-reforming (IIR) stack, reforming takes place in separate "reformer plates" adjacent to the "fuel cell plates" and in good contact with the anodes of the cells.

The *DIR concept* has the advantages of direct product and heat exchange. The kinetics of the reforming reaction (see Section 9.2.5) are rapid because of

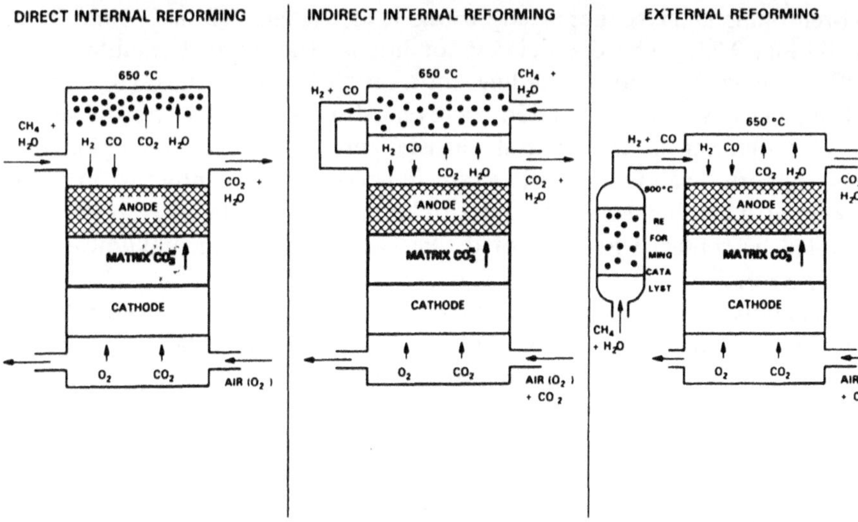

Figure 9.72. Types of reforming configuration. (From Murahashi.[15])

this direct exchange. Conversion therefore, approaches the equilibrium limit. However, in this configuration the catalyst particles are placed in the anode gas channel, close to the anode itself. In the DIR stack catalyst is embedded in the corrugated plate, with catalyst concentrated in certain areas (e.g., increasing in downstream direction of fuel gas or in the downstream compartment of a return-flow configuration, as in Fig. 9.75b). The objective is to reduce the pressure drop and protect the catalyst from electrolyte attack, as well as ensure temperature control. Because of their proximity to the anode, the catalyst particles are exposed to electrolyte vapor which permeates the porous anode and condenses in cooler parts of the channel. Electrolyte films which form by creepage (Section 9.3.9) can also cover the catalyst and its support. This leads to a gradual loss of activity.

The rate of activity loss by the catalyst is a major factor affecting lifetime and, therefore, economic viability of the IR-MCFC. As discussed in Section 9.4.7, the mechanism of activity loss is a key area requiring further work to improve long-term performance.

In the *IIR concept,* the fuel gas flows through a separate reformer chamber in thermal contact with the fuel cells before it contacts the electrolyte. The catalyst therefore is not affected by electrolyte vapor, and catalyst poisoning does not occur. However, the reforming reaction is not driven by the local H_2O partial pressure, as in the DIR, and the local temperature gradient supplying heat to the reformer chamber is not so steep. Methane conversion therefore is less.

Since the basic IIR configuration is inferior to the DIR, with respect to methane conversion, a cascaded IIR type is considered. In such a configuration, residual CH_4 in the first stack is converted in the second stack, so the overall conversion approaches that in a DIR stack. Another option is a hybrid type of DIR and IIR, in which the reforming load of the DIR is kept intentionally small to stretch catalyst lifetime.

The major challenges to IR-MCFC stack development are catalyst degradation and the need to optimize the temperature distribution.

a. Catalyst Degradation. The decay of the CH_4 conversion rate for the nickel-reforming catalyst supported on MgO, in a single cell of 272 cm^2 area, is shown in Fig. 9.73, under open-circuit conditions. The thermal equilibrium level at 650°C is shown by a horizontal line, for a ratio of steam to CH_4 (S/C) of 2.0.

The mechanism of deactivation, as revealed by posttest analyses,[128] seems to be in part sintering of the catalyst particles and in part contamination of the catalyst support. The rate of particle growth exceeds significantly that in external reformers at much higher temperature. The particle growth rate appears to be accelerated by contact with electrolyte. Perhaps other, as yet unknown, factors also play a role.[15]

Contaminants such as alkaline compounds and sulfur (in concentrations far below those tolerated by the MCFC anode) attack the catalyst support and change the interaction between support and catalyst, with consequent degradation of catalyst activity. The fuel gas should contain less than 0.1 ppm sulfur to maintain quality.[129] This implies that natural gas used in IR-MCFCs should first be stripped of any H_2S, mercaptans, etc., as well as any odorants added for consumer safety.

Carbonate condensation onto the catalyst via the anode gas leads to

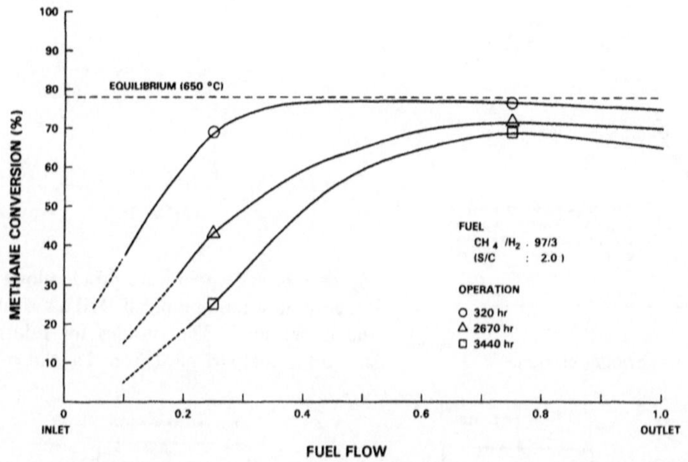

Figure 9.73. Decay of methane conversion rate with time, along the anode flow direction of a IR-MCFC cell. (From Tanaka *et al.*[127])

deactivation when the carbonate deposit exceeds 5% of the catalyst mass.[130] The carbonate resistance of the catalyst support is also of great importance. MgO and $LiAlO_2$ appear to be adequate; MgO is currently considered the best choice. Even these materials need to be protected from direct exposure to carbonate vapor. Strategies followed to accomplish this include modification of the cell design to avoid condensation, and use of an absorbent to scavenge carbonate vapor before its contacts the catalyst. Alternative support materials are also being explored. A broader choice of support materials would be very desirable, since it could be used to modify hardware surfaces (e.g., of separator, current collector, flow guides) so as to make them catalytically active.

By proper choice of electrolyte-resistant catalyst support and other improvements, cell life has now been extended to several thousand hour. Beyond 5000 h, other life-limiting factors appear to influence the cell performance; therefore it is difficult to predict lifetime based on catalyst degradation only. Significantly longer catalyst lifetime is expected in the IIR mode of operation, when only thermal degradation occurs.

b. Temperature Distribution. Thermal management has different requirements for the DIR stacks than for IIR stacks. As discussed above, the reforming reaction is strongly endothermic and in the cells of a freshly loaded and active DIR stack, this causes a pronounced minimum in temperature at the cell inlet. This is illustrated in Fig. 9.74 for an early DIR stack; the behavior of reformate gas is shown for comparison, and the heating effect of current load is also evident.

Such a steep temperature gradient at the stack inlet is not advisable, and several ideas have been implemented to avoid it. One method is to use cathode gas recycling and/or change the gas flow configuration. In the 3-kW DIR stack of MELCO, a fuel return manifold is used to achieve better thermal balance (Fig. 9.75); the improved distribution of temperature, conversion, and current density is shown in Fig. 9.76. The other method is to change the way in which the catalyst is loaded into the reformer plate. In the latter case, the idea is to suppress the

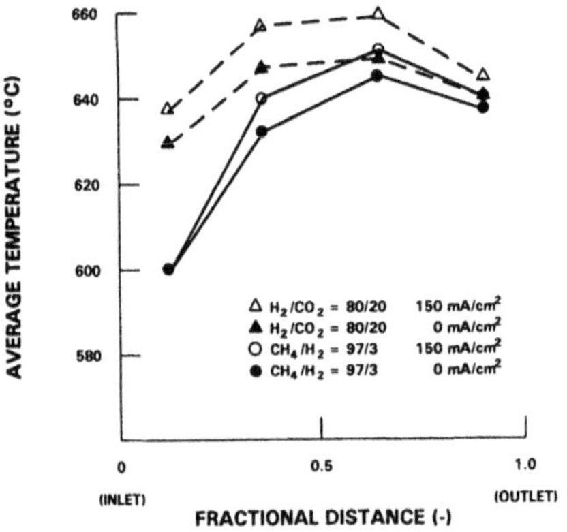

Figure 9.74. Temperature profile along the fuel flow direction in a nonoptimized DIR stack at open circuit and under load. The profiles for reformate gas are shown for comparison. (From Tanaka *et al.*[127])

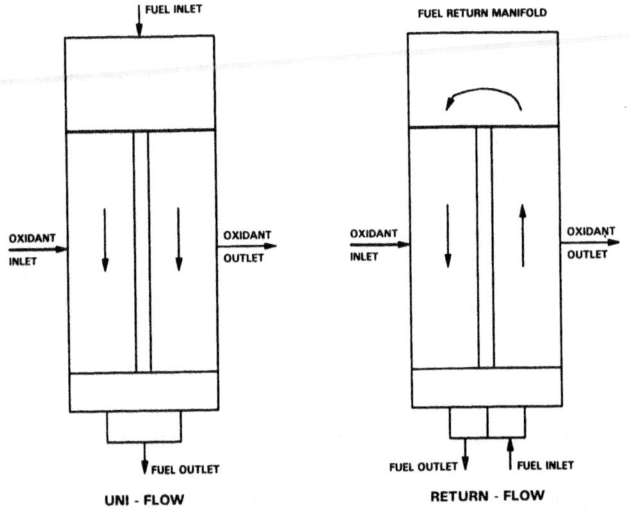

Figure 9.75. Flow configuration of 3-kW DIR stack (MELCO). (From Tanaka *et al.*[131])

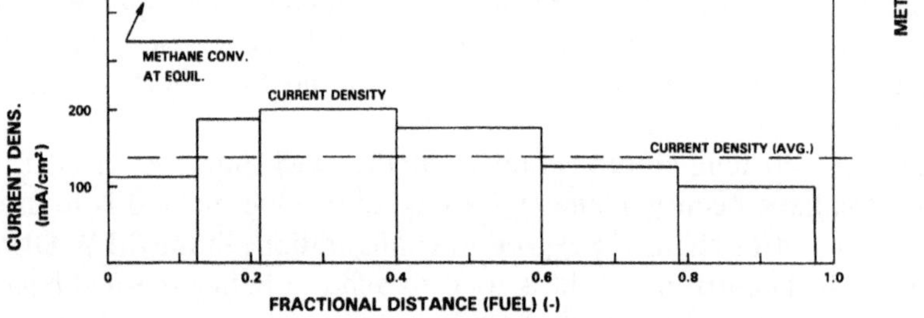

Figure 9.76. Profiles of temperature, CH_4 conversion, and current density along the flow direction of cells in 3-kW DIR stack. (From Matsumura *et al.*[132])

conversion of the reforming reaction at the inlet of the plate by reducing the catalyst loading.

Compared with the DIR, the IIR stack presents greater difficulties with respect to design for temperature uniformity.[133] The IIR typically has an internal reformer plate for every five or six cells; this plate includes the catalyst. Fuel (with a higher S/C ratio than in DIR stacks) is fed directly to the reformer plates via a specially designed manifold. Reformed hydrogen gas is fed into the anodes of cells adjacent to the reformer plates.

The endothermic effect of CH_4 conversion is locally large and causes especially low temperatures at the anode inlet area. In the case of the DIR this heat sink is present in every cell and can be corrected by fuel gas recycling or by a flow configuration as shown in Fig. 9.75. In the IIR the heat sink leads to gradients in the stacking direction, which must be controlled by three-dimensional stack design features.

Figure 9.77 shows the stack structure of the IIR stack of MELCO, applied successively to 10-kW and 30-kW stacks. Figure 9.78a represents a horizontal section of the reformer and illustrates the gas flow directions. Fuel gas mixture flows into the reforming unit through a dead-ended "bypass channel" without catalyst filling, turns to flow into the channels filled with catalyst, and turns again to exit through a bypass channel into the "anode gas" (reformed hydrogen-rich gas) inlet manifold. This manifold distributes the anode gas vertically among adjacent cells. The anode gas–oxidant flow pattern in the cells is crossflow. The cathode gas entrance and exit areas function as intensive heat exchangers so that the corresponding areas in the reformer unit are left empty of catalyst (bypass channel). The corresponding temperature distributions and heat loads, calculated from a numerical model, are shown in Fig. 9.78b,c. The actual temperature difference in cell 28 of the 30-kW IIR stack is shown in Fig. 9.79. The maximal temperature difference in this stack has been reduced to 80°C.

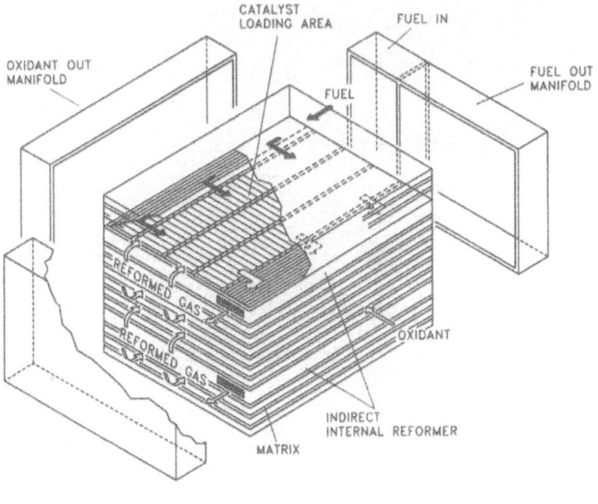

Figure 9.77. Stack structure of an indirect internal reforming MCFC, according to MELCO.[131,134]

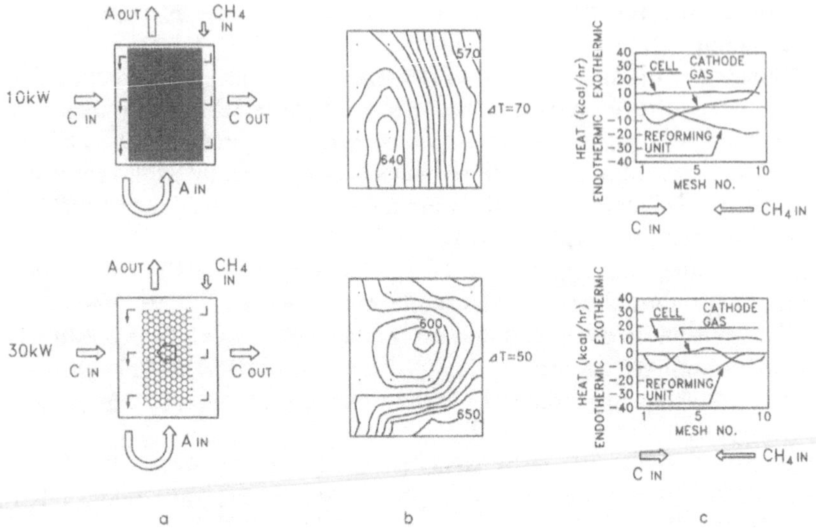

Figure 9.78. Thermal design of large IIR stacks: (a) schematic of flow configuration; (b) predicted temperature distribution in cell 1; (c) predicted heat balance in the stack. (From Ide *et al.*[135])

9.4.4.3. *Internal-Reforming Stack Performance*

At present two companies, Energy Research Corporation (ERC) in the United States and Mitsubishi Electric Corp. (MELCO) in Japan are leading the development of scaled-up internal reforming stacks, with the aim of commercialization. Energy Research Corporation, which pioneered the internal-reforming stack (direct fuel cell or DFC concept), has recently been concentrating on life-time improvement as well as scale-up, with performance tests mainly on reformate fuel. All ERC's current stacks have DIR–IIR combined configuration. MELCO, concentrating on scale-up and the internal reforming function, has pursued the IIR approach discussed above.

In the 1990–1992 period,[136] ERC built a 2-kW stack with 4000-cm^2

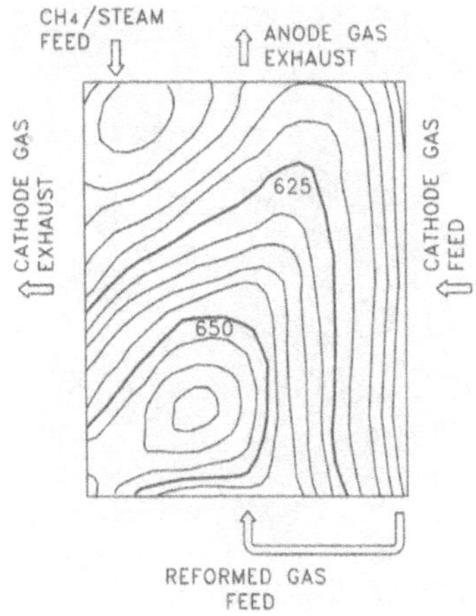

Figure 9.79. Temperature profile in cell 28 or 30-kW IIR stack. (From Ohtsuki *et al.*[134])

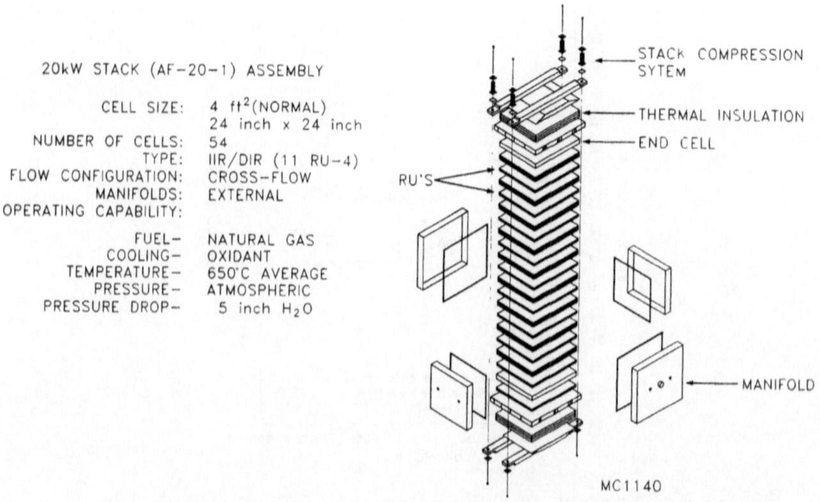

Figure 9.80. Schematic of 20-kW stack assembly (AF20-1) by ERC. (From Maru.[137])

components and tested it for 10,000 h. An accelerated subscale stack test was conducted for 7000 h. This test served to validate ERC's manifold seal design. Since ERC's (as well as MELCO's) stack design is based on external manifolding, it is necessary to minimize the electrolyte redistribution caused by migration, as discussed above. The performance of these stacks will be discussed in Section 9.4.7.

In 1991, ERC assembled a 20-kW stack (AF-20-1), using 4000-cm^2 components, as shown schematically in Fig. 9.80. Initial performance data for this stack are shown in Fig. 9.81. They show good cell-to-cell uniformity between the

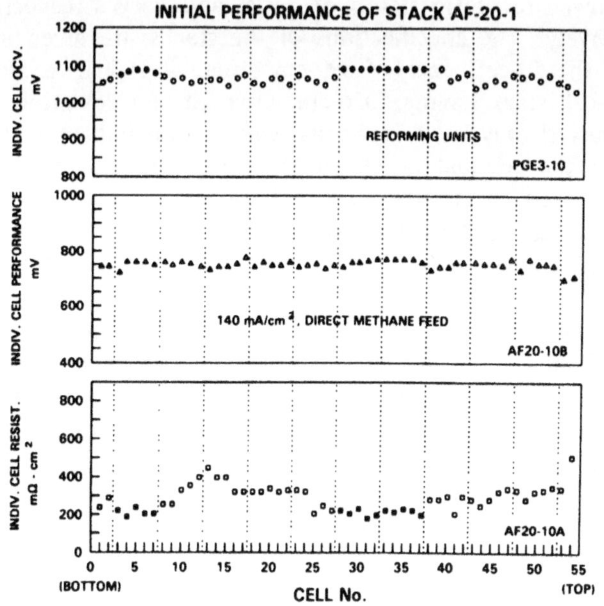

Figure 9.81. Initial performance of stack AF-20-1 (ERC). (From Maru.[137])

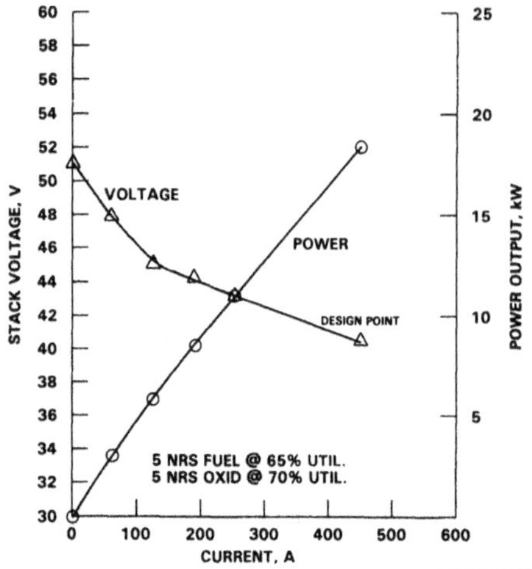

Figure 9.82. 20-kW stack performance (ERC). (From Maru.[137])

54 cells. The stack was first tested for 400 h at ERC (Danbury, CT) and then again, at low power density, after transport to Pacific Gas & Electric Co., San Ramon, California. The stack was successfully restarted and achieved stable operation comparable to that before shipping. The performance curve at partial power is shown in Fig. 9.82. The stack was operated for 1000 h and thermally cycled several times before shutdown.

Based on the experience gained with the 20-kW stack, a full-height stack (AF-100-1), containing 234 cells of 4000-cm^2 area (nominal rating 70 kW) has recently been assembled by ERC and is undergoing initial testing (Fig. 9.83).

Performance figures for MELCO's recent large-scale DIR stacks (Matsumura *et al.*[132] are shown, for DIR stacks of 3 kW and 5 kW, respectively, in Fig. 9.84a,b, respectively. The specifications of the stacks are given in Table 9.12. Performance of the 50-cell stack of 899-cm^2 area, at 1850 h, is almost equal to that of the five-cell stack having 5016-cm^2 area. Strong deviations of individual cell voltage occurred only for a few cells, which is an indication that migration effects seem to be adequately ctronolled. End-cell effects occurred; these appeared to be caused by gas flow maldistribution.

In MELCO's stacks, the electrolyte loss rate by vaporization was found to depend on fuel gas composition,[138] but a typical rate of 4.3×10^{-9} g/std-cc of fuel gas has been stated.[138]

The performance of scaled-up IIR stacks (Ohtsuki *et al.*[134]) is shown in Fig. 9.85, for a 30-kW stack. The specifications of these and some recent and planned stacks are shown in Table 9.13. The 30-kW stack itself is shown in Fig. 9.86, and the system configuration in Fig. 9.87. Exhaust gas from the cathode was cooled to 250°C before being recycled.

In the future carbon dioxide recycling will be required. Pressure swing adsorption was used in the 30-kW stack test to separate carbon dioxide from the anode exhaust gas and recycle carbon dioxide. The purity of the separated CO_2 was better than 99%, and the separation factor was 0.85. The incorporation of this separation scheme required a relatively small pressure adjustment.

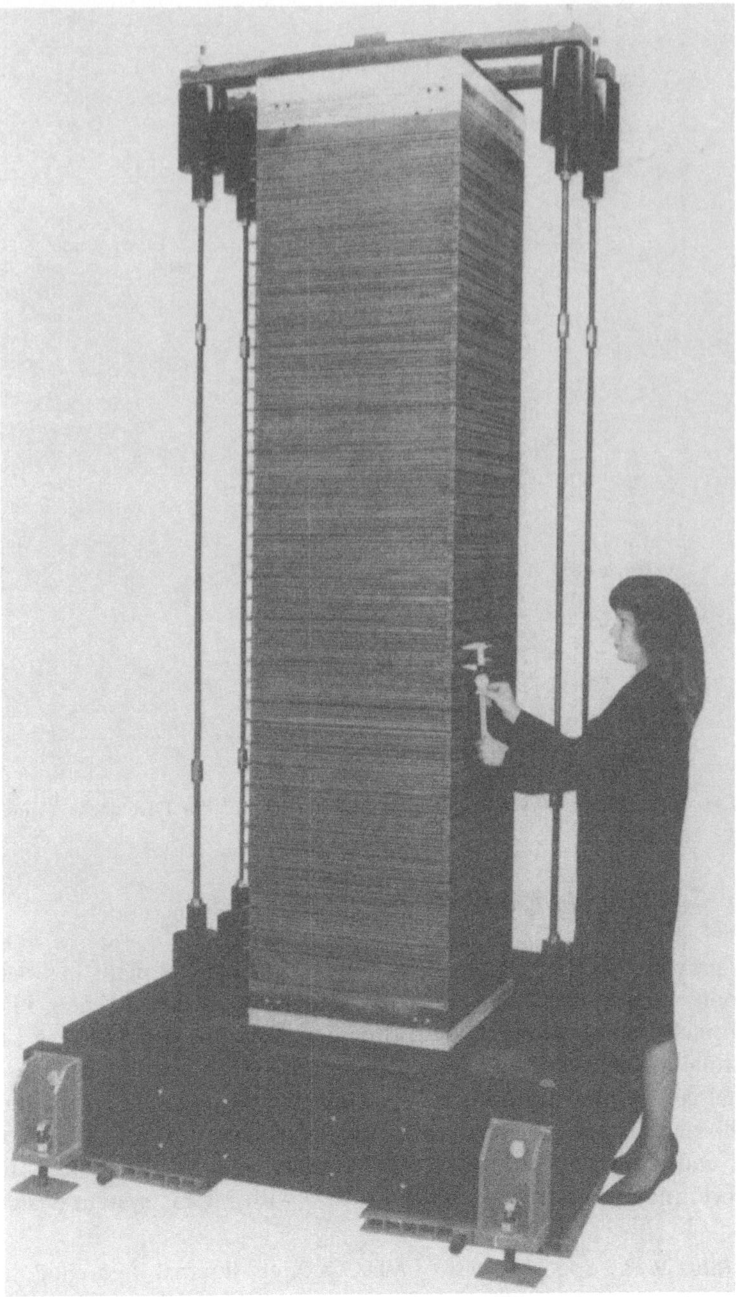

Figure 9.83. Assembled 70-kW stack (AF-100-1) of ERC. [From Maru.[137] (Courtesy ERC.)]

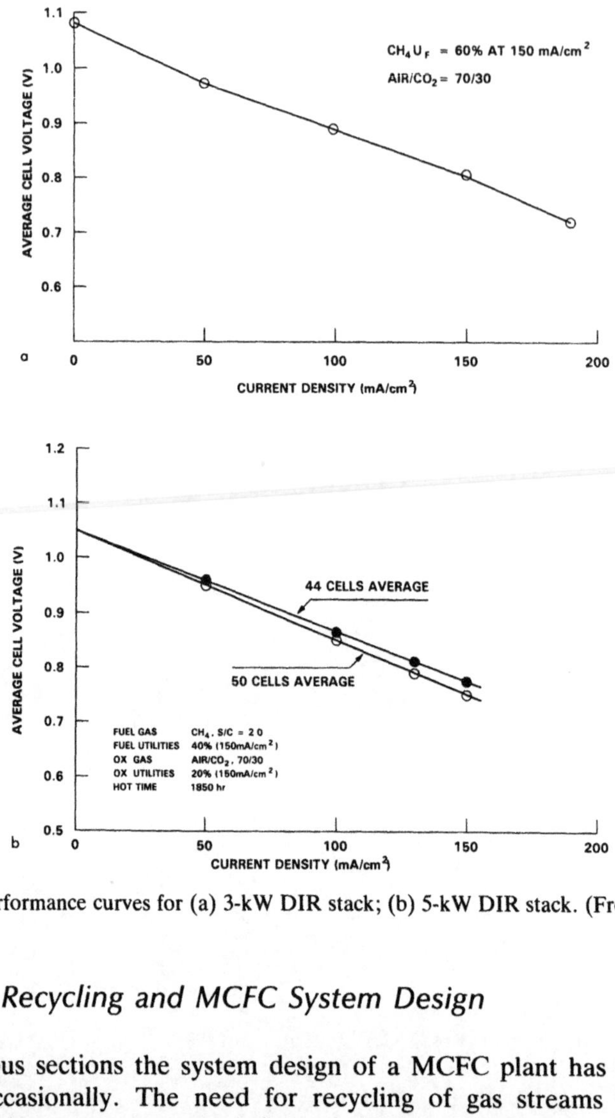

Figure 9.84. Performance curves for (a) 3-kW DIR stack; (b) 5-kW DIR stack. (From Matsumura *et al.*[132])

9.4.5. Gas Recycling and MCFC System Design

In previous sections the system design of a MCFC plant has been touched upon only occasionally. The need for recycling of gas streams in large-scale MCFC stacks has come up several times, in different contexts, e.g., as a device for thermal management and smoothing temperature distributions.

Fuel cell system design is to a certain extent independent of the details of fuel cell stacks. Several chapters in this book are devoted to fuel cell system design, and several recent articles have dealt with MCFC system design.[139–147] Therefore, the present section does not deal with MCFC system design as such.

Table 9.12. Specifications of MELCO Direct Internal Reforming Stacks

Rated power	3 kW	5 kW
Output power	3 kW	5 kW
Number of cells	5	50
Cell effective area	5,016 cm^2	899 cm^2
IR design	Direct	Direct
Catalyst loading	0.12 g/cm^2 (uniform)	0.05–0.20 g/cm^2 (graded)

From Ref. 134.

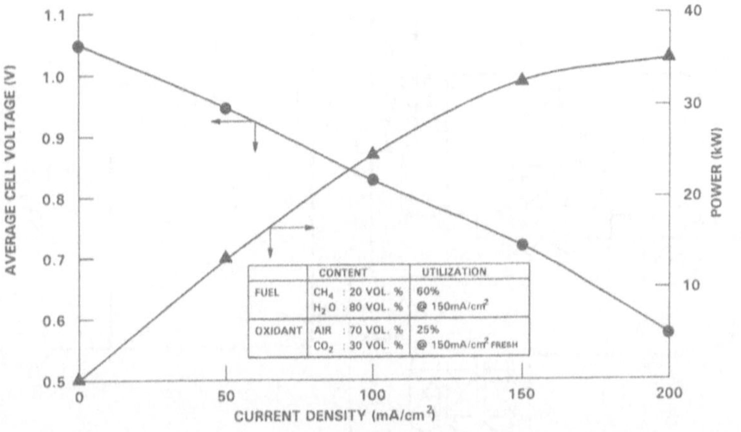

Figure 9.85. Peformance curve for 30-kW IIR stack. (From Ohtsuki *et al.*[134])

Table 9.13. Specifications of MELCO Indirect Internal Reforming Stacks

Rated power	10 kW	30 kW	100 kW (planned)
Output power	11 kW	35.1 kW	
Number of cells	20	62	192
Cell effective area	5,016 cm^2	4,864 cm^2	4,864 cm^2
Reformer plates	3	10	32
Time operated	2,052 h	2,131 h	2–5000 h

From Ref. 132.

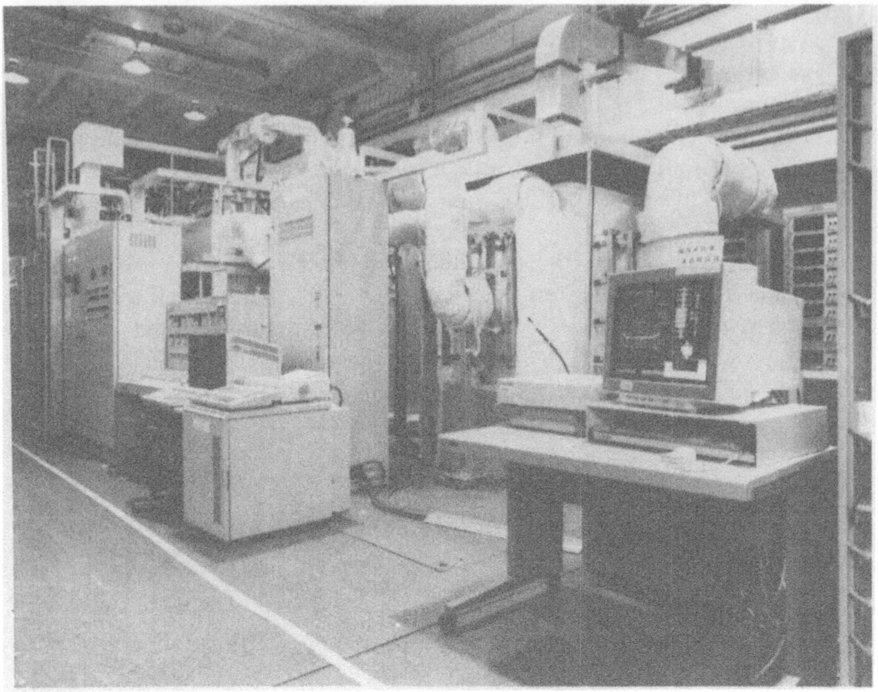

Figure 9.86. 30-kW indirect internal reforming stack of Mitsubishi Electric Company (MELCO). (From Ohtsuki *et al.*[134])

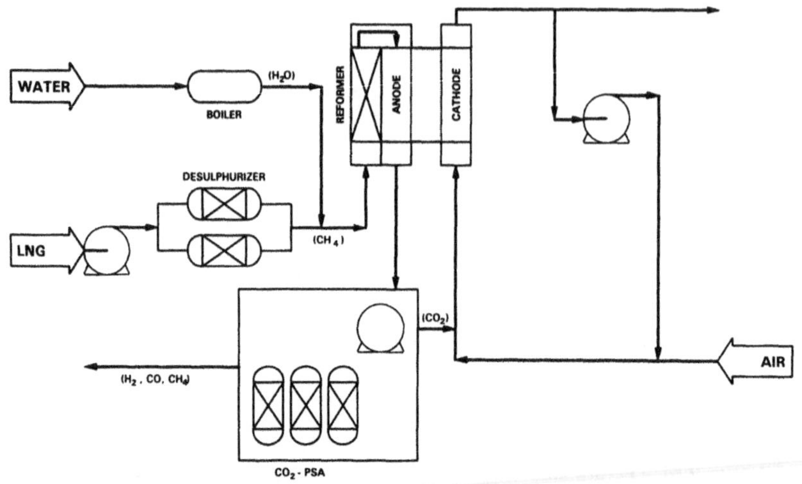

Figure 9.87. System configuration for 30-kW IIR-MCFC stack testing. (From Ohtsuki *et al.*[134])

Instead, we concentrate here briefly on a few aspects of system design which are specific to the MCFC:

1. The effect of gas recycling on stack performance
2. Simplified stack performance prediction for system design
3. Dynamic behavior of stacks

9.4.5.1. Gas Recycling in MCFC Stacks

The performances required of the MCFC in power plants using LNG as fuel include, according to Japanese standards:[147]

1. Average cell voltage of 0.8 V at 150 mA/cm^2 for a fuel utilization of 80%
2. Decay rate of 8 mV/kh or less in long-term operation
3. Net plant efficiency of 45% based on HHV, in pressurized operation

Under these conditions, the cell efficiency is 60%, so 40% of the chemical energy of the fuel is converted to heat. The cell therefore must be cooled efficiently, and usually anode or cathode gas recycling is adopted as a cooling

Table 9.14. Specifications of Stack Used for Recycling Test

Rated output	6 kW
Electrode area	3,120 cm^2
Number of cells	18
Manifolding/flow config.	Internal/parallel flow
Anode	Ni–Cr alloy (tape cast) with bubble barrier
Cathode	NiO (tape-cast Ni, oxid. *in situ*)
Electrolyte	(62 Li–38 K) CO_3
Matrix support material	$LiAlO_2$ (tape-cast)
Heating system	Electric heating
Compression system	Pressurized air bellows
Separator material	SS Type 310–Ni bimetal, Al on wet-seal areas

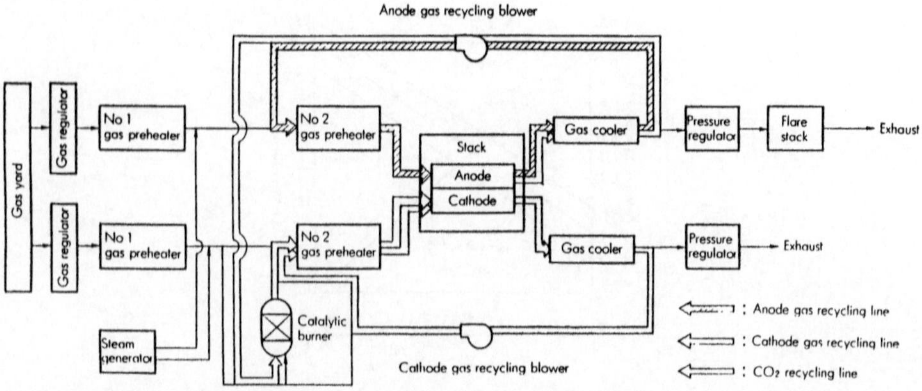

Figure 9.88. Schematic of test facility for 10-kW stack testing. (From Ogoshi *et al.*[148])

strategy. Anode exhaust gas recycling to the cathode inlet is also necessary to close the CO_2 balance of the MCFC, if not completely, then at least to an acceptable extent. These recycling operations must be carried out under pressurized conditions. Experimental data are available for pressurized cells (see Section 9.3.8) but their applicability to stacks is not *a priori* certain and must be confirmed.

A systematic series of such recycling operations under pressure were carried out recently[149] on a 6-kW IHI stack with the specifications given in Table 9.14. These measurements were carried out in cooperation between IHI and the Central Research Institute of Electric Power Industry (CRIEPI), Yokosuka, Japan. The flow diagram of the recirculation testing facility is shown in Fig. 9.88. Figure 9.89 shows the initial performance of the stack under atmospheric pressure, and Fig. 9.90 the voltage in pressurized operation, at various current densities. The result showed that the increase in cell voltage with increase of pressure from P_1 to P_2, at 650°C was well approximated, regardless of utilization, by the expression

$$\Delta V = 95 \sim 105\ {}^{10}\log(P_2/P_1). \tag{9.46}$$

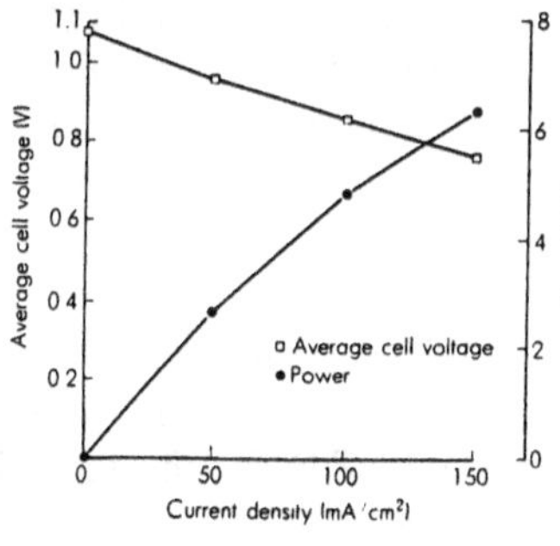

(Note)
Anode gas: $H_2/CO_2/H_2O = 66/17/17$ Oxygen utilization: 16%
Cathode gas: $Air/CO_2 = 70/30$ Hot time: 256 h
Fuel utilization: 79%

Figure 9.89. Cell performance of 6-kW stack under atmospheric operation. (From Ogoshi *et al.*[148])

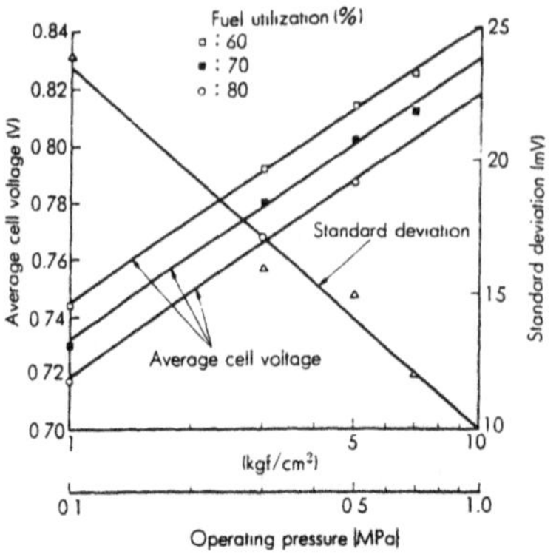

Figure 9.90. Dependence of average cell voltage of 6-kW stack on operating pressure. (From Ogoshi *et al.*[148])

The results also showed that methane formation is very slow, also under pressure, and that the increase in H_2O partial pressure in the fuel gas has the effect of decreasing the methanation rate.

Anode gas recycling is one method to cool the stack. The recycling ratio is defined as the feedback flow rate to the inlet of the stack divided by the stack outlet flow rate. Recycling reduces the average voltage. At constant fuel utilization the anode outlet gas composition is nearly constant regardless of the recycling ratio, but the inlet hydrogen concentration decreases as the recycling ratio increases. It is also found that the variation in individual cell voltages decreases, probably because the higher flow rate causes a more stable and uniform distribution of the flow.

With LNG reformed gas as fuel, the heat required for the reforming reaction must be supplied by combustion of anode exhaust gas, therefore, the limit of fuel utilization is 80%, or perhaps 85%, depending on the efficiency of heat exchange at the reformer burner. When coal gas is used, the stack can be operated at much higher fuel utilization. These high utilizations are made possible by recycling under pressure, as shown for 7 bar in Fig. 9.91. Operating at 70% anode gas recirculation and 90% overall fuel utilization, an average cell voltage of 0.738 V at 150 mA/cm^2 was obtained, equivalent to an efficiency of 53%.

CO_2 gas recirculation was investigated by burning the H_2 and CO of the anode exhaust gas in a catalytic burner and supplying the combustion product to the cathode inlet as CO_2 source. This operation was investigated by itself and in combination with cathode recycling for heat recovery. Figure 9.92 illustrates this schematically and also provides data for operation at 7 bar, at 70% fuel utilization and 30% oxygen utilization.

When the stack is cooled by recycling the cathode gas, the recycling ratio must be changed with changing fuel utilization, to keep the stack temperature constant. In the tests performed by CRIEPI, the recycling ratio was controlled so as to keep stack temperature constant (at 632.7°C in the center of the stack). The results are shown in Fig. 9.93. As the fuel utilization increases, the cathode gas recycling ratio increases. The average voltage, which was 0.81 V when the cathode gas was 70% air–30% CO_2 and the fuel utilization 70% (Fig. 9.90), decreases to 0.75 V in the combined operation. If the oxygen utilization is kept

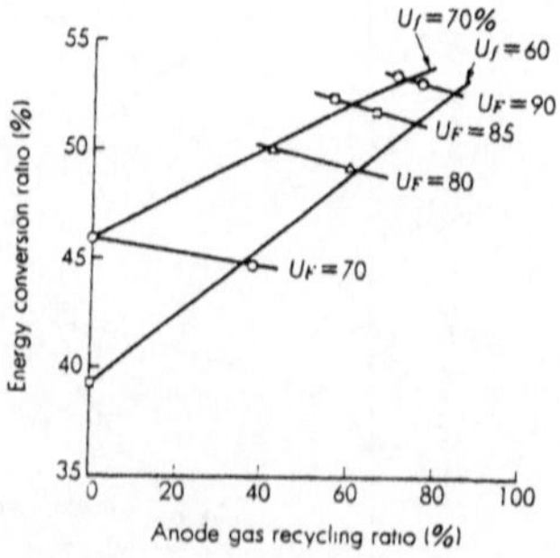

(Note)
U_f: One path fuel utilization
U_F: Overall fuel utilization
Anode gas: $H_2/CO_2/H_2O$=66/17/17
Cathode gas: Air/CO_2=70/30
Oxygen utilization: 23~45%
Current density: 150 mA/cm^2
Operating pressure: 7 kgf/cm^2 (0.69 MPa)
Hot time: About 1,250 h

Figure 9.91. Relationship between anode gas recycling ratio and energy conversion efficiency. (From Ogoshi *et al.*[148])

constant, increased fuel utilization causes a slightly higher O_2 concentration in the cathode gas, but the cell voltage decreases because the CO_2 concentration decreases.

The 6-kW stack was operated for 2000 h, and the average cell decay rate was 12 mV/kh, with extremes of 1 and 56 mV/kh.

Recently, CRIEPI has also operated a 10-kW IHI stack, consisting of ten 1000-cm^2 area cells. This stack, tested for long-term performance (see Section 9.5.8), was also operated at various pressures and a range of anode recycling ratios. The results[148] are in good agreement with those described above.

As discussed in Section 9.4.4.3 in connection with performance testing of internal-reforming stacks, alternative methods of CO_2 separation and recycling are being explored, such as pressure swing adsorption.[134]

Direct transfer of CO_2 from anode exhaust to cathode inlet would make fuel utilizations of close to 100% possible, with corresponding gains in overall efficiency. Earlier, several devices were proposed which are essentially fuel cells

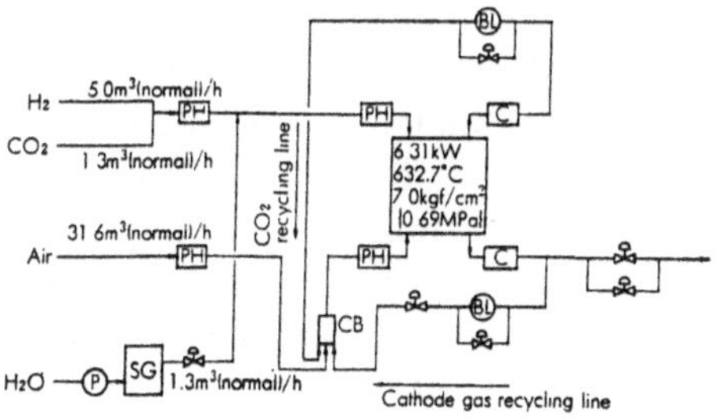

(Note)
BL: Blower
C: Cooler
CB: Catalytic burner
SG: Steam generator
P: Pump
PH: Preheater

Figure 9.92. Schematic of CO_2 and cathode gas recycling operation. (From Ogoshi *et al.*[148])

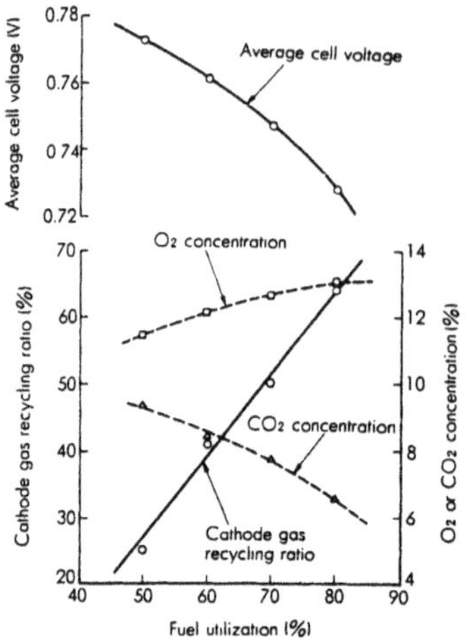

(Note)
Operating pressure: 7 kgf/cm² (0.69 MPa)
Operating temperature: 630°C
Oxygen utilization: 30%

Figure 9.93. Relationship between fuel utilization and cathode gas recycling ratio, as functions of fuel utilization. (From Ogoshi *et al.*[148])

operated in reverse to accomplish a transfer from one gas stream to another. For example, a phosphoric acid fuel cell can be used to transfer hydrogen from the anode exhaust stream to an inert carrier stream, at the expense of some electrical energy.[150] A MCFC-like cell can also, in principle, be used as a CO_2 transfer-device;[151] however, the operation under high H_2O partial pressure is not straightforward.

These ideas for "active" transfer devices have apparently not been developed beyond the conceptual stage. An all-solid-state "passive" device for selective transfer of CO_2 would be highly desirable, and an intensified search for such a device can be foreseen as the MCFC enters the early commercialization stage.

9.4.5.2. Stack Performance Prediction for System Design

System designers have a particularly strong interest in reliable data for the pressure dependence and utilization dependence of the stack voltage, within prescribed boundaries of temperature. Such data can be provided by a complete stack model of the type described in Section 9.4.3. However, the parameters in such a model computation are so numerous and, in part, so uncertain that a much more compact description would be desirable as a system design tool.

Performance correlations on a statistical–empirical basis can fulfill the role of predictors for design purposes. It should be kept in mind that such empirical correlations, of which Eq. (9.46) is an example, are valid only for the type of stack in question and within the operational range covered by the data they are based on. The uncertainty (statistical deviation or fitting error) must also be considered.

Using the data gathered for the 6-kW internally manifolded stack described above, one such empirical correlation ("voltage characteristic equation") was derived as follows:[148]

$$V = A_0 + \sum A_i \log P_i + A_A \log Q_A + A_C \log Q_c, \tag{9.47}$$

where

$$A_i = \frac{RT}{2F}\left(\pm 1 + \frac{a_i}{\alpha}\right) \quad \text{for non-O}_2 \text{ species,}$$
$$A_i = \frac{RT}{2F}\left(0.5 + \frac{a_i}{\alpha}\right) \quad \text{for O}_2. \tag{9.48}$$

Here a_i is the reaction order, and α is the transfer coefficient of the reaction. Furthermore

$$P_i = \frac{(P_i)_{in} + (P_i)_{out}}{2}, \tag{9.49}$$

where P_i is the partial pressure of the gas, and

$$Q_{A,C} = \frac{(9Q_{A,C})_{in} + (Q_{A,C})}{2}. \tag{9.50}$$

The constants have the following values:

$$A_0 = 801.0$$
$$A_{H_2O} = -65.2$$
$$A_{H_2} = 111.7$$
$$A_{CO_2,a} = -112.0$$
$$A_{O_2} = 77.2$$
$$A_{CO_2,c} = 90.5$$
$$A_A = 118.2$$
$$A_C = 57.6$$

The reaction orders measured were, on the anode side, for H_2 0.35, for CO_2 −0.35, for H_2O 0.39, at $\alpha = 0.7$; on the cathode side, for O_2 0.72, for CO_2 0.02, also for $\alpha = 0.5$. The accuracy of predictions by such an equation is suggested by Fig. 9.94.

9.4.5.3. Dynamic Behvaior of MCFC Stacks

As MCFC stacks of 10–30 kW capacity have undergone testing and 100-kW tests are scheduled by 1995, the response of large stacks to interruptions of gas flow, changes in gas composition, and changes in electric load is becoming an important issue. Initial data on the dynamic behavior of stacks are being collected, and modeling studies of transient stack performance are undertaken. These will be very important for developing *process control* procedures applied to MCFC stacks and systems.

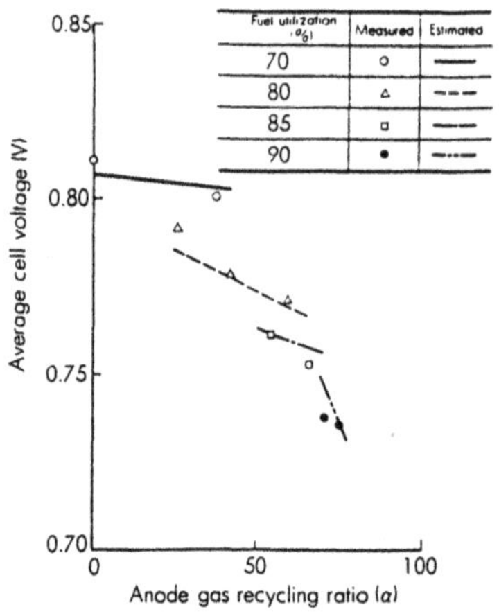

Anode gas: $H_2/CO_2/H_2O = 66/17/17$
Cathode gas: $Air/CO_2 = 70/30$
Oxygen utilization: 23~45%
Current density: 150 mA/cm²
Operating pressure: 7 kgf/cm² (0.69 MPa)
Hot time: About 1,250 hours

Figure 9.94. Predicted and measured dependence of cell voltage on anode gas recycling ratio, in a 6-kW stack. (From Ogoshi *et al.*[148])

Watanabe *et al.*[152,153] investigated the transient response to load of a 10-kW stack and showed that the response to small variations in current density has a short-range time constant of a few tens of seconds, and a long-range time constant of several thousand seconds. The response can be predicted very well from bench-scale cell performance data.

9.4.6. Electrolyte Migration and Loss

9.4.6.1. Electrolyte Migration

In externally manifolded stacks, the cells are ionically connected via the manifold gasket. This is usually a ceramic tile or paste; zirconia is frequently used for the latter. The few volume percent void fraction in such a gasket fill up with electrolyte by capillary equilibrium (Section 9.2.4). An electrolyte column is thereby formed which, though thin and fairly resistant, conducts current from the positive end plate of the stack to the negative endplate. This is a "bipolar leak current" or "shunt current" of the type commonly found in batteries or electrolyzers with a bipolar arrangement of cells if there is a common electrolyte pool. Such a common electrolyte creates an ionic path between individual cells (Fig. 9.96a).

In a MCFC stack there is in principle no common electrolyte reservoir, and each cell contains its own amount of electrolyte between adjoining separator plates. However, it is necessary to seal each cell against escape of gas. This is ensured by the wet seal, i.e., by clamping the semisolid electrolyte matrix layer between the edges of the separator plates (Fig. 9.4). Consequently, electrolyte motion in or out of the cell is possible, and this may cause electrolyte paths to

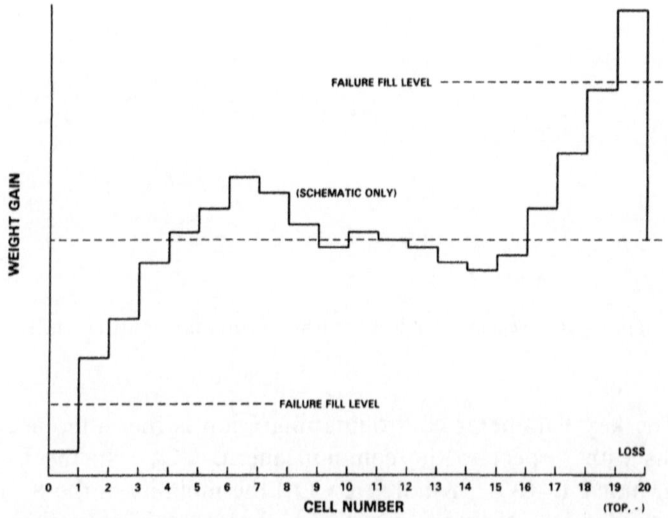

Figure 9.95. Schematic of electrolyte distribution in UTC stack No. 6, posttest. (From Ref. 154.)

form between cells via hardware surfaces (creepage) or via pores in the manifold gasket if an external manifold is used.

Though creepage of electrolyte is not yet fully understood, it appears likely to be driven by a combination of electrocapillary forces and corrosion. As discussed in Section 9.3.9, electrolyte can slowly move out of the cell via the wet seal and spread as a film onto the outside surface of the cell hardware. If the spreading continues, an ionic contact is formed between cells, and sustained electrolyte movement from one cell to another, via the outer hardware surface, may result. The ensuing electrolyte displacement and hardware corrosion can be a cause of performance decline of some cells or of the stack as a whole.

Electrolyte creepage onto internal manifold hardware surfaces can conceivably also cause cell-to-cell ionic contact. However, the effect seems to be relatively unimportant. Electrolyte movement via porous connections (such as external manifold gaskets) appears to be more effective than via surface films.

In externally manifolded stacks the electrolyte contained in the wet seal of a cell contacts each manifold gasket in two places, and the manifold gasket forms an uninterrupted ionic path from one end plate to the other. A stack typically contains hundreds of cells, and the full stack voltage is applied across the high-ohmic resistance formed by this electrolyte column. The result is a leakage current which, though by itself not very significant, causes a slow, continuous one-way movement of the electrolyte. This "ion pumping" is the net result of migration, i.e, motion of ions at different speeds, due to the electric field.

The electrolyte migration effect became evident immediately during operation of the first stacks (externally manifolded), since it caused severe flooding of the cells near the negative end of the stack and electrolyte draining from cells near the positive end (Fig. 9.95). The performance of the stack deteriorated drastically almost from the beginning, and a net electrolyte loss of almost 60% in 1000 h was found to have taken place.[154]

The migration phenomena responsible for performance deterioration are fairly complicated. A comprehensive review has been presented by Kunz and

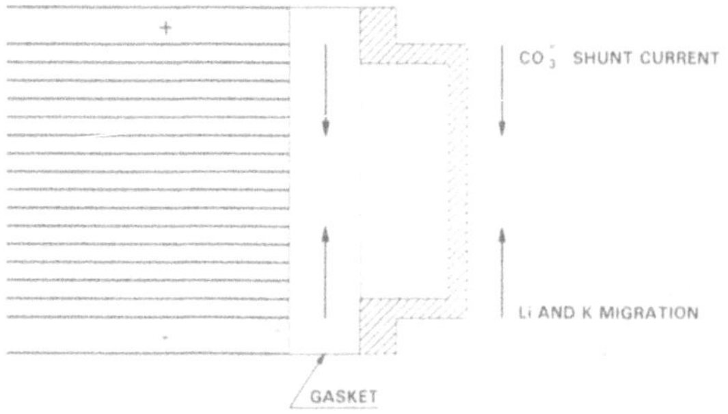

Figure 9.96. Schematic of leakage (shunt) current in a MCFC stack.

Bregoli.[155] The key parameter controlling migration is the difference in mobility of the cations with respect to the common anion, CO_3^{2-}. In the Li–K eutectic carbonate (62 mol% Li_2CO_3, 38 mol% K_2CO_3) the mobility of the K^+ ion is larger than that of the Li^+ ion. Near the negative end plate, migration therefore causes enrichment in K^+ ions and depletion in Li^+ ions; the opposite occurs near the positive end plate. Such accumulations are indeed observed (Fig. 9.97) and because the partial volumes of Li and K carbonate are different, this change in composition causes a displacement (net velocity) of the carbonate melt from the positive end to the negative end. The cells communicating with the low-pressure positive end will lose electrolyte, while those communicating with the high-pressure negative end will be flooded, in both cases causing poor performance. In addition, electrolysis of carbonate is likely to occur at the end plates, since the stack voltage far exceeds that necessary for decomposition of carbonate. This decomposition causes net loss of electrolyte.

It can be appreciated that the overall patterns of electrolyte movement, electrolyte loss, and stack performance decay are extremely complex. The electrolyte displacement by migration has been modeled mathematically and verified experimentally, using dummy cells.[156,157] A leakage current of 1 mA was found to be equivalent to an electrolyte displacement of the order of 1 mg/h.[158]

Attempts have been made to counter the migration effect by built-in electrolyte reservoirs, although these are obviously difficult to control once a large-scale stack has started operating. It would seem most desirable to apply a

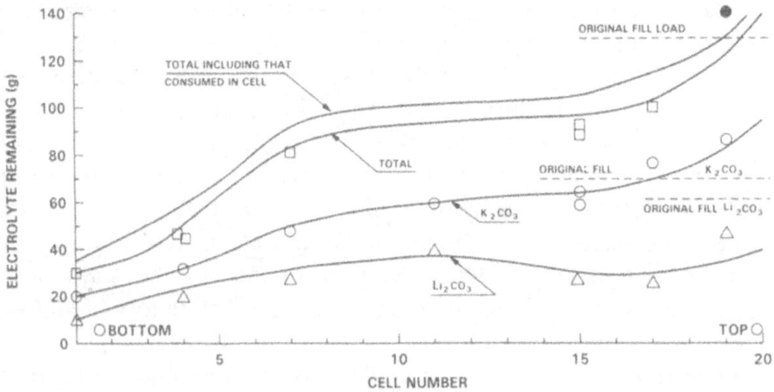

Figure 9.97. Variation of electrolyte composition along a 20-cell stack after 650 h operation (UTC). (After Kunz and Bregoli.[155])

completely dense dielectric as a manifold gasket. However, even if such a material (most closely approximated by α-alumina) would be available, a rigid gasket would be impractical for structural reasons as discussed in Section 9.4.1.

The only practical solution therefore is to *minimize* migration. Two strategies can be followed.

1. Minimize bipolar leak currents by minimizing the electric field in the leakage path, e.g., by superimposing an electric field of opposite sign over the entire height of the manifold gasket This is a classical leakage control device which has proved effective in bipolar battery technology.[159]
2. Apply an electrolyte composition such that the main cations have equal mobility. In Li–K carbonate this isomobility composition is rather high in lithium; this may be attractive from the viewpoint of ohmic resistance but not from that of polarization (due to decreasing gas solubility and inferior kinetics).[160] In Li–Na or Li–Na–K carbonate melts a more favorable compromise might be found. This is a subject of continuing research, with important implications also for corrosion.

The first method has proved fairly effective in reducing migration. Additional, undisclosed methods of preventing migration appear to have further improved electrolyte stability. Recent data on individual cell performance in externally manifolded stacks have shown impressive advances. IFC has presented data which indicate that the 40,000-h goal is well within the capability of present technology (Fig. 9.98). ERC has presented performance data for a 20-kW stack (Fig. 9.99) which show good cell-to-cell uniformity in the initial phase of operation, when migration problems tend to manifest themselves in the form of rapidly deteriorating performance of end cells. Figure 9.99 shows that this uniformity holds after 900 h of operation and relocation of the stack. ERC has

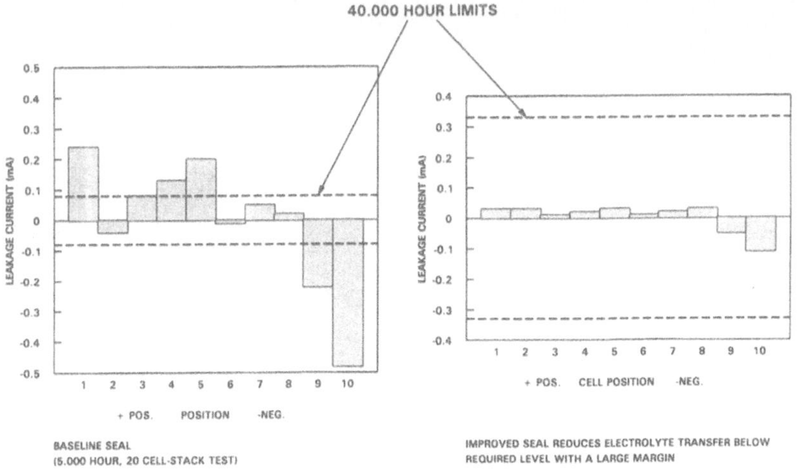

Figure 9.98. Reduced electrolyte transfer due to improved manifold gasket, tested in a rig simulating conditions in a 20-cell 25-kW stack of 8-ft^2 cells. The leakage current is proportional to electrolyte transfer at a rate of 1 g/A · h. (From Ref. 158.)

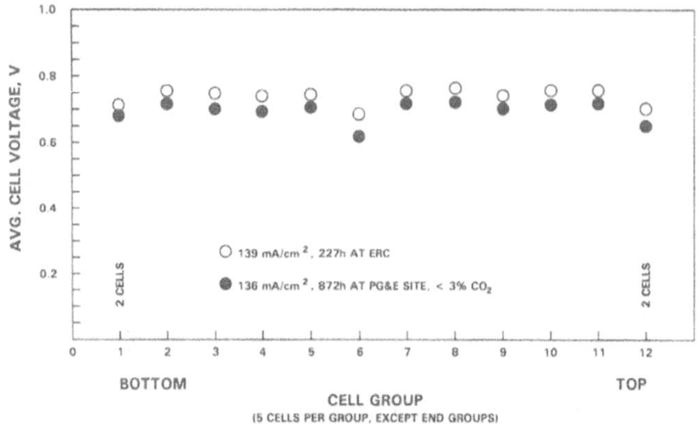

Figure 9.99. Cell performance of a 18-kW, 54-cell ERC stack of 4000-cm^2 area components (in groups of five cells except for two end-cells). (From Maru *et al.*[161])

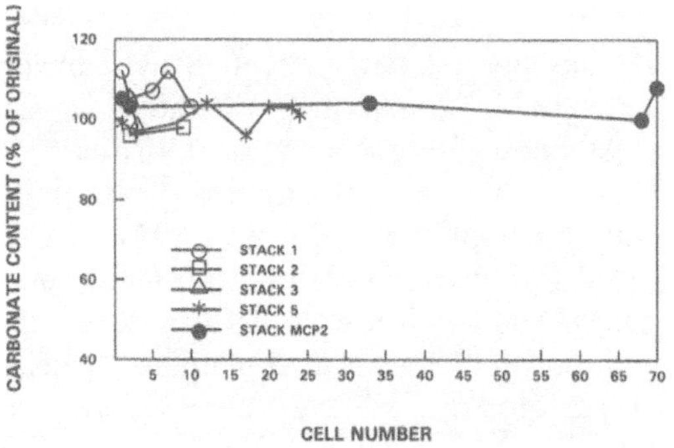

Figure 9.100. Electrolye distribution in internally manifolded stacks, according to M-C Power Corp. (From Marianowski *et al.*[120])

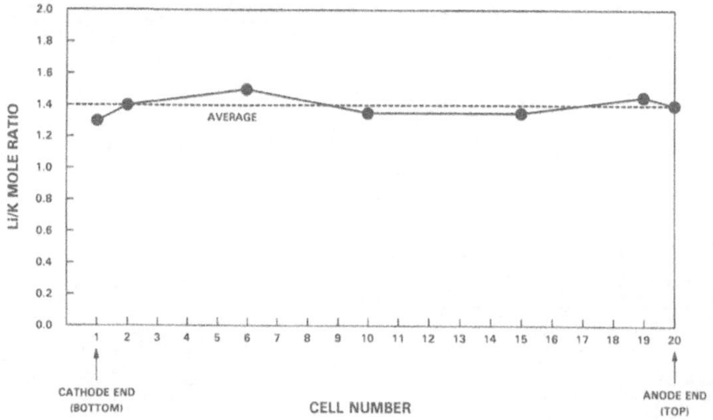

Figure 9.101. Distribution of Li/K ratio in an internally manifolded stack of twenty 0.1-m^2 cells. (From Sato.[95])

also reported performance data for 7300-h operation of a 250-cm^2 active area stack, which showed no significant migrational effect up to the end of testing.

For comparison, the electrolyte distribution in an internally manifolded stack is shown in Fig. 9.100 (M-C Power Corp.). The Li–K distribution, which is a sensitive diagnostic of electrolyte migration, is shown for another internally manifolded stack in Fig. 9.101 (IHI Corp.).

9.4.6.2. Electrolyte

If electrolyte displacement due to migration is minimized and no catastrophic corrosion occurs (usually due to unprotected wet-seal areas), a stable level of stack performance is established fairly rapidly, usually within 1000 h. The electrolyte loss in stacks thereafter would be expected to follow approximately the same pattern as in cells (Section 9.3.12). As shown in Fig. 9.48, it appears to be linear with time, after an initial corrosion-related loss.

Recently, electrolyte loss in stacks has been analyzed more widely and in more detail, since it is the main cause of performance decay. The corrosion-related loss has been determined by Urushibata and Murahashi,[162] using tape-cast electrolyte, for a variety of gas compositions, in cells operated at 150 mA/cm^2. Their data indicate that Li_2CO_3 loss due to corrosion is proportional with the square root of time, as one would expect from their data for metal loss by corrosion ($t^{1/2}$ dependence in agreement with earlier data illustrated in Figs. 9.39 and 9.40). K_2CO_3 loss, independent of time, is approximately 2 μmol · cm^{-2} (Fig. 9.102). Significantly, the corrosion-related loss of Li_2CO_3, which follows the correlation

$$\text{rate} = 0.38\ \mu\text{mol} \cdot \text{cm}^{-2} \cdot \text{h}^{-1/2} \tag{9.51}$$

continues to increase well beyond 10,000 h. The overall electrolyte loss due to corrosion is illustrated in Fig. 9.103, as a percentage of the initial electrolyte loading. In a 5000-cm^2 type of cell (or stack), gas crossover starts to affect performance when the electrolyte level falls below 65% of initial loading, which occurs at approximately 20,000 h. For comparison, IHI reports that electrolyte content of 1000-cm^2 and 25000-cm^2 cells in stacks was 74% of initial loading after 10,000 h.[163]

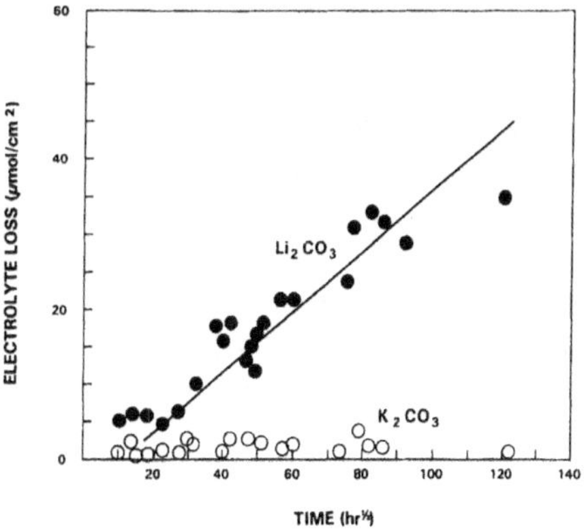

Figure 9.102. Electrolyte loss due to corrosion of Type 316 stainless steel current collector. (From Urushibata and Murahashi.[162])

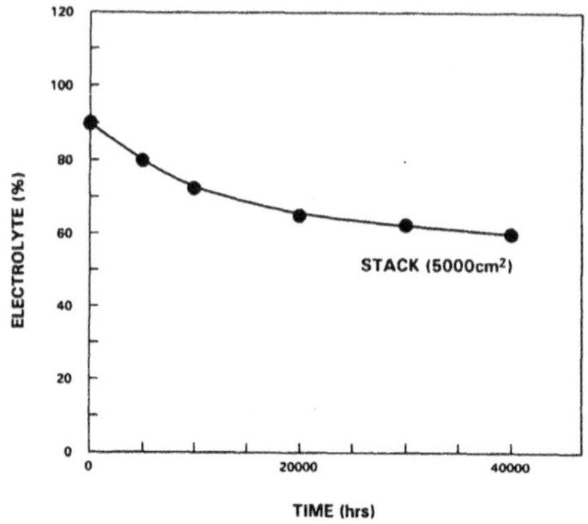

Figure 9.103. Electrolyte loss due to corrosion, in percent of initial loading. (From Urushibata and Murahashi.[162])

The electrolyte loss rate by vaporization, measured by Hirai *et al.*,[164] depends on fuel gas composition. The data indicate that the electrolyte vaporization rate decreases with increasing pressure, according to

$$\text{rate} \sim (P)^{-1}(x_{H_2O}x_{CO_2})^{1/2}. \tag{9.52}$$

In out-of-cell experiments, a typical rate of 4.3×10^{-9} g electrolyte/std-cc of fuel gas was found for reformed natural gas at 1 atm.[165] However, in-cell measurements of vaporization loss yield a much lower value (by a factor of 0.25). This discrepancy has not yet been explained satisfactorily.

9.4.7. Stack Performance Data

The first MCFC stacks, designed as stacks (as distinct from bicells or triple cells) and housed in a pressure vessel, were operated in the United States by United Technologies Corporation (UTC) and, later, its fuel cell subsidiary International Fuel Cells Corporation (IFC) in the early 1980s. These stacks were initially 0.1 ft^2 (93 cm^2) in area. They were externally manifolded and could be operated under pressure. Figure 9.104 shows the "learning curve" of stacks operated under 65 psia (4.4 ata) at 650°C. The performance at 160 ASF (172 MA/cm^2) of this type of stack can be expected to approach 0.8 V, i.e., 0.135 W/cm^2, at moderate fuel utilization. The figure also demonstrates that cell-to-cell variation can be reduced to less than 30 mV and that long-term performance depends strongly on control of electrolyte movement (especially by migration) and electrolyte loss, as well as improved cell components.

From 1984 on, larger stacks, starting from 0.3 m^2, have been operated by UTC–IFC and other companies. IFC has operated externally manifolded stacks of up to 20 cells of 0.74-m^2 (8-ft^2) active area, with a capacity of up to 25 kW. The long-term performance of such stacks has been tested for up to 2000 h.[166] Recently IFC has scaled-up to stacks of 1.932-m^2 (20.8 ft^2) cells, of a design termed *thin cell.* A 25-kW stack of this type[167,168] (Fig. 9.105) has been tested for 2200 h. The performance of this stack, shown in Fig. 9.106, was equivalent to

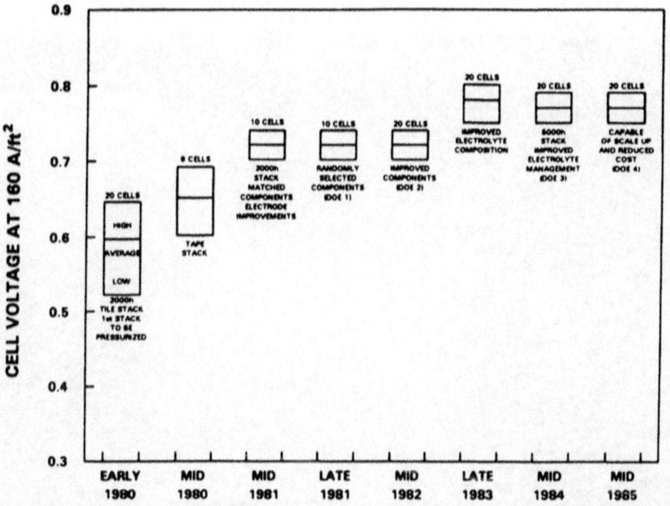

Figure 9.104. Performance of early 93-cm^2 stacks at UTC (now International Fuel Cells Inc., IFC) on low-Btu fuel. Stacks 1–5 used Li/K atomic ratio of 62/38, whereas stack 6 had close to 70/30 ratio. Stack 7 operated under reducted temperature, to improve long-term performance. The final stack of this type operated for 5000 h. (Adapted from Ref. 166.)

0.114 W/cm^2 (106 WSF) at 150 mA/cm^2 (160 ASF) and 0.164 mW/cm^2 (152 WSF) at 269 mA/cm^2 (250 ASF), with standard gas conditions and utilizations. The 160-WSF range is within the regime that can be considered for application to commercial power plants in the United States. The scaled-up stack of Fig. 9.105 forms part of IFC's integrated stack unit (ISU), which is being developed as a self-contained power generator with a rating of 100 kW to 1 MW or greater.[170] IFC is cooperating with the Toshiba Corporation in developing and demonstrating this stack technology.

Since 1989 IFC has assumed a less prominent role in stack and system development under Department of Energy (DOE) sponsorship, but remains involved with emphasis on development of stack components. Other U.S. developers are constructing and testing stacks of 0.4-m^2 to 0.6-m^2 cells as prototypes for commercial megawatt-level systems, under sponsorship of DOE, the Electric Power Research Institute (EPRI), and the Gas Research Institute (GRI). The principal developers of these stacks, as of 1992, are Energy Research Corporation (ERC) and M-C Power Corporation (M-CP). ERC has been the main proponent of the "direct fuel cell" approach, i.e., the internal-reforming fuel cell. The principles and design of the IR-MCFC have been extensively discussed in Section 9.4.4. ERC has emphasized both scale-up and endurance (long-term performance) since it started operating stacks in the early 1980s.

During 1990 and 1991, ERC built and tested several stacks using 0.4-m^2 (4-ft^2) cells.[171] The 20-kW and 70-kW stacks were discussed in Section 9.4.4, which also contains performance figures for the 20-kW stack (AF-20-1; see Figs. 9.80 and 9.81). The initial performance of the full-height 70-kW stack (AF-100-1), which contains 234 cells with 39 reforming units at 6-cell intervals, was obtained at part load (due to facility limitations at ERC), producing 42 kW with 59% efficiency (based on LHV) at 0.78 fuel utilization. The appearance of this

Figure 9.105. IFC 25-kW molten carbonate stack. (Courtesy International Fuel Cells Inc.)

stack was shown in Fig. 9.83. Further scale-up of cell size to 0.6 m^2 (6 ft^2, i.e., 2 ft × 3 ft) is now under way in short stacks.

ERC has recently operated a 2-kW stack of five 4000-cm^2 cells for a duration of 10,000 h. This considerably exceeded the design period (7000 h), in spite of the older electrolyte management technology used in this stack. The performance decay in this stack averaged 3.125 mV/1000 h, for a total extrapolated loss of

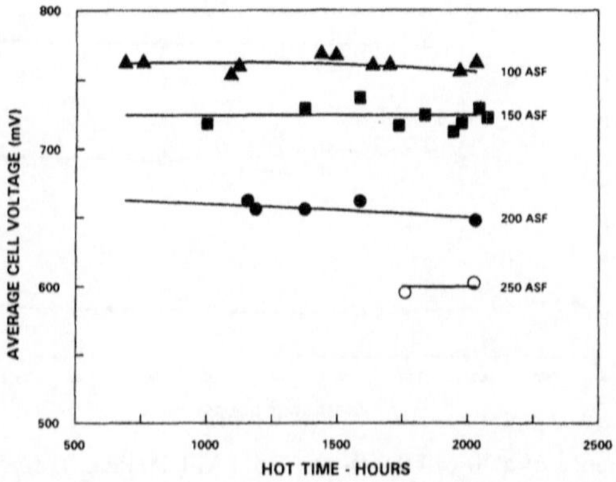

Figure 9.106. Average voltage of center cells of the stack of Fig. 9.105. (From Ref. 169, p. 263.)

125 mV in 40,000 h. This may be considered very encouraging. A small five-cell stack (300 cm^2 cell area) showed even lower loss in accelerated testing (i.e., large seal-to-active-area ratio) during the design period of 7000 h (Fig. 9.107). Most of the loss is due to increase of ohmic resistance. The decay in 40,000 h was projected to be 75 mV.

Tests of ERC's 100-kW stacks are scheduled for 1992–1993. A 2-MW power plant demonstration test is scheduled in 1994 at Santa Clara, CA.

M-C Power Corporation, which is commercializing IGT's fuel cell technology, has tested numerous subscale (1-ft^2) stacks containing from 3 to 70 cells. IGT and M-CP pioneered the internal manifolding concept applied to MCFC stacks (IMHEX stack design concept), and these tests demonstrated the viability of internal manifolding, from the viewpoint of manufacturing (ease of assembly), operation (gas sealing), and performance (electrolyte management). Figure 9.108 shows the performance of a 70-cell stack of 0.091-m^2 (1-ft^2) cells, operated for a total of 1580 h. The power density achieved was 0.124 W/cm^2 (115 WSF). In 1992 a full-scale 0.929-m^2 (10-ft^2) 19-cell stack was tested, under sponsorship of the

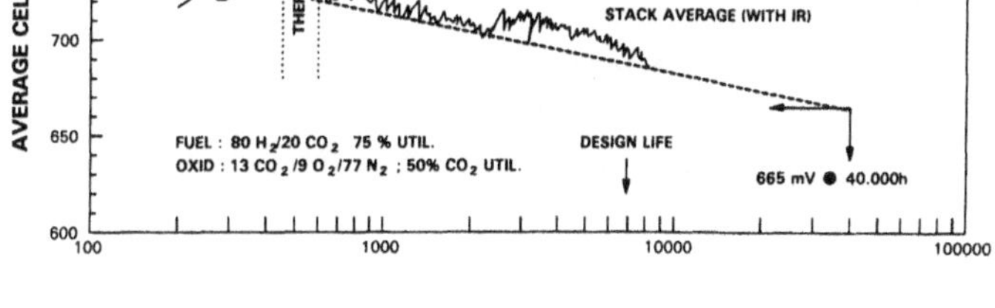

Figure 9.107. Life-graph of accelerated stack test and performance projection for a five-cell 300-cm^2 ERC stack, operated at 160 mA/cm^2 under integrated system conditions. (From Maru *et al.*[171])

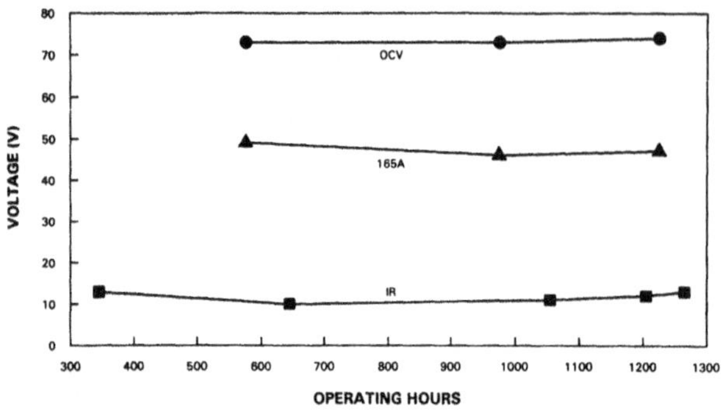

Figure 9.108. Performance of a 70-cell 0.091-m^2 (1 ft^2) M-CP stack. (From Marianowski *et al.*[172])

Gas Research Institute (Fig. 9.109). Longer term performance tests (4000, 8000, and 12,000 h) are being planned.

The M-CP near-term goal is to operate a full-scale stack of 250 10-ft^2 cells, for a total capacity of 250 kW, in 1993. This demonstration of a small MCFC system will be carried out at San Diego, in cooperation with San Diego Gas & Electric as well as other members of the IMHEX commercialization alliance (ACCT).

Among the Japanese companies building and testing MCFC stacks, Ishikawajima-Harima Heavy Industries Ltd (IHI) also has opted for internal manifolding. The MCFC development program of this company is illustrated in Fig. 9.110. IHI aims to demonstrate power generation by a 100-kW stack of 1-m^2 electrode area by 1993, and to demonstrate a 1-MW system in 1997. The performance of IHI's 6-kW and 10-kW stacks was discussed in Section 9.4.5 in connection with system design and gas recycling.

IHI's present effort, after successful demonstration of large-scale separator manufacture (see Section 9.4.4. and Fig. 9.57), is concentrated on tall stacking and thermal cycling. A 50-cell stack of 0.3-m^2 area was operated for almost 2000 h, generating 16.1 kW, showing both uniform performance and high sealing efficiency. The performance of this stack is shown in Fig. 9.111, and the 50-cell 1.015-m^2 50-kW stack (Fig. 9.112) is just starting to operate, with an initial cell performance of 0.72 V at 150 mA/cm^2, on simulated reformed gas (steam/carbon = 3) at fuel utilization of 0.70.

As for endurance and thermal cycling, a five-cell stack of 1 m^2 has operated for 4300 h and experienced six thermal cycles successfully. A smaller 2.5-kW stack operated to test endurance has functioned well for more than 10,000 h, with performance decay less than 4 mV/1000 h (Fig. 9.113).

As discussed earlier (Sections 9.4.2 and 9.4.4), among the Japanese MCFC developers only Mitsubishi and Sanyo are pursuing mainly the internal-reforming MCFC concept (IR-MCFC). IHI, like Hitachi and Toshiba, make use of external reforming, although, as discussed in Section 9.4.2, IHI has developed a plate-type reformer for application in MCFC systems.

The Hitachi Corporation was among the first Japanese companies to start MCFC development and is at present constructing and testing large-scale (1-m^2) stacks with the goal of demonstrating a 100-kW plant by 1993 and a 1-MW system in 1997. To make construction of a large cell possible without utilizing meter-size

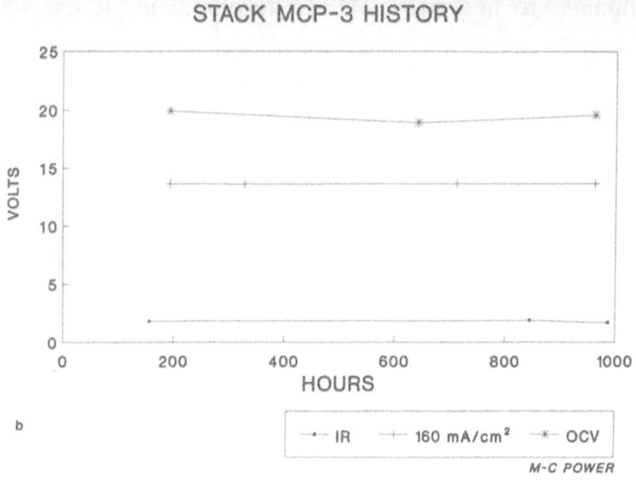

Figure 9.109. (a) Close-up view of 19-cell, 1-m^2 stack MCP-3 of M-C Power Corp.; (b) Performance-life graph, using simulated reformed natural gas at 75% utilization; 1 atm pressure. (Courtesy M-C Power Corp.)

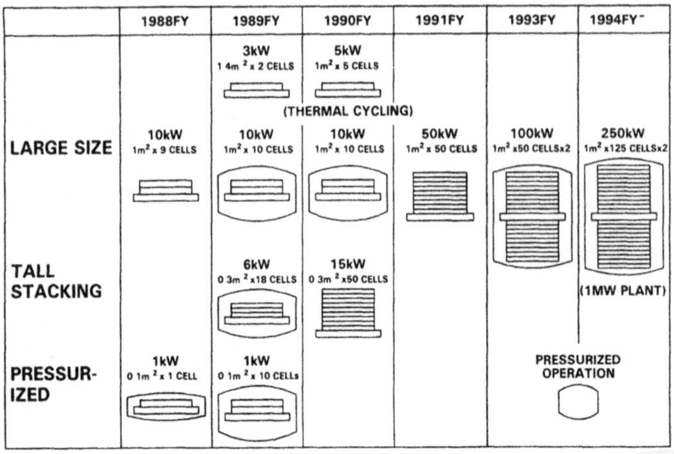

Figure 9.110. MCFC development and demonstration program of IHI. (From Suzuki *et al.*[173])

tape casting and sintering equipment, Hitachi developed the multiple large-capacity-type cell consisting of four 0.36 m^2 subcells, as discussed in Section 9.4.1 and illustrated in Fig. 9.59. The advantages cited for this type of composite-separator stack are small gas pressure drop, therefore more uniform gas flow and temperature distribution, as well as the ease of handling smaller components during assembly.

In recent years, two 25-kW stacks have been constructed and tested, the second for over 4000 h, in atmospheric as well as pressurized operation (6 ata). Figure 9.114 shows the performance of the more recent 25-kW stack (Fig. 9.115). A schematic of the 100-kW stack is shown in Fig. 9.116. To improve the long-term performance of MCFC stacks, Hitachi is now giving much systematic attention to electrode structure and preparation procedures.

Mitsubishi Electric Company (MELCO) was, with Hitachi, among the first Japanese companies to undertake MCFC development. It has since chosen to

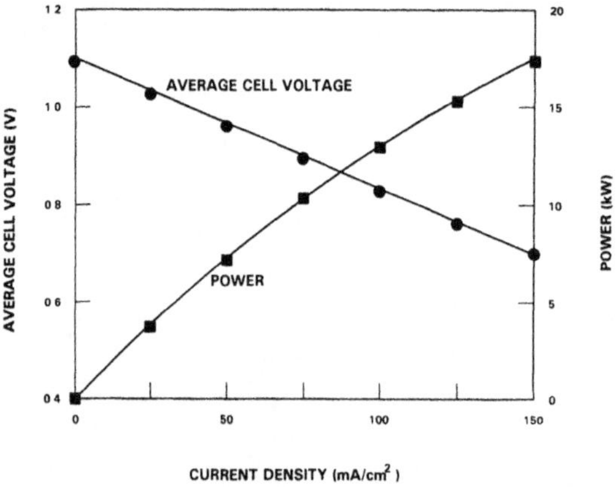

Figure 9.111. Performance of a 50-cell 16-kW IHI stack of 0.3-m^2 electrode area. (From Suzuki *et al.*[173])

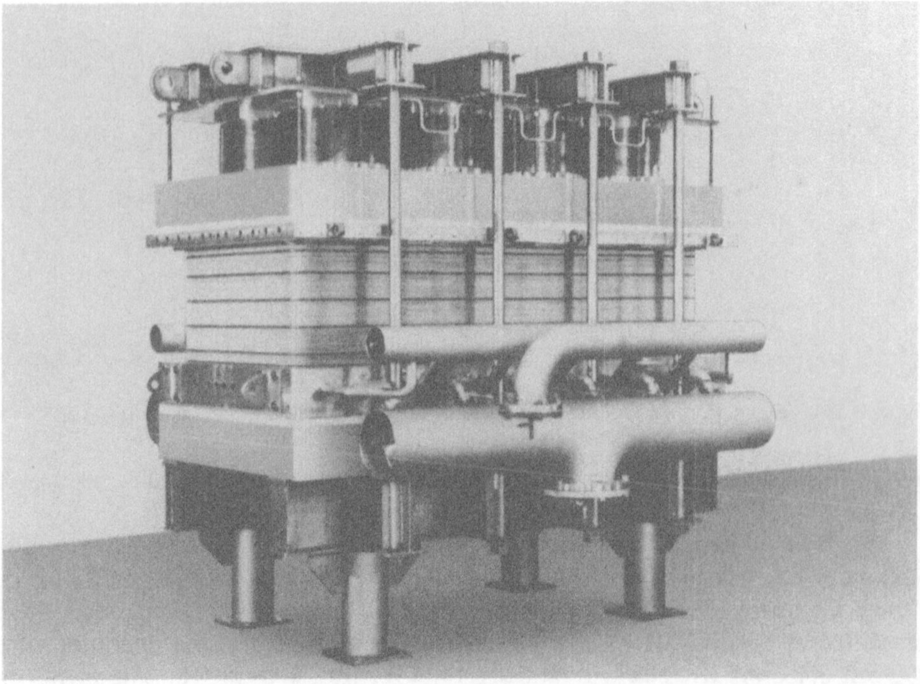

Figure 9.112. 50-kW IHI stack (50-cell, 1.015-m^2 separator area.) (Courtesy of IHI, Inc.)

focus almost exclusively on the IR-MCFC, in cooperation with ERC. Several important aspects of Mitsubishi's MCFC development, in particular the DIR and IIR concepts, have been discussed in Section 9.4.4, where performance figures of recent stacks were also given.

MELCO is pursuing DIR development under government sponsorship (NEDO) with the intention of operating a 30-kW (62-cell, 5000-cm^2) cell for a long-term test during 1992–1993. This is a continuation of the tall (50-cell) stack

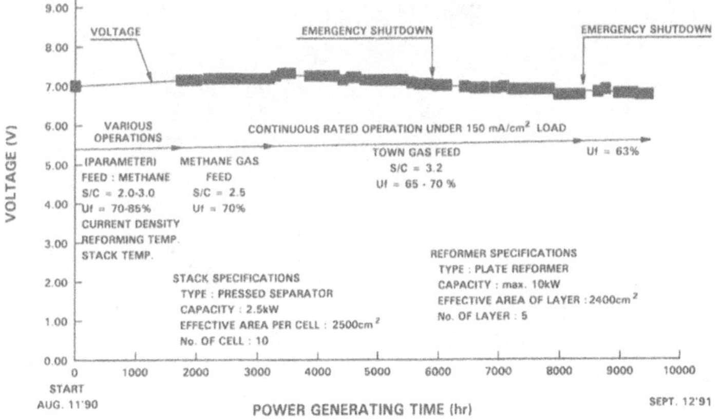

Figure 9.113. 9500-h life-graph of 2.5-kW MCFC system operated by IHI. (Courtesy of IHI, Inc.)

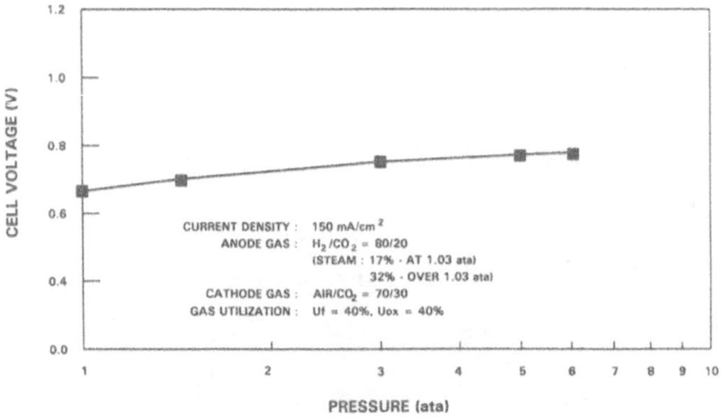

Figure 9.114. Cell performance of Hitachi 25-kW stack (No. 12 cell). (From Takashima *et al.*[174])

design tested in a 5-kW stack (see Section 9.4.4) but using full-scale (5000-cm^2) cell area.

Meanwhile the indirect internal-reforming (IIR-MCFC) concept, discussed in Section 9.4.2, has been further advanced in cooperation with Kansai Electric Power Company (KEPCO). Details of the IIR construction and tests were presented in Section 9.4.4. The goal of this development is the operation of a 192-cell 0.486-m^2 IIR stack, nominally 100 kW, for 2000–5000 h. The layout of this system is shown in Figs 9.117 and 9.118; one of the 30-kW (nominally 35.1-kW) stacks which make up the 100-kW system was shown in Fig. 9.86.

Sanyo Electric Co. has only fairly recently started MCFC development, under sponsorhsip of the Petroleum Energy Center (PEC) and using petroleum

Figure 9.115. Appearance of Hitachi 25-kW stack. (From Takashima *et al.*[174] Courtesy Hitachi Ltd.)

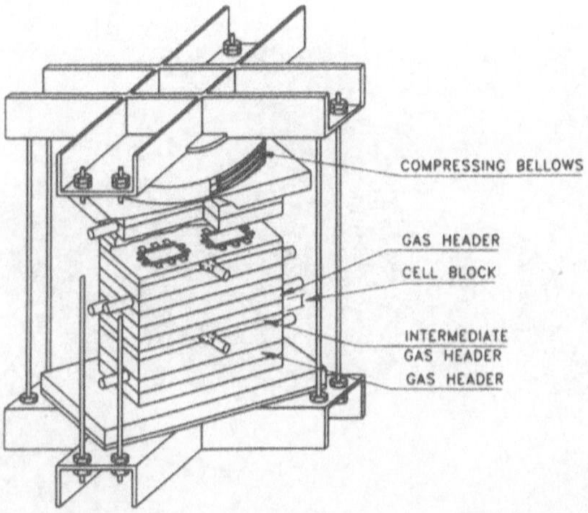

Figure 9.116. Schematic of 100-kW stack design according to Hitachi. (From Takashima *et al.*[174])

light fraction as a fuel. A 10-kW stack has been constructed in 1990 (40 cells of 0.24-m^2 area) and has been operated for more than 2000 h. The performance of the average cell was 0.124 W/cm^2, at a relatively modest fuel utilization (40%). An example of the performance is shown in Fig. 9.119.

The MCFC was invented and first developed in Europe (in the 1950s), but sustained MCFC development programs have been underway in Europe only since 1986 (see Section 9.5). Most programs are still in the basic research or component development stage, with the exception of the Dutch program as noted below. However, in the future more extensive stack testing can be expected in Europe.

The Netherlands Energy Research Center (ECN) has been constructing and testing small-scale stacks of internally manifolded design since 1990. Tests of 1000-cm^2 cell stacks at the 1-kW level are underway, with a 10-kW test by 1993 as a goal. The ECN stack is constructed with a FLEXSEP separator (see Section 9.4.1) and has a design area of 0.335 m^2 (composites of 1-m^2 area may be used for later stacks, up to 250 kW). Figure 9.120 shows an example of recent stack construction and performance.

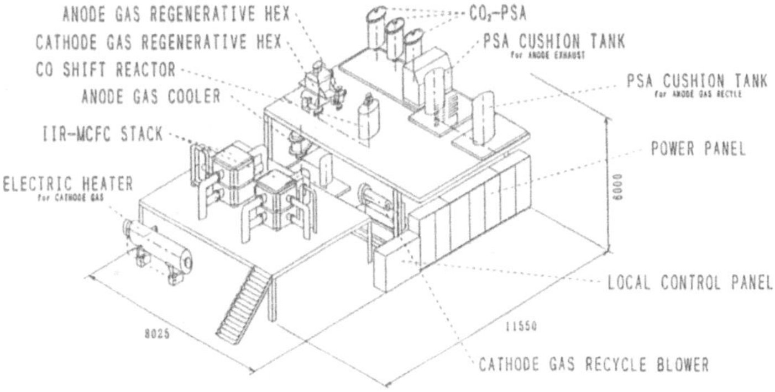

Figure 9.117. Layout of 100-kW IIR-MCFC plant according to Mitsubishi Electric Co.[175]

Figure 9.118. View of 100-kW MCFC plant at Kobe Works of Mitsubishi Electric Corp. (Courtesy MELCO.[175])

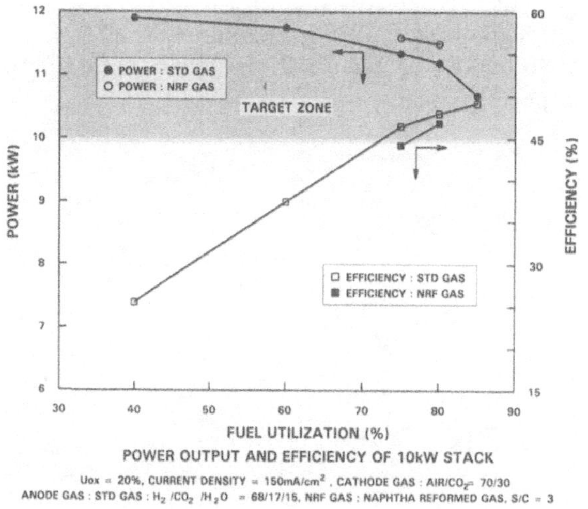

Figure 9.119. Performance of Sanyo (PEC) 10-kW stack operating on petroleum naphtha. (From Aoyagi *et al.*,[176] Fig. 4.)

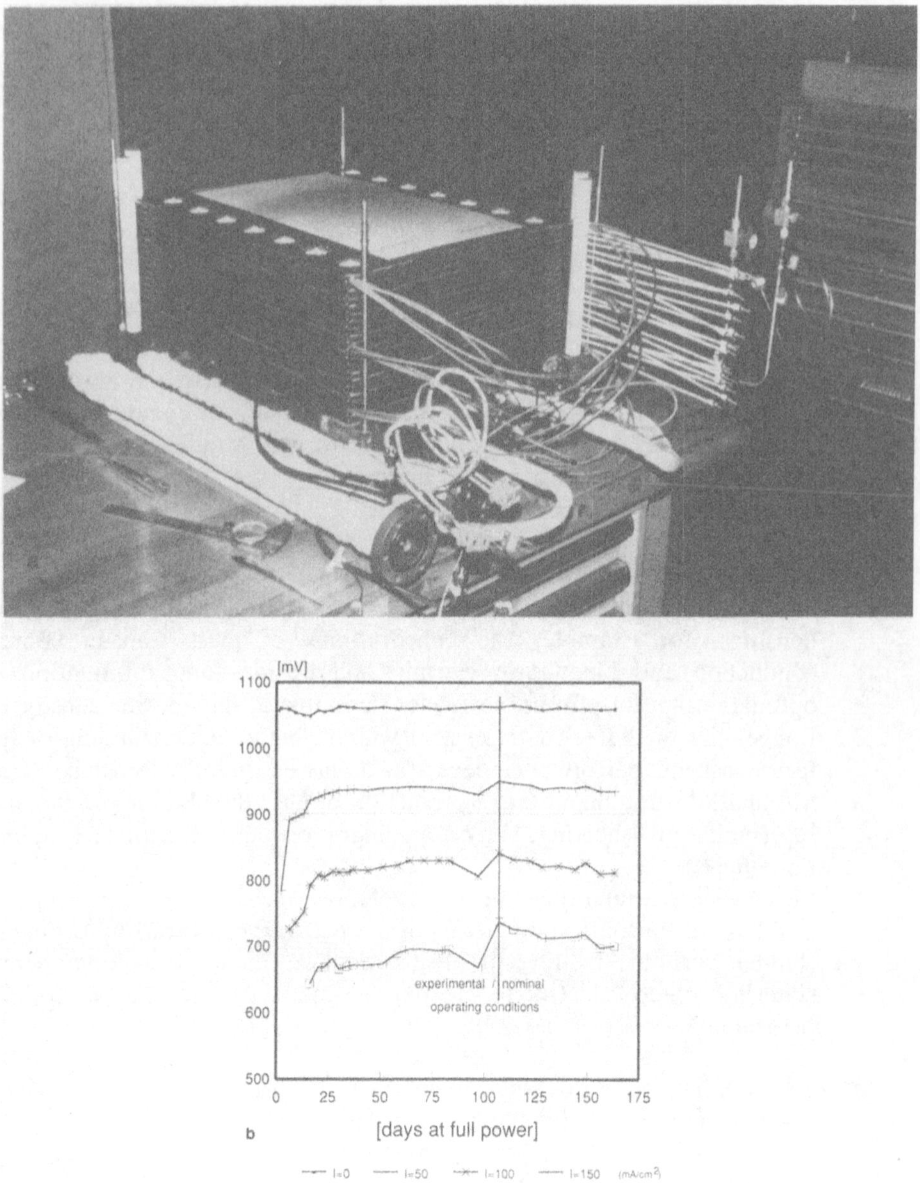

Figure 9.120. (a) View of 20-cell, 1000-cm^2 stack of ECN; (b) performance-life graphs. (Courtesy ECN, The Netherlands.)

9.5. STATUS

9.5.1. Research Goals

The MCFC, which as a concept is approximately 50 years old, has now reached the early demonstration stage of precommercial stacks. This necessarily brings with it a certain convergence of designs and materials, which is helpful in the assessment of commercial viability, as the increasing size of the demo plants requires more and more funds.

As a consequence, the current *research goals* of MCFC R&D projects are technological and focus largely on currently used materials. However, this does not imply that basic research plays no further role in MCFC development. There are still many aspects of the MCFC which require deeper physicochemical insight, and this insight can lead to significant improvements. In this section the most important research directions are briefly summarized.

The *overall goals* are

- Higher power density of the electrodes
- Longer life of the cell (stack) as a whole
- Less expensive materials and manufacturing procedures

At present, most research is directed at state-of-the-art materials, such as Ni–alloy anodes (Ni–Al or Ni–Cr), NiO cathodes, alkali–carbonate electrolyte (with alkaline–earth additions), and common alloys (with modified compositions or special surface treatment) as structural materials.

Specifically, the above overall goals require

1. Higher power density of the electrodes—better understanding of the mechanism of operation of porous electrodes, especially at high current densities, for example, the contributions of mass transfer, ohmic conduction and kinetics in causing polarization—and formulation of optimal design for porous electrodes operating at high power density.
2. Longer life of the cell (stack) as a whole—better understanding of the factors causing performance decay (as discussed, e.g., by Urushibata and Murahashi[177] and quantitative prediction of NiO dissolution and metallic reprecipitation, shorting, corrosion, migration, and electrolyte evaporation effects.
3. Less expensive materials and manufacturing procedures—better understanding of the controlling factors in the occurrence of start-up problems (binder burnout, melting of electrolyte, wetting of cell components, sintering, creep, electrolyte creepage)—and active search for optimal programming of stack assembly.

The research defined in this way is indeed very focused, but such *technology-based research* is far from routine and can stimulate further development.

In addition, *alternative materials and processes* are actively being pursued. Alternative cathode materials (such as $LiFeO_2$ and $LiCoO_2$) in the place of nickel are being explored, and preliminary tests have given rise to optimism (Plomp *et al.*[178]). Alternative separator plate materials (and alternative structural materials) may result from work on alloy corrosion currently being conducted in Japan and Europe, as well as the United States.

9.5.2. R&D Programs and Demonstration Projects

This section intends to give only a very brief overview of MCFC R&D programs. Since there are constant changes of focus and direction, the reader is referred to the proceedings of the International Fuel Cell Conferences, held with regular intervals since 1986 (the latest one in February 1992 in Makuhari, Japan), and the biennial Fuel Cell Seminar held in the United States (most recently, the 1990 Fuel Cell Seminar at Tucson, Arizona).

In the United States,[179] MCFC development is part of the Department of Energy–Fossil Energy program directed by the Morgantown Energy Technology Center (METC), Morgantown, West Virginia. Its objective is to establish the technology base for cost-effective fuel cell power generation which can be commercialized in the 1990s. The primary focus is on coal-fueled large-size plants with natural gas backup, and on U.S. technology competitiveness. MCFC funding in FY 1992 was \$28.6 million (out of a total of \$50.8 million). The department plans to award two product development test contracts in FY 1992, of 0.5–1 MW size. An important precondition for placing these contracts is that prototype manufacturing facilities must be operating and commercialization plans must be implemented. ERC and M-CP have been designated as such.

In addition to DOE, EPRI[180] supports MCFC development and demonstration in the United States, based on environmental benefits and efficiency for electric utility generation. The fuels considered are natural gas and liquid fuels, as well as coal gas. MCFC research, development, and demonstration constitutes the major part of EPRI's fuel cell budget, which in 1991 was \$80 million. EPRI supports both ERC's demonstration project with Pacific Gas & Electric (San Ramon, California) and M-CP's demonstration project referred to in Section 9.4.7.

GRI[181] also supports MCFC development, through the M-CP 250-kW demonstration project (single stack) at Unocal in California and, following that, the 250-kW gas-fired system test by San Diego Gas & Electric, in 1993–1994.

In Japan,[182] the New Energy Development Organization (NEDO) of the Ministry of International Trade and Industry (MITI) supports MCFC development at a level of 2.5 billion yen (\$20 million) in 1992. The Japanese interest in the MCFC started in 1981 and concerns R&D of two types: on fuel cell stacks, and on peripheral devices and system technology (balance of plant). MCFC stack development is carried out mainly by two developers (Hitachi and IHI) for MCFC systems with external reformer, and by one (Mitsubishi) for systems with internal reforming.

Peripheral and system research is being carried out by the Technology Research Association for MCFC Power Generation System (MCFC Research Association). This organization, in which all major developers and many equipment suppliers participate, plans to conduct the design, construction, and operational tests of a 1-MW MCFC system, to be completed in 1997, with an efficiency of 45% or higher based on the HHV of natural gas as fuel.

The electric utilities in Japan are actively cooperating in MCFC system development. The Central Research Institute of Electric Power Industry (CRIEPI) thus far has taken a lead role in testing stacks, up to the 1-MW stage. Kansai Electric Power Industry is cooperating with Mitsubishi Electric Corporation in developing the IIR-MCFC (see Section 9.4.4).

Tonen Corporation, Sanyo Electric Corporation, and Toyo Engineering Corporation are conducting tests on 10-kW external-reforming-type MCFC stacks, funded by the Petroleum Energy Center (see also Section 9.4.7). In Europe[183] the MCFC is being developed as part of several national R&D programs. The Netherlands and Italy started MCFC programs in 1986. Germany and Spain started in 1988 and 1989, respectively. Sweden has shown some interest recently. The Commission of the European Communities (CEC) is taking a very active part in stimulating and partially funding these programs since 1988.

In the Netherlands, ECN under the national fuel cell program made use of a technology transfer program with IGT to accelerate its own well-structured and dynamic approach, starting in 1987. This led to the first operation of a 1-kW (externally reformed) stack in 1989. The construction of two 250-kW units is planned for 1995, in cooperation with industrial partners. Thus far, approximately 30 MECU has been spent on MCFC R&D in the Netherlands.

In Germany, Messerschmitt Bölkow Blohm GmbH (MBB) has started a collaboration with ERC in the United States, and a technology transfer program has been set up. The German federal budget provides for 70 MECU over a period of nine years. By 1994, 250-kW units are expected to be operational. An MCFC production facility in Germany is to be established by 1998.

In Italy, Ansaldo has built a 1-kW MCFC stack and intends to construct a 300-kW plant with external reforming in 1994. This development is being carried out in cooperation with IFC. The overall Italian fuel cell program is managed by the governmental energy department, ENEA, and several universities and public research institutes participate in it. Ansaldo spent 7 MECU on MCFC research in 1991.

Spain started a 13-MECU five-year MCFC program last year. In Denmark the utility Elkraft has purchased one 7-kW MCFC stack from ERC for its operation. In Sweden the government is stimulating the interest of utilities in the MCFC, and a program involving technical universities has started.

The CEC initially took a basic research interest in the MCFC, but since 1989 it has funded major development projects as well. These include

- Development of a 1-kW IRMCFC in the Netherlands (ECN) by 1992
- Development of a 20-kW MCFC pilot plant for coal gas in Germany (MBB) by 1992

The CEC contributes 40% of the total cost of these programs, which, including basic research, amounted to 8 MECU in 1991.

9.5.3. Commercialization

The MCFC is expected to position itself favorably in the power generation market. The technical implications of fuel cell commercialization have been discussed extensively in a 1989 monography by Appleby and Foulkes.[76] The position of fuel cells with respect to other advanced power generation systems, such as the combined-cycle gas turbine, is a theme of the present volume (see Chapter 12).

Successful demonstration of the MCFC will require concentrated effort to remove the few remaining barriers. Commercial success, however, will also require further long-term research, as discussed earlier in this section.

9.6. CONCLUDING REMARKS

This chapter has aimed to provide most of the necessary technical information to understand the design and operation of molten carbonate fuel cells and fuel cell stacks. It is intended as a guide and reference for those working in the field or wanting to familiarize themselves with the MCFC.

When a technology rapidly expands, as is the case for advanced fuel cells and for the MCFC in particular, it is practically impossible to stay on top of all new developments continuously. The author is grateful to friends and colleagues who reviewed the text, and accepts responsibility for whatever has escaped their scrutiny.

Advanced fuel cells have an important role to fulfill in the energy economy of the twenty-first century, and the molten carbonate fuel cell will be a key element in their market penetration. From a chemical engineer's viewpoint, the MCFC is a remarkable and unique type of reactor. There remains a lot to understand, to play with, and to enjoy in designing and operating these devices.

REFERENCES

1. G. H. J. Broers, *High Temperature Galvanic Fuel Cells,* Doctoral thesis, University of Amsterdam, 1958.
2. C. A. Reiser and C. R. Schroll, *Abstracts National Fuel Cell Seminar* (Norfolk, VA, 1981), p. 144.
3. H. A. Liebhafsky and E. J. Cairns, *Fuel Cells and Fuel Batteries* (Wiley, New York, 1968), ch. 3, and 4.
4. J. R. Selman and H. C. Maru, in *Advanced Molten Salt Chemistry* (G. Mamantov and J. Braunstein, eds.), Vol. 4, (Plenum, NY, 1981), p. 159.
5. J. R. Selman and L. G. Marianowski, in *Molten Salt Technology* (D. G. Lovering, ed.) (Plenum, New York, 1982), p. 323.
6. A. J. Appleby and J. P. Ackerman, eds., *Proc. DOE/ERPI Workshop on Molten Carbonate Fuel Cells* EPRI WS-78-135, 1979.
7. J. R. Selman and T. D. Claar, eds., *Molten Carbonate Fuel Cell Technology* (The Electrochemical Society, Pennington, NJ, 1984), PV84-13.
8. J. R. Selman, *Energy* **11,** 153 (1986).
9. (a) N. Q. Minh, *J. Power Sources* **24,** 1–19 (1988); (b) *Chemtech* **Jan**., 32–37 (1991).
10. A. Pigeaud, H. C. Maru, L. Paetsch, J. Doyon, and R. Bernard, in *Proc. Symp. Porous Electrodes: Theory and Practice* (H. C. Maru, T. Katan, and M. G. Klein, eds.), *PV84-8* (The Electrochemical Society, Pennington, NJ, 1984), p. 398.
11. J. R. Selman, *Energy* **11,** 153 (1986).
12. H. C. Maru and B. S. Baker, *Prog. Batt. Solar Cells* **5,** 264 (1984).
13. H. C. Maru, *Quarterly Progress Report,* DOE Project DE-AC03-76 (ET-11304) (Energy Research Corporation, Danbury, CT, 1985).
14. T. Murahashi, in *Proc. 2nd Molten Carbonate Fuel Cell Symp.*, PV90-16 (The Electrochemical Society, Pennington, NJ, 1990), p. 100.
15. T. Murahashi, in *Proc. 2nd Molten Carbonate Fuel Cell Symposium,* PV90-16 (The Electrochemical Society, Pennington, NJ, 1990), p. 100.
16. P. S. Patel, *Assessment of a 6500-Btu/kWh Heat Rate Dispersed Generator,* (Energy Research Corporation, Nov. 1983), EM-3307, Final Report, EPRI-EM-3307.
17. T. J. George and M. J. Mayfield, *Fuel Cells: Technology Status Report,* DOE/METC-90/0268, 1990.
18. S. H. Lu and J. R. Selman, in *Proc. Symp. MCFC Technology* (J. R. Selman and T. D. Claar, eds.), PV84-13 (The Electrochemical Society, Pennington, NJ, 1984), p. 385.
19. K. Kinoshita, F. R. McLarnon, and E. J. Cairns, *Fuel Cells, A Handbook* DOE/METC-88/6096, 1988, p. 26.
20. L. A. H. Machielse, in *Proc. Symp. Modeling Batteries and Fuel Cells* (R. E. White, M. W. Verbrugee, and J. F. Stockel, eds.), PV91-10 (The Electrochemical Society, Pennington, NJ, 1991), p. 166.
21. B. S. Baker, in *Proc Symp. Molten Carbonate Fuel Cell Technology* (J. R. Selman and T. D. Claar, eds.), PV84-13 (The Electrochemical Society, Pennington, NJ, 1984), p. 15.

22. S. N. Simons, R. B. King, and P. R. Prokopius, in *Symp. Proc. Fuel Cells Technology: Status and Applications* (E. H. Camara, ed.) (Institute of Gas Technology, Chicago, 1982), p. 45.
23. *Development of Improved Molten Carbonate Fuel Cell Technology,* Final Report Project RP-1085-4 to Electric Power Research Institute (United Technologies Corp., 1983), Fig. 5-10.
24. R. J. Boersma, *Energiespectrum* **14,** 260 (1990).
25. H. C. Maru, L. Paetsch, and A. Pigeaud, in *Proc. Symp. Molten Carbonate Fuel Cell Technology* (J. R. Selman and T. D. Claar, eds.), PV84-13 (The Electrochemical Society, Pennington, NJ, 1984), p. 20.
26. R. J. Petri and T. G. Benjamin, in *Proc. 21st Intersoc. Energy Conversion Engineering Conf.*, Vol. 2 (American Chemical Society, Washington, D.C., 1986), p. 1156.
27. R. D. Pierce, in *Fuel Cells: Technology Status and Applications* (Institute of Gas Technology, Chicago, 1982), p. 67.
28. K. Kinoshita, in *Proc. DOE/EPRI Workshop on Molten Carbonate Fuel Cells* (EPRI-WS-78-135, 1979), p. 4.
29. Y. Mugikura, T. Watanabe, Y. Izaki, T. Hamamatus, T. Abe, H. Ishikawa, N. Kusunose, Y. Shundo, S. Maruyama, K. Koseki, and T. Nakanishi, *Performance of Bench-Scale MCFC with Electrolyte Plates Made by Paper-Making Method,* CRIEPI Report EW91001, 1991.
30. C. E. Baumgartner, V. J. DeCarlo, P. G. Glugla, and J. J. Grimaldi, *J. Electrochem. Soc.* **132,** 57 (1985).
31. P. G. Glugla and V. J. DeCarlo, *J. Electrochem. Soc.* **129,** 1745 (1982).
32. C. D. Iacovangelo and B. R. Karas, *J. Electrochem. Soc.,* **133,** 1395 (1986).
33. *Development of Molten Carbonate Fuel Cell Power Plant,* DOE/ET/17019-20, Final Report Contract DE-AC02-80ET 17019, (2 vols.) (General Electric Corp., Schenectady, NY, 1985).
34. L. G. Marianowski, E. T. Ong, R. J. Petri, and R. J. Remick, *Development of Internal Manifold Heat Exchanger (IMHEX®) Molten Carbonate Fuel Cell Stacks* (42nd Meeting Int. Soc. Electrochem., Montreux, Switzerland, 1991).
35. H. C. Maru, A. Pigeaud, R. Chamberlin, and G. Wilemski, in *Proc. Symp. Electrochemical Modeling of Battery, Fuel Cell and Photoenergy Conversion Systems* (J. R. Selman and H. C. Maru, eds.) (The Electrochemical Society, Pennington, NJ, 1986), p. 398.
36. H. R. Kunz, *J. Electrochem. Soc.* **134,** 105 (1987).
37. (a) Y. Itoh, Y. Tonoike, Y. Akiyama, M. Nishioka, T. Saito, and N. Furukawa, in *Proc. 2nd Symp. MCFC Technology* (J. R. Selman, H. C. Maru, D. A. Shores, and I. Uchida, eds.), PV 90-16 (The Electrochemical Society, Pennington, NJ, 1990), p. 169. (b) T. Saito, Y. Itoh, Y. Akiyama, K. Okudo, M. Nishioka, S. Murakami and N. Furukawa *J. Power Sources* **36,** 529 (1991).
38. R. D. Pierce, J. L. Smith, and R. B. Poeppel, *Proc. Symp. Molten Carbonate Fuel Cell Technology* (J. R. Selman, and T. D. Claar, eds.), PV84-13 (The Electrochemical Society, Pennington, NJ, 1984), p. 147.
39. L. J. Bregoli and H. R. Kunz, *J. Electrochem. Soc.* **129,** 2711 (1982).
40. C. Y. Yuh and J. R. Selman, *J. Electrochem. Soc.* **131,** 2062 (1984).
41. H. R. Kunz, L. J. Bregoli, and S. T. Szymanski, *J. Electrochem. Soc.* **131,** 2815 (1984).
42. J. R. Selman, in *Tutorial Lectures in Electrochemical Engineering and Technology* (R. Alkire and T. Beck, eds.), *A.I.Ch.E. Symp. Ser.* 204, Vol. 77, 138 (1981).
43. C. Y. Yuh and J. R. Selman, *J. Electrochemical Soc.* **138,** 3542 (1991).
44. S. Takashima, K. Ohtsuka, N. Kobayashi, and H. Fujimura, in *Proc. Second Int. Symp. MCFC Technology* (J. R. Selman, H. C. Maru, D. A. Shores, and I. Uchida, eds.), PV90-16 (The Electrochemical Society, Pennington, NJ, 1990), p. 378.
45. G. Wilemski and T. L. Wolf, in *Proc. Symp. Electrochemical and Thermal Modeling of Battery, Fuel Cell, and Photoenergy Conversion Systems* (J. R. Selman and H. C. Maru, eds.), PV86-12 (The Electrochemical Society, Pennington, NJ, 1986), p. 334.
46. J. R. Huff, in *1986 Fuel Cell Seminar* (Tucson, AZ).
47. H. C. Maru (ERC), private communication, 1985.
48. H. R. Kunz and L. A. Murphy, in *Proc. Symp. Electrochemical and Thermal Modeling of Battery, Fuel Cell, and Photoenergy Conversion Systems* (J. R. Selman and H. C. Maru, eds.), PV86-12 (The Electrochemical Society, Pennington, NJ, 1986), 395.
49. B. S. Baker, in *Proc. Symp. MCFC Technology* (J. R. Selman and T. D. Claar, eds), PV84-13 (The Electrochemical Society, Pennington, NJ, 1984), p. 15.
50. Y. Mugikura, T. Abe, T. Watanabe, and Y. Izaki, *Development of a Correlation Equation for the Performance of MCFC* CRIEPI-EW91002, 1991.

51. (a) E. J. Cairns and A. D. Tevebaugh, *J. Chem. Eng. Data* **9,** 453 (1964); (b) A. Pigeuad and J. Klinger, *Study of the Effects of Soots, Particulate and Other Contaminants on Molten Carbonate Fuel Cells Fueled by Coal Gas,* Final report to U.S. DOE under contract no. DE-AC21-84MC21154, 1987.
52. *Development of Improved Molten Carbonate Fuel Cell Technology,* Final Report to Electric Power Research Institute (United Technologies Corp., 1983), EPRI-RP-1085.
53. G. H. J. Broers and B. W. Treijtel, *Adv. Energy Conver.* **5,** 365 (1965).
54. B. S. Baker, S. Gionfriddo, A. Leonida, H. Maru, and P. Patel, *Internal Reforming Natural Gas Fueled Carbonate Fuel Cell Stack,* Final Report to Gas Research Institute (Energy Research Corporation, 1984), GRI Contract 5081-244-0545.
55. K. Ota, S. Mitsushima, K. Kato, and N. Kamiya, in *Proc. 2nd Symp. Molten Carbonate Fuel Cell Technology* (J. R. Selman, H. C. Maru, D. A. Shores, and I. Uchida, eds.), PV90-16 (The Electrochemical Society, Pennington, NJ, 1990), p. 318.
56. A. J. Appleby and F. R. Foulkes, *Fuel Cell Handbook* (Van Nostrand Reinhold, New York, 1989), p. 560.
57. L. G. Marianowski, private communication (Institute of Gas Technology, 1984).
58. D. A. Shores and P. Singh, in *Proc. Symp. Molten Carbonate Fuel Cell Technology* (J. R. Selman and T. D. Claar, eds.), *PV84-13* (The Electrochemical Society, Pennington, NJ, 1984), p. 271.
59. C. Y. Yuh, in *Proc. 2nd Symp. Molten Carbonate Fuel Cell Technology* (J. R. Selman and T. D. Claar, eds.), PV90-16 (The Electrochemical Society Inc., Pennington, NJ, 1990), p. 368.
60. R. A. Donado, L. G. Marianowski, H. C. Maru, and J. R. Selman, *J. Electrochem. Soc.* **131,** 2535 (1984).
61. S. Sato, in *Proc. 2nd Symp. Molten Carbonate Fuel Cell Technology* (J. R. Selman, D. A. Shores, H. C. Maru, and I. Uchida, eds.), PV90-16 (The Electrochemical Society, Pennington, NJ, 1990), p. 137.
62. K. Kinoshita, F. R. McLarnon, and E. J. Cairns, *Fuel Cells, A Handbook* (DOE/METC-88/6096, May 1988), p. 78.
63. G. L. Anderson and P. C. Garrigan, in *Proc. Symp Molten Carbonate Fuel Cell Technology* (J. R. Selman and T. D. Claar, eds), PV84-13 (The Electrochemical Society, Pennington, NJ, 1984), p. 299.
64. A. Pigeaud, C. Y. Yuh, and S. F. Hon, in *Proc. First Ann. Fuel Cells Contributors Rev. Meeting* (W. J. Huber, ed.) DOE/METC-89/6105, 1989, p. 214.
65. W. M. Vogel and S. W. Smith, *J. Electrochem. Soc.* **129,** 1441 (1982).
66. S. W. Smith, H. R. Kunz, W. M. Vogel, and S. J. Szymanski, in *Proc. Symp. Molten Carbonate Fuel Cell Technology* (J. R. Selman and T. D. Claar, eds.), PV84-13 (The Electrochemical Society, Pennington, NJ, 1984), p. 246.
67. L. G. Marianowski, *Progr. Batt. Solar Cells* **5,** 283 (1984).
68. R. J. Remick, *Effects of H_2S on Molten Carbonate Fuel Cells,* Final report, (Institute of Gas Technology, Chicago, May 1986), DOE/MC/20212-2039.
69. R. J. Remick, *Effects of H_2S on Molten Carbonate Fuel Cells,* Final Report, section 3, (Institute of Gas Technology, Chicago, May 1986), DOE/MC/20212-2039.
70. R. J. Remick, J. R. Jewulski, T. L. Osif, and R. Donelsen, *Contaminant Resistant Molten Carbonate Fuel Cell,* Final Report, (Institute of Gas Technology, Chicago, 1988), Contract DE-AC21-86MC23023.
71. R. J. Remick, T. L. Osif, and M. G. Lawson, *Sulfur-Tolerant Anode Materials,* Final report, (Institute of Gas Technology, Chicago, 1986), Contract DE-AC21-86MC23267.
72. S. H. Lu, in *Proc. 2nd Symp. Molten Carbonate Fuel Cell Technology* (J. R. Selman, H. C. Maru, D. A. Shores, and I. Uchida, eds.), PV90-16 (The Electrochemical Society, Pennington, NJ, 1990), p. 251.
73. A. Pigeaud, in *Proc. Sixth Ann. Contractors Meeting on Contaminant Control in Coal-Derived Gas Streams* (K. E. Markel and D. C. Cicero, eds.), DOE/METC-86/6042, 1986.
74. A. Pigeaud, *Progress Report by Energy Research Corporation to U.S. Department of Energy,* Contract DE-AC21-84MC21154 (Morgantown, WV, 1987).
75. *Fuel Cell Research on Second-Generation Molten Carbonate Systems,* Project 9105, Final technical report (Institute of Gas Technology, Chicago, 1978), SAN-1735-4.
76. A. J. Appleby and F. R. Foulkes, *Fuel Cell Handbook* (Van Nostrand Reinhold, New York, 1989).

77. I. Uchida, in *Proc. 2nd Symp. MCFC Technology* (J. R. Selman, H. C. Maru, D. A. Shores, and I. Uchida, eds), PV90-16 (The Electrochemical Society, Pennington, NJ, 1990), p. 206.
78. J. R. Selman, in *Proc. 2nd Symp. MCFC Technology* (J. R. Selman, H. C. Maru, D. A. Shores, and I. Uchida, eds), PV90-16 (The Electrochemical Society, Pennington, NJ, 1990), p. 187.
79. C. Y. Yuh and J. R. Selman, *J. Electrochem. Soc.* **131,** 2062 (1984).
80. C. Y. Yuh and J. R. Selman, *J. Electrochem. Soc.* **139,** 1373 (1992).
81. G. L. Lee, *Dynamic Analysis of MCFC Porous Electrodes,* thesis (Illinois Institute of Technology, Chicago, May 1992).
82. C. Y. Yuh and A. Pigeaud, *Determination of Optimum Electrolyte Composition for Molten Carbonate Fuel Cells,* Final report (Energy Research Corporation, Danbury, CT, 1989), DOE/MC/23264-2756.
83. K. Ota, B. Kim, S. Asano, H. Yoshitake, and N. Kamiya, in *Proc. Int. Fuel Cell Conf.* (Makuhari, Japan, 1992), p. 165.
84. K. Tanimoto, Y. Miyazaki, M. Yanagida, S. Tanase, T. Kojima, H. Okuyama, and T. Kodama, in *Proc. 2nd Symp. Molten Carbonate Fuel Cell Technology* (J. R. Selman, H. C. Maru, D. A. Shores, and I. Uchida, eds.), PV90-16 (The Electrochemical Society, Pennington, NJ, 1990), p. 357.
85. K. Tanimoto, Y. Miyazaki, M. Yanagida, S. Tanse, T. Kojima, N. Ohtori, H. Okuyama, and T. Kodama, in *Proc. Int. Fuel Cell Conf.* (Makuhari, Japan, 1992), p. 185.
86. L. G. Marianowski, private communication (Institute of Gas Technology, Chicago, 1991).
87. L. G. Marianowski and J. B. O'Sullivan, *Status of MCFC Technology* (8th Ann. Energy Technology Conf. Exp., Washington D.C., 1981).
88. E. T. Ong and T. D. Claar, in *Proc. Symp. Molten Carbonate Fuel Cell Technology* (J. R. Selman and T. D. Claar, eds.), PV84-13 (The Electrochemical Society, Pennington, NJ, 1984), p. 54.
89. H. Urushibata and T. Murahashi, in *Proc. Int. Fuel Cell Conf.* (Makuhari, Japan, 1992), p. 223.
90. H. Urushibata and T. Murahashi, in *Proc. 32nd Battery Symp.* (Kyoto, Japan 1991), p. 17.
91. M. C. Williams and T. J. George, in *Proc. 26th IECEC, Am. Nucl. Soc.* (LaGrange Park, IL, 1991), p. 577.
92. H. R. Kunz and J. W. Pandolfo, *J. Electrochem. Soc.* **138,** 1549 (1992).
93. R. J. Boersma, *Energie Spectrum* **10,** 260 (1990).
94. J. M. King, A. P. Meyer, C. A. Reiser, and C. R. Schroll, *Molten Carbonate Fuel Cell Verification and Scale-Up* (EPRI, Palo Alto, CA, 1985), EM-4129.
95. S. Sato, in *Proc. 2nd Symp. Molten Carbonate Fuel Cell Technology* (J. R. Selman, H. C. Maru, D. A. Shores, and I. Uchida, eds.) PV90-16 (The Electrochemical Society, Pennington, NJ, 1990), p. 137.
96. L. G. Marianowski, E. T. Ong, R. J. Petri, and R. J. Remick, *Development of Internal Manifold Heat Exchanger (IMHEX®) Molten Carbonate Fuel Cell Stacks* (42nd Meeting Int. Soc. Electrochem., Montreux, Switzerland, 1991).
97. M. Ohtsubo, Y. Kato, N. Zaima, S. Kasa, T. Shima, and A. Tezuka, *IHI Eng. Rev.* **24,** 90 (1991).
98. R. J. Boersma, L. A. H. Machielse, and R. Ijpelaan, in *Proc. Int. Fuel Cell Conf.* (Makuhari, Japan, 1992), p. 255.
99. L. A. H. Machielse, R. J. Boersma, C. Croon, W. M. A. Klerks, and G. Rietveld, in *Proc. Int. Fuel Cell Conf.* (Makuhari, Japan, 1992), p. 269.
100. M. Hosaka, Y. Yamamasu, M. Tooi, N. Zaima, and T. Matsuyama, *IHI Eng. Rev.* **31,** 414 (1991).
101. S. Takashima, K. Ohtsuka, T. Kahara, M. Takeuchi, Y. Fukui, and H. Fujimura, in *Proc. Int. Fuel Cell Conf.* (Makuhari, Japan, 1992), p. 265.
102. A. J. Appleby and J. R. Selman, in *Electrochemical Hydrogen Technologies* (H. Wendt, ed.) (Elsevier, New York, 1990), p. 456.
103. M. Ohtsubo, Y. Kato, N. Zaima, S. Kasa, T. Shima, and A. Tezuka, *IHI Eng. Rev.* **24,** 90 (1991).
104. M. Koga, T. Kamata, S. Kawakami, and K. Tanigawa, *IHI Eng. Rev.* **31,** 421 (1991).
105. L. A. H. Machielse, R. J. Boersma, C. Croon, W. M. A. Klerks, and G. Rietveld, in *Proc. Int. Fuel Cell Conf.* (Makuhari, Japan, 1992), p. 269.
106. Y. Yamamasu, T. Kakihara, E. Kasai, and T. Morita, in *Proc. Int. Fuel Cell Conf.* (Makuhari, Japan, 1992), p. 161.

107. E. J. Vesely, *Corrosion of Materials in Molten Carbonate Fuel Cells*, Final report DE-AC21-86MC23265 (IIT Research Institute, Chicago, 1990).
108. Y. Miyazaki, M. Yanagida, K. Tanomoto, S. Tanase, T. Kodama, H. Itoh, C. Nagai, and K. Morimoto, in *Abstracts 1988 Fuel Cell Seminar*, p. 304.
109. S. van der Molen, private communication (ECN, Petten, Netherlands, 1991).
110. T. J. George and M. J. Mayfield, *Fuel Cells*, Technology status report, DOE/METC-90/0268, 1990.
111. H. C. Maru, M. Farooque, and A. Pigeaud, in *Proc. 2nd Symp. Molten Carbonate Fuel Cell Technology*, PV90-16 (The Electrochemical Society, Pennington, NJ, 1990), p. 121.
112. *Proc. 3rd Ann. Fuel Cells Contractors Rev. Meeting* (W. J. Huber, ed.), DOE/METC-91/6120, 1991).
113. A. Suzuki, M. Tooi, M. Hosaka, T. Matsuyama, Y. Masuda, T. Nakane, and T. Osato, in *Proc. Int. Fuel Cell Conf.* (Makuhari, Japan, 1992), p. 273.
114. Y. Yamamasu, T. Kakihara, E. Kasai, and T. Morita, in *Proc. Int. Fuel Cell Conf.* (Makuhari, Japan, 1992), p. 161.
115. S. Takashima, K. Ohtsuka, T. Kahara, M. Takeuchi, Y. Fukui, and H. Fujimura, in *Proc Int. Fuel Cell Conf.* (Makuhari, Japan, 1992), p. 265.
116. *Research and Development on Fuel Cell Power Generation Technology*, FY 1990 annual report (NEDO, Tokyo, Japan, 1991), p. 81.
117. H. Ozu, T. Akasaka, K. Nakagawa, H. Tateishi, and K. Tada, in *Proc. 32nd Battery Symp.* (Japan, 1991), p. 25.
118. T. Nishimura, K. Sato, and T. Murahashi, in *Proc. 32nd Battery Symp.* (Japan, 1991), p. 15.
119. T. Kakihara, C. Shindou, M. Koga, and Y. Yamamasu, in *Proc. 32nd Battery Symp.* (Japan, 1991), p. 119.
120. L. G. Marianowski, E. T. Ong, R. J. Petri, and R. J. Remick, *Development of Internal Manifold Heat Exchanger (IMHEX®) Molten Carbonate Fuel Cell Stacks* (42nd Meeting Int. Soc. Electrochem., Montreux, Switzerland, 1991).
121. R. J. Boersma, *Energie Spectrum* **10,** 260 (1990).
122. M. Yamamoto and S. Takahashi, in *Proc. Int. Fuel Cell Conf.* (Makuhari, Japan, 1992), p. 181.
123. T. Watanabe, M. Koga, and S. Morishima, in *Abstracts 1988 Fuel Cell Seminar* (Long Beach, CA, 1988), p. 56.
124. T. Watanabe, T. Hirata, and M. Mizusawa, *IHI Eng. Rev.* **31,** 430 (1991).
125. S. Takashima, K. Ohtsuka, N. Kobayashi, and H. Fujimura, in *Proc. 2nd Molten Carbonate Fuel Cell Symp*, PV90-16 (The Electrochemial Society, Pennington, NJ, 1990), p. 378.
126. N. Kobayashi, H. Fujimura, and K. Ohtsuka, *JSMA Intn. J. Ser. II* **32,** 420 (1989).
127. T. Tanaka, T. Murahashi, and T. Nishimura, in *Proc. 23rd Intersoc. Energy Conv. Eng. Conf.*, 1988, p. 245.
128. T. Tanaka, M. Matsumura, Y. Gonjio, C. Hirai, T. Okada, and M. Miyazaki, in *Proc. 25th Intersoc. Energy Conv. Eng. Conf.*, 1990, p. 201.
129. A. J. Appleby, *Ann. Rev. Energy* 267 (1988).
130. E. H. Camara and E. T. Ong, *National Fuel Cell Seminar Abstracts*, 1983, p. 46.
131. T. Tanaka, M. Matsumura, T. Gonjo, M. Miyazaki, A. Sasaki, K. Sato, H. Urushibata, and T. Murahashi, in *Proc. 26th Intersoc. Energy Conv. Eng. Conf.*, 1991, p. 583.
132. M. Matsumura, Y. Gonjyo, C. Hirai, and T. Tanaka, in *Proc. Int. Fuel Cell Conf.* (Makuhari, Japan, 1992), p. 247.
133. T. Okada, H. Ide, M. Miyazaki, T. Tanaka, S. Narita, and J. Ohtsuki, in *Proc. 25th Intersoc. Energy Conv. Eng. Conf.*, 1990, p. 207.
134. J. Ohtsuki, A. Kusunoki, T. Murahashi, T. Tanaka, and E. Nishiyama, in *Proc. Int. Fuel Cell Conf.* (Makuhari, Japan, 1992), p. 251.
135. H. Ide, T. Okada, M. Miyazaki, T. Tanaka, and J. Ohtsuki, *Proc. 32nd Battery Symp.* (Kyoto, Japan, 1991), p. 19.
136. H. C. Maru, M. Farooque, L. Paetsch, C. Y. Yuh, P. Patel, J. Doyon, R. Bernard, and A. Skok, in *Proc. Int. Fuel Cell Conf.* (Makuhari, Japan, 1992), p. 145.
137. H. C. Maru, private communication (Energy Research Corporation, 1992).
138. C. Hirai, M. Matsumura, and T. Tanaka, in *Proc. 32nd Battery Symp.* (Kyoto, Japan, 1991), p. 21.
139. M. N. Mugerwa, L. J. M. J. Blomen, and K. G. Staller, *Design Optimisation and Environmental Aspects of Fuel Cell Systems* (Report to PEO, Utrecht, Netherlands, 19.65-010.10 (KTI BV, Zoetermeer, Netherlands, 1988).

140. H. C. Healy, in *Proc. First Ann. Fuel Cell Contractors Rev. Mtg* (W. H. Huber, ed.), DOE/METC-89/6105 1989, p. 112.
141. M. Farooque, G. Steinfeld, H. Maru, S. Kremenik, and G. McCleary, in *Proc. 25th Intersoc. Energy Conv. Eng. Conf.*, 1990, p. 207.
142. K. A. Trimble, in *Proc. 2nd Symp. Molten Carbonate Fuel Cell Technology*, PV90-16 (The Electrochemical Society, Pennington, NJ, 1990), p. 36.
143. E. J. Daniels, C. B. Dennis, M. Krumpelt, and V. Minkov, in *Abstracts 1988 Fuel Cell Seminar*, p. 41.
144. V. Minkov, E. Daniels, C. Dennis, and M. Krumpelt, *Abstracts 1986 Fuel Cell Seminar* (Tucson, 1986).
145. K. Kinoshita, F. P. McLarnon, and E. J. Cairns, *Fuel Cells: A Handbook* DOE/METC-88/6096, p. 127ff.
146. M. Ogoshi, T. Yoshida, K. Mochizuki, T. Inoue, M. Tanaka, S. Ohmoto, and T. Ishikawa, *IHI Eng. Rev.* **31,** 435 (1991).
147. L. J. Christiansen and K. Aasberg-Petersen, in *Proc. Int. Fuel Cell Conf.* (Makuhari, Japan, 1992), p. 231.
148. M. Ogoshi, T. Shimizu, S. Sato, T. Matsuyama, H. Saito, T. Abe, T. Watanabe, Y. Izaki, and Y. Mugikura, *IHI Eng. Rev.* **24,** 1 (1991).
149. Y. Izaki, T. Watanabe, Y. Mugikura, H. Kinoshita, E. Kouda, T. Abe, T. Matsuyama, T. Shimizu, and S. Sato, in *Proc. Int. Fuel Cell Conf.* (Makuhari, Japan, 1992), p. 243.
150. B. S. Baker and H. G. Ghezel-Ayagh, U.S. patent 4,532,192 (July 30, 1985).
151. M. P. Kang and J. Winnick, *J. Appl. Electrochem.* **15,** 431 (1985).
152. T. Watanabe, E. Koda, Y. Mugikura, and Y. Izaki, in *Proc. 32nd Battery Symp.* (Kyoto, Japan, 1991), p. 113.
153. Y. Mugikura, T. Abe, T. Watanabe, Y. Izaki, E. Koda, and H. Kinoshita, in *Proc. 32nd Battery Symp.* (Kyoto, Japan, 1991), p. 115.
154. *Molten Carbonate Fuel Cell System Verification and Scale-Up,* Project 1273-1 final report, (United Technologies Corp., S. Windsor, CT, 1985), EPRI EM-4129.
155. H. R. Kunz and L. J. Bregoli, in *Proc. 2nd Symp. Molten Carbonate Fuel Cell Technology* (J. R. Selman, H. C. Maru, D. A. Shores, and I. Uchida, eds), PV90-16 (The Electrochemical Society, Pennington, NJ, 1990), p. 157.
156. H. R. Kunz, *J. Electrochem. Soc.* **134,** 105 (1987).
157. S. Kuroe, M. Takeuchi, S. Nishimura, and K. Ohtsuka, in *Proc. Int. Fuel Cell Conf.* (Makuhari, Japan, 1992), p. 205.
158. In *Proc. 3rd Ann. Fuel Cell Contractors Rev. Meeting* (W. J. Huber, ed.), DOE/METC-91/6120, 1991, p. 444.
159. P. Grimes, R. Bellows, and M. Zahn, in *Electrochemical Cell Design* (R. E. White, ed.) (Plenum, New York, 1984); P. Grimes and R. Bellows, in *Electrochemical Cell Design* (R. E. White, ed.) (Plenum, New York; 1984), pp. 277–292.
160. C. Y. Yuh and A. Pigeaud, *Determination of Optimum Electrolyte Composition for Molten Carbonate Fuel Cells* (Energy Research Corp., Danbury, CT, 1989), DOE/MC/23264-2756.
161. H. C. Maru, M. Farooque, L. Paetsch, C. Y. Yuh, P. Patel, J. Doyon, R. Bernard, and A. Skok, in *Proc. Int. Fuel Cell Conf.* (Makuhari, Japan, 1992), p. 145 (Fig. 3).
162. H. Urushibata and T. Murahashi, in *Proc. Int. Fuel Cell Conf.* (Makuhari, Japan, 1992), p. 223.
163. Y. Yamamasu, T. Kakihara, E. Kasai, and T. Morita, in *Proc. Int. Fuel Cell Conf.* (Makuhari, Japan, 1992), p. 161.
164. C. Hirai, M. Matsumura, and T. Tanaka, in *Proc. 32nd Battery Symp.* (Kyoto, Japan, 1991), p. 21.
165. M. Matsumura, Y. Gonjyo, C. Hirai, and T. Tanaka, in *Proc. Int. Fuel Cell Conf.* (Makuhari, Japan, 1992), p. 247.
166. F. Gmeindl, in *Proc. 21st Intersoc. Energy Conv. Eng. Conf.*, Vol. 2 (American Chemical Society, Washington, D.C., 1986), p. 1129, Fig. 3.
167. W. H. Johnson, in *Proc. Second Ann. Fuel Cell Contractors Rev. Meeting* (W. J. Huber, ed.), DOE/METC-90/6112, 1990, p. 66.
168. In *Proc. 3rd Annual Fuel Cell Contractors Rev. Meeting* (W. J. Huber, ed.), DOE/METC-91/6120, 1991, p. 444.
169. H. C. Healy, W. H. Johnson, and C. A. Reiser, in *Proc. Int. Fuel Cell Conf.* (Makuhari, Japan, 1992), p. 261.

170. W. H. Johnson, in *Proc. First Ann. Fuel Cell Contractors Rev. Meeting* (W. J. Huber, ed.), DOE/METC-89/6105, 1989, p. 105.
171. H. C. Maru, M. Farooque, L. Paetsch, C. Y. Yuh, P. Patel, J. Doyon, R. Bernard, and A. Skok, in *Proc. Int. Fuel Cell Conf.* (Makuhari, Japan, 1992), p. 145 (Fig. 3).
172. L. G. Marianowski, E. T. Ong, R. J. Petri, and R. J. Remick, *Development of Internal Manifold Heat Exchanger (IMHEX®) Molten Carbonate Fuel Cell Stacks* (42nd Meeting Int. Soc. Electrochm., Montreux, Switzerland, 1991).
173. A. Suzuki, M. Tooi, M. Hosaka, T. Matsuyama, Y. Masuda, T. Nakane, and T. Osato, in *Proc. Int. Fuel Cell Conf.* (Makuhari, Japan, 1992), p. 273.
174. S. Takashima, K. Ohtsuka, T. Kahara, M. Takeuchi, Y. Fukui, and H. Fujimura, in *Proc. Int. Fuel Cell Conf.* (Makuhari, Japan, 1992), p. 265.
175. T. Murahashi, private communication (Mitsubishi Electric Co., 1992).
176. Y. Aoyagi, T. Hashimoto, T. Nishimoto, A. Saiai, Y. Miyake, T. Nakajima, K. Harima, T. Saitoh, H. Yanaru, and H. Fukuyama, in *Proc. Int. Fuel Cell Conf.* (Makuhari, Japan, 1992), p. 235.
177. H. Urushibata and T. Murahashi, in *Proc. Int. Fuel Cell Conf.* (Makuhari, Japan, 1992), p. 223.
178. L. Plomp, J. B. J. Veldhuis, E. F. Sitters, F. P. F. van Berkel, and S. B. van der Molen, in *Proc. Int. Fuel Cell Conf.* (Makuhari, Japan, 1992), p. 157.
179. M. Mayfield, in *Proc. Int. Fuel Cell Conf.* (Makuhari, Japan, 1992).
180. E. A. Gillis, in *Proc. Int. Fuel Cell Conf.* (Makuhari, Japan, 1992), p. 17.
181. M. P. Whelan, in *Proc. Int. Fuel Cell Conf.* (Makuhari, Japan, 1992), p. 21.
182. K. Hirose, in *Proc. Int. Fuel Cell Conf.* (Makuhari, Japan, 1992), p. 11.
183. P. Zegers, in *Proc. Int. Fuel Cell Conf.* (Makuhari, Japan, 1992), p. 3.

10

Research, Development, and Demonstration of Solid Oxide Fuel Cell Systems

K. A. Murugesamoorthi, S. Srinivasan, and A. J. Appleby

10.1. INTRODUCTION

10.1.1. Principles: Technological Merits

The solid oxide fuel cell (SOFC) is an all-solid-state power system which operates at high temperatures to ensure adequate ionic and electronic conductivity of its components ($T = 1000°C$ for the state-of-the-art system). The electrolyte is an oxide ion conductor. Since it is a solid-state system, it has advantages over other types of fuel cells.[1] It is simpler in concept than both the molten carbonate fuel cell (MCFC) and the phosphoric acid fuel cell (PAFC), because these require the presence of three separate phases, compared with two in the SOFC. Two-phase (gas–solid) contact reduces corrosion and eliminates any problems of electrolyte management. Because of its high operating temperature, activation overpotentials are low during cell operation, and noble metal electrocatalysts are not necessary.[2] However, this advantage is largely offset by the high fuel cell operating temperature, at which the free energy of formation is less negative than at room temperature, so that the open-circuit potential at practical gas utilizations is about 0.9 V. In spite of this, thermal efficiencies greater than 50% (corresponding to operation at 0.75 V, at least at modest current densities) can be obtained for chemical-to-electrical energy conversion in SOFCs. Because of its high operating temperature, the system provides for internal reforming of methane

K. A. Murugesamoorthi, S. Srinivasan, and A. J. Appleby • Center for Electrochemical Systems and Hydrogen Research, Texas Engineering Experiment Station, The Texas A&M University System, College Station, Texas 77843-3402. *Present address for K. A. Murugesamoorthi*: AT&T Bell Laboratories, Energy Systems, Mesquite, Texas 75749.

Fuel Cell Systems, edited by Leo J. M. J. Blomen and Michael N. Mugerwa. Plenum Press, New York, 1993.

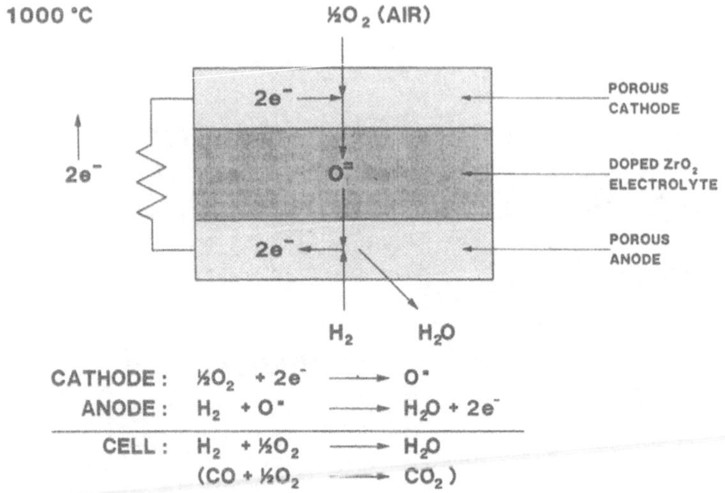

Figure 10.1. Principle of operation of solid oxide fuel cell.

and other hydrocarbons to produce CO and H_2, which is even more effective than in lower temperature MCFCs. Alternatively, even with external reforming the SOFC fuel feedstock stream does not require the extensive steam reforming with shift conversion as do low-temperature fuel cell systems. Finally, the SOFC can provide high-quality waste heat for (i) cogeneration applications and (ii) bottoming cycles utilizing conventional steam turbines for additional electric power generation.

Figure 10.1 shows the operating principle of the SOFC. External work is done by the electrons moving from the anode through the external circuit to the cathode through the voltage difference between these electrodes. The electrons transfer their charge at the cathode–electrolyte interface to oxygen molecules to produce the oxide (O^{2-}) ion. Oxide ion transport occurs through the electrolyte to the anode–electrolyte interface. Here, these ions combine with H_2 and CO to produce H_2O and CO_2 respectively, together with electrons, which complete the circuit. See Chapter 3 for the general electrochemistry of fuel cell systems.

10.1.2. History

The first solid-state oxygen ion conductor, which had the composition $(ZrO_2)_{0.85}(Y_2O_3)_{0.15}$, was described by Nernst in 1899.[3] It was accidently discovered because Nernst observed oxygen evolution at the anode with his "glower," which he developed as a commercial light source. In 1937 Baur and Preis[4] used this material to construct the first SOFC. This cell, with a coke anode and an iron oxide cathode, was operated at 1050°C with hydrogen and air as reactants, and yielded a current density of 1 mA/cm^2 at 650 mV. Later Weissbart and Ruka demonstrated the operation of SOFC employing calcia-stabilized zirconia as the electrolyte, using both hydrogen and hydrocarbon fuels.[5] The output voltage was essentially limited by the resistance of the thick electrolyte layer used. A multicell SOFC generator was first demonstrated by Archer *et al.*;[6] a 100 W solid electrolyte fuel cell employing calcia-stabilized zirconia as the electrolyte and sintered platinum as electrodes was developed in 1965. The expensive platinum electrodes used at first were subsequently replaced by a nickel–zirconia cermet, and an electronically conducting oxide for the fuel and air electrodes respectively.[7,8] Rohr[9] fabricated cylindrical SOFC multicells with different anode and cathode materials deposited by plasma spraying on cylindrical electrolytes.

10.1.3. Recent Advances in SOFC Technology

In recent years, considerable advances have been made in the development and fabrication of cylindrical SOFCs. A number of 3-kW systems with tubular cells have been manufactured and tested by the Westinghouse Electric Corporation.[10] Monolithic[11] and planar[12] solid oxide fuel cells are also being developed to attain higher power densities. However, both of these are only in an early stage of development, and only short-term performance data on systems of limited size are available. Developments made on electrodes, electrolytes, and interconnection materials for SOFCs, together with alternative cell designs, along with the progress made in their performance, are presented in detail in this chapter.

10.2. CATHODE DEVELOPMENT

10.2.1. Materials Selection Criteria

High operating temperatures limit the choice of materials suitable for SOFCs. The main functions for the cathode material are (i) to exhibit high electrocatalytic activity for oxygen reduction and (ii) to serve as a current collector. Two other requirements of the cathode material are that it should be stable in the electrochemical and thermal environments and should be inexpensive. The first requirement arises as a consequence of the physical and chemical environment in which the cathode operates.

The combination of a high temperature and an oxidizing atmosphere leads to severe materials problem for the cathode. This electrode should be stable over a wide range of partial pressures of oxygen. It should not react with other cell components, either at the high cell operating temperature or during the fabrication of ther cell components. The volatility of the oxide should be very low, so that the cathode is stable for long periods of operation. In addition, it should not undergo any destructive phase changes at the operating temperature of the cell. The cathode should adhere to the electrolyte layer so that there is no appreciable contact resistance and, hence, negligible ohmic voltage loss at the cathode–electrolyte interface. A further requirement is that the cathode should have a thermal expansion coefficient which is equal to (or at least very close) that of the electrolyte layer, to avoid mechanical failure of the cell during thermal cycling. The electronic and ionic conductivities of all components should be high to reduce ohmic losses to a minimum. For example, a high ionic conductivity is necessary for oxygen ions to pass through the thickness of the electrolyte, and a high electronic conductivity is required for good current collection along the length of the electrode. Since the electronic and ionic conductivities are inversely related to the porosity and thickness of an electrode containing both components as separate solid phases, a low ionic conductivity may be acceptable if it is compensated by enhanced electronic conductivity of the other component, or vice versa.

The electrode material must exhibit low kinetic overpotential losses, as well as high thermodynamic phase stability over its operating temperature range. It should be inert in regard to chemical reactivity with the electrolyte, current collector, and interconnection materials. Most of the cathode materials investigated to date fail to satisfy one or more of the above criteria. The required properties of the anode, which operates in a reducing atmosphere, are not particularly stringent. In contrast, the cathode must possess certain necessary mechanical characteristics, since it must maintain physical and electrical contact with the electrolyte and interconnection during cell operation, as well as the porosity required for efficient oxygen transport.

The wide variety of materials proposed for cathodes may be divided into four categories: (i) metals, (ii) oxides with current collector grids, (iii) electronically conducting oxides without current collector grids, and (iv) mixed conducting oxides without a separate current collector.

10.2.2. Metal Cathodes

During the early stages of SOFC development, metals were considered as possible electrode materials, particularly for the cathode. In this application, the metal acts as an electrocatalyst as well as a current collector. Only noble metals such as Pt, Pd, Au, and Ag could be considered because of the highly oxidizing environment. Of these, Ag and Au cannot be used, due to melting point and sintering conderations. Palladium has a considerable vapor pressure at the SOFC operating temperature, which limits long-term cell performance. Platinum is the most suitable metallic cathode material, since it has a thermal expansion coefficient similar to that of the yttria-stabilized zirconia electrolyte, to which it can be easily applied by sputtering or as a paste. The use of noble metals in electrodes or interconnections may be acceptable for fundamental investigations. However, they would have a significant, if not prohibitive, economic penalty for use in a practical power plant.

10.2.3. Oxide Current Collector Cathodes

Iron oxide was used as a cathode material in the first solid oxide fuel cell in 1937.[4] Since then, a number of other oxide materials have been investigated. One of these is a composite system consisting of a matrix of an oxide material, which provides the path for oxygen transport either because of its porosity or its oxide ion conduction (or both), used along with a metallic wire current collector grid to provide electronic conductivity. The oxide material also serves as the electrocatalyst. A platinum wire grid, embedded in porous zirconia is a good example of this type of electrode. Porous zirconia is a good cathode material, since it is very compatible with the electrolyte layer in terms of thermal expansion coefficient and chemical stability. However, it has low electronic conductivity, which can, however, be increased by doping with multivalent cations. Fuel cells with an electrode of this type have been built and tested, and satisfactory performance has been obtained. The use of platinum or palladium wire grids may fail to satisfy the necessary economic criteria, so this type of cathode is often considered not practical for operation in large systems.

10.2.4. Electronically Conducting Oxide Cathodes

Electronically conducting oxides represents one of the more interesting classes of materials which have been proposed for SOFCs. The electrode consists of a porous oxide material with a sufficiently high electronic conductivity, so that a metallic current collector grid is not required. The electrode should be porous to reduce mass transport overpotentials by allowing gas-phase diffusion, even though this may not increase the effective surface area for charge transfer, represented by the electrode–electrolyte contact area, because electronically conducting oxides do not exhibit high oxide ion conductivity. Materials considered for cathodes of this type have been Li-doped NiO, Sr-doped $LaCoO_3$, Sn-doped In_2O_3, doped ZnO_2, doped SnO_2, and $PrCoO_3$. Conductivity data for several oxide compounds are shown in Fig. 10.2, which shows that the resistivity criterion may be satisfied.

However, all the above materials fail to meet one or more of the other requirements discussed earlier. Li-doped NiO rapidly loses Li, due to the high vapor pressure of LiOH at the cell operating temperatures, which makes this material unsuitable for cathode applications. Sr-doped $LaCoO_3$ was found to have a thermal expansion incompatibility with the electrolyte layer ($2.8 \times 10^{-5}\,C^{-1}$ versus $1.0 \times 10^{-5}\,C^{-1}$ at 1000°C, which resulted in cell failure).[8] After about 500 h of operation, $LaCoO_3$ electrodes showed significant reaction with the electrolyte layer. During short-term operation, satisfactory results were obtained with ZnO cathodes doped with Zr or Al over the temperature range 800–1000°C.[13] However, at about 1000°C ZnO reacted with the electrolyte, resulting in rapid failure of the cell. Sn-doped In_2O_3 appeared to be more promising, but exhibited high gas-phase polarization.[7]

Among the oxides proposed for cathode materials, *p*-type perovskites (ABO_3) were the most promising. Substitution of donor- or acceptor-type cations

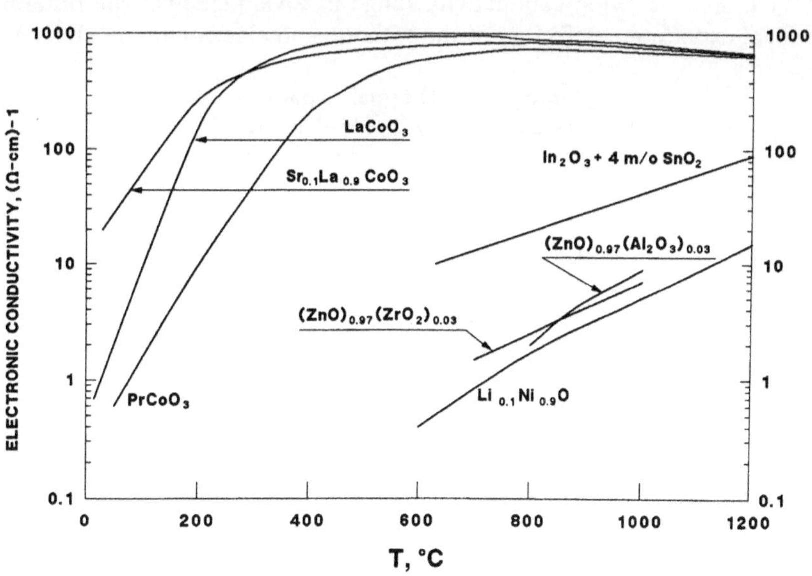

Figure 10.2. Electrical conductivity as a function of temperature for various oxides proposed for use in cathodes.[8]

in the A or B sites increases their electrical conductivity, depending on the degree of substitution and the intrinsic defect structure of the parent oxide. However, below a critical oxygen pressure, their electrical conductivity can decrease, rendering these materials unsuitable for electrode applications. The electrical conductivity of $LaCrO_3$ and $LaMnO_3$ results from the electronic 3*d* band of Cr or Mn ions;[14] thus, an increase in conductivity may be expected when lower valence ions are substituted in the La^{3+}, Cr^{3+}, or Mn^{3+} sites, resulting in the formation of Cr^{4+} or Mn^{4+}. However, if this substitution is compensated by the formation at oxide ion vacancies, little contribution to electronic conductivity may be anticipated. The conditions under which charge equilibration in $LaCrO_3$ and $LaMnO_3$ takes place will determine whether electronic or ionic compensation will be favored. Lanthanum cobaltite has a high electronic conductivity range varying from 10^2 to $10^4\ S \cdot cm^{-1}$ at 1120 K, depending on the dopant level.[15,16] However, its thermal expansion coefficient is twice that of the yttria-stabilized zirconia electrolyte, so spalling of the electrode film occurs, making it unsuitable as a cathode material. Undoped $PrCoO_3$ perovskite materials are good electronic conductors, but react with the electrolyte layer in a complex manner;[8] in addition, they do not survive thermal cycling, since the thermal expansion coefficient of $PrCoO_3$ is about two and one half to three times larger than that of yttria-stabilized zirconia.

The state-of-the-art cathode material for SOFCs is strontium-doped lanthanum manganite, a *p*-type semiconductor, which obeys the "small polaron" mechanism* only up to about 1000°C; above this temperature metallic behavior is observed.[14] Carrier mobilities for this oxide range from 0.05 to $0.1\ cm^2/V \cdot sec$. Thermal expansion coefficients of Sr-doped lanthanum manganite materials, for a range of Sr concentration, have been determined from 25 to 1000°C[17] and are given in Table 10.1. While the thermal expansion coefficients of Sr-doped lanthanum manganites are greater than that of the yttria-stabilized zirconia electrolyte, particularly in the 10 and 20 mol% Sr range, this material is preferred for SOFCs, since its wide conductivity range is advantageous. The difference in the thermal expansion coefficient of the cathode and electrolyte materials must be

Table 10.1. Thermal Expansion of $La_{.99-x}Sr_xMnO_3$ (25–1,100°C)

Composition	$\alpha \times 10^6/°C$
$La_{.99}MnO_3$	11.2 ± 0.3
$La_{.94}Sr_{.05}MnO_3$	11.7
$La_{.89}Sr_{.10}MnO_3$	12.0
$La_{.79}Sr_{.20}MnO_3$	12.4
$La_{.69}Sr_{.30}MnO_3$	12.8

From Ref. 17.

*The polaron theory of mobility is based on the polar optical mode scattering. The conductivity is dominated by the mobility rather than by the carrier concentration. The expression for conductivity derived for the small polaron mechanism is

$$\sigma = (A/T)\exp(-E/kT),$$

where σ is the conductivity, A is the preexponential factor, k is Boltzmann's constant, E is the activation energy, and T is the absolute temperature.

reduced, and work in this direction is in progress.[18] The conductivities and electrical properties of the oxides vary greatly with method of preparation, firing conditions, oxygen partial pressure, stoichiometry, and bulk- and thin-film morphology.

10.2.5. Oxides with Mixed Conduction

Cathodes using oxides with mixed conduction are the most promising for the SOFC. Charge transfer reaction occurs over the entire three-dimensional electrode structure when these materials are used, because they possess both holes and oxide ion mobility. Mixed conductivity can enhance charge transfer and gas adsorption by effectively increasing the interfacial area for the electrolyte–electrode reactions, so that polarization losses at the electrode–electrolyte interface may be significantly reduced due to the large increase in the charge transfer reaction area. Application of this class of materials may reduce the requirement of fabricating and maintaining the optimized porous cathode structure used in the state-of-the-art cells. Strontium-doped lanthanum manganite, which is widely used in present-day SOFCs, is the best available material for high-temperature fuel cell cathodes. It is an electronic conductor, but under high cathodic polarization conditions it becomes partially reduced, creating oxygen vacancies, hence mixed conduction.[19] Ionic conduction in known O^{2-} conductors may be explained by the migration of oxygen vacancies between octahedral edge sites, as is shown schematically in Fig. 10.3. The charge transfer processes in the

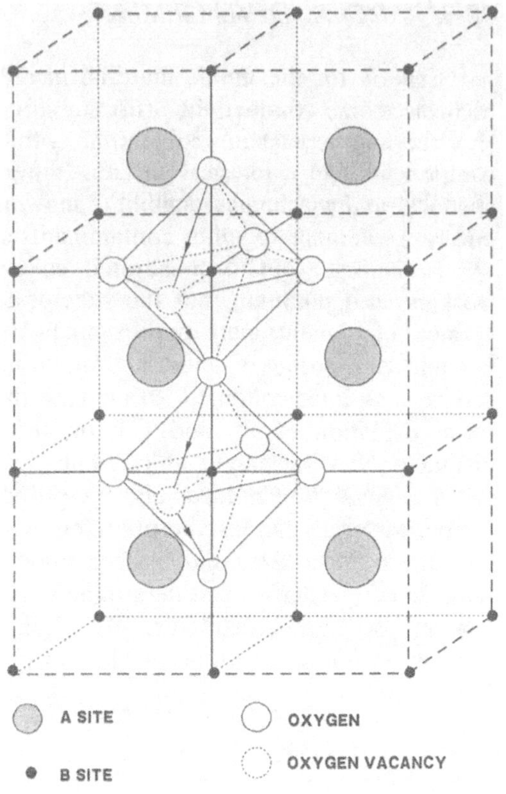

Figure 10.3. Ionic conduction pathway through perovskite ABO_3 via oxygen octahedral edge sites.

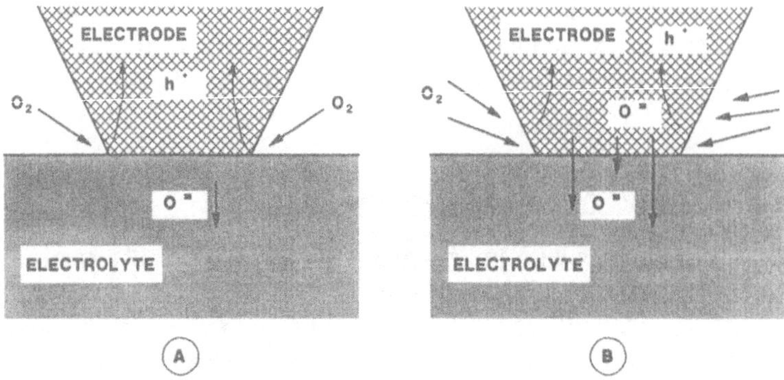

Figure 10.4. Charge transfer process occurring in (a) electronically conducting and (b) mixed conducting cathodes.[19]

third and fourth types of cathode materials discussed above are given in Fig. 10.4. While the bulk conductivity of Sr-doped lanthanum manganite is more than $100\,S \cdot cm^{-1}$ at 1000°C, its effective conductivity as a cathode material is much lower, due to the required porosity and pore geometry. Other promising perovskite-type cathode materials of this class have been proposed.[20,21] The ideal air electrode for SOFCs has yet to appear, but a search for more effective mixed conductors is presently being conducted.[22,23]

10.3. ANODE DEVELOPMENT

Criteria for the anode material include (i) effective oxidation catalysis, (ii) high electronic conductivity, (iii) stability in the reducing anodic environment, (iv) thermal expansion compatible with the electrolyte and other fuel cell components, (v) a physical structure offering low fuel transport resistance, (vi) chemical and mechanical stability, showing no high-temperature phase change, and (vii) tolerance to sulfur contaminants and hydrocarbon fuels.

In the first SOFC demonstrated, coke was used as the anode material.[4] Later workers used platinum, but the state-of-art anode material is a nickel–zirconia cermet, i.e., an intimate porous mixture of Ni and ZrO_2–Y_2O_3.[24] The doped zirconia is a porous support and sintering inhibitor for the nickel phase, and serves as an ionic conductor to increase the effective surface area for electrochemical oxidation of H_2 and CO on the nickel electrocatalyst and electronic conductor. Several papers and patents describe its preparation, using slurry and vapor deposition techniques and the properties of the cermet thus prepared.[25–28] Similar materials can also be prepared by tape casting,[29] and the effects of nickel content, particle size, and surface area on the electronic conductivity of the reduced cermet have been determined. This is strongly dependent on the nickel content and its microstructure.[30–32] Cermet resistivities have been observed to vary by several orders of magnitude for only a 10% change in metal content.[33–37] Table 10.2 shows the porosities and conductivities at 1000°C of tape-cast Ni/ZrO_2–Y_2O_3 cermets investigated by Dees *et al.*[29] Materials with nickel contents greater than 30 vol% of total solids are electronically conducting due to improved particle-to-particle contact. Below this concentration, conductivity is

Table 10.2. Tubular Solid Oxide Fuel Cell—Summary of Component Thicknesses, Materials, and Fabrication Processes

Component	Thickness	Material	Fabrication process
Support tube	1.2	Calcia-stabilized zirconia	Extrusion, sintering
Air electrode	1.4	Modified lanthanum manganite	Slurry coating/sintering
Electrolyte	0.04	Yttria-stabilized zirconia	Electrochemical vapor deposition
Interconnection	0.04	Modified lanthanum chromite	Electrochemical vapor deposition
Fuel electrode	0.1	Nickel-zirconia cermet	Slurry coating/electrochemical vapor deposition

From Ref. 74.

considerably reduced and is dominated by ionic conduction. Increasing zirconia particle size considerable improves conductivity. Since nickel has the highest thermal expansion coefficient of all SOFC materials, increasing the amount of nickel to enhance conductivity also increases the thermal expansion coefficient of the anode material.[33] About 35 vol% Ni is required to maintain the required level of electronic conductivity in the anode layer, so this concentration is used in SOFCs to minimize the thermal expansion coefficient mismatch with other cell components.

Mixed oxides with different compositions of Gd-doped ceria with Zr and Y oxides have been investigated[23] as anode materials for oxidation of methane in high-temperature fuel cells. While satisfactory oxidation rates were observed, cracking of the electrodes occurred during thermal cycling.

10.4. ELECTROLYTE DEVELOPMENT

Criteria for SOFC electrolytes are (i) high oxygen ion conductivity and minimum electronic conductivity, i.e., an oxide ion transport number close to unity; (ii) good chemical stability with respect to electrodes and/or substrates; (iii) high density, to avoid fuel crossover to the cathode; (iv) high stability in both oxidizing and reducing atmospheres; and (v) thermal expansion compatibility with the other cell components.

Yttria-stabilized zirconia was the electrolyte used in the first SOFC to be demonstrated. It is still the most effective of the available electrolytes for high-temperature fuel cells, although several others, including Bi_2O_3,[38] CeO_2,[39,40] and Ta_2O_5[41,42] have been investigated. The oxide ion transport number for 12% yttria-stabilized zirconia is close to unity, and it is highly stable in reducing and oxidizing environments. Yttria-stabilized zirconia samples with greater than 92–93% of the theoretical density have a hydrogen permeability of less than 10^{-8} cm^2/s which meets the electrolyte requirements for SOFC.

Zirconia electrolytes are stabilized in the fluorite structure, which is responsible for oxide ion conductivity, by the addition of stable aliovalent cations such as Y^{3+}. Oxide ion conduction occurs between vacant tetrahedral oxygen

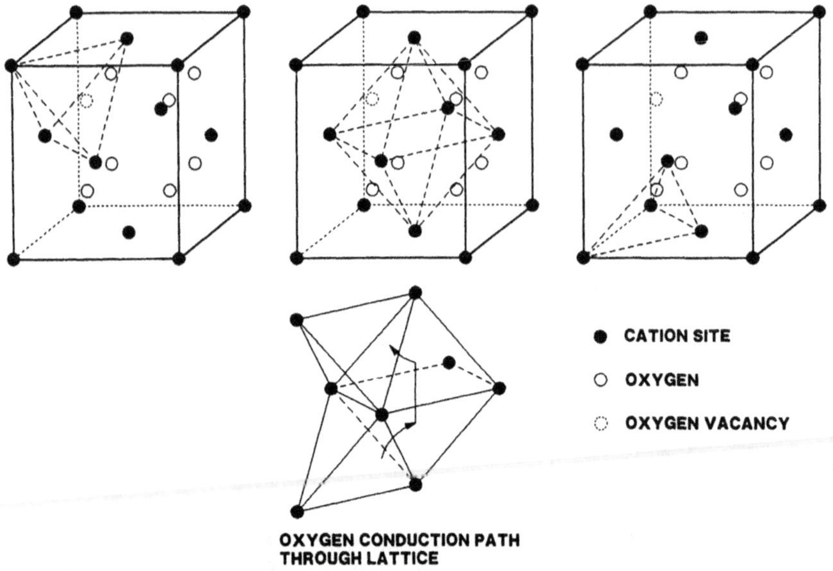

Figure 10.5. Oxygen vacancy conduction path through stabilized zirconia (fluorite structure) between tetrahedral sites via unoccupied sites.

sites via an unoccupied octahedral site, as is shown in Fig. 10.5. The electrolytes have anion vacancy concentrations proportional to the dopant concentration, yielding reasonably high ionic conductivities at high temperature (about $1\ S \cdot cm^{-1}$ at 1000°C). Materials such as Bi_2O_3 and doped CeO_2 show higher ionic conductivities than yttria-stabilized zirconia; however they are more susceptible to reduction at the anode, producing defect oxides with increased electronic conductivity. Such electrolytes will reduce fuel cell potentials and, hence, energy conversion efficiencies.[43] The ionic conductivity constraint necessitates a high SOFC operating temperature. Many SOFC materials and fabrication problems would be reduced if a lower temperature oxide ion conductor were available. Operation at intermediate temperatures (about 600°C) will still retain the advantages of improved electrode kinetics and the ability to internally reform hydrocarbons. Perovskite (ABO_3) compounds, similar to those used for the cathode materials, are being investigated as solid electrolytes for operation at intermediate temperatures.[44,45]

10.5. INTERCONNECT DEVELOPMENT

The "interconnect" is the means of electronic connection between two neighboring cells. It should have (i) high electronic conductivity with no ionic conductivity, (ii) chemical stability in both anode and cathode environments, (iii) mechanical stability, and (iv) thermal expansion compatibility with other cell components. Among the metals, the choice is limited to the noble metals and certain nickel alloys (e.g., Inconels). The cost of the noble metals is prohibitive, and metallic materials possess larger thermal expansion coefficients than the other cell components. An alternative is the use of inexpensive perovskite-structured oxide (ABO_3) layers. Lanthanum chromite is the material of choice for the SOFC interconnect. Its electronic conductivity is greatly enhanced when Mg or an alkaline earth cation (Ca, Sr, or Ba) is substituted for La in relatively small amounts. This enhancement results from charge compensation at the B lattice sites, where a transition from Cr^{3+} to Cr^{4+} creates localized excess positive charges on an equal fraction of the chromium cations. The conduction is by a hopping mechanism.[46,47] Substitution of La by 16% Sr yields the conductivity range required. The doped material has the required chemical stability in both air and in a highly reducing atmosphere at SOFC operating temperatures; however, the thermal expansion coefficients of Sr-doped lanthanum chromites are reported to be about 10% less than that of the electrolyte layer.[48] It is not possible to attain 100% density of the material by sintering at a temperature which is compatible with the properties of the other fuel components (cathode, anode, and electrolyte). Under compatible sintering conditions, the porosity of the interconnect remains high, and crossover of the fuel and/or oxidant will occur, lowering cell efficiency. To obtain dense lanthanum chromite, it is necessary to sinter at high temperatures (1625°C) in a highly reducing atmosphere ($P[O_2] = 10^{-10}$ to 10^{-12} atm).[49] The air electrode material, lanthanum manganite, is unstable under reducing conditions. If the entire cell has to be sintered in one operation, these two materials will appear to be incompatible from a fabrication standpoint.

The above problem can be solved by fabricating the fuel cell components individually, using electrochemical vapor deposition (EVD).[50] The EVD technique, however, is complex, slow, costly, and limits the range of available chromite compositions. Westinghouse presently uses Mg (rather than Ca or Sr) as the dopant in the EVD deposition of the interconnect, because $MgCl_2$ has a higher vapor pressure than, say, $SrCl_2$. Unfortunately magnesium substitutes for chromium as well as for lanthanum, resulting in lower electrical conductivities.[51] Several substituted lanthanum chromite powders, produced using the glycine–nitrate method,[52] are highly reactive and can be sintered to nearly a 100% density at 1550°C in air. Furthermore, addition of Al into the Sr-doped lanthanum chromite structure, as a substitute for a fraction of the chromium, results in enhanced sintering ability and closed porosity. $La_{0.82}Sr_{0.15}Cr_{0.85}Al_{0.15}$ has been found to have a high electrical conductivity and a thermal expansion coefficient matching that of the electrolyte. However, the long-term stability of this compound in an operating SOFC requires verification. The challenge of sintering the interconnect material under conditions which do not adversely affect the other SOFC materials therefore still remains.

10.6. DESIGN AND DEVELOPMENT OF MULTICELL STACKS

10.6.1. SOFC Design

The original SOFC design was the so-called bell-and-spigot design, in which many small cells with the air electrode on the outside of a tubular electrolyte, and the anode on the inside, were mechanically connected in a linear series by a press-fit interconnect. This type of cell, in which the current is collected at the edge of the cathode so that cell length is limited to about 1 cm, is now obsolete.

Three types of SOFC designs are being developed today: (i) tubular, (ii) monolithic, and (iii) planar. The tubular design is the most advanced. Monolithic and planar designs are still in the research and development stage.

10.6.2. Tubular Design

The tubular concept of the SOFC was first introduced by Rohr[9] at the Westinghouse Electric Corporation Research Laboratories. The design used a porous 15% calcia-stabilized zirconia support tube, 1–2 mm thick, on which the cylindrical cermet anodes of individual cells were deposited by filtration and sintered. The anodes were separated by bands of bare support tube, produced by electrochemical etching. After masking, a band of interconnect material was then deposited by EVD over the end of each anode. After further masking, the electrolyte was deposited by EVD in such a way that a cylindrical band of this interconnect material was still exposed. The end of each electrolyte later next to the interconnect was then masked, a porous layer of tin-doped indium oxide was deposited by chemical vapor deposition (CVD). Finally, this layer was impregnated with praeseodymium nitrate, which was heated to form a catalytic cathode. The cells were edge-connected in series, the current being carried along the length of the anode of one cell to the interconnect, and then along the length of the tin-doped indium oxide cathode current collector of the next cell. The whole system was a solid-state thin-film version of the bell-and-spigot structure, bearing the same relationship to it as a printed circuit board does to a soldered electronic device. Indeed, the procedures used were inspired by those in electronics. At 1000°C, maximum power densities of 0.3 W/cm^2 with hydrogen and oxygen and 0.2 W/cm^2 with hydrogen and CO–air were attained with this complex structure of small cells. The air electrode and current-collector structure was later abandoned because of the high vapor pressure of indium oxide at 1000°C.

The design was modified at Westinghouse by Isenberg in 1980–82[53] by turning it inside out so that the anode was external, which allowed a simpler form of current collection in the anode atmosphere by using a nickel felt between the anode of one cell and the interconnect of the next. It uses a closed-end tubular support upon which the cathode, interconnection, electrolyte, and anode are sequentially deposited (Fig. 10.6). Air is fed into the tube via a concentric alumina air injection tube (not shown in Fig. 10.6). The air-injection tube is integrated with a heat exchanger, to allow maintenance of the correct cell operating temperatures as a function of power output. Figure 10.7 shows the cell-to-cell connection in the tubular SOFC bundle configuration. The series–parallel design protects the system against complete stack failure in the event of individual cell failure.

As in the original concept, the support tube is made of calcia-stabilized zirconia, which accounts for 50% of the materials cost, because it is much higher than that of the cell itself. It therefore has a significant effect on the specific power densities (power/weight and power/volume). An advantage of this design is that the support tube provides a structural support for the cell components so that their structures can be made very thin. This reduces the ohmic losses in the electrolyte layer, but current is collected peripherally in this system, which means that the layer thickness cannot be significantly reduced nor can tube diameter be significantly increased. At present, the support tube has a diameter of about

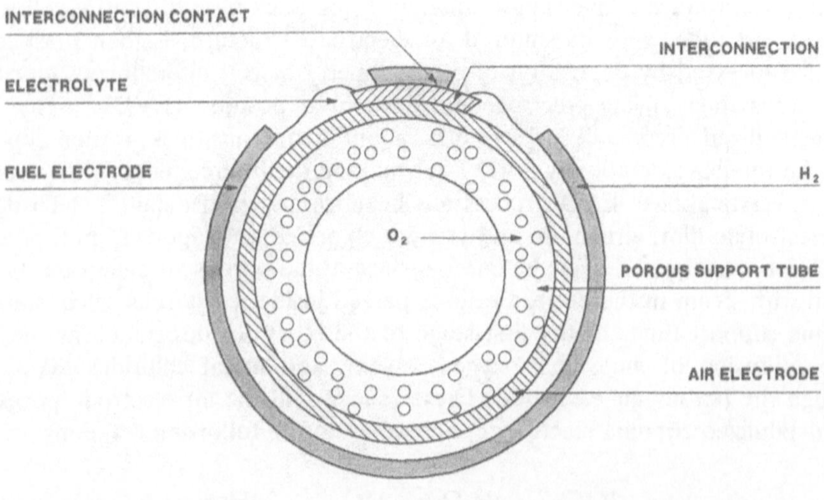

Figure 10.6. Cross section of the Westinghouse tubular solid oxide fuel cell.[74]

1.5 cm, a wall thickness of 1.5 mm, and a porosity of about 50% to allow oxygen to diffuse to the cathode. Its length varies from 30 to 60 cm, and it has an effective surface area of 100–200 cm^2,[10] since the upper (open) part of the tube is inoperative as a fuel cell, serving as a manifold in a ceramic header. This is described below.

Since the support tube consists of stabilized zirconia, its thermal expansion coefficient matches those of the electrolyte, cathode, anode, and interconnection. It is fabricated by extruding a cylindrical section, inserting a plug to close one end, and sintering. The porous air electrode is fabricated by filtering a slurry of Sr-doped $LaMnO_3$ power over the support tube and sintering the composite

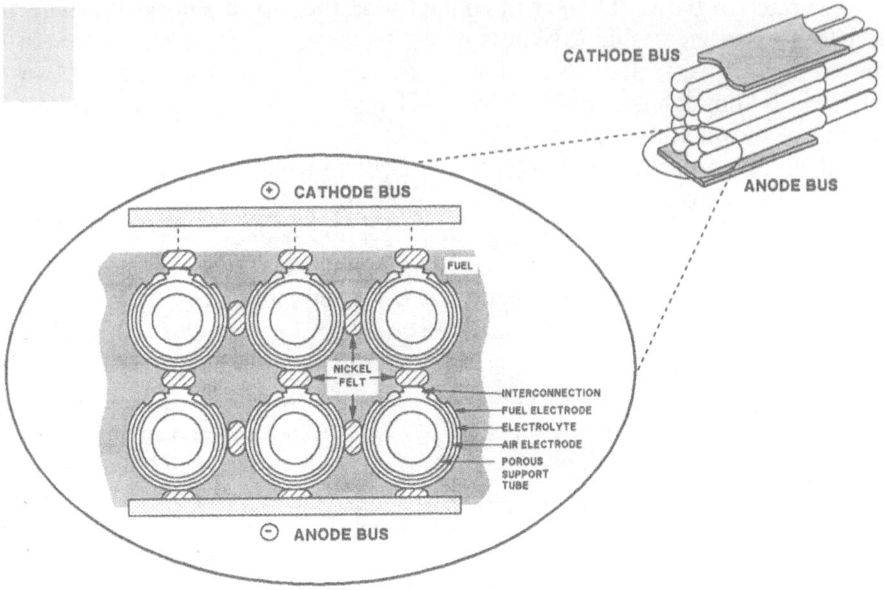

Figure 10.7. Tubular bundle configuration of SOFCs.[74]

structure to produce a layer approximately 1 mm thick over the entire cell length. The support tube, with its sintered air electrode structure, is then masked and further processed by deposition of a Mg-doped $LaCrO_3$ interconnect film about 30 microns thick, using electrochemical vapor deposition (EVD).[54] The dense yttria-stabilized zirconia electrolyte film, about 30 microns thick, is then deposited over the tubular assembly by EVD after making the interconnect. The procedure for the Westinghouse EVD process has been discussed in detail.[55–58] To deposit the electrolyte film, zirconium and yttrium chlorides are vaporized in a predetermined ratio and passed over the outer surface of the porous air electrode. Oxygen mixed with steam in the desired ratio is passed inside the porous calcia-stabilized zirconia support tube. In the first stage of the reaction, designated as the CVD stage, diffusion of molecular oxygen, steam, and metal chlorides takes place through the porous air electrode. These react to fill the air electrode pores with yttria-stabilized zirconia electrolyte, according to the following reactions:

$$2MCl_y + yH_2O = 2MO_{y/2} + 2yHCl, \quad (10.1)$$

$$4MCl_y + yO_2 = 4MO_{y/2} + 2yCl_2, \quad (10.2)$$

where M is the cation species (zirconium and/or yttrium) and y is the vacancy associated with the cation. The temperature, pressure, and flow rates of the gases are chosen so that the above reactions are thermodynamically and kinetically favored.

The electrolyte film formed over the air electrode in this way is uniform in thickness and provides a gastight barrier during the exposure of the cell to the oxidant and the fuel atmosphere. During this stage the pores in the air electrodes are filled with the yttria-stabilized zirconia.

In the second stage of the reaction, electrochemical reduction of oxygen occurs at the air electrode–CVD electrolyte layer interface, and oxide ions pass to the surface of the CVD layer in contact with the mixed chlorides, where they react to give an increasing thickness of dense doped oxide and chlorine. In this part of the process, the oxide ion is transported electrochemically, so the technique is thus labeled EVD. The CVD and EVD stages in electrolyte film growth are shown schematically in Fig. 10.8.[54]

Finally, a porous nickel–zirconia cermet fuel electrode layer with a thickness of about 100 microns is deposited on the electrolyte by the application of nickel powder slurry. This is followed by its impregnation with yttria-stabilized zirconia by EVD. The cells are then checked for gas tightness before electrical testing. Table 10.3 lists the tubular solid oxide fuel cell components and their dimensions, along with the fabrication techniques.[54] Each 30-cm-long tubular cell has a design power output of 18 W; typical performance on reformed methane at 90% utilization was 12 W at 160 mA/cm^2.

Since both hydrogen and oxygen utilization rise along the length of the tubular cell, its current density is higher at the closed end. Under normal operating conditions, the fuel utilization is 85%, and the unused 15% of the fuel leaks through the holes in the tube sheet, which loosely retains the tubular cells. The depleted fuel is burned with excess air in the plenum chamber into which the open ends of the tubular cells project. This serves as a heat exchanger to heat the incoming process air in the alumina entry tubes, as is shown in Fig. 10.9. Figure

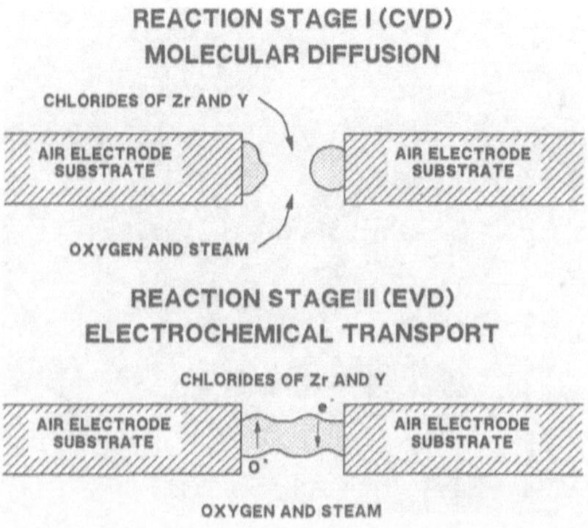

Figure 10.8. Two stages of reaction occurring during the deposition of yttria-stabilized zirconia.[54]

10.10 shows the potential of state-of-the-art cells as a function of current density at 85% H_2 fuel utilization and 25% air utilization. Figure 10.11 shows cell potentials as a function of time at a current density of 250 mA/cm^2, which includes prototype cells from 1986–1987, and shows the progressive increase in cell performance and stability through 1988 and 1989. By 1990, cells produced in 1988 had demonstrated 15,000 h of operation.

A 400-W SOFC generator was designed and constructed by Westinghouse Electric (USA) and was delivered to Tennessee Valley Authority (USA) in 1986. It was successfully operated for 1760 h. This demonstrator showed the capability of the system for automatic unattended operation. In 1987 two 3-kW systems were built by Westinghouse Electric and installed and tested at Osaka Gas Company and at Tokyo Gas Company with reformed, desulfurized pipeline natural gas as fuel. The Tokyo Gas system was operated continuously for 4900 h, while that at Osaka Gas Company was operated for 3700 h. Both units have met the target specifications, with availabilities greater than 98%, at a higher heating value (HHV) efficiency of 50%. They also substantiated the expected low

Table 10.3. Porosities and Conductivities at 1,000°C of Ni/ZrO_2–Y_2O_3 Cermet Samples

v/o Ni of total solids	v/o porosity		Zirconia powder	Conductivity (Ω^{-1} cm^{-1})
	Air-fired	Reduced		
25	25	36	Toyo Soda	0.162
30	30	42	Toyo Soda	0.307
32	19	34	Toya Soda	526
40	26	42	Toyo Soda	1100
15	5	14	Zircar	0.0753
25	4	18	Zircar	0.1141
30	6	23	Zircar	13.1
40	22	39	Zircar	268
50	23	43	Zircar	524

From Ref. 29.

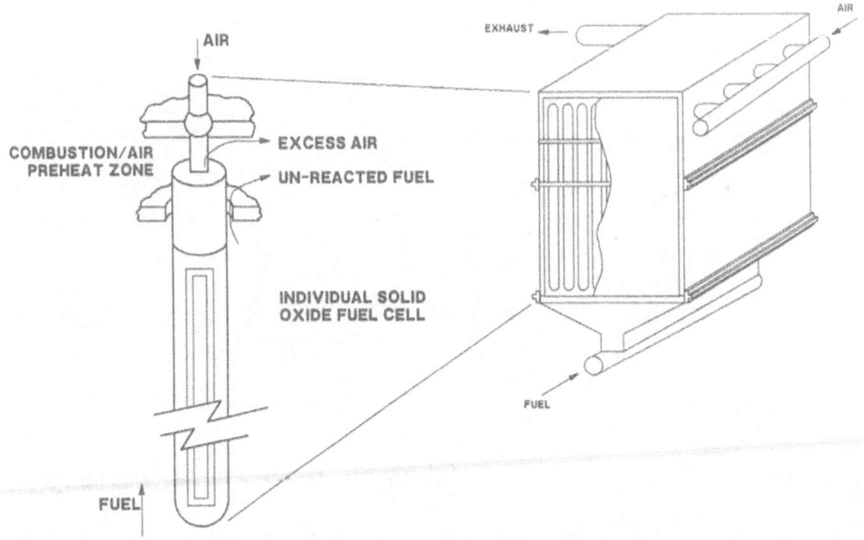

Figure 10.9. Tubular solid oxide fuel cell module.[74]

environmental impact of SOFC systems. Easy installation, start-up, operation, and maintenance were demonstrated. For start-up, air is preheated up to 700°C and then to 500°C during cell operation, using a subsystem which can deliver up to 600 L of electrically heated air per minute. The exhaust gas mixture contains about 77% N_2, 17% O_2, 4% H_2O, and 2% CO_2 and is at a temperature of 700°C.

A 3-kW system was built with financial support from Gas Research Institute (GRI), and was tested using reformed, desulfurized natural gas. After 5500 h of continuous operation and five complete thermal cycles, the generator was shut down as scheduled. At the time of writing, 25-kW systems are being constructed using cells produced on Westinghouse's pilot production line, which was installed in 1989 and has a capacity of about 2 MW/day.[59] Three to five 25-kW units will be

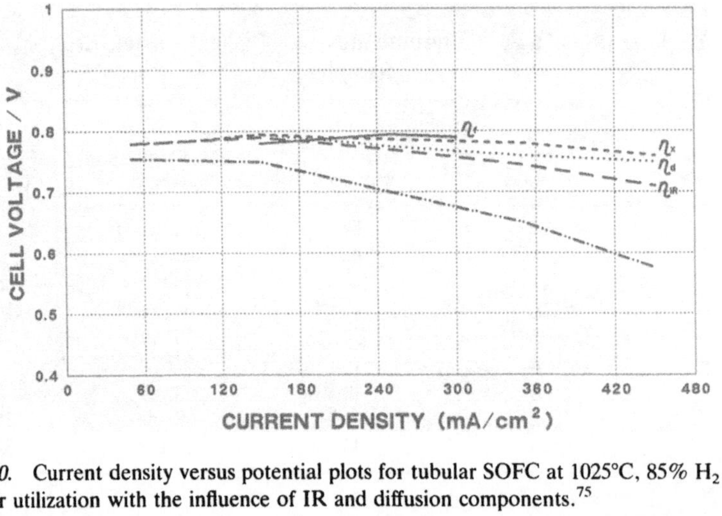

Figure 10.10. Current density versus potential plots for tubular SOFC at 1025°C, 85% H_2 utilization, and 25% air utilization with the influence of IR and diffusion components.[75]

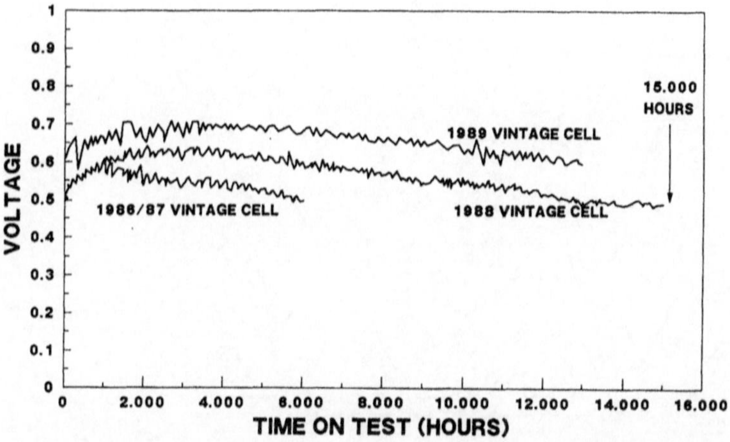

Figure 10.11. Cell test data verifying improved cell performance and life.[10]

fabricated and tested during the 1990s. Westinghouse also has plans for constructing larger units in the 100–200 kW and 10–50 MW classes, which are aimed at the gas and electric utilise market.

10.6.3. Monolithic Design

The monolithic solid oxide fuel cell (MSOFC), first proposed by Argonne National Laboratory[60] in 1983, is a planar cell design in which the ceramic cell components mutually support each other. Thus, no supporting material is used, and all components are active, allowing enhancement of the specific power density (in terms of both power/weight and power/volume). It is projected that this design may offer power densities of 8.08×10^3 W/kg and 4×10^3 kW/m^3, compared with 100 W/kg and 140 kW/m^3 for the Westinghouse tubular system. MSOFCs are still in an early stage of development,[61] and much more R&D is required before the system becomes truly practical. Figure 10.12 shows a schematic cross section of a MSOFC in a coflow configuration, indicating the component materials.

Presently, MSOFCs are fabricated by the following steps: (i) slurries of electrode, electrolyte and the interconnect material powder are made; (ii) tape-casting or tape-calendaring processes are used for the production of thin green ceramic layers; (iii) these layers are laminated, corrugated, and stacked; (iv) the electrochemical cell stack is sintered. Tape casting is a common fabrication process used for forming large-area, thin, flat ceramic layers. The process uses a slurry of ceramic material dispersed in a dilute solution of organic binders and plasticizers in a suitable solvent or solvents (referred to as a "slip"). This is spread on a flat surface from a linear hopper moving perpendicular to the direction of the film to a controlled thickness using the knife edge of a doctor blade moving with the hopper. The solvents are then allowed to dry, giving a flexible tape containing the ceramic powders and plastic binders, which can be stripped from the casting surface, cut to size, and corrugated before firing. The anode, electrolyte, and cathode layers are fabricated by sequentially casting one layer on top of another.

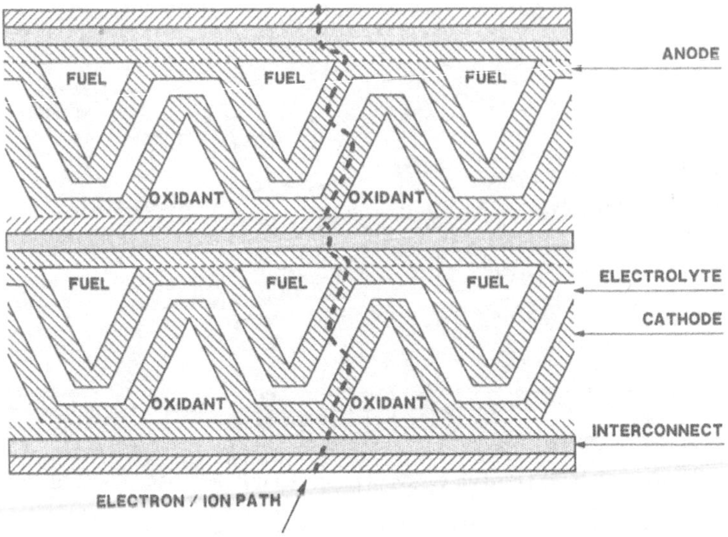

Figure 10.12. Cross section of the coflow type monolithic solid oxide fuel cell.

A tape-calendering technique, similar to tape casting, can also be used to produce thin, flexible green ceramic layers. In this process, the ceramic powder, binder, and plasticizer are mixed in a high-intensity mixer, as shown in Fig. 10.13. The friction resulting from the mixing heats the batch to form a plastic mass. This is then calendered (i.e., rolled) into a thin flat tape in a two-roll mill. To fabricate composite tapes, individual tapes are laminated in a second rolling operation. These are corrugated using a two-piece mold designed to yield the desired corrugation pitch and amplitude. The tape to be corrugated is laid on the bottom half of the mold, and the top half is lowered to engage the first corrugation. This process is then advanced to the remaining corrugations successively until the

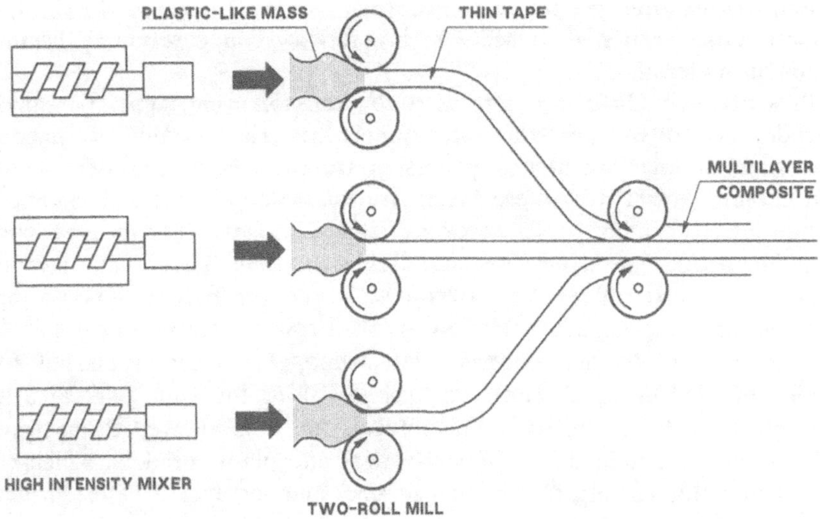

Figure 10.13. Tape calendering process for monolithic solid oxide fuel cell.[11]

entire tape has been formed on the bottom mold.[18] Corrugation has also been carried out by compression molding.[62] The corrugation step is followed by cutting, then stacking in the proper sequence and orientation to form the fuel cell stack. Following the corrugation process, the laminated layers may be compressed into a honeycomb-like structure, in which the layers are stacked to give a structure as in Fig. 10.12. This stack is then sintered at the desired temperature. This design is of the coflow type; i.e., fuel flows and airflows are in the same direction. Several coflow stacks have been fabricated for demonstration purposes. Gas supply to and from coflow stacks requires complex geometry if external manifolding is to be used. Internally manifolded designs may be less complex.

Another MSOFC stack design utilizes crossflow of fuel and air, which can be manifolded. Stacks with two, three, and four cells in a crossflow configuration have been successfully fabricated and tested. A single cell, with an active surface area of 14.5 cm^2, operated for over 100 h and achieved a power of 4.4 W.[18]

10.6.4. Planar Design

Whereas the MSOFC is made up of components which are sintered in one piece, planar designs which superficially resemble it are made by building up separately fired components. Planar designs of this type may therefore have the advantage of posing less technological risk than monolithic structures, even though their assembly may be more complex and costly. Planar SOFCs may also have some advantages over tubular designs, particularly in their potential for allowing a higher power density, and simpler, less costly methods of fabrication. They may be regarded as a compromise between tubular and monolithic SOFCs. Planar cells may be more suitable than tubular SOFCs for large scale (10–100 MW) on-site power plants, because they contain a lower mass of raw material, reducing cost, and since they are bipolar rather than having edge collection of current, the ohmic resistance of tubular cells will generally be larger than that of the planar cells.[63] In addition, large power plants with a tubular design will require a very large number of separate cells and electronic connections, which will increase cost compared with that of plants consisting of a small number of bipolar stacks. Planar cells are being developed at a number of R&D facilities.[64–69]

Planar cells have normally been fabricated by tape casting. Characteristic methods of preparation have been reported by Iwata *et al.*[70] Electrolyte layer slurries were prepared by mixing the organic binder (polyvinyl butyral), a solvent (toluene-2-propanol), a dispersant (fish oil), a plasticizer (dibutyl phthalate), a homogenizer (the surfactant Triton X-100), and yttria-stabilized zirconic powder in a ball mill. Using the doctor-blade method, the slurry was cast into tapes of thicknesses between 500 and 700 microns. These tapes were then cut into the sizes required. The green samples were first burnt out at 350°C to remove the organic compounds, and then calcined at 1500°C for about 2 h. The final samples were 300–500 microns thick.

To form the anode layer, ethanol was added to a mixture of nickel oxide and yttria-stabilized zirconia, which was ball-milled and calcined at 1500°C for 2 h to give a uniform powder. Ethanol slurries of this crushed Ni (35 vol%)–yttria-stabilized zirconia cermet powder were then prepared. The slurry was then coated on the electrolyte layer and sintered at 1500°C for 2 h. The final thickness of the anode layer was about 150–200 microns.

For the cathode layer, lanthanum oxide, strontium carbonate, and manganese carbonate powders were mixed in proportions to give the composition $(La_{0.7}Sr_{0.3})_{0.9}MnO_3$ with ethanol, ball-milled and calcined at 1000°C for 2 h. Lanthanum strontium manganite slurries were then prepared from the crushed powders. These slurries were coated on the electrolyte plates and calcined at 950°C for 2 h. Slurry coating was also carried out by a screen-printing process. The final thicknesses of the cathode layers were between 150–200 microns. The interconnector material (Sr-doped lanthanum chromite) can be sintered in air and usually tape-casted and fired separately in a reducing atmosphere and attached to the cell for planar SOFCs.

The method described above is general, and conditions of sintering temperature have been varied by different developers. In some cases, electrolyte layers have been prepared by CVD instead of tape casting. Similarly, the electrodes have been prepared by plasma-arc-spray techniques.[66] Single cells fabricated by the above methods have been mounted in alumina holders for electrical testing. Small-diameter cells have generally been found to show better performance than those which were larger. Singer *et al.*[66] achieved power densities at 1000°C of about 3.5 kW/m^2 using hydrogen and air as reactants, and up to 9 kW/m^2 with hydrogen and oxygen. Figure 10.14 shows two planar SOFC configurations with self-supported and substrate structures for multicell stacks.[71] In the self-supported structure, each side of the interconnect had channels for fuel and air, and the self-supported planar SOFCs were stacked between the interconnects. In the substrate structure, the cathodes and the anodes contained the gas channels, and flat interconnect and electrolyte layers were sandwiched between the anodes and cathodes (Fig. 10.14). Figure 10.15 shows the *I–V* characteristics of planar single cells with both types of configuration. The self-supported design was found to

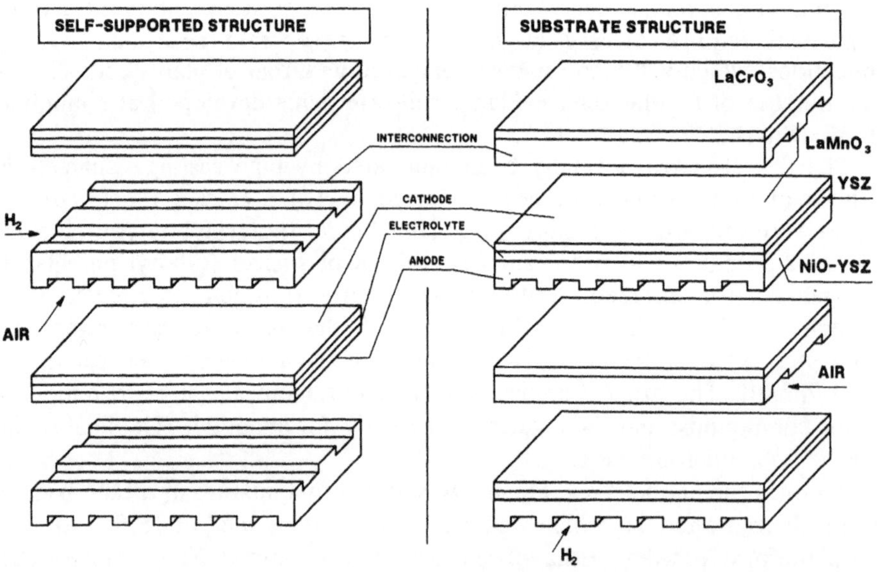

Figure 10.14. Planar SOFC components, showing self-supported and substrate structures.[71]

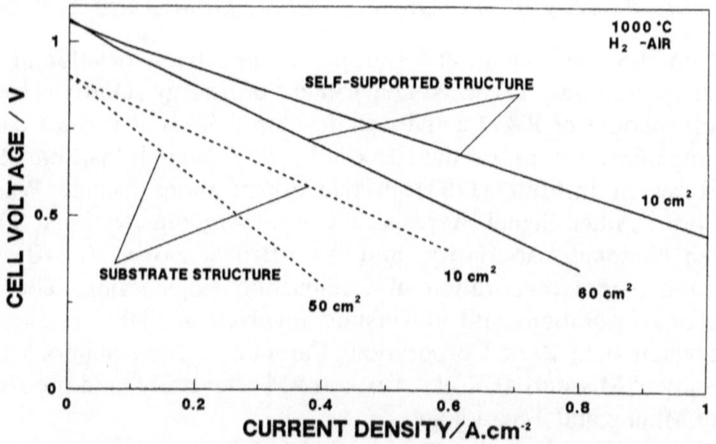

Figure 10.15. Current density versus cell voltage of single planar cells.[71]

have better performance since its *IR* drop was lower. The maximum power density for this design with hydrogen and air reactants was 0.48 W/cm^2 and 0.3 W/cm^2 at 0.6 A/cm^2 for active cell areas of 10 and 60 cm^2, respectively.

10.7. SYSTEM DEVELOPMENT

10.7.1. Cell Cooling

Cell cooling is normally carried out by using an excess flow of process air. The sensible heat contained in the effluent gases, combined with combusted anode tail-gas, is used to preheat the air entering the cell. In effect, this configuration integrates a high-temperature heat exchanger and the fuel cell stack, without the need for gastight seals in the case of the Westinghouse tubular design.

10.7.2. Contaminant Levels

SOFCs operate equally well on dry or humidified hydrogen and/or carbon monoxide fuel. Carbon dioxide has no effect on cell performance. Because of its high operating temperature, it requires no catalysis for internal reforming of steam–methane (i.e., steam–natural gas) or steam–methanol mixtures, and it can probably internally reform heavier hydrocarbons provided that steam-to-carbon ratios are sufficiently high to prevent fuel cracking and/or carbon deposition. Because of the latter problems, it cannot directly oxidize pure carbonaceous fuels. Moreover, its high operating temperature minimizes catalyst poisoning. A concentration of 50 ppm of hydrogen sulfide in the fuel lowers the operating cell potential by only about 5%; the cell potential reverts to the original voltage when the contaminant is removed from the fuel mixture. The sulfur tolerances are about one to two orders of magnitude higher than for other fuel cell types. SOFCs are thermodynamically more tolerant to S-containing fuel streams (H_2S or COS) than MCFC and PAFC systems.[72]

10.7.3. R&D Activity in the United States, Japan, and Europe

The United States, Japan, and Europe are all actively developing SOFC for commercial applications. The U.S. Department of Energy (DOE) is sponsoring a considerable amount of R&D activity to develop SOFCs at various corporations and other organizations, as are the Electric Power Research Institute (EPRI) and the Gas Research Institute (GRI). Active organizations include Westinghouse Electric Corp., Allied-Signal Aerospace Corp., Argonne National Laboratory, Brookhaven National Laboratory, and Asea-Brown Boveri (ABB) Corporate Research, the parent corporation of Combustion Engineering. There are also other smaller corporations and universities involved in SOFC research, namely Eltron Research Inc., Ztek Corporation, Ceramatec, Inc., Stanford University, the University of Missouri at Rolla, Texas A&M University, and the New Mexico Institute of Mining and Technology.

Westinghouse has been involved in SOFC R&D in order to commercialize tubular SOFCs for nearly three decades. Presently, tubular SOFC technology development is being supported at Westinghouse by DOE and GRI. Several 3-kW tubular SOFC power units have been constructed and tested for extended periods of time. Westinghouse has plans for the future fabrication and testing of 25-kW units in 1991, 100–200 kW units during 1992–1994, and 1–2 MW units in 1994–1996. Its plans also include development of a 10–50 MW power plant during the mid-1990s for the electric utility market. Allied-Signal Aerospace Company and Argonne National Laboratory are actively involved in the development of MSOFCs, concentrating on the fabrication and testing of larger multicell stacks and related materials research. Allied-Signal Aerospace R&D is supported by DOE, GRI, and Southern California Edison Company. Cerametec Inc. and Ztek Corporation are developing planar SOFCs. Other organizations are largely involved in materials research for the various varieties of SOFC.

Under Japan's "Moonlight Project," an SOFC feasibility study started during 1989. The Electrotechnical Laboratory and the National Chemical Laboratory for Industry are the national institutes leading Japanese SOFC R&D. Japanese universities involved include Yokohama National University, together with those at Mie, Kyoto, Tottori, and Kyushu. Manufacturing companies include Mitsubishi Heavy Industries and Mitsui Engineering and Shipbuilding, and potential users supporting the program are the Tokyo Electric Power Company (TEPCO), Tokyo Gas, and Osaka Gas.

European programs include those at Imperial College, London (electrode research, investigation of methanol oxidation), Twente University, the TNO Institute and the Delft University of Technology in Holland (development of EVD and CVD techniques for the deposition of the electrode and the electrolyte layers), ECN–Energy Center Netherlands, Siemens, ABB Corporate Research in Germany (planar SOFCs), and Denmark's Risø University National Laboratory (investigation of the oxidation of dry methane).

10.8. APPLICATIONS AND ECONOMICS

10.8.1. Potential Applications

SOFCs are certainly attractive for electric utility and industrial applications. The high operating temperature and tolerance to impure fuel streams make the

SOFC systems ideal for utilizing H_2 and CO from natural gas steam reforming and coal gasification plants, in which its high-quality waste heat can be used either in bottoming cycles and/or to provide all or part of the process heat for gasification. In principle, the latter could result in coal generating plants with unprecedented efficiencies in the 65% range. For large central station applications, the target is to reduce the capital cost of the entire fuel cell power plant to \$1000 to \$1500 per kilowatt, depending on the fuel used (natural or coal gas). Of this, the cost of the SOFC stack must be in the \$300 per kilowatt range.

Future application of the system must be seen in the context of the economics versus its potential competition, namely gas turbine combined-cycle systems and systems using other fuel cells, in particular MCFC. The latter are likely to be less efficient (perhaps 45% when combined with a coal gasifier), but they are likely to have lower capital cost than a unit using SOFCs. In consequence, the trade-off will involve the comparative effects of capital recovery and fuel cost on the overall cost of electricity. At least in first-generation plants, it is almost certain that efficiency of the SOFC stack will be sacrificed to reduce its capital by operation at a high current density so that the lowest cost of electricity, rather than the highest fuel savings, result. The MCFC may have a higher electrochemical efficiency than the SOFC, although its waste heat will be of a lower quality, meaning that it will be less valuable in providing process heat for coal gasification or natural gas reforming, and will result in a somewhat lower efficiency when used in a bottoming cycle. Thus, the efficiency of many SOFC and MCFC large generating plant designs may be comparable, the lower electrochemical efficiency of the former being offset by the higher efficiency of its bottoming cycle. The deciding factors will therefore be the capital cost, reliability, and lifetime of the competing technologies.

10.8.2. Research and Development Funding Structures

Recognition of the need for financial support of a national U.S. SOFC program gained momentum in the 1980s. The most advanced U.S. project (at Westinghouse) is cost-shared by the DOE, GRI, Westinghouse Electric Corporation itself, and two Japanese consortia. One of the latter consists of the New Energy Industrial Technology Development Organization (NEDO), Tokyo Gas, Osaka Gas, and Toho Gas, whose financial support to the developer is sent via GRI. The second consortium is composed of the NEDO, the Tokoku, Chugoku, and Kyushu Electric Power Companies, and the Electric Power Development Corporation. EPRI is responsible for managing the financial support of the second consortium. The goals of the Westinghouse project are (i) development and conceptual design of natural-gas-fueled SOFC cogeneration systems and (ii) evaluation of the technology for utility applications.

The U.S. Department of Energy and the Gas Research Institute also provide financial support, though at a considerably lower level, for MSOFC Development at Allied Signal–Aerospace Company and for the planar SOFC Development at Cerametec and Ztek Corporation. The monolithic and planar technologies are still at the bench-scale development stage. Financial support by the U.S. Department of Energy for materials R&D is still at a lower level. It includes advanced materials research at Argonne National Laboratory and at the University of Missouri at Rolla, studies on alternative materials at the Pacific

Northwest Laboratories, and the effect of fuel contaminants at Westinghouse Electric Corporation.

The NEDO in Japan supports R&D activities in universities and in industrial laboratories under cost-share arrangements. This includes work at the Electrotechnical Laboratory, Mitsubishi Heavy Industries, Chubu Electric Power Company, Tonen Corporation, and Hitachi. The R&D tasks focus on alternative materials, novel methods of fabrication, and new designs.

The European Economic Community is sponsoring the development of planar SOFC systems, ceramic component development, and testing at ECN in the Netherlands, stack design and development at Siemens in Erlangen, Germany, methods for the deposition of a thin electrolyte layer on supported air electrodes at the University of Twente, the Netherlands. GEC Alsthom Engineering Research Center and Imperial College, both in England, are program subcontractors.

10.8.3. Estimates of Research, Development, and Demonstration

The applications of SOFC power plants are most likely to be for power generation by electric and gas utilities and cogeneration by chemical and other industries. A less probable application is as a power source for transportation vehicles, in which its high operating temperature will be a disadvantage in vehicles with intermittent duty cycles. However, one may certainly envisage megawatt-class power plants for trains and ships and possibly units in the 100-kW class for intensively used buses and trucks, if these are sufficiently compact and lightweight. Thus, the most probable rated power levels for SOFCs will in the megawatt range. The U.S. Department of Energy's target for the capital cost of fuel cell power plants was \$500–\$1000 per kilowatt in 1989.[73] The fabrication processes for SOFCs are sophisticated compared with those for MCFCs and PAFCs; a cost of \$2850 per kilowatt for 200-kW PAFC system prototype units is currently being quoted by International Fuel Cells. In view of this, it can be estimated that a 1-MW prototype SOFC power plant will cost at least \$6–\$10 million. This projection may even be low, since the first 3-kW Westinghouse units using tubes from the pilot production line cost about \$250,000 each.

Compared with other fuel cells, the SOFC is unique in that it is a two-phase solid–gas system. This overcomes the complexities of flooding, drying out, electrolyte replenishment, or maintaining a three-phase zone within the electrode, which should make it particularly reliable. However, challenges still to be faced in its development are (i) cost-effective improved fabrication techniques for the thin cell component layers, (ii) alternative materials for interconnects and other cell components which will allow operation at lower temperatures (perhaps 700°C) to permit less performance degradation and a longer lifetime, equal to that of a normal industrial plant; essential criteria for such materials are the correct physical and chemical properties (electronic and ionic conductivity, electrocatalytic activity, thermal compatibility, and inertness to fuel contaminants); (iii) alternative designs such as the monolithic and planar SOFCs, which use less ceramic material per kW and offer *IR* drop; and (iv) methods for increasing the reliability of multicell stacks, e.g., by series–parallel connections.

Financial support is thus required to cover the spectrum of (i) materials selection and evaluation, (ii) development of cost-effective fabrication techniques, and (iii) engineering design and development of multicell stacks and power plants. A realistic estimate for the cost of a concentrated program for the development, demonstration, and performance evaluation of advanced 1-MW SOFC prototype power plants, with alternate stack designs to those presently used, will be \$100 to \$300 million over 10 years.

10.8.4. Projected Capital Cost and Cost of Electricity

Major advantages of high-temperature fuel cell power plants based on MCFC and SOFC systems are (i) the fuel processing and dc power generation steps can be efficiently integrated; (ii) noble metal electrocatalysts are not required; (iii) CO is a reactant, not a catalyst poison; and (iv) high-grade waste heat can be utilized for additional electricity generation or for thermal energy requirements. As semiconductor technology has advanced, the potential serial production cost of power conditioning subsystems has effectively decreased from about \$500 to \$100 per kilowatt during the last 10 years. The elimination of the conventional fuel processor in high-temperature fuel cell power plants and the lower cost of power conditioners make it reasonable to assume an ultimate capital cost of \$1000 per kilowatt for the SOFC power plant.

Assuming a capital recovery factor (CRF) of 15% per year and a power plant efficiency of 50% using natural gas at \$2.00/MBtu, then the electricity cost based on 7000 h per year availability will be about 35 mils/kW-h. Doubling the cost of natural gas will raise this to 49 mils/kW-h. To these production costs, the plant operating and maintenance (O&M) expenses must be added. These will be considerably less for SOFC power plants than for conventional power plants, which are about 30 mils/kW-h, because SOFC plants operating on natural gas will run unattended on a remote dispatch basis, which has already been proven by the PAFC. A conservative estimate for the cost of electricity from SOFC power plants may thus become about 55–70 mils/kW-h, which will be highly competitive with other technologies.

ACKNOWLEDGMENTS

The authors wish to express their appreciation to the Electric Power Research Institute and the Center for Space Power (CSP), Texas A&M University for sponsoring the solid oxide fuel cell project in our Center from July 1988 to August 1991.

REFERENCES

1. A. J. Appleby and F. R. Foulkes, *Fuel Cell Handbook* (Van Nostrand Reinold, New York, 1989).
2. J. T. Brown, *Assessment of Research needs for Advanced Fuel Cells*, DOE/AFCWG, 1985, p. 209.
3. W. Z. Nernst, *Electrochemistry* **6,** 141 (1900).
4. E. Baur and H. Z. Preis, *Electrochemistry* **43,** 727 (1937).
5. J. Weissbart and R. J. Ruka, *J. Electrochem. Soc.* **109,** 723 (1962).
6. D. H. Archer, R. L. Zahrandnik, E. F. Sverdrup, W. A. English, L. Elikan, and J. J. Alles, *Proc. Power Sources Conf.* 1965, p. 36.

7. E. V. Sverdrup, D. H. Archer, and A. D. Glasser, in *Fuel Cell Systems II* (R. F. Gould, ed.) Advances in Chemistry Series 90 (American Chemistry Society, Washington, D.C., 1969).
8. C. S. Tedmon, H. S. Spacil, and S. P. Mitoff, *J. Electrochem. Soc.* **116,** 1170 (1969).
9. F. J. Rohr, in *Proc. Workshop on High Temperature Solid Oxide Fuel Cells* (H. S. Isaacs, S. Srinivasan, and I. L. Harry, eds) 1977, p. 122.
10. W. G. Parker, in *Proc. 25th Intersociety Energy Conversion Engineering Conf.*, Vol. 3 (P. A. Nelson, W. W. Schertz, and R. H. Till, eds) (American Institute of Chemical Engineers, New York, 1990), p. 213.
11. N. Q. Minh, C. R. Home, F. S. Liu, D. M. Moffatt, P. R. Staszak, T. L. Stillwagon, in *Proc. 25th Intersociety Energy Conversion Engineering Conf.*, Vol. 3 (P. A. Nelson, W. W. Schertz, and R. H. Till, eds.) (American Institute of Chemical Engineers, New York, 1990), p. 230.
12. M. Dokiya, N. Sakai, T. Kawada, I. Yokoawa, T. Iwata, and M. Mori, in *Proc. First Int. Symp. on Solid Oxide Fuel Cells* (S. C. Singhal, ed.) (Electrochemical Society Inc. 1989), p. 325.
13. T. Takahashi, Y. Suzuki, K. Ito, and H. Hasegawa (American Ceramic Society, Washington D.C., May 1966).
14. H. U. Anderson, J. H. Kuo, and D. M. Sparlin, in *Proc. First Int. Symp. on Solid Oxide Fuel Cells* (S. C. Singhal, ed.) (Electrochemical Society Inc. 1989), p. 111.
15. B. L. Kuzin, A. N. Vlazov, *Sov. Electrochem.* **20,** 1510 (1984).
16. Y. Ohano *et al.*, *Solid State Ionics* **9&10,** 1001 (1983).
17. S. Srilomsak, D. P. Schilling, and H. U. Anderson, *Proc. First Int. Symp. on Solid Oxide Fuel Cells* (S. C. Singhal, ed.) (Electrochemical Society Inc. 1989), p. 129.
18. U. Balachandran, S. E. Dorris, J. J. Picciolo, R. B. Poeppel, C. C. McPheeters, and N. Q. Minh, *Proc. 24th Intersociety Energy Conversion Engineering Conf.*, Vol. 3 (P. A. Nelson, W. W. Schertz, and R. H. Till, eds.) (American Institute of Chemical Engineers, New York, 1989), p. 1341.
19. A Hammouche, E. Siebert, M. Kleitz, and A. Hammou, *Proc. First Int. Symp. on Solid Oxide Fuel Cells* (S. C. Singhal, ed.) (Electrochemical Society Inc., 1989), p. 265.
20. J. Lambert, W. J. Weber, and C. W. Griffin, *Proc. First Int. Symp. on Solid Oxide Fuel Cells* (S. C. Singhal, ed.) (Electrochemical Society, Inc., 1989), p. 141.
21. M. A. Priestnall and B. C. H. Steele, *Proc. First Int. Symp. on Solid Oxide Fuel Cells* (S. C. Singhal, ed.) (Electrochemical Society, Inc., 1989), p. 157.
22. S. S. Liou and W. L. Worrell, *Proc. First Int. Symp. on Solid Oxide Fuel Cells* (S. C. Singhal, ed.) (Electrochemical Society, Inc., 1989), p. 81.
23. M. Mogensen and J. J. Bentzen, *Proc. First Int. Symp. on Solid Oxide Fuel Cells* (S. C. Singhal, ed.) (Electrochemical Society, Inc., 1989), p. 99.
24. D. C. Fee and J. P. Ackerman, *Abstracts of the 1983 Fuel Cell Seminar* (Orlando, FL, 1983), p. 11.
25. R. Schmidberger, German patent 2,747,647/B/.22, (1979).
26. A. O. Isenberg and G. E. Zymboly, U.S. patent 4,582,766. 15 (1986).
27. A. O. Isenberg, U.S. patent 4,597,170. 1, (1986).
28. A. O. Isenberg, U.S. patent 4,702,971. 27, (1987).
29. D. W. Dees, T. D. Claar, T. E. Easler, D. C. Fee, and F. C. Mrazek, *J. Electrochem. Soc.* **134,** 2141 (1987).
30. G. E. Pike and C. H. Seager, *J. Appl. Phys.* **48,** 5152 (1977).
31. G. Ondracek and B. Schulz, *Rev. Int. Hautes Temp. Refract.* **7,** 397 (1970).
32. G. E. Pike and C. H. Seager, Report No. SAND-76-0558, (Sandia National Laboratory, 1977).
33. G. E. Pike, Report No. SAND-77-1042C, (Sandia National Laboratory, 1979).
34. C. A. Neugebauer, *Thin Solid Films* **6,** 443 (1970).
35. J. Gasperic and B. Navensik, *Thin Solid Films* **36,** 353 (1976).
36. J. C. Garland and D. B. Tanner, ed., *Electrical Transport and Optical Properties of Inhomogeneous Media,* (AIP Conf. Proceedings, no. 40, American Institute of Physics, New York, 1978).
37. T. E. Easler, B. K. Flandermeyer, T. D. Claar, D. E. Busch, R. J. Fousek, J. J. Picciolo, and R. B. Poppel, *1986 Fuel Cell Seminar Abstracts* (Courtesy Associates, 1986), p. 72.
38. T. Takahashi and H. Iwahara, *Mater. Res. Bull.* **13,** 1447 (1978).
39. H. L. Tuller and A. S. Nowick, *J. Electrochem. Soc.* **122,** 255 (1975).
40. T. Kudo and H. Obayashi, *J. Electrochem. Soc.* **123,** 415 (1976).
41. A. E. McHale and H. L. Tuller, *Solid State Ionics* **5,** 515 (1984).
42. G. M. Choi, H. L. Tuller, and J. S. Haggerty, *J. Electrochem. Soc.* **136,** 835 (1989).
43. H. L. Tuller, in *Non-Stoichiometric Oxides* (O. T. Sorensen, ed.) (Academic Press, New York, 1981), p. 271.

44. A. F. Sammells, R. L. Cook, D. J. Kuchynka, *Abstracts 1990 Fuel Cell Seminar,* 1990, p. 119.
45. T. Kodo and H. Obayashi, *J. Electrochem. Soc.* **123,** 415 (1976).
46. J. B. Webb, M. Sayer, and A. Mansingh, *Can. J. Phys.* **55,** 1725 (1977).
47. D. P. Karim and A. T. Aldred, *Phys. Rev. B* **20,** 2255 (1979).
48. J. Jacobs, K. M. Castelliz, W. Manuel, and H. W. King, in *Proc. Conf. on High Temperature Sciences Related to Open-Cycle, Coal-Fired MDH Systems* (Argonne National Laboratory, Argonne, IL, 1977), ANL-77-21, p. 148.
49. J. W. Halloran and H. U. Anderson, *J. Am. Ceram. Soc.* **57,** 150 (1974).
50. S. C. Singhal, R. J. Ruka, and S. Sinharoy, DOE/MC/21184-1 (Westinghouse R&D Center, Pittsburgh, PA, 1985).
51. B. K. Flandermeyer, M. M. Nasrallah, D. H. Sparlin, and H. U. Anderson, in *Transport in Nonstoichiometric Compounds* (G. Simkovich and V. S. Stubican, eds.) (Plenum, New York, 1984), p. 87.
52. L. A. Chick, J. L. Bates, L. R. Pederson, and H. E. Kissinger, in *Proc. First Int. Symp. on Solid Oxide Fuel Cells,* (S. C. Singhal, ed.) (Electrochemical Society Inc., 1989), p. 170.
53. A. O. Isenberg, *National Fuel Cell Seminar Abstracts* 1982, p. 154.
54. U. B. Pal and S. C. Singhal, *National Fuel Cell Seminar Abstracts* 1989, p. 41.
55. A. O. Isenberg, in *Proc. Symp. on Electrode Materials and Processes for Energy Conversion and Storage* (The Electrochemical Society Inc., 1977), p. 572.
56. A. O. Isenberg, *Abstracts, National Fuel Cell Seminar,* (Sandiego, CA, 1980), p. 135.
57. A. O. Isenberg, *Solid State Ionics* **3,** 431 (1981).
58. U. B. Pal and S. C. Singhal, *Proc. Sixth Int. Conf. on High Temperatures—Chemistry of Organic Materials* (Gaithersburg, MD, 1989).
59. A. O. Isenberg, *Abstracts, National Fuel Cell Seminar,* 1983, p. 78.
60. D. C. Fee, R. K. Steunenberg, T. D. Claar, R. B. Poppel, and J. P. Ackerman, *Abstracts 1983 Fuel Cell Seminar,* 1983, p. 74.
61. K. A. Murugesamoorthi, S. Srinivasan, D. L. Cocke, and A. J. Appleby, *Proc. Intersociety Energy Conversion Engineering Conf.* (P. A. Nelson, W. W. Schertz, and R. H. Till, eds.) (American Institute of Chemical Engineers, New York, 1990), p. 246.
62. N. Minh, F. Liu, P. Staszak, T. Stillwagon, and J. Van Ackeren, *Abstracts 1988 Fuel Seminar,* 1988, p. 105.
63. M. Dokiya, N. Sakai, T. Kawada, H. Yokokawa, T. Iwata, and M. Mori, *Proc. 24th Intersociety Energy Conversion Engineering Conf.* (P. A. Nelson, W. W. Schertz, and R. H. Till, eds.) (American Institute of Chemical Engineers, New York, 1989), p. 1547.
64. T. Hoshina, T. Yoshida, and S. Sakurada, *Abstracts 1990 Fuel Cell Seminar,* 1990, p. 516.
65. J. P. P. Huijsmans, E. J. Siewers, and S. B. van der Molen, *Abstracts 1990 Fuel Cell Seminar,* 1990, p. 512.
66. R. F. Singer, F. J. Rohr, and A. Belzner, *Abstracts 1990 Fuel Cell Seminar*, 1990, p. 111.
67. M. Hsu and H. Tai, *Abstracts 1990 Fuel Cell Seminar,* 1991, p. 115.
68. E. Ivers-Tiffee, W. Wersing, and B. Reichelt, *Abstracts 1990 Fuel Cell Seminar,* 1990, p. 137.
69. A. C. Khandkar, K. L. Stuffle, and S. Elangovan, *Proc. First Int. Symp. on Solid Oxide Fuel Cell* (S. C. Singhai, ed.) (Electrochemical Society Inc., 1989), p. 377.
70. T. Iwata, N. Sakai, T. Kawada, H. Yokokawa, and M. Dokiya, *Proc. 25th Intersociety Energy Conversion Engineering Conf.* (P. A. Nelson, W. W. Schertz, and R. H. Till, eds.) (American Institute of Chemical Engineers, New York, 1990), p. 235.
71. K. Koseki, N. Kusunose, K. Shimizu, H. Shundo, T. Iwata, S. Maruyama, and T. Nakanishi, *Abstracts 1990 Fuel Cell Seminar,* 1990, p. 107.
72. L. G. Marianowski, Presentation made at a meeting of the DOE/AFCWG, July 27, 1984.
73. G. Hagey, *Proc. 1st Int. Fuel Cell Workshop on Fuel Cell Technology Research and Development* (M. Watanabe, P. Stonehart, and K. Ota, eds.) (1989), p. 3.
74. T. J. George, M. J. Mayfield, DOE/METC-90/0268 (1990).
75. N. J. Maskalick, *Proc. First Int. Symp. on Solid Oxide Fuel Cell* (S. C. Singhal, ed.) (Electrochemical Society Inc., 1989), p. 279.

11

Research, Development, and Demonstration of Solid Polymer Fuel Cell Systems

David S. Watkins

11.1. INTRODUCTION

Solid polymer fuel cells (SPFC) go by many names. General Electric (GE), the original developers, referred to ion exchange membrane (IEM) Fuel Cells in the early stages. Later they were given the GE (now United Technologies Corporation/Hamilton Standard) trademark name SPE for solid polymer electrolyte. Recently, the name proton exchange membrane (PEM) fuel cells has been used by many developers. This chapter will use SPFC.

By any name, these fuel cells all use a solid polymer membrane as their electrolyte. Figure 11.1 shows the basic operation of these acid type fuel cells. The solid membrane, from 50 to 250 microns thick, provides an electrolyte capable of withstanding high-pressure differentials with no free corrosive liquids. Besides functioning as an acid electrolyte, the membrane separates the fuel gas (hydrogen or a hydrogen rich mixture) from the oxidant gas (oxygen or air).

Today's membranes are solid, hydrated sheets of a sulfonated fluoropolymer similar to Teflon. The acid concentration of the membrane is fixed and cannot be diluted by product or process water. The acid concentration of a particular membrane is characterized by equivalent weight, EW (g dry polymer/mole of ion exchange sites). This number is the reciprocal of the ion exchange capacity in moles per gram. Generally, lower EW and thinner membranes result in higher cell performance. However, thinner membranes result in higher parasitic cross-diffusion of reactant gas.

Typical cell hardware consists of the membrane with electrodes attached and, cell, cooling, and end plates in a plate and frame or filter press arrangement. The

David S. Watkins • Ballard Power Systems Inc., 980 West 1st Street, North Vancouver, B.C., Canada V7P 3N4.

Fuel Cell Systems, edited by Leo J. M. J. Blomen and Michael N. Mugerwa. Plenum Press, New York, 1993.

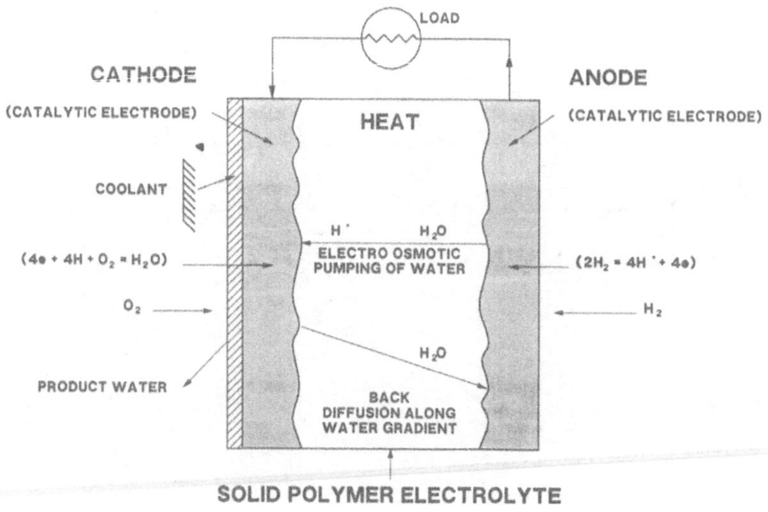

Figure 11.1. SPFC schematic.

cell plates are metal, graphite, or graphite composites. Metals have included titanium, zirconium, and niobium. The plates have embossed, machined or molded grooves which distribute gas over the electrode surfaces from an internal perimeter manifold and, in some designs, provide paths for the removal of the liquid product water. The water is liquid because of the low operating temperature (<100°C) of the SPFC. Other designs have used wicking arrangements to remove the liquid water.

Effective product water removal is required to prevent flooding of the electrode which prevents access of gas to the catalyst–membrane interface where the reactions take place. At the cathode or air electrode, this process is a countercurrent and competing process where oxygen moves toward the interface and product water moves away. Flooding hampers the rate of mass transfer and results in poor cell performance, characterized by the inability to maintain high current at a given cell voltage.

Heat removal is accomplished using cooling plates with air or liquid coolants passed through the cell hardware. Some low-power cell designs have used evaporation of the product water and removal of the resulting vapor to control temperature.

Advantages and disadvantages of SPFC can be summarized as follows:

Advantages	*Disadvantages*
High efficiency	Low-quality waste heat
Solid noncorrosive electrolyte	Sensitive to CO*
Low temperature, rapid start-up	High catalyst loading*
High power density	High membrane cost*
Insensitive to differential pressures	Few membrane suppliers*
No carbonate formation	
Long life	
Potable liquid product water	

*Historically. Now being addressed successfully.

Versatility of application
Wide range of power covered
Simple design
Ease of manufacture
Many construction materials compatible
Low-temperature operation
H_2 and reformed fuels

11.2. HISTORY

11.2.1. 1959–1982

The biggest effort in SPFC was conducted by GE during 1959–1982.[1] However, Engelhard produced some small cells of about 30 W during the late 1970s and early 1980s.

Performance ranges on H_2–O_2 for five different membranes between 1959 and 1980 are shown in Table 11.1.[2] In 21 years, cell performance had jumped by 160-fold from 5 to 800 W/ft^2 (5.4–861 mW/cm^2). Cell sizes had been increased from a few square centimeters to 1000 cm^2 (1.1 ft^2). Stack sizes had gone from watts to kilowatts.

The first SPFC used in an operational system was the GE-built 1-kW Gemini power plant shown in Fig. 11.2.[1] Two, 1-kW modules provided primary power for each of the seven Gemini spacecraft during the early 1960s. Each module could provide the full mission power requirements. Figure 11.3[1] shows the two power plants installed in the Gemini spacecraft. Performance and life of the Gemini fuel cells was limited due to the polystyrene sulfonic acid membrane used at that time.

The second GE SPFC unit was the 350-W module which powered the Biosatellite spacecraft in 1969. An improved Nafion membrane manufactured by DuPont was used as the electrolyte. Figure 11.4 shows the structural characteristics of Nafion.[2] Performance and life has significantly improved since Nafion's introduction in 1968. Demonstrated cell life ranges with four different membranes are shown in Fig. 11.5.[2] With commercial Nafion 120, lifetimes of over 50,000 h have been achieved. The generally accepted degradation mechanism, determined by GE, is shown in Fig. 11.6.[2] The result is a HO_2^- radical which attacks the membrane. As shown in Fig. 11.6, metal ions such as Fe^{2+} catalyze the effect.

Table 11.1. Power Densities Attained for SPFC on H_2–O_2

Time	Membrane	Power density (W/ft^2)
1959–1961	Phenol sulfonic	5–10
1962–1965	Polystyrene sulfonic	40–60
1966–1967	Polytrifluorostyrene sulfonic	75–80
1968–1970	Nafion experimental	80–100
1971–1980	Nafion production	600–800

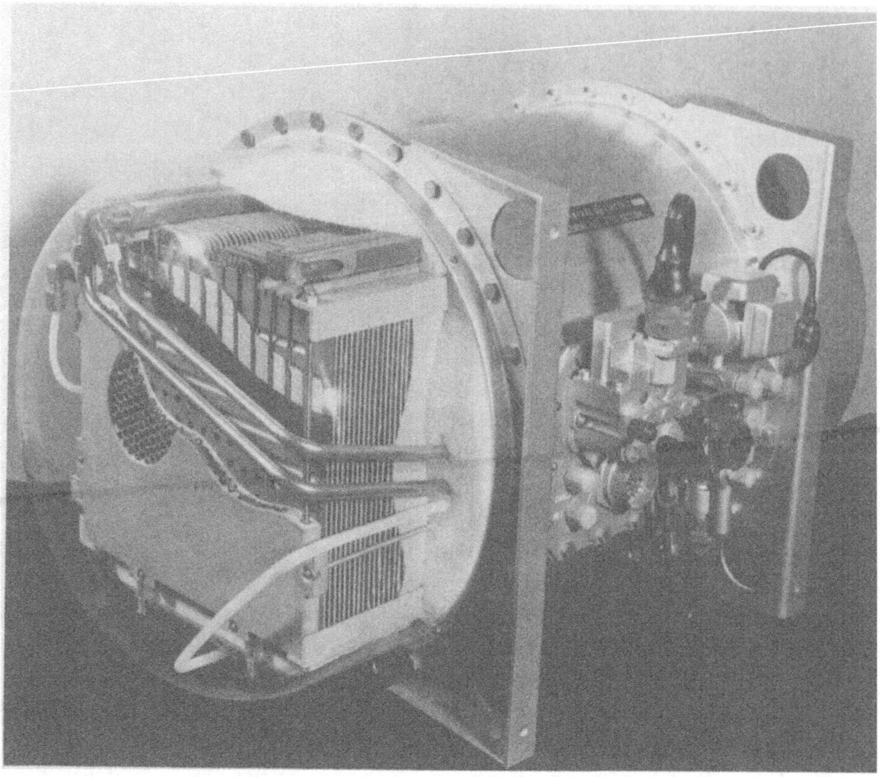

Figure 11.2. GE 1-kW Gemini SPFC module.

Figure 11.7[1] shows a 3-kW flight-qualified module developed by GE for a Navy high-altitude balloon program. This unit exhibited an energy density more than five times that of the Gemini fuel cells. More advanced H_2–O_2 regenerative fuel cell–electrolyzer breadboard hardware was delivered to NASA by GE in early 1983.

Besides systems using H_2–O_2, a 44-kWh (5 W continuous with short 500-W peaks) self-contained sonobuoy SPFC power supply was built by GE for the U.S. Navy. This system used sodium–aluminum hydride as the source of H_2 and sodium chlorate as the source of O_2. Another solid-fueled 12-W device was built by GE for an army transceiver. Figure 11.8[1] shows a picture of the hardware. The unit used pellets of LiH or CaH_2 reacted with water to provide H_2. Ambient air was used as the oxidant. Engelhard also developed cells fueled by chemical hydrides.[3]

With respect to hydrocarbon fuels, GE developed a prototype 1.5-kW power plant in the 1966–1968 period which was designed to run on JP-4 jet aircraft fuel. Further funding for the system was not available, and the unit was not pursued. A photograph of the power plant is shown in Fig. 11.9.[1]

In 1981 GE did an important study on an indirect (steam-reformed) methanol–air power plant for transportation applications for Los Alamos National Laboratory (LANL) in the United States.[4] This study concluded that SPFC was a candidate to power a compact passenger car. The GE–LANL study identified economics as the deterring factor. The key areas for cost

Figure 11.3. Gemini space capsule.

NAFION

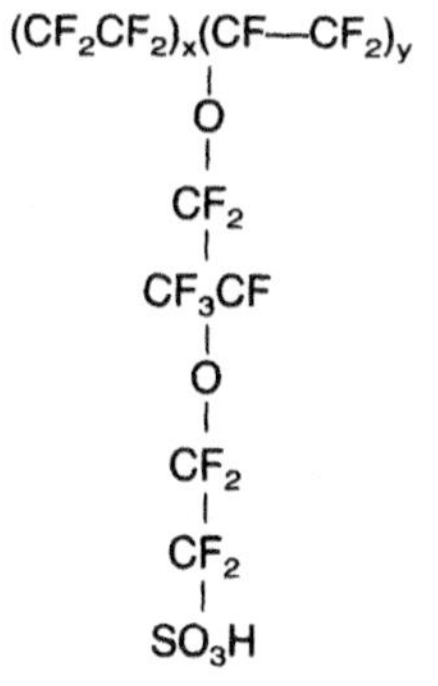

Figure 11.4. Structural characteristics of DuPont's Nafion membrane.

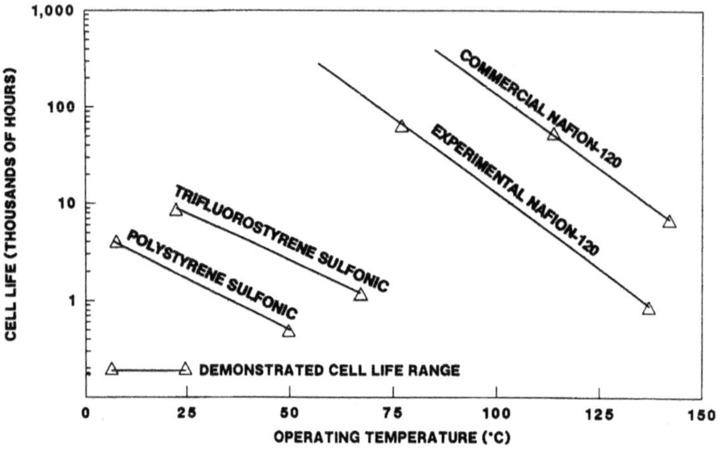

Figure 11.5. SPFC life capability.

saving (in dollars per kilowatt) were catalyst loading and membrane cost. A relatively small follow-on contract to pursue reduction of catalyst and membrane cost was completed,[5] but the aggressive targets laid out in the 1981 study were not attained.

11.2.2. 1982 to Present

In the mid-1980s, GE SPE technology (fuel cells and electrolyzer) was transferred to UTC–Hamilton Standard, where it resides now.[6] Before this, a license for the basic know-how for SPFC was transferred from GE to Siemens AG in Germany during 1983–1984.[7] The Siemens application was a power source for an air-independent submarine.

Besides Siemens and UTC–Hamilton Standard, other organizations began work on SPFC. In late 1983 Ballard Technologies Corporation in Canada (now

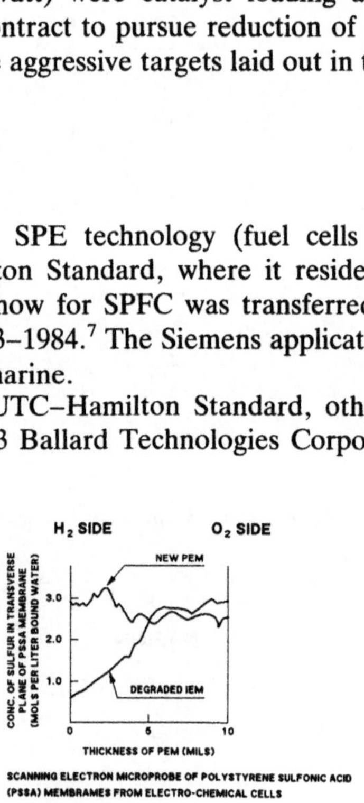

DEGRADATION MECHANISM

STEP 1: $H_2 \xrightarrow{Pt} 2H\cdot$

STEP 2: $H\cdot + O_2$ (DIFFUSED THRU PEM) $\rightarrow HO_2\cdot$

STEP 3: $HO_2\cdot + H\cdot \rightarrow H_2O_{2(l)}$ (WHICH CAN DIFFUSE INTO PEM) AT LOCUS NEAR DEGRADED FRONT IN PEM.

STEP 4: $H_2O_2 + M^{++}$ (Fe^{++} & Cu^{++} ALWAYS FOUND IN PEM'S) $\rightarrow H^{+}M^{+3} + H_2O + \cdot OH$

STEP 5: $\cdot OH + H_2O_2 \rightarrow H_2O + HO_2\cdot$ (ATTACKS PEM)

Figure 11.6. SPFC membrane degradation mechanism.

Figure 11.7. GE 3-kW SPFC module.

Ballard Power Systems) began developing SPFC technology with funding support from the Canadian Department of National Defense. Shortly thereafter Ergenics Power Systems Inc. in the United States began using SPFC technology acquired from A. F. Sammer Corp., which used technology developed at Engelhard. Treadwell in the US also began work. Fundamental SPFC research and studies at LANL accelerated, with LANL being the greatest advocate of SPFC during the years. In addition, Dow Chemical was developing a membrane for chloroalkaline cells which could be used in an SPFC.

In the mid- to late 1980s organizations such as International Fuel Cells (IFC, a joint venture between UTC and Toshiba), General Motors (Delco Remy, GM Research), Giner Inc., and Texas A&M (all in the United States) and, DeNora (in Italy) began work. DeNora is developing a 10-kW (50-V, 200-A) SPFC for delivery in 1991 with support from ENEA.[8] In 1990 Analytic Power of the United

Figure 11.8. GE 12-W LiH SPFC power unit.

States was added to the list. At the time of this writing (March, 1991) the author has heard of low-level efforts in Japan, but information is scarce. Billings in the U.S. reports developing SPFC hardware, but no technical papers reviewing their work have been published.

Significant developments after 1982 have included the demonstration of the ability to run SPFC on CO-containing fuel gas by using a selective oxidation process after the fuel reformer to convert the CO to CO_2 external to the cell (Ballard, 1985),[9] a two-and-one-half- to threefold improvement in maximum power density from 800 to 2000 and 2500 W/ft^2 by using a new Dow membrane (Ballard, 1987 and 1988),[10–12] the attainment of low catalyst loadings from 8 mg Pt/cm^2 total (both anode and cathode) to $<$1 mg Pt/cm^2 by using improved electrode structures (LANL, 1987–1988),[13–18] the development of advanced passive product water removal (wicking) SPFC hardware (Ergenics–International Fuel Cells, 1987–1988),[19,20] demonstration of a fuel-cell-powered commercial submarine (Ballard–Perry Energy Systems, 1988–1989),[21] and the demonstration of "methanol in" and "electricity out" with an independent methanol–air SPFC brassboard power plant (Ballard, 1990).

11.3. RECENT RESEARCH

The main areas of research on SPFC have been in CO conversion processes and CO-tolerant catalysts to allow low-temperature SPFC operation on reformed

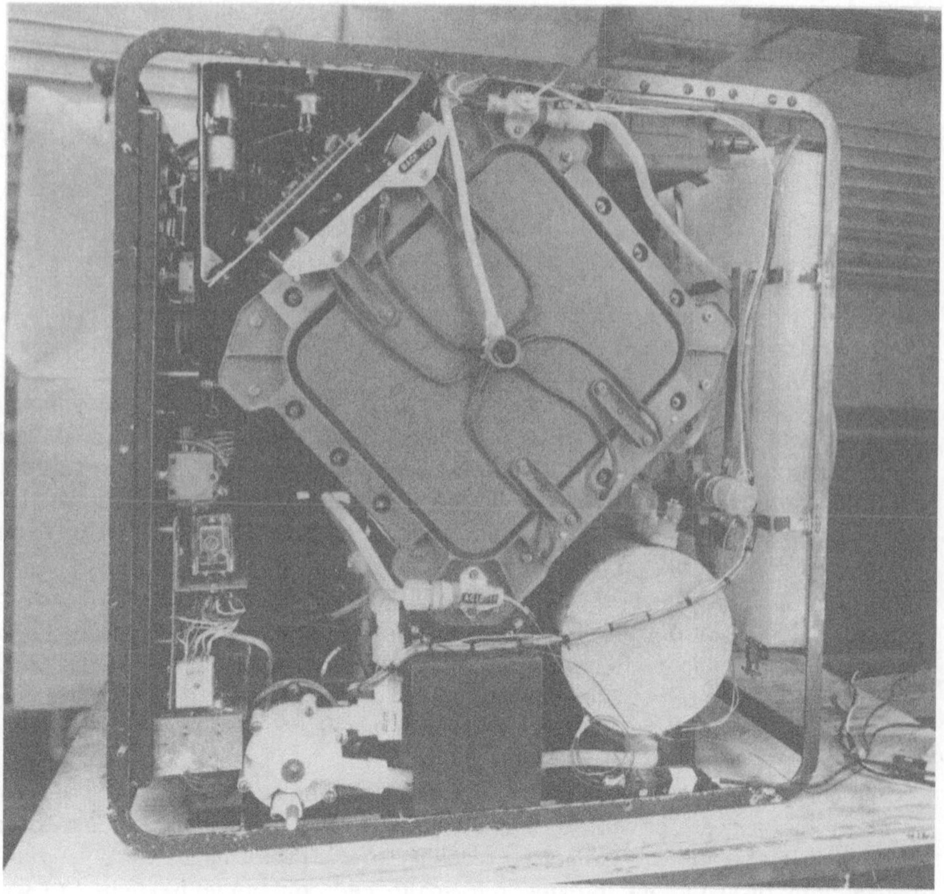

Figure 11.9. GE 1.5-kW hydrocarbon–Air SPFC power plant.

fuels, membranes for high power and low cost, and low catalyst loadings for low cost.

11.3.1. Tolerance to Reformed Fuels

The SPFC is being developed for transportation applications such as automobiles and buses. Liquid methanol would likely be a good fuel for this application. When methanol is steam-reformed a gas containing about 75% H_2, 24% CO_2, and 0.3 to 1% CO results. This fuel gas, when fed to the low-temperature (<100°C) SPFC, "poisons" the Pt anode catalyst, causing dramatic decreases in performance. Alloy catalysts, such as that developed by GE–Hamilton Standard[5] and Giner,[22] tend to reduce the effect. For instance, Giner catalysts have shown good tolerance to levels as high as 60 ppm CO with polarizations of about 20 mV compared to pure H_2 at 200 mA/cm^2 in H_2–H_2 test cells. However, performance approaching pure H_2 on reformate-containing fractions of a percent CO has not been obtained with these catalysts.

In 1985 Ballard used an external selective oxidation process to convert the CO to CO_2.[9] Since then the process has been adopted by LANL and GM.[23,24] Briefly, reformate is conditioned by mixing it with a small amount of air and then

Table 11.2. Ballard SPFC Reformate–Air 12-Cell Stack Performance (30 psig, 85°C, Nafion 117 membrane, 400 A/ft^2)

Fuel gas	Voltage (V)	Current (A)	Percent of H_2–air
H_2	8.22	20.6	100
H_2–25% CO_2	8.05	20.6	98
Conditioned Reformate	7.94	20.6	97

passing it through a column packed with 0.5% Pt on alumina. The selective oxidation can yield residual CO in the 2–100 ppm range. In the first Ballard demonstration, effluent from the column, containing from 2 to 10 ppm CO was fed directly to a single cell operating at about 55°C and atmospheric pressure. The resulting cell performance was about 95% of that on pure H_2. Feed to the column was 0.3% CO, 25% CO_2, balance H_2.

Table 11.2 shows more recent 12 cell (50 cm^2/cell) stack test results by Ballard. Air utilization was 50% and H_2 utilization was just over 90% (10% excess). In the table conditioned reformate means reformate which has been passed through a catalytic column designed to convert CO to CO_2 prior to being fed to the cell. Again the feed to the column was 0.3% CO, 25% CO_2, balance H_2.

In cells without alloy catalysts, CO at the 100 ppm level in the fuel gas causes severe loss of cell performance. An alternative to alloy catalysts has been developed by LANL.[25,26] Complete recovery of H_2 performance by "bleeding" 2% O_2 into the CO-contaminated reformate feed containing 100 ppm CO was achieved. Results are shown in Fig. 11.10.[25] This technique is seen to enhance the use of selective oxidation similar to CO-tolerant alloy catalysts. For instance, selective oxidation can be used to convert the bulk of the CO, and bleeding is used to convert the residual amounts preventing low-level CO anode polarization. Alternatively, alloy catalysts can be used to prevent the low-level CO polarization

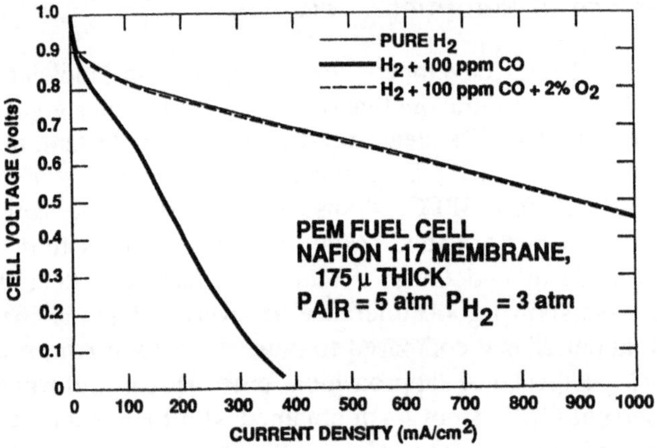

Figure 11.10. LANL SPFC performance on H_2 vs. H_2/100 ppm CO.

effects. If the entire CO oxidation process were done in the cell at the membrane–electrode assembly, the heat produced would place undue stress on the membrane, resulting in cell failure.

11.3.2. Membranes

The most significant membrane development since the introduction of Nafion by DuPont in 1968 has been a new perfluorosulfonic acid membrane, with a shorter side chain, made by Dow Chemical.[27] The structural characteristics of the Nafion and Dow membranes are compared in Fig. 11.11.[2]

Dow has developed a family of functional ionomers that can be synthesized to lower equivalent weights than the DuPont product and still maintain physical strength. The Dow material is prepared by the copolymerization of tetrafluoroethylene with a vinyl ether monomer. The polymer can be described as having a Teflon-like backbone structure with a side chain attached via an ether group. The side chain is characterized by a terminal sulfonate functional group. Properties of the material relating to fuel cell performance are believed to be related to side-chain length.[28,29] As shown in Fig. 11.11, the Dow side chain is shorter.

Fuel cell research at Dow Chemical is directed toward optimizing Dow's perfluorosulfonic acid membrane for use in both O_2 and air-breathing systems. The primary focus has been to concentrate on correlating physical and mechanical properties with polymer synthesis and membrane fabrication. Characteristics such as ion exchange capacity (EW), hydration, and the effects on such properties as mechanical strength and ionic conductivity are of concern if a material is to be suited for application. To date, research and development efforts have produced membranes which have uniaxial and biaxial strength as demonstrated by tensile and dynamic mechanical analysis, as well as ionic conductivities in some cases a factor of 2 greater than existing art.[28] Other pertinent parameters useful in understanding fuel cell performance are also under investigation, including water dynamics, fuel and oxidant humidification, electrode hydrophobicity, operating temperature and pressure, and flow rates and are being related to membrane hydration characteristics.

Testing of the Dow membrane at Ballard in 1987 and 1988 resulted in

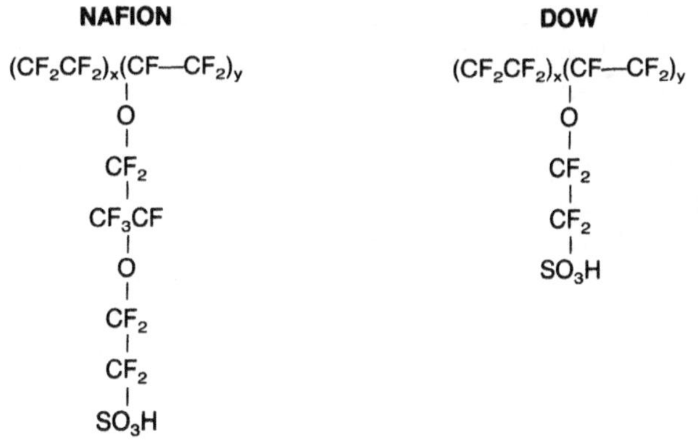

Figure 11.11. Structural characteristics of DuPont's Nafion and Dow membranes.

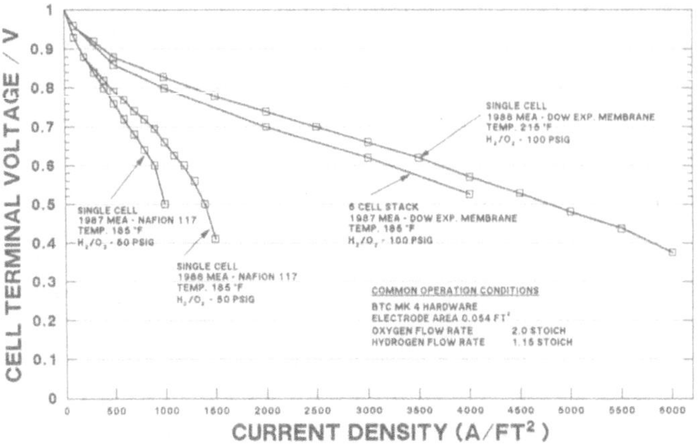

Figure 11.12. Ballard performance improvements on H_2–O_2.

dramatic increases in SPFC performance levels.[10–12] Power densities as high as 2.5 and 1.4 kW/ft^2 were achieved on H_2–O_2 and H_2–air, respectively.

Figure 11.12 compares 1987–1988 best single-cell performance of Nafion 117 in the MK IV Ballard hardware versus that of the Dow experimental membrane (DEM) both in a six-cell stack and in a single cell. The Nafion membrane has an EW of 1100, whereas the DEM had an EW of about 800. Wet thickness of the DEM membrane was about 100 microns, whereas the Nafion was about 230 microns. This DEM is the best performer of a number of versions tested. Other variations include different thickness and EW.

The first curve in Fig. 11.12 shows that performance of Nafion 117 under 50 psig H_2/O_2 at 85°C coolant outlet was 0.5 V at 1000 A/ft^2 (ASF) of active electrode area. Cell active area was 0.054 ft^2 (50 cm^2) for the MK IV. The second curve shows that, under similar conditions, best 1988 performance had increased to 0.5 V at 1400 ASF. In comparison, the third curve shows that 1987 DEM performance in a six-cell MK IV stack was an average of 0.53 V/cell at 4000 ASF. This stack was run for 20 h at 4000 ASF before the polarization curve data were taken. Recently, these results have been obtained by Siemens with similar cell size under similar operating conditions.[7] Previous to the six-cell Ballard stack, a single cell had been operated for 115 h at 3900 ASF with no signs of performance decay. The fourth curve in Fig. 11.12 shows increased performance of the same DEM under elevated temperatures and pressures. Performance as high as 0.5 V at 5000 ASF was achieved.

Figure 11.13 shows H_2–air performance of the DEM under different pressures and temperatures in the Ballard MK IV hardware. The first curve shows that under 50 psig and 71°C, performance was 0.69 V at 1000 ASF and 0.50 V at 1900 ASF. Under similar conditions, average Nafion 117 performance is about 0.7 V at 500 ASF. The second curve shows that under 100 psig and about 100°C, DEM performance was 0.71 V at 1200 ASF and 0.50 V at 2800 ASF.

Figure 11.14[30] shows work by Siemens AG with three membranes. Two are versions of Nafion which differ in thickness but not EW, and the other is an unspecified version of the Dow membrane. Reactants were H_2–O_2 at about 2 bar. Cell size was large at 1180 cm^2 (1.3 ft^2). At the reference voltage of 0.684 V,

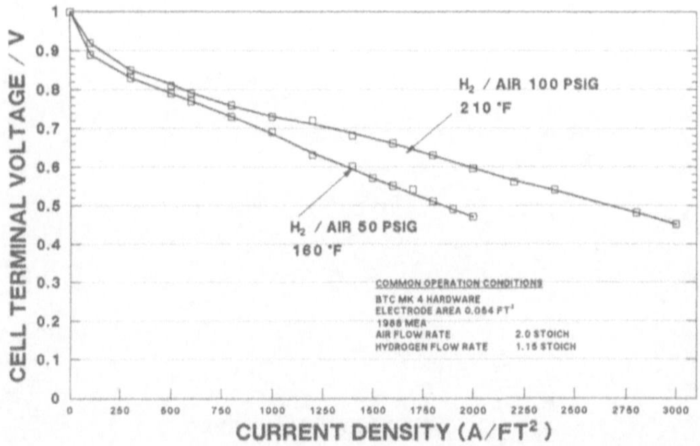

Figure 11.13. Ballard H_2–air performance using Dow membrane.

Nafion 115 increased current capability by over 40% compared to Nafion 117. The Dow membrane increased current capability by twofold.

Figure 11.15 shows testing by Ballard with three membranes on H_2–air. As with the Siemens work, two are versions of Nafion which differ in thickness but not EW, the other is the current "commercial" version of the Dow membrane. H_2–air reactants were at 50 psig. Cell size is 232 cm^2 (0.25 ft^2).

Testing of different membranes has also been done by LANL and Texas A&M. Several Nafion membranes with 1100 EW but with varying thickness from 175 to 50 microns are compared to a 125-micron lower EW Dow in Fig. 11.16.[31] The cell with the Dow membrane showed the best result. The cell with the 50-micron Nafion showed the next best result. However, the open-circuit potential in this cell was 50–100 mV lower than in the cells with the others. It is thought that this effect may be due to excessive cross-diffusion of reactant gases.

Fundamental studies of both the Dow and DuPont membranes are being

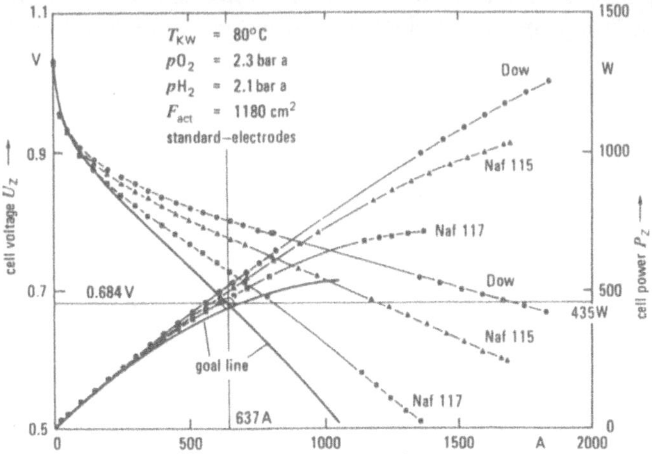

Figure 11.14. Siemens H_2–O_2 performance using Nafion and Dow membranes.

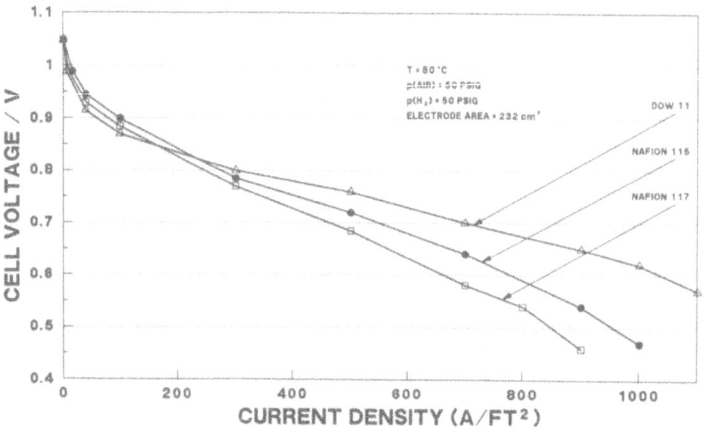

Figure 11.15. Ballard H_2–air performance using Nafion and Dow membranes.

undertaken by organizations such as General Motors Research Laboratories, Los Alamos National Laboratories, and Texas A&M. Membrane properties are being determined and mathematical models are being developed. A proper evaluation of this work is beyond the scope of this chapter. Instead, a sampling of this type of work is listed in the references.[14,32–34]

Besides the U.S. membranes, membranes made in Japan have been evaluated in SPFC. For instance, a membrane made by Chlorine Engineers has been tested by LANL.[14] Performance of a 900-EW, 125-micron membrane appears similar to DuPont's Nafion.

11.3.3. Catalyst Reduction

High energy efficiencies and high power densities are necessary to minimize weights, volumes, and capital and operating costs of fuel cells for marine, space, and terrestrial applications, such as cited and transportation-type power plants. Until recently, SPFC have used noble metal electrodes with loadings 10 times higher (4 mg Pt/cm^2) than those for PAFC. Methods resulting in tenfold catalyst reductions by using improved electrode structures were developed by LANL in

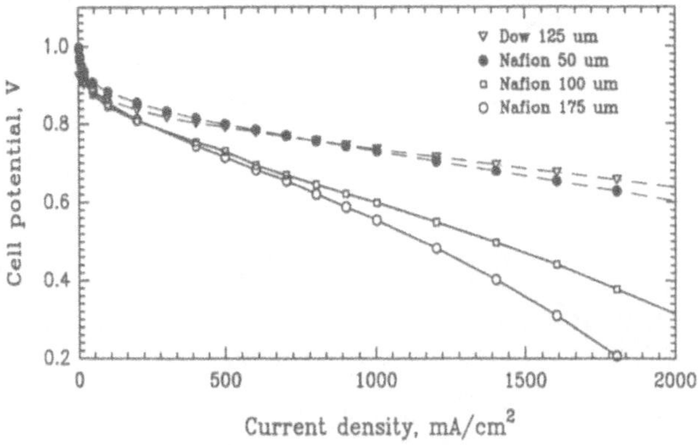

Figure 11.16. H_2–O_2 performance using various thickness membranes (95°C, 4–5 atm).

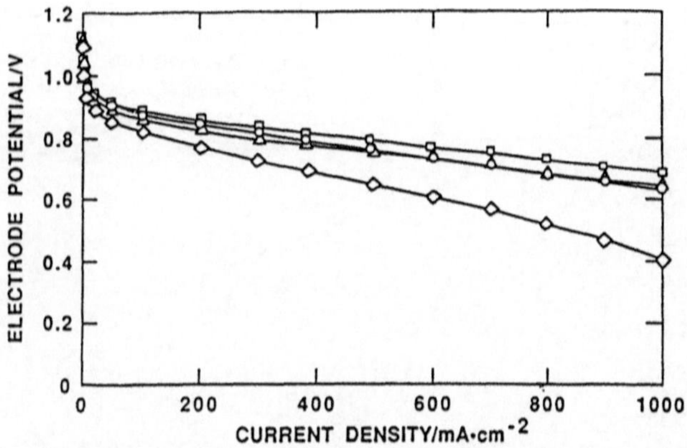

Figure 11.17. Effect of localization of Pt near front surface of Prototech electrodes on performance of oxygen electrode in H_2–O_2 single cell at 80°C and at 3–5 atm. 40 wt% Pt/C plus 50 nm sputtered film of Pt (○); 20 wt% Pt/C plus sputtered film of Pt (□); 20 wt% Pt/C (△); and 10 wt% Pt/C (◇).

the U.S. based on electrodes from Prototech.[14–18] Since then others, such as Texas A&M, have become active in the field.[13,35]

At LANL, structures which used ionomer-impregnated (liquid Nafion) gas-diffusion electrodes hot-pressed onto the membrane were developed. The effect has been said to extend the three-dimensional reaction zone, resulting in a considerable improvement in O_2 reduction kinetics.

LANL utilized carbon cloth custom-made electrodes from Prototech with 20 or 40% Pt on carbon, while maintaining the same Pt loading (0.4 mg/cm^2) and sputter-deposited a thin layer of Pt (0.05 mg/cm^2) on the front surface. Figure 11.17[31] shows the effect of localization of Pt near the front surface of Prototech electrodes on performance of the O_2 electrode in a H_2–O_2 single cell at 80°C and 3–5 atm. The H_2 counterelectrode in each case was the same as the working electrodes. Figure 11.18[31] compares performances using H_2–O_2 and H_2–air as

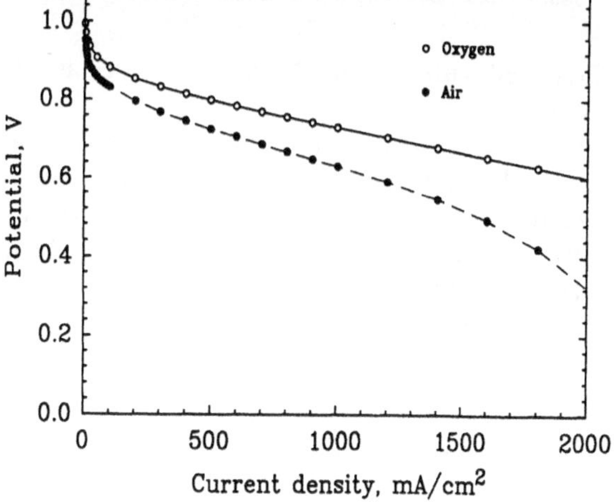

Figure 11.18. Comparison of performance using H_2–O_2 and H_2–air as reactants in single cell with 0.40 mg Pt/cm^2 electrodes. Temperature 95°C and pressure 5 atm.

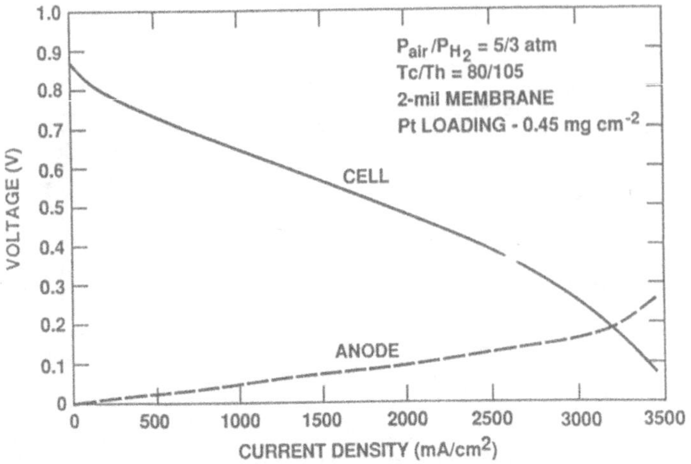

Figure 11.19. Best LANL performance on H_2–air.

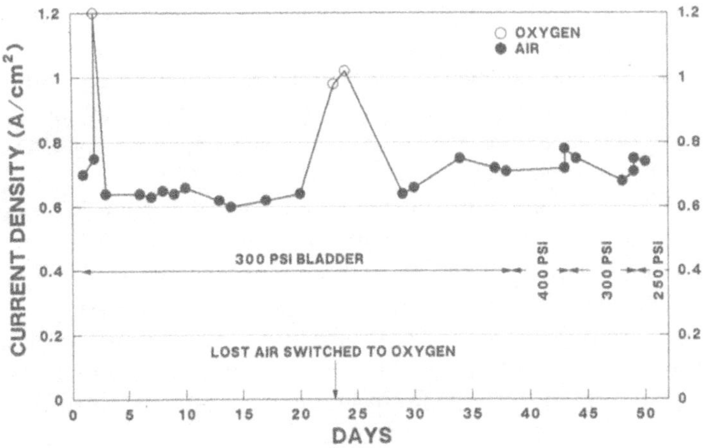

Figure 11.20. LANL life test of low Pt loading single SPFC.

reactants in a single cell with 0.40 mg Pt/cm^2 electrodes at a temperature of 95°C and a pressure of 5 atm.

The best LANL performance to date on pressurized air and H_2 in cells using Prototech carbon cloth electrodes with 0.45 mg Pt/cm^2 has been achieved with thin Nafion 112 membranes. Figure 11.19[14] shows the results.

LANL lifetime results with a 5-cm^2 cell with low Pt loadings on H_2–air are shown in Fig. 11.20.[14] During the test, cell voltage was held at 0.5 V. No significant loss of performance was seen after 2200 h.

11.4. DEVELOPMENT AND DEMONSTRATION

The following section describes development and demonstration activities worldwide. Present systems under development and applications are discussed.

11.4.1. Systems under Development

Systems can be broadly broken down according to the fuel and oxidant used. Included are those using H_2 as the fuel and O_2 as the oxidant, H_2 as the fuel and air as the oxidant, methanol as the fuel and air or oxygen as the oxidant, and natural gas as the fuel and air as the oxidant. Systems may use passive or dynamic product water removal for the stacks. System heat management techniques include liquid and air cooling for the stacks. Besides power generators, some systems include a fuel cell and a solar panel or other source of power in combination with an electrolyzer and reactant storage to give a regenerative-type energy storage system.

H_2–O_2 systems usually apply to submersible or space applications. Those using methanol and O_2 apply to submersibles. H_2 is stored in the form of a gas, a liquid, a reversible metal hybride, a chemical hydride such as LiH, $LiAlH_4$, or CaH_2, or methanol which is steam-reformed to yield H_2 and CO_2 (indirect methanol system). Oxygen is stored as a gas, a liquid, or a chemical, such as sodium chlorate or potassium superoxide.

H_2–air systems apply to applications where a source of H_2 is available, such as off-gas from a chemical process, or in a load-leveling application where off-peak (cheaper) electricity is used to generate H_2 with an electrolyzer. The H_2 is stored and later converted to electricity during peak demand by using a fuel cell. In this application O_2 could also be stored and used during the peak. H_2–air systems also apply in low-power applications as a substitute when batteries are too heavy or bulky to provide the required mission duration. In this case H_2 can come from low-weight composite pressure vessels, reversible metal hydrides, or chemical hydrides. If air is used, it is ambient air obtained for the stack by diffusion (low power) or from blowers or compressors–turbochargers.

Methanol–air systems apply to mobile power generators and transportation applications such as automobiles, buses, trucks, or locomotives. The methanol is steam-reformed to H_2 and CO_2 which is fed as fuel gas to the stack. Again the air is ambient from blowers or compressors–turbochargers.

Natural gas–air systems usually apply to sited or fixed power plants in a central or distributed power configuration. The natural gas is steam-reformed to give a H_2-rich fuel gas for the stack.

Systems which use passive product water removal were first used by GE in the Gemini fuel cells. Structures which wicked the water out of the stack were installed adjacent to the cathode. Later GE developed a dynamic water removal system in which the water was "pushed" out of the cell with excess O_2. The liquid product water in the effluent O_2–water mixture was separated in a drum and the O_2 gas recirculated back to the inlet. In the case of air, excess air "pushed" the water out of the stack where the liquid was separated in a drum and the spent air vented.

An example of an advanced passive water removal system is that developed by Ergenics Power Systems Inc (EPSI).[19] The principle of the cell is shown in Fig. 11.21.[36] The EPSI technology uses a porous hydrophilic collector structure separated by gas-impervious members. These structures provide a wicking path for product water transport within the cell and facilitate reactant prehumidification and passive product water removal. In systems with dead-ended reactant supply (H_2–O_2), product water is wicked by the porous collector to a suitable

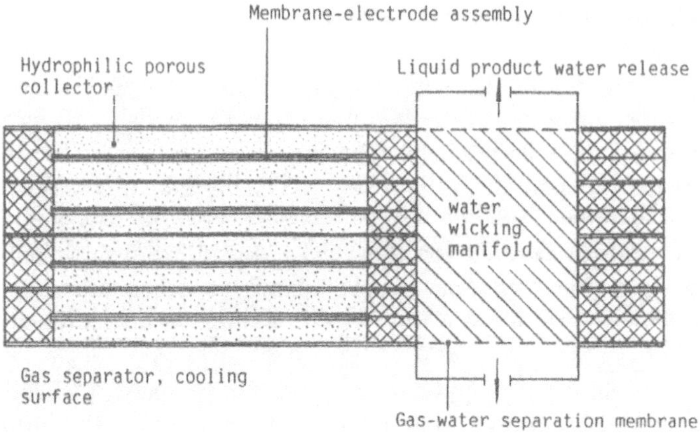

Figure 11.21. EPSI SPFC construction with hydrophilic collector structures and water-gas separation.

Figure 11.22. EPSI 30-V, 1-kW SPFC power supply.

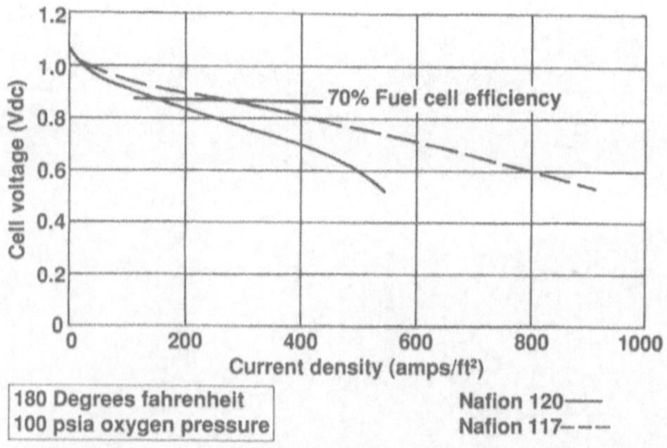

Figure 11.23. Hamilton standard performance curves for 0.78-ft^2 passive water removal SPFC.

conduit and rejected as liquid by gas–water separation using a microporous membrane. This cell technology is being considered for applications requiring watts to kilowatts. A 1-kW, 30-V H_2–O_2 system is shown in Fig. 11.22.[36]

Passive water removal systems have also been developed by United Technologies–Hamilton Standard.[6,20,37,38] Figure 11.23[37] shows performance of a 0.78-ft^2 passive water removal metal cell under 100 psia H_2–O_2 and 85°C using Nafion 120 and 117.

Dynamic water removal systems are used by Siemens, Treadwell, and Ballard. A 35-cell (0.25 ft^2 per cell), 5-kW Ballard graphite stack employing dynamic water removal and using a version of the Dow membrane is shown in Fig. 11.24. Excess air which pushes the product water out of the cells is fed to the

Figure 11.24. Ballard 5-kW H_2–air graphite stack using dynamic water removal.

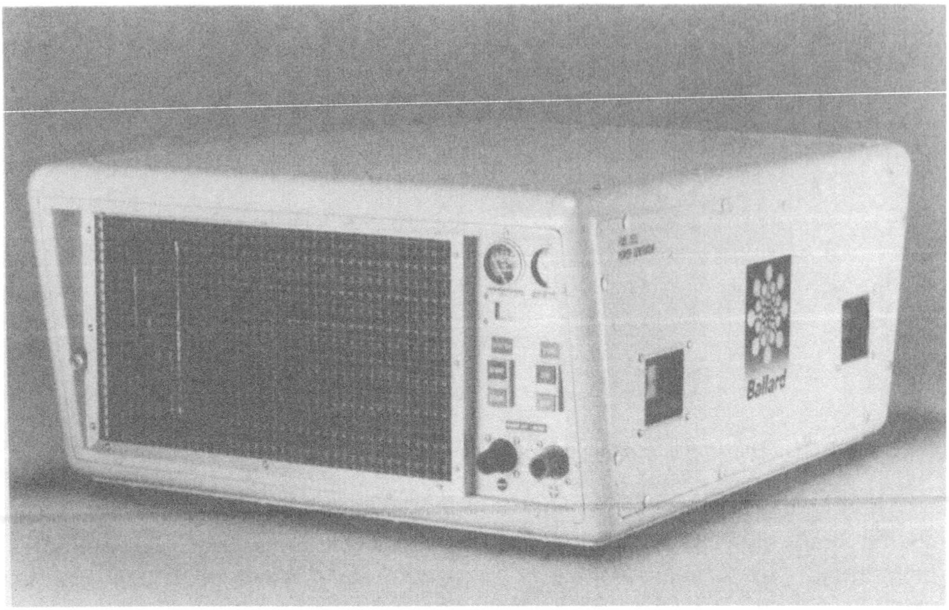

Figure 11.25. Ballard 5-kW H_2–air power plant.

Figure 11.26. Ballard 5-kW H_2–air power plant with top removed.

stack. Air utilization for the stack is 50%. A 5-kW H_2–air power system using this stack is shown in Fig. 11.25. The same system with top removed is shown in Fig. 11.26. Electrical and control, fuel cell, and heat and water management subsystems are visible. The stack is water-cooled with the cooling water pumped to an air-cooled radiator and then circulated back through the stack. Inputs are process H_2 and air at 30–50 psig and cooling air. Outputs are spent process air, cooling air, product water, and unregulated dc power.

11.4.2. Current Applications

Applications can be broadly grouped into three categories: space and military, motive, and utilities. The following sections describe current activity in these categories.

11.4.2.1. Space and Military

Space applications include regenerative fuel cells (RFC) and power units for the extramobility unit (EMU). Specific military applications include submersibles, surface ships, low power units to replace batteries and portable–mobile field generators.

a. Regenerative Fuel Cells. NASA is evaluating RFC under the surface power program of Project Pathfinder. The objective of the surface power program is to develop power technology to a level of readiness sufficient to enable or enhance extraterrestrial surface missions to the Earth's moon or Mars. RFC energy storage subsystems and photovoltaic power generation subsystems appear to satisfy mission and system requirements.[39]

Hamilton Standard is developing hardware for an RFC system.[37] Figure 11.27[37] shows Hamilton Standard's RFC schematic. A 25-kW fuel cell together with a 35-kW electrolyzer are being considered. Cell sizes for a 150-Vdc bus for both the electrolyzer (0.23 ft^2, 214 cm^2) and the fuel cell (0.78 ft^2, 725 cm^2) have been developed by Hamilton Standard.

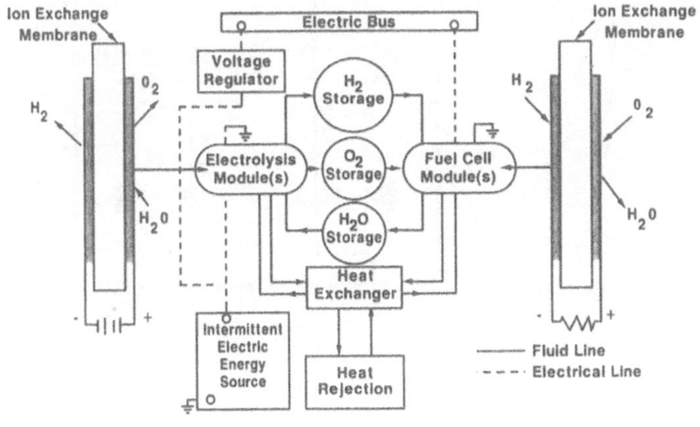

Figure 11.27. Hamilton standard's RFC schematic.

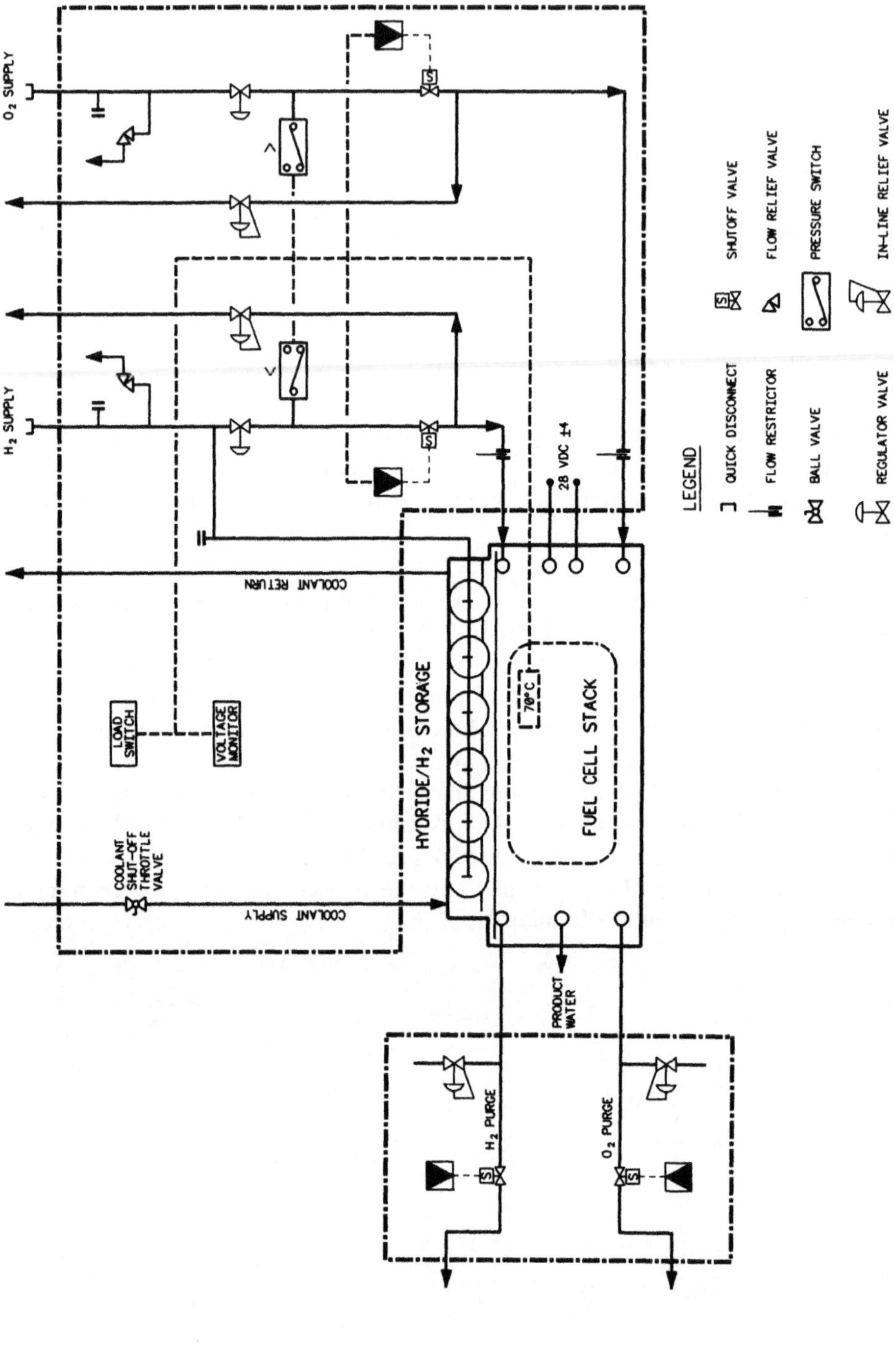

Figure 11.28. EPSI EMU systems schematic.

For the fuel cell a voltage efficiency of 65% during 10,000 h is required. SPFC metal hardware stacks appear to be the only candidate to meet these targets.[39]

Treadwell Corporation is currently under contract to the U.S. Air Force to design, fabricate, and test an SPFC stack for an RFC system. The use of SPFC technology to satisfy the continually growing power requirements of Air Force satellites is a natural application of the distinct advantages the fuel cells provide over other power sources.

The Treadwell design will supply 12 kW of power at 28 V for a satellite application. For purposes of survivability, the system will consist of six separate stacks with 35 cells in each. All of the stacks have integral humidifiers and cooling sections. Each stack is about 10 in. (25 cm) long, 6 in. (15 cm) wide, 6 in. (15 cm) high, and weighs 10 lb (4.5 kg). The reactants are H_2 and O_2 stored at 500 psi. Selection of the component materials was dictated by the demanding outer space environment and aerospace volume and weight constraints.

b. Extramobility Unit. Ergenics Power Systems Inc. is under contract to NASA for the development of a 200-W, 1500-W · h fuel cell energy storage subsystem using SPFC technology and H_2 stored in reversible metal hydrides.[40–43] The device is to replace short-life zinc–silver oxide batteries currently used in the EMU. The SPFC units compare favorably with secondary batteries based on cycle life, recharge speed, capacity retention, and energy density. A systems schematic is shown in Fig. 11.28.[36] The breadboard 28-V, 7-A unit is shown in Fig. 11.29.[36]

c. Submersibles. IFC, Siemens, Treadwell, Vickers, Perry Energy, and Ballard are currently involved in submersible applications. IFC is developing SPFC power systems for air-independent supply of submarine auxiliary power[44] through DARPA and the SUBTEC program. However, no public information is available at this time.

Marine applications of fuel cells have been evaluated by V. W. Adams of the Procurement Executive, Ministry of Defence, United Kingdom.[45] For submarines, Adams concluded that "in terms of power density, fuel consumption, fastest start-up time, shortest development time and least complexity SPFCs are seen as having the greatest potential for marine applications, although the reformation of fuels would be required to obtain the greatest benefits."

Siemens. Fuel cell systems for use in submarines have been under development in Germany since 1980. Companies involved include Howaldtswerke-Deutsche Werft AG (HDW), Ingenieurkontor Lubeck (IKL), and Ferrostaal (FS). The fuel cells are being provided by Siemens. IKL has been in charge of the overall design.[46]

In 1988–1989 a 100-kW alkaline fuel cell power plant was tested in the German submarine U1.[47] However, the target power plant is a SPFC. In the early 1980s GE transferred the basic SPFC know how to Siemens. SPFC was chosen for high power density, high overload capability, low power degradation, long lifetime, and absence of a liquid, corrosive electrolyte.

The Siemens SPFC power plant consists of the fuel cell module, the gas supply, the waste heat, product water, and rest gas remover, and the electrical

Figure 11.29. EPSI 28-V, 7-A breadboard FCESS for EMU.

control and monitoring component shown in Fig. 11.30.[7] The main components of the fuel cell system are connected to each other through the electrical and media (fluid) interface. The system may contain one or several modules, depending on the required voltage and power output.

The Siemens fuel cell module consists of a stack with about 70 cells,

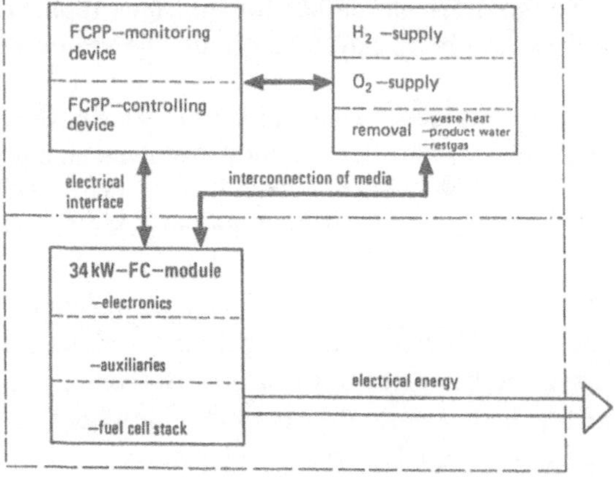

Figure 11.30. Siemens SPFC power plant schematic.

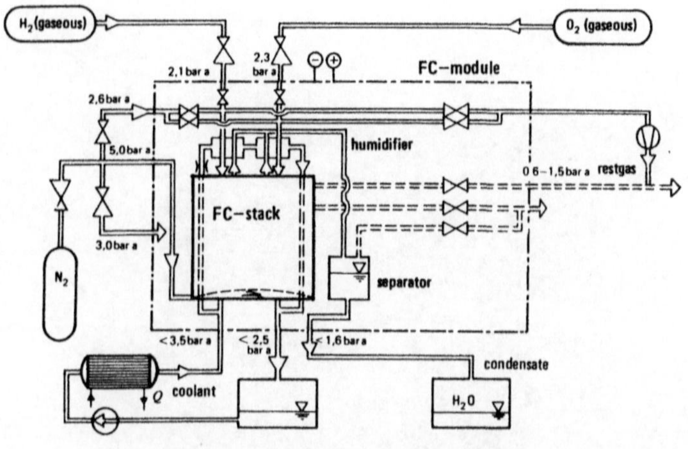

Figure 11.31. Siemens fuel cell module schematic.

humidifiers, product water separator, and electromechanical and electronic support hardware. The fuel cell stack and ancillaries constitute the module block, which is installed in a pressure vessel filled with a nitrogen blanket at 3.5 bar. A schematic of the fuel cell module is shown in Fig. 11.31.[7]

The FC module power output is specified as 34 kW at 53 V. Cell outer dimensions are 400 mm × 400 mm. Many of these cells have been tested. Long-term behavior at constant load has been investigated for over 10,000 h. A 20-cell stack with about 10-kW rated power has been constructed and tested.

The next step in the Siemens program is the construction and testing of the first 34-kW laboratory module in 1991. The project will be finished at the end of 1993 after qualifying a 34-kW prototype module ready for series production.

Treadwell. The U.S. Navy is evaluating the use of Treadwell Corporation's SPFC technology as a power source for underwater vehicles. Treadwell has designed and constructed a closed-loop 1-kW SPFC for installation into an unmanned underwater vehicle (UUV).

Mission requirements demand a minimal weight–displacement vehicle design with the capacity for extended mission durations. The vehicle's power source must provide high energy per unit volume, high power capability, and good standby performance.

The SPFC provides efficient, quiet, and reliable power. The system can be rapidly recharged by refilling gas storage containers.

Treadwell's 1-kW, 34-cell SPFC stack for the UUV application is shown in Fig. 11.32.[48] The stack was designed to produce 40 A nominally with a sprint capability of 60 A. The voltage regulation limits were set at 28 ± 4 Vdc. The voltage design point set the active cell area at about 0.05 ft^2 (46 cm^2). The stack is about 6 in. (15.3 cm) in length, 4 in. (10.2 cm) wide, 4 in. (10.2 cm) high, and weighs about 7 lb (3.2 kg).

The support system for the stack is minimal. The major functions are supplying the stack with reactants, removing product water and waste heat, and safely controlling the power plant. A system schematic is shown in Fig. 11.33.[48]

A microprocessor controller governs start-up and shutdown, oxygen recir-

Figure 11.32. **Treadwell 1-kW SPFC stack.**

culation, sensor monitoring, and fault detection. If an operational fault is detected, the system safely shuts down.

The UUV power plant provides 80 A · h, occupies 0.66 ft^3 (0.02 m^3) and weighs less than 14 kg fully charged. The complete system is shown during bench testing in Fig. 11.34.[48]

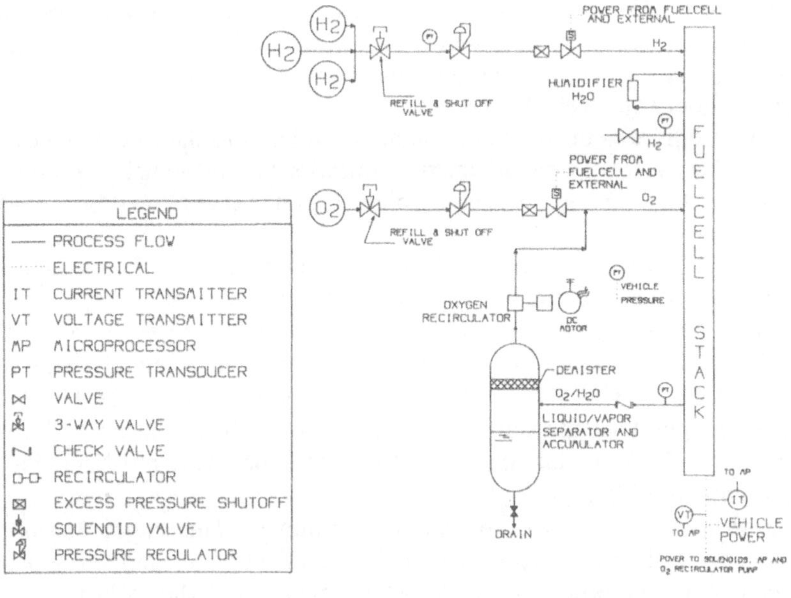

Figure 11.33. **Treadwell UUV SPFC system schematic.**

Figure 11.34. Treadwell 1-kW UUV power source.

Vickers–Construction–John Brown (CJB) Developments–Ballard. Vickers Shipbuilding and Engineering Limited (VSEL), a major submarine builder, and CJBD have embarked upon a project aimed ultimately at the production of power generation systems based upon SPFC technology. VSEL and CJBD have decided to base systems upon the Ballard SPFC stack and concentrate their efforts on the development of supporting systems and overall package. Working with Ballard in this way and feeding back system performance requirements allows the fuel cell to be further developed to more closely match the applications.[49]

The current phase of the project is aimed at producing a 20-kW system demonstrator based on methanol fuel. The system will be generic at this time, not being designed for a specific application. It will, however, serve to demonstrate the potential attractions to a variety of markets, both military and industrial. Among the military aspects, a potential fit into ocean going submarines is being considered.

Perry Energy–Ballard. In July of 1989 Perry Energy of the United States successfully tested the first commercial fuel-cell-powered submarine. The fuel cell was a Ballard SPFC. The boat was a two-man observation submersible shown in Fig. 11.35. The fuel cell system and gas bottles which supply the H_2 and O_2 are about half the size of the existing battery pod on the sub, but they supply three times the energy. The submarine, which was limited to about 2 h at top speed on batteries, can now travel 6 h with the fuel cell system. By adding two more gas bottles, this can be increased to 12 h. Recharging the system is as quick and simple as changing the gas bottles. Previously, the system required an 8-h battery recharge.[21]

Figure 11.35. Perry SPFC powered PC-1401.

d. Surface Ships. In a Phase I U.S. small business innovative research (SBIR) program for the U.S. Office of Naval Technology monitored through DTRC, Analytic Power (AP) determined that SPFCs were the optimum power approach for several surface ship missions.[50] In the Phase II program now underway, AP is constructing a 10-kW SPFC power plant producing dc power from diesel fuel and air. The purpose of the program is to verify AP's claims as to the performance attainable on reformed fuels and air.

The water-cooled fuel cell stack under construction will produce 36 V at 1000 A/ft^2 (1.1 A/cm^2) of active area for an areal power density of 700 W/ft^2 (0.75 W/cm^2). The stack contains 56 cells and 28 coolers. The coolers can operate with either single- or two-phase coolant. It will weigh about 23 kg and be 23.5 cm in diameter and 28 cm high. The plastic unitized cells will use Dow membrane and Pt-catalyzed electrodes. AP is also evaluating conventional Pt black and low-load supported Pt electrocatalysts. The bipolar plates are titanium with injection-molded seal rings and silicone rubber seals. The seal design has evolved from AP's electrochemical compressor programs which operate at pressures up to 1000 psi. The cell stack is designed to operate up to 100°C and 150 psi cathode air pressure. The anode pressure will be significantly lower.

AP is constructing a modular test stand to house the fuel cell stack, the turbocharged air supply, and the thermal and water management systems. The instrumentation integral to the test facility is computer-linked for easy data storage and analysis. The diesel fuel processor will be housed in a separate module.

e. *Low-Power units.* Low-power units of less than or equal to 1 kW are being discussed. The units would be useful for commercial as well as military applications. They are mainly designed to replace batteries (as with the EPSI EMU for the space application, discussed above) or to fulfill missions which batteries cannot. Such units are being developed by EPSI and AP (for the U.S. company H-Power) and are being studied by Ballard for the Canadian Department of National Defence. Development efforts by EPSI and AP are described below.

EPSI. EPSI is developing SPFC devices for energy storage applications and high-energy-density power generation. Cells are cooled by air, liquid, or by conduction to a heat sink. Reactants are stored as compresed gases, or, if volume

Figure 11.36. EPSI 28-V, 400-W H_2–O_2 SPFC stack with heat transfer fins for passive cooling.

is critical, H_2 is stored in metal hydrides. The devices offer improved cycle life, speed of recharge, capacity retention, and energy density when compared to batteries. Figure 11.36[36] shows a 28-V, 400-W H_2–O_2 SPFC stack with heat transfer fins for passive stack cooling. Figure 11.37[36] shows a 17-V, 7-A H_2–O_2 stack with integrated metal hydride hydrogen storage vessel. H_2 release is accomplished by direct heat transfer from the cell stack. Figure 11.38[36] shows 12- or 24-V, 500- and 1000-W H_2–O_2 systems which operate at 35 psig and use air cooling.

Figure 11.37. EPSI 17-V, 7-A, H_2–O_2 SPFC with integrated hydride hydrogen storage vessel. Hydrogen release is accomplished by direct heat transfer from cell stack.

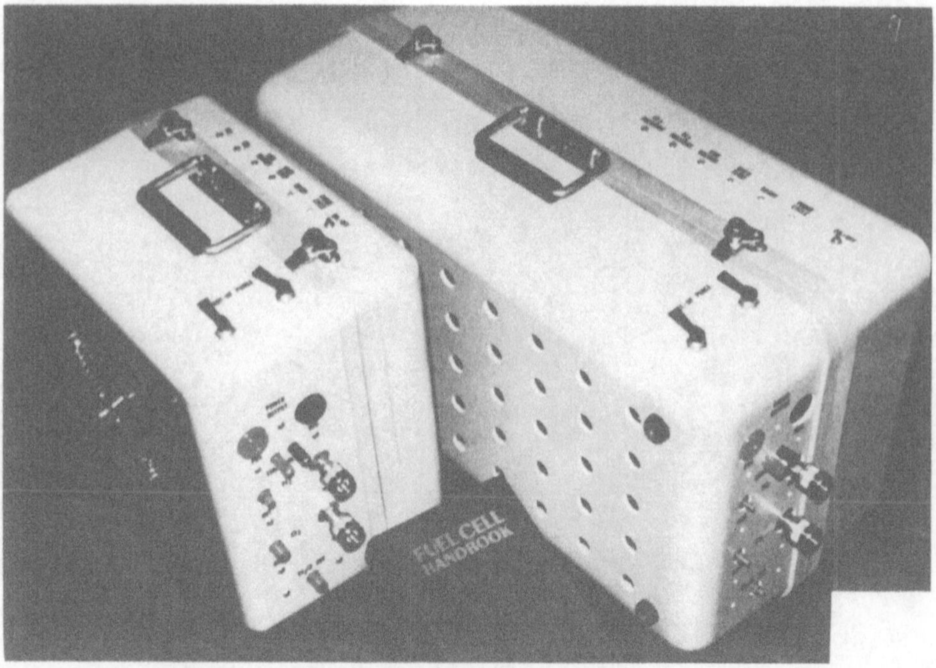

Figure 11.38. EPSI H_2–O_2, 12–24 V, 0.5–1 kW power supplies.

Analytic Power. Analytic Power has a program with H-Power Corp. to develop a 25-W H_2–air power supply. The program has successfully tested a passive water management system for this unit. The power supply is focused at replacing NiCd batteries in the communications industry.

AP claims the units are rugged and reliable. They have been operated in intermittent fashion for over 3000 h, drop-tested, and tested at open and short circuits. AP claims the stacks to be oblivious to mistreatment and continue to produce power from ambient pressure, room-temperature reactants. The present lab-scale design is now being revised for commercial manufacture.

f. Portable–Mobile Field Generators. Ballard has developed a 4-kW, 28-Vdc methanol–air power plant to the brassboard stage for the Canadian Department of National Defence. The target application is a portable field generator. In December 1990, Ballard demonstrated "methanol in" and regulated dc "electricity out" in the independent brassboard configuration. Included were the fuel cell, air supply, fuel gas supply, heat rejection, water recovery, and electrical and control subsystems. An artist's rendition of the packaged unit based on the brassboard design is shown in Fig. 11.39.

11.4.2.2. Motive

Two major SPFC motive programs involving development targeted for an automobile and for a bus are described. The automobile program[51,52] is a government–cost-shared funded effort between the Department of Energy (DOE), the South Coast Air Quality Management District, GM Allison, General

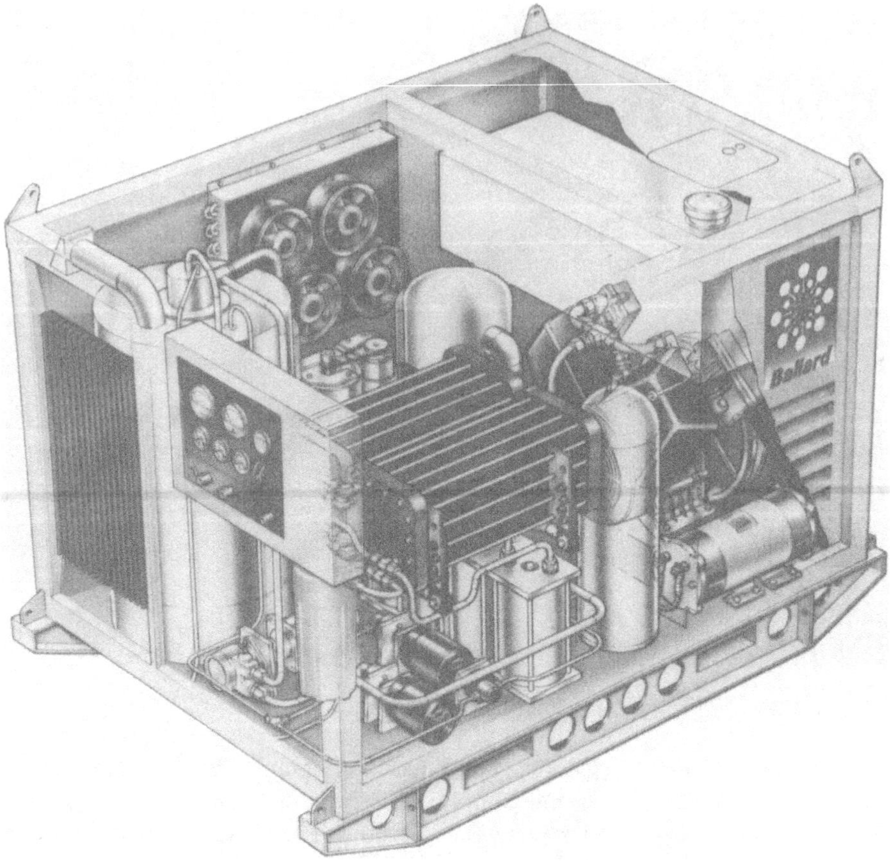

Figure 11.39. Ballard 4-kW, 28-Vdc methanol–air power plant.

Motors Technical Staffs, Los Alamos National Laboratory, and Dow Chemical Company in the United States and Ballard Power Systems in Canada. The bus program is a government–cost-shared effort between the Canadian Province of British Columbia, the Federal Department of Energy Mines and Resources, and Ballard Power as the prime contractor.

a. *Automobile.* The objective of the program is to develop an advanced indirect methanol–air SPFC propulsion system for vehicular applications. GM, the prime contractor, views an initial two-year effort as the first phase of a six-and-one-half year program that develops an advanced power plant and culminates in an actual fuel cell–hybrid vehicle demonstration. The preliminary milestone schedule for the multiyear program is presented in Fig. 11.40.[52]

GM sees a four-phase program. The first phase was initiated with the award of the two-year initial contract by September 1, 1990. The objectives of the first phase are the development of a 10-kW power source system evaluator, shown in Fig. 11.41,[52] and delivery of a conceptual design study (RPD in Fig. 11.40).

The second phase of the program begins in the third year following program go-ahead and runs for two more years. Its objectives are the development of a 40-kW brassboard power plant, its integration with vehicle drive train com-

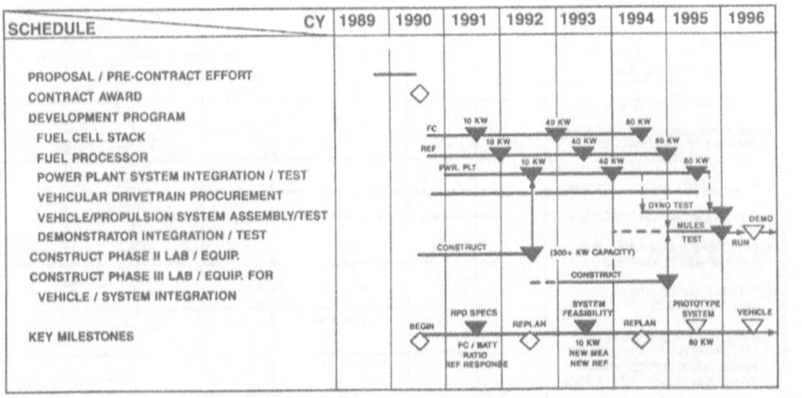

Figure 11.40. GM–DOE multiyear program preliminary milestone schedule.

ponents, and the initiation of brassboard power plant dynamometer testing. Other activities include system optimization of the 10-kW unit and continuing work on the vehicle propulsion system. Control capability will be developed for the 40-kW unit based on optimization of the 10-kW model. A "mule," or laboratory vehicle, ready to accept the 40-kW unit will be prepared.

The third phase, beginning in the fifth year, extends for 18 months. The

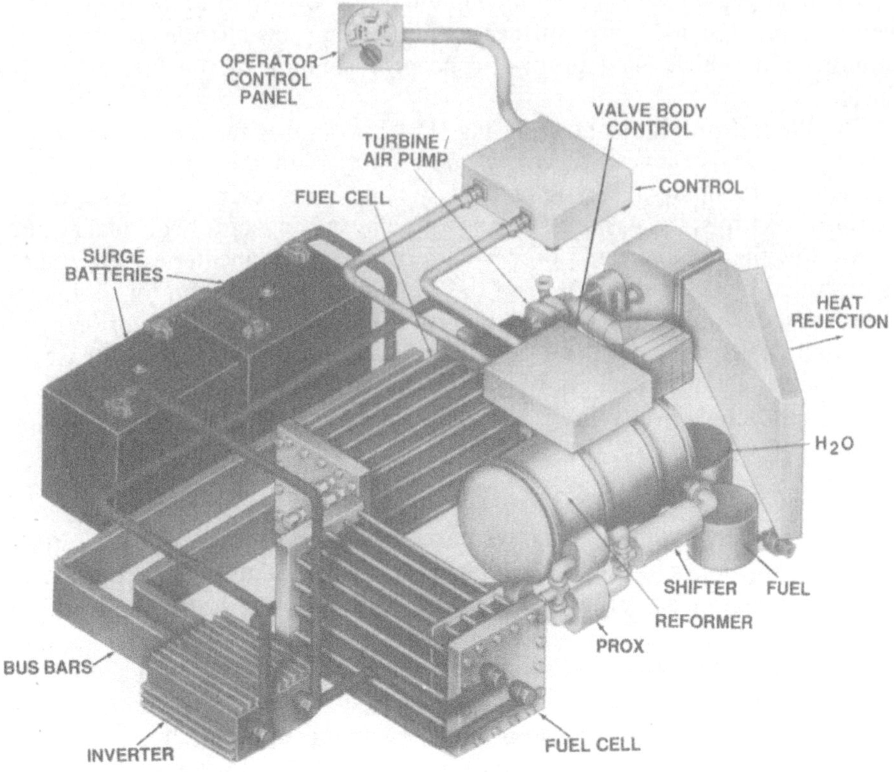

Figure 11.41. GM–DOE 10-kW conceptual power source.

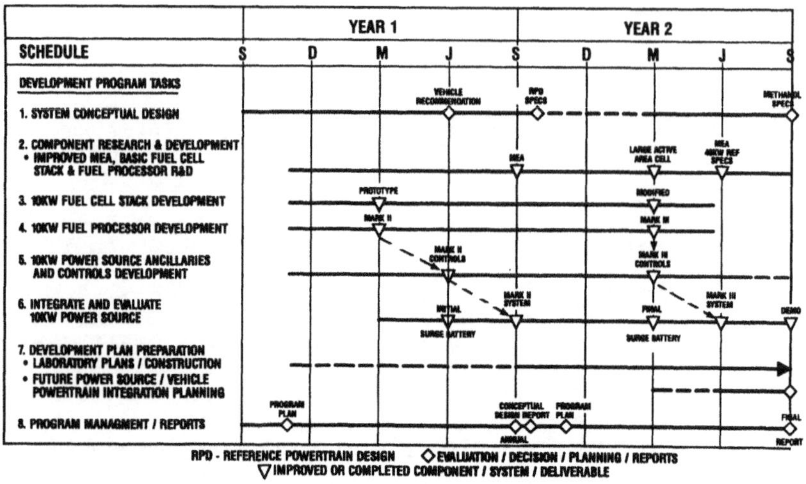

Figure 11.42. GM–DOE first-phase preliminary milestone schedule.

objective is the development of an 80-kW fuel cell power source prototype, its dynamometer testing and initial installation in the mule vehicle. As presently envisioned, the 80-kW fuel cell stack will be composed of two 40-kW stacks. A 40-kW power source will be evaluated first on a dynamometer and then in the mule. Following development of an 80-kW reformer, the 80-kW unit will be integrated, dynamometer-tested, and installed in the mule.

The final phase extends from the middle of the sixth year to the middle of the seventh year. Included are further mule testing, installation in the actual demonstration vehicle, and proof-of-concept evaluation of the fuel cell–hybrid vehicle.

The Phase I program, shown in Fig. 11.42,[52] emphasizes the development of technologies GM believes are critical. The purpose of the 10-kW system is to demonstrate feasibility. Key elements are the fuel cell, a fuel processor, electronic controls, a gas-pressurizing system, water and heat management systems, low thermal inertia, a proper battery, and other ancillaries. Feasibility is demonstrated by delivery to DOE of the complete 10-kW system for evaluation.

In addition to the R&D efforts on critical components, a propulsion system conceptual design study will be conducted during the first year of Phase I to define the actual reference power train design (RPD) specifications for the full-scale fuel cell–hybrid vehicle. Finally, a detailed multiyear program plan will be developed.

b. Bus. The ultimate objective of the BC bus program is to demonstrate a full-sized SPFC-powered transit but that meets all transit authority requirements. Phase I of a multiphase program to achieve this goal was started in November 1990.

The Phase I objective is to integrate a 75–100 kW SPFC H_2–air power plant into a small 30–35 ft transit but and demonstrate "driver acceptability." Driver acceptability is perceived as a transit authority driver, using the bus in his or her normal daily environment, being satisfied that the fuel-cell-powered bus is in no way less responsive than the normally assigned diesel bus. Phase I will not attempt to address all of the mission requirements of a bus suitable for transit authority use. For example, Phase I may demonstrate driver acceptability in a small demonstration bus that operates for a 2-h duty cycle and does not have a transit-authority-acceptable refueling system. The limited duty cycle will result from the use of bottled H_2 as the initial fuel. Later phases will address alternative fuels such as methanol, fuel handling, refueling, and safety, both on the street and in the depot.

BC Transit of Vancouver, British Columbia, is the immediate customer. The systems integration and project management have been subcontracted by Ballard to Science Applications International Corporation (SAIC). Program management and the SPFC stacks will be supplied by Ballard. Major decisions will be made by an internationally constituted steering committee chaired by BC Transit. Committee members also include Ballard, the BC Government, the University of Victoria, Georgetown University, Discovery Foundation, and Texas A&M.

Phase I is scheduled to complete a brassboard system by October 1991 and to have the Phase I bus available for testing in the late fall of 1992.

11.4.2.3. Utilities

Stationary and/or distributed power generating systems using natural gas as the indirect source of H_2 for the SPFC have not yet been demonstrated as they have with PAFC. Ballard, at the time of this writing, March, 1991, is in the planning and proposal stage to obtain such a demonstration program. However, SPFC power generation at a chemical plant using H_2 by-product gas as the fuel is being demonstrated by Dow and Ballard in Canada.

Dow Chemical Canada Inc. became interested in SPFC for industrial power generating applications at sites with abundant H_2 supplies from chloroalkaline plants, when the technology showed signs of becoming viable and reliable. This, combined with Dow's role as a supplier of solid polymer membranes, made this application of special interest to Dow Canada. In 1988, Dow Chemical Canada Inc. and the Ontario Ministry of Energy agreed on a jointly funded research project to install and test a 10-kW fuel cell system using Ballard's SPFC technology and Dow's proton exchange membrane. The Ontario Government's portion of the funding came through the Ontario Ministry of Energy's EnerSearch Program. The goal of this work was to define the operating parameters and limitations associated with operating such a facility in an industrial environment, using industrial-grade gases.

The project was engineered and installed directly in Dow's chloroalkaline process plant in the Sarnia Division. The system was designed to be fully automated and operate under computer control. Data collection and process is also done by the microVax control computer. The unit was started up in a 5-kW mode for the first time in May 1990 and has since gone through a number of experimental programs. Data collection have pointed the way to improvements in the operating system, cell design, and membrane itself. These changes have seen the unit go from relatively short operational times without maintenance intervention in the beginning to progressively longer periods. Since its installation, the system has been a significant factor in development of other joint membrane–fuel cell research projects for a number of applications.

Figure 11.43. Dow–Ballard 10-kW chloroalkaline installation.

The system was upgraded in February 1991 to the full 10-kW size and testing will continue for the foreseeable future. Figure 11.43 shows the 10-kW installation. Dow and Ballard are continuing to work together on improvements to the fuel cell system and expect to operate this facility to achieve high performance with minimal maintenance in the industrial environment.

ACKNOWLEDGMENTS

The author wishes to thank the following people who provided information for this chapter.

Victor W. Adams, Ministry of Defence, UK
Otto Adlhart, Ergenics Power Systems Inc., USA
David P. Bloomfield, Analytic Power Corporation, USA
Donald C. Brant, Senior Development Programs Manager, DuPont, USA
Glenn A. Eisman, Dow Chemical Company, USA
Shimshon Gottesfeld, Project Leader, Los Alamos National Laboratory, USA
A. B. LaConti, Vice President, Giner Inc., USA
Richard J. Lawrence, Director, Research & Development. Treadwell Corporation, USA
James F. McElroy, Chief, Advanced Systems/Electrochemical Products, Hamilton Standard, USA
Pandit Patil, U.S. Department of Energy

Ian Robson, Dow Chemical Canada Inc., Canada
A. E. Steck, President & CEO, Ballard Advanced Materials, Canada
C. M. Seymour, Projective Developments Manager, Vickers Shipbuilding and Engineering LTD, UK
Karl Strasser, Deputy Director, Senior Project Manager Fuel Cell, Siemens AG, Germany
Supramanian Srinivasan, Deputy Director, Center for Electrochemical Systems and Hydrogen Research, Texas A&M University, USA
Robert D. Sutton, Technical Coordinator, Electrochemical Engine Project, GM Allison, US
Nicholas E. Vanderborgh, Principal Investigator, Los Alamos National Laboratory, US

REFERENCES

1. L. J. Nuttall, *Development Progress on Solid Polymer Electrolyte Fuel Cells with Various Reactant Sources* (General Electric Company, Electrochemical Energy Conversion Programs, 1982).
2. A. B. LaConti, in *Power Systems for Small Underwater Vehicles* (Massachusetts Institute of Technology, Marine Industry Collegium, 1988).
3. O. Adlhart, Engelhard Minerals and Chemicals Corporation, *Design and Development of a 30 W Solid Polymer Electrolyte Fuel Cell Power Source Fueled with Calcium Hydride,* Final Technical Program Report, Contract DAAK 70-77C-0222 (U.S. Army Mobility Equipment Research and Development Command, Fort Belvoir, VA, 1978).
4. General Electric, Aircraft Equipment Division, Direct Energy Conversion Programs, *Feasibility Study of SPE Fuel Cell Power Plants for Automotive Applications, Final Study Report* (University of California, Los Alamos National Laboratory, Los Alamos, New Mexico, 1981).
5. Electro-Chem Products, Space & Sea Systems Department, Hamilton Standard, Division of United Technologies Corporation, *Final Report New Membrane-Catalyst for Solid Polymer Electrolyte Systems* (University of California, Los Alamos National Laboratory, Los Alamos, New Mexico, 1985).
6. K. Strasser, *Ber. Bunsenges. Phys. Chem.* **94,** 1000–1005 (1990) (VCH Verlags gesellschaft mbH, Weinheim, 1990).
7. R. Vellone, in *Program and Abstracts 1988 Fuel Cell Seminar* (Long Beach, CA, 1988), p. 168.
8. D. S. Watkins *et al., Abstracts 32rd Int. Power Sources Symp.* (The Electrochemical Society, Pennington, NJ, 1986), p. 590.
9. *Chemical Week,* p. 13, March 18, 1987.
10. G. A. Eisman, *Abstracts Fuel Cell Technology and Applications* (Int. Seminar, The Netherlands, 1987), p. 287.
11. D. S. Watkins *et al., Abstracts 33rd Int. Power Sources Symp.* (The Electrochemical Society, Pennington, NJ, 1988), p. 782.
12. S. Srinivasan *et al., Program and Abstracts 1988 Fuel Cell Seminar* (Long Beach, CA, 1988) p. 324.
13. O. J. Adlhart *et al., Ion Exchange Fuel Cell Assembly with Improved Water and Thermal Management,* U.S. Patent 4,826,741 (1989).
14. C. A. Reiser, U.S. patent 4,826,742 (1989).
15. Perry Oceanographics, *Press Release,* 7/28/89.
16. J. Kosek *et al., Proc. 24th Intersociety Energy Conversion Engineering Conf.* (Washington, D.C., 1989).
17. N. E. Vanderborgh *et al., Abstracts Fuel Cell Technology and Applications* (Int. Seminar, The Netherlands, 1987), p. 253.
18. N. E. Vanderborgh *et al., Program and Abstracts 1988 Fuel Cell Seminar* (Long Beach, CA, 1988), p. 52.
19. M. T. Paffett *et al., Program and Abstracts 1990 Fuel Cell Seminar* (Phoenix, AZ, 1990), p. 441.
20. U.S. patent 4,910,099 (1990).

21. U.S. patents: 4,358,545; 4,417,969; 4,478,695; 4,578,512; 4,804,727.
22. G. A. Eisman, in *Proc. Symp. Diaphragms, Separators, and Ion-Exchange Membranes* (The Electrochemical Society, 1986), p. 156.
23. M. R. Trant *et al.*, *Multi-phase Polymers: Blends and Ionomers* ACS Symposium Series 395 (American Chemical Society, Washington D.C., 1989), pp. 370–400.
24. K. Strasser, Siemens AG, Germany, 1990.
25. S. Srinivasan, Texas A&M University, 1990.
26. M. W. Verbrugge *et al.*, *J. Electrochem. Soc.* **137,** 3770–3777 (1990).
27. D. M. Bernardi, *J. Electrochem. Soc.* **137,** 3344–3350 (1990).
28. C. Derouin *et al.*, *Program and Abstracts 1990 Fuel Cell Seminar* (Phoenix, AZ, 1990), p. 437.
29. I. D. Raistrick, *Proc. Symp. on Diaphragms, Separators and Ion Exchange Membranes* (The Electrochemical Society, 1986), p. 172.
30. I. D. Raistrick, U.S. patent 4,876,115 (1989).
31. E. A. Ticianelli *et al.*, *J. Electrochem. Soc.* **135,** 2209 (1988).
32. E. A. Ticianelli *et al.*, *J. Electroanal. Chem.* **251,** 275 (1988).
33. T. Zawodzinski, *Program and Abstracts 1990 Fuel Cell Seminar* (Phoenix, AZ, 1990) p. 445.
34. M. J. Rosso *et al.*, *Proc. 33rd Int. Power Sources Symp.*, pp. 792–798.
35. O. J. Adlhart *et al.*, in *19th SAE/Intersociety Conf. on Environmental Systems,* 1989.
36. M. J. Rosso *et al.*, in *18th SAE/Intersociety Conf. on Environmental Systems,* 1988.
37. O. J. Adlhart *et al.*, in *20th SAE/Intersociety Conf. on Environmental Systems,* 1990.
38. Ergenics Power Systems Inc., *Advanced Extravehicular Mobility Unit (AEMU) Fuel Cell Energy Storage System (FCESS)*, Final Report, NAS9-17775, 1990.
39. J. F. McElroy, *Program and Abstracts 1990 Fuel Cell Seminar* (Phoenix, AZ, 1990), p. 282.
40. J. R. Huff *et al.*, *Program and Abstracts 1990 Fuel Cell Seminar,* (Phoenix, AZ, 1990), p. 286.
41. A. P. Meyer *et al.*, in *Proc. 33rd Int. Power Sources Symp.* 1988, p. 799.
42. A. Leonida, in *Proc. 33rd Int. Power Sources Symp.* (1988), p. 776.
43. A. P. Meyer, *Program and Abstracts 1990 Fuel Cell Seminar,* (Phoenix, AZ, 1990) (presentation only, no abstract submitted).
44. V. W. Adams, *Program and Abstracts 1990 Fuel Cell Seminar* (Phoenix, AZ, 1990), p. 273.
45. K. Ledjeff *et al.*, *Program and Abstracts 1990 Fuel Cell Seminar* (Phoenix, AZ, 1990), p. 449.
46. K. Stasser, in *Proc. Grove Anniversary Fuel Cell Symp.* (London), *J. Power Sources* **29**(1/2) (1989).
47. C. M. Seymour *et al.*, *Program and Abstracts 1990 Fuel Cell Seminar* (Phoenix, AZ, 1990), p. 379.
48. D. P. Bloomfield *et al.*, *Program and Abtracts 1990 Fuel Cell Seminar* (Phoenix, AZ, 1990), p. 278.
49. P. G. Patil *et al.*, in *Symp. Proc. 10th Int. Electric Vehicle Symp.* (Hong Kong, 1990), p. 657.
50. H. F. Creveling, *Program and Abstracts 1990 Fuel Cell Seminar* (Phoenix, AZ, 1990), p. 57.
51. S. Srinivasan *et al.*, in *Proc. 1st Int. Fuel Cell Workshop on Fuel Cell Technology Research and Development* (Ohokayama, Tokyo, 1989), pp. 119–150.
52. O. Adlhart, Ergenics Power Systems Inc.
53. R. J. Lawrence, Director Research & Development, Treadwell Corporation.

12

Fuel Cell System Economics

Michael N. Mugerwa and Leo J. M. J. Blomen

12.1. INTRODUCTION

Chapter 6 on the design and optimization of fuel cell systems has quite clearly shown the importance of understanding how various components of a fuel cell system interact. However, the technology issues concerning fuel systems cannot be considered in isolation, and a detailed analysis of the economics of fuel cell systems must also be made. In essence, this chapter will report the results of an extensive set of economic analyses that has been used to assess the future commercial prospects of fuel cell systems.[1]

Fuel cells, with their very attractive characteristics of high efficiency, modular construction, fast load response, low emissions, and cogeneration potential, have been shown to have a very large market potential in several studies.[2,3] (see also Chapter 13). However, the extent of energy market penetration by fuel cell systems is strongly dependent upon their projected price. Obviously, with a technology such as the fuel cell, which is still undergoing intensive research and development efforts, the initial costs are high. This is a natural precursor to the onset of commercialization of any technology;[4–8] however, major capital cost reductions can be expected when discounting, order duplication, direct labor and material cost reduction, and the learning curve effects of mass production are considered. For a technology such as fuel cells, which has already generated very considerable interest from fuel cell developers and, even more importantly, from potential market users, such as the electric and gas utilities, the capital cost reductions necessary for economically acceptable costs of electricity are expected to be achievable.

In this chapter installed-cost estimates for entire phosphoric acid fuel cell (PAFC) and molten carbonate fuel cell (MCFC) power plants are presented. These estimates were generated using the process flow diagrams and equipment

---

Michael N. Mugerwa • Kinetics Technology International S.p.A., Via Monte Carmelo 5, 00166 Rome, Italy. Leo J. M. J. Blomen • Mannesmann Plant Construction, 2700 AB Zoetermeer, The Netherlands. *Present address*: Blomenco B. V., Achtemonde 31, 4156 AD Rumpt, The Netherlands.

Fuel Cell Systems, edited by Leo J. M. J. Blomen and Michael N. Mugerwa. Plenum Press, New York, 1993.

summary sheets of eight base-case flowsheets based either on a once-off engineering premise or on a first series of five fuel cell power plants. To these estimates, cost reduction factors have been applied whose magnitude is dependent upon the capacity of the system under study and its production volume. Levelized capital and operating costs have been calculated according to the Electric Power Research Institute's (EPRI) *Technical Assessment Guide* (TAG) methodology,[9] and have been presented for all the fuel cell power plants under study. The TAG software made possible the calculation of levelized costs of electricity (COE) and enabled extensive parametric sensitivity analyses to be carried out. In addition, this chapter will present some economic comparisons of fuel cell system technology with existing conventional power generation technologies, both with and without cogeneration. Finally, an overview of expected technology improvements to fuel cell system design is given.

12.2. INSTALLED-COST ESTIMATING AND COST REDUCTION

Based upon optimized designs, the costs of the major equipment items in fuel cell systems can be developed from manufacturer's quotations, cost estimating techniques used in engineering offices, and database information. Piping, instrumentation, and other construction costs can be developed on the basis of process flow diagrams, process control schemes, and expected plot plans.

The philosophy of the estimating and cost reduction reported here is based upon evaluating the expected fuel cell system costs as the number of systems produced increases. The produced number of units is obviously a function of the size of the system, so, for example, for the small 25-kW_e systems, the production volume is assumed to range up to 2000 units in a robotized production scenario, whereas at a 100-MW_e utility plant scale, the number of produced power plants ranges up to 20. Table 12.1 details the number of units assumed to be produced for the different capacities of fuel cell systems. Please note that a list of acronyms used in the graphs and tables is given at the end of this chapter.

Table 12.1 reflects the expectation that the on-site market (25 and 250 kW_e) will be at least 50 MW_e during the advanced stages of commercialization, whereas the corresponding market for the industrial and utility scale systems is expected to be about 300 MW_e and 2000 MW_e, respectively (see Chapter 13). In any case, the figures have been structured to enable readers to adjust the cost estimates of the fuel cell systems according to their own market projections.

From an engineering contractor's point of view, a fuel cell system may be considered as a hydrogen (or synthesis gas) production plant integrated with fuel

Table 12.1. Number of Units

Capacity (kW_e)	First unit (FU)	First series (FS)	Low-volume production (LVP)	High-volume production (HVP)	Robotized production (RP)
25	—	5	20	200	2,000
250	—	5	20	200	—
3,250	—	5	20	100	—
100,000	1	5	20	—	—

Table 12.2. PAFC Stack Costs (\$/kW$_e$)

Stack size (kW$_e$)	FS	LVP	HVP	RP
25	1,000	500	350	200
250	750	350	250	—

cell and inverter electrical systems, with supplemental utilities that include, among others, water treatment plant, inert gas units, instrument air packages. The hydrogen production plant consists of well-known, proven plant components (hydrogen plant technology is very mature with respect to reliability and safety, due to the crucial role of hydrogen in most refinery and petrochemical plant complexes), whereas the fuel cell stack, and to a lesser extent the inverter, components are the major development items. Consequently, stack prices exhibit a strong learning curve cost reduction dependency which will have a major impact on total system prices during the early stages of commercialisation. Moreover, the differing fuel cell types are all at different stages of R&D and/or (pre)production level, which makes reasonable price comparisons difficult[2] (see Chapter 14). Therefore, for the sake of simplicity, the following costing philosophy has been assumed:

1. Fuel cell stack and inverter prices are based upon literature and manufacturers' information on target costs for different series and represent consensus values of such target costs. They tend to be considered as optimistic so far as current pricing and near-future expectations are concerned.
2. The remainder of the systems and plants have been costed using well-proven methods widely used in engineering, and that have a reasonable degree of accuracy (±30% on an absolute basis; or ±5–10% on a relative basis, i.e., when comparing different system designs.) As before, this approach permits the reader to adapt the cost estimates to include his or her own price expectations.
3. All prices in this chapter (unless stated otherwise) are based on summer 1988 cost estimates (European market), and have been converted to U.S. dollars at the exchange rate of December 8, 1989 (i.e., \$1 U.S. = 2.00 Dfl).

The assumed fuel cell stack costs for the differing production scenarios are given in Tables 12.2 and 12.3. It has been assumed that fuel cell manufacturers will produce a range of stack sizes, 25-kW$_e$, 250-kW$_e$, and 1000-kW$_e$. The 25-kW$_e$ stack size will be used only in the smallest power plants, whereas the 1-MW$_e$ stack is for utility-scale use. Note that the present costs for fuel cells in development

Table 12.3. MCFC Stack Costs (\$/kW$_e$)

Stack size (kW$_e$)	FU	FS	LVP	HVP	RP
25	—	900	400	250	150
250	—	650	250	150	—
1,000	350	150	100	—	—

Table 12.4. Power Conditioner Costs ($\$/kW_e$)

Number	Capacity (kW_e) 25	250	3,200	10,000[a]
5	850	450	175	100
50	500	325	125	85
500	400	250	100	—

[a]For the 100-MW_e plant, 10 units of 10 MW_e are assumed.

projects are in some cases several times higher, but increasing production volumes will result in substantial cost reductions. Note also that MCFC stacks are assumed to become cheaper than PAFC (which will not be the case with current materials selection,[2] but is still considered to be an achievable target by most manufacturers).

Table 12.4 details the cost estimates of the power conditioner systems (inverters) for use in the different fuel cell systems, and shows that significant cost reductions may be expected.[10]

Table 12.5 details the number of engineering hours per plant required for the design and construction of the various capacities and production volumes of fuel cell power plants.

A total of eight flowsheets ("base cases") have been subject to systematic estimating and cost reduction analyses, all based on the same assumptions, as follows (all natural-gas-fueled, unless otherwise noted):

25-kW_e PAFC system
25-kW_e MCFC system
250-kW_e PAFC system
250-kW_e PAFC alternative flow scheme (different design)
250–kW_e MCFC system
3.25-mW_e MCFC system
100-mW_e MCFC system

From the first unit (FU)–first series (FS) basis, cost reductions have been applied with increasing production volume. The basis of the cost reduction includes discounting, order duplication, and the benefits of mass production such as the learning curve effect. An 85% learning curve is applied for the reduction of direct labor cost components in the construction and installation of the power plants. This learning curve principle means that for each doubling in the number

Table 12.5. Engineering Hours (65 \$/h)

Capacity (kW_e)	FU	FS	LVP	HVP	RP
25	—	2,000	400	50	(500 \$/unit)
250	—	3,500	500	80	—
3,250	—	7,000	1,700	450	—
100,000	60,000	28,000	22,000	—	—

of systems produced, a 15% reduction in direct labor cost is applied. This is a standard procedure in any cost reduction effort and has been historically proven in industries as diverse as automobiles and personal computers.[8,11]

The greatest cost reductions are possible with the multi-kW_e-capacity plants, because one is dealing with larger volumes of smaller components. This means that direct labor costs can be reduced substantially, and the material usage can be more easily economized. In addition, the smaller components may be more easily manufactured by a far greater number of vendors such that a greater degree of competitive pricing is likely. Table 12.6 details the overall cost reduction factors that have been applied to the total fuel processor equipment, piping, instrumentation, electrical, insulation, and painting requirements for all the fuel cell power plants under consideration.

Table 12.7 gives a further breakdown of the cost reduction factors for the total fuel processor equipment.[12]

For each base case, costing tables were generated containing details of the cost estimates from the first unit and series production volume level through to mass production volumes, incorporating the cost reduction factors outlined in Table 12.6. In addition to the total cost of equipment and installation, costs that would normally be charged by any engineering contractor working to design and construct these systems are included. These are called contingency, process guarantee provision (PGP), and profit and are essentially charged as a percentage of the total installed plant cost. For the purpose of this work, this figure is varied from 10% for the first series of units through to 0.5% per unit at a robotized

Table 12.6. Cost Reduction Factors

Capacity (kW_e)[a]	Total fuel processor equipment[b]		Piping		Instrumentation		Electrical/insulation/painting	
	M	E	M	E	M	E	M	E
25[c]								
LVP	0.5	0.5	0.6	0.5	0.8[d]	0.4	0.75	0.5
HVP	0.16	0.16	0.2	0.125	0.4[e]	0.2	0.5	0.2
RP	0.1	0.05	0.1	0.05	0.1	0.05	0.1	0.05
250[c]								
LVP	0.76	0.9	0.8	0.723	0.9[d]	0.723	0.9	0.723
HVP	0.53	0.8	0.6	0.42	0.8[e]	0.42	0.75	0.42
3,250[c]								
LVP	0.845	0.9	0.8	0.723	0.9	0.723	0.9	0.723
HVP	0.69	0.8	0.6	0.5	0.8	0.5	0.75	0.5
100,000[f]								
FS	0.85	0.9	0.8	0.686	0.9	0.686	0.9	0.686
LVP	0.7	0.8	0.6	0.5	0.8	0.5	0.75	0.5

[a]The production rates relate to the first unit for the 100-MW_e case, and to the first series for the others.
[b]M = materials; E = erection.
[c]With respect to FS.
[d]Remove gas analyzers.
[e]Remove PC as well.
[f]With respect to FU.

Table 12.7. Breakdown of the Cost Reduction Factor for the Total Fuel Processor Equipment at High-Volume Production

Cost reduction factor (CRF)[12]

Item	Fraction of original cost: $25\ kW_e$[a]	Fraction of original cost: $250\ kW_e$–$100\ MW_e$
Heat exchangers	0.5	0.5
Catalysts	0.5	0.7
Rotating equipment	0.5	0.8
Custom-made	—	0.03–0.07[b]
turbocompressors		~0.65[c]
Vessels + structural steel	0.5	~0.7[c]
Reformer	0.5	

Sample calculation: $3.25\ MW_e$

Item	CRF	Fraction of total fuel processor cost (%)	CRF × fraction/100
Heat exchangers	0.5	29.9	0.1495
Catalysts	0.7	0.2	0.0014
Rotating equipment	0.8	50.9	0.4072
Vessels + steel	0.66	8.3	0.0548
Reformer	0.73	10.7	0.0781
Total		100%	0.691

Thus the weighted average CRF ≈ 0.69.

[a]Because of the very high numbers of units, $25\ kW_e$ is considered a special case (also, turbocompressors included in rotating equipment).
[b]The high initial cost is due mainly to the large amount of engineering needed to develop the first unit, hence the large cost reduction.
[c]These factors vary slightly since the learning curves are applied to fraction of the reformer cost attributable to labor (typically 30%).

production level. Table 12.8 is an example of the costing table generated for the 250-kW_e PAFC power plant and clearly shows that installed costs decrease sharply with volume production. In practice, this table was generated by abstracting more detailed cost estimates within each of the categories indicated.

Table 12.9 contains a summary of the installed cost estimates for all the base cases outlined earlier, and enables the reader to deduce the effect of capacity and fuel cell type on the magnitude of cost reduction possible. Note once again that all designs have been individually optimized, and therefore all have different flowsheets, pressures, temperatures, rotating equipment, heat exchanger networks, and process control schemes.

Finally, Table 12.10 gives a breakdown of the total fuel processor equipment costs for all the base cases, and it is worth noting the significant proportion of the total cost attributable to rotating equipment.

12.3. COSTS OF POWER PRODUCTION

Using the standard EPRI TAG methodology[1,9,13] for calculating the COE, and based on the costing tables of the previous section (Tables 12.8 and 12.9),

Table 12.8. Cost Structure Breakdown for 250-kW_e Unit (cell type = PAFC unit size = 250 kW_e nominal)

	First series (5th unit) of 5		LVP (20th unit) of 100		HVP (200th unit) of 100	
Cost basis: $1991	Mat.	Erect.	Mat.	Erect.	Mat.	Erect.
Total fuel proc. equip.	184,750	13,000	140,400	11,700	98,350	10,400
Fuel cell stacks	187,500	13,000	87,500	11,700	62,500	10,400
Piping	52,500	75,000	42,000	54,190	31,500	31,500
Instrumentation	149,750	38,500	89,775	27,800	55,800	16,150
Power conditioner	112,500	7,875	81,250	5,690	62,500	4,375
Elect./insul./paint	40,000	44,500	36,000	32,150	30,000	18,700
Engineering	—	227,500	—	32,500	—	5,200
Subtotal	*a* 727,000	*b* 419,375	476,925	175,730	340,650	96,725
Total of *a* + *b*	1,146,375		652,655		437,375	
Conting/PGP/profit	114,640		32,635		8,750	
Grand total (installed plant cost)	1,261,015		685,290		446,125	
Installed cost per kW_e	4,490		2,440		1,590	
Additional information						
actual capacity = 281 kW_e						
overall efficiency = 47.1%						

systematic calculations for the various fuel cell systems have been carried out. Table 12.11 contains the assumptions, which vary as a function of system capacity, used in these calculations. A sample calculation of the cost of electricity generated by a high-volume-production 250-kW_e PAFC power plant is given in Table 12.12. In this table, the annual levelized capital, operating and maintenance, as well as fuel cost contributions are computed separately. This cost figure is then divided by the total number of kilowatt-hours (kW·h) of electricity produced by the plant in one year to yield the levelized costs of electricity. This is the price that would have to be charged per unit of electricity if all costs are to be covered over the life of the project. Moreover, the price remains constant (hence

Table 12.9. Installed Plant Costs $/$kW_e$ (1990) for Fuel Cell Power Plants as a Function of Production Volume

	LHV[a] efficiency	FU	FS	LVP	HVP	RP
25-kW_e PAFC	41.2%	NA	22,415	9,095	3,295	1,300
25-kW_e MCFC	51.1%	NA	23,460	9,696	3,465	1,355
250-kW_e PAFC (nominal)	47.1%	NA	4,490	2,440	1,590	NA
250-kW_e PAFC (alternative)	36.9%	NA	4,550	2,435	1,580	NA
250-kW_e MCFC (nominal)	49.1%	NA	4,820	2,670	1,740	NA
250-kW_e MCFC (naphtha feed)	50.1%	NA	5,550	3,135	2,070	NA
3.25-MW_e MCFC	53.4%	NA	2,810	1,835	1,330	NA
100-MW_e MFC	56.6%	1,300	780	600	NA	NA

[a]Efficiencies are *net* electrical efficiencies: $\frac{\text{kW ac after inverter} - \text{parasitic power use and losses}}{\text{LHV feed + fuel to plant}}$

Table 12.10. Further Breakdown of Total Fuel Processor Equipment Cost

	Fraction of total cost (%)							
	25 kW$_e$		250-kW$_e$				3.25-MW$_e$	100-MW$_e$
			PAFC		MCFC			
Plant items	PAFC	MCFC	base	alt	base	naphtha	MCFC	MCFC
Heat exchangers	17.0	5.3	26.9	15.1	39.4	35.0	29.9	19.6
Catalysts	9.3	—[a]	4.5	4.1	0.6	4.2	0.2	0.5
Rotating equipment	37.2	72.8	40.0	48.2	42.9	41.5	50.9	48.3
Vessels and structural steel	14.9	4.9	13.7	15.9	2.5	6.5	8.3	22.4[b]
Reformer	21.6	17.0	14.9	16.7	14.5	12.8	10.7	9.2

[a]Included in vessel costs.
[b]Includes CO_2 removal equipment.

levelized) over the life of the project, and so is adjusted to take into account future inflation, escalation, and interest on loans. This means that the levelized cost of electricity is a useful way of comparing the economics of various generator systems, but may not be suitable for direct comparisons with the present prices charged for electricity by the public utilities (who often use lower real interest rates, longer depreciation periods, capitalized investment based on lower prices for "older" plants, etc.).

Figures 12.1 and 12.2 detail how the total costs of electricity change with the production volumes for all the base cases outlined in the previous section. The figures clearly demonstrate the effect of increasing production volume upon the costs of electricity. This effect becomes more pronounced for the smaller system capacities. The curves show how, with increasing system size, the costs of producing electricity decreases. This is a conclusion consistent for any given production volume.

Figure 12.3 is a more detailed breakdown of the cost of electricity produced from the 25-kW$_e$ PAFC and MCFC power plants as a function of production volume. The cost of electricity is subdivided into fuel, operating and maintenance, and capital contributions. It can be seen that at the onset and early stages of commercialization the capital contribution will be the greatest determinant of the final price of electricity. At the higher production volumes the fuel cost takes over as the major contributor to the total cost of electricity. The operating and maintenance contributions to the electricity cost decline with increasing production volume. The gas price used for the 25-kW$_e$ electricity cost calculations is that that would be incurred by a small gas user in the European Community (averaged over all EC countries), which is significantly higher than that incurred by large industrial users. The very high number of service hours (8592 h/year) indicate that these power plants are expected to run in a fully automated and unattended mode.* The 25-kW$_e$ MCFC power plant was optimized using an oxygen-enrichment delivery system and has significantly

*As a general remark, decreasing the number of service hours does not significantly change the relative position of the bars compared to each other (this is true for Fig. 12.4 but also most other bar charts of this chapter); it does, however, increase the height of all bars almost proportionally (see also Section 12.4).

Table 12.11. Economic Basis

i = Inflation rate = 3%
j = Interest rate = 10% (reflects the current money market)
c = Weighted cost of capital required for a constant dollar analysis (i.e., in the absence of inflation) = 6.8%
where $j = (1 + c)(1 + i) - 1$
Annual escalation rate assumed to be 0% for base cases
Income tax rate (NL) 42%
Property tax and insurance rate 2%
Investment tax credit 0%
Straight-line depreciation assumed

Book life years	Levelized carrying charge, %Total capital requirement/ year (TCR)
15	14.35
20	12.85
30	11.55

Plant lifetimes

25 kW_e	15 years
250 kW_e	20 years
3.25 MW_e	20 years
100 MW_e	30 years

Levelizing factor for fuel and operating and maintenance costs at

10 years	1.154
15 years	1.21 (interpolated)
20 years	1.264
30 years	1.344

Fuel cost natural gas

System size, kW_e	gas price based on LHV with no value-added tax, \$/GJ[14]
25	7.9
250	6.35
3,250	5.85
100,000	5.5

Fuel cost
Naphtha

System size, kW_e	Naphtha price, \$/GJ[15]
25	9

Maintenance costs
Stack replacement every 5 years, therefore 20% of installed fuel cell per annum.
Catalyst replacement every 10 years, therefore 10% of catalyst cost per annum.
Maintenance cost FS/FU = 3% of (TCR − fuel cell stacks)
LVP = 2% of (TCR − fuel cell stacks)
HVP = 1% of (TCR − fuel cell stacks)
RP = 0.5% of (TCR − fuel cell stacks)
(TCR = total capital requirement)
Labour costs: rate = 20 \$/hr

25-kW_e–250-kW_e power plants
Service = 8592 h/annum
Annual maintenance[a] = 1 week, 2 men, day shift only, 1 operator for 1 h each week operation.
Overhead at 30%.

3.25-MW_e power plant
Service = 8000 h/annum
1 person (on average), 3 shifts per day
Overheads at 30%

100-MW_e power plant
Service = 8000 h/annum
3 people, 3 shifts per day
Utility costs:

Boiler feed water	1.375 \$/ton
Cooling water	0.025 \$/ton

[a]Reduced by half for the highly automated systems made by robotized production.

Table 12.12. Cost of Electricity Calculation

PAFC 250 kW_e high-volume production
Net output = 281-kW_e capacity factor = 100% levelizing factor = 1.264

Capital costs

Section name	Equipment	Installed cost
	1000 US$	1000 US$
TTL F.P. equipment	99	109
Fuel cell stack	63	73
Piping	32	63
Instrumentation	56	72
Power conditioner	63	67
Elect/insulation/paint	30	49
Engng/contin/PGP/profit	0	14
Total	343	447

Annual carrying charge (at 12.8%) = 57 × 1000 US$/year

Operating and maintenance cost

	1000 US$/year
Normal O + M costs	28
Utility costs	5
Total	33
Fuel costs	140

Operating credits 1000 US$/year

Hot water	0
Steam	0
Gas	0
Total	0

Total O + M 1000 US$/year 173

Cost of electricity (US$/kW·h)

Capital contribution	0.024
O + M contribution	0.014
Fuel contribution	0.058
Contribution from credits	0.000
Total cost of electricity	0.096

higher efficiency than the equivalent PAFC power plant. The difference in system efficiencies between the PAFC and MCFC systems is shown up by the relative fuel contributions to the cost of electricity. There is an approximately $0.01 (U.S.)/kW·h difference in the price of electricity.

Figure 12.4 details the cost of electricity contributions for the 250-kW_e PAFC power plants, for both the optimized base case and the alternative design, as a function of the production volume scenarios. As before, increasing the production volume reduces the capital and the operating and maintenance contributions to the total costs of electricity. In fact, these contributions more than halve from the first series through to high-volume production units and, when compared with those for the 25-kW_e systems, are a significantly lower percentage of the total

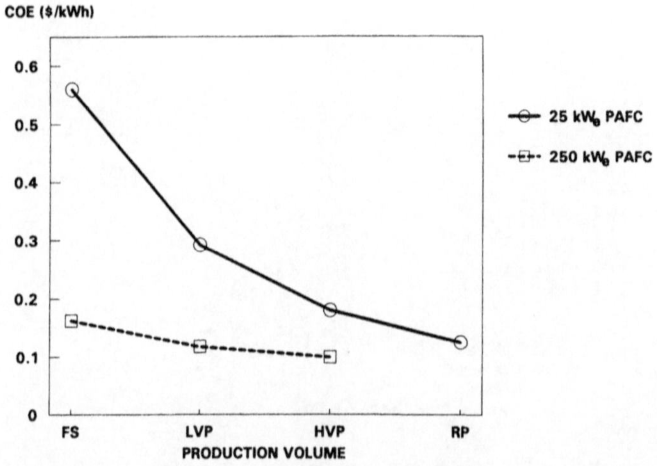

Figure 12.1. COE vs. production volume (PAFC power plants).

costs of electricity. The fuel contribution to the total cost of electricity is lower because the 250-kW$_e$ system has a higher efficiency and the gas price for this size of user is about 20% cheaper. In the alternative design a steam-from-convection-bank approach is used (see Chapter 6) and the waste heat from the fuel cell is barely utilized in the process flow scheme. It is quite noticeable that the fuel contribution to the cost of electricity is significantly higher than in the base case and results in a \$0.02 (U.S.)/kW·h price difference.

Figure 12.5 details the cost of electricity calculations versus production volume for the 250-kW$_e$ MCFC power plant base case and for a similar capacity MCFC power plant optimized to accept naphtha as feedstock (with the high-volume production of the base case 250-kW$_e$ PAFC power plant for comparison). As noted previously, there are reductions in cost due to decreased capital, and operating and maintenance contributions with increasing production volume. It is evident that the naphtha case has significantly higher capital,

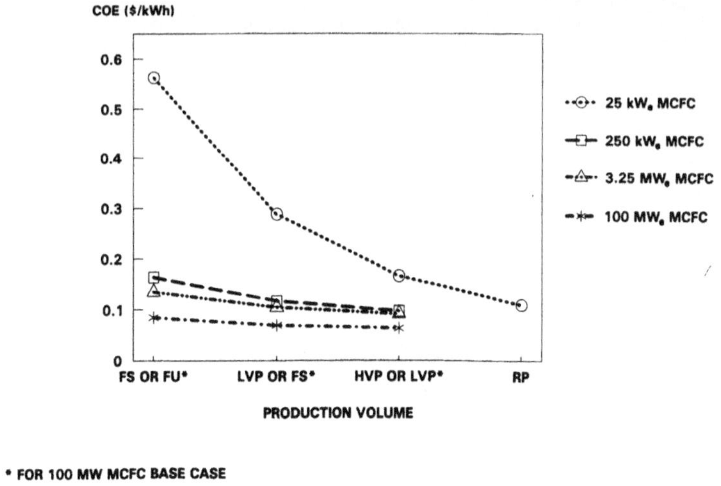

Figure 12.2. COE vs. production volume (MCFC power plants).

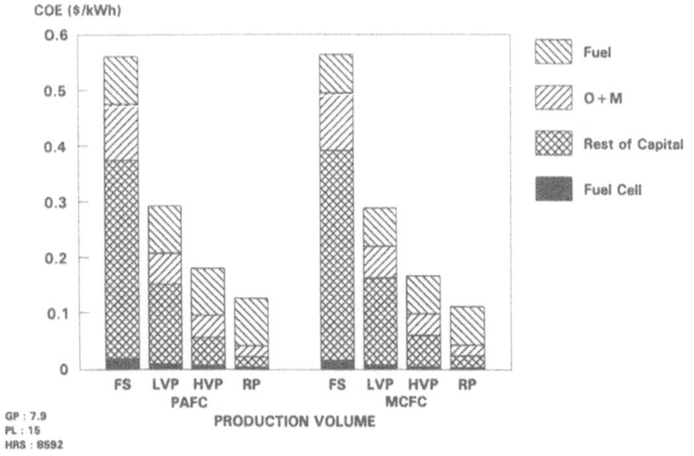

Figure 12.3. 25-kW$_e$ power plants.

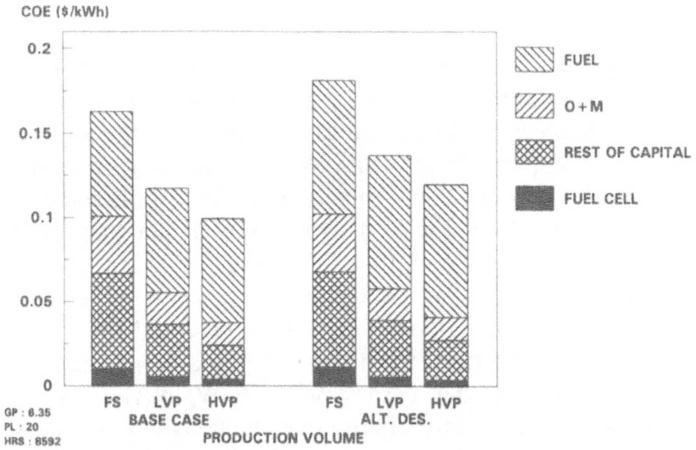

Figure 12.4. 250-kW$_e$ PAFC power plants.

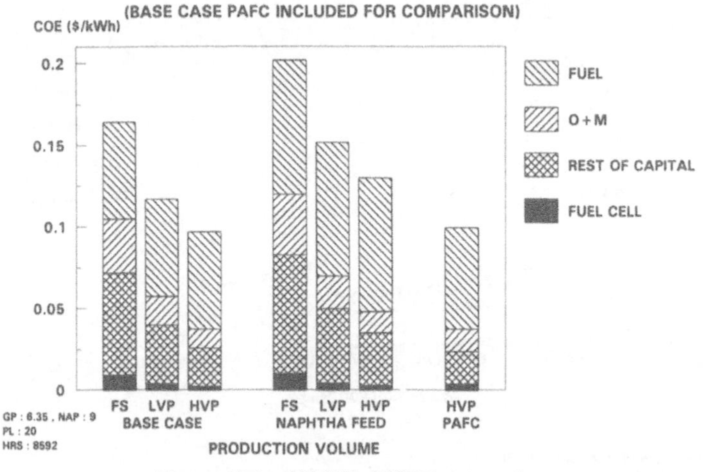

Figure 12.5. 250-kW$_e$ MCFC power plants.

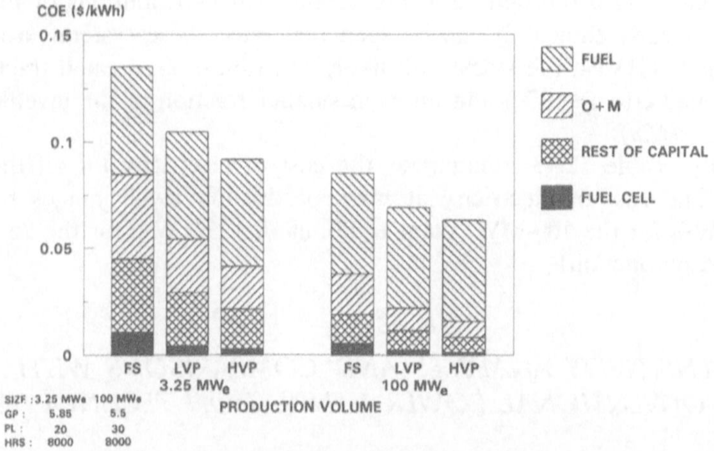

Figure 12.6. 3.25-MW$_e$ and 100-MW$_e$ MCFC power plants.

operating and maintenance, and fuel contributions to the overall cost of electricity which causes a \$0.03/kW·h difference with the base case. This is despite the higher net system efficiency in the naphtha case, and is in a large part due to the higher cost of the naphtha feedstock. The total cost of electricity of the natural-gas-fueled 250-kW$_e$ PAFC and MCFC base cases is very similar. The only difference is that the MCFC power plant has slightly higher capital, and operating and maintenance contributions combined with a slightly lower fuel contribution to the total cost of electricity. The reason for the higher capital cost of the MCFC plant is the higher pressure at which it has to operate to achieve higher fuel efficiency.

Figure 12.6 details the cost of electricity calculations for the 3.25-MW$_e$ and 100-MW$_e$ MCFC power plants. It is most noticeable that the proportion of the fuel contribution is much higher in the 100-MW$_e$ case. In addition, both the capital, and operating and maintenance contribute only \$0.016/kW·h to the cost of electricity from the 100-MW$_e$ plant, compared to \$0.041/kW·h from the 3.25-MW$_e$ plant.

Table 12.13. Summary of Costs of Electricity for All Fuel Cell Power Plants Studied

Electricity production	Type of plant	Cost of electricity (\$/kWh) at various production volumes FU	FS	LVP	HVP	RP
25 kW$_e$	PAFC	—	0.560	0.292	0.180	0.125
	MCFC	—	0.563	0.288	0.166	0.110
250 kW$_e$	PAFC	—	0.163	0.117	0.096	—
	PAFC[a]	—	0.181	0.137	0.120	—
	MCFC	—	0.164	0.117	0.097	—
	MCFC[b]	—	0.202	0.152	0.130	—
3.25 MW$_e$	MCFC	—	0.135	0.105	0.092	—
100 MW$_e$	MCFC	0.081	0.069	0.059	—	—

[a]Alternative design.
[b]Naphtha feed.

In Figures 12.3 through 12.6, the capital cost contributions of the fuel cell stacks are clearly shown. It can be seen that once these systems are produced commercially (HVP), the stacks will likely contribute but a small fraction of the total installed cost (≤18%) and an even smaller fraction of the levelized cost of electricity (≤4%).

Finally, Table 12.13 summarizes the cost of electricity for all the fuel cell systems. The cost of electricity at mass production levels ranges from about \$0.059/kW·h for the 100-MW_e plant to about \$0.125/kW·h for the 25-kW_e plant (robotized production).

12.4. SENSITIVITY ANALYSES AND COMPARISONS WITH CONVENTIONAL POWER GENERATION TECHNOLOGIES

In this section the results of sensitivity analyses are presented, and the effect of seven variables on the total cost of electricity is given. The variables are capital cost, fuel cost, operating and maintenance costs, capacity, plant efficiency, levelizing factor, and levelized carrying charge rate. Also included in this section is a brief comparison of the competitiveness of the fuel cell systems with conventional power generation technologies. The sensitivity analyses are applied to all four sizes of MCFC (25 kW_e–100 MW_e) with natural gas as fuel. The PAFC systems exhibit comparable sensitivity behavior since the breakdown of contributions to the cost of electricity follows quite similar patterns.

To cover uncertainties in capital cost estimation and the cost reduction effort, it is necessary to carry out sensitivity analyses to evaluate the effect of this uncertainty on the cost of electricity. At the first series production volume a variation of ±30% is used to cover the uncertainty in the budget estimating. In addition, the following variations are used to cover cost reduction uncertainties: at low volume production (LVP), ±45%, at high-volume production (HVP), ±60%; and for robotized production (RP), ±75%. Thereafter, the results of sensitivity analyses are given, which evaluate the effect of the remaining six variables on the cost of electricity for the HVP cases (or RP case for the 25-kW_e power plant) only.

The natural gas price is varied according to the differences in cost of natural gas throughout the EC. The variations in price level differ depending on the capacity of the fuel cell system, and are varied in the range +70% and −60%. The price of natural gas can vary by almost a factor 5, from as low as \$2.65 (U.S.)/GJ for the 100-MW_e case in West Germany to \$12.8(U.S.)/GJ for the 25-kW_e case in Italy. In addition, the operating and maintenance requirements are varied ±50% to cover uncertainties in these estimates.

The capacity of a system under study is varied from −50% to +10% to cover part-load and a 110% load operation, and, indirectly, possible variations in the total operating hours per year. Since fuel cell systems are assumed to operate at the same net system efficiency during part-load operation down to 50% of nominal capacity,* this is equivalent to halving the number of service hours per annum. The net system efficiencies were varied from 80% of the base-case value up to the base case value, to cover possible shortfalls in plant performance.

*This is an approximation; see Fig. 6.23.

Finally, the effect on the cost of electricity from these fuel cells systems of two important economic factors is also evaluated. These factors include the levelizing factor, which takes into account the effect of inflation and escalation rates on the operating and maintenance costs of a plant, and the levelized carrying charge rate, which effectively takes into account the cost of repayment of capital over the plant life. These factors were varied between 80% and 120%, and 40% and 160%, of their base-case values, respectively.

As briefly indicated earlier, the uncertainties in relative comparisons between different fuel cell systems are much less than the above figures would indicate, since the same assumptions were used throughout. The data for conventional technologies, however, was gathered from a wide variety of sources,[16–20] so exact comparisons are more difficult. The values used to generate the cost of electricity for these technologies are given in Table 12.14 to establish the basis for calculation. Table 12.15 summarizes the costs of electricity obtained using these data.

The 25-kW_e MCFC is considered first. Figure 12.7 shows the sensitivity of cost of electricity to deviations from the original capital cost estimate for all production rates. Figure 12.8 plots variations in the other factors in the same way, but applies them only to robotized production plants.

The sensitivity to variations in the capital decreases rapidly with increasing production rates as the contribution of capital to the total cost of electricity diminishes. At the robotized production rate the price of natural gas becomes so important that a 10% increase causes a 6% increase in the electricity cost. A reduction of 10% in electrical efficiency or capacity (electricity produced per year) leads to increases in electricity cost of 8% or 4%, respectively. These figures are helpful in setting limits on the mode of operation and acceptable

Table 12.14. Basis of COE Calculations for Conventional Technologies

Plant type	Nominal power generated	Total installed capital (1000$)	Installed capital per kW (1000 $/kW)	Electrical efficiency (%)	O + M (100 $/yr)	Utilities (1000 $/yr)	Hours operation (h/yr)	Plant life (yr)
Gas engine	25 kW_e	32	1.3[a]	27	7	0[b]	8,100	4
	250 kW_e	375	1.5	30	31	0[b]	8,650	20
	3.25 Mw_e	4,098	1.25	33	397	123	8,650	20
Gas turbine	250 kW_e	435	1.75	18	9	20	8,060	20
	3.25 MW_e	2,500	0.75	25	100	168	8,060	20
Combined cycle	100 MW_e	50,000	0.5	50	2,000	1,725	8,060	30
Natural-gas-fired conventional-al power plant (steam turbine)	100 MW_e	80,000	0.8	41	3,200	2,159	8,060	30
Uninterruptable power supply (UPS)	25 kW_e	31	1.2	82[c]	2	0	8,760	20

[a]Low cost due to scaling-up data from high-volume production, commercial 15-kW_e unit.
[b]*Air cooling (other cases have water cooling).*
[c]Electricity from grid is the "fuel" in this case priced at 0.11 $/kWh.

Table 12.15. Cost of Electricity for Alternative Technologies

Power output	Type of plant	Levelized cost of electricity (COE)($/kWh)
25 kW_e	Gas engine	0.198
	battery-backed uninterruptable power supply	0.200
250 kW_e	Gas engine	0.136
	Gas turbine	0.208
3.25 MW_e	Gas engine	0.123
	Gas turbine	0.132
100 MW_e	Combined cycle	0.066
	Conventional (gas-fired)	0.086

deterioration in performance for a given market niche. The operating and maintenance contribution exhibits similar sensitivity behavior to that of capital. The most sensitive factor, however, is also the most uncertain: the levelizing factor. A 10% increase in the levelizing factor yields a 9% increase in the cost of electricity. The importance of this is more difficult to assess, since the costs of electricity from all the technologies under study will be affected by the financial factors used to calculate the levelizing factor (see Table 12.11).

Some possible alternative technologies at the 25-kW_e scale are a gas engine for independent power generation, and a rechargeable battery backup to public utility electricity for an uninterruptable power supply (UPS). Figure 12.9 compares their costs to the fuel cell systems. The fuel cells are more economical than both alternatives even at high-volume production. Robotized production of fuel cell systems yields costs of electricity of only about 60% of the costs of the alternative technologies. Note that the "fuel cost" for the UPS is the cost of buying electricity from a public utility at $0.11(U.S.)/kW·h.

Similar sets of graphs may be obtained for each of the other plant sizes; see Figures 12.10–12.15. They all display the same trends as the 25-kW_e case. However, as the size of the plant increases the capital, operation and main-

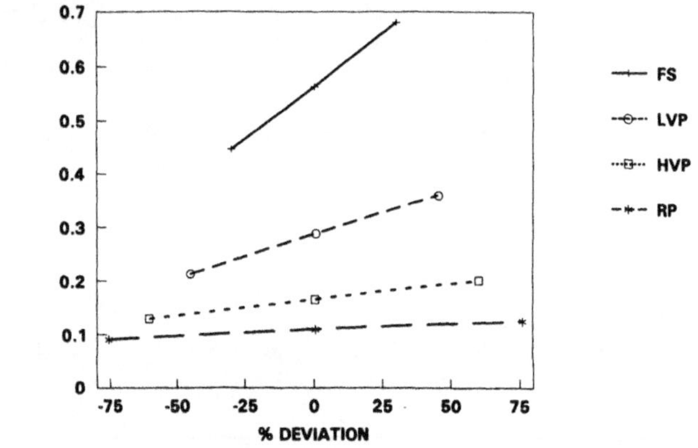

Figure 12.7. Sensitivity of cost of electricity to variations in capital costs at various production rates, 25-kW_e MCFC.

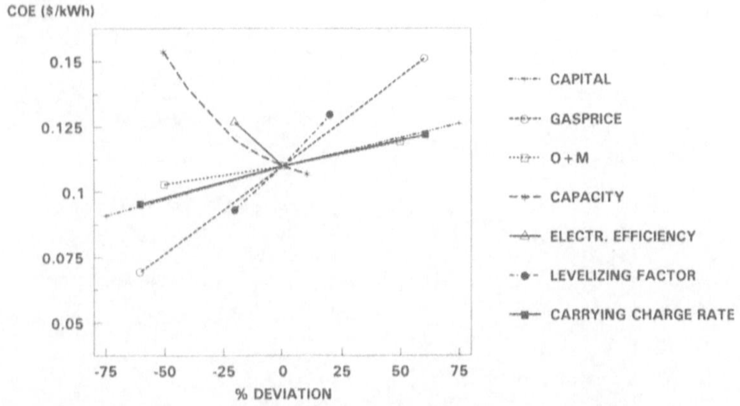

Figure 12.8. Sensitivity of cost of electricity to variations in various parameters at robotized production, 25-kW$_e$ MCFC.

tenance costs constitute a reduced fraction of the total electricity cost while the fuel cost contribution increases, despite lower gas prices and higher efficiencies. Thus a 10% increase in the gas price for a 100-MW$_e$ plant gives a 7.5% increase in the cost of electricity (it was 6% for the 25-kW$_e$ case). The effect of reduced efficiency or capacity is slightly less, and the levelizing factor gives almost the same effect as for the 25-kW$_e$ case.

Fuel cell systems always compare favorably with alternative technologies in terms of cost of electricity (see Figures 12.9, 12.11, 12.13, 12.15). This applies even with only high-volume production for the 25-kW$_e$ plants and low-volume production for the 250-kW$_e$ and 3.25-MW$_e$ plants. The main advantage of fuel cell systems over alternative technologies is not in the capital employed but in their higher overall electrical efficiencies. This advantage is eroded for the 100-MW$_e$ power plants for which the combined cycle gas plus steam turbine systems have electrical efficiencies approaching 50%. However, unlike the combined cycle the fuel cell system can be operated at half-load with little

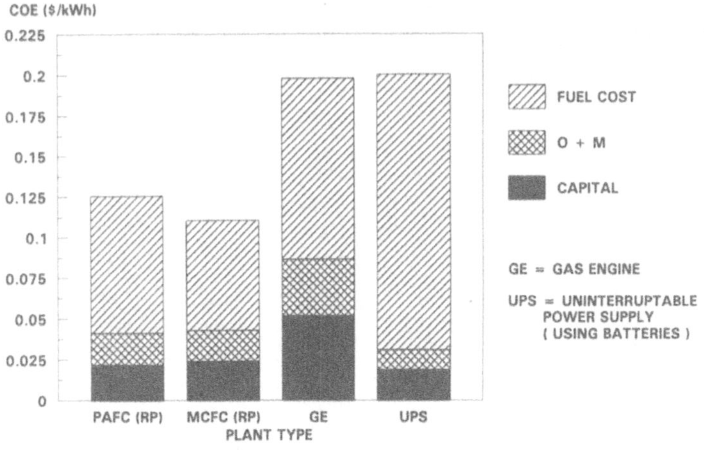

Figure 12.9. Comparing fuel cells with other technologies, 25-kW$_e$ power plants.

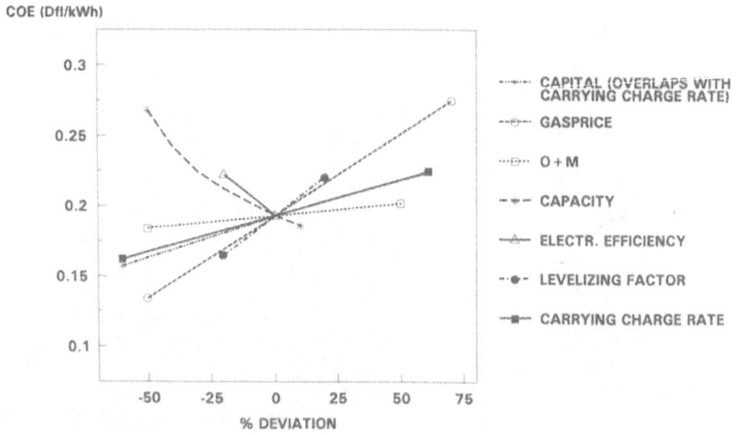

Figure 12.10. Sensitivity of cost of electricity to variations in various parameters at high-volume production, 250-kW$_e$ MCFC.

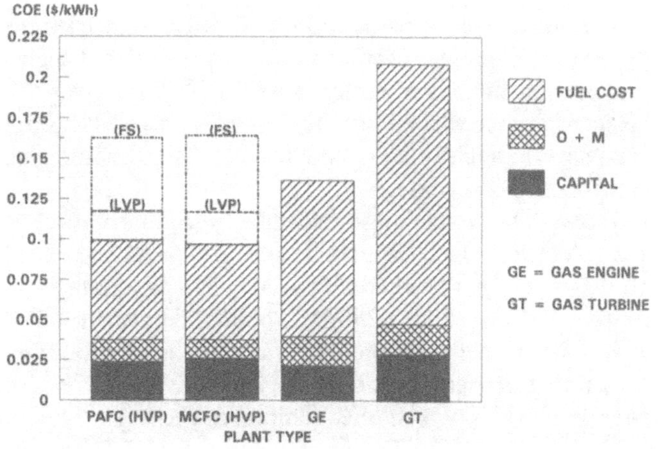

Figure 12.11. Comparing fuel cells with other technologies, 250-kW$_e$ power plants.

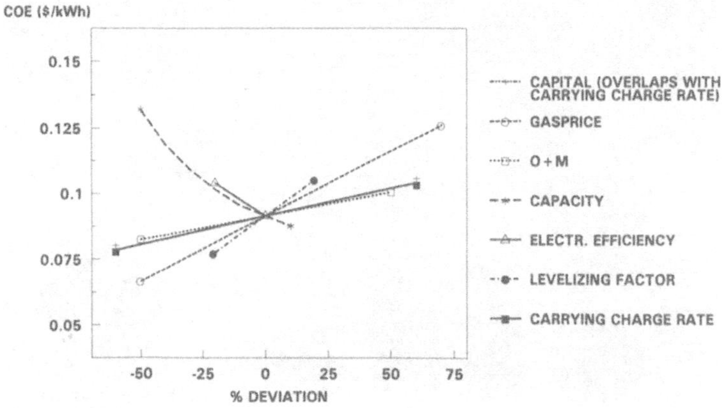

Figure 12.12. Sensitivity of cost of electricity to variations in various parameters at high-volume production, 3.25-MW$_e$ MCFC.

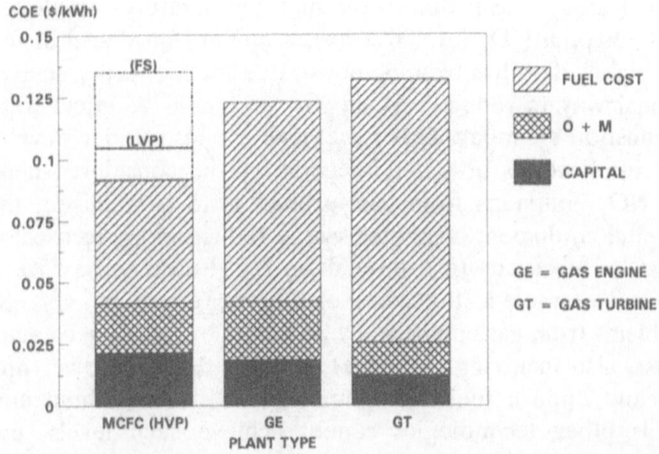

Figure 12.13. Comparing fuel cells with other technologies, 3.25-MW_e power plants.

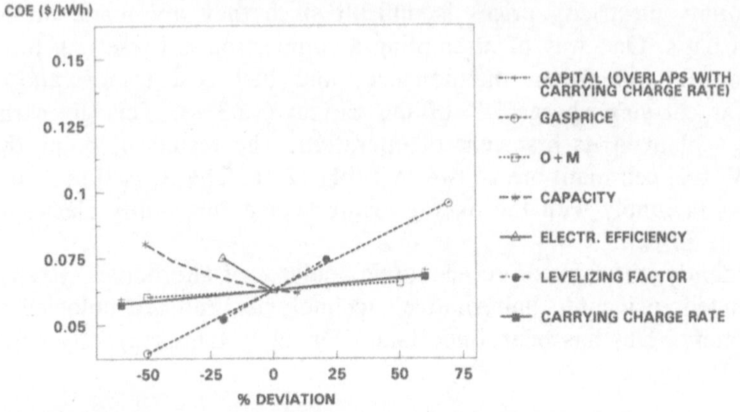

Figure 12.14. Sensitivity of cost of electricity to variations in various parameters at high-volume production, 100-MW_e MCFC.

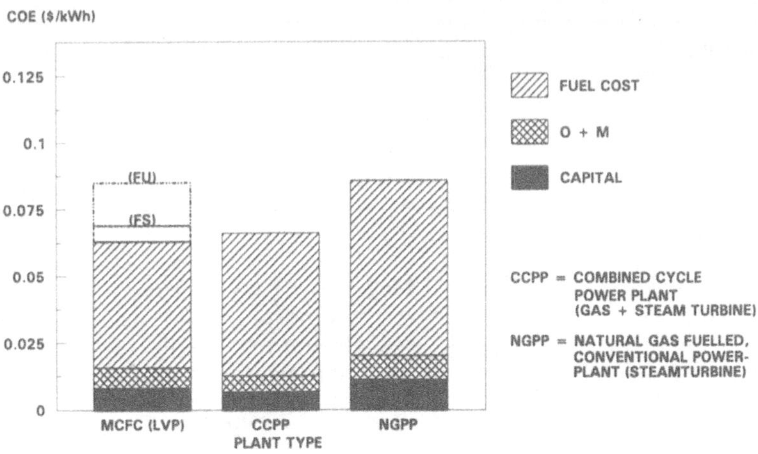

Figure 12.15. Comparing fuel cells with other technologies, 100-MW_e power plants.

reduction in efficiency. Also, due to the high temperatures employed in the gas turbine about 130 ppm NO_x (at 400% excess air) are produced in the flue gas.[20] This is much higher than has been demonstrated for fuel cell systems (7 ppm).

The normal way to reduce NO_x in gas turbines is to inject water or steam into the combustion chamber. Other methods are also under development, but only the cost implications of water injection will be considered here. If a 50% reduction in NO_x emissions from gas turbines is to be realized, the necessary additional capital equipment combined with a reduction in electrical efficiencies[20] will cause costs of electricity to rise by about 2%. Further, if a 75% reduction in NO_x emissions is targeted, the costs of electricity rise by 4%. Similar NO_x emission problems from gas engines may be solved by adding a catalytic converter to the exhaust, also incurring extra cost. Even with much lower limits on NO_x emissions (below 7 ppm), fuel cell technology will cope without any additional costs, whereas other technologies cannot achieve such levels, even at high additional cost (gas turbines may eventually achieve 10–15 ppm NO_x emission levels, or perhaps even lower, though at substantial extra cost).

Comparing the levelized costs of electricity calculated in this chapter with public utility electricity prices is difficult since they are made on a different financial basis. One way of attempting a comparison is by setting the levelizing factor on the operating, maintenance, and fuel cost to 1.0 and using the "first-year carrying charge"[9,13] of the capital (16.3%). This gives the cost of running a plant in its first year of operation. The results of doing this for the 100-MW_e fuel cell plant are shown in Table 12.16. The revised cost of electricity compares favorably with the average European public utility electricity prices[21] (see Table 12.17).

Evidently, the heretofore economic analyses of alternative systems can also be extended to include "alternative" technologies and technologies still under development. This has been undertaken for all systems currently known to the authors.[3]

Table 12.16. Adjusting the 100-MW_e Fuel Cell Power Plant's Cost of Electricity to a First-Year Cost basis

Cost contribution	Levelized COE[a]	First-year cost
Capital	0.008	0.011
O + M	0.007	0.006
Fuel cost	0.044	0.035
Total	0.059 $/kWh	0.052 $/kWh

[a]Levelizing factor = 1.344.

Table 12.17. Public Utility Prices Averaged over European Countries

User power consumption	Electricity price ($/kWh)
25 kW_e	0.110
250 kW_e	0.079
3.25 MW_e	0.064

From Ref. 21.

12.5. COGENERATION

12.5.1. Introduction

In previous sections, attention was directed primarily at achieving the highest possible electrical efficiency from the fuel cell power plants. However, in many applications heat in the form of hot water or steam is also desired. In these cases the waste heat from the plants can have significant value, since the alternative is to generate this heat separately at cost.

In general, electricity has more economic value than heat, for which higher temperature (e.g., 500°C) heat tends to have higher value than low-temperature heat (e.g., 100°C). As a consequence, energy production systems which can generate both electricity (E) and heat (H) should first aim for maximum electrical efficiency and then try to maximize the total (E + H) efficiency. This will generally allow the most economical use of the valuable fossil fuel under consideration. However, in practice, these conditions are often not fulfilled—for example, in countries where electricity is cheap and abundantly available (hydroelectric power) but heat is expensive. Legislation and monopoly situations may be—and still are—unfavorable in many countries, and very often excess electricity cannot be sold back to the grid (or only at unattractively low prices). This creates situations in which electricity loses its economic value advantage over heat, and, as a consequence, many, if not most, cogeneration installations worldwide have been chosen to utilize all waste heat rather than to optimally produce clean power. Nevertheless it seems reasonable to expect that fossil-fuel shortages and environmental and cost constraints will increase the relative value of electricity to heat.

Table 12.18 shows that the relative demand for heat and electricity varies

Table 12.18. The Relative Amounts of Heat and Electricity Used in Various Applications

$$R = \frac{\text{electricity consumption}}{\text{total energy consumption}}$$

Commercial application in Japan[22]	R_{ave}
Supermarket	0.88
Office	0.66
School	0.55
Restaurant	0.34
Hotel	0.30
Hospital	0.24
Food catering	0.11
Laundry	0.08
Industrial applications[23a]	
Sodium	0.78
Acrylic acid	0.60
Chlorine + caustic soda	0.42
Low-density polyethylene	0.28
Flat glass	0.17
Petroleum refining	0.035

[a]Heat required at up to 1000°C in some cases.

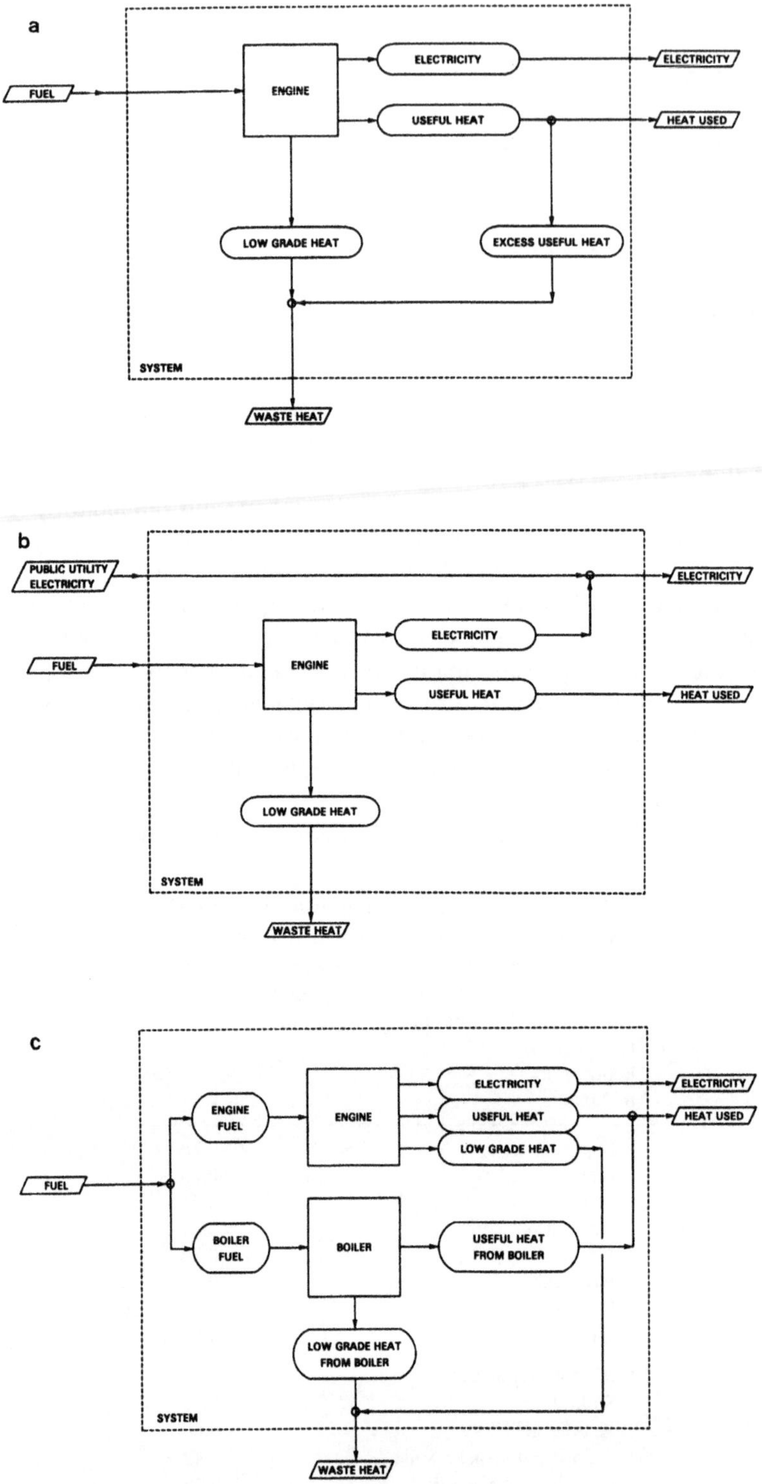

Figure 12.16. Diagrammatic representation of the heat flows in three possible cogeneration scenarios: (a) only part of useful heat is needed; (b) alternative solution to case (a) with all useful heat recovered; (c) supplementary boiler installed to provide extra heat.

widely between applications. The amount and quality of heat available from a power plant may not always match the demand, so some useful heat may have to be lost or, at the other extreme, a supplementary boiler may have to be installed (or in a fuel cell system the reformer convection section may be increased in size). Figures 12.16a–c show the heat flows in various scenarios.

Consideration of cogeneration possibilities is important because alternatives to fuel cell systems which have lower electrical efficiencies may be more cost-effective in applications where high heat-to-electricity ratios are needed. This is especially important since most competitive systems are now used in such applications and tend to have relatively low capital costs; for example, mature gas turbines having low electrical efficiencies are usually selected on the basis of heat output.

12.5.2. Basis of Calculation

The cogenerated heat is assumed to be in the form of hot water heated from 60 to 80°C in a closed-cycle system, e.g., central heating. This means that it has been assumed that heat should not be extracted from process streams below 100°C. Any further cooling gives only lower-grade heat. This sets the limits on how much of the fuel energy may be usefully employed.

The main problem in assessing the benefits of cogenerating hot water is the assignment of value. The approach adopted here is to determine the levelized cost of producing water at a range of heat flow rates. The figures used are given in Table 12.19. These levelized costs of heat are subsequently used as economic *credits* to all electric power generation technologies that cogenerate heat, thereby reducing the levelized costs of electricity.

Table 12.19. Valuing Hot Water Produced by Boilers

Installed capital cost of hot-water generator[15]

Power (kW)	Capital ($/W)
10	0.220
500	0.180
>500	~0.175

For values between 10 and 500 kW, interpolate.

Lifetime: Same as plant for which the hot-water value is to be calculated.
Availability: ~8600 h/yr
Heat efficiency: 95%
Fuel price: Same as plant for which the hot water is to be calculated.
Operating and maintenance: No continuous supervision needed, only annual overhaul by a man for five days on day shift only: 1050 $/yr; no major repair costs expected.
Utilities: None, except perhaps very small makeup to the closed hot-water system.

Resulting cost generated

Heat duty (kW)	Levelized cost of heat ($/kWh)
10	0.061
50	0.037
500	0.034
1000	0.033

For intermediate values, interpolate.
A value of 0.031 $/kWh was used for all calculations at the 3.25-MW_e scale.

Enabling a power plant to cogenerate heat often necessitates the addition or replacement of heat exchangers. The FS cost of these exchangers is estimated on the same basis as the original cost estimates[24] (see Section 12.2). For fuel cell systems, the piping on the plant is not expected to change much, so only a 7% installation charge is added. However, for alternative technologies, extra piping and instrumentation is required, so an installation charge of 120% is necessary. In both cases, the costs are discounted for high-volume or robotized production, as appropriate. The operating and maintenance costs are increased in proportion to the increased installed capital cost, and cooling-water costs are reduced if appropriate. Adding a boiler for extra heat generation is costed on the same basis as the hot-water credit. Finally, the pumping and distribution facilities for the hot water are assumed to be in place and, therefore, are not included in the costing calculations.

The average relative electricity to total energy usage ratio (R) (see Table 12.18) is used throughout and is defined as

$$R = \frac{\text{Electricity consumption}}{\text{Electricity plus heat consumption}}.$$

Hence, $R = 1$ implies only electricity is used, while $R \approx 0$, implies hot-water usage is very high compared to electricity usage. Low-grade heat is not included in the ratio. The relative amounts of heat and electricity needed are assumed constant for a particular application. The fraction of the lower heating value of the fuel fed to the plant that ends up as low-grade or wasted heat (W) can be calculated from R and the overall electricity efficiency (e, which includes the fuel used in a supplementary boiler, if present) as follows:

$$W = 1 - \frac{e}{R}.$$

12.5.3. *Effect of Cogeneration on the Cost of Electricity for Fuel Cell and Conventional Technologies*

The effect on the levelized cost of electricity of recovering as much cogenerated heat as possible from the fuel cell systems is given in Table 12.20 and plotted in Figure 12.17. At a capacity of 25 kW_e, cogeneration levels out the difference in cost of electricity between the two types of fuel cells, because the less integrated PAFC system design has more heat available for cogeneration. At a capacity of 250 kW_e, the MCFC system gains most from cogeneration because it produces less low-grade heat than the PAFC system (due to condensing water from process gas at a higher pressure). Deliberately "deoptimizing" the heat integration, as in the alternative PAFC system design, produces more hot water but does not produce a lower overall cost of electricity than the optimally integrated base case with cogeneration credits. At the 3.25-KW_e and 100-MW_e scales the high degree of heat integration, designed to maximize electrical efficiency, means that the majority of waste heat is produced at a temperature too low to cogenerate significant quantities of hot water. In fact, the assumption not to

Table 12.20. The Effect of Maximum Use of Cogenerated Hot Water On the Levelized Cost of Electricity of the Fuel Cell Systems (Robotized or High-Volume Production)

Power output	Type of plant[a]	COE($/kWh) No cog. ($R = 1$)	COE($/kWh) max. cogen. ($R = R_{min}$)	R_{min}[b]	W[c]
25 kW_e	P	0.125	0.101	0.713	0.422
	M	0.110	0.101	0.853	0.401
250 kW_e	P	0.099	0.084	0.710	0.337
	P (alt. design)	0.120	0.090	0.535	0.310
	M	0.097	0.073	0.617	0.204
3.25 MW_e	M	0.092	0.083	0.800	0.332
100 MW_e	M	0.063	0.063	1.000[d]	0.434

[a] P = phosphoric acid fuel cells; M = molten carbonate fuel cells.

[b] $R_{min} = \dfrac{\text{electricity output}}{\text{max. useful energy output}}$.

[c] W = low-grade heat total fuel energy (LHV) input.

[d] No cogeneration is possible for the 100-MW_e plant.

recover heat from any process stream below 100°C means that in most fuel cell systems optimized for electric power generation, between 20 and 43% of the fuel energy will be lost as low-grade heat.

A more detailed look at the cost of electricity at various heat cogeneration rates and for a range of technologies generating 250 kW_e and 3.25 MW_e of electricity is given in Figures 12.18 and 12.19. The value of hot water, size of heat exchangers, etc., was calculated for each data point separately, so the plots represent a series of plants operating at a given heat recovery duty rather than one plant operating at a range of duties. However, the difference in the curves would be small.

The sloping portion of the curves represents the situation depicted in Figure 12.16a, where some of the useful heat is not required. The relatively flat curves at

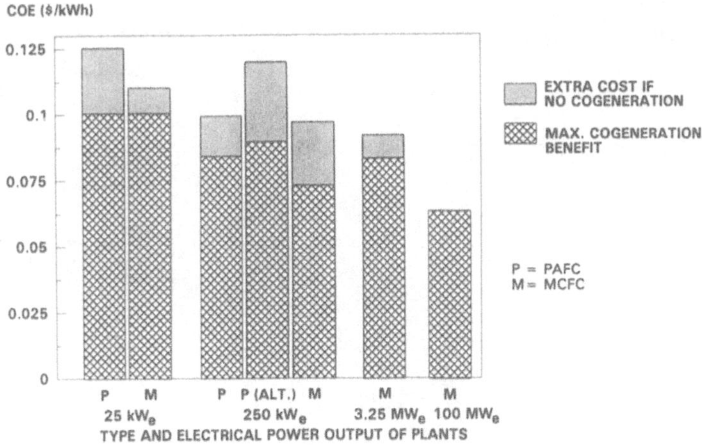

Figure 12.17. Effect of cogeneration on the cost of electricity from fuel cell systems.

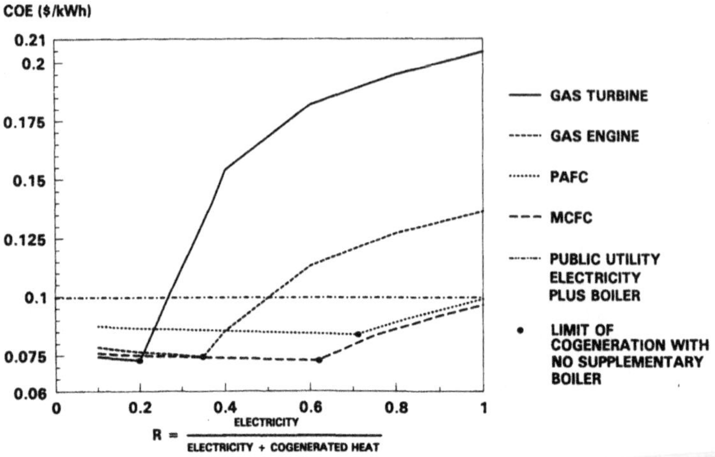

Figure 12.18. Effect of cogeneration on COE for various technologies at 250-kW_e electrical power generation.

low R represent the situation in Figure 12.16c, where all possible cogenerated heat has been extracted from the power plant and a supplementary water heater had to be installed to meet the demand. The relatively small effect of the heater on the cost of electricity is due to the basis of calculation, which dictates that the unit cost of producing extra hot water is only slightly greater than the unit cost of using a larger heater covering the whole heating duty.

At 250 kW_e, all technologies become equally competitive at very high heat demands ($R = 0.1$), except for the PAFC system, which has less useful heat available. Where less heat is needed ($R > 0.5$), only the fuel cell systems can yield electricity at a lower cost than public-utility electricity combined with a conventional boiler for hot water.

At 3.25 MW_e, the MCFC system has a reduced advantage in electrical efficiency over the alternative technologies. It is therefore less economical at high

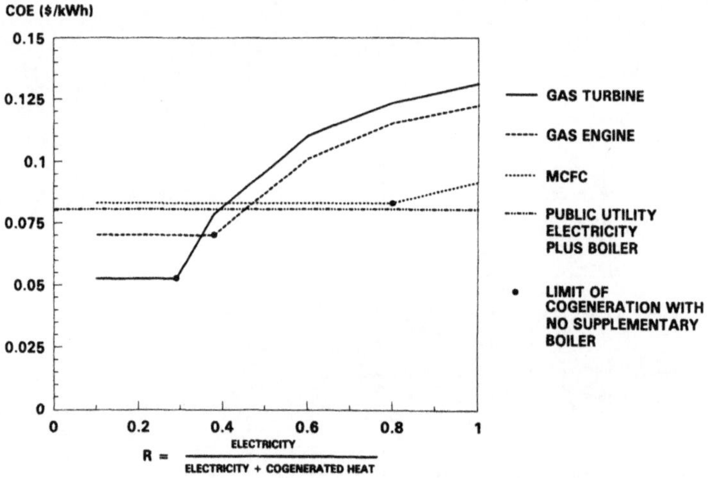

Figure 12.19. Effect of cogeneration on COE for various technologies at 3.25-MW_e electrical power generation.

heat generation rates ($R < 0.4$). Also, the lower price of public-utility electricity at this scale means that the MCFC is slightly more expensive. However, this would depend on the electricity-to-gas price ratio of the country in question.

12.5.4. Discussion

The calculations described above clearly show that obtaining high overall electrical efficiencies and generating large amounts of cogenerated heat are incompatible goals. Fuel cells, due to their inherently high electrical efficiencies, are best used where high electrical-power-to-heat ratios are required. Deintegrating the heat exchanger system allows more heat to be generated up to a limit and results in a small penalty on the overall costs. When large amounts of heat and relatively low amounts of electricity are required, conventional technologies are no longer at a cost disadvantage. This implies that fuel cell systems need to arrive at lower capital costs to compete effectively in these market segments.

The calculations have been based on hot-water production (at 80°C), employing the fuel cell system's waste heat above 100°C only. For many applications low or high-pressure steam is desired. The fully integrated PAFC and MCFC system designs have heat available at temperatures up to about 170°C. Higher temperatures would require redesign of the system so that the reformer and fuel cell could be used as boilers. On the other hand, gas engine and gas turbine technologies can produce cogenerated heat up to about 500°C.

Figure 12.20 illustrates the ranges in electricity-to-heat ratio and temperature covered by each technology with a supplementary boiler added where needed. The choice of the most economical technology clearly depends on the application. For instance, the gas turbine is evidently not suited to producing low-pressure steam (at 150°C) with relatively large amounts of electricity ($R > 0.3$) when compared with a molten carbonate fuel cell system unless expensive public-utility electricity is bought in. Thus, in several cogeneration applications fuel cell

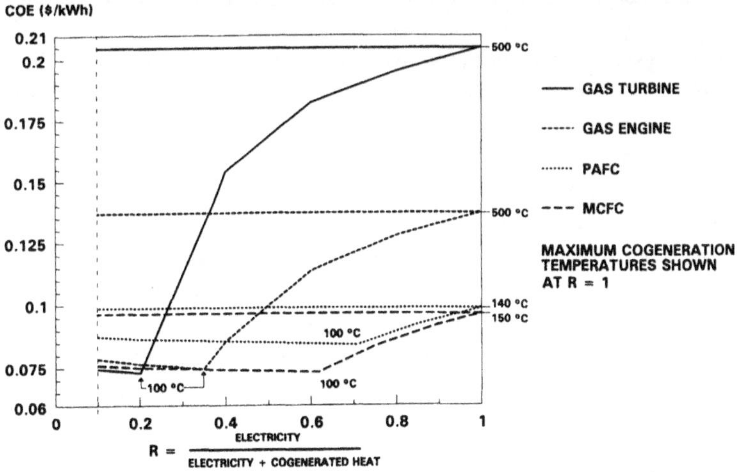

Figure 12.20. Effect of increasing the temperature of the cogenerated heat for the 250-kW$_e$ power plants.

systems offer the best solution. Legislation would be the major factor in determining the number of these applications (see Chapter 14).

12.6. COMPOSITE SYSTEMS

The systems evaluated for this chapter are all based on "traditional" designs of fuel cell power plants (see Chapter 6) that are expected to constitute the early commercial generation of fuel cell systems. As briefly referred to in Chapter 6, new systems are under development in which advanced rotating equipment, particularly gas turbines, are being combined with fuel cells. Apart from yielding higher efficiencies, these systems will tend to cost less per kW_e for the following reasons:

1. More electric power is produced from essentially thc same sizes of equipment and piping.
2. Rotating equipment, constituting a relatively large proportion of the fuel processor costs (see Table 12.7), is replaced by "more standard" gas turbine equipment, which can be up to a factor of 2 cheaper since it is an established serial product.

However, early composite systems may not demonstrate this cost advantage, since they may require more control equipment to guarantee reliability, to combine generator and inverter (or otherwise to combine the power output from both power generating devices), and to overcome the learning curve cost threshold of the totally different flowsheets involved.

12.7. EXPECTED TECHNOLOGY IMPROVEMENTS

Evidently, major improvements are needed to reduce fuel cell stack and inverter costs to commercially competitive levels, especially for the "later-generation" fuel cell types, but also, to a lesser extent, for the "early-generation" fuel cells.

However, system improvements should not be underrated in importance. Flowsheets have been developed[3] that reduce the investment cost of some of the systems described in this chapter by a further 20–40% at all production volumes simply by combining existing unit operations into new configurations. Reformers are being simplified, impurity-tolerant catalysts are being developed, material costs are decreasing, and safety philosophy is being adapted to allow lower cost instrumentation and electrical components. Compactness of design and layout miniaturization have already yielded some very promising results. The data in this chapter can therefore be considered conservative.

For some fuel cell types internal reforming systems are being developed (MCFC and SOFC). This eliminates part of the fuel processor costs, but other systems are still necessary, including water treatment, purification, desulfurization, preheating and other heat exchange, control system, instrument air, inert gas evaporation, and electrical systems. Moreover, the fuel cell stacks themselves are not likely to be cheaper than "simple" stacks operating on

Table 12.21. Relative System Cost Comparison of External Reforming (ER) and Internal Reforming (IR) High-Temperature Fuel Cell Systems (data are approximate only)

	ER	IR	
		Optimistic	Pessimistic
Hydrogen plant	1.0	0.3	0.6
FC skid	1.0	1.1	1.2
Control + El	1.0	1.0	1.0
Σ	3.0	2.4	2.8
		| −20%	| −7%

synthesis gas or pure hydrogen, and a smaller proportion of the plant will benefit from economy of scale. This makes internal reforming more attractive for relatively small plants, and external reforming for larger plants.

A typical power plant of, say, a few hundred kW_e capacity in high-volume production would have relative costs[3] distributed as in Table 12.21. Overall, internal reforming systems are expected to be up to 20% cheaper. The data are indicative only.

Stack replacement has a considerable effect on cost. In their most cost-competitive scenarios fuel cell systems run continuously for 8000 h/annum or more, and stack replacement at five-year intervals has been assumed (as part of the operating and maintenance cost) in this chapter. These target values are only now being realized for the PAFC. It is expected that lifetime (if defined as the time at which the stack voltage has fallen to 90% of its originally rated voltage) can be extended to 100,000 h, perhaps before the end of this decade. This issue will reduce the cost of electricity, especially in earlier series, and imposes rather severe constraints on the development targets for the second-generation fuel cells, in particularly the MCFC and SOFC. Present target values for several MCFC developers range from 10,000 to 20,000 h, but these will not enable achievement of the same COE values as for PAFC systems. MCFC lifetime must at least approach the target values now being set for the PAFC.

Stack cost is another important issue; however, due to the different level of maturity of each cell type, a relative cost comparison is difficult, and publications on this topic have been limited.[2] Platinum-containing cells are reputed to be most expensive, but analysis of the current material price of construction[2] for each fuel cell type suggests that current MCFC and SOFC technologies are using perhaps the most expensive raw materials per kW_e. In order to decrease the cost of these high-temperature fuel cell stacks, emphasis should be given to the following issues:

Less weight per kW_e generator capacity must be accomplished (e.g., thin-layer SOFC).
A high degree of recycling should be applied (to reduce net nickel and stainless steel cost per kW_e in MCFC).
Noble metal loadings should be lowered (e.g., $5 \rightarrow 3$ g/W of platinum in PAFC).

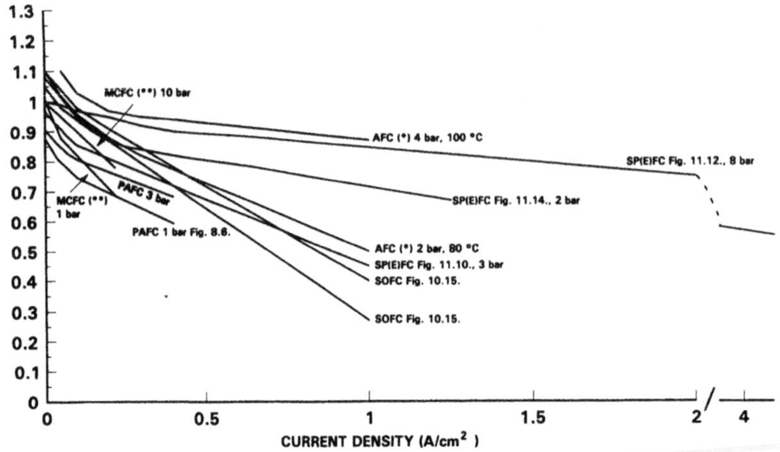

Figure 12.21. Examples of voltage–current density relations for different types of fuel cells.

PAFC costs are approaching "reasonable" limits (relatively close to competitive levels), and in the future up to 95% of platinum may be recycled, further reducing the costs of stack replacement. In AFC, as well as SP(E)FC, raw material costs can be reduced by applying lower loadings of noble metal catalysts (which is readily accomplished at lower current densities, e.g., for stationary applications) and by mass production of membrane materials. Some alternative materials to reduce platinum usage have been employed, but in state-of-the-art fuel cells no real cost reductions have as yet been achieved. For example, Raney nickel with silver in AFC has approximately the *same* raw metal costs (due to higher loading requirements) as the platinum-based version.

System improvements are also possible by way of design simplifications. One way of simplifying an integrated fuel cell system is by feeding high-purity hydrogen to the fuel cell anodes (see Chapter 6). Fuel utilization may be increased to 95 + %, and the lower efficiency of the losses in hydrogen purification are largely offset by the higher fuel cell efficiency (see Chapter 6). In fact, this enables one plant to operate with basically more than one fuel cell type. It also gives almost similar efficiencies, since the difference between the efficiencies of all fuel cell types is not very large if they all would run (at similar current densities) on pure hydrogen. This is illustrated in Figure 12.21, where some voltage–current density plots for all fuel cell types (from the data in Chapters 7–12 and the literature) have been plotted. The variations in performance for different operating conditions (and materials' selections) with one fuel cell type is almost as large as the difference between the curves of the various types of fuel cells. Therefore, the difference between the fuel cell types tends not to be significant, and approximately the same system efficiencies on pure hydrogen result (at a chosen current density).*

In general, hydrogen purification is expensive. In standard hydrogen plants, it is accomplished with multibed molecular sieve adsorption–desorption cycles in

*Evidently this statement does not deny that *practical* purity limitations in fuel cell systems operating *on fossil fuels* do give rise to significant differences between the fuel cell *system* types. It does, however, imply that in a future hydrogen economy, or if cheaper hydrogen separation technologies become available, the applicability of certain fuel cell types may be strongly affected.

pressure swing adsorption (PSA) systems (see Chapter 6). This technology is particularly expensive for low capacities, and alternative, cheaper systems, using either molecular sieves or selective membranes, are being developed. If such a system can be produced at low cost, with sufficient recovery and minimal losses, fuel cell systems may readily adopt them, and the reliability of the systems will easily approach state-of-the-art hydrogen plant technology.

An issue that has not been considered in any detail here has been the cost of extracting carbon dioxide from power plant flue gases using state-of-the-art carbon dioxide removal technology combined with disposal of the recovered carbon dioxide in aquifiers, depleted gas fields, deep sea injection, chemical storage, or any other means. We consider evaluation of such costs as outside the scope of this book. However, carbon dioxide removal from the gases for larger power plants (100 MW_e and above) may increase plant investment and costs of electricity by several tens of percent.

For smaller plants carbon dioxide removal will be even more costly, unless selective (preferably high-temperature) membranes (or other devices) come into the market at sufficiently low prices (this could well occur this decade). However, other power generation technologies (most of them with lower efficiencies, and therefore with higher carbon dioxide emissions per kW_e produced) will find it even more difficult to achieve similar levels of carbon dioxide removal. The relative competitive position of fuel cells can only improve, the more so against technologies based on "heavier" carbon-containing fuels (oil, coal).

Finally, biogas-fueled cell systems will also achieve commercial status. As shown in Chapter 6, natural gas PAFC systems have been adapted for landfill–biogas usage, and acceptable results have been obtained (albeit for a short test period only). Standard systems have only to be adapted with simple off-the-shelf components for removal of impurities, and the larger carbon dioxide content of the biogas favors the reforming equilibria.

12.8. CONCLUSIONS

The calculations reported in this chapter give a sound basis for evaluating the economic viability of optimized fuel cell systems. Clearly, great reductions in cost can be expected as production rates increase, but some aspects of the size of the reductions will remain under discussion. It is important to note that the fuel cells themselves, given current assumptions for their price, will constitute less than 20% of the total capital cost. However, it is perhaps equally important that many fuel cells still have a long way to go before they will approach these projected target prices, whereas the costs of other plant components can be more readily predicted.

Bearing this in mind, the most promising economic advantages of fuel cell systems over other technologies occur in the 250-kW_e and 3.25-MW_e electrical output range, even at low-volume production. This region lies between the small sizes, where the greater electrical efficiencies of the fuel cell systems have difficulty in compensating for their greater capital costs, and the very large power plants, where high efficiencies can be achieved using the most modern competitive systems as well, such as combined cycle power plants.

For 25-kW_e and 250-kW_e sizes, the total costs of electricity are similar for

PAFC and MCFC systems. The slightly higher capital cost for MCFC systems are offset by higher efficiency and therefore lower fuel cost contributions. For application in cogeneration, MCFC systems offer a slightly greater benefit as regards availability of low-temperature heat, but can only generate more high level heat after redesign of the system, which will in itself imply lower electrical efficiencies.

A key advantage of fuel cell systems–high electrical efficiency—is of less interest when large amounts of cogenerated heat are required, especially at high temperatures. However, in cogeneration applications which require low-temperature cogenerated heat (<150°C) and relatively high electricity-to-heat ratios (>0.7), fuel cells exhibit an economic advantage over other technologies. System improvements will further improve this economic advantage.

It therefore seems reasonable to expect that fuel cell systems will become the most economic and environmentally friendly technology for a wide range of power and heat generation applications, provided the initial cost barriers can be absorbed by the markets.

LIST OF ACRONYMS

CCPP	combined cycle power plant (gas + steam turbine)
COE	cost of electricity
FS	first series
FU	first unit
GE	gas engine
GP	gas price
GT	gas turbine
HVP	high-volume production
LVP	low-volume production
NAP	naphtha price
NGPP	natural-gas-fired conventional power plant (steam turbine)
O + M	operating and maintenance
PL	plant life
RP	robotized production
UPS	uninterruptable power supply

REFERENCES

1. M. N. Mugerwa *et al.*, *Fuel Cell Systems, Design Optimization and Environmental Aspects of Fuel Cell Systems,* KTI BV report (Dutch Management Office for Energy Research (PEO), Utrecht, 1988).
2. Arthur D. Little, confidential report to the Platinum Association, 1991.
3. Mannesmann confidential information, 1990.
4. S. L. Mullick and D. W. Warren, *System and Component Cost Evaluation for a Utility Volume Manufactured Westinghouse 7.5 MW Fuel Cell Power Plant,* KTI Corp. report (Southern California Edison Company, 1987).
5. G. G. Venture and S. Chandra, *Solar Energy in Florida* (Florida Solar Energy Center, 1980).
6. American Solar Energy Society, Inc., *Solar Age Magazine,* Boulder, CO, 1984–1985.
7. D. Marier and P. Gipe, *ASE Mag.* (83), Aug./Sept. (1986).
8. Cost of Personal Computers, *Byte Magazine,* McGraw-Hill, New York, 1981, 1984, 1986.

9. Electric Power Research Institute (EPRI), *TAG-Technical Assessment Guide* P-2410 SR 1982.
10. Holec Nederland BV, private communication, 1987.
11. *Solar Cost Reduction through Technical Improvements: The Concepts of Learning and Experience*, Report No. SERI/RR-52-173 (Solar Energy Research Institute, 1979).
12. KTI Corp., *Plant Cost Potential of Phosphoric Acid Fuel Cell Systems*, Report No. GM 605041-3, 1987.
13. R. E. Billings, *Hydrogen from Coal: A Cost Estimation Guide Book* (Pen Well Books, Tulsa, 1983).
14. Gas Prices, 1980–1986, *Eurostat*, April 1986.
15. WEBCI/WUBO, Dutch Association of Cost Engineers, *Prijzenboekje*, 12th Ed., August 1986.
16. Chloride, private communication, January 1988, and brochures concerning uninterruptable power supply Products.
17. KTI Corp., *Electricity from Waste Fuel Gases: A Handbook for Evaluation and System Design* (KTI Corp., Monrovia, CA, 1983).
18. *Nieuwe Ontwikkelingen op het gebied van WK-systemen: Gasturbine en STEG* (Centrum voor Energiebesparing, 1986).
19. *Nieuwe Ontwikkelingen op het gebied van WK-systemen: de Gas-, Diesel-, en Dual-fuel-motor* (Centrum voor Energiebesparing, 1986).
20. L. Blomen *et al.*, *Molten Carbonate Fuel Cell Power Generation: A Comparison of a 3.25 MW Molten Carbonate Fuel Cell Power Plant with an Existing Gas Turbine Cogeneration Facility* (KTI B.V., The Netherlands, 1985).
21. Electricity Prices, 1980–1986, *Eurostat* (1986).
22. *On-site Fuel Cells*, Tokyo Gas Co. Ltd., Lecture, May 20, 1987.
23. *Fuel Cell Support Studies—On-site Molten Carbonate Systems* (Institute of Gas Technology, Chicago, 1979), Project No. 30514, final report.
24. *A Guide to Capital Cost Estimating*, 3rd ed. (Institute of Chemical Engineers, 1988).

13

Market

Diane Traub Hooie

13.1. INTRODUCTION

An international fuel cell market perspective was developed by studying the available market projections and overlaying these projects on the growth of various market segments as well as fuel cell availability. Market penetrations were developed for the period 1990 to 2000. The international fuel cell market is a difficult market to assess since there are five basic fuel cell types which in the future may supply virtually all of the world's energy needs. For this study, applications for these basic fuel cell system types were considered for key market regions: Europe, Japan, and the United States. Overlaid on these applications are the specific energy needs for each segment by country. This chapter will review the markets, based on the results of this study. The likelihood of the fuel cell manufacturers supplying sufficient numbers of power plants to meet these needs has not been assessed here.

13.2. MARKET SEGMENTS

Currently, developing nations use only one sixth as much purchased energy as developed nations. However, on a per capita basis in 1990 primary energy use averaged 6.3 kW in industrialized countries but only about 1 kW in developing nations, including 0.4 kW of noncommercial energy use (in many areas, less than half of the housing units have access to electricity). If developing countries follow the pattern of the developed countries, global energy use would need to double or quadruple over the next four to five decades in order for the world's impoverished majority to achieve a decent standard of living. Supplying that much additional energy would present enormous economic and environmental problems (see also the introduction to this book).

Throughout the past two decades, per capita commercial energy use in developing countries grew at a vigorous average annual rate of 3.6%. If that pace

---

Diane Traub Hooie • NORA Management, Inc., P.O. Box 525, Matawan, New Jersey 07747.

Fuel Cell Systems, edited by Leo J. M. J. Blomen and Michael N. Mugerwa. Plenum Press, New York, 1993.

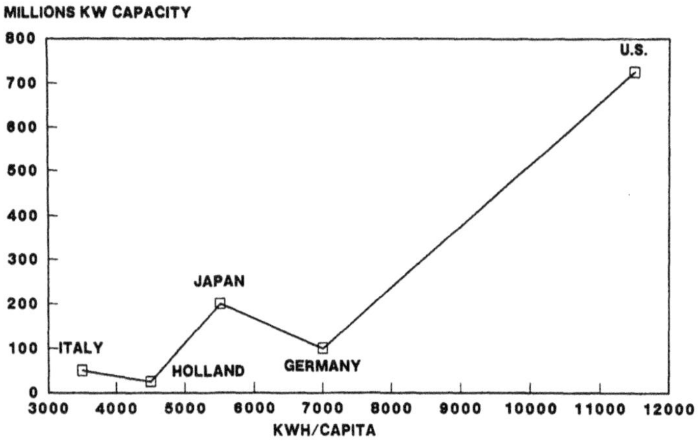

Figure 13.1. Energy consumption by country, 1987.

of the energy use continues, then by the year 2020 developing countries could equal 1.3 times the 1980 world energy use. The relationship between per capita use and energy production is shown in Fig. 13.1. With the exception of Middle Eastern and North African countries, the increase in demand in the rest of the world cannot be met with domestically produced oil and gas. Because of the high efficiencies of fuel cells and the low emissions, they have been targeted as an ideal solution to the projected energy shortfall.

The most desirable market segments for fuel cells are those which can take advantage of the inherent properties and features of fuel cells. Ideal fuel cell markets are those which can exploit the fuel cell's economic viability. These markets can be broken down into the following categories: electric utility, on-site (cogeneration and generation), portable or mobile, and vehicular. The subcategories are shown in Table 13.1.

Table 13.1. Market Segments and Subcategories

Generation	100 kW to 250 MW
Electric utility	
Remote	
Dispersed (stationary) generation	
On-site	4 kW to 10 MW
Residential	
Multifamily residential	
Commercial	
Light industrial	
Heavy industrial	
Portable/mobile	1 to 500 kW
Commercial, portable	
Underground mining	
Vehicular	1 kW to 2.5 MW
Automotive	
Buses, trucks	
Railway	
Space	
Submersibles	

Desirable markets to penetrate for each of these applications vary by country as well as by region within each country. However, since the generation and cogeneration markets are expected to have the highest volume market penetration for the next decade, the major focus of this chapter will be on these segments. Reference is also made to the relevant sections of the review chapters on each fuel cell type (Chapters 7–11).

13.2.1. Generation Industry

The electric power generating equipment industry is international and involves a limited number of manufacturers worldwide. Equipment required to meet this industry's needs are considered "heavy electrical equipment" and are multimegawatt in size. Thus, for the purpose of this study, power plants between 1 and 250 MW are included in this category. Fuels of choice for this market include natural gas and coal. To effectively compete, fuel cells would have to be available for perhaps $1000 per kilowatt or less and have an operating cost of $0.04 per kilowatt hour. Higher prices are possible, but will limit market penetration, lower prices will increase penetration.

In general, the world capacity to produce heavy electrical equipment exceeds demand. U.S. manufacturers were the first to exhaust their order backlogs; Japanese and Western European manufacturers followed the U.S. example closely. Worldwide, manufacturers continue to consolidate operations and shutdown marginal facilities in response to weak demand from utilities, which buy virtually all of the equipment.

Types of generating equipment currently competing in this market include geothermal, hydroelectric, nuclear, supercooled generators, and turbines. These will be briefly discussed here.

13.2.1.1. Turbines

Overall turbine productivity in major European factories appears to be close to that of the United States. Major heavy electrical producers in Germany and Switzerland factories have been operating on a single-shift basis and, along with their raw material and component suppliers, can easily triple their output by going to additional shifts. They also have continuing apprenticeship programs to develop the required skilled labor force.

In the United States, shipments of electric-utility-size turbine generators in 1987 were 2215 MW, a 35% decline from 1986 and less than 5% of the 46,550 MW averaged in the peak years of 1972–1974. One of the main contributions to this decreased demand is the conservation efforts that began after the oil price war in 1974. As a result, projected additional capacity needs have not been realized.

In addition, for economic reasons, some U.S. utilities have contracted for electricity from unregulated qualifying facilities and deferred building new capacity of their own. Between 1979 and 1986, 12,500 MW of energy was installed by qualifying facilities. This is roughly equivalent (in capacity) to 22 nuclear facilities. The nonutility growth rate has run about 15%, compared to the utilities installed capacity growth of 2.4% annually. As long as this trend continues, the product mix will favor gas turbines, waste-heat boilers, small steam turbines, small boilers designed to handle a broad range of fuels, and on-site fuel cells. As a result, in the last decade, two major U.S. turbine manufacturers, Westinghouse and General Electric, closed their manufacturing facilities for large turbines.

13.2.1.2. Supercooled Generators

European firms have the technological lead in the area of supercooled generators. They expect to make significant gains in efficiency and reductions in weight in generators that operate at temperatures near absolute zero.

In 1985, the United States abandoned their program in this area on the grounds that the need for it did not exist. Instead, the United States has focused on more flexible, reliable, and efficient fossil steam power plants. In fact, these programs involved the adoption, adaption, and demonstration of existing foreign technology in U.S. power plants. This raises the possibility that European and Japanese producers could position themselves to serve the returning U.S. market better than domestic producers.

13.2.1.3. Hydroelectric Power

In 1981, the total world hydroelectric power potential was estimated at more than 2.5 million MW. Only 19% of this total has been developed. Asia has about 30% of the total world capacity, while North America has 16%. The most intensive development of hydroelectric energy has occurred in Europe and North America. These two regions have developed 44% and 32% of their respective capacities and over 56% of the world's developed capacity.

Presently, hydroelectric power production accounts for 7% of the world's primary energy production. This figure represents about a 2% increase from 1973. Unfortunately, the amount of hydroelectric energy is limited by environmental and site-specific concerns. As a result, relatively little increase in hydroelectric production facilities is expected in the next decades.

13.2.1.4. Geothermal

The world's natural interior heat is enormous, but difficult to harness with present technologies. However, world geothermal capacity grew by 8.3% per year from 1920 through 1978. After 1978, capacity grew at about 16.5% per year. In 1985, an estimated 4.763 MW of geothermal energy was produced worldwide. By 1990, this worldwide output is expected to have increased to 6.398 MW (with capacity expected to have reached 6.398 MW).

Tapping geothermal resources requires tools and procedures similar to those employed for oil exploration. Drilling costs are high, and the lack of scientific surveys of potential thermal areas is also retarding rapid geothermal development. In addition, geothermal energy generation may cause pollution of surface waters and groundwater by nontoxic chlorides, sulfates, carbonates, and/or silica. Experimental systems are being tested whereby closed-loop hot dry rocks are used. System tests have been somewhat successful and appear to be free of the pollutants normally associated with conventional geothermal recovery systems.

13.2.1.5. Nuclear Power

One of the most controversial generation systems of the century has been nuclear power. The 1986 accident at Chernobyl in the USSR has affected much of this industry. Austria has abandoned all efforts to operate its nuclear power plant at Zwentendorf, which was completed but never used. Instead it has begun to dismantle the plant and sell marketable components in order to recover a small portion of the $645 million invested.

The United States had nuclear power plants totaling 8000 MW come on stream during 1986. Although schedules called for significantly more to enter the full-power stage during 1987 and 1988, legal restraints and extracautious utilities restricted outputs to the 1986 levels. No new domestic orders have been placed for nuclear steam supply systems since 1978.

13.2.2. On-site Cogeneration and Generation

On-site generation and cogeneration usually consists of equipment in the size range of 4 to 1000 kW. Some sample market segments and their plant sizes are shown in Table 13.2. In general, the lower end of this market is not economically feasible since it represents the residential and specialty segments. However, between 40 and 1000 kw a significant market opportunity exists in the multifamily residential, commercial, and light industrial segments.

During the past decade, the worldwide cogeneration market has grown continuously. By 1990, it is projected that over 27,000 MW of cogeneration will be installed in the United States alone (projected in 1989). Of this amount, about 74% or 20,000 MW, will be in the industrial sectors. Between 1981 and the first quarter of 1989, there were 1315 small natural-gas-fueled cogeneration system installations operating in the United States. Of these, 244 (or 18.5%) were packaged cogenerators, that is, small units (650 kW or less) delivered as one unit. Although many of these units have turbine drivers, the majority of them are driven by internal combustion engines.

It is also interesting that 49 of these units were fuel cell plants! These units were delivered to a total of 310 sites. Six manufacturers provided 80% of the site installations. These are shown in Fig. 13.2.

In Japan, on the other hand, it was projected in 1988 that over 15,000 MW of cogeneration equipment will be installed by 1990. Again, the industrial sector makes up the largest segment, accounting for 92% or 13,800 MW.

Table 13.2. Sample Market Needs

		Fuels[a]						Costs ($/kW)	
Market	Plant size range (kW)	H_2	CH_3OH	CH_4	LNG	Coal	Other	Mature	Operating
Residential	<10			X				<$500	<Grid
Multifamily	10–200			X				<$1000	<Grid
Commercial	40–1000			X				<$1500	<$0.04
Light industrial	40–1000	X		X	X	X	X	<$1000	<$0.04
Industrial	500–1000	X		X	X	X	X	<$1000	<$0.04

[a]Fuels include: H_2 = hydrogen; CH_3OH = methanol; CH_4 = methane; LNG = liquid natural gas; Coal = coal; Other = other fuels such as biomass.

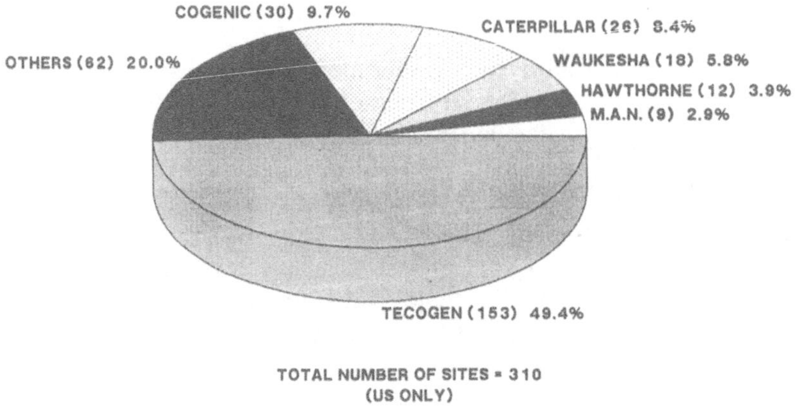

Figure 13.2. Distribution of packaged cogeneration systems by manufacturer (Purpa QCF & non-purpa QCF).

In Europe, cogeneration is expected to have a smaller impact in nonindustrialized countries and a heavy impact in industrialized areas such as Germany and Holland, during the remainder of this century.

Worldwide, the primary industries projected to use cogeneration are mainly chemical, paper, and pulp, and iron and steel industries. Together, these energy-intensive industries account for about 85% of the projected industrial cogeneration usage.

13.2.3. Portable or Mobile

The portable or mobile segment of the market is a relatively small, high-priced market segment. Units in this segment can range from 1 to 500 kW in size and up to $30,000 per kilowatt. Fuels for these applications are generally hydrogen or methanol, although the use of logistic fuels (such as diesel and jet-A) are also desired. Since this market is so small in size compared with the generation and on-site markets, it is insignificant in the total numbers in megawatts, and will not be discussed further here. Because of the high allowable cost, it is, however, an interesting market for introducing (even rather simple) fuel cell systems and has played an important role to date in system introduction of several fuel cell types, such as AFC (Chapter 7) and SP(E)FC (Chapter 11).

13.2.4. Vehicular

In addition to the potential energy savings advantage, the vehicular market is a key market worldwide because of environmental considerations such as acid rain and the greenhouse effect. In the United States alone, about 81% of the total air emissions (in 1986) were from the transportation segment. This is true even for the most pollution-conscious segment of the United States—the South Coast Air Basin around Los Angeles.

Also in the United States, transportation accounts for over 60% of petroleum usage, an amount comparable to the total domestic production. Since fuel cells can provide approximately twice the energy efficiency of internal combustion engines with comparable range and performance, they are expected to replace internal combustion engines in applications such as buses, trucks, trains, and submersibles.

In the long term, they are also expected to replace the internal combustion automobile engine. This is not projected for the near term because of the high initial cost of the power plant as well as the rigorous consumer requirements that would be put on power plant performance.

It will be difficult for fuel cells to compete with internal combustion engines on an initial-cost basis. For example, in the automotive industry, they would have to be commercially available for less than \$150 per kilowatt. Likewise, for the truck and bus market segment, they would have to be available for less than \$300 per kilowatt. As shown in the previous chapters, it is highly unlikely that these types of prices will be realized.

Programs are currently underway in the United States, Italy, and Japan to develop and test city buses powered by fuel cells (see Chapters 7, 8, and 11 of this book). These systems are expected to be between 25 and 250 kW in size and operate on fuels such as methanol, LPG, and propane. Systems have been designed and are expected to be tested, under actual operating conditions, within the next year or so.

A joint program between the United States and Japan to develop and test a fuel-cell-powered forklift truck (see Chapter 8) has been slowed down because of the high cost of the power plant system. These systems would range up to 10 kW and could be readily installed in the current electric-powered forklift trucks in the 1–3 ton range.

13.3. FUEL CELL APPLICATIONS

To determine the market opportunities for each system, the type of fuel cell must be compared to the market requirements. Each type of power plant has its

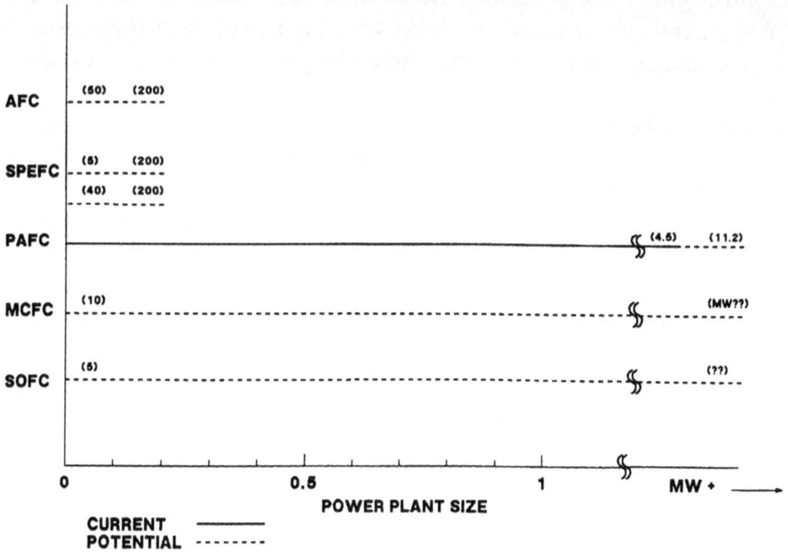

Figure 13.3. Economically feasible plant sizes.

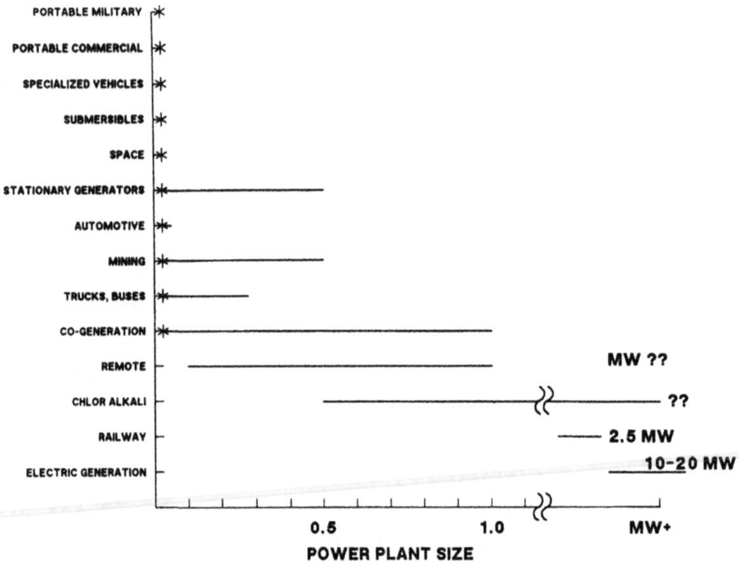

Figure 13.4. Plant size by market.

economically and technologically feasible plant sizes. These are generally believed to be as shown in Fig. 13.3.

This has been overlaid with the output requirements for each market application, as shown in Fig. 13.4. For example, markets such as the portable military, portable commercial, specialized vehicles, submersibles, space, etc., all require small systems (and generally a small thermal footprint). Thus, systems with high operating temperatures, such as the molten carbonate fuel cell and the solid oxide fuel cell, would not meet this market requirement.

Likewise, some systems require quite large units, such as the generation and industrial segments. As a result, systems with high manufacturing costs, such as the proton exchange membrane (at currently perceived prices) would not be desirable.

The ideal markets for each power plant system, based on size and footprint (weight, thermal, and physical), are shown in Table 13.3. Note that the potential mature market price has not been considered in this initial analysis. As can be seen from this table, the alkaline fuel cells and proton exchange membranes are the most desirable cell types for portable or mobile systems.

Phosphoric acid systems have the potential of addressing the intermediate markets such as multifamily residential, commercial, and light industrial, as well as the early-entry electric utility markets.

Molten carbonate systems are expected to compete with phosphoric acid systems in the commercial, industrial, and electrical generation markets. However, each system is expected to have its own market niche.

Lastly, because of their immature development stage, it is not known which market segments the solid oxide systems can adequately address. However, it is currently being studied for specialized vehicle uses (such as subways, buses, portable), and also for cogeneration and stationary applications.

Table 13.3. Comparison of Technological Market Applications by Fuel Cell Type

Market	Fuel cell types[b]				
	AFC	SPEFC	PAFC	MCFC	SOFC
Generation					
Electric utility			X	X	
Remote	X		X	X	
Dispersed (stationary)	X	X	X	X	X
On-site cogeneration		X	X	X	X
Portable/mobile					
Commercial, portable	X				
Military, portable		X			X
Underground mining	X	X			
Vehicular		X			
Automotive	X		X		
Buses, trucks	X	X	X		X
Railway	X	X			
Space	X	X	X		
Submersibles		X			

[a]Fuel cell types: AFC = alkaline fuel cell; SPEFC = proton exchange membrane; PAFC = phosphoric acid fuel cell; MCFC = molten carbonate fuel cell; SOFC = solid oxide fuel cell.

13.4. MARKET OPPORTUNITY

The worldwide fuel cell opportunity has been evaluated based on a proprietary market model which was developed to compare fuel cells to other cogeneration equipment. This model based the allowable cost on a life cycle costing analysis where the system was sized to take advantage of the fuel cell's output.

Obviously, since this model is designed to optimize the fuel cell with respect to the application, the bias is in favor of the fuel cell system. However, since the overall market opportunity is so large, the effect of this bias is expected to be insignificant.

In addition, as in any modeling program, the output is only as good as the projections and assumptions that go into it. Therefore, any market model should be used only as a guideline for business decisions. It should be used to answer the following business strategy questions:

1. How difficult will the business be to satisfy corporate goals?
2. Is the market structured for a "safe" penetration strategy?

13.4.1. Market Model Algorithms

As stated previously, this model based the allowable cost on a life-cycle-cost basis where the system was sized to take advantage of the fuel cell's output. That is, it was sized to optimize the use of the thermal and electrical energy of the output. Thus, in many instances, the comparative equipment required supplemental thermal and electrical energy. This assumption is actually equal to the normally applied procedure in designing cogeneration units based on gas turbines. In most countries they are designed on the available waste heat (since excess power production is often still badly reimbursed by the grid operating utility), and in many studies fuel cells have to match this perfect gas turbine heat-to-power ratio. Fuel cells will be designed differently, since less waste heat is produced. So in this chapter, we have optimized the fuel cell heat-to-power ratio instead. In a few instances, the thermal energy from the fuel cells was supplemented by the use of a gas boiler. Thus, the following relationships were used:

$$\text{Cogen } a = \text{Cogen } b, \tag{13.1}$$

where

$$\text{Cogen a,b} = [\$\text{elec} + \text{Sum(op + maint)life}] + [\$\text{therm} + \text{Sum(op + maint)life}], \tag{13.2}$$

where

a,b = two technologies being compared
Cogen a,b = allowable life-cycle cost* of cogeneration equipment
$elec = capital cost of electrical generating equipment for the grid, this is zero)
thermal = capital cost of the thermal generating equipment (usually the cost of a gas-fired boiler)
Sum = summation of operating (op) and maintenance (maint) costs of respective capital equipment over life of the equipment (includes cost of fuel, regular maintenance, and repairs)

In the case of the fuel cell system,

$$\text{Cogen} = \text{fuel cell system capital cost} + [\text{Sum(op + maint)}_{\text{life}}] + \text{makeup thermal costs (where necessary).} \tag{13.3}$$

The allowable cost of the fuel cell is then compared to the total market opportunity. This market opportunity is dependent upon the price of the system as well as the market environment, that is, the ecological, regulatory, and political environment of the specific geographic market areas. In addition, other subjective factors, such as the industrial growth and innovativeness of the market environment must be considered. These subjective factors have been assigned a weight and projected value by country and market in the model. Because of the complexity of the factors and the proprietory nature of the model, they are not included here.

* Life-cycle cost is defined as the total, cumulative cost of purchasing, maintaining, and operating the equipment over the (economic) lifetime, and is thus related via the load factor to the COE (cost of electricity). See Chapter 12 for further clarification.

13.4.2. Equipment and Market Readiness

The readiness of the fuel cell to meet the market has also been considered in this analysis. For example, during the next decade it is highly unlikely to see a significant number of fuel cell systems in the vehicular market segment because it requires additional development by the fuel cell manufacturer as well as the vehicle manufacturer. In the case of the vehicle manufacturer, it may require the design of a new vehicle as well as a complete redesign of the manufacturing facilities.

In addition, for the on-site systems, in many instances, the building codes and regulations must be addressed to meet the special needs of the fuel cell. For instance, a water-cooled fuel cell system, during heat-up, superheated boiling water is circulated through the system. That is, the same water loop that is used to cool the system is used to preheat the system. For the larger systems, because of the volume of water circulating, they may meet the current definition of a boiler and, as such, could be subject to the requirements of licensed boiler personnel on-site. However, *the fuel cell is not a boiler and is designed to operate in an unattended mode.* Thus, the requirement that boiler personnel be on-site should not be applicable to the fuel cells. The codes should be modified to require boiler personnel on-site *only* during the initial system heat-up cycle.

From a similar point of view, the risk of hydrogen operation should be considered. In many countries, hydrogen is seen in the same category of flammable and explosive gases as is natural gas, but adequate safety protection of the extremely light hydrogen is much easier and should lead to simpler electrical safety and controls requirements. Too strict rules have a negative impact on market penetration of a technology which is already expensive.

The readiness of the fuel cell systems and manufacturing status of fuel cells should also be taken into consideration in assessing the market opportunity. Some systems, such as the molten carbonate and solid oxide fuel cells, will begin field testing in the 1990s. As a result, they are not expected to make a significant impact on the market until the next century. This will be discussed in more detail later.

13.4.3. Market–Price Relationship

The price of the fuel cell system significantly affects the market opportunity. As shown in Fig. 13.5, the market significantly increases as the price of the fuel cell system drops below \$2000 per kilowatt. However, it should be noted that a significant market still exists at \$3000 per kilowatt. Thus, the manufacturers should not have to wait to reduce the fuel cell price before entering the market.

However, as seen in Fig. 13.6, the fuel cell system price is not really a true indication of the market value, that is, the dollar value of the unit sales (price times number of units). This really is just simple mathematics. If the selling price of the fuel cell system is reduced, say, by 50%, the market opportunity must double that of the \$1000 per kilowatt to equal the same market value (i.e., x units at a = $2x$ units at $0.5a$). However, the market does not double for fuel cells, and it is apparent that the optimal market price is around \$1000 per kilowatt. In addition, it can be seen that the market value differential between \$2,000 and \$3,000 per kilowatt is not very significant. Therefore, once again, it is suggested

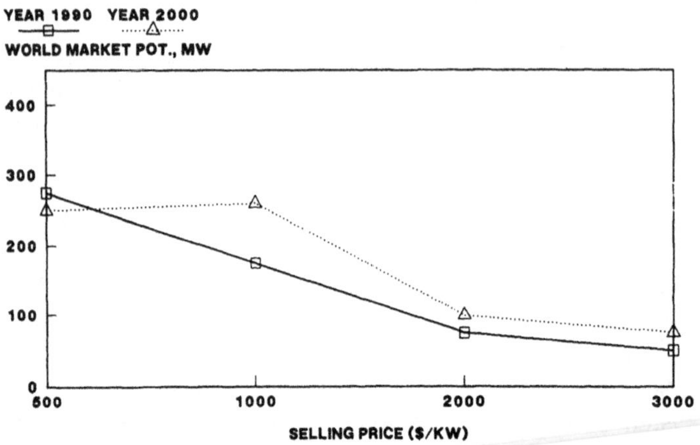

Figure 13.5. Effect of fuel cell system price on market opportunity ($ million US).

that the manufacturers need not wait for the selling price to drop to $1000 per kilowatt before entering the market.

13.4.4. *Worldwide Market Potential*

The worldwide market potential is dependent upon the selling price of the fuel cell and the year under discussion. Most manufacturers have indicated a selling price goal of $1,000 per kilowatt for market penetration. Therefore, the following analysis will be conducted using that figure as the baseline price. The overall market will increase or decrease according to the relationships set out in the previous section as the selling price changes (see Figs. 13.5 and 13.6).

For 1990, the total worldwide market opportunity for all fuel cell technologies at $1000 per kilowatt was 161,048 MW (per year). That is, they can readily compete with other technologies at these prices. About 91% of this is in the commercial sector, while about 8% is in the electric utility sector. By the year 2000, the market opportunity will have increased, by about 55%, to about 250,000 MW (per year). However, the sector segmentation does not change significantly. This is shown in Fig. 13.7.

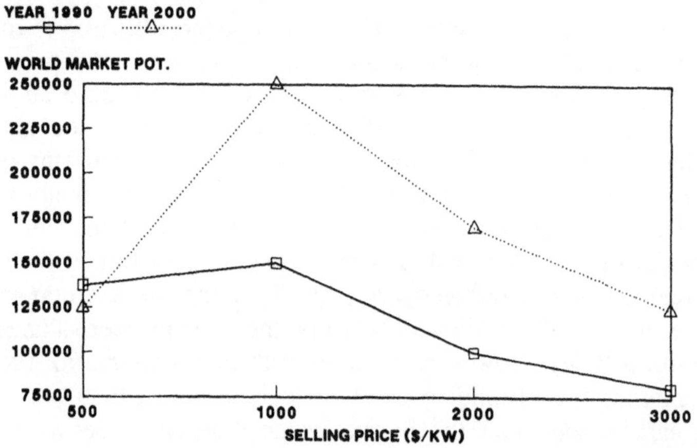

Figure 13.6. Effect of fuel cell system market and price on market value ($ million US).

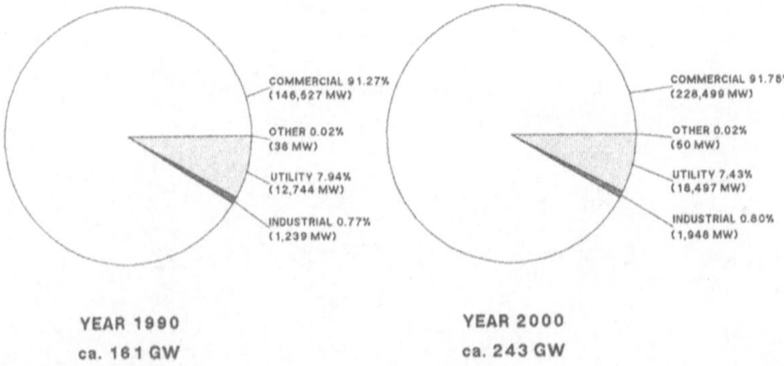

Figure 13.7. Market sectors for total worldwide market by year at $1000 per kilowatt.

As can be seen from Fig. 13.8, the bulk of the commercial opportunity is in the United States, with Europe second. This is because the United States market has already been exposed to fuel cells, and many of the primary customers, the gas utilities, have identified a major business opportunity in selling and maintaining fuel cells. Studies conducted by the various gas utilities, in conjunction with the Gas Research Institute and Department of Energy Field Test Program, have indicated a profit potential of over 70% for the gas utilities. That is, gas utilities, if they install and operate the fuel cells for their customers, will realize revenues of 70%, compared to delivery of natural gas to the end users, through the energy service business. This is in addition to the increased gas sales. End users (gas utility customers) are also afforded a 15–20% savings in their total energy bill. One U.S. gas utility has contracts to install, maintain, and operate 10 units for its end users. They have contractually guaranteed the end users a 15% savings on their energy use. Again, this should not change significantly by the year 2000.

As can be seen from Fig. 13.9, for the three major country market segments

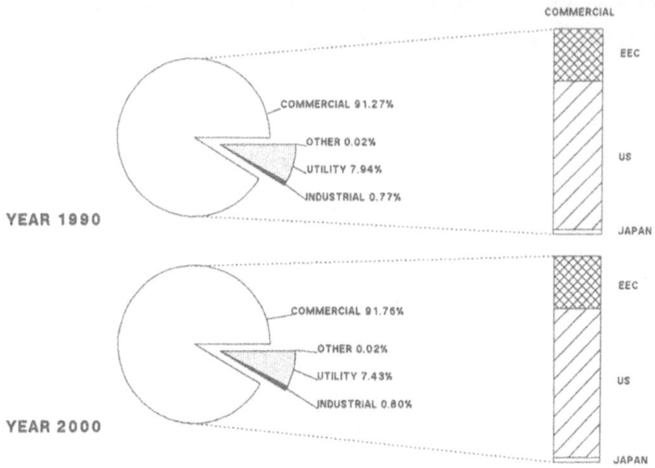

Figure 13.8. Market sectors for total worldwide market by year at $1000 per kilowatt.

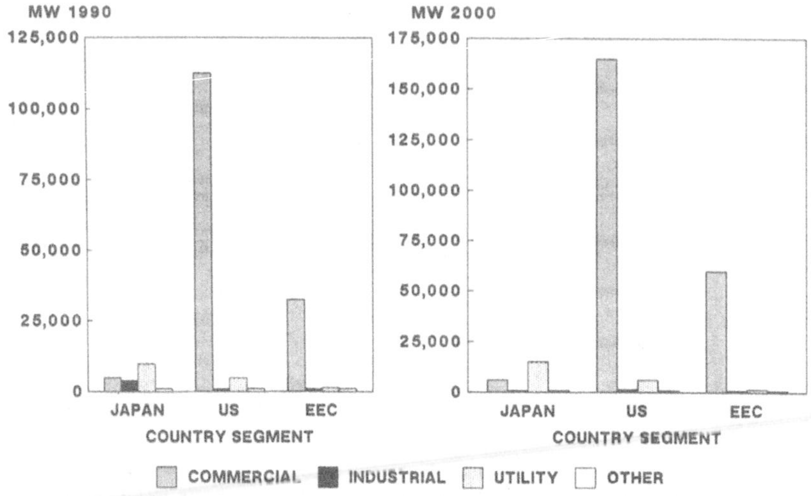

Figure 13.9. Market opportunity by country and segment at $1000 per kilowatt.

the primary market opportunity for both 1990 and 2000 is in the commercial sector. However, in countries other than the United States, this may be a difficult segment to penetrate since it involves selling to many different customers. If the average fuel cell size was 100 kW, this would mean that there is the potential of selling to 371,700 customers in Europe alone during 1990. This will increase to 579,100 by the year 2000.

For the utility industry, on the other hand, it is estimated that the desired size would be about 10 MW in 1990. Thus, to capture this market, the manufacturer would have to sell about 21 units in Europe, 59 units in the United States, and 44 units in Japan in 1990.

By the year 2000, however, the average unit size is expected to increase to over 50 MW. Thus, to capture the market in Europe, the manufacturer would have to sell 32 units, in the United States 92, and in Japan 74, all in the year 2000.

In many respects it is easier to sell to fewer customers since it requires fewer personnel and a smaller customer base. However, in general, these customers are much more sophisticated and would be more difficult to "close" during a sale.

13.4.5. Market Penetration

Obviously, the market opportunity appears to be gigantic. However, this represents the opportunity for fuel cells as well as all of the competing equipment. In practice, the fuel cell share will only be a small portion of this market.

If fuel cells were to capture only 1% of this market, this would already represent a worldwide potential opportunity of $1605 million in 1990 and $2490 million in the year 2000 (at $1000 per kilowatt). This is shown in Table 13.4.

If the fuel cells were to capture 25% of this market, the opportunity would increase to $40,137 million in 1990 and $62,257 million by the year 2000. Again, this is at $1000 per kilowatt.

Table 13.4. Market Penetration at $1000/kW and Given Market Share

Mkt Pot ($MM US)	At 1% market share				At 25% market share			
	Japan	U.S.	EEC	Total	Japan	U.S.	EEC	Total
Year 1990								
Commercial	32	1,062	372	1,465	789	26,550	9,293	36,632
Industrial	4	6	2	12	111	148	52	310
Utility	80	35	12	127	1,991	885	310	3,186
Other	0	0	0	0	4	3	3	9
Total	116	1,103	386	1,605	2,895	27,585	9,657	40,137
Year 2000								
Commercial	51	1,655	579	2,285	1,283	41,364	14,477	57,125
Industrial	7	9	3	20	186	230	80	496
Utility	133	38	13	185	3,330	959	336	4,624
Other	0	0	0	0	5	4	4	12
Total	192	1,702	596	2,490	4,804	42,556	14,897	62,257

Little opportunity exists in the vehicular (other) segment during the decade analyzed. This is not true after 2000, since it is believed that the market environment will be much more receptive to vehicular-powered systems by that time.

13.5. FUEL CELL READINESS

Although the market is ready for fuel cells at this moment, it appears that fuel cells are not ready for the market. Water-cooled, 40-kW phosphoric acid fuel cells were successfully field-tested in the United States during the early- to mid-1980s. However, the manufacturer later decided to enter the market with 200-kW units. This decision was based on their internal market assessment and manufacturing cost analysis.

Other phosphoric acid manufacturers are not yet at the stage of development where they can sell larger numbers of their units, so they sell demonstration and test plants and have begun producing the very first series (see Chapter 8 on PAFC).

For the carbonate systems, stacks are being tested. Thus, the stacks must still be integrated with the system and field-tested before they can be sold commercially. Therefore, these systems are not expected to be available until the mid-1990s. This is shown in Fig. 13.10.

Solid oxide fuel cells, on the other hand, are farther behind the carbonate system. Many technical and scale-up issues still need to be addressed before they can be commercialized. Thus, these units are not expected to be available until the late 1990s. This is shown in Fig. 13.11.

13.6. CONCLUSIONS

The market opportunity for fuel cells is significantly large for all of the fuel cell system manufacturers as well as the cogeneration manufacturers. In general,

Figure 13.10. Carbonate fuel cell commercialization schedule.

the fuel cell needs to capture only a small percentage of the market to provide a significant return to the manufacturer. However, each manufacturer must independently assess whether the opportunity is large enough to meet the corporate goals for a new business opportunity.

We have just celebrated the 150th birthday of the fuel cell during the period that this book was being produced. And yet, they still are not commercialized. They must be commercialized since there is a window of market opportunity for fuel cells which could close once again. If fuel cells do not pass out of the development stage into commercialization, customer interest may be lost. In addition, the credibility of the technology may be questioned. This may make

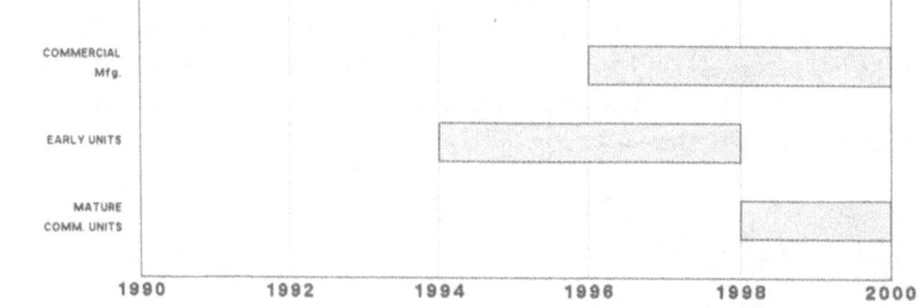

Figure 13.11. Solid oxide fuel cell commercialization schedule.

market penetration at a later date more difficult. We are close to commercialization of PAFC now, and have good chances on other types of cells. The window is still open, with the strong environmental awareness and boom of the late 1980s and early 1990s. The opportunity must be taken now!

REFERENCES

1. Yusuke Narimatsu and Noboru Higuchi, *Abstracts 1985 National Fuel Cell Seminar,* Tucson, AZ, May 1985, pp. 6–9.
2. Noboru Itoh, Minoru Simuzu, and Masaki Nagashima, *Abstracts 1986 National Fuel Seminar,* Tucson, AZ, October 1986, pp. 2–5.
3. Diane Traub Hooie, *The Future of Fuel Cells and Its Impact on the Precious Metals Industry* (IPMI Meeting, Boston, MA, 1988).
4. M. Shioiri and T. Satomi, *Abstracts 1986 National Fuel Cell Seminar,* Tucson, AZ, October 1986.
5. Fuel Cell Development Information Center, *Fuel Cell in Japan* (June 1988).
6. Sidney Gross and Thomas Murray, *Abstracts 1985 National Fuel Cell Seminar,* Tucson, AZ, May 1985, pp. 2-3A–2-3C.
7. W. Meyer, R. Shangraw, M. Brolin, S. Bretschneider, and D. Wilemon, *Abstracts 1985 National Fuel Cell Seminar,* Tucson, AZ, May 1985, pp. 2-10A–2-10D.
8. Meirios Moechtar and T. M. Soelaiman, *Abstracts 1986 National Fuel Cell Seminar,* Tucson, AZ, October 1986, pp. 364–368.
9. J. D. Pandya, *Abstracts 1986 National Fuel Cell Seminar,* Tucson, AZ, October 1986, pp. 361–363.
10. Claire Gibbs, *Fuel Cell News,* **5,** 6–7 (1988).
11. John B. O'Sullivan, *The Role of Fuel Cells in Industrial Cogeneration* (1981 National Fuel Cell Seminar), pp. 243–293.
12. Takehiko Takahashi, *In Expectation of the Future of Fuel Cells* (Fuel Cell Symp. Tokyo, 1988).
13. R. H. Goldstein, *Molten Carbonate Fuel Cell Status* (Fuel Cell Symp., Tokyo, 1988).
14. J. Gilbert, H. Hirschenhofer, and J. L. Humes, *Abstracts 1985 National Fuel Cell Seminar,* Tucson, AZ, May 1985, pp. 90–94.
15. S. Penner, ed., *Advanced Fuel Cell Working Group Report.*
16. Office of Technology Assessment, *New Electric Power Technologies: Problems and Prospects for the 1990's,* (U.S. Congress, 1985).
17. Keith B. Prater and Paul F. Howard, *Overcoming Impediments to Fuel Cell Commercialization,* (National Fuel Cell Seminar, 1981).
18. Robert R. Barthelemy, *Defense Applications of Fuel Cells* (National Fuel Cell Seminar, 1981), pp. 109–115.
19. J. H. Altsermer, J. F. Roach, M. C. Krupka, and J. M. Anderson, *Abstracts 1986 National Fuel Cell Seminar,* Tucson, AZ, October 1986, pp. 354–357.
20. J. Michael Torrey and John D. Leeper, *Abstracts 1986 National Fuel Cell Seminar,* Tucson, AZ, October 1986, pp. 346–349.
21. A. R. Landgrebe and P. G. Patin, *Fuel Cells for Transportation* (National Fuel Cell Seminar, Tucson, AZ, November 1988).
22. F. Baron, *Abstracts 1988 National Fuel Cell Seminar,* Tucson, AZ, November 1988, pp. 255–258.
23. W. G. Parker, S. M. Knable, Charles M. Zeh, *Abstracts 1988 National Fuel Cell Seminar,* Tucson, AZ, November 1988, pp. 248–253.
24. Jack Kleib, ed., *Fuel Cells Workshop Proceedings: Potential Opportunities in Canada* (Toronto, Canada, 1988). (a) N. R. Beck, W. A. Adams, and D. D. McLeod, *Canadian Background Paper*; (b) E. Gillis, *Commercialization of Large Scale Fuel Cells*; (c) R. J. Petri, *Fuel Cell Technology for the Gas Industry*; (d) F. W. Spillers, *Fuel Cells from the Viewpoint of Industrial Users*; (e) Halina Wroblowa, *Fuel Cells for Vehicular Applications*; (f) G. Belanger, D. Klavana, C. Leigh, and R. Breault, *Fuel Cells at Hydro Quebec.*
25. M. N. Mugerwa *et al., Fuel Cell Systems: Design Optimisation and Environmental Aspects of Fuel Cell Systems,* Final Report to Ministry of VROM (Kinetics Technology International Group, 1988)

Epilogue

Leo J. M. J. Blomen and Michael N. Mugerwa

During the four years it has taken to complete this book, the world has changed considerably. Economic relations have shifted: the dollar–German mark exchange rate varied by over a factor 2, as did the oil price. Major changes in East–West European relations exposed new pollution and energy problems and are now paving the way for a trans-European energy treaty. The two Germanies merged. Russia opened its borders to foreign investors, allowing the export of profits. A concerted international effort was geared toward extinguishing the major oil fires in the Middle East following the Gulf War. A recent major disaster in Bangladesh generated a wave of publicity that showed the increased likelihood of repetition of such disasters if fossil fuel use and deforestation are allowed to continue.

Over the last four years many countries have accepted legislative penalties on carbon dioxide, nitrogen oxides, and other emissions. Many states of the United States have adopted least-cost utility planning (LCP), which takes into account environmental factors. Recent investigations indicated that it is essentially concerns for the environment that are forcing us to change our legislative conditions, such that we spontaneously create optimal penetration conditions for clean, highly efficient technologies. Moreover, most energy demand growth will come from the developing nations, where most population growth and probably also increase in relative affluence will occur, but which are acutely short of financial resources. It is now generally accepted that this is prompting an increasing trend toward decentralization of electric power generation. Without any doubt, this will be the most important factor determining market penetration of fuel cell and other systems with similar (or comparable) environmental characteristics and performance. Despite the present situation in which most affluent countries have a considerable proportion of central power generation, the

Leo J. M. J. Blomen • Mannesmann Plant Construction, 2700 AB Zoetermeer, The Netherlands. *Present address*: Blomenco B. V., Achtermonde 31, 4156 AD Rumpt, The Netherlands. Michael N. Mugerwa • Kinetics Technology International S.p.A., Via Monte Carmelo 5, 00166 Rome, Italy.

Fuel Cell Systems, edited by Leo J. M. J. Blomen and Michael N. Mugerwa. Plenum Press, New York, 1993.

decentralization trend has some undeniable advantages:

1. Lower investment costs, among others because of reduced infrastructure costs (transmission and distribution).
2. Shorter planning and construction times, which can be up to a factor of 4 less (a major factor, also with respect to financing of new capacity).
3. Simultaneous use of electricity and waste heat (production near end user).
4. Grid flexibility, both for the owner and/or power producer, because of the possibility to "manage" many small units rather than few large ones (this reduces the excess capacity to be reserved for peak power).
5. Reduced transmission energy losses (which can amount to approximately 8% of the centrally produced power).

Relatively small units are especially important for the less affluent nations and enables a step-by-step low-cost entry into an economically viable and environmentally acceptable energy scenario (the more so if biogas can be used in the decentral plants), with lower investments, as well as shorter delivery times. Such an approach will lower the risk of bad investments and requires less long-term economic and political stability guarantees.

Major driving forces for fuel cell development and commercialization will increasingly come from the environmental and fossil-fuel-saving pressures. The easiest way to facilitate this process is to create conditions that favor superior economics to both the end user and the producer. To this end, many legislative factors should perhaps change, such as:

1. Measures that yield incentives for the delivery of *excess* power from a clean, decentral unit to the grid, for a guaranteed, reasonable price.
2. Establishing low natural gas tariffs for high-efficiency electric power producers (the same or even better rates than for central users: for example, charges that are inversely proportional to the efficiency and proportional to the pollution caused, *not* inversely proportional to capacity installed).
3. Subsidies for clean technologies dependent on installed *electric* capacity.
4. Standardization of contracts between decentral and central producers.
5. More attractive purchasing prices for extra electricity *demands* of decentral producers.
6. The possibility to sell decentrally produced electricity directly to end users.
7. A stiffening of emission requirements aimed at the cleanest technologies that can be made available (all other technologies pay increased penalties proportional to the amount of pollution caused).
8. The evaluation of total environmental effects, including distribution losses, optical pollution of high-voltage lines, electromagnetic field effects, etc.
9. Long-term legislation planning for emissions and tariffs, in order to provide long-term certainty for the potential decentral customer (planning; depreciation period) (in many countries only large utilities can

afford to plan 20–30 years ahead and depreciate their, sometimes "ancient," plants over very long periods).

10. Tax effects on subsidies and cost should be carefully recalculated to give a logical benefit to the cleanest production technology (in many countries, central utilities do not pay any corporate taxes and thus have a strong advantage over the decentralized producer).
11. Control of unfair competition against traders in cogeneration installations.

For some of these issues, the current situation is undergoing rapid change. In several countries, authorities are now convinced of the need to review some of these issues. However, from a global point of view, most or all of these issues should be addressed and appropriate legislation implemented, and the rate at which such change takes place will have a large effect upon potential market and market penetration of high-efficiency clean systems and, consequently, fuel cell systems.

However, market introduction of fuel cell systems on a commercial scale has only just begun. During the 1990s considerable progress on certain major issues must be made in order to guarantee a successful breakthrough, and due to their advanced technology status PAFC systems will probably play the major role here. The issues that need to be tackled may be summarized as

reliability
price
lifetime

of which the first factor is most important.

Potential customers too often perceive fuel cells as nonattractive, since they have been in existence for so long, have cost so much money to develop, and still have not been commercialized on a wide scale. Certainly part of this perception is wrong (the amount of R&D spent on fuel cells is a disproportionally small part of total energy R&D in almost all countries), but the fact that the technology is not yet commercial is real, being especially due to the limited reliability and availability of the systems demonstrated so far. Moreover, many decision makers, environmental groups, and energy experts are hardly aware of fuel cells and their potential, or just consider them as laboratory curiosa. These perceptions should, and will, change soon.

Even the many fuel cell experts themselves do not fully realize what promises this technology has in store. As an example, it is not widely known that if all fuel cell types were to operate at similar conditions on pure hydrogen at similar current densities (in the range of 100–300 mA/cm^2) then their efficiencies would vary less between the different types than between different varieties of one type. Or, in other words, the efficiency differences between the fuel cell types are hardly significant, but the practical purity limitations and application-governed operating conditions cause the major differences in fossil-fuel-based system efficiencies. This implies that the AFC and SP(E)FC may become very competitive if some improvements in gas treatment technologies (carbon dioxide removal, hydrogen separation) can be cheaply realized, for instance with selective ceramic or polymer membranes. Similarly, combinations of the various fuel cell types with gas turbines may yield, albeit at a cost, very high system efficiencies. Due to the issue of reliability, these so-called composite systems are not likely to become the first generation of fuel cell products, but they may largely constitute the second generation. Then the rotating equipment will once more determine the system reliability and performance rather than the fuel cell stacks themselves.

Another widely spread misunderstanding concerns the "Christmas tree lights" effect: many worry that if one cell fails, the plant fails. While this may be true for very small systems, stack and plant design enable (especially for larger plants) a very flexible layout with cell groups, substacks, variable stack and cell surface area sizes, different stack configurations and combinations, and electrical switching and connection options, ensuring guaranteed operation of the (mature technology) plant close to, or at, 100% capacity even if one cell or stack fails. This can all be achieved at marginally increased expense of some extra piping, valves, and control equipment. This flexibility is an inherent advantage of the fuel cell system over any other power generation technology.

Evidently, fuel cells themselves are being continually developed, and many problems still have to be solved, as outlined in this book. However, in the distant future, fuel cell systems will not only have to compete against "established" and "other advanced" technologies, but also against other fuel cell system types. Over the next few years the PAFC will set the scene against which the other technologies will be judged, from the point of view of cost, lifetime, availability, and other factors. In this scenario, the total system behavior and economics will be governing. For this purpose, detailed analyses of the production (as well as the dismantling) cost of systems, including stack production costs, should be made to enable comprehensive comparisons of economics. The trade-off of platinum cost (in the low-temperature cells) and (almost) established lifetime of the PAFC (40,000 h with 10% decay in performance) should be compared with costs for the production of other fuel cell types. The importance of such analyses cannot be overstressed, since they may lead to resetting some goals for R&D. For instance, MCFC lifetime targets of 1% decay per 1000 h (corresponding to 15,000 h lifetime) are hardly sufficient to develop a competitive MCFC system, since frequent stack replacement costs would become prohibitive. Similarly, the need to lower SOFC material costs may favor the commercial application of thin-layer SOFC technology.

Similar economic criteria affect the transportation market. In California, it is expected that 200,000 electric vehicles or zero-emission vehicles (ZEV) will be on the road by 1997, and 200,000 *per year* may be introduced in 2003. The automotive application requires the lowest acceptable cost, with a target of \$150–\$300 kW_e. This excludes all technologies with major stack production and raw material cost. It favors SP(E)FC, even though this is one of the "youngest" fuel cell types. Because of its mere size, the vehicle market may have the strongest effect on volume production, which is needed to enable cost reduction to economically acceptable levels.

The availability of raw materials is also an issue to be studied. The markets for stationary applications are expected to permit a fuel cell system penetration of a few thousand megawatts per annum by the end of this decade. There will be no platinum supply problem for those capacities involved, and platinum cost will not play a major role in the total system's cost reduction problem. With the possibility of recycling platinum from used fuel cells, it is not expected that platinum cost and/or availability will limit fuel cell commercialization. Only a massive acceptance of the fuel cell in the world automotive market might pose a threat to platinum availability and cost, but before that stage will ever be reached fuel cells will be globally considered as the major electric power and propulsion technology*.

In this epilogue, we have only highlighted a few points that are of influence to the exciting development and commercialization of fuel cells, particularly this decade. Still within this context, this book is testimony to the current status of interdisciplinary work, being carried out by some 2000 scientists and engineers worldwide, on a 150-year-old invention, of which the inventor foresaw the use of the technology as a commercial electricity source, especially if hydrogen as a fuel could be replaced by coal, wood, etc. (see Section 1.2.1).

In June 1991, Arthur D. Little expressed the following vision: "Fuel cells may be one of the critical technologies which will allow for the expanded use of energy services needed for continued worldwide economic expansion compatible with maintaining the environmental integrity of the planet." It is now in our hands to realize this vision.

INCENTIVES TO PRODUCE

Worldwide demand for platinum is dominated by two applications: the automotive exhaust catalyst (42%) and jewelry (37%). The remaining 21% includes applications in the chemical, electrical, glass, and petroleum industries as well as metal absorbed by investment demand. For the present, the fuel cell is the

*The following is based on information provided by Impala Platinum, S.A.: Despite the popular perception of platinum as a "scarce" or "precious" metal, in reality only a small proportion of the reserves of platinum-bearing ore in the world has been recovered. A recent authoritative study of the opportunities available to platinum producers and fabricators from the commercialization of fuel cells (Arthur D. Little, Inc., May 1991), concluded that the annual platinum demand to be generated by the fuel cell could be as much as 440,000–580,000 ounces (13–17% of present annual platinum demand) by the year 2000. This finding and the widely held notion that the fuel cell will make an increasingly valid contribution to the need for energy sources after 2000 invite the question as to whether platinum will be available in quantities that will be sufficient to sustain the growth and development of the fuel cell industry. In answering this question, two critical aspects of platinum supply conditions need to be considered, namely, whether sufficient incentives exist for the platinum industry to meet the demand that will emanate from the new market for the metal and whether platinum reserves are available in quantities sufficient to ensure that platinum-bearing ore is accessible for economic production.

World Platinum Reserves

Some 89% of the world's known reserves of platinum is found in South Africa's Bushveld Igneous Complex, a saucer-shaped geologic structure that extends some 400 km across the Transvaal province. Outside South Africa, approximately 9% of world reserves is located in the northwest part of Siberia, near Norilsk, in an area mined primarily for nickel and copper but in which platinum group metals make a considerable contribution to the profitability of the mining operations.

Stillwater (USA), Sudbury (Canada), and Hartley (Zimbabwe) represent 1.6% of known ore reserves.

Currently, proven recoverable reserves of platinum to a depth of 2500 m total 543×10^6 ounces, sufficient to satisfy the 1990 level of demand, at 3.66×10^6 ounces, for another 148 years. The above projection represents a conservative view, as it is more than likely that new deposits will be found. In addition, there is growing evidence that an increasing percentage of platinum supplies will be recovered from spent autocatalysts and fuel cells. This recycling of platinum will effectively extend the life of the currently identified platinum reserves.

only significant new opportunity for the industrial application of platinum. Understandably, platinum producers and fabricators have a vested interest in the commercialization of the fuel cell.

In addition, the supply of the metal to the Western World is dominated by South African producers, who account for 75% of it, and Russia, which has a 19% market share and also meets "local" requirements for platinum. Both countries require sustained economic growth to satisfy the needs and expectations of their respective populations, and platinum production will play an increasingly meaningful role in achieving those objectives. These factors will provide the major incentives to the producers to supply the metal required by the fuel cell industry.

Index

GPSR Compliance

The European Union's (EU) General Product Safety Regulation (GPSR) is a set of rules that requires consumer products to be safe and our obligations to ensure this.

If you have any concerns about our products, you can contact us on ProductSafety@springernature.com

In case Publisher is established outside the EU, the EU authorized representative is:

Springer Nature Customer Service Center GmbH
Europaplatz 3
69115 Heidelberg, Germany

Batch number: 10762209

Printed by Printforce, the Netherlands